Online Ocean Studies

The Southern California coast, near Santa Barbara. [Photo: Edward J. Hopkins]

Online Ocean Studies

Major Contributors:

M. Grant Gross
Washington College

Elizabeth Gross
Scientific Committee on Oceanic Research

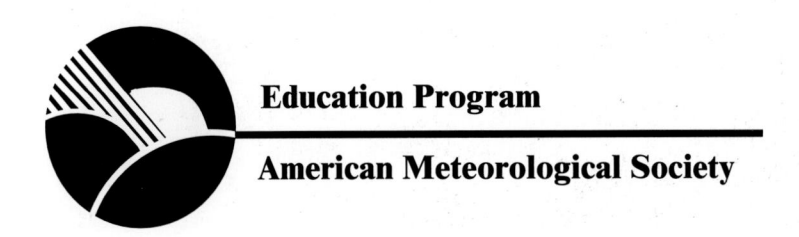

Education Program
American Meteorological Society

The American Meteorological Society Education Program

The American Meteorological Society (AMS), founded in 1919, is a **scientific and professional** society. Interdisciplinary in its scope, the Society actively promotes the development and dissemination of information on the atmospheric and related oceanic and hydrologic sciences. AMS has more than 10,000 professional members from more than 100 countries and over 135 corporate and institutional members representing 40 countries.

The Education Program is the initiative of the American Meteorological Society fostering the teaching of the atmospheric and related oceanic and hydrologic sciences at the precollege level and in community college, college and university programs. It is a unique partnership between scientists and educators at all levels with the ultimate goals of (1) attracting young people to further studies in science, mathematics and technology, and (2) promoting public scientific literacy. This is done via the development and dissemination of scientifically authentic, up-to-date, and instructionally sound learning and resource materials for teachers and students.

Online Ocean Studies, a new component of the AMS education initiative, is an introductory undergraduate oceanography course offered partially via the Internet in partnership with college and university faculty. **Online Ocean Studies** provides students with a comprehensive study of the principles of oceanography while simultaneously providing pedagogically appropriate investigations and applications focusing on web-delivered real-world current data. It provides experiences that demonstrate the value of computers and electronic access to time-sensitive data and information.

ISBN 1-878220-67-5

Copyright © 2005 by the American Meteorological Society

All rights reserved. No part of this publication may be reproduced, stored in a retrieval system, or transmitted, in any form or by any means, electronic, mechanical, photocopying, recording or otherwise, without the prior written permission of the publisher.

Published by the American Meteorological Society
45 Beacon Street, Boston, MA 02108

Printed in the United States of America

Cover Photograph © Mark Conlin/Larry Ulrich Stock

BRIEF CONTENTS

PREFACE

CHAPTER 1:	OCEAN IN THE EARTH SYSTEM	1
CHAPTER 2:	OCEAN BASINS AND PLATE TECTONICS	29
CHAPTER 3:	PROPERTIES OF OCEAN WATER	53
CHAPTER 4:	MARINE SEDIMENTS	77
CHAPTER 5:	THE ATMOSPHERE AND OCEAN	97
CHAPTER 6:	OCEAN CURRENTS	125
CHAPTER 7:	OCEAN WAVES AND TIDES	151
CHAPTER 8:	THE DYNAMIC COAST	173
CHAPTER 9:	MARINE ECOSYSTEMS	203
CHAPTER 10:	LIFE IN THE OCEAN	229
CHAPTER 11:	THE OCEAN, ATMOSPHERE, AND CLIMATE VARIABILITY	257
CHAPTER 12:	THE OCEAN AND CLIMATE CHANGE	279
CHAPTER A:	THE FUTURE OF OCEAN SCIENCE	309
CHAPTER B:	OCEAN STEWARDSHIP	325
CHAPTER C:	OCEAN PROBLEMS AND POLICY	341
APPENDIX I:	CONVERSION FACTORS	357
APPENDIX II:	OCEAN TIMELINE	361
GLOSSARY		369
INDEX		387

CONTENTS

CHAPTER ONE: OCEAN IN THE EARTH SYSTEM — 1

Case-in-Point 1
Driving Question 2
Earth as a System 3
    Hydrosphere 3
    Atmosphere 5
    Geosphere 7
    Biosphere 9
    Biogeochemical Cycles 11
The Ocean in the Global Water Cycle 12
Observing the Ocean 15
    Space-Based Observations 15
    Monitoring the Ocean's Depths 18
Modeling the Ocean 18
Human Impact on the Ocean 19
Conclusions 20
Basic Understandings 20
ESSAY: What Killed the Dinosaurs? 23
ESSAY: Origin of the Ocean and Earth System 25

CHAPTER TWO: OCEAN BASINS AND PLATE TECTONICS — 29

Case-in-Point 29
Driving Question 30
Distribution of the World Ocean 31
Oceanic Crust and Continental Crust 32
    Earth Materials 32
    Rock Cycle 33
Ocean Bottom Profile 34
    Continental Margins 34
    The Ocean Basin 35
Plate Tectonics and Ocean Basin Features 36
    Evidence for Plate Tectonics 36
    Divergent Plate Boundaries 38
    Convergent Plate Boundaries 40
    Transform Plate Boundaries 42
    Marine Volcanism 42
    Hydrothermal Vents 43
Spreading and Closing Cycles 44
Conclusions 46
Basic Understandings 47
ESSAY: Investigating the Ocean Bottom 48
ESSAY: Hotspot Volcanism and the Hawaiian Islands 50

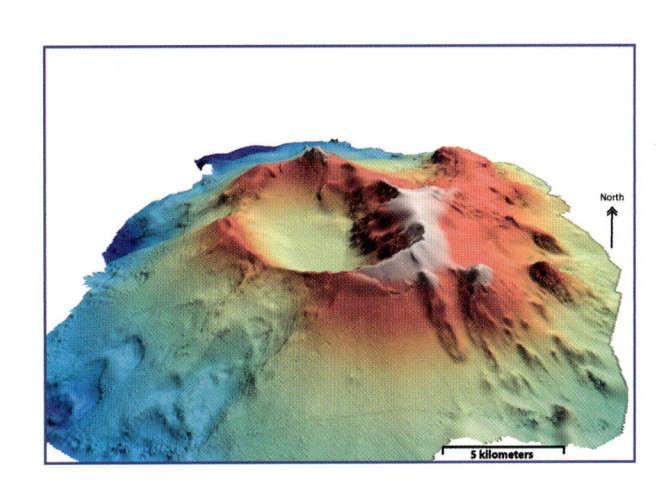

CHAPTER THREE: PROPERTIES OF OCEAN WATER 53

Case-in-Point 53
Driving Question 54
The Water Molecule 54
Water as Ice, Liquid, and Vapor 55
Temperature and Heat 56
Changes in Phase of Water 57
Specific Heat of Water 58
Maritime Influence on Climate 59
Chemical Properties of Seawater 60
    Water as a Solvent 60
    Sea Salts 60
    Dissolved Gases 63
    Alkalinity of Seawater 64
Physical Properties of Seawater 65
    Water Density and Temperature 66
    Pressure 67
    Sea Ice 68
    Sound Transmission 68
Conclusions 70
Basic Understandings 70
ESSAY: Desalination 73
ESSAY: Sea Ice Terminology 74

CHAPTER FOUR: MARINE SEDIMENTS 77

Case-in-Point 77
Driving Question 78
Sediment Size and Accumulation 78
    Size Classification 78
    Terminal Velocity 79
Classification of Marine Sediments 80
    Lithogenous Sediment 80
    Biogenous Sediment 83
    Hydrogenous Sediment 85
    Cosmogenous Sediment 86
Marine Sedimentary Deposits 86
    Continental-Margin Deposits 86
    Deep-Ocean Deposits 88
Marine Sedimentary Rock 89
Resources of the Seafloor 90
    Oil and Natural Gas 90
    Mineral Resources 90
    Exclusive Economic Zone 91
Conclusions 92
Basic Understandings 92
ESSAY: Heinrich Events 94
ESSAY: Burgess Shale: A Glimpse into Ancient Marine Life 95

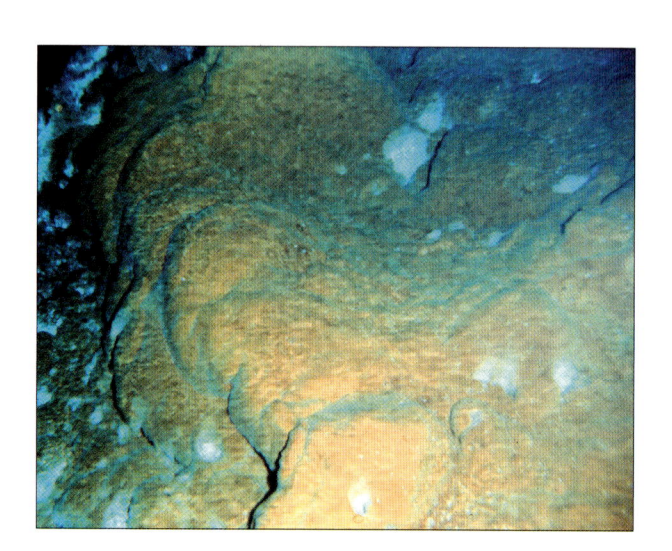

CHAPTER FIVE: THE ATMOSPHERE AND OCEAN 97

Case-in-Point 97
Driving Question 98
Weather and Climate 98
Heating and Cooling Earth's Surface 99
 Solar Radiation 99
 Solar Radiation Budget 102
 Solar Radiation and the Ocean 104
 Infrared Radiation and the Greenhouse Effect 105
Heating Imbalances: Earth's Surface versus Atmosphere 106
 Latent Heating 108
 Sensible Heating 108
Heating Imbalances: Tropics versus High Latitudes 110
 Heat Transport by Air Mass Exchange 111
 Heat Transport by Storms 112
 Heat Transport by Ocean Circulation 112
Circulation of the Atmosphere: The Forces 112
 Pressure Gradient Force 112
 Coriolis Deflection 113
Circulation of the Atmosphere: Patterns of Motion 114
 Planetary-Scale Circulation 114
 Synoptic-Scale Weather Systems 116
Conclusions 118
Basic Understandings 118
ESSAY: Location at Sea, An Historical Perspective 120
ESSAY: The Stratospheric Ozone Shield and Marine Life 121

CHAPTER SIX: OCEAN CURRENTS 125

Case-in-Point 125
Driving Question 127
Ocean's Vertical Structure 127
Ocean in Motion: The Forces 128
 Wind-Driven Currents and Ekman Transport 128
 Geostrophic Flow 130
Wind-Driven Surface Currents 130
 Gyres 131
 Equatorial Currents 132
 Western Boundary Currents 133
 Rings 134
 Upwelling and Downwelling 135
Thermohaline Circulation 137
 Monitoring the Deep Ocean 137
 Water Masses 139
Oceanic Conveyor Belt 141
Conclusions 142
Basic Understandings 142
ESSAY: Predicting Oil-Spill Trajectories 145
ESSAY: The Global Positioning System 146
ESSAY: Profiling the Ocean Depths 148

CHAPTER SEVEN: OCEAN WAVES AND TIDES 151

Case-in-Point 151
Driving Question 152
Wind-Driven Waves 152
    Wind-Wave Generation 153
    Deep-Water and Shallow-Water Waves 154
    Seiche 156
    Atmosphere-Ocean Transfer 157
Ocean Tides 157
    Tide-Generating Forces 158
    Types of Tides 159
    Tides in Ocean Basins 160
    Tidal Currents 162
    Observing and Predicting Tides 162
    Open-Ocean Tides 163
Internal Waves 164
Tsunamis 164
Conclusions 166
Basic Understandings 166
ESSAY: The State of the Sea 168
ESSAY: Monitoring Sea Level from Space 170
ESSAY: Tidal Power 171

CHAPTER EIGHT: THE DYNAMIC COAST 173

Case-in-Point 173
Driving Question 174
Coastline Formation 175
Coastline Features 177
    Beaches 177
    Barrier Islands 179
    Deltas and Salt Marshes 180
    Human Alterations 181
Estuaries 183
Coastal Storms and Storm Surge 185
Tropical Cyclones 186
    Where and When 187
    Hurricane Life Cycle 189
    Hurricane Hazards 189
    Evacuation 191
Extratropical Cyclones 192
Coastal Zone Management 193
Conclusions 194
Basic Understandings 194
ESSAY: Moving the Cape Hatteras Lighthouse 196
ESSAY: Restoring Salt Marshes 198
ESSAY: The Great Lakes and the Ocean 199

CHAPTER NINE: MARINE ECOSYSTEMS — 203

Case-in-Point 203
Driving Question 204
Requirements for Marine Life 204
Structure of Marine Ecosystems 205
    Producers 205
    Consumers 207
    Decomposers 208
    Trophic Structure of Ecosystems 208
    Bioaccumulation 209
Ecosystem Processes 210
    Energy for Growth and Reproduction 210
    Production in the Photic Zone 211
    Nutrients and Trace Elements as Limiting Factors 213
    Microbial Marine Ecosystems 213
Ocean's Role in the Global Carbon Cycle 215
    Physical Pump 215
    Biological Pump 215
Ecosystem Observations and Models 217
Conclusions 218
Basic Understandings 218
ESSAY: Iron Fertilization and Climate Change 221
ESSAY: Gas Hydrates, A Future Energy Source 223
ESSAY: Ocean Color and Marine Productivity 226

CHAPTER TEN: LIFE IN THE OCEAN — 229

Case-in-Point 229
Driving Question 230
Marine Habitats 230
    Oceanic Life Zones 231
    Plankton in the Pelagic Zone 232
    Nekton in the Pelagic Zone 233
Life Strategies and Adaptations 234
    Vertical Migration 234
    Light and Vision 234
    Sound 235
    Feeding Strategies 235
Life at the Ocean's Edge 236
    Intertidal Zone 236
    Sea Grass Beds and Salt Marshes 238
    Kelp Forests 240
    Coral Reefs 240
    Benthic Feeding Habits 243
    Life on the Deep-Sea Floor 244
Marine Animals 244
    Fishes 244
    Marine Mammals 247
    Marine Reptiles 248
    Sea Birds 249

Conclusions 250
Basic Understandings 250
ESSAY: Marine Sanctuaries and Reserves 253
ESSAY: Ecosystem Approach to Fisheries Management 255

CHAPTER ELEVEN: THE OCEAN, ATMOSPHERE, AND CLIMATE VARIABILITY 257

Case-in-Point 257
Driving Question 258
Earth's Climate System 258
    Climate Controls 259
    Role of the Ocean 259
The Tropical Pacific Ocean/Atmosphere 261
Historical Perspective 261
Neutral Conditions in the Tropical Pacific 262
El Niño, The Warm Phase 264
    The 1997-98 El Niño 266
La Niña, The Cold Phase 266
Predicting and Monitoring El Niño and La Niña 268
Frequency of El Niño and La Niña 270
North Atlantic Oscillation 271
Arctic Oscillation 271
Pacific Decadal Oscillation 272
Conclusions 272
Basic Understandings 272
ESSAY: Sea Surface Temperature and Drought in Sub-Saharan Africa 275
ESSAY: El Niño in the Past 277

CHAPTER TWELVE: THE OCEAN AND CLIMATE CHANGE 279

Case-in-Point 279
Driving Question 280
The Climate Record 280
    Marine Sediments and Climate 280
    Other Proxy Climatic Data Sources 281
    Geologic Time 282
    Past Two Million Years 284
    Instrument-Based Temperature Trends 286
Lessons of the Climatic Past 288
Possible Causes of Climate Change 288
    Climate and Plate Tectonics 289
    Climate and Solar Variability 290
    Climate and Volcanoes 292
    Climate and Earth's Surface Properties 294
    Climate and Human Activity 294
The Climate Future 295
    Global Climate Models 295
    Enhanced Greenhouse Effect and Global Warming 296
Impact of Climate Change on the Ocean 297

  Sea Level Fluctuations 297
  Arctic Sea Ice Cover 300
  Marine Life 301
Conclusions 301
Basic Understandings 302
ESSAY: Climate Rhythms in Glacial Ice Cores 304
ESSAY: The Drying of the Mediterranean Sea 306
ESSAY: Sea Level Rise and Saltwater Intrusion 308

CHAPTER A: THE FUTURE OF OCEAN SCIENCE   309

Investigating the Ocean 310
  Voyages of Exploration 310
  Challenger Expedition (1872-1876) 312
Modern Ocean Studies 313
  Technological Innovations 313
  Remote Sensing 315
  Scientific Ocean Drilling 316
Emerging Ocean-Sensing Technologies 317
  Autonomous Instrumented Platforms and Vehicles 317
  Ocean Floor Observatories 319
  Animal-Borne Instruments 320
  Computers and Numerical Models 320
Challenges in Ocean-Sensing Technologies 321
Conclusions 322
Basic Understandings 323

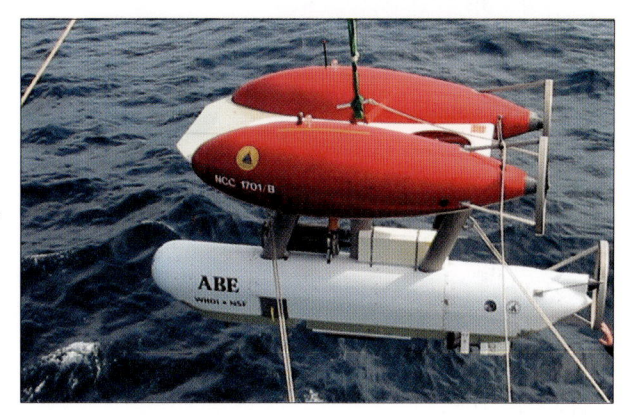

CHAPTER B: OCEAN STEWARDSHIP   325

Stewardship of Ocean Life 325
Fisheries and Sustainable Exploitation 326
  Overfishing 326
  Maximum Sustainable Yield 328
  Ecologically Sustainable Yield 328
  Bycatch 330
  Restoring Fisheries 330
  Habitat Destruction and Restoration 331
  Recreational Fisheries 332
Protecting Endangered Marine Species 333
  Sea Turtles 333
  Whales 334
  Water Birds 336
Mariculture 336
Marine Exotic Species 337
Conclusions 338
Basic Understanding 338

CHAPTER C: OCEAN PROBLEMS AND POLICY   341

Milestones in Ocean Governance 342
    Freedom of the Seas 342
    Antarctic Treaty 342
    Exclusive Economic Zones 342
Human Impact in the Coastal Zone 343
    Population Trends 343
    Environmental Pollution 344
Oil Spills 345
Dams and Marine Habitats 347
Restoring Chesapeake Bay 349
Waste Disposal in the Ocean 350
Deep-Ocean Carbon Storage 352
Obstacles to Ocean Policy Making 353
Conclusions 353
Basic Understandings 354

Appendix I: Conversions Factors   357

Appendix II: Ocean Timeline   361

Glossary   369

Index   387

Preface

Welcome to Online Ocean Studies! You are about to embark on a study of the world ocean and the role of the ocean in the Earth system. Online Ocean Studies is a major initiative of the American Meteorological Society (AMS) that explores the ocean in the Earth system with special emphasis on (1) the flow and transformations of water and energy into and out of the ocean, (2) the physical and chemical properties of seawater, (3) ocean circulation, (4) marine life and its adaptations, (5) interactions between the ocean and the other components of the Earth system (i.e., hydrosphere, atmosphere, geosphere, and biosphere), and (6) the human/societal impacts on and response to those interactions. The purpose of this book is to provide you with background information on the physical, chemical, geological, and biological foundations of oceanography. This information will assist your work on twice-weekly Benchmark Investigations whose components are available in the companion Study Guide and delivered via the Online Ocean Studies Homepage.

You will explore twelve principal themes plus three optional themes that are arranged by chapter, each corresponding to one week of the Online Ocean Studies course. Themes are organized so that concepts build logically upon one another as the ocean and its role in the Earth system are demonstrated to follow patterns described by physical laws. Benchmark Investigations, partially delivered via the Internet and focused on some aspect of the ocean in the Earth system, are tied directly to each chapter. The first chapter introduces the Earth system and examines the ocean's place in the Earth system. The second chapter focuses on the characteristics of the ocean basin that are largely the products of plate tectonics. Chapter 3 deals with the unique physical and chemical properties of seawater and Chapter 4 covers the origin and distribution of sediment in the ocean with special emphasis on ocean resources. Chapter 5 investigates air-sea interactions and the flow of heat energy in the Earth-atmosphere-ocean system. The next three chapters cover the dynamic ocean: surface and deep-ocean currents (Chapter 6), waves and tides (Chapter 7), and shoreline processes (Chapter 8). Chapters 9 and 10 examine marine ecosystems and life in the ocean. The following two chapters explore the ocean's role in and short-term climate variability (Chapter 11) and long-term climate change (Chapter 12). Optional chapters summarize the future of ocean science (Chapter A), ocean stewardship (Chapter B), and public policy and the ocean (Chapter C).

Each of the first twelve chapters opens with a Case-in-Point, an authentic event or issue that highlights or applies one or more of the main concepts introduced in the chapter. In essence, the Case-in-Point previews the chapter and is intended to engage reader interest in the topic early on. Chapter 2, for example, opens with a discussion of the impacts of massive volcanic eruptions on the Earth system. The Case-in-Point is followed by a sample Driving Question, a broad-based query that links chapter concepts and provides a central focus from the beginning of the chapter. Content is science-rich and informs additional driving questions. Each chapter closes with a list of Basic Understandings. Essays at the end of each chapter address in some depth a specific topic that builds on a concept introduced in the narrative. Examples include Origin of the Ocean and Earth System, Desalination, Profiling the Ocean Depths, The State of the Sea, Moving the Cape Hatteras Lighthouse, and Marine Sanctuaries and Reserves. All bold-faced terms are defined in the Glossary at the back of the book.

Online Ocean Studies is pedagogically guided by a teaching approach that seeks to engage learners in exploring their world by investigating meaningful questions. Online Ocean Studies incorporates driving questions, investigations, collaboration, technology, and artifacts. The course offers one driving question per chapter but each chapter plus investigations inspire additional driving questions. Each Benchmark Investigation has printed and electronic components that make use of environmental data available on the Internet. Investigations promote critical thinking as participants engage in observation, data analysis, inference, and prediction. The course presents opportunities for students to collaborate with their instructor and fellow students as together they negotiate understanding. Application of information-age technology provides the student with experience in retrieving and analyzing real-world data (some in real-time) and sharing interpretations. Throughout the course, students assemble learning materials (artifacts) for assessment purposes.

Online Ocean Studies learning materials are the products of collaboration among many individuals with extensive educational backgrounds and teaching experience. This textbook was derived from a manuscript co-authored by M. Grant Gross of Washington College (MD) and Elizabeth Gross of the Scientific Committee on Oceanic Research. Joseph M. Moran of the AMS education program served as managing editor for the writing project. Textbook development benefited greatly from suggestions, constructive criticisms and editing provided by Ira W. Geer, Bernard A. Blair, and Elizabeth W. Mills of the AMS education program, Robert S. Weinbeck of SUNY College at Brockport and the AMS education program, David R. Smith of the U.S. Naval Academy, Edward J. Hopkins of the University of Wisconsin-Madison, James A. Brey of the University of Wisconsin-Fox Valley, H. Joseph Niebauer of the University of Wisconsin-Madison, Ronald D. Stieglitz of the University of Wisconsin-Green Bay, and many NOAA scientific personnel. Norman J. Frisch of Brockport, NY did an excellent job of turning line drawings into final art. Bernard Blair of the AMS education program met the numerous technical challenges in converting the original manuscript into this book with his usual skill, attention to detail, dedication, and perseverance. Special thanks are extended to the group of outstanding K-12 teachers who serve as AMS Educational Resource Agents (AERAs) for providing valuable advice and encouragement during the development and testing of orean related learning materials. Thanks also to Ashley L. Boivin of the University of Wisconsin-Green Bay for her thorough and dedicated library research.

A special note concerns the use of units in Online Ocean Studies. Generally, the International System of Units (abbreviated SI, for Systèm Internationale d'Unitès) is employed with equivalent English or other units following in parentheses. Exceptions are units used by convention or convenience in oceanography (e.g., pressure is given in units of decibars). Also, the equivalence between units is given in context; that is, where general estimates are given, approximate values are shown for all units. Conversion factors are in Appendix I.

Earlier work completed by the AMS education program in projects funded by the National Science Foundation (NSF) and the National Oceanic and Atmospheric Administration (NOAA) contributed to the development of Online Ocean Studies. The goals of all these efforts have been to enhance public understanding of the fluid Earth system emphasizing the atmospheric, oceanic, and hydrologic sciences and to promote activity that will contribute to greater human resource diversity in the nation's scientific workforce. Particularly, the goals of Online Ocean Studies are in alignment with the AMS/NOAA Cooperative Program for Earth System Education (CPESE).

Through CPESE, the AMS assists NOAA in the advancement of its goals directed toward environmental assessment and prediction, protection of life and property, and the fostering of global environmental stewardship. NOAA's success in meeting its mission objectives is highly dependent upon synergistic relationships between it and the users of its products and services. CPESE nurtures this synergy through precollege teacher, introductory undergraduate, and general educational activity. Fundamental to CPESE are (1) breadth, demonstrating the comprehensive need for describing and predicting changes in the Earth's environment and conserving and wisely managing the nation's coastal and marine resources; (2) visibility, increasing pubic awareness of the ways environmental assessment, prediction, and stewardship touch the lives of all Americans every day; and (3) diversity, promoting educational activity and outreach to attract members of groups underrepresented in science, technology, engineering and mathematics to study and consider careers in those fields, including those for which NOAA has employment needs and opportunities. For additional information on CPESE and the AMS education program, go to: http://www.ametsoc.org/amsedu.

Ira W. Geer
AMS Education Program

CHAPTER 1

OCEAN IN THE EARTH SYSTEM

Case-in-Point
Driving Question
Earth as a System
    Hydrosphere
    Atmosphere
    Geosphere
    Biosphere
    Biogeochemical Cycles
The Ocean in the Global Water Cycle
Observing the Ocean
    Space-Based Observations
    Monitoring the Ocean's Depths
Modeling the Ocean
Human Impact on the Ocean
Conclusions
Basic Understandings
ESSAY: What Killed the Dinosaurs?
ESSAY: Origin of the Ocean and Earth System

A rain shaft pierces a tropical sunset as seen from Man-of-War Bay, Tobago, Caribbean Sea. [Courtesy of NOAA.]

Case-in-Point

Earth is unique in the solar system in that ocean waters cover much of its surface. But as scientists continue to explore the moon and Earth's neighboring planets, they are finding that water, as either liquid or ice, is more common in the solar system than once believed.

    Mars has long been suspected of having water in its past. As far back as 1877 astronomers observed features on the Martian surface that they interpreted as canals that at one time transported water. Today we know that these "canals" were actually defects in the early telescopes. In recent decades, Mars-orbiting satellites and robots placed on the planet have mapped its surface, now dry and covered with windblown dust (Figure 1.1). Mariner satellite missions of the 1960s and Viking orbiter missions of the 1970s photographed vast flood channels and networks of valleys on the Martian surface, presumably cut by running water when temperatures were higher than they are now. About two decades later, the National Aeronautics and Space Administration's (NASA's) Pathfinder mission to Mars confirmed these findings. On 4 July 1997, the Pathfinder spacecraft landed on an ancient floodplain that had been scoured by catastrophic flooding billions of years ago. Also, Pathfinder's tiny rover, *Sojourner*, photographed pebbles and rocks on the Martian surface having physical characteristics indicative of transport by running water. More recently, NASA's Mars Global Surveyor and Mars Odyssey spacecraft detected ancient lakebeds, gullies, and shorelines. And in 2004, two Mars Exploration Rovers (MERs) landed on the Martian surface in search of more evidence that water once flowed on Mars.

    Polar ice caps form during the Martian winter and mostly disappear in spring. Unlike their counterparts on Earth, the Martian ice caps consist of mostly solid carbon

FIGURE 1.1
The surface of Mars viewed from space is marked by numerous craters produced by meteorite impacts over billions of years. Canyons up to 8 km (5 mi) deep were apparently cut by running water, which flowed into a large basin in the upper right of the image. Volcanoes (dark spots) are visible on the left side of the image. [Courtesy of NASA and the U.S. Geological Survey]

dioxide (i.e., dry ice) and some water ice. The carbon dioxide (CO_2) in the ice caps is derived from the very thin Martian atmosphere, which is chiefly carbon dioxide. The small amount of water in Mars' polar ice caps would scarcely fill a large municipal swimming pool.

Liquid water no longer exists on Mars' surface because the planet's temperature is much too low. Average temperatures on the Martian surface range from about −60°C (−76°F) at the equator to as low as −123°C (−189°F) at the poles. Sensors on Mars-orbiting satellites have found evidence of water ice, buried under several meters of dry soil. The ice layer is estimated to be roughly equivalent to twice the volume of water in Lake Superior, the largest of the North American Great Lakes.

Water also occurs elsewhere in the solar system. Three of Jupiter's largest moons, Europa (about the size of Earth's moon), Ganymede, and Callisto apparently have deep layers of salty water, which may be considered subsurface oceans. In the 1990s, images taken by the Galileo spacecraft revealed a fractured 20-km (12-mi) thick layer of ice covering the surface of Europa. A liquid layer of salty water underlies this icy shell. Frictional heat generated by tides induced by Jupiter's gravitational attraction keeps the water in a liquid state. Tidal motions on Europa resemble ocean tides on Earth, which are caused by the combined gravitational attraction of the moon and sun (Chapter 7).

Jupiter is about 1300 times more massive than Earth and has an enormous gravitational field. Europa orbits Jupiter at approximately the planet's rotation rate so that the same side of Europa faces Jupiter for hundreds of years. In one Europan day (85-hours) Jupiter's gravitational attraction pushes and pulls Europa's icy crust causing it to rise and fall about 30 m (100 ft), generating frictional heat. Radiation of heat from Europa's surface to space keeps the outer shell frozen. This icy crust insulates the inner portion, which becomes sufficiently warm to maintain a salty water layer about 150 km (95 mi) thick. Some scientists speculate that these salty waters could harbor some forms of life, perhaps in niches associated with cracks in the ice.

Discovery of water on other bodies in the solar system will help scientists to better understand the origins of Earth's ocean and life. An abundance of liquid water is thought to be essential for the evolution of life.

Driving Question:

What is the relationship between the ocean and Earth system?

This book focuses on Earth's ocean and the energy and processes responsible for formation of its basins, the supply of seawater and dissolved salts, the circulation of its waters, and the abundant life it harbors. We also consider the ocean's interactions with the continents, atmosphere, the rocks and sediments that form the ocean floor, and the life on Earth. By studying the ocean in this way, we are adopting an Earth system perspective, viewing the planet as a set of interacting subsystems. Thus, we examine the hydrosphere (Earth's water and ice), atmosphere (its gaseous envelope), geosphere (Earth's solid surface), and biosphere (all living organisms). In this opening chapter we introduce the Earth system and briefly describe Earth's various interacting subsystems with special emphasis on the ocean.

In short, we use the Earth system perspective to guide our investigation of the ocean. Space explorers also use this systems approach to study the other planets of the solar system. We also examine the flow and transformations of matter and energy within and among Earth's subsystems. We stress the transport and cycling of water, salts, carbon, and oxygen as we seek to understand how

Earth's ocean functions and how it influences our lives. Furthermore, the systems approach is valuable in studying how the planet responds to large-scale environmental change. An example is presented in the first Essay at the end of this chapter.

In this chapter, we also survey some of the advances in direct (in situ) and remote sensing of the ocean made possible by modern technology (e.g., Earth-orbiting satellites, floats). We introduce the contributions that electronic computers and numerical modeling have made in the collection, analysis, and interpretation of environmental data. Finally, we briefly examine how human activity influences the ocean.

Earth as a System

What is the Earth system and more fundamentally, what is a system? A **system** is an interacting set of components that behave in an orderly way according to the laws of physics, chemistry, geology, and biology. One familiar example of a system is the human body, which consists of various subsystems including the nervous, respiratory, and reproductive systems plus energy and matter input/output. In a healthy person, these subsystems function internally and interact with one another in regular and predictable ways. Based on extensive observations and understanding of a system, scientists can predict how the system and its components are likely to respond to changing conditions. This predictive ability is especially important in dealing with the complexities of global climate change and its potential impacts on Earth's subsystems (Chapter 12).

The Earth system consists of four major interacting subsystems: hydrosphere, atmosphere, geosphere, and biosphere. Here we briefly examine each subsystem, its composition, basic properties, and some of its interactions with other components of the Earth system. The view of Planet Earth in Figure 1.2, resembling a "blue marble," shows all the major subsystems of the Earth system. The ocean, the most prominent feature, appears blue; clouds mostly obscure the ice sheets that cover much of Greenland and Antarctica; and the atmosphere is made visible by swirling storm clouds over the Pacific near Mexico and the middle of the Atlantic Ocean. Viewed edgewise, the atmosphere appears as a thin, bluish layer. Land (part of the geosphere) is mostly green because of vegetative cover (biosphere). The dominant color of Earth is blue because the ocean covers more than two-thirds of its surface; in fact, often Earth is referred to as the "blue planet" or "water planet."

HYDROSPHERE

The **hydrosphere** includes water in all three phases (ice, liquid, and vapor) that continually cycles from one reservoir to another within the Earth system. (We discuss the global water cycle in more detail later in this chapter.) Water is unique in the Earth system in that it is the only naturally occurring substance that co-exists in all three phases at normal temperatures and pressures at Earth's surface. The ocean, by far the largest reservoir in the hydrosphere, covers about 70.8% of the planet's surface and has an average depth of about 3.8 km (2.4 mi) but accounts for only 0.02% of Earth's mass. About 97% of the hydrosphere is ocean salt water. The next largest reservoir in the hydrosphere is glacial ice, most of which covers much of Antarctica and Greenland. Considerably smaller quantities of water occur on the land surface (lakes, rivers), in the subsurface (soil moisture, groundwater), the atmosphere (water vapor, clouds, precipitation), and biosphere (plants, animals). The possible origins of water on Earth are described in this chapter's second Essay.

The hydrosphere is dynamic. Earth's waters are continually moving although at different rates through different parts of the Earth system. Nonetheless, the ocean is the ultimate destination of all water moving on or beneath the land surface. Water flowing in river or stream channels may take a few weeks to reach the ocean. Groundwater typically moves at a very slow pace through fractures and tiny openings in subsurface rock and sediment and feeds into rivers, lakes, or directly into the ocean. Water in large, deep lakes also moves slowly, in some cases taking centuries to reach the ocean. Water frozen into mountain glaciers takes thousands of years to be released to the sea whereas ice in the Greenland and Antarctic ice sheets may be sequestered for hundreds of thousands of years.

The ocean and atmosphere are coupled such that winds drive surface ocean currents. Wind-driven currents are restricted to a surface ocean layer typically about 100 m (300 ft) deep and require a few months to years to cross an ocean basin (Chapter 6). Deep-ocean currents, at depths greater than 100 m (300 ft), are much more sluggish and more challenging to study than surface currents because of the considerable difficulties in making measurements at great depths. Movements of deep-ocean waters are caused primarily by small differences in water density (mass per unit volume) arising from small differences in water temperature and salinity (a measure of dissolved salt content). Cold water, being denser than warm water, tends to sink whereas warm water, being less dense, is buoyed upward

4 Chapter 1 OCEAN IN THE EARTH SYSTEM

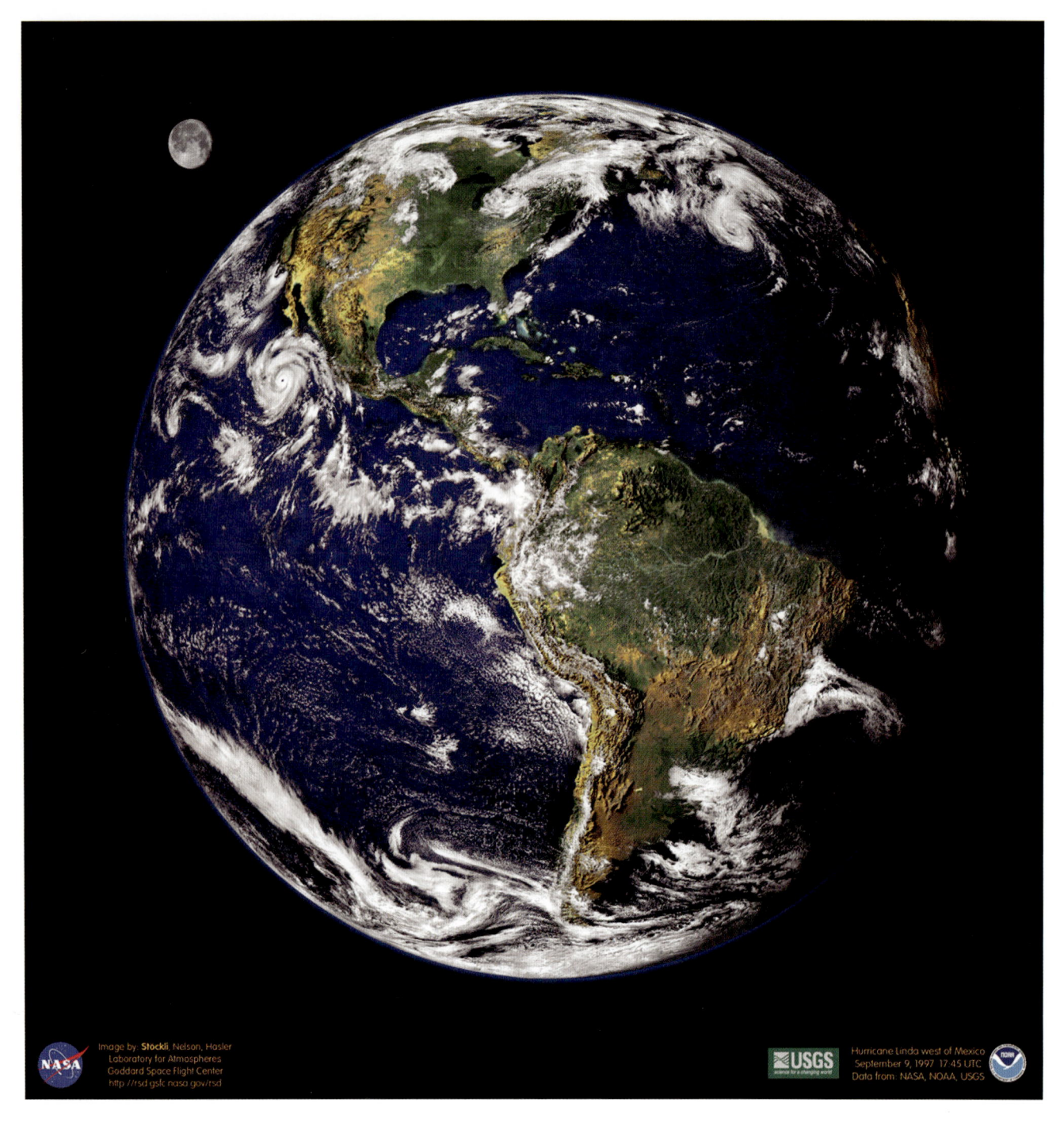

FIGURE 1.2
Planet Earth, viewed from space, appears as a "Blue Marble," with its surface mostly ocean water and partially obscured by swirling cloud masses. The moon, appearing in the upper left limb of the Earth, is the planet's only natural satellite and exerts a major influence on the ocean through tides. [Courtesy of NASA, Goddard Space Flight Center]

by (or floats on) colder water. Likewise, saltier water is denser than less salty water and tends to sink whereas less salty water is buoyed upward. The combination of temperature and salinity determines whether a water mass remains at its original depth or sinks to the bottom. Even though deep currents are relatively slow, they keep ocean waters well mixed so that the ocean has a nearly uniform chemical composition (Chapter 3).

The densest ocean waters form in polar or nearby subpolar regions. Salty waters become even saltier where sea ice forms at high latitudes because growing ice crystals exclude dissolved salts. Chilling of this salty water near Greenland and Iceland and in the Norwegian and Labrador Seas further raises its density so that it sinks and forms a bottom current that flows southward under equatorial surface waters and into the South Atlantic as far south as Antarctica. Here, deep water from the North Atlantic mixes with deep water around Antarctica (Chapter 6). Branches of that cold bottom current then spread northward into the Atlantic, Indian, and Pacific basins. Eventually, the water slowly diffuses to the surface, mainly in the Pacific, and then starts its journey on the surface through the islands of Indonesia, across the Indian Ocean, and around South Africa and into the tropical Atlantic. There, intense heating and evaporation make the water hot and salty. This surface water is then transported northward in the Gulf Stream thereby completing the cycle. This so-called ocean *conveyor belt system* and its transport of heat energy and salts is an important control of climate (Chapter 6).

The frozen portion of the hydrosphere, known as the **cryosphere**, encompasses massive glacial ice sheets, mountain glaciers, ice in permanently frozen ground (*permafrost*), and the pack ice and bergs floating at sea. All of these ice types except pack ice (frozen sea water) are freshwater. A **glacier** is a mass of ice that flows internally under the influence of gravity. Huge glacial ice sheets, in places up to 3 km (1.8 mi) thick, cover nearly all of Greenland and Antarctica. The Antarctic ice sheet contains 90% of the ice in the Earth system. Much smaller glaciers (tens to hundreds of meters thick) primarily occupy the highest mountain valleys on all continents. At present, glacial ice covers about 10% of the planet's land area but at times during the past 1.7 million years, glacial ice expanded over as much as 30% of the land surface, primarily in the Northern Hemisphere. At the peak of the last glacial advance, about 20,000 to 18,000 years ago, the Laurentide ice sheet covered much of Canada and the northern tier states of the United States. At the same time, a smaller ice sheet buried the British Isles and portions of northwest Europe. Meanwhile, mountain glaciers worldwide thickened and expanded.

Glaciers form where annual snowfall exceeds annual snowmelt. As snow accumulates, the pressure of the overlying snow converts snow to ice. As the ice forms, it preserves traces of the original seasonal layers of snow and traps gas bubbles. Chemical analysis of the ice layers and air bubbles in the ice provides clues to climatic conditions at the time the original snow fell (Chapter 12). Ice cores extracted from the Greenland and Antarctic ice sheets yield information on changes in Earth's climate and atmospheric composition extending back hundreds of thousands of years—to about 450,000 years ago in Antarctica.

Under the weight of the overlying snow and ice, glacial ice flows slowly from sources at higher latitudes or higher elevations (where some winter snow survives the summer) to lower latitudes or lower elevations, where the ice either melts or flows into the nearby ocean. Glaciers normally expand (thicken and advance) and shrink (thin and retreat) slowly in response to changes in climate (long-term average temperature and snowfall). Mountain glaciers take decades to centuries to form or totally disappear. On the other hand, continental ice sheets change on time scales of tens of thousands to hundreds of thousands of years.

Around Antarctica, streams of glacial ice flow out to the ocean. Ice, being less dense than seawater, floats, forming ice shelves (typically about 500 m or 1600 ft thick). Thick masses of ice break off the shelf edge, forming flat-topped icebergs that are carried by surface ocean currents around Antarctica (Figure 1.3). Likewise, irregularly shaped icebergs break off the glacial ice streams of Greenland and flow out into the North Atlantic. In 1912, the newly launched luxury liner, *RMS Titanic*, struck a Greenland iceberg southeast of Newfoundland and sank with the loss of more than 1500 lives. Most sea ice surrounding Antarctica forms each winter through freezing of surface seawater. During summer most of the sea ice around Antarctica melts whereas in the Arctic Ocean sea ice can persist for several years before flowing out through Fram Strait into the Greenland Sea and eventually melting. This "multi-year" ice loses salt content with age, so that Eskimos harvest this older, less salty ice for drinking water.

ATMOSPHERE

Earth's **atmosphere** is a relatively thin envelope of gases and suspended particles surrounding the planet. Compared to Earth's diameter, the atmosphere is like the thin skin of an apple and accounts for only about 0.07% of the mass of the Earth system. But the thin atmospheric skin is essential for life and the orderly functioning of physical and biological processes on Earth. Unlike the nearly uncompressible ocean water, air is compressible, so that air density decreases with increasing altitude above Earth's surface. About half of the atmosphere's mass is concentrated within about 5500 m (18,000 ft) of Earth's surface and 99% of its mass occurs below an altitude of 32 km (20 mi). At an altitude of about 1000 km (620 mi), the atmosphere

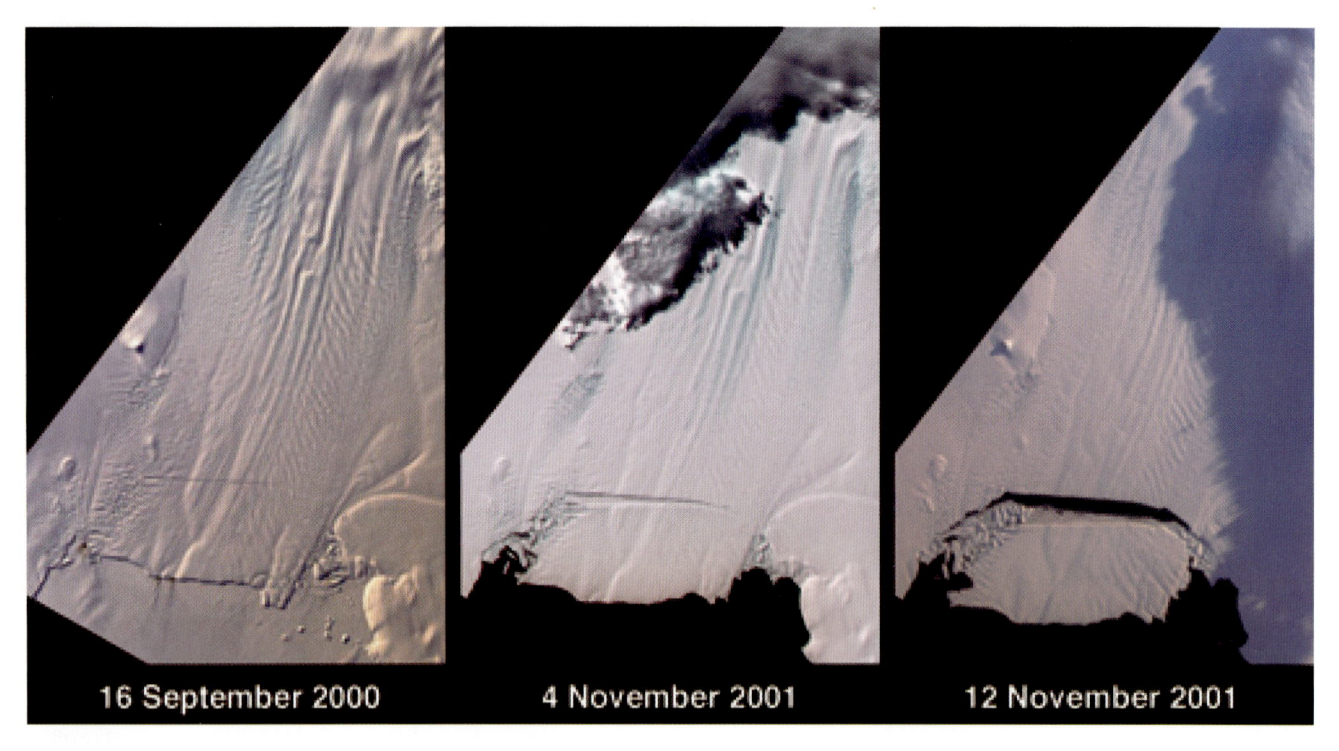

6 Chapter 1 OCEAN IN THE EARTH SYSTEM

FIGURE 1.3
A massive iceberg (42 km by 17 km or 26 mi by 10.5 mi) is shown breaking off Pine Island Glacier, West Antarctica (75 degrees S, 102 degrees W) in early November 2001 along a large fracture that formed across the glacier in mid 2000. Images of the glacier were obtained by the Multi-angle Imaging SpectroRadiometer (MISR) instrument aboard NASA's Terra spacecraft. Pine Island Glacier is the largest discharger of ice in Antarctica and the continent's fastest moving glacier. [Courtesy of NASA]

merges with the highly rarefied interplanetary gases, hydrogen and helium.

Based on the average vertical temperature profile, the atmosphere is divided into four layers (Figure 1.4). The **troposphere** (averaging about 10 km or 6 mi thick) is where the atmosphere interfaces with the hydrosphere, geosphere, and biosphere and where most weather takes place. In the troposphere, the average air temperature drops with increasing altitude so that it is usually colder on mountaintops than in lowlands. The troposphere contains 75% of the atmosphere's mass and 99% of its water. The *stratosphere* (10 to 50 km, 6 to 30 mi above Earth's surface) contains the ozone shield, which protects organisms from exposure to potentially lethal levels of solar ultraviolet radiation. Above the stratosphere is the *mesosphere* where average temperatures generally decrease with altitude and above that is the *thermosphere* where average temperatures increase with altitude but are particularly sensitive to variations in incoming solar radiation.

Nitrogen (N_2) and oxygen (O_2), the chief atmospheric gases, are mixed in uniform proportions up to an altitude of about 80 km (50 mi). Not counting water vapor (which has a highly variable concentration), nitrogen occupies 78.08% by volume of the lower atmosphere, and oxy-

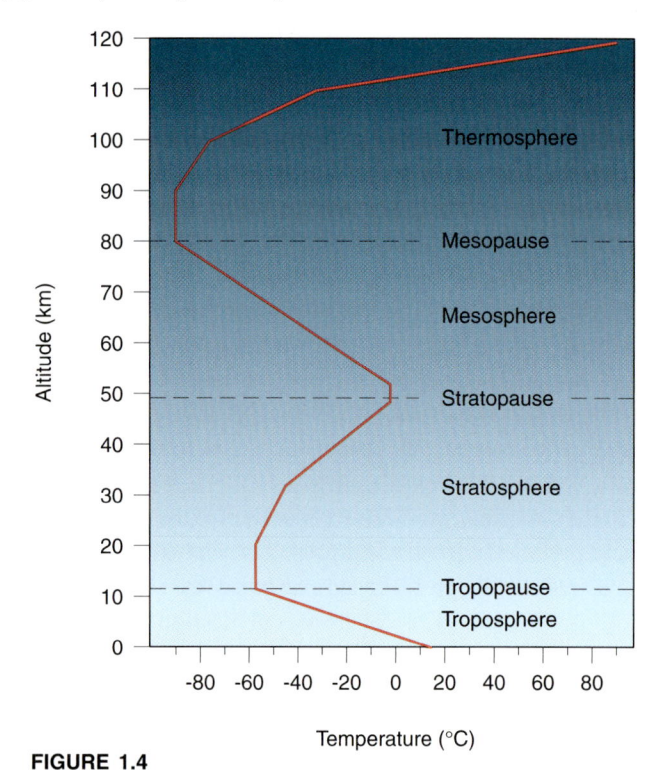

FIGURE 1.4
Based on variations in average air temperature with altitude, the atmosphere is divided into the troposphere, stratosphere, mesosphere, and thermosphere.

TABLE 1.1
Gases Composing Dry Air in the Lower Atmosphere (below 80 km)

Gas	% by Volume	Parts per Million
Nitrogen (N_2)	78.08	780,840.0
Oxygen (O_2)	20.95	209,460.0
Argon (A)	0.93	9,340.0
Carbon dioxide (CO_2)	0.03694	369.4
Neon (Ne)	0.0018	18.0
Helium (He)	0.00052	5.2
Methane (CH_4)	0.00014	1.4
Krypton (Kr)	0.00010	1.0
Nitrous oxide (N_2O)	0.00005	0.5
Hydrogen (H)	0.00005	0.5
Xenon (Xe)	0.000009	0.09
Ozone (O_3)	0.000007	0.07

gen is 20.95% by volume. The next most abundant gases are argon (0.93%) and carbon dioxide (0.037%). Many other gases occur in the atmosphere in trace concentrations, including ozone (O_3) and methane (CH_4) (Table 1.1). Unlike nitrogen and oxygen, the percent volume of some of these trace gases varies with time and location.

In addition to gases, minute solid and liquid particles, collectively called **aerosols**, are suspended in the atmosphere. A flashlight beam in a darkened room reveals an abundance of tiny dust particles floating in the air. Individually, most atmospheric aerosols are too small to see but in aggregates, such as water droplets and ice crystals composing clouds, they may be visible. Most aerosols occur in the lower atmosphere, near their sources on Earth's surface; they are derived from wind erosion of soil, ocean spray, forest fires, volcanic eruptions, industrial chimneys, and the exhaust of motor vehicles. Although the concentration of aerosols in the atmosphere is relatively small, they participate in some important processes. Aerosols function as nuclei that promote the formation of clouds, essential for the global water cycle. And some aerosols (e.g., volcanic dust, sulfurous particles) influence air temperatures by interacting with incoming solar radiation.

The significance of an atmospheric gas is not necessarily related to its concentration. Some atmospheric components that are essential for life occur in very low concentrations. For example, most water vapor is confined to the lowest kilometer or so of the atmosphere and is never more than about 4% by volume even in the most humid places on Earth (e.g., over tropical rainforests and seas). Without water vapor, the planet would have no water cycle, no rain or snow, no ocean, and no fresh water. Also, without water vapor, Earth would be much too cold for most forms of life. Although comprising only about 0.037% of the lower atmosphere, carbon dioxide (CO_2) is essential for photosynthesis. Without carbon dioxide, green plants and the food webs they support could not exist. Although the atmospheric concentration of ozone (O_3) is minute, the chemical reactions responsible for its formation (from oxygen) and dissociation (to oxygen) in the stratosphere (mostly at altitudes between 30 and 50 km) shield organisms on Earth's surface from potentially lethal intensities of solar ultraviolet radiation.

The atmosphere is dynamic; that is, the atmosphere is always in motion in response to differences in rates of heating and cooling within the Earth system. On an average annual basis, Earth's surface experiences net radiational heating (from the sun) and the atmosphere undergoes net radiational cooling (to space). Also, net radiational heating occurs in the tropics, while net radiational cooling characterizes higher latitudes. Differential heating and cooling give rise to *temperature gradients*, that is, changes in temperature from one location to another. In response to temperature gradients, the atmosphere (and ocean) circulates and redistributes heat within the Earth system. Heat is conveyed from where it is warmer to where it is colder: from the Earth's surface to atmosphere and from the tropics to higher latitudes. As discussed in Chapter 5, the global water cycle and accompanying phase changes of water play an important role in this planetary-scale transport of heat energy.

GEOSPHERE

The geosphere is the solid portion of the planet composed of rocks, minerals, and sediments. Most of Earth's interior cannot be observed directly—the deepest mines and oil wells penetrate to a depth of only about 12 km (7 mi) below the surface. Most of what we know about the composition and physical properties of Earth's interior comes from studying seismic waves generated by earthquakes and explosions. In addition, meteorites provide valuable clues regarding the chemical composition of Earth's interior. From study of the behavior of vibrations that penetrate the planet, geologists have determined that Earth's interior consists of four spherical shells: crust, mantle, and outer and inner core (Figure 1.5). Earth's interior is mostly solid and accounts for much of the mass of the Earth system. The outermost solid skin of the planet, called the *crust*, ranges in thickness from only about 8 km (5 mi) under the ocean to about 70 km (45 mi) in some mountain belts. We live on the crust and it is the source of almost all rock, mineral, and fuel (e.g., coal, oil, and natural gas) resources that are essential for our industrial-based economy. The rigid uppermost portion of the *mantle* plus the overlying crust constitutes Earth's **lithosphere**, averaging about 100 km (62 mi) thick. Two sets of geological processes continually modify the lithosphere: surface geological processes and internal geological processes.

Surface geological processes encompass weathering and erosion taking place at the interface between the lithosphere (mainly the crust) and the other Earth subsystems. **Weathering** entails the physical disintegration, chemical decomposition, or solution of exposed rock. Rock fragments produced by weathering are known as *sediments*. Water plays an important role in weathering by dissolving soluble rock and minerals and participating in chemical reactions that decompose rock. Also water's unusual physical property of expanding when freezing can fragment rock. Water seeps into the tiny cracks and crevices that are present in all rock and when the temperature drops below 0°C (32°F), water freezes and expands, forcing the cracks open. Repeated freeze-thaw cycles cause cracks to eventually penetrate the rock, reducing the rock mass to progressively smaller fragments.

The ultimate weathering product is *soil*, a mixture of organic (humus) and inorganic matter (sediment) on Earth's land surface that supports rooted plants and supplies mineral nutrients and water for plants. Soils derive from the weathering of bedrock or sediment and vary widely in texture (particle size). A typical soil is 50% open space (pores) that is occupied by air and water in roughly equal proportions. Plants also participate in weathering through the physical action of their growing roots and the carbon dioxide they release to the soil.

Erosion involves the removal and transport of sediments by gravity, moving water, glaciers, and wind. Running water and glaciers are pathways in the global water cycle. Erosive agents transport sediments from source regions (usually highlands) to low-lying depositional environments (e.g., ocean, lakes). Weathering aids erosion by reducing massive rock to particles that are sufficiently small to be transported by agents of erosion. Erosion aids weathering by removing sediment and exposing fresh surfaces of rock to the atmosphere and weathering processes. Together, weathering and erosion reduce the elevation of the land.

Internal geological processes counter surface geological processes by uplifting the land through tectonic activity, including volcanism and mountain building. Most tectonic activity occurs at the boundaries between lithospheric plates. The lithosphere is broken into 12 massive plates (and many smaller ones) that are slowly driven (typically less than 20 cm per year) across the face of the globe by huge convection currents in Earth's mantle. Continents are carried on the moving plates and, as we will see in Chapter 2, ocean basins are formed by seafloor spreading.

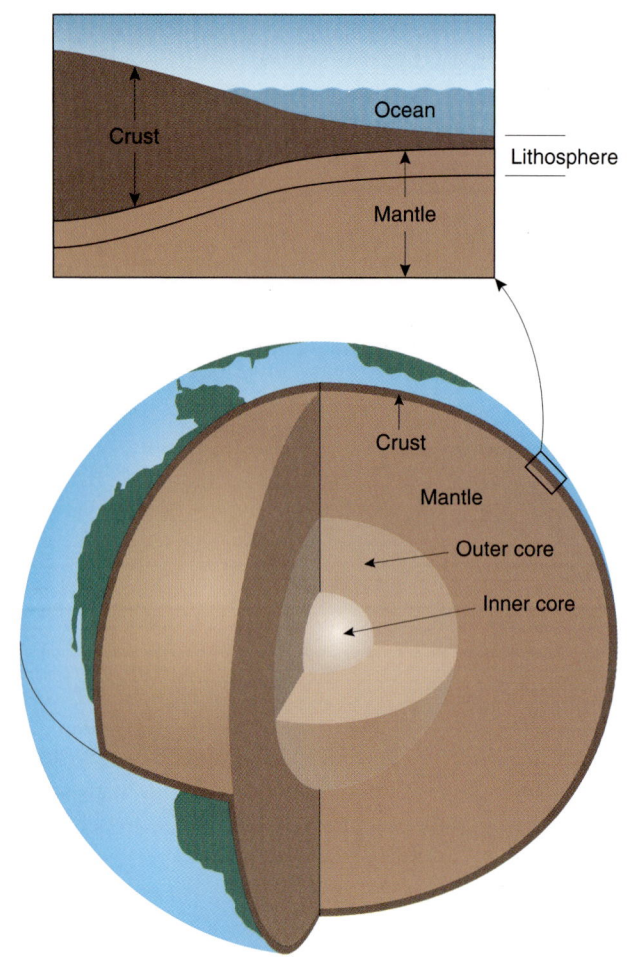

FIGURE 1.5
Earth's interior is divided into the crust, mantle, and outer and inner core. The lithosphere is the rigid upper portion of the mantle plus the overlying crust. (Drawing is not to scale.)

Plate tectonics probably has operated on the planet for at least 3 billion years, with continents periodically assembling into supercontinents and then splitting into smaller segments (Chapter 2). The last supercontinent, called *Pangaea* (Greek for "all land"), broke apart about 200 million years ago and its constituent landmasses, the continents of today, slowly moved to their present locations. Plate tectonics explains such seemingly strange discoveries as glacial sediments in the Sahara Desert and fossil coral reefs indicative of tropical climates in Wisconsin. Such discoveries reflect climatic conditions hundreds of millions of years ago when the continents were at different latitudes than they are today.

Geological processes occurring at boundaries between plates produce large-scale landscape and ocean bottom features, including mountain ranges, volcanoes, deep-sea trenches, and the ocean basins themselves. Enormous stresses develop at plate boundaries, bending and fracturing bedrock over broad areas. Hot molten rock material, known as **magma**, wells up from deep in the crust or upper mantle and migrates along rock fractures. Some magma pushes into the upper portion of the crust where it cools and solidifies into massive bodies of rock that form the core of mountain ranges (e.g., Sierra Nevada). Some magma feeds volcanoes or flows through fractures and spreads over Earth's surface as lava flows that cool and solidify (e.g., Columbia Plateau in the Pacific Northwest). At spreading plate boundaries on the sea floor, magma solidifies into new oceanic crust (Chapter 2). Plate tectonics and associated volcanism are important in geochemical cycling and release water vapor, carbon dioxide, and other gases to the atmosphere.

Volcanic activity is not confined to plate boundaries. Some volcanic activity occurs at great distances from plate boundaries and is due to so-called hot spots in the mantle. A **hot spot** is a long-lived source of magma caused by rising plumes of hot material originating deep in the mantle (mantle plumes). As a plate moves over a hot spot (assumed to be essentially stationary), magma may break through the crust and form a volcano. The big island of Hawaii is volcanically active because it sits over a hot spot located in the mantle under the Pacific plate. A hot spot underlying Yellowstone National Park is the source of heat for geyser eruptions (including Old Faithful).

BIOSPHERE

All the living organisms on Earth are components of the biosphere (Figure 1.6). Organisms range in size from microscopic, single-celled bacteria to the very largest plants and animals that ever lived (e.g., redwood trees and blue whales). Bacteria and other one-celled organisms dominate the biosphere, both on land and in the ocean. The

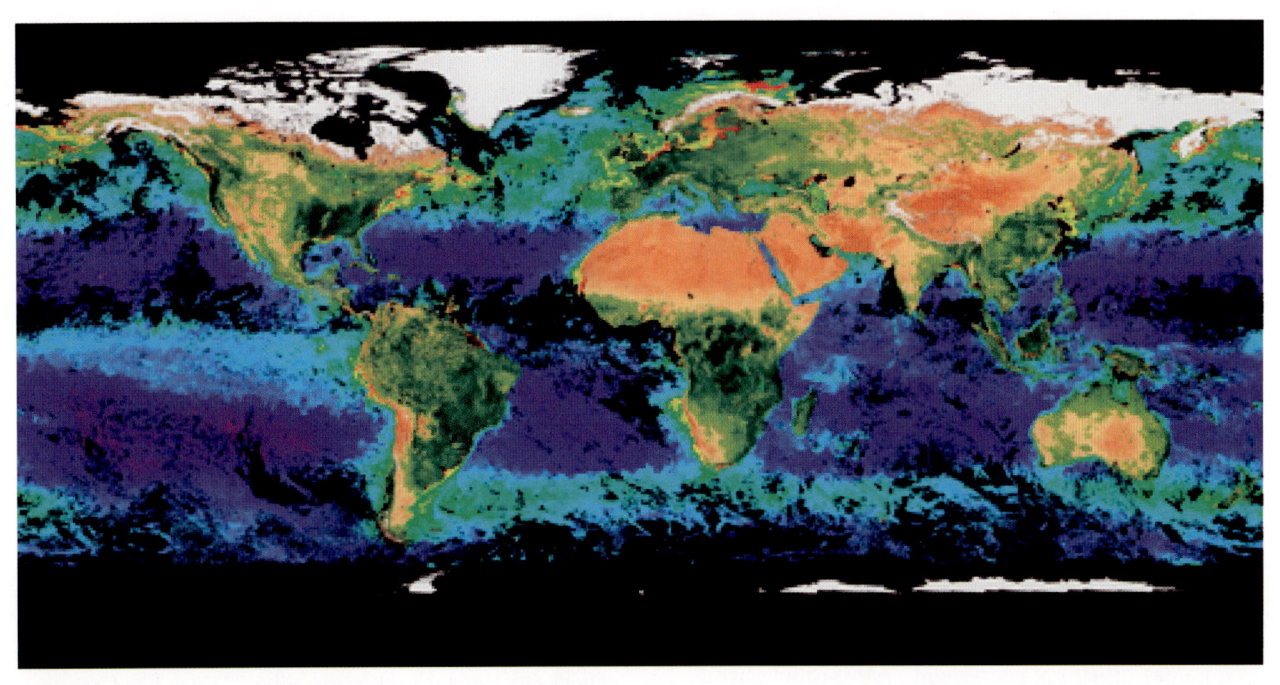

FIGURE 1.6
Earth's biosphere viewed by instruments flown aboard NASA's SeaWiFS (Sea-viewing Wide Field-of-view Sensor) on the SeaStar satellite launched in August 1997. Biological productivity is color-coded and highest where it is green and, and lowest where it is violet. White indicates snow and ice covered areas. [Provided by the SeaWiFS Project, NASA/Goddard Space Flight Center and ORBIMAGE.]

average animal in the ocean is the size of a mosquito. Large, multi-cellular organisms (including humans) are relatively rare on Earth.

Organisms on land or in the atmosphere live close to Earth's surface. However, marine organisms occur throughout the ocean's depths and even inhabit rock fractures, volcanic vents, and mud on the ocean floor. Certain organisms live in so-called *extreme environments* at temperatures and pressures previously thought impossible for life. In fact, some scientists estimate that the mass of organisms living in fractured rocks below Earth's solid surface may vastly exceed the mass of organisms living on or above it. Most of these deep-dwelling organisms remain a mystery.

Photosynthesis and cellular respiration are essential for life and exemplify the interaction of the biosphere with the other subsystems of the Earth system. **Photosynthesis** is the process whereby green plants use light energy from the sun to combine carbon dioxide from the atmosphere with water to produce sugars, a form of carbohydrate containing a relatively large amount of energy. (An important byproduct of photosynthesis is oxygen (O_2).) Animals that consume plants harvest some of that energy. Other animals consume those animals and energy progresses up the food chain. Through **cellular respiration**, an organism processes food and liberates energy in a form that can be used for maintenance, growth, and reproduction. Thereby, carbon dioxide, water, and heat energy are released (cycled) to the environment.

The primary source of energy for most living organisms is sunlight. However, for specialized organisms inhabiting the darkness of the deep ocean and rock fractures on the ocean floor, chemical energy rather than light drives biological processes. Through chemosynthesis, these marine organisms derive energy from substances such as hydrogen sulfide (H_2S) that originates in the Earth's interior (Chapter 9).

Dependency of organisms on one another (e.g., as a source of food) and on their physical and chemical environment (e.g., for water, oxygen, carbon dioxide and habitat) is embodied in the concept of ecosystem. The biosphere is composed of **ecosystems**, communities of plants and animals that interact with one another, together with the physical conditions and chemical substances in a specific geographical area. Deserts, tropical rain forests, tundra, estuaries, marshes, lakes, streams and coral reefs are examples of natural ecosystems. Most people live in highly modified terrestrial ecosystems such as cities, towns, farms, or ranches.

An ecosystem is home to producers (plants), consumers (animals), and decomposers (bacteria, fungi). *Producers* (also called *autotrophs* for "self-nourishing") form the base of most ecosystems. Using solar energy, their photosynthetic pigments form energy-rich carbohydrates. Consumers that depend directly or indirectly on plants for their food are called *heterotrophs*. Animals feeding directly on plants are called *herbivores*; those that prey on other animals are called *carnivores* and animals that consume both plants and animals are *omnivores*. After death, organisms are broken down by decomposers, usually bacteria and fungi, which cycle nutrients back to the environment.

Feeding relationships, called **food chains**, can be quite simple. In a hypothetical food chain on land, plants (primary producers) are eaten by deer (herbivores), which in turn are eaten by wolves (carnivores) or humans (omnivores). In a food chain, each stage is called a trophic level (or feeding level). On average, only about 10% of the energy available at one trophic level is transferred to the next higher trophic level. *Biomass*, the total weight or mass of organisms, is more readily measured than energy so that scientists describe the transfer of energy in food chains in terms of so many grams or kilograms of biomass. Thus 100 g of plants are required to produce 10 g of deer, which in turn produces 1 g of wolf. Simple food chains occur on land, in some lakes, but rarely in the ocean. Marine organisms usually eat many different kinds of food, and in turn, are eaten by a host of other consumers. These more complicated feeding relationships are called *food webs* (Figure 1.7).

An important ecosystem in the coastal zone is an estuary, where the biosphere, hydrosphere, and lithosphere interact in a special way. An **estuary** forms where fresh and salt water mix usually at the mouths of rivers and in tidal marshes and bays. Water in an estuary undergoes daily tidal oscillations (i.e., periodic rise and fall of sea level in response to the gravitational attraction of the moon and sun). Organisms living in an estuary are adapted to frequent fluctuations in currents, temperature, salinity, and concentrations of suspended sediment. Among the nation's best-known estuaries are Chesapeake Bay (largest in the U.S.), San Francisco Bay, and Puget Sound (Washington).

Estuaries are among the most productive ecosystems on the planet because of a special combination of biological and physical characteristics. They receive a continual supply of nutrients and organic matter delivered by both river water and ocean tides. Water circulates in an estuary in a way that traps and re-circulates nutrients and detritus (dead or partially decomposed remains of plants

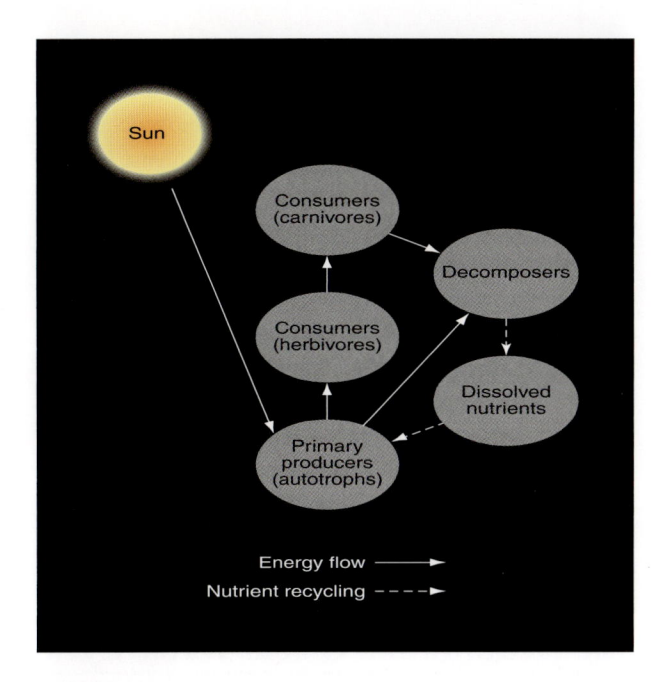

FIGURE 1.7
A simple marine food web in which solar radiation powers photosynthesis and energy is conveyed to successively higher trophic (feeding) levels.

and animals). These conditions support luxuriant plant growth (e.g., phytoplankton, marsh grass) as well as large populations of animals that feed on detritus. The abundant supply of food coupled with the unique environmental conditions of an estuary favor development and protection of juvenile fish. Low salinity and shallow water deter many ocean predators that would feed on juvenile fish. Estuaries provide favorable habitat for oysters, mussels, clams, lobsters, shrimp, and snails. These organisms, in turn, are an abundant food source for fish, birds, and humans. Estuaries are also home to juvenile anadromous fish (fish that swim from the sea to freshwater upriver to spawn) such as striped bass and salmon.

BIOGEOCHEMICAL CYCLES

Biogeochemical cycles are pathways along which solid, liquid, and gaseous materials flow among the various reservoirs of the Earth system. Accompanying the flow of materials is the movement and transformation of energy. Examples of biogeochemical cycles are the water cycle, carbon cycle, rock cycle, and oxygen cycle. The Earth system is an open (or flow-through) system for energy. (*Energy* is defined as the capacity for doing work.) The Earth system receives energy from the sun and from Earth's interior while emitting energy in the form of infrared radiation to space. Within the Earth system, energy is neither created nor destroyed although it is converted from one form to another. This is the **law of energy conservation** (also known as the *first law of thermodynamics*).

The Earth system is essentially closed for matter, that is, it neither gains nor loses matter over time (except for meteorites and asteroids). All biogeochemical cycles obey the *law of conservation of matter*, which states that matter can neither be created nor destroyed, but can change in chemical or physical form. When a log burns in a fireplace, a portion of the log is converted to ash (and heat energy), and the rest goes up the chimney as carbon dioxide, water vapor, creosote and heat. In terms of accountability, all losses from one reservoir in a cycle can be accounted for as gains in other reservoirs of the cycle. Stated succinctly, for any reservoir:

$$\text{Input} = \text{Output} + \text{Storage}$$

The amount of some substance within a biogeochemical reservoir depends on the rates at which the material is cycled into and out of the reservoir. *Cycling rate* is defined as the amount of material that moves from one reservoir to another within a specified period of time. If the input rate exceeds the output rate, the amount in the reservoir (storage) increases. If the input rate is less than the output rate, the amount decreases. Over the long term, the cycling rates of materials among the various global reservoirs have been relatively stable; that is, equilibrium tends to prevail between the rates of input and output.

Closely related to cycling rate is residence time. *Residence time* is the average length of time for a substance in a reservoir to be completely replaced, that is,

$$\text{Residence time} = \frac{\text{(amount in reservoir)}}{\text{(rate of addition or removal)}}$$

The residence time of a water molecule in the various reservoirs of the hydrosphere varies widely from only about 10 days in the atmosphere to tens of thousands of years or longer in glacial ice sheets. Residence time of dissolved constituents of seawater range from 100 years for aluminum to 260 million years for sodium.

Consider the global cycling of carbon as an illustration of a biogeochemical cycle. Photosynthesis cycles carbon dioxide (CO_2) from the atmosphere into green plants where carbon is incorporated into sugar ($C_6H_{12}O_6$). Plants use sugar to manufacture other organic compounds including fats, proteins, and other carbohydrates. As a byproduct of cellular respiration, plants and animals transform a portion of the carbon in these organic compounds back into carbon dioxide that is released to the atmosphere. The same cycle occurs in the ocean when carbon dioxide is cycled into and out of marine organisms through photosynthesis and respiration. In addition to the uptake of CO_2 in photosynthesis, marine organisms also use carbon for calcium carbonate ($CaCO_3$) to make hard, protective shells. Furthermore, decomposer organisms (e.g., bacteria) both on land and in the ocean act on the remains of dead plants and animals and through respiration release CO_2 to the atmosphere and ocean.

When marine organisms die their shells and skeletons settle to the ocean bottom. Over long periods of time, these organic remains accumulate, compress under their own weight and the weight of other sediments, and gradually transform into solid, carbonate rock. Common carbonate rocks are limestone ($CaCO_3$) and dolostone ($CaMgCO_3$). Subsequently, tectonic processes uplift these rocks and expose them to chemical and physical weathering processes. For example, atmospheric CO_2 dissolves in rainwater producing carbonic acid (H_2CO_3) that, in turn, dissolves carbonate rock (e.g., limestone) releasing CO_2. As part of the global water cycle, rivers and streams transport these weathering products to the sea where they settle out of suspension or precipitate as sediments that accumulate on the ocean floor. Over the millions of years that constitute geologic time, weathering and erosion of carbon-containing rocks have significantly altered the atmospheric concentration of CO_2.

About 280 to 345 million years ago, during the geologic time interval known as the Carboniferous Period, trillions of metric tons of organic remains (detritus) accumulated at the bottom of the ocean and in low-lying swampy terrain. The supply of detritus was so great that decomposer organisms could not keep pace. In some marine environments, plant and animal remains were converted to oil and natural gas. In swampy terrain, heat and pressure from accumulating organic debris concentrated carbon, converting the remains of luxuriant swamp forests into thick layers of coal. Today, when we burn coal, oil, and natural gas, collectively called *fossil fuels*, we are tapping energy that was originally locked in vegetation through photosynthesis hundreds of millions of years ago. During combustion, carbon from these fossil fuels combines with oxygen in the air to form carbon dioxide and water vapor, both of which escape to the atmosphere.

Another important biogeochemical cycle operating in the Earth system is the water cycle, which is closely linked to all other biogeochemical cycles. Reservoirs in the water cycle (ocean, atmosphere, lithosphere, living organisms) are also reservoirs in the other cycles. Furthermore, water is an important mode of transport in biogeochemical cycles. In the nitrogen cycle, for example, intense heating of air associated with lightning combines atmospheric nitrogen (N_2), oxygen (O_2), and moisture to form droplets of extremely dilute nitric acid (HNO_3) that are washed by rain to the soil. In the process, nitric acid converts to nitrate (NO_3^-), which can be taken up by plant roots and is an important plant nutrient. Plants convert nitrate to ammonia (NH_3), which is incorporated into a variety of compounds including amino acids, proteins, and DNA. On the other hand, both nitrate and ammonia readily dissolve in water so that heavy rains can deplete soil of these important nutrients and wash them into waterways.

The Ocean in the Global Water Cycle

We can reasonably assume that the total amount of water in the Earth system is neither increasing nor decreasing although natural processes continually generate and breakdown water. Water vapor accounts for perhaps half of all gases emitted during a volcanic eruption; at least some of this water was originally sequestered in magma and solid rock. Volcanic activity is more or less continuous on Earth and adds to the supply of water. Also, a minute amount of water is contributed to Earth by meteorites and other extraterrestrial debris continually bombarding the upper atmosphere. At the same time, intense solar radiation entering the upper atmosphere converts (*photodissociates*) a small amount of water vapor into its constituent hydrogen and oxygen atoms, which may escape to space. Also, water chemically reacts with other substances and is thereby locked up in various compounds. Annually, additions of water from volcanic eruptions roughly equal losses of water through photodissociation of water vapor and chemical reactions. This balance of give and take has prevailed on Earth for perhaps hundreds of millions of years.

As noted earlier, the fixed quantity of water in the Earth's system is distributed in its three phases among oceanic, terrestrial (land-based), atmospheric, and biospheric reservoirs. The ceaseless movement of water among its various reservoirs on a planetary scale is known as the **global water cycle** (Figure 1.8). In brief, water vaporizes from ocean and land to the atmosphere where winds can transport water vapor thousands of kilometers. Clouds form and rain, snow and other forms of precipitation fall from clouds to Earth's surface and recharge the ocean and the

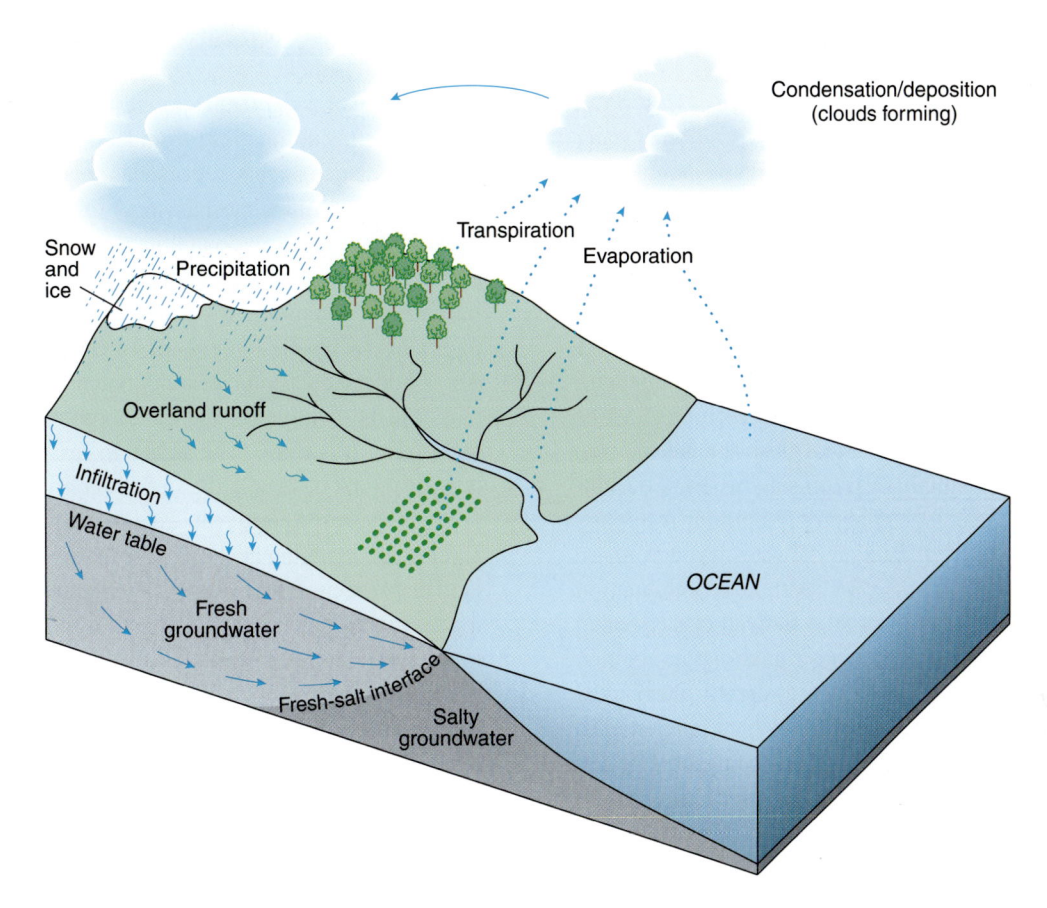

FIGURE 1.8
The global water cycle.

terrestrial reservoirs of water. From terrestrial reservoirs, water seeps and flows back to the ocean. The continuity of the global water cycle is captured in a verse from Ecclesiastes: "Every river flows into the sea, but the sea is not yet full. The waters return to where the rivers began and starts all over again."

Transfer of water between Earth's surface and the atmosphere is key to the existence of the global water cycle. Water cycles from Earth's surface to the atmosphere via evaporation, sublimation, and transpiration. Ultimately, solar radiation supplies the heat energy for these phase changes and powers the global water cycle. **Evaporation** is the process whereby water changes from a liquid to a vapor. Water evaporates from the surface of bodies of water including the ocean, lakes and rivers as well as from soil and the damp surfaces of plant leaves and stems. About 85% of the total annual evaporation in the Earth system takes place at the ocean surface so that the ocean is the principal source of water in the atmosphere. During **sublimation** water changes directly from a solid to a vapor without first becoming liquid. Through sublimation, snow banks shrink and patches of ice on sidewalks and roads disappear even while the air temperature remains below freezing.

Transpiration is the process whereby water that is taken up from the soil by plant roots eventually escapes as vapor through the tiny pores (called stomates) on the surface of green leaves. On land during the growing season, transpiration is considerable and is often more important than direct evaporation of water in delivering water vapor to the atmosphere. A single hectare (2.5 acres) of corn typically transpires 34,000 liters (8800 gallons) of water per day. Measurements of direct evaporation from Earth's surface plus transpiration are often combined as *evapotranspiration*.

Water cycles from the atmosphere to land and ocean via condensation, deposition, and precipitation. **Condensation** is the process whereby water changes phase from vapor to liquid (in the form of small droplets). Water droplets forming on the outside surface of a cold can of soda on a warm and humid day is an example of condensation of water vapor. **Deposition** is the process whereby water changes directly from vapor to solid (ice crystals) without first becoming a liquid. Appearance of frost on an automobile windshield is an example of deposition of water vapor. Condensation or deposition within the atmosphere produces clouds. **Precipitation** is water in frozen or unfrozen forms (e.g., rain, drizzle, snow, ice pellets, hail) that falls from clouds and reaches Earth's surface.

Significantly, evaporation (or sublimation) followed by condensation (or deposition) purifies water. As water vaporizes (evaporation, transpiration, or sublimation), all suspended and dissolved substances such as sea salts are left behind. Through this natural cleansing mechanism, ocean water that originally was too salty to drink eventually falls as freshwater precipitation, replenishing reservoirs on Earth's surface. Purification of water through phase changes is known as *distillation*. Although precipitation is fresh water, it is not free of all dissolved or suspended materials. Condensation and deposition take place on tiny aerosols suspended in the atmosphere, slightly altering the chemical composition of cloud and precipitation particles. Also, as precipitation falls through the atmosphere, it dissolves or captures gases or suspended particles that also change its composition (e.g., acid rain).

Return of water from the atmosphere to the land and ocean via condensation, deposition, and precipitation completes an essential subcycle in the global water cycle. To learn more about the atmosphere-Earth's surface subcycle of the global water cycle, we compare the movement of water between the continents and the atmosphere with that between the ocean and atmosphere. The balance sheet for inputs and outputs of water to and from the various global reservoirs is called the *global water budget* (Table 1.2).

Over the course of a year, the volume of precipitation (rain plus melted snow) that falls on land exceeds the total volume of water that vaporizes (via evaporation, transpiration, and sublimation) from land by about one-third. (This imbalance occurs because landmasses favor certain precipitation-forming mechanisms.) Over the same period, the volume of precipitation falling on the ocean is less than the volume of water that evaporates from the ocean. (Evaporation is more important because the ocean surface supplies a nearly limitless supply of water vapor.) The global water budget thus indicates an annual net gain of water mass on the continents and an annual net loss of water mass from the ocean. The annual excess of water on the continents about equals the deficit from the ocean. But year after year the continents do not get any wetter and sea level is not changing appreciably because excess water on land drips, seeps, and flows back to the sea, thus completing the global water cycle. The net flow of water from land to sea also implies a net flow of water in the atmosphere from sea to land.

The flow of excess surface and subsurface water from the continents to the ocean has important implications for the composition of seawater and the global distribution of water pollutants. The ocean is the ultimate destination for all substances dissolved or suspended in surface or subsurface waters. This is one (but not the only) source of salt in seawater. The ocean is also the ultimate destination of contaminants dumped into waterways (e.g., rivers, streams, canals). These contaminants primarily end up in estuaries and other coastal waters, which are also the most productive portion of the ocean. Not surprisingly, considerable debate centers on the capacity of the ocean to assimilate waste—especially toxic and hazardous industrial waste.

TABLE 1.2
Global Water Budget

Source	Cubic meters per yr	Gallons per yr
Precipitation on the ocean	$+3.24 \times 10^{14}$	$+85.5 \times 10^{15}$
Evaporation from the ocean	-3.60×10^{14}	-95.2×10^{15}
Net loss from the ocean	-0.36×10^{14}	-09.7×10^{15}
Precipitation on land	$+0.98 \times 10^{14}$	$+26.1 \times 10^{15}$
Evapotranspiration from land	-0.62×10^{14}	-16.4×10^{15}
Net gain on land	$+0.36 \times 10^{14}$	$+09.7 \times 10^{15}$

Precipitation that reaches the land surface can follow one of several pathways. Some precipitated water vaporizes (evaporates or sublimates) directly back into the atmosphere and some is temporarily stored in lakes, snow and ice fields, or glaciers. Some either flows on the surface as rivers or streams (*runoff component*) or seeps into the ground as soil moisture or groundwater (*infiltration component*). The ratio of the portion of water that infiltrates the ground to the portion that runs off depends on rainfall intensity, vegetation, topography, and physical properties of the land surface. For example, rain falling on frozen ground or city streets mostly runs off whereas rain falling on unfrozen sandy soil readily soaks into the ground.

During the growing season, plant roots take up some soil moisture and almost all of that water (98%) transpires back to the atmosphere. Water that seeps to greater depths may completely fill the spaces between particles of soil or sediment or the fractures within rock, constituting *groundwater*. In most environments, groundwater flows very slowly in the subsurface from recharge areas at Earth's surface toward discharge areas including wells, springs, rivers, lakes, and the ocean. Where the groundwater reservoir is located well below surface water reservoirs, water may seep out of rivers and lakes and recharge the groundwater reservoir.

Rivers and streams plus their tributaries drain a fixed geographical area known as a *drainage basin* (or watershed). The quantity and quality of water flowing in a river depends on the climate, topography, geology, and land use in the drainage basin. A drainage basin may also include lakes or glaciers, temporary impoundments of surface water.

Observing the Ocean

Scientists observe the ocean not only out of curiosity but also to better understand the ocean's role in the Earth system. Scientists want to know, for example, the physical and biological factors governing the distribution of marine life and how the ocean influences climate variability. Answers to these and other questions require observational data on the ocean's properties and processes. One important source of such data is Earth-orbiting satellites. Earth-orbiting satellites now routinely monitor the ocean's surface waters, the atmosphere, and other components of the Earth system. Powerful computers collect, process, and analyze enormous quantities of environmental data. In only minutes, the instruments on an ocean-observing satellite collect as much data as an ocean-research vessel operating at sea continuously for a decade or longer. While data collected via space platforms are extremely valuable, oceanographic ships and instrumented buoys are still essential for surveys and field experiments. A great deal of ocean research has been and is still done from ships. It is the only way to get out on and stay at sea to investigate what cannot be detected remotely. As we will see in subsequent chapters, a variety of instruments and techniques have contributed significantly to our understanding of the ocean.

SPACE-BASED OBSERVATIONS

Instruments observing Earth from orbiting spacecraft measure selected wavelengths (or frequencies) of **electromagnetic radiation** reflected or emitted by the various components of the Earth system. Electromagnetic radiation (or simply *radiation*) describes both a form of energy and a means of energy transfer. Forms of radiation include gamma rays, X-rays, ultraviolet (UV) radiation, visible light, infrared (IR) radiation, microwaves, and radio waves. Together, these forms of radiation make up the **electromagnetic spectrum** (Figure 1.9). All types of radiation travel as waves at the speed of light and the different segments of the electromagnetic spectrum are differentiated by wavelength or frequency. *Wavelength* is the distance between successive wave crests or troughs whereas *frequency* is the number of wave crests or troughs passing a point during a period of time, usually a second (Figure 1.10).

The clear atmosphere is essentially transparent to visible light and is selectively transparent (by wavelength) to other types of radiation such as infrared. Satellite-borne sensors monitor these forms of radiation to gather information on atmospheric processes and properties of Earth's ocean and surface. Ocean water is much less transparent to electromagnetic radiation than is the atmosphere so that remote sensing by satellite is essentially limited to obtaining data on surface or near-surface ocean properties.

The earliest ocean observations from space came from sensors on meteorological satellites orbited in the 1960s. Beginning in 1978, special satellites were launched specifically to monitor the ocean. Now instruments flown on Earth-orbiting satellites routinely provide global images of ocean conditions, which are typically updated every few days. Satellite-borne instruments fall in one of two categories: passive or active. Passive instruments include those that measure radiation emanating from the ocean surface, that is, visible solar radiation reflected by the ocean surface and invisible infrared radiation emitted by the ocean surface. Among the ocean properties measured by passive

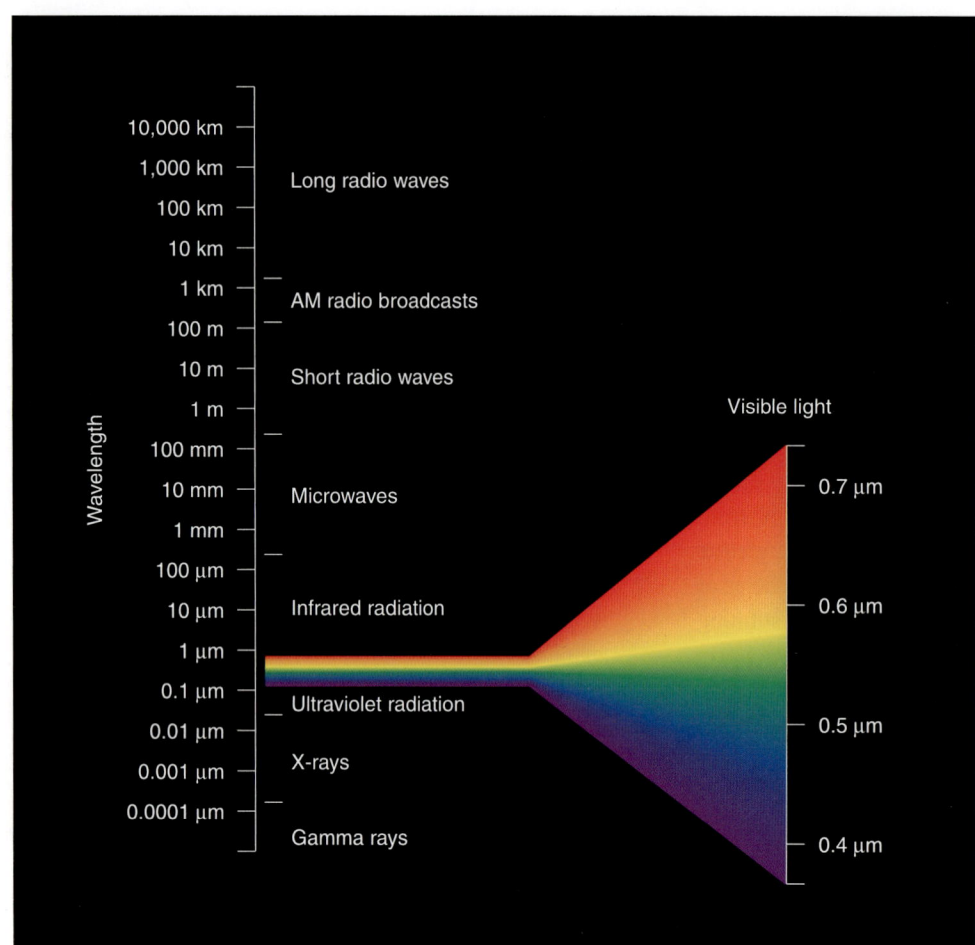

FIGURE 1.9
The electromagnetic spectrum. The various forms of electromagnetic radiation are distinguished by wavelength in micrometers (µm), millimeters (mm), meters (m), and kilometers (km).

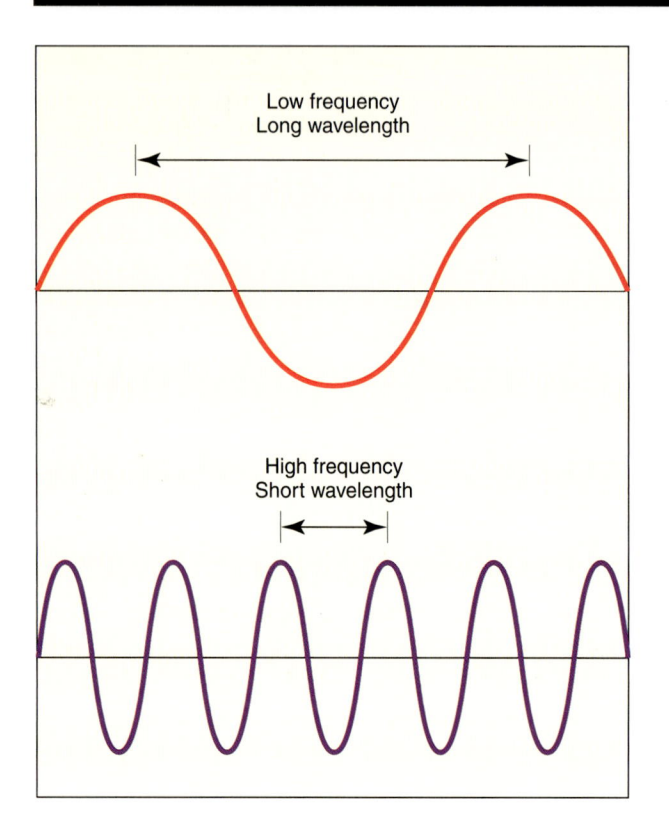

FIGURE 1.10
Wavelength is the distance between two successive crests or, equivalently, the distance between two successive troughs. Wavelength is inversely related to wave frequency.

radiation sensors are sea-surface temperature (SST) and water color, a measure of marine productivity or sediment concentration. Radar instruments mounted on satellites are active sensors; they emit pulses of microwave radiation and then record the reflected signal to measure surface roughness (an indicator of surface wind speeds and wave heights) as well as ocean-surface elevations, which are used to map bottom topography and surface currents. Techniques to measure salinity and other ocean properties remotely are under development. Such measurements permit studies of ocean surface "weather" (week-to-week variations) as well as ocean "climate" (variability over decades to centuries).

Satellites that monitor the ocean are in either geostationary or polar orbits. We are probably most familiar with images provided by geostationary satellites, which are launched into relatively high orbits (36,000 km or 22,300 mi) (Figure 1.11). A **geostationary satellite** revolves around Earth at the same rate and in the same direction as the planet rotates so that the satellite always remains over the same point on the equator and its sensors monitor the same portion of Earth's surface. For this reason, these satellites are sometimes described as geosynchronous. Each geostationary satellite views about one-third of Earth's surface and five satellites are needed to provide complete and overlapping coverage of the globe between about 60 degrees N and 60 degrees S. Considerable distortion sets in poleward of about 60 degrees latitude.

Polar-orbiting satellites are in low-altitude (800 to 1000 km, 500 to 600 mi) orbits that pass near the north and south poles (Figure 1.12). Earth rotates eastward under the satellite whose orbit remains fixed in space so that sensors monitor successive strips of Earth. A polar-orbiting satellite that follows the sun (sun-synchronous) passes over the same area twice each 24-hr day. Other polar-orbiting satellites are positioned so that they require several days before passing over the same point on Earth.

The flood of data from satellite-based instruments is analyzed, stored, and manipulated by computers. Computer speed doubles about every 18 months so that, ordinary laptop computers, now widely and cheaply available, far exceed the power of the most powerful, room-sized supercomputers available only a few decades ago. Web sites now make readily available details of ocean surface properties that were impossible to obtain only a decade ago.

An example of the utility of remotely sensed data is the determination of the distribution and abundance of chlorophyll (plant pigments). Satellite-derived images of "ocean color" locate much of the "food" used in marine

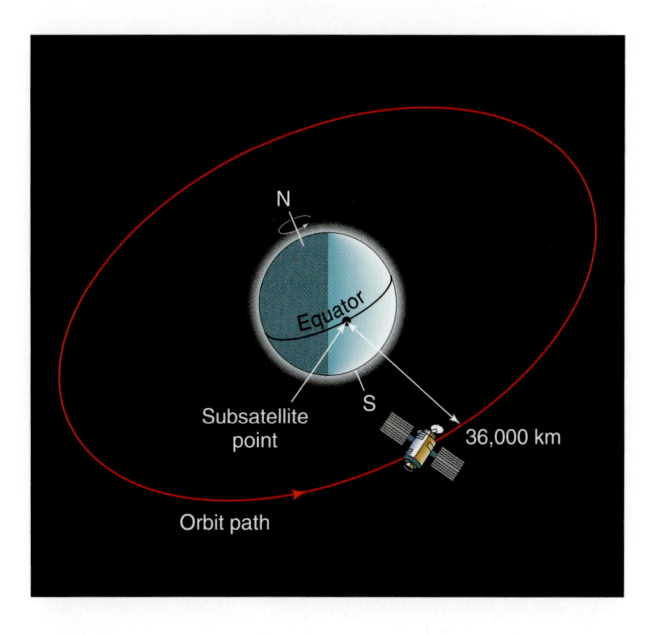

FIGURE 1.11
A satellite in geostationary orbit about the Earth.

ecosystems and indicate how its concentration varies with time (Chapter 9). (These same sensors are also used to map the distribution of vegetation on land.) Thus large areas of Earth's marine ecosystems can be routinely studied without risking injury or seasickness onboard research ships. (Refer back to Figure 1.6 for a sample image.)

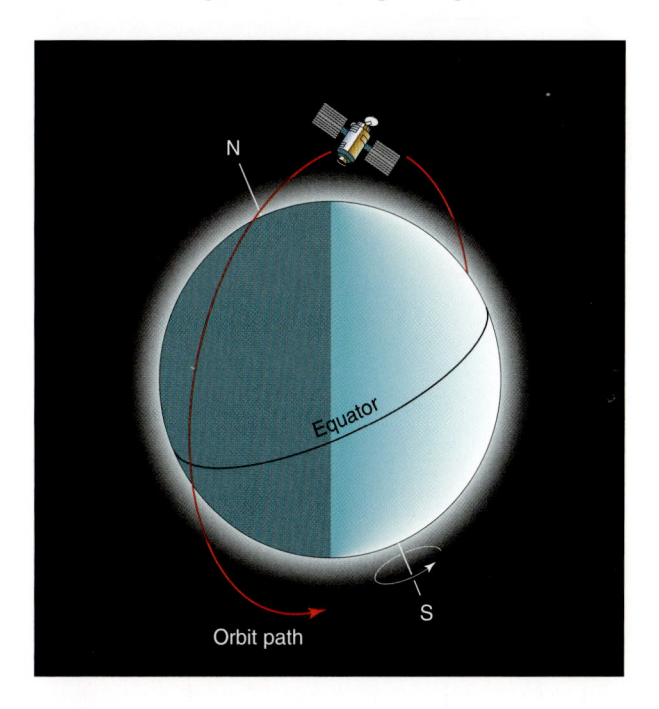

FIGURE 1.12
A satellite in polar orbit about the Earth.

Another example of remotely sensed data used in ocean studies is sea-surface temperature (SST) patterns obtained by infrared (IR) sensors onboard satellites. The intensity of radiation emitted by an object increases rapidly as the surface temperature of the object rises. By calibrating temperature against IR-emission, a satellite sensor measures SST and can distinguish between warm and cold ocean currents. For example, Figure 1.13 is an IR satellite image of the western North Atlantic Ocean. SST is color coded so that oranges and reds represent the highest temperatures. In this depiction the warm Gulf Stream is clearly visible as it parallels the coast from Florida north to the Carolinas and then turns northeastward away from the coast.

Studies of the ocean in the Earth system also require satellite-based communications systems. With communications satellites, scientists working onboard research ships at sea can communicate nearly instantaneously with labs, computers or colleagues anywhere in the world. They can also work with schools to provide a virtual shipboard experience for students in their classroom. Remotely operated instruments on drifting buoys, or on unmanned autonomous underwater vehicles, can send the data they collect to scientists on shore via communications satellites. These communications capabilities are now routinely combined as one system for observing both the ocean and atmosphere.

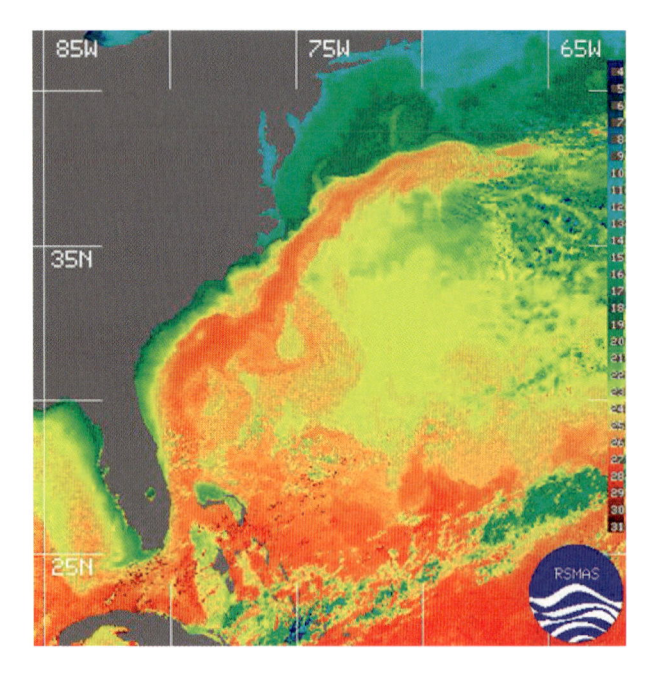

FIGURE 1.13
Satellite-derived sea-surface temperatures (SSTs) over the western North Atlantic Ocean; highest temperatures are shown in orange and red. Dark blue and purple patches represent cloud cover. [Source: NOAA]

MONITORING THE OCEAN'S DEPTHS

Satellite-borne monitoring instruments, such as radar, can directly observe only the ocean's surface because water is nearly opaque to electromagnetic radiation. Therefore, other techniques are necessary to examine the ocean's interior. Instruments lowered from ships are used to sample ocean water at depth. And instruments moored on the sea floor monitor properties of the ocean beneath the surface. One technique uses the speed of sound waves in water to measure ocean-water temperature and its variation through time. Ocean water is highly transparent to sound, just as the atmosphere is nearly transparent to visible light. A sound pulse from a transmitter on one side of the ocean can be detected by a sensitive receiver thousands of kilometers away on the other side of the ocean. Whales, for example, take advantage of this property of seawater to communicate long distances across ocean basins, to attract mates and to locate food.

Sound travels faster in warm water than in cold water. Thus measuring the speed of sound in seawater can be used to determine the average water temperature between the transmitter and the receiver. Furthermore, variations in sound speed between many transmitters and receivers can be combined to obtain a three-dimensional representation of seawater temperature. This technique, called *acoustic tomography*, is similar in concept to the X-ray scans used to examine the internal organs of humans. We have more to say about acoustic tomography in Chapter 3.

A **float** is an instrument package that obtains vertical profiles of water temperature and salinity and is used to study deep-ocean currents. Floats are designed to submerge to a preset depth, typically about 2000 m (6600 ft), where they drift with the current. After a pre-determined time, the float returns to the surface, collecting temperature and conductivity (salinity) data while rising. At the surface, the float radios its data and location to a satellite, which in turn transmits this information to labs ashore for analysis and storage. After uploading its data to the satellites, the float submerges and begins the cycle again. By the end of 2005, some 3000 floats will be deployed throughout the ocean. This approach is greatly expanding our knowledge of the mean state of the upper two kilometers of the ocean.

Modeling the Ocean

A model is an approximate representation or simulation of a real system. It incorporates only the essential features (or variables) of a system while omitting details considered non-essential. Models have been widely used in investigating the Earth system and its components, including the ocean. Models are conceptual, physical, or numerical.

A *conceptual model* is a statement of a fundamental law or relationship. Such a model is used to organize data or describe the interaction between components of a system. For example, conceptual models are used to explore linkages among physical and biological subsystems in the ocean and they enable us to understand why and how ocean water circulates. We use conceptual models throughout this book.

A *physical model* is a small-scale (miniaturized) representation of a system. For example, engineers may assess the aerodynamic properties of a proposed new aircraft by testing a scale model in a wind tunnel. Before powerful electronic computers were widely available, physical models were used to study the flow of water in river channels, harbors, and estuaries, such as Puget Sound in Washington. Physical models were used to predict floods on the Mississippi River or changes in currents that would result from deepening the shipping channels leading into New York Harbor. Although the results of physical modeling were generally useful and easily understood, the models were expensive to build, maintain and operate.

Today, because of the availability of powerful electronic computers, numerical models have essentially replaced physical models. Weather forecasts are probably the most familiar products of numerical models; in fact, numerical models of the Earth-atmosphere system have been used to forecast weather since the 1950s. A *numerical model* consists of one or more mathematical statements that approximate the process(es) under study. Observational data are necessary for use as initial and boundary conditions as well as to guide and verify model predictions. Very powerful computers are needed to handle the complex mathematics and huge data sets required for ocean, atmosphere, or coupled ocean-atmosphere numerical models. A single set of complex calculations may require hours to a few days to run. Equations may be altered or new data provided to simulate different situations. For example, a numerical model of New York Harbor is used to predict movements of water and sediment in response to changing wind patterns. Numerical models are powerful tools with which to study the response of systems to long-term environmental change.

A model is an approximation of a real system and there may be advantages to comparing different models and model runs. Climate change studies, for example, routinely compare the output of multiple numerical models.

Interacting atmospheric and oceanic processes are simulated using coupled models of the ocean, atmosphere, and biosphere for predicting responses to global climate change. Numerical models can also be used to simulate currents in ancient ocean basins based on reconstructions of global climate and the shapes of ocean basins at the time. Time scales for these models of ancient oceans can be millions of years.

Human Impact on the Ocean

Humans have altered the Earth system to such an extent that we must now consider human activities as equal in magnitude to many geological processes. Human activities impact the orderly functioning of the ocean (and other components of the Earth system) by altering cycling rates and disrupting the equilibrium of biogeochemical cycles. Agriculture, mining, and electrical power generation are examples of widespread human activities that affect the ocean and other components of the Earth system. Most of the waste byproducts of these activities are widely distributed by winds and/or rivers that discharge into the ocean. Consider an example.

Runoff from agricultural activities has impacted the ocean, especially in coastal areas. Nitrogen compounds (e.g., nitrates), essential for terrestrial and marine plant growth, are manufactured as fertilizer, and routinely applied to agricultural fields in spring and summer. Excessive nitrogen runoff from fields enters rivers, and eventually reaches bays and coastal waters causing major pollution problems. Excess nitrogen stimulates explosive growth of algal populations. While alive, these algae shade out sea grasses growing on the bottom, which soon die, depriving other organisms of habitat for reproduction and growth. After they die, algae settle to the bottom and decompose, depleting the oxygen dissolved in the near-bottom waters. The result is large volumes of water that can support only bacteria that grow in the absence of oxygen (anaerobic bacteria). Most commercially valuable fish and shellfish (e.g., crabs, oysters) die.

Each summer a large (New Jersey-size) area of the ocean on the continental shelf off the mouth of the Mississippi River in the Gulf of Mexico becomes devoid of dissolved oxygen (Figure 1.14). Seasonal changes in weather cause replenishment of dissolved oxygen by late fall. In late 2001, G.F. McIsaac, an environmental scientist at the University of Illinois at Urbana-Champaign, reported that this "dead zone" in the Gulf of Mexico could be alleviated

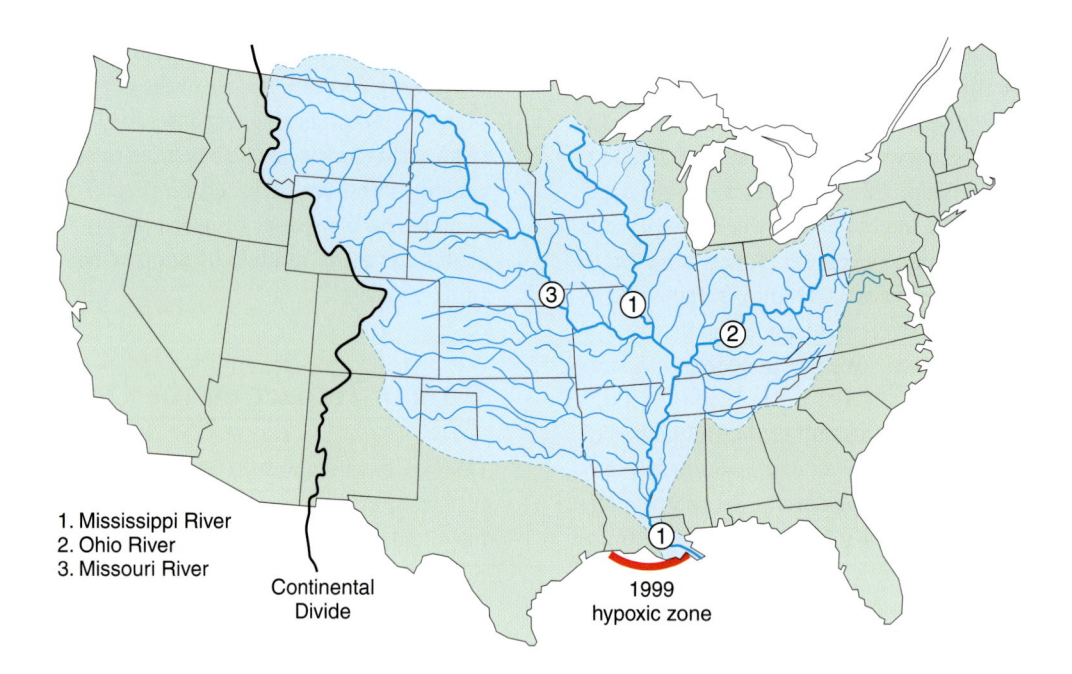

FIGURE 1.14
By mid-summer, a large volume of water significantly depleted in dissolved oxygen develops in the Gulf of Mexico just offshore of Louisiana. These hypoxic conditions eliminate many bottom-dwelling organisms (hence the name "dead zone") and are caused by excessive amounts of nitrogen fertilizer washed from agricultural land in the Mississippi River drainage basin. [Courtesy of the U.S. Geological Survey]

significantly if farmers in the Mississippi River drainage basin cut their application of nitrogen fertilizer by 12%. McIsaac argued that such a reduction would not necessarily reduce crop yields but would decrease the amount of nitrates reaching the Gulf of Mexico by one-third by the year 2010.

Conclusions

The Earth system consists of several subsystems: hydrosphere (including the cryosphere), atmosphere, geosphere, and biosphere. These subsystems interact globally over various time scales ranging from seconds to hundreds of millions of years. Interactions adhere to fundamental principles, such as the laws of physics, chemistry, geology and biology. Matter and energy cycle through these systems in a variety of global biogeochemical cycles. Matter cycles in closed systems that obey the law of conservation of matter whereas energy cycles in open systems, as the Earth system continually receives energy from the sun and from Earth's interior while emitting infrared radiation to space.

Application of natural laws combined with monitoring of the system functioning permits subsystem behavior to be understood, modeled and future conditions to be predicted. Each subsystem is studied using a variety of observation techniques. Remote sensing by satellite of Earth's subsystems has greatly improved our ability to predict their behavior as a whole and on global scales over long periods of time. The huge quantity of environmental data that is collected via Earth-orbiting satellites is fed into computer models that generate forecasts of the future state of components of the Earth system.

In the next chapter, we consider plate tectonics with special emphasis on the ocean basin and its interaction with ocean waters.

Basic Understandings

- A system is an interacting set of components that behave in an orderly way according to the principles of physics, chemistry, geology, and biology. The Earth system consists of four major interacting subsystems: hydrosphere (including the cryosphere), atmosphere, geosphere, and biosphere.
- The hydrosphere encompasses water in all three phases (ice, liquid, and vapor) that continually cycles from one reservoir to another in the Earth system. The ocean is the largest reservoir in the hydrosphere, con-

- taining 97.2% of all water on the planet and covering 70.8% of Earth's surface. The next largest reservoirs of water in the hydrosphere are the glacial ice sheets that cover much of Antarctica and Greenland.
- The hydrosphere is dynamic with water flowing at different rates through and between different reservoirs. The time required for water to reach the ocean varies from days to weeks in river channels to hundreds of thousands of years for water locked in ice sheets. The ocean features wind-driven surface currents and density-driven deep-ocean currents caused by small differences in temperature and salinity.
- In addition to the Antarctic and Greenland ice sheets, the frozen portion of Earth's hydrosphere (called the cryosphere) encompasses mountain glaciers, permafrost, sea ice (frozen sea water), and bergs.
- Earth's atmosphere is a relatively thin envelope of gases and tiny suspended solid and liquid particles (aerosols) surrounding the planet. Based on the average vertical temperature profile, the atmosphere is divided into the troposphere, stratosphere, mesosphere, and thermosphere. The lowest layer, the troposphere, is where most weather occurs and where the atmosphere interfaces with the other subsystems of the Earth system.
- Nitrogen (N_2) and oxygen (O_2), the principal atmospheric gases, are mixed in uniform proportions up to an altitude of about 80 km (50 mi). Not counting water vapor (which has a highly variable concentration), nitrogen occupies 78.08% by volume of the lower atmosphere and oxygen is 20.95% by volume.
- The significance of atmospheric gases and aerosols is not necessarily related to their concentration. Some atmospheric components that are essential for life occur in very low concentrations. Examples are water vapor (needed for the water cycle), carbon dioxide (for photosynthesis), and stratospheric ozone (protection from ultraviolet radiation). The atmosphere is dynamic and circulates in response to different rates of radiational heating and cooling.
- The geosphere is the solid portion of the planet composed of rocks, minerals, and sediments. The rigid uppermost portion of Earth's mantle plus the overlying crust constitutes Earth's lithosphere. Surface geological processes (i.e., weathering and erosion) and internal geological processes (i.e., mountain building, volcanic eruptions) continually modify the lithosphere. Weathering refers to the physical and chemical breakdown of rock into sediments. Erosive agents (i.e., rivers, glaciers, wind) remove, transport, and subsequently deposit sediments.
- The biosphere encompasses all life on Earth and is dominated on land and in the ocean by bacteria and one-celled plants and animals. Photosynthesis and cellular respiration are essential for life and exemplify the interaction of the biosphere with the other subsystems of the Earth system. The biosphere is composed of ecosystems, communities of plants and animals that interact with one another, together with the physical conditions and chemical substances in a specific geographical area. Ecosystems consist of producers (plants), consumers (animals), and decomposers (bacteria, fungi). These organisms occupy trophic levels in food chains.
- Biogeochemical cycles are pathways along which solid, liquid, or gaseous materials flow among the various reservoirs within subsystems of the Earth system. Biogeochemical cycles follow the law of conservation of matter, which states that matter can neither be created nor destroyed, but can change chemical or physical form. The time required for a unit mass of some substance to cycle into and out of a reservoir is the residence time of the substance in the reservoir.
- Powered by sunlight, the ceaseless movement of water among the various reservoirs in the hydrosphere on a planetary scale is known as the global water cycle. The ocean is by far the largest reservoir in the global water cycle. Key to the existence of the global water cycle is the transfer of water between the Earth's surface and atmosphere. Water cycles into the atmosphere from the continents and ocean via evaporation, sublimation, and transpiration. Water cycles from the atmosphere to land and sea via condensation, deposition, and precipitation. To balance the global water budget, the net flow of water is from land to ocean.
- Satellites orbiting the planet are platforms for sensors that monitor the Earth system, measuring the flux of electromagnetic radiation reflected or emitted by the system. Electromagnetic radiation is both a form of energy and a means of energy transfer. The various forms of radiation are distinguished on the basis of wavelength and frequency.
- Passive and active remote sensing by satellite is a valuable tool for observing the ocean and other components of the Earth system. A geostationary satellite orbits the planet at a rate that matches Earth's rotation and in the same direction so that it is always positioned above the same spot on Earth's equator. A po-

- lar-orbiting satellite travels along relatively low north-south trajectories that take it over polar areas.
- Satellite-based sensors can directly observe only the ocean's surface waters so that other techniques such as acoustic tomography and instrumented floats, ships, and buoys are used to monitor ocean properties at depth.
- A model is a simulation of a real system and includes only those variables considered to be essential to the system. Conceptual, physical, and numerical models have been used to investigate the Earth system and its ocean. Numerical models are increasingly important in studying the ocean; they consist of one or more mathematical equations that describe the relationship among variables in a system and are programmed into an electronic computer.
- Human activities impact the orderly functioning of the ocean (and other components of the Earth system) by altering cycling rates and disrupting the equilibrium of biogeochemical cycles. Often such impacts are the unintended consequences of other activities.

Chapter 1 OCEAN IN THE EARTH SYSTEM 23

ESSAY: What Killed the Dinosaurs?

Large asteroids striking Earth have repeatedly caused mass extinctions of life, in the ocean and on the continents. Such extinctions have occurred on Earth every few tens of millions to hundreds of millions of years. Some scientists believe the potential for a large asteroid hitting Earth in the centuries ahead is real and could devastate modern society. A well-preserved, small asteroid crater in Arizona is shown in Figure 1.

FIGURE 1
The Barringer Crater, Arizona, formed by a meteorite striking Earth about 49,000 years ago, is more than 1 km (0.6 mi) across. [Courtesy of NASA]

The most recent and best-known large asteroid impact on Earth occurred about 65 million years ago and produced a 180-km (112-mi) wide crater on the floor of the ancient Caribbean Sea (Figure 2). Marine sediments gradually filled the crater and geological forces later elevated a portion of the crater above sea level. Today, what remains of the Chicxulub crater forms part of Mexico's Yucatán peninsula. Radar images obtained by the space shuttle Endeavour in 2000 revealed a 5-m (16-ft) deep, 5-km (3-mi) wide trough on the Yucatán peninsula that may mark the outer rim of the crater. Drilling through the layers of sediment on the floor of the nearby Gulf of Mexico has recovered cores of fractured and melted rock from the impact zone.

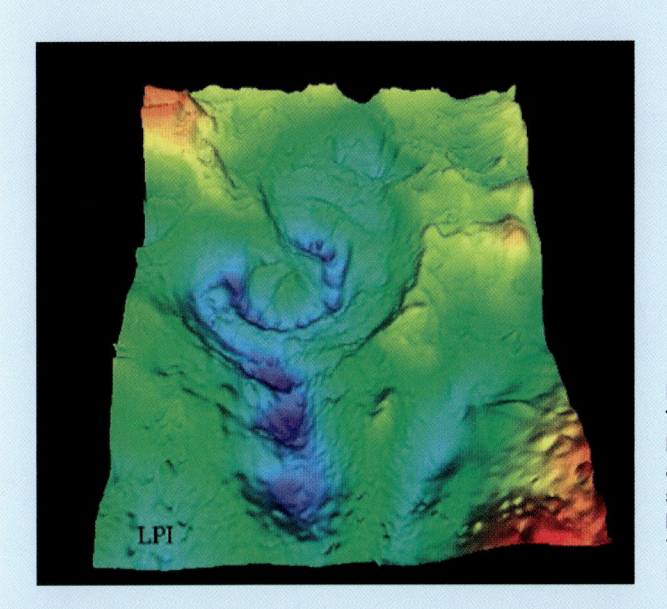

FIGURE 2
The Chicxulub Crater, Yucatan Peninsula, Mexico, is about 180 km (112 mi) across. It formed about 65 million years ago when a mountain-size asteroid (10 to 20 km or 6 to 12 mi across) struck Earth's surface. The effects of the impact are thought to be responsible for the extinction of the dinosaurs and about 70% of all species then present on the planet. [Courtesy NASA, Lunar Planetary Institute, V.L. Sharpton]

Other evidence of the impact consists of bits of glassy rock, which are chilled droplets of molten rock blasted into the atmosphere by the impact. These droplets cooled as they fell through the atmosphere and onto the land or into the ocean. They were recovered from nearby deep-ocean sediments (Chapter 4). Many rocks on land contain mineral grains deformed by the extreme heat and pressure from the impact. Other unusual sediment deposits were produced by enormous waves (tsunamis) generated when the asteroid (about 10 km or 6 mi in diameter) impacted the ocean surface.

The effects of the asteroid impact on life on Earth were catastrophic. Best known is the extinction of the dinosaurs, which had dominated Earth for more than 250 million years. Dinosaurs were not the only victims, however. The asteroid impact destroyed more than 70% of the other life forms then existing on the planet. Small mammals evolved rapidly to take the place of the dinosaurs. The impact also caused major extinctions among many groups of marine organisms.

What precisely caused the demise of the dinosaurs? One widely accepted theory is that the asteroid impact vaporized large amounts of sulfur-containing deep-sea sediments. This sulfur was blown into the atmosphere where it generated enormous clouds of tiny sulfate particles. These clouds prevented sunlight from reaching Earth's surface for long periods and most plants died because they could not photosynthesize. Dinosaurs dependent on eating plants starved and the carnivores that fed on them soon met the same fate. Only small animals (like mammals) could survive by eating the dead plants and animals until the sulfate particles eventually settled out of the atmosphere so that photosynthesis could resume and new plants replaced those lost after the impact. Another theory is that red-hot, impact-generated particles rained through the atmosphere making it so hot that most plants and animals were killed directly.

ESSAY: Origin of the Ocean and Earth System

Earth is called the water planet because ocean waters currently cover almost 71% of its surface. From a different perspective, Earth is relatively dry in that ocean water accounts for only about 0.02% of the planet's total mass. Yet in view of how the solar system is believed to have formed, it is surprising that even that much water is present on Earth. Where did the water come from?

According to astronomers, Earth, the sun, and the entire solar system evolved from an immense rotating cloud of dust, ice, and gases (called a *nebula*) more than 4.5 billion years ago. Variations in the physical properties across the nebula affected the distribution of water within the evolving solar system. Temperature, density, and pressure were highest at the center of the nebula and gradually decreased toward its outer limits. Extreme conditions at the nebula's center vaporized ice and light elements and drove them toward the outer reaches of the nebula. Consequently, dry rocky masses formed the inner planets (including Earth). Farther out, meteorites (carbonaceous chondrites) and the giant outer planets Saturn and Jupiter formed. These meteorites are about 10% ice by mass and the giant planets also contain some ice, but most of the water in the nebula condensed in comets at distances beyond Saturn and Jupiter.

Spectral studies show that comets are about half ice. This prompted scientists to hypothesize that during the latter stages of Earth's formation, comets bombarded the planet's surface delivering a veneer of water. A key factor leading to comets colliding with Earth is thought to be the growth of Jupiter by accretion and the consequent strengthening of its gravitational attraction. Jupiter's gravitation may have drawn a multitude of ice-rich comets from the outer to the inner reaches of the solar system and toward Earth. This hypothesis remained popular until the past decade when scientists discovered that the chemistry of water on Earth differs from that of ice in comets. Spectral analyses of three comets that came close to Earth in recent years (Halley in 1986; Hyakutake in 1996; and Hale-Bopp in 1997) revealed that comet ice contains about twice as much deuterium as that in water on Earth. *Deuterium (D)* is an isotope of hydrogen whose nucleus consists of one proton plus one neutron and is very rare on Earth. Common hydrogen's nucleus consists of one proton only. Deuterium behaves chemically like ordinary hydrogen and can form water (D_2O). Based on this finding, some scientists have estimated that bombardment by comets could account for no more than half of the water on Earth and perhaps much less. If correct, where did the rest of the water come from? What are other possible sources of water for the planet?

One potential source is the impact on Earth's surface of planetesimals (large meteorites a few kilometers across) containing water having less deuterium than comet ice. Planetesimal impacts may have occurred toward the end of Earth's formation. However, the ratio of some other chemical components of planetesimals is not the same as the ratio of those same components in the Earth system. Another possibility is that Earth's water is indigenous; that is, the center of the nebula may have been cooler than previously assumed and some of the materials present in the inner solar system that went into the early formation of Earth were water-rich. These materials may have included silicate rock with encapsulated ice.

Whatever the ultimate source of water on Earth, the ocean co-evolved with Earth's atmosphere. At the very beginning, Earth's atmosphere was probably mostly hydrogen (H) and helium (He) plus some hydrogen compounds including methane (CH_4) and ammonia (NH_3). Because these atoms and molecules are relatively light, this earliest atmosphere eventually escaped to space. In time, Earth congealed and volcanic activity began spewing huge quantities of lava, ash, and gases. By about 4.4 billion years ago, the planet's gravitational field was sufficiently strong to retain a thin gaseous envelope of volcanic origin, Earth's primeval atmosphere.

The principal source of Earth's atmosphere was *outgassing*, the release of gases from rock through volcanic eruptions and the impact of meteorites on the planet's rocky surface. Perhaps as much as 85% of all outgassing took place within a million or so years of the planet's formation, but outgassing continues to this day although at a slower pace. Outgassing produced a primeval atmosphere that was mostly carbon dioxide (CO_2), with some nitrogen (N_2) and water vapor (H_2O), along with trace amounts of methane, ammonia, sulfur dioxide (SO_2), and hydrochloric acid (HCl). Radioactive decay of an isotope of potassium (potassium-40) in the planet's bedrock added argon (Ar), an inert (chemically non-reactive) gas, to the evolving atmosphere. Dissociation of water vapor into its constituent atoms, hydrogen and oxygen, by solar ultraviolet radiation contributed a small amount of free oxygen to the primeval atmosphere. (The lighter hydrogen, having relatively high molecular speeds, escaped to space.) Also, some oxygen combined with other elements in various chemical compounds such as carbon dioxide.

Scientists suggest that between 4.5 and 2.5 billion years ago, the sun was about 30% fainter than it is today. This did not mean a cooler planet, however, because of the abundance of carbon dioxide. Earth's CO_2-rich atmosphere was perhaps 10 to 20 times denser than the present atmosphere. Carbon dioxide slows the escape of Earth's heat to space, so that average surface temperatures may have been as high as 85 °C to 110 °C (185 °F to 230 °F).

By perhaps 4 billion years ago, the planet began to cool and the Earth system underwent major changes. Cooling caused atmospheric water vapor to condense into clouds that produced rain. Precipitation plus runoff from landmasses gave rise to the ocean that eventually covered as much as 95% of the planet's surface. The global water cycle (which helped further cool Earth's surface through evaporation) and its largest reservoir (the ocean) were in place. Rains also led to a substantial decline in the concentration of atmospheric CO_2. Carbon dioxide dissolves in rainwater producing weak carbonic acid that reacts chemically with bedrock. The net effect of this large-scale geochemical process was increasing amounts of carbon chemically locked in rocks and minerals while less and less CO_2 remained in the atmosphere. Weathering and erosion on land delivered some carbon-containing sediments to the ocean. Also, rains washed dissolved carbon dioxide directly into the sea and some atmospheric CO_2 dissolved in ocean waters as sea surface temperatures fell. The ocean began to emerge as a major reservoir in the global carbon cycle.

Although carbon dioxide has been a *minor* component of the atmosphere for at least 3.5 billion years, its concentration fluctuated from time to time in the geologic past with important implications for global climate. All other factors being equal, more CO_2 in the atmosphere means higher temperatures at Earth's surface. (As noted earlier, carbon dioxide slows the loss of Earth's heat to space.) For example, geologic evidence points to a burst of volcanic activity on the Pacific Ocean floor about 100 to 120 million years ago. Some CO_2 released during that activity eventually found its way to the atmosphere and elevated the global mean temperature by as much as 10 Celsius degrees (18 Fahrenheit degrees). During the Pleistocene Ice Age (1.7 million to 10,500 years ago), atmospheric carbon dioxide levels fluctuated, decreasing during episodes of glacial expansion and increasing during episodes of glacial recession (although it is not clear whether variations in atmospheric CO_2 were the cause or effect of global-scale climate variations).

Living organisms also played an important role in Earth's evolving atmosphere, primarily through photosynthesis. As noted elsewhere in this chapter, *photosynthesis* is the process whereby plants use sunlight, water, and carbon dioxide to manufacture their food. A byproduct of photosynthesis is oxygen (O_2). (Although vegetation is also a *sink* for carbon dioxide, photosynthesis probably was not as important as the geochemical processes just described in removing carbon dioxide from the atmosphere.) Photosynthesis dates to at least 2.5 billion years ago when the first primitive forms of life (blue-green algae) appeared in the ocean. During the subsequent 500 million years, most oxygen generated via photosynthesis combined with marine sediments while very little entered the atmosphere. By about 2.3 billion years ago, however, oxidation of ocean sediments had tapered off. Oxygen generated photosynthetically by marine plants initially dissolved in ocean water. Eventually, oxygen saturated surface ocean waters and began cycling into the atmosphere, ultimately building to near current levels. With the concurrent decline in CO_2, within 500 million years, oxygen became the second most abundant atmospheric gas after nitrogen.

With oxygen emerging as a major component of Earth's atmosphere, the *ozone shield* developed. In the portion of the upper atmosphere known as the stratosphere, incoming solar ultraviolet (UV) radiation powers reactions that convert oxygen to ozone (O_3) and ozone to oxygen. Absorption of UV radiation in these reactions prevents potentially lethal intensities of UV radiation from reaching Earth's surface. By about 1 billion years ago, formation of the *stratospheric ozone shield* made it possible for organisms to live and evolve on land. UV radiation does not penetrate ocean water to any great depth so that marine life was able to exist in the ocean depths prior to the formation of the ozone shield. But with the ozone shield, marine life was able to thrive in surface waters.

Nitrogen (N_2), a product of outgassing, became the most abundant atmospheric gas because it is relatively inert chemically and its molecular speeds are too slow for this gas to readily escape to space. Furthermore, compared to other atmospheric gases, such as oxygen and carbon dioxide, nitrogen is less soluble in water. All these factors greatly limit the rate at which nitrogen cycles out of the atmosphere. While nitrogen continues to be generated as a minor component of volcanic eruptions, today the principal source of free nitrogen entering the atmosphere is the denitrification process involved in bacterial decay of the remains of plants and animals, both in the ocean and on land. But this input is countered by an amount removed from the atmosphere by *biological fixation* (i.e., direct nitrogen uptake by leguminous plants such as clover and soybeans) and *atmospheric fixation* (i.e., the process whereby the high temperatures associated with lightning causes nitrogen to combine with oxygen to form nitrates).

Earth is the water planet but scientists are unsure as to the origins of that water. Nonetheless, scientists have developed an emerging picture of how the Earth system formed and it is evident that Earth's ocean, atmosphere, and biosphere co-evolved through billions of years.

CHAPTER 2

OCEAN BASINS AND PLATE TECTONICS

Case-in-Point
Driving Question
Distribution of the World Ocean
Oceanic Crust and Continental Crust
    Earth Materials
    Rock Cycle
Ocean Bottom Profile
    Continental Margins
    The Ocean Basin
Plate Tectonics and Ocean Basin Features
    Evidence for Plate Tectonics
    Divergent Plate Boundaries
    Convergent Plate Boundaries
    Transform Plate Boundaries
    Marine Volcanism
    Hydrothermal Vents
Spreading and Closing Cycles
Conclusions
Basic Understandings
ESSAY: Investigating the Ocean Bottom
ESSAY: Hotspot Volcanism and the Hawaiian Islands

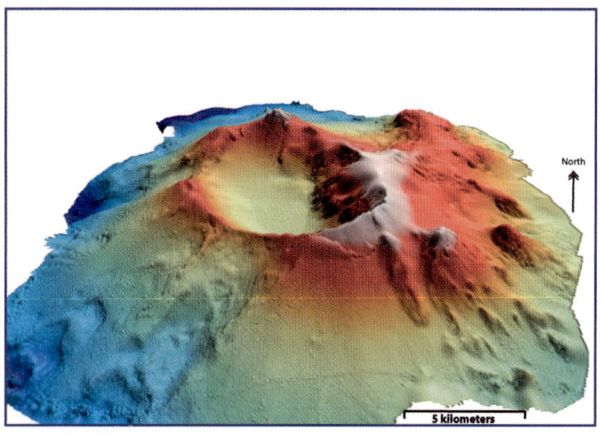

EM300 bathymetry of West Rota submarine volcano. The data were collected using the EM300 multibeam system mounted on the hull of the R/V *Thompson*. The image is 2 times vertically exaggerated. [Courtesy of NOAA.]

Case-in-Point

Tectonic activity plays a central role in the evolution of the basins that are occupied by the ocean. In addition, through the years the volcanic eruptions and earthquakes that are products of tectonic activity both at sea and on land have claimed many lives and caused considerable property damage. Study of the Earth system has greatly improved scientific understanding of tectonic processes and the conditions that contribute to volcanic eruptions, earthquakes, and their after-effects (e.g., tsunamis). This basic understanding is enabling scientists to develop techniques to predict volcanic eruptions and earthquakes, thereby saving many lives.

One of the greatest natural disasters of the past few centuries was the eruption of Krakatoa, a volcano on a small island in the Sunda strait between Java and Sumatra in Indonesia. In 1883, after 200 years of inactivity, Krakatoa came to life and remained moderately active for several weeks. On 26-27 August, the volcano erupted and violent explosions were heard as far away as Rodriguez Island, 4653 km (2885 mi) across the Indian Ocean, and the atmospheric pressure waves generated were recorded around the world. Two days of gigantic explosions obliterated two-thirds of the volcanic island and triggered enormous ocean waves, called *tsunamis*, which reached estimated heights of 40 m (130 ft). Tsunamis swept inland 16 km (10 mi) on the densely populated islands of Java and Sumatra, killing at least 36,000 people and destroying 165 coastal villages.

Thick clouds of ash carried by winds settled out of the atmosphere producing deep deposits over a region of about 780,000 square km (300,000 square mi), roughly equivalent to the area of Chile. Thinner deposits covered 3.9 million square km (1.5 million square mi), equal to nearly half the area of the United States. Thick beds of ash also accumulated on the surrounding ocean floor. In all, an estimated 21 cubic km (5 cubic mi) of ash and other larger ejecta were blown into the air, with significant amounts of ash rising 50 km (30 mi) into the stratosphere. Ash remained suspended in the upper atmosphere for several years and was transported by winds around the world, causing brilliantly colored sunsets.

Prior to Krakatoa, in 1815, another Indonesian volcano, Tambora, erupted violently and impacted the climate in many parts of the world. It blasted a considerable amount of sulfur dioxide into the stratosphere where the gas combined with moisture to produce plumes of tiny sulfurous acid droplets and sulfate particles (collectively called *sulfurous aerosols*). Winds transported sulfurous aerosol plumes around the globe. In the stratosphere, sulfurous aerosols reflect some incoming solar radiation back to space and absorb some solar radiation. The net effect was less solar radiation reaching Earth's surface and cooling of the lower troposphere. Cooling was particularly notable during 1816, referred to in New England as the "year without a summer." Late spring snows and midsummer freezes destroyed crops in the northeast United States and across eastern Canada, bringing considerable hardship to many people.

The largest volcanic explosion in the past 10,000 years took place about 3650 years ago on the Greek island of Santorini in the eastern Mediterranean. This violent eruption and the gigantic ocean waves it caused nearly destroyed the island, giving rise to the legend of the "lost city" of Atlantis. About 30 cubic km (7 cubic mi) of ash fell downwind over the eastern Mediterranean Sea and Turkey. Widespread destruction throughout the region apparently hastened the decline of the Minoan civilization on the nearby island of Crete and the subsequent spread of Greek colonies.

More recently, in June 1991, Mount Pinatubo, located on Luzon Island in the Philippines, erupted violently, ejecting huge amounts of ash and sulfurous aerosols into the stratosphere (Figure 2.1). Sulfurous aerosols in the

FIGURE 2.1
Sulfurous aerosols emitted during the spectacular eruption of Mount Pinatubo in the Philippines in June 1991 likely impacted global climate for a year or two. [NOAA National Geophysical Data Center, Boulder, CO]

stratosphere reduced the amount of solar radiation reaching Earth's surface. The Mount Pinatubo eruption temporarily interrupted the post-1970s global warming trend and was likely responsible for the relatively cool summer of 1992 over continental areas of the Northern Hemisphere. The death toll was relatively low in this eruption, 350 people, in large part due to advance warning and better communications and transportation systems.

Several major volcanic eruptions have occurred in North America, but fortunately none since the continent became heavily populated. Even the spectacular eruption of Mount St. Helens in Washington on 18 May 1980 was relatively small compared to the many massive eruptions that have shaped Earth and impacted its inhabitants in the past. Processes that take place in Earth's solid interior, which we discuss in this chapter, are responsible for volcanic eruptions.

Driving Question:

What processes shape the lithosphere and how do those processes affect ocean basins?

Many forces work together to shape and reshape Earth's lithosphere, creating and destroying both ocean basins and continents. These forces arise primarily from movement of lithospheric plates and geological processes that occur at plate boundaries. In this chapter we learn about these forces, their origins, and the features they produce on the ocean floor. We compare the properties of oceanic crust with those of continental crust in terms of rock type, density, and thickness. The ocean bottom profile shows the transition from continental crust to oceanic crust and is divided into the continental shelf, slope, and rise. Although the shelf, slope, and rise are submerged, they are part of the continental crust whereas the abyssal plains, trenches, and ridges beyond are portions of the oceanic crust. Once thought to be flat and featureless, modern technology has revealed many features on the ocean bottom that provide clues as to the origin of the ocean basins; features include prominent volcanic mountain ranges and deep trenches. These and other characteristics of the ocean bottom are associated with processes that take place at divergent, convergent, and transform plate boundaries. We begin by examining the geographical distribution of the world ocean.

Distribution of the World Ocean

Ocean basins and continents are unevenly distributed over Earth's surface (Figure 2.2). The ocean covers about 71% of the planet with the remaining 29% land. However, most land is in the Northern Hemisphere whereas ocean dominates the Southern Hemisphere. The Northern Hemisphere is 39.3% land and 60.7% ocean whereas the Southern Hemisphere is 19.1% land and 80.9% ocean. The world's ocean is composed primarily of the separate, but connected oceans called the Pacific, Atlantic, and Indian Oceans as the largest basins, with the smaller Arctic Ocean basin connected to the far north Atlantic. By international agreement in 2000, the Southern Ocean was defined as the waters between 60 degrees S and the Antarctic continent. In the Southern Ocean, the Antarctic Circumpolar Current flows eastward around the Antarctic continent virtually unimpeded by landmasses. The only exception is the northward deflection of currents caused by the relatively narrow Drake Passage (about 650 km or 400 mi across) between the southern tip of South America and Antarctica. A comparison of the polar regions shows the Arctic as an ocean surrounded by continents whereas the Antarctic is a continent surrounded by the Southern Ocean.

Where no convenient continental boundaries exist, arbitrary lines delineate ocean basins. For example, the 150 degrees E longitude line from Australia to Antarctica separates the Pacific and Indian Ocean basins. North of Australia, the islands of Indonesia divide the Pacific and Indian Ocean basins. Far to the north, the Bering Strait (between Alaska and Siberia) separates the Pacific and the

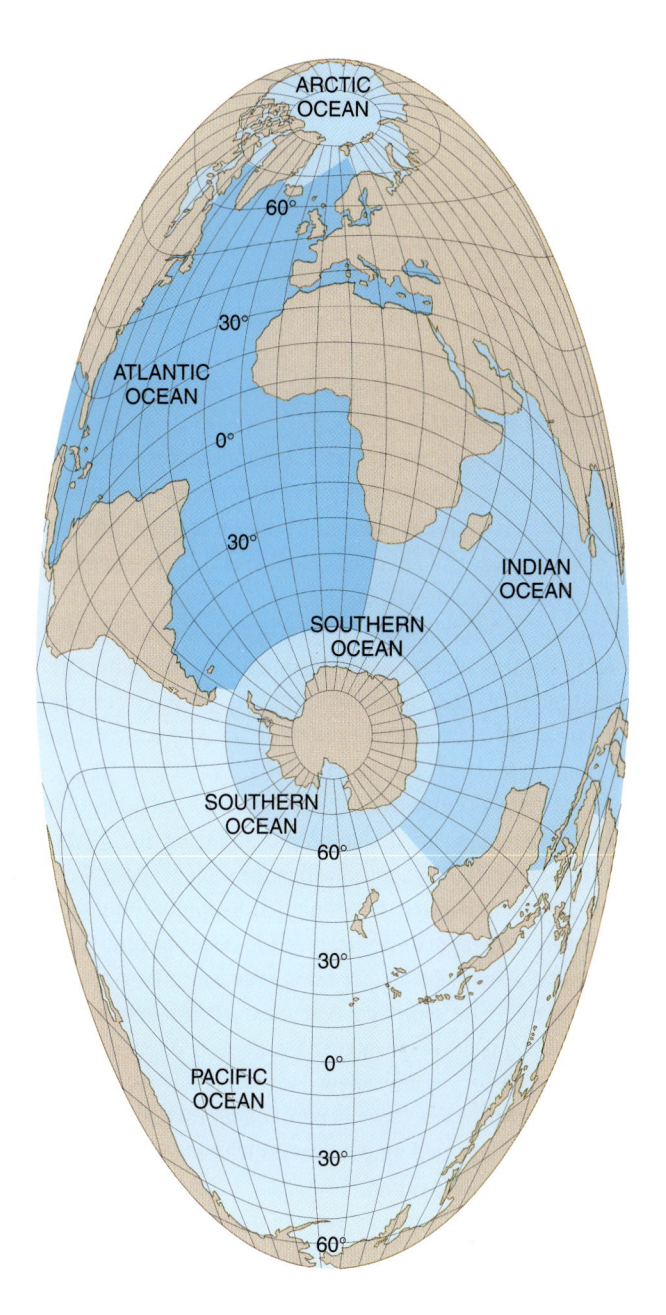

FIGURE 2.2
Ocean basins and continents are unevenly distributed on Earth's surface. Ocean dominates the Southern Hemisphere and most land is in the Northern Hemisphere.

Arctic Ocean. The 70 degrees W longitude line between Cape Horn (the southern tip of South America) and Antarctica can be considered the boundary between the Pacific and Atlantic Ocean. Still another north-south meridian (20 degrees E) from the Cape of Good Hope (southern tip of Africa) to Antarctica separates the Atlantic and Indian Ocean. In addition, numerous seas and gulfs occur adjacent to and connected to the main ocean basins.

Oceanic Crust and Continental Crust

Earth's relatively thin solid **crust** is the portion of the lithosphere that interfaces with the ocean, atmosphere, cryosphere, and biosphere. The crust beneath the ocean differs from the crust of the continents in composition, density, and thickness. In this section, we compare oceanic and continental crust beginning with the basic distinction among rock types.

EARTH MATERIALS

As noted in Chapter 1, internal and surface geological processes continually shape and reshape the lithosphere. These same processes produce a multitude of rock types that compose the crust. Rocks, in turn, are made up of one or more minerals. A *mineral* is a naturally occurring inorganic solid characterized by an orderly internal arrangement of atoms (the basic structural units of all matter) and fixed physical and chemical properties. Some rocks and minerals on land and on the sea floor are economically important resources (Chapter 4).

Rocks composing the crust are classified as igneous, sedimentary, or metamorphic based on the general environmental conditions in which the rock formed. Cooling and crystallization of hot molten *magma* produces *igneous rock*. Magma originates in the lower portion of the lithosphere or in the upper mantle and migrates upward towards the Earth's surface. Magma may remain within the crust and cool slowly forming coarse-grained igneous rock such as granite or it may spew onto Earth's surface as *lava* through vents or fractures in bedrock and solidify rapidly forming fine-grained igneous rock such as basalt or glassy material such as obsidian.

Sedimentary rock may be composed of any one or a combination of compacted and cemented fragments of rock and mineral grains, partially decomposed remains of dead plants and animals (e.g. shells, skeletons), and minerals precipitated from solution. Sediments form as rocks undergo physical disintegration and chemical decomposition when exposed to rain, atmospheric gases, and fluctuating temperatures at or near Earth's surface. These are *weathering processes*. Sediments are washed into rivers that transport them to the sea and other standing bodies of water where they settle out of suspension, accumulate on the bottom, and eventually compact into layers of solid sedimentary rock. Sediments are also transported and deposited by wind, glaciers, and icebergs at sea. Most sedimentary rocks have a granular texture; that is, they consist of individual grains that are compressed or cemented together (e.g., sandstone, shale), although some consist of precipitated minerals and are crystalline (e.g., limestone, rock salt).

Like many sedimentary rocks, *metamorphic rocks* are derived from other rocks. A rock is metamorphosed (changed in form) when exposed to high pressure, intense heat, and chemically active fluids—conditions that exist in geologically active mountain belts or near sources of volcanic heat. Like igneous rocks, metamorphic rocks are crystalline; that is, they are composed of crystals that interlock like the pieces of a jigsaw puzzle. Marble is a common metamorphic rock formed by the metamorphism of limestone ($CaCO_3$) and quartzite is a very durable metamorphic rock formed by metamorphism of sandstone (mostly SiO_2).

Most of the bedrock composing Earth's crust is igneous with some metamorphic rock locally. In many places, thick layers of sedimentary rock and unconsolidated sediments overlie crystalline igneous and metamorphic rocks. Unconsolidated sediments include soil (on land) and clay, silt, sand, or gravel (on land and ocean bottom) and deposits of sediment vary widely in thickness, from a thin veneer to thousands of meters.

Continental crust is mostly granite, a coarse-grained rock rich in minerals containing silica and aluminum. **Oceanic crust**, on the other hand, is mostly basalt, a fine-grained rock rich in minerals containing iron and magnesium. Continental crust is thicker (20 to 90 km or 12 to 56 mi) and less dense than oceanic crust (only 5 to 10 km or 3 to 6 mi thick). As noted in Chapter 1, both continental and oceanic lithosphere includes the crust and the rigid upper portion of the mantle (Figure 1.5). Oceanic lithosphere has a maximum thickness of about 100 km (62 mi) whereas continental lithosphere ranges in thickness from 100 to 150 km (62 to 93 mi). The lithosphere floats on the underlying **asthenosphere**, a deformable region of the upper mantle. With temperatures at the base of the lithosphere ranging from 450 °C to 750 °C (840 °F to 1380 °F), the asthenosphere exhibits plastic-like behavior; that is, it readily deforms in response to stress. As we will see later in this chapter, the lithosphere is continually generated in the oceanic ridge system and drawn down into the mantle at subduction zones.

Granite is less dense than basalt so that continental lithosphere floats higher and extends deeper into the mantle than oceanic lithosphere. In fact, most of the continental lithosphere is below sea level. The continents float in the upper mantle similar to icebergs floating in the ocean. (This form of buoyancy is known as *isostacy*.) Figure 2.3 is a comparison between land elevation and ocean depth.

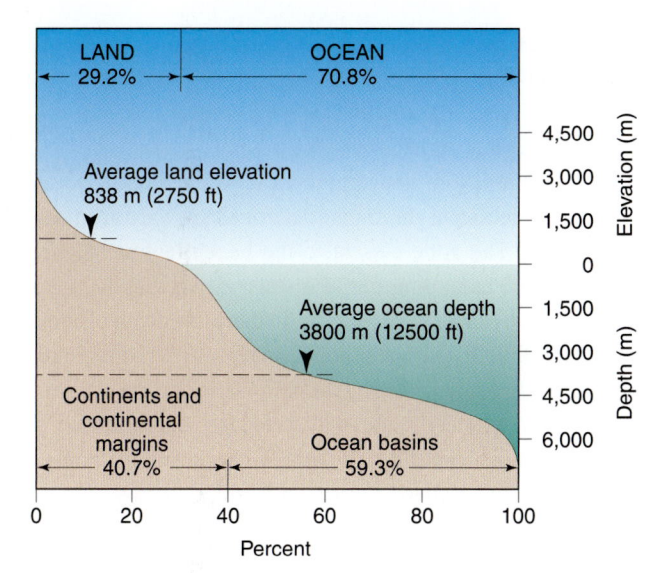

FIGURE 2.3
Hypsographic curve of Earth's surface, showing the relative distribution of ocean and land.

Where ocean waters are less than 1000 m (3300 ft) deep, the ocean bottom is usually a submerged portion of the continental crust (the *continental margin*). Where the ocean is deeper than 4000 m (13,000 ft), the bottom is usually oceanic crust. Mountains higher than 6000 m (20,000 ft) are rare on land, as are portions of ocean basins deeper than 6000 m (20,000 ft). These extreme heights and depths occupy less than 1% of Earth's surface.

ROCK CYCLE

Surface and internal geological processes transform rock from one type to another in the **rock cycle** (Figure 2.4). Through the rock cycle, rocks and their component minerals are continually regenerated. Consider an example. Physical and chemical weathering fragment an igneous rock mass that is exposed to the atmosphere into sediments which are subsequently transported by running water and deposited in a low-lying basin. In time, the accumulated sediments compact and are cemented together as they gradually convert to sedimentary rock. The mounting weight of the continually accumulating sediments forces the sedimentary rock to greater depths within the crust. As we will see later in this chapter, subduction also transports crustal material to great depths in the mantle. Temperature and confining pressure increase with depth so that the rock is eventually metamorphosed, that is, re-crystallized into metamorphic rock. At some depth in the subsurface, the temperature is so high that the metamorphic rock melts into magma. The magma may subsequently migrate upward along

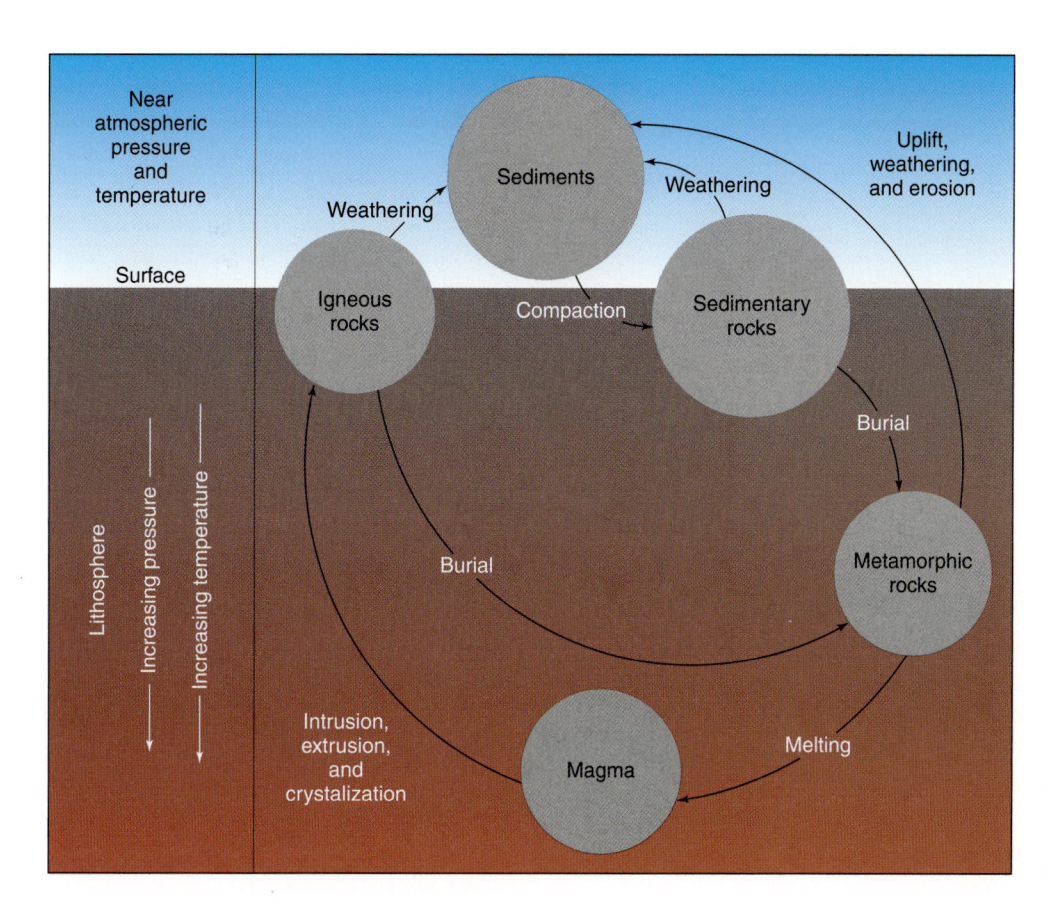

FIGURE 2.4
Through the rock cycle, geological processes (black arrows) convert rock from one type to another.

fractures within the crust, and then cool and crystallize into igneous rock, thereby completing the rock cycle. As we will see later in this chapter, movements of lithospheric plates play an important role in the rock cycle.

Rock transformations involved in the rock cycle are extremely slow; regeneration of rocks and minerals may take many millions of years. Hence, in the time frame of a human lifetime or even of civilization, the rate of regeneration of rock, mineral, and fuel resources (e.g., coal, natural gas) is so slow that for all practical purposes the supply is fixed and finite. For this reason, the resources found in the Earth's crust are essentially *nonrenewable*. We have more to say about resources from the ocean bottom in Chapter 4.

Ocean Bottom Profile

Ocean depth varies markedly from one location to another. Over large areas water depth is less than 200 m (650 ft); other areas are as deep as 11,000 m (36,000 ft); and the average ocean depth is about 3800 m (12,500 ft). Here, we examine the vertical cross-sectional profile of the ocean bottom, including the continental margin and ocean basin (Figure 2.5). In places the ocean bottom is nearly flat and essentially featureless whereas in other places the ocean floor exhibits considerable topographic relief. Parts of the ocean bottom are volcanically active with lava interacting chemically with seawater. But over vast areas of the ocean floor, the only significant geological process operating is a very gentle rain of particles that resembles dust settling in a schoolroom.

CONTINENTAL MARGINS

Based on measurements of water depth, ocean scientists delineate three distinct zones seaward from the coastline. The zone closest to the beach features a very gentle slope extending out to a water depth that averages about 130 m (430 ft). Seaward from there to a depth of about 3000 m (9800 ft) the water depth increases much more rapidly with distance. Then a relatively narrow zone is transitional from the steep slope of the previous zone to the more-or-less flat ocean basin. The initial, gently sloping zone is the **continental shelf**; the second, more steeply sloping zone is the **continental slope**; and the third transition region is the **continental rise**.

The continental shelf, slope, and rise together comprise the *continental margin*. This is not a misnomer because the bedrock of the continental margin is the same as the continental crust. From a geological perspective, the continents do not end at the beach, but at the continental rise. At its outer edge, the continental margin merges with the deep-sea floor or descends into an oceanic trench (Figure 2.5).

The continental shelf is nearly flat, sloping less than 1 degree seaward. Although water depth generally increases with distance offshore, the rate of increase is quite small—averaging about 2 m per km (10 ft per mi). Shelf width generally ranges from a few tens of meters to 1000 km

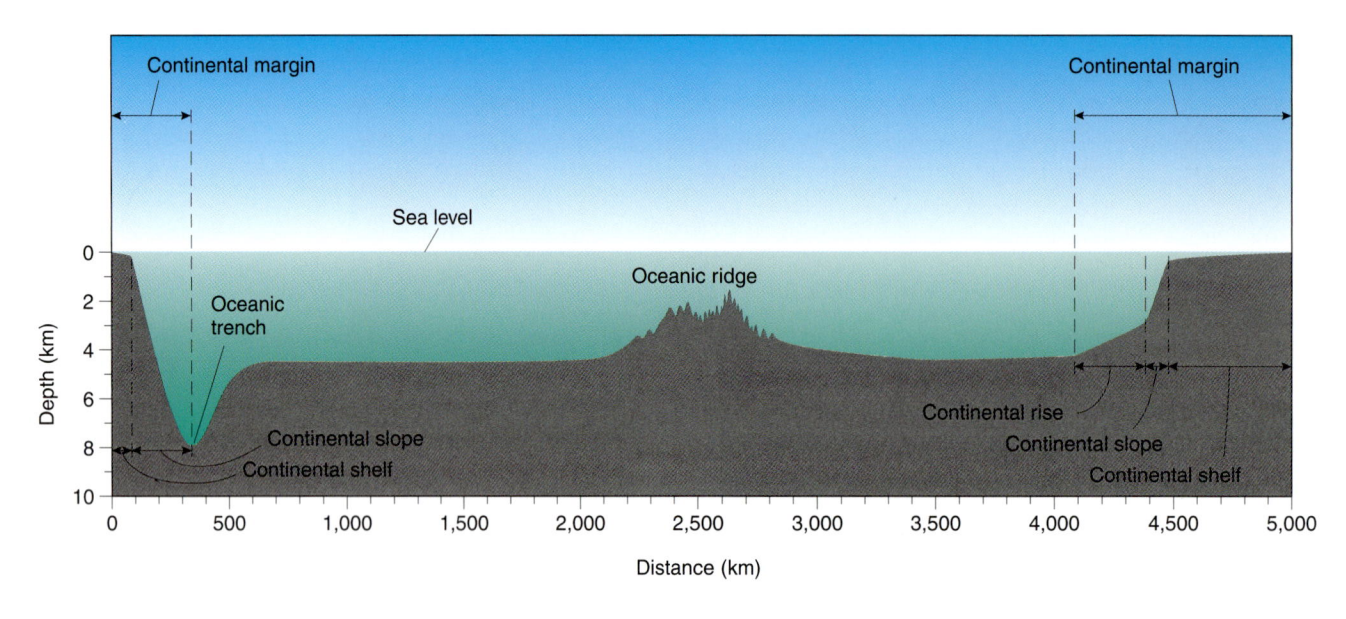

FIGURE 2.5
Cross-sectional profile of the continental margin and ocean bottom with the vertical scale greatly exaggerated.

(620 mi). As a general rule, the shelf is narrowest where the continental margin is tectonically active (e.g., subduction zones) and is widest where the continental margin is passive (i.e., no plate boundary nearby). Hence, the North American shelf is wider along the *passive* East Coast (as much as several hundred kilometers) than the *tectonically active* West Coast (a few kilometers). About 7.5% of the total area of the ocean overlies the continental shelf.

The inclination of the continental slope averages about 4 degrees but ranges between 1 and 25 degrees; water depth typically increases by about 50 m per km (265 ft per mi) in a seaward direction, a significantly greater rate of increase than in the shelf zone. In many places such as around the margin of the Atlantic, the continental slope merges with the more gently sloping continental rise. (The term "rise" here means sloping less steeply to the flat ocean bottom.) In this passive continental margin, sediment spreads over the ocean floor forming vast, flat **abyssal plains** seaward of the continental rise. Along tectonically active continental margins, such as that surrounding nearly the entire Pacific Ocean basin, the slope descends directly into deep ocean trenches. Land-derived sediment flows into trenches and is not available to form either a continental rise or abyssal plains. About 15% of the total area of the ocean is situated over the continental slope and rise.

Large numbers of deep, steep-sided **submarine canyons** slice into the continental slope and some run up onto the continental shelf; many of these canyons cut into solid rock (Figure 2.6). The most probable explanation for most submarine canyons is erosion by turbidity currents. A **turbidity current** is a down-slope flow of water heavily laden with suspended sediment and denser than normal seawater. In many respects, turbidity currents resemble underwater avalanches. Sediment delivered to the ocean by rivers accumulates on the continental shelf. Eventually the pile of sediments builds to an unstable height and suddenly (perhaps triggered by an earthquake) moves downhill as a unit, scouring the ocean bottom in the process. These sediments accumulate at the base of the continental rise as a series of overlapping **submarine fans**. Turbidity currents have been clocked at speeds as fast as 100 km (60 mi) per hr, so it is easy to imagine these flows gouging the ocean bottom and clearing submarine canyons of accumulated sediments.

Many submarine canyons appear to be the natural extensions of existing rivers (e.g., the Hudson Canyon off the New York Bight), lending further support to the tur-

FIGURE 2.6
The darker water in the upper right of this NASA Space Shuttle image is the southern portion of the Tongue of the Ocean in the Bahamas, one of two main branches of the Grand Bahama Canyon. This submarine canyon's nearly vertical walls rise 4285 m (14,000 ft) above the canyon floor. The Grand Bahama Canyon has a length of more than 225 km (140 mi) and a width of 37 km (23 mi) at its deepest point. The lighter blue water covers the shallow Bahama shelf, much of which was above sea level during the last glacial maximum. At the time water draining into the canyon eroded the gullies that are visible at the shelf edge. [NASA]

bidity current explanation. For submarine canyons having no obvious association with existing rivers, sediment sources may be located upstream of longshore currents. As discussed in Chapter 8, longshore currents transport sand and other sediment along the shoreline. Also, some submarine canyons may be valleys cut by rivers during the Pleistocene Ice Age (1.7 million to 10,500 years ago). At times during the Pleistocene, sea level was as much as 130 m (425 ft) lower than today (because so much water was locked up in glaciers) and much of the continental shelf was above sea level. With this increased gradient, rivers flowed across these exposed shelves and eroded deep canyons that were subsequently flooded at the close of the Pleistocene when glaciers melted and sea level rose.

THE OCEAN BASIN

Fringed by continental margins, **ocean basins** encompass the remaining portion of the oceanic area. Once thought to be largely flat and monotonous plains, the ocean basin actually has a varied topography featuring deep trenches, seamounts, and submarine mountain ranges. Indeed, undersea terrain is just as diverse as terrestrial terrain and exhibits even greater relief. Nevertheless, much of the ocean bottom (about 42%) is comprised of plains and low hills, usually rising no more than about 100 m (330 ft) above the plain. Blanketed with sediments that tend to smooth out any irregularities in the bedrock below, abyssal plains and abyssal hills are typical of ocean basin topography—apart from the 23% of the ocean bottom that is covered by ridge systems.

Ocean scientists use many different instruments and techniques to investigate the properties and composition of the ocean bottom. For more on this topic, refer to this chapter's first Essay.

Plate Tectonics and Ocean Basin Features

The most prominent features of the ocean basin, including trenches and mid-ocean ridge systems, are products of tectonic stresses. Earth's lithosphere is divided into many plates that move very slowly over the face of the globe (Figure 2.7). All plates, except the Pacific, include parts of continents as well as ocean basins. Mountain building and most volcanic activity and earthquakes take place at boundaries between plates, many of which occur either within the ocean basins or along basin boundaries.

EVIDENCE FOR PLATE TECTONICS

Plate tectonics is a unifying concept that combines continental drift plus sea floor spreading to describe the generation, movement, and destruction of Earth's lithosphere and the formation of ocean basins. The historical evolution of this concept illustrates the **scientific method**, a systematic form of inquiry involving observation, interpretation, speculation, and reasoning—components of *critical thinking*.

Evidence for the continental drift part of plate tectonics dates back at least to the time of the British Philosopher Francis Bacon (1561-1626) who noted how the eastward bulge of South America closely fit the configuration of a portion of the west coast of Africa. (Comparing the edge of the continental shelves provides an even better fit than the coastlines.) By the middle of the 19th century, scientists and others proposed that the Atlantic Ocean

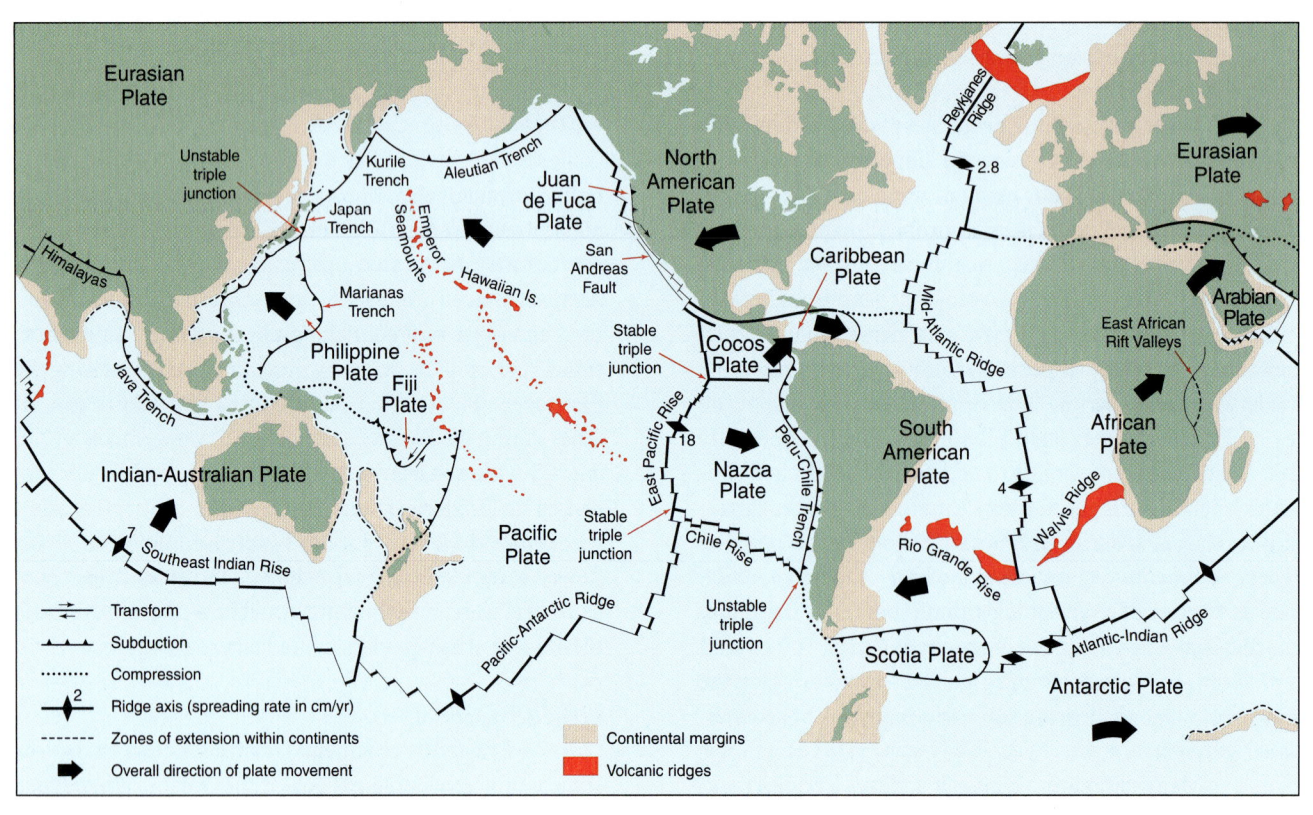

FIGURE 2.7
Divergent, convergent, and transform boundaries of major lithospheric plates. [U.S. Geological Survey]

formed when landmasses separated during some catastrophic event. Early in the 20th century, the Austrian geologist Eduard Suess (1831-1914) hypothesized that at one time the individual continents of the Southern Hemisphere were together as one huge continent, which he called *Gondwanaland*. Around 1910, the German meteorologist Alfred Wegener (1880-1930) and the American geologist Frank B. Taylor (1860-1939) independently proposed **continental drift**, the hypothesis that the continents move over the surface of the planet. Wegener proposed that the supercontinent *Pangaea* split into *Laurasia* (encompassing present day North America and Eurasia) and *Gondwanaland* (encompassing South America, Africa, India, Australia, and Antarctica). He based his idea on the observed close fit between continental margins, similarities in fossil plants and animals, and the continuity of rock formations and mountain ranges between continents on either side of the Atlantic.

Wegener's hypothesis was ridiculed and mostly ignored by geologists on the grounds that he did not provide a scientifically sound mechanism for driving the continents. At the time, most scientists believed that continents and ocean basins were fixed in place so it was hard to understand how Wegener's continents could plow through solid rock on the bottom of the ocean. Little was then known about the ocean floor or Earth's interior. All this began to change in the 1950s and 1960s when investigations of the deep sea floor revealed prominent mountain ranges and deep trenches. In the early 1960s, these findings inspired Harry Hess (1906-69) of Princeton University and the geological oceanographer Robert S. Dietz (1914-95) to resurrect an idea first presented in 1929 by the British geologist Arthur Holmes (1890-1965). Hess and Dietz proposed that convection in the Earth's mantle was the driving force behind continental drift.

According to Hess and Dietz, rock in the lower mantle is heated by the decay of certain radioactive elements. This warmer less dense material gradually rises within the mantle and then flows along the base of the cool, rigid lithosphere. Cooling of the mantle material increases its density and it sinks back toward the lower mantle where it is reheated thus completing the convective circulation. Convection currents drag the lithosphere, and are responsible for **sea-floor spreading**, the divergence of adjacent plates on the ocean bottom. They supply heat and magma to the ocean ridges, and cause **subduction**, the descent of lithospheric plates into the mantle. Today, some scientists argue that two zones of convection operate in the mantle: deep convection (below about 700 km or 435 mi) and shallow convection (above 700 km) that interfaces with the lithosphere.

In subsequent years, evidence accumulated in support of the Hess-Dietz model and sea-floor spreading. Much of this evidence came from deep-sea drilling from specially outfitted ships (discussed in this chapter's first Essay) and includes: (1) the concentration of earthquake activity along plate boundaries, with the deepest earthquakes generally associated with subduction zones (Figure 2.8), (2) the enhanced heat flow from Earth's interior along mid-ocean ridges, (3) increasing age of the ocean crust with increasing distance on either side of the mid-ocean ridges, and (4) increasing thickness of sea-floor sediment with increasing distance from the mid-ocean ridges.

Perhaps the most convincing argument for plate tectonics was the discovery of a distinctive pattern of anomalies in the Earth's ancient magnetic field on the sea floor. We can visualize Earth's magnetic field as emanating from a huge bar magnet centered in the Earth's core with a north and south magnetic pole. Earth's magnetic poles, however, do not coincide with the planet's geographical poles. Currently, the north magnetic pole is located in northwest Canada while the south magnetic pole is in the Southern Ocean. (Precise locations vary over time.) Magnetic lines of force converge toward the two magnetic poles so that the needle on a compass aligns with the magnetic lines of force and points toward the magnetic north.

Oceanic crust contains the mineral magnetite, an iron oxide, which acts like a compass needle in aligning with the planet's magnetic lines of force. This alignment is frozen into the rock when the temperature of cooling lava drops below the *Curie temperature* of 580 °C or 1076 °F, named for its discoverer Pierre Curie (1859-1906). Hence, information on the direction of the planet's magnetic field at the time lava cooled and solidified is locked in the rocks.

Studies of the same types of rock on land reveal times in the past when the polarity of the Earth's magnetic field reversed; that is, the north magnetic pole became the south magnetic pole and vice versa. This rock record indicates 170 reversals over the past 76 million years. Beginning in the 1950s, research vessels and aircraft mapped magnetic polarity reversals over much of the ocean floor. Instruments towed behind ships measured the intensity of the magnetic field and revealed ribbon-like patterns of magnetic anomalies (Figure 2.9). Each anomaly strip was a few kilometers to tens of kilometers wide and thousands of kilometers long and oriented roughly parallel to the mid-ocean ridges. Most intriguing was the discovery that the magnetic anomaly pattern on one side of an ocean ridge was the mirror image of the pattern on the other side of the ridge.

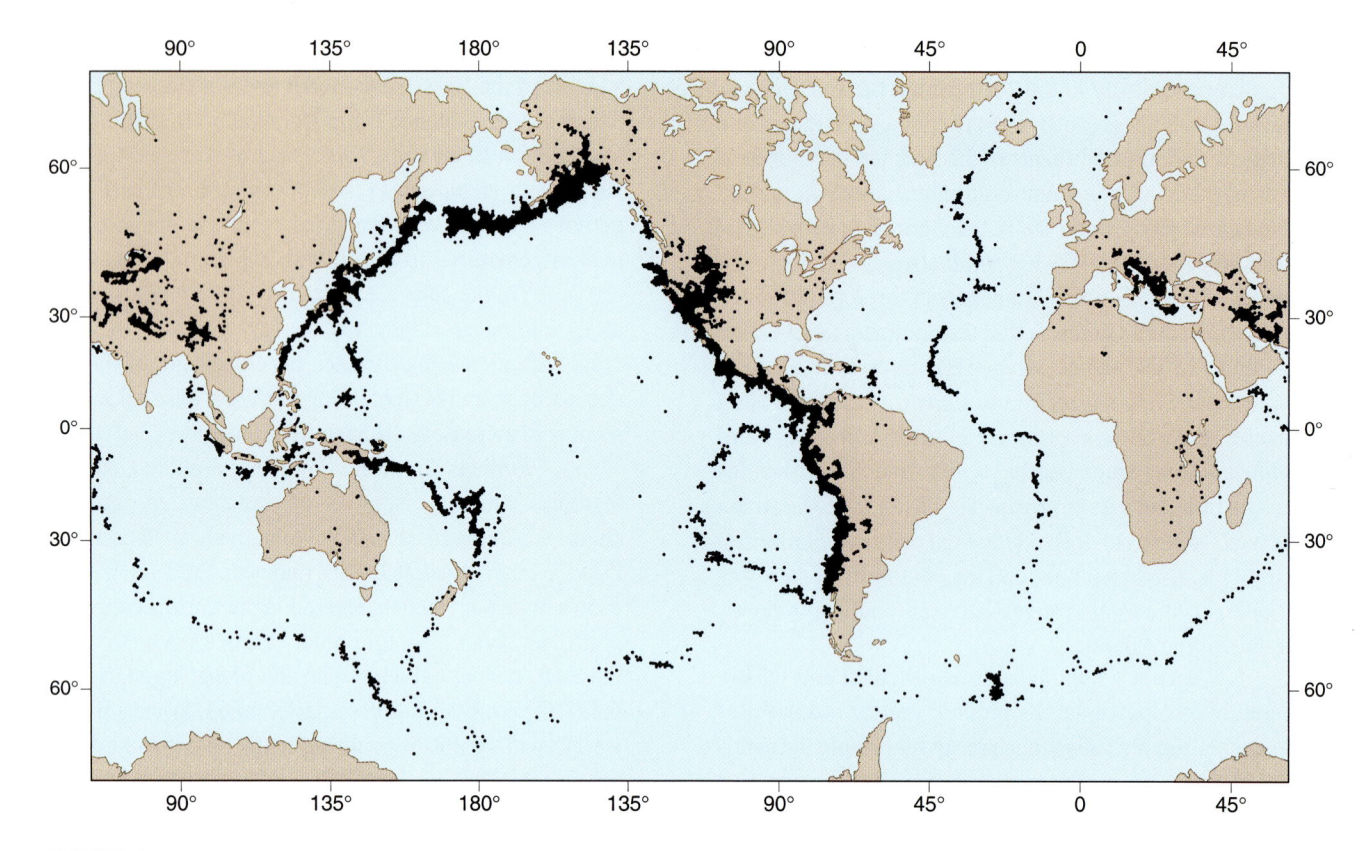

FIGURE 2.8
Most earthquakes occur along or near the boundaries of major lithospheric plates. [U.S. Geological Survey]

In 1963, F.J. Vine and D.H. Matthews of Cambridge University interpreted these anomaly patterns as a record of past magnetic reversals indicating that newly formed crust moved away from the mid-ocean ridges. That is, magnetic anomaly patterns confirmed sea-floor spreading and continental drift and the unifying theory of plate tectonics soon became widely accepted.

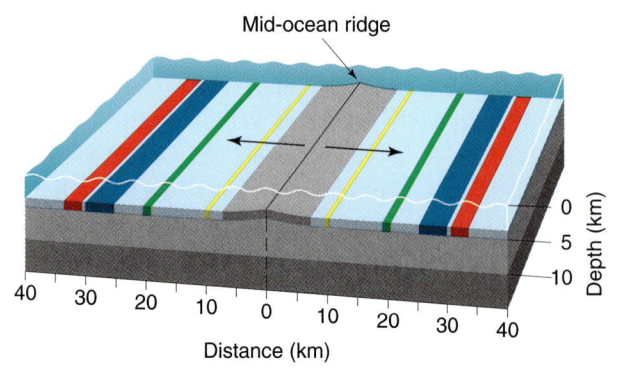

FIGURE 2.9
This block diagram shows the regular pattern of magnetic anomalies on either side of the mid-ocean ridge system. The age of the ocean crust increases with increasing distance from the ridge axis. Light blue stripes represent normal polarity; other colored stripes indicate reversed polarity.

Boundaries between the rigid lithospheric plates are where the action is geologically. Depending on relative movement, plate boundaries are designated divergent, convergent, or transform. Movements along plate boundaries give rise to a variety of geographic and geological features on the ocean floor and on land.

DIVERGENT PLATE BOUNDARIES

Along a **divergent plate boundary**, adjacent plates move apart producing rifts (fractures) in crustal bedrock along which magma wells up from below (Figure 2.10). Magma (called *lava* when it reaches Earth's surface) has a temperature of about 1200 °C (2200 °F) and cools and solidifies rapidly when it comes in contact with much colder seawater (at 2 °C to 5 °C or 36 °F to 41 °F). Through this process, new oceanic lithosphere is generated at divergent plate boundaries. In fact, more than 18 cubic km (4 cubic mi) of new lithosphere is produced in this way every year.

On the seafloor, the outer surfaces of lava flows exhibit characteristic tube-like and pillow-shaped structures (Figure 2.11). These volcanic eruptions at spreading centers are generally tranquil events, unlike the violent explo-

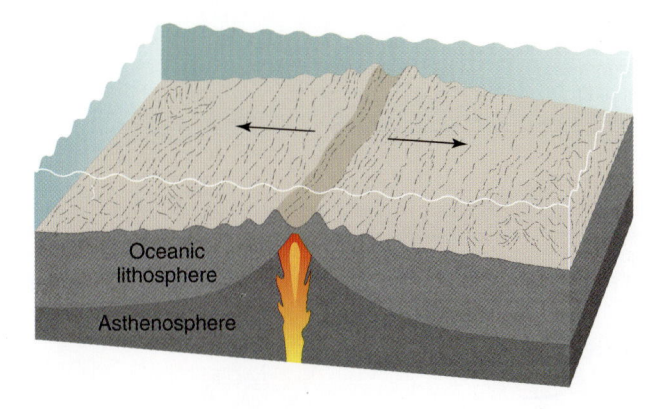

FIGURE 2.10
Divergent plate boundary at a mid-ocean ridge.

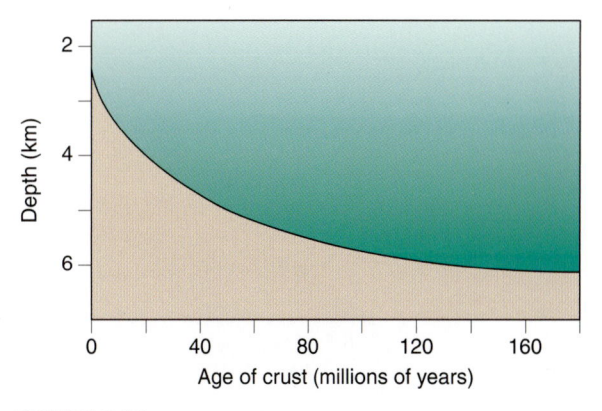

FIGURE 2.12
When formed at a mid-ocean ridge, oceanic crust is about 2.5 km (1.6 mi) below the sea surface. As the oceanic crust ages, it cools and becomes denser and moves away from the mid-ocean ridge, accumulating sediment and sinking deeper into the upper mantle. At an age of 160 million years, the ocean crust is at a depth greater than 6 km (3.7 mi).

sions that characterize volcanoes associated with subduction zones (discussed below). Also, shallow earthquakes (centered a few kilometers to a few tens of kilometers below the ocean floor) are often associated with these volcanic eruptions.

Newly formed oceanic crust stands higher than nearby older oceanic crust because recently formed warm rock is less dense than older and colder rock. As oceanic crust ages and moves away from plate boundaries, it cools and becomes denser. Also, as the crust is slowly loaded with sediments, it sinks deeper into the underlying asthenosphere (not unlike a boat as it is loaded). Consequently, the older the crust, the deeper is the water above it (Figure 2.12). At divergent plate boundaries (mid-ocean ridge), the ocean bottom is about 2500 m (8200 ft) deep, but away from the boundary where the crust is 160 million years old, the oceanic crust is about 6000 m (19,500 ft) below sea level.

Thus, the ocean floor not only contains information on magnetic reversals and plate movements, but functions as conveyor belts on either side of the mid-ocean ridge system, moving newly formed crust away from divergent plate boundaries.

A prominent submerged volcanic mountain range marks divergent plate boundaries on the ocean floor. The Mid-Atlantic Ridge, a segment of this ridge system, divides the Atlantic Ocean into two roughly symmetrical eastern and western basins and is more than 2000 km (1240 mi) wide and rises about 2500 m (8200 ft) above the adjacent sea floor (Figure 2.13). The Mid-Atlantic Ridge occurs along a boundary between oceanic plates that have been diverging for perhaps 200 million years. That divergence split apart the ancient supercontinent *Pangaea*, creating the Atlantic Ocean basin. At present, the Mid-Atlantic Ridge is spreading apart at a rate of about 3 cm per year (about as fast as fingernails grow) and rifting is evident in the gaping fractures that dissect solidified lava flows in Iceland, a volcanic island formed at a hot spot along the ridge (Figure 2.14). Other examples of divergent plate boundaries include the East African Rift Valleys and the rifts occupied by the Red Sea and Gulf of Aden in North Africa.

The ocean ridge system—of which the Mid-Atlantic Ridge is a part—winds its way from the Arctic Ocean down the middle of the Atlantic, curves around South Africa and into the Indian Ocean (*Mid-Indian Ridge*), and then into the Pacific Ocean (*East Pacific Rise*). Its total length exceeds 65,000 km (40,000 mi), much longer than any mountain range on land. A major feature of most segments

FIGURE 2.11
The pillow shape exhibited by these basaltic flows on the slope off Hawaii is characteristic of lava that cools rapidly on the seafloor. [OAR/National Undersea Research Program (NURP); NOAA]

40 Chapter 2 OCEAN BASINS AND PLATE TECTONICS

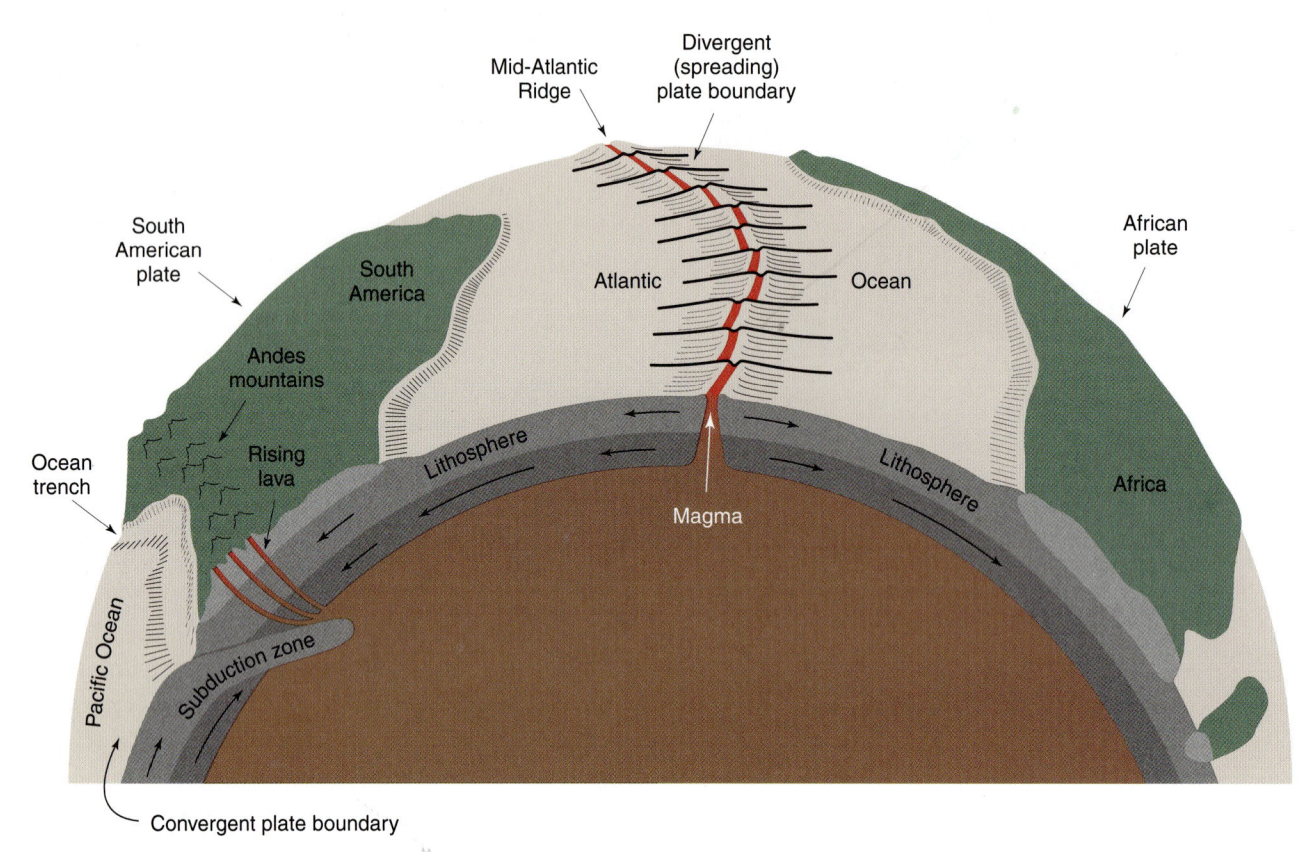

FIGURE 2.13
The Mid-Atlantic Ridge occurs where adjacent plates diverge and is part of a lengthy ridge system that winds through the ocean basins.

of the ocean ridge system is a narrow, steep-sided *rift valley* with low volcanoes along its margins. Where formation of new oceanic crust is relatively slow (with low spreading rates typically less than 6 cm or 2.4 in. per year), ocean ridges have rugged rift valleys, 1 to 2 km (0.6 to 1.2 mi) deep and a few tens of kilometers across. The Mid-Atlantic Ridge is an example. More rapid spreading results in less rugged, broader dome-like rises, usually with no rift valleys such as the East Pacific Rise, located in the eastern Pacific Ocean, near South America.

Lengthy, nearly straight, parallel fractures are oriented perpendicular to the mid-ocean ridge system and offset segments of the ridge. These fractures allow the spreading motion to adjust to the curvature of the Earth. Rugged sea bottom topography as well as lines of submarine volcanoes or islands mark some fracture zones. The Cape Mendocino Fracture Zone intersects the California coast at Cape Mendocino and extends westward thousands of kilometers on the North Pacific Ocean floor.

FIGURE 2.14
Iceland is a volcanic island formed at a hot spot along the Mid-Atlantic ridge. The volcanic rocks exposed in this scene were fractured as lava flows cooled, congealed, and contracted. [Photo by J.M. Moran]

CONVERGENT PLATE BOUNDARIES

Two plates move toward one another along a **convergent plate boundary**. There are three possibilities here: (1) two oceanic plates collide (e.g., the Pacific and Phillipine plates), (2) an oceanic and continental plate collide (e.g., the Nazca and South American plates), and (3) two continental plates collide (e.g., the Indian and Eurasian plates).

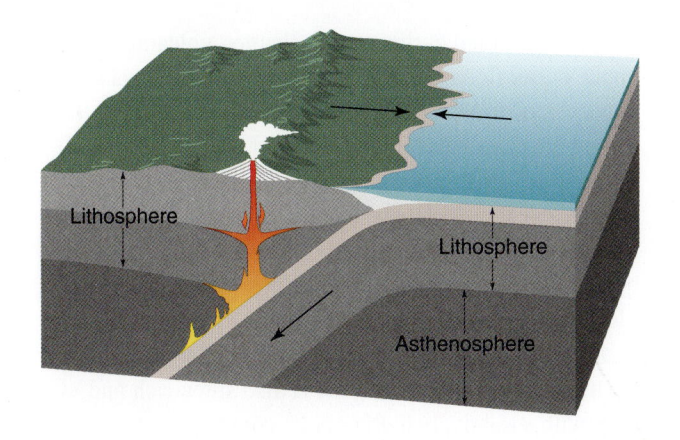

FIGURE 2.15
A convergent plate boundary where an oceanic plate subducts under a continental plate.

A **subduction zone** forms where two oceanic plates converge or where an oceanic plate converges with a continental plate (Figure 2.15). In a subduction zone, the denser oceanic plate slips under the other plate and descends into the mantle. With downward transport of rock and sediment, a deep elongated depression (trench) forms on the ocean floor. Temperature and confining pressure increase with depth within the planet so that a plate is heated and compressed as it descends into the mantle. Also, the friction of plates grinding past one another generates heat and earthquakes. At depths approaching 80 km (50 mi), the subducting plate begins to melt into magma, which migrates upward toward Earth's surface and contributes to mountain building and explosive volcanic eruptions. (The eruptions of Mount St. Helens and Pinatubo are recent examples.) Whereas new lithosphere forms at divergent plate boundaries, lithosphere is incorporated into the mantle in subduction zones at convergent plate boundaries.

In oceanic-to-oceanic plate collisions, volcanic activity associated with subduction forms **island arcs**, curved chains of volcanic islands. Trenches usually lie on the seaward sides of island arcs with relatively shallow seas near continents. For example, the islands of Japan are part of a volcanic island arc with the Sea of Japan on the continental side and a trench on the Pacific Ocean side. An **ocean trench** is relatively narrow, typically 50 to 100 km (30 to 60 mi) wide, and may be thousands of kilometers long. In cross-section, a trench is slightly asymmetrical with the steeper slope on the continental side. Depths greater than 8000 m (26,250 ft) are not uncommon and some trenches have deep holes called *deeps* that exceed 10,000 m (32,800 ft) in depth. The deepest place on the ocean floor is probably the Marianas Deep, in the Marianas Trench in the Pacific, with a maximum depth around 11,000 m (36,000 ft). For all their importance in plate tectonics, trenches make up only 1 to 2% of the ocean basin floor. Trenches surround most of the Pacific Ocean basin but volcanic island arcs are mostly on the western and northern sides of the Pacific. Besides Japan, Alaska's Aleutian Islands/Aleutian Trench is another example of a volcanic island arc/trench system.

At the convergent plate boundary just off the west coast of South America, the Nazca plate subducts under the South American plate, producing the offshore Peru-Chile Trench, which is 5900 km (3660 mi) long, 100 km (62 mi) wide, and more than 8000 m (26,300 ft) deep. Associated with subduction of the Nazca plate is the Andes, a prominent mountain range that forms the backbone of South America.

The only gaps in the circum-Pacific subduction/trench system, also known as the *Pacific Ring of Fire*, are in Antarctica and in western North America. The Pacific Ring of Fire includes the U.S. West Coast, an active continental margin, whereas the East Coast is not a plate boundary and is passive. The action in the Atlantic is down the middle (along the Mid-Atlantic Ridge) whereas the action in the Pacific is around the rim. This difference is related to the fact that the Atlantic is spreading with little subduction, while the Pacific is shrinking at the subduction zones.

Deep earthquakes (originating at depths greater than 100 km or 60 mi) are usually associated with subduction zones. Earthquakes occur where adjacent crustal blocks move past each other, releasing energy stored in rocks deformed during plate movements when descending into the mantle at subduction zones. This energy travels as waves both along Earth's surface as well as through its interior. The depth to earthquake energy release generally increases with distance inland with the deepest earthquakes originating at depths of 700 km (430 mi).

Plate movements also cause continents to collide, but such convergence does not form trenches or island arcs. Continents are much thicker and less dense than oceanic crust so they do not subduct into the mantle. A conspicuous continental convergence is taking place in the Mediterranean region, where Africa and Europe are colliding as evidenced by volcano and earthquake activity in the eastern Mediterranean. The Himalayan Mountains are the result of the continuing collision of the Asian and Indian plates. Over the past 40 million years, the plate containing the Indian subcontinent has moved thousands of kilometers northward through the Indian Ocean and is now colliding with Asia. The Himalayan Mountains occur where the leading edge of the Indian plate thrust under the Asian plate. The Rocky Mountains of North America were formed in a similar manner when the Pacific plate thrust beneath the North American plate. Their present height results from the buoyant, relatively young Pacific plate material lying beneath the North American continental crust.

TRANSFORM PLATE BOUNDARIES

Adjacent plates slide laterally past one another along a **transform plate boundary** (Figure 2.16). Although crust is neither created nor destroyed and these boundaries are generally free of volcanic activity, slippage can deform rock and trigger earthquakes. The San Andreas Fault of California, site of frequent earthquakes, occurs along a transform plate boundary where the Pacific plate (carrying a piece of California) slides toward the northwest, past the North American plate. The Anatolian transform fault in Turkey has produced some particularly disastrous earthquakes.

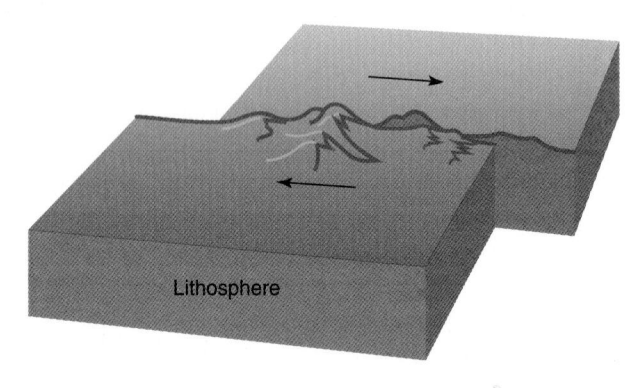

FIGURE 2.16
A transform plate boundary is often the site of shallow earthquakes.

MARINE VOLCANISM

Marine volcanic activity occurs where magma erupts through vents or fissures at Earth's surface, that is, along divergent and convergent plate boundaries and over hot spots. Products of volcanic eruptions include materials that are molten (e.g., lava), solid (e.g., ash, cinders, rock blocks), and gaseous (e.g., water vapor). Viscosity (internal resistance of a substance to flow), gas content, and rate of extrusion of molten material govern volcanic activity and the shape of a volcano built by an eruption. Volcanoes range from small cinder cones to massive shield-like accumulations of solidified lava. Whereas about 20 major volcanic eruptions occur annually on the continents, many more eruptions occur on the ocean bottom and almost always are unobserved. Only shallow submarine eruptions and those occurring on islands have been extensively studied, although this is changing because of the development of new observing techniques.

As noted earlier in this chapter, volcanic activity on the ocean floor consists of mostly quiescent lava flows. However, continental volcanoes often erupt violently. This difference in volcanic activity stems from the silica (SiO_2) composition and gas content (volatiles) of the magma that feeds a volcano. Magmas that are silica-rich are relatively viscous and may block volcanic vents. Pressure produced by volatile components (mostly water vapor) beneath the block increases and eventually the volcano explodes. Continental crust is mostly silica-rich granite so that magma formed from melting this crust is also silica-rich and viscous with a relatively high volatile content.

A high volatile content magma develops when wet oceanic crust subducts under continental crust and melts at depth. This magma migrates upward carrying volatiles, melting and mixing with continental crust, and forming andesitic to granitic magmas that are 60-70% silica and 3-6% volatiles. Such magmas are responsible for violent eruptions characteristic of volcanoes of the Cascade and Andes mountain ranges. On the other hand, oceanic crust is mostly silica-poor basalt derived from the upper mantle. It is also relatively dry with only 1-2% volatile content. Hence, the magma that composes the oceanic crust is volatile-poor, silica-poor, has low viscosity, and flows readily as lava. This explains why mid-ocean spreading centers exhibit quiescent volcanic activity (e.g., Iceland). Intermediate in activity are volcanic island arcs (e.g., the Aleutian Islands) where one oceanic plate subducts under another oceanic plate. Wet oceanic crust melts but there is no silica-rich granite to mix with. The volatile-rich magma is silica-poor so that volcanic activity tends to be intermediate between violent and quiescent.

A significant volcanic eruption, usually persisting over a long period of time, builds **seamounts**, underwater volcanic peaks rising more than 1000 m (3300 ft) above the ocean floor. Examples include the Emperor Seamounts in the North Pacific Ocean basin. Massive eruptions may cause the top of the volcano to break the water surface, rising above sea level and forming a volcanic island. The Hawaiian Islands, Iceland, and most other oceanic islands are products of such submarine volcanic activity. With the passage of time and the decline and eventual end of volcanic activity, weathering and erosion rapidly dissect and greatly reduce the elevation of volcanic islands. Meanwhile, the subsurface void produced by the discharge of lava during island formation may start to collapse, causing the island to sink back into the sea. Movement of the oceanic crust away from divergent plate boundaries also transports volcanoes into deeper waters. Such a deep-water seamount featuring a wave-eroded flat top is called a **guyot**.

Chapter 2 OCEAN BASINS AND PLATE TECTONICS 43

Another example of an oceanic feature caused by volcanic activity followed by subsidence, is the coral atoll. A **coral atoll** is a ring-shaped island surrounding a seawater lagoon (Figure 2.17). While serving as a naturalist on the British research ship *HMS Beagle*, Charles Darwin (1809-82) first proposed that atolls were produced by coral growing upward around the edge of the island at the same rate that the island gradually sank into the sea. Modern drilling of atolls extracts coral remains continuously down to the underlying volcanic rock. Coral live in warm near-surface waters, so these drilling data indicate that the volcanic material was once at the surface, just as Darwin had proposed.

Lithospheric plates are rigid and little volcanic activity takes place away from plate boundaries. An important exception is hotspot volcanism, which is responsible for the Hawaiian Islands, as described in the second Essay of this chapter.

HYDROTHERMAL VENTS

At locations near divergent plate boundaries, superheated waters (with temperatures as high as 400 °C or 750 °F) circulate through rock fractures on the seafloor and discharge through spectacular **hydrothermal vents**. A few regions of extremely hot water have been known to exist for many years, but it was not until the Woods Hole submersible *Alvin* descended into the Galapagos Rift in 1979 that scientists observed hydrothermal vents directly.

Hydrothermal vent waters are mineral-rich. Materials, including iron and sulfur, precipitate out of solution as the hot water mixes with the surrounding cold, oxygen-containing ocean water. Precipitates may appear as dark clouds, called *black smokers*, and accumulate in the form of conical chimneys 10 m (33 ft) or more in height (Figure 2.18). The largest hydrothermal chimney observed to date is located near the top of a seamount in the Atlantic about 2500 km (1550 mi) east of Bermuda; it rises to the height of

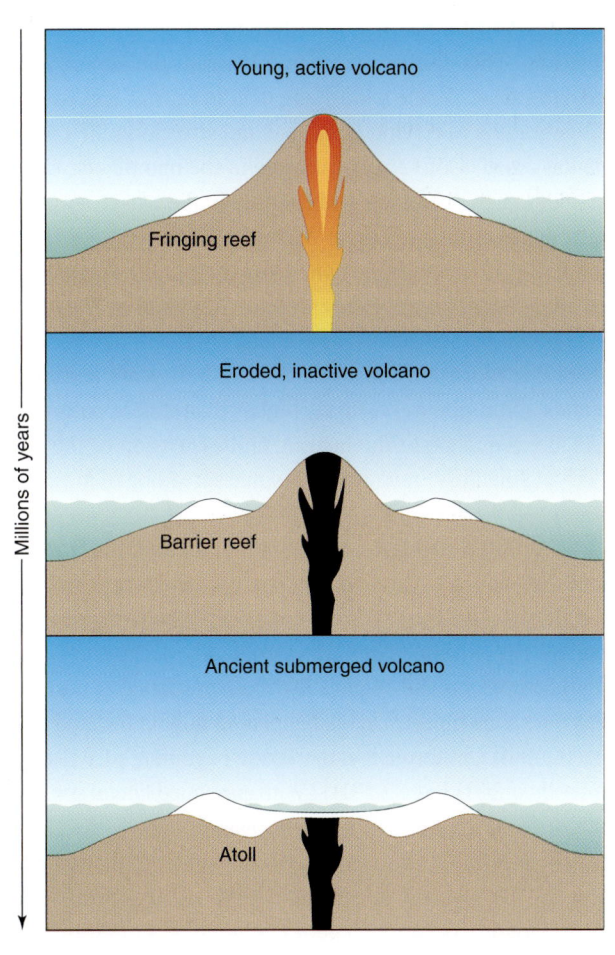

FIGURE 2.17
Stages in the development of a coral atoll.

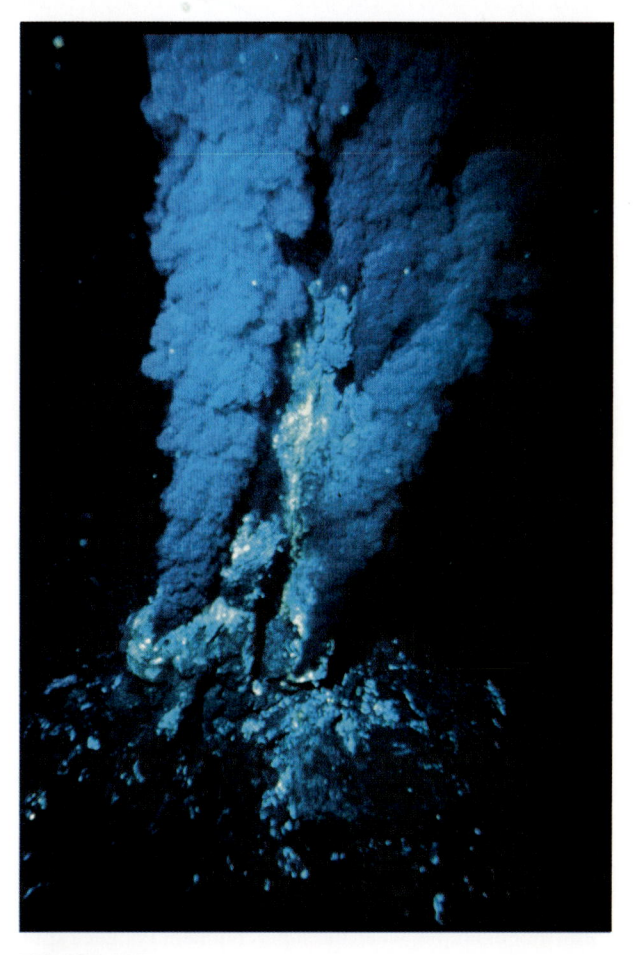

FIGURE 2.18
Black smoker emitted at a submarine hydrothermal vent on the mid-Atlantic ridge. [OAR/National Undersea Research Program (NURP); NOAA]

an 18 story building. Whereas most hydrothermal vents occur along the ocean ridge system, this one is located about 15 km (9 mi) from the Mid-Atlantic Ridge.

Hydrothermal activity on the ocean bottom is another source of the dissolved constituents of seawater, in addition to suspended and dissolved substances delivered to the ocean by rivers. This is not an insignificant control of seawater composition—the entire volume of the ocean is recycled about every 10 million years. Two processes cool newly formed ocean crust. First, heat conducted through the ocean floor removes about one-third of the heat escaping from recently erupted lavas near ocean ridges. In the second process, seawater circulating through fractures in the seafloor crust (to depths of 600 m or 1,950 ft) removes the other two-thirds. These waters are heated as they flow through fissures in the oceanic crust and may penetrate well below the sea floor. Fissures form when lava cools, solidifies, and contracts in contact with cold seawater. This *hydrothermal circulation* through recently formed crust is an important source of dissolved chemical constituents of seawater.

Communities of organisms, including bacteria, crabs, mussels, starfish, and tubeworms live near hydrothermal vents. In most marine communities, photosynthesizing organisms form the base of food chains. In the deep water where hydrothermal vents occur, no sunlight is available for photosynthesis, and the base of the food chain consists of bacteria that practice *chemosynthesis* (Chapters 1 and 9). That is, they oxidize sulfur compounds from hot vent waters and thereby derive energy for themselves and other members of marine food chains.

Spreading and Closing Cycles

By about 2.5 billion years ago, Earth's lithosphere had much of the same make up as it does today, and cycles of ocean basin formation and closing were underway. Since then, the planet probably has undergone three or more cycles of supercontinent formation and break-up (Figure 2.19). That is, during at least three long-term cycles, all the continental pieces were brought together as one huge continent and subsequently split apart into smaller continents. Associated with supercontinents are super-oceans so that when a supercontinent breaks up, one or more new ocean basins form. For example, when the supercontinent Pangaea split into Laurasia (to the north) and Gondwanaland (to the south), Tethys, a narrow seaway formed between the two. New ocean basins first expand and later contract as a new supercontinent forms. The entire cycle from start to finish may take 500 to 600 million years. Cycles of ocean basin spreading and closing are called **Wilson cycles** after the Canadian geologist J. Tuzo Wilson (1908-1993) who first recognized and described the stages in the life span of an ocean basin.

We know the most about ocean basin changes over the past 500 to 600 million years, the most recent cycle. During this time, marine animals formed durable shells, which as fossils can be used to date the rocks containing them fairly precisely. Dating older rock devoid of fossils requires analysis of radioactive elements. These techniques have become more accurate and now permit reconstruction of earlier Wilson cycles. Here we describe the stages in the most recent Wilson cycle.

A Wilson cycle consists of six stages: embryonic, juvenile, mature, declining, terminal, and suturing. The cycle begins because thick continental crust does not conduct heat as readily as thinner oceanic crust. A supercontinent that remains in one location for hundreds of millions of years acts like a blanket, retarding heat flow from Earth's interior. This causes the mantle beneath the supercontinent to warm. As the underlying mantle warms, it expands, elevating the overlying continent and stretching the continental crust. Convection currents in the mantle also contribute to this stretching and eventually the crust fractures, forming a rift valley. This is Wilson's *embryonic stage* (Figure 2.19A). As noted earlier in this chapter, a *rift valley* is an elongated topographic depression bordered by fractures (faults) in the bedrock. The African continent has been in its present location for about 200 million years and now stands about 400 m (1300 ft) higher than other continents as a consequence of the heating and expansion of the underlying mantle. Fracturing and some spreading of the crust produced the rift valleys of East Africa.

With rifting of the continental crust, the broken sides rise about a kilometer enclosing a valley that fills with fresh water. In East Africa, long, deep lakes now occupy narrow rift valleys. Rift valleys gradually widen and eventually connect to the ocean and the freshwater lakes become narrow saline gulfs. This is happening now in the Red Sea and Gulf of California and marks the beginning of Wilson's *juvenile stage* (Figure 2.19B). With continued lateral spreading of the rift valley a divergent plate boundary develops and new oceanic crust is generated signaling the *mature stage* of the Wilson cycle (Figure 2.19C). Today, the Atlantic is a mature ocean with geologically passive margins.

Subduction becomes more widespread around the border of the ocean basin during the *declining stage* of the

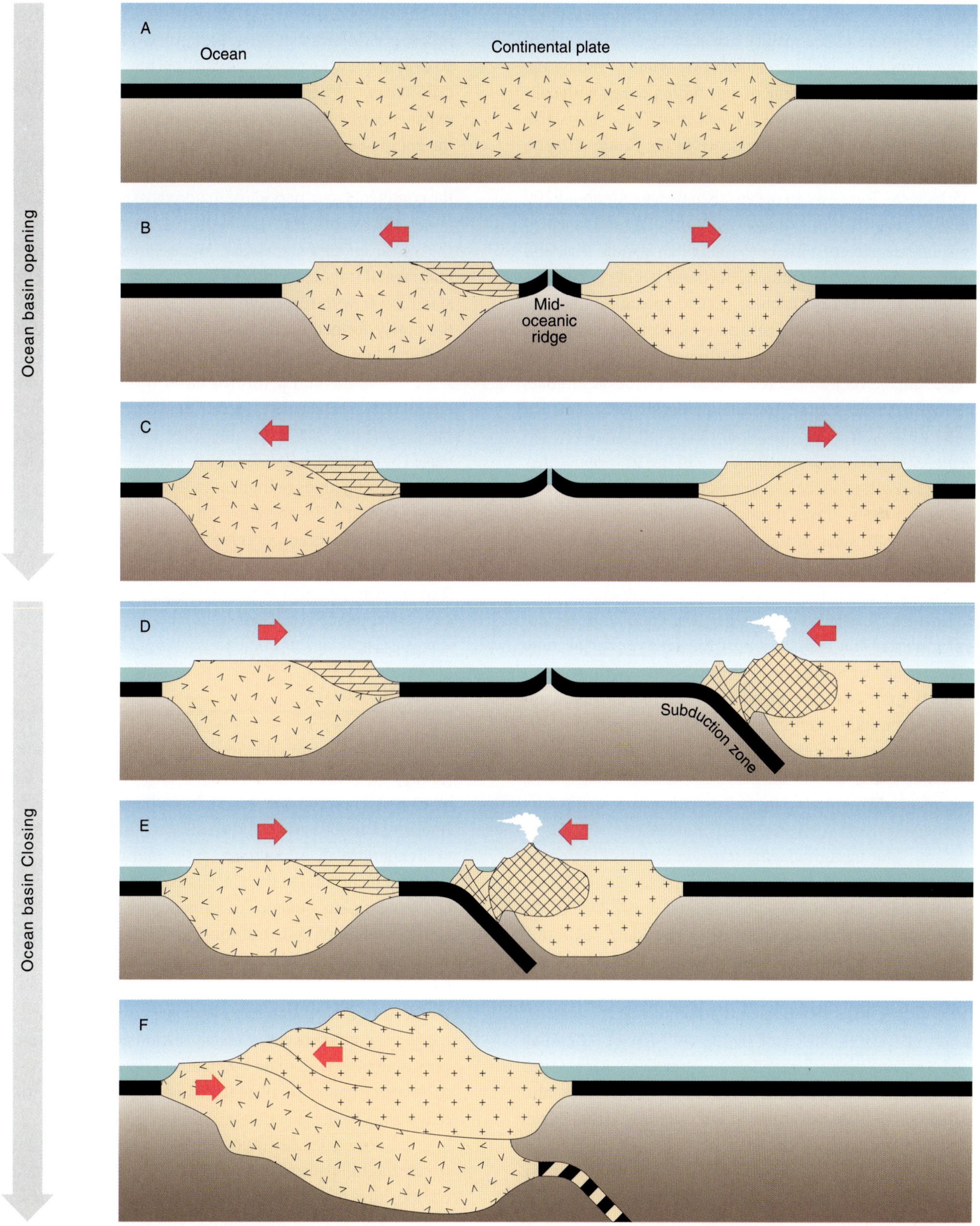

FIGURE 2.19
A model of the Wilson cycle, showing the formation and closing of ocean basins.

Wilson cycle (Figure 2.19D). Convection in the mantle drags the lithosphere of the mature ocean away from divergent plate boundaries and into subduction zones. Today, the Pacific is the best example of a declining ocean with subduction zones forming the *Pacific Ring of Fire*. This is also now happening on a smaller scale in the South Atlantic (Scotia Arc near Antarctica) and in the West Indies (near Barbados). Typically an ocean basin widens for about 200 million years before subduction begins. As evidence of this, no oceanic crust has been found that is more than 200 million years old. Eventually, the basin begins to close as subduction rates (at trenches) exceed spreading rates (at mid-ocean ridges). For Earth as a whole, spreading must equal subduction—otherwise Earth would be shrinking or expanding. Over the next 200 million years, the ocean basin continues to close and its sediment deposits are deformed and uplifted, creating mountain ranges on the newly assembled supercontintent on the site of the former ocean.

Following the *declining stage* of the Wilson cycle, the ocean basin closes through subduction as continents from opposite sides of the ocean basin bear down on one another and eventually collide. These events signal the final two stages of the Wilson cycle: the terminal and suturing stages. In the *terminal stage*, the continents are not yet touching but subduction of the intervening oceanic crust causes a narrowing of the sea separating the continents (Figure 2.19E). Volcanic eruptions, earthquakes, uplift, and mountain building accompany subduction of the oceanic crust. An example of this today is the African continent converging with the European continent producing the intervening Mediterranean Sea with nearby volcanism, earthquakes, and young rugged mountains (e.g., the Alps).

During the *suturing stage*, collision of the continents is complete and the intervening sea is gone (Figure 2.19F). The two colliding continental crusts, being less dense than the oceanic crust, do not subduct but rather override one another causing uplift and mountain building. Collision of the continents squeezes out the intervening ocean and causes subduction of oceanic crust. Today, the suturing stage is illustrated by the collision of the Indian and Eurasian plates generating the Himalayan Mountains. The Appalachian Mountains of eastern North America also formed in this way at the end of the previous Wilson cycle, about 450 million years ago.

The Pacific began as a vast super-ocean, known as Panthalassa, surrounding the supercontinent Pangaea. When other ocean basins such as the Atlantic started to form, the Pacific Ocean entered the declining stage of the Wilson cycle with concomitant changes in its size and shape.

As an ocean basin narrows, subduction accommodates the lithospheric plates that are colliding at or along its margins. Except in a few locations, subduction zones border the Pacific basin. When the Atlantic basin begins closing by expansion of subduction along its margins, subduction in the Pacific basin will diminish or cease.

Wilson cycles also influence sea level relative to the continents over periods of hundreds of millions of years. When supercontinents dominate, continents stand high as the mantle under them warms. When spreading begins, the continental fragments move off the heated mantle and thus stand lower with respect to the sea surface. Furthermore, the newly formed ocean basin floor is relatively shallow, and thus low-lying continental areas are flooded.

As the newly formed basin widens and its crust ages, cools, and deepens (under the weight of accumulating sediments), sea level falls. Sea level has apparently varied by as much as 250 m (820 ft) over the past 100 million years. During times of higher sea level, shallow seas, similar to Canada's Hudson Bay, covered large areas of continents. At such times, about 80% of Earth's surface was ocean water. During episodes of lower sea level, shorelines moved seaward to the continental margins; the ocean then covered about 65% of Earth's surface.

Conclusions

In a geological sense, the present ocean floor is relatively young—the oldest oceanic crust is only about 200 million years old whereas some bedrock on the continents is billions of years old. Oceanic crust is continually generated at divergent plate boundaries (at the mid-ocean ridge system) and incorporated into the mantle as it is drawn down at convergent plate boundaries (subduction zone). Over periods of perhaps 500 million years, ocean basins have opened and closed (the Wilson cycle). In this way, plate tectonics explains the origin of the large-scale features of the deep ocean bottom (e.g., trenches, ocean ridges) and the geological processes (e.g., volcanism, earthquakes) operating mostly at plate boundaries. Furthermore, as we will see in Chapter 12, plate tectonics helps to explain some long-term changes in climate.

Processes operating at the interface between the ocean and lithosphere (e.g., hydrothermal vents) affect the composition of seawater. Earth materials dissolve in seawater and influence its salinity. The salinity of seawater coupled with its temperature control its density, which in turn affects its vertical motion (thermohaline circulation).

Water is a substance with unique physical and chemical properties that not only influence the salinity of seawater but also the exchange of heat energy and moisture between the ocean and atmosphere. In the next chapter, we examine the uniqueness of water and its implications for the ocean in the Earth system.

Basic Understandings

- Ocean basins and continents are unevenly distributed over Earth's surface with ocean covering about 71% of the Earth and the remaining 29% land. The Northern Hemisphere is about 39.3% land and 60.7% ocean whereas the Southern Hemisphere is 19.1% land and 80.9% ocean.
- Earth's relatively thin solid crust is the part of the geosphere that interfaces with the ocean, atmosphere, cryosphere, and biosphere. Rocks composing the crust are classified as igneous, sedimentary, or metamorphic based on the general environmental conditions in which they formed. The rock cycle is the gradual transformation of one rock type to another. Earth's crust is composed of mostly igneous rock with some metamorphic rock locally. In most areas, sediments and sedimentary rocks of varying thickness overlie the igneous and metamorphic rock of the crust.
- Oceanic crust is chiefly basaltic (volcanic) rock that is denser and thinner than the mostly granitic rock of continental crust. The lithosphere includes the crust and the rigid upper portion of the mantle and is thicker under the land surface than under the ocean. The lithosphere floats on the underlying asthenosphere, a deformable (plastic-behaving) region of the mantle.
- The continental margin (a submerged part of a continent) consists of a relatively shallow and gently sloping continental shelf, bordered by the continental slope that extends to the deeper continental rise. At geologically inactive continental margins where there are no plate boundaries (e.g., surrounding the Atlantic basin), sediment spreads over the ocean floor forming flat abyssal plains adjacent to the continental rise. At geologically active continental margins with convergent plate boundaries (e.g., surrounding much of the Pacific basin), the continental slope descends into deep ocean trenches. Turbidity currents—downslope flows of water heavily laden with suspended sediments—have cut submarine canyons into the continental slope and shelf. The deep ocean basin has a varied topography featuring mountain ranges, trenches, and seamounts.
- Plate tectonics is a unifying concept that combines continental drift and sea floor spreading to describe the fragmentation and movement of Earth's lithospheric plates and the formation of ocean basins. Evidence for plate tectonics consists of cross-Atlantic similarities in continental margins, fossils, rock formations, and mountain ranges. Sea-floor spreading, a key aspect of plate tectonics, occurs at divergent plate boundaries and is confirmed by many lines of evidence including the systematic pattern of magnetic anomalies on the sea floor.
- Convection currents in the mantle are proposed as the driving force for plate movement, sea-floor spreading, and subduction.
- New oceanic crust forms at divergent plate boundaries, at the mid-ocean ridge system. This volcanic mountain range winds through the world ocean, forming one of the deep-ocean's most conspicuous features. A narrow, steep-sided rift valley occurs along most segments of the ridge system.
- Where two oceanic plates converge or where an oceanic plate converges with a continental plate, a subduction zone forms. In a subduction zone, the denser oceanic plate slips under the other plate and descends into the mantle. Hence, lithosphere is incorporated into the mantle in subduction zones. Subduction zones produce trenches on the deep ocean floor and are associated with shallow to deep earthquakes and violent volcanic eruptions.
- A transform plate boundary occurs where adjacent plates slide laterally past one another. They are often the site of earthquake activity but not volcanoes.
- Hot, newly formed oceanic crust is cooled by contact with cold seawater circulating through fractures on the seafloor. This hydrothermal circulation also extracts minerals and salts from rock. Minerals precipitate out of the hot waters and build spectacular chimneys, tens of meters high, on the ocean floor.
- The Wilson cycle, with a period of about 500 million years, describes the successive opening and closing of ocean basins through plate tectonics.

ESSAY: Investigating the Ocean Bottom

Ocean scientists employ a variety of methods to investigate the ocean bottom. Some are quite simple such as devices that recover physical samples from the ocean bottom, whereas others, such as acoustic and gravimetric instruments, are more complicated. Some are *in situ* (direct) measurements and other methods involve remote sensing techniques.

Sampling the ocean bottom is straightforward. If a sample of bottom material is desired, a *grab sampler* may be used. This mechanical device operates much as a human hand when it grasps some sand. If scientists require a vertical section of the ocean bottom showing sediment layering, a weighted coring tube is lowered to within a few meters of the bottom and then allowed to fall with the weight of the system driving it into the bottom sediments. The sediment core, retrieved when the tube is brought to the surface, is a record over time of local sedimentation, and it is even possible to date the various layers. Where the ocean bottom is bare rock or covered by thick accumulations of sediment, deep-sea drilling is used to obtain bottom samples much like sediment cores.

Our understanding of the deep ocean floor was greatly advanced by study of sediment and rock cores extracted by specially outfitted research vessels. The first of these, the *Glomar Challenger*, operated from 1968 until 1983, as part of the *Deep Sea Drilling Program (DSDP)*. The Glomar Challenger drilled 1092 holes at 624 deep ocean sites; laid end-to-end, the recovered rock and sediment cores would have a total length of 96 km (60 mi). The success of the DSDP inspired the *Ocean Drilling Program (ODP)*, an international partnership of scientists and research institutions organized to explore the evolution and structure of Earth through study of deep ocean cores. Sponsorship of ODP was by the National Science Foundation (NSF) and agencies in 20 other nations under the management of *JOIDES* (*Joint Oceanographic Institutions for Deep Earth Sampling*).

The 143-m (469-ft) long drill ship *JOIDES* (pronounced *joy-deez*) *Resolution*, was the centerpiece of the Ocean Drilling Program (see Figure). (It was named after the *HMS Resolution*, commanded by Captain James Cook during his exploration of the Pacific Ocean more than 200 years ago.) With this ship, ODP scientists extracted sediment and rock cores in water depths as great as 5980 m (19,620 ft). Built in 1978 in Halifax, Nova Scotia, the ship was originally a conventional oil-exploration vessel. She was converted for scientific research in 1984, equipped with some of the world's finest ship-board laboratories, and went into service for ODP in January 1985. Annually the *JOIDES Resolution* embarked on six scientific expeditions, each lasting about two months. The research ship has drilled in all the world's ocean basins, including north of the Arctic Circle (to 85.5 degrees N) and south of the Antarctic Circle in the Weddell Sea (to 70.8 degrees S). As of September 2003, ODP had recovered cores that end-to-end total more than 226 km (140 mi) in length.

After a core is brought to the surface, it is marked with its original location on the seafloor, coded to distinguish top from bottom, measured, and cut into smaller sections for study and storage. Each section of the core is sliced lengthwise. One half is used for non-destructive analysis before being stored. Scientists seeking to reconstruct a chapter in Earth's history analyze samples from the other half of the core. No aspect of the core is overlooked. Paleontologists examine microfossils in the cores to determine the age of the rock or sediment. Other scientists measure physical properties such as density, strength, and ability to conduct heat. Paleomagnetists use state-of-the-art equipment to read the record of Earth's magnetic field changes, information that helps determine when and where specific rocks were formed. At the end of an expedition,

FIGURE
The JOIDES Resolution is a drill ship that was the centerpiece of the Ocean Drilling Project (ODP) from 1985 to 2003, extracting sediment and rock cores from the ocean bottom around the world. It is currently performing the same service for the Integrated Ocean Drilling Program (IODP). [Courtesy of IODP]

cores are transported to one of four repositories for storage and future research. Scientists are able to access these repositories much as the general public uses a library.

In October 2003, the *Integrated Ocean Drilling Program (IODP)* succeeded the ODP. Sponsored initially by the U.S. and Japan, the IODP may eventually involve as many as 20 other nations. The goal of the new drilling program is to use two drill ships plus specialized drilling platforms rather than a single general-purpose ship and drill more and deeper holes on the ocean floor. The original IODP plan called for an upgraded *JOIDES Resolution* or a new similar drill ship but because of budget constraints, the U.S. sponsor (the National Science Foundation) will continue to use an unmodified *JOIDES Resolution* at least for the foreseeable future.

By late 2006, the *Chikyu*, a new drill ship built and operated by Japan, will join the U.S. ship. The *Chikyu* is much larger than earlier drill ships, technologically more advanced, and equipped with specialized equipment that will minimize environmental hazards (e.g., petroleum blowouts). Lacking such capabilities, DSDP and ODP drilling sites were restricted to the deep-ocean floor far from shore where the chance of encountering oil or natural gas deposits is minimal. *Chikyu's* first project will be to investigate earthquake mechanisms by boring into the Nankai Trough subduction zone located offshore of Honshu, Japan—a location that was off limits to the *JOIDES Resolution* because of the potential for a petroleum blowout. Another option with IODP is the use of specialized drilling platforms for areas where drill ships could not safely or efficiently operate, such as in the Arctic multi-year sea ice.

In addition to deep-sea drilling, other methods are available for study of the ocean bottom. One of these techniques involves the use of a sound source, such as an underwater explosion, coupled with a number of strategically positioned acoustic receiving devices. Knowing that sound travels at different speeds through different materials, and different acoustic components of the initial sound signal travel at different speeds through the same material, much can be learned about the thickness, extent, and composition of the oceanic crust. This information is then combined with gravimetric data, obtained using a gravimeter that measures the acceleration of gravity. From knowledge of the local gravitational acceleration, information about local crustal mass, and thus composition, can be inferred.

Probably the oldest type of measurement involving the ocean bottom is determination of depth. Classically this was done with a sounding line, a weighted rope marked off in *fathoms*, where 1.0 fathom equals 1.8 m or 6 ft. The line was thrown over the side, allowing the weight to reach the bottom. Today, an *echo sounder* (earlier called a *fathometer*) is commonly used, where a sound pulse is directed toward the bottom and travel time to the bottom and back of an acoustic pulse is recorded. Knowing the speed of sound in seawater, the depth is calculated. Acoustic sounders first came into common use on U.S. Navy ships during World War II, and these data led to the discovery of the variability of ocean bottom topography.

Depth data are needed for preparation of navigation charts, but acquiring these data by acoustic sounder can be time consuming because ship speeds average only about 30 km per hr (15 knots). An alternative is a system called *laser bathymetry*. A downward facing laser gun is mounted on the underside of an aircraft so that light pulses can be aimed at the ocean bottom while the aircraft is in flight. A computer measures and records the time for a light pulse to reach the bottom and return. An aircraft can travel nearly ten times faster than a ship, so that close to 1000 square km (386 square mi) can be surveyed in a single day. The only limitation to this system is how far the laser light can penetrate into a given water column, limiting the technique to relatively clean and shallow waters. For depths seaward of the continental shelf, ocean scientists depend on acoustic sounding.

50 Chapter 2 OCEAN BASINS AND PLATE TECTONICS

ESSAY: Hotspot Volcanism and the Hawaiian Islands

Most volcanic activity on Earth takes place along convergent plate boundaries and is readily explained by plate tectonics. However, volcanic activity also occurs on the ocean floor and continental interiors at great distances from plate boundaries and at locations along spreading centers (divergent plate boundaries). These regions of anomalously high volcanic activity are known as *hot spots*. A prominent product of hotspot volcanism is the Hawaiian Island chain (see Figure) and its submerged northward continuation, the Emperor Seamounts. This 6000-km (3700-mi) long chain of volcanoes, extending from the "Big Island" of Hawaii to the Aleutian Trench off Alaska, is thousands of kilometers from the nearest plate boundary. Iceland is a site of hot spot volcanism along the Mid-Atlantic Ridge (a spreading center) and Yellowstone National Park is near a mid-continental hot spot.

What is the origin of hotspot volcanism? In 1971, only a few years after widespread acceptance of the theory of plate tectonics, W.J. Morgan proposed that narrow plumes of hot material ascending within the mantle delivered copious amounts of melt to the lithosphere where it feeds volcanic eruptions (i.e., hotspots). According to Morgan, plumes originate at the mantle/core boundary. The core is much hotter than the mantle so that heat is conducted to the base of the mantle. Morgan and other scientists assumed that plumes, rooted deep in the mantle, would anchor hot spots at fixed locations relative to one another. Over many millions of years, as an oceanic plate passed over a mantle plume/hot spot, submarine volcanic eruptions built massive accumulations of basaltic lava that eventually emerged above sea level as a volcanic island. Then as the plate carried the island beyond its lava source, volcanic activity died out. With hot spots stationary and plates moving, age-progressive chains of volcanoes developed. This was proposed to be the origin of the Hawaiian-Emperor Seamount chain.

Today, active volcanism is confined to the southeast end of the Hawaiian-Emperor Seamount chain, on and near the "Big Island," which is composed of five overlapping volcanoes. Kilauea, the Island's youngest volcano, has erupted

FIGURE
True-color Terra MODIS image of the Hawaiian Islands. View is towards the north with the Big Island (Hawaii) to the lower right. From the youngest, Big Island, located over the "Hawaiian hotspot," progressively older vlocanic islands stretch towards the northwest. The underlying tectonic plate is moving northwest at the rate of 5 to 10 cm per year, carrying the islands with it. [NASA]

continuously since the early 1980s. The rate of lava emission per unit area at this hot spot is greater than at any other place on the planet. A new, still submerged volcano called Loihi is forming on the Pacific Ocean floor southeast of Hawaii. The other Hawaiian Islands are eroded remnants of now extinct volcanoes. The oldest and most extensively eroded volcanoes, marked by coral atolls, lie at the northwestern end of the Hawaiian chain (i.e., Kauai). Beyond that a chain of deeply submerged seamounts, the Emperor Seamounts, stretches to the northwest on the Pacific Ocean floor. This southeast-to-northwest trending chain of progressively older volcanic islands and seamounts was assumed to mark the track of the Pacific plate over a stationary mantle plume/hot spot. The bend in the chain (near 170 degrees E longitude) toward a more northerly heading was attributed to a change in direction of the Pacific plate about 50 million years ago. At the northern end of the Emperor Seamount chain, extinct submarine volcanoes are subducting into the Aleutian Trench.

In recent years, geoscientists have begun to question whether deep mantle plumes are associated with all hot spots. Also, the scientific community never completely accepted the hypothesis that hotspots are fixed in location. In 2003, scientists at Princeton University reported on the results of a new technique to investigate Earth's mantle using seismic (earthquake) waves that penetrate Earth's interior. Similar to tomography used to examine the internal organs of humans, this technique enabled scientists to search for and locate mantle plumes. They found that plumes originating near the mantle/core boundary feed some hotspots (e.g., Hawaii) but other hotspots (e.g., Iceland, Galápagos) have relatively shallow plumes (originating at depths near 660 km or 410 mi), and some hotspots (e.g., Yellowstone) have no plumes at all.

Evidence is also mounting that hot spots do not remain fixed relative to one another. Convection in the mantle may cause plumes to tilt thereby shifting the location of hotspots. It now appears likely that the Hawaiian-Emperor Seamount chain was the product of motion of both the Pacific plate (moving northward) and the hot spot (moving southward). In addition, evidence is not convincing that the Pacific plate actually changed direction 50 million years ago. Movement of the Hawaiian hot spot relative to the Pacific plate may be a more plausible explanation for the abrupt change in the trend of the Hawaiian-Emperor Seamount chain.

As of this writing, the scientific community has yet to reach a consensus on the origins of hotspot volcanism. Many question the validity of the use of seismic tomography to investigate mantle plumes. Better scientific criteria are needed to identify mantle plumes and hot spots. While Morgan's mantle plume model may account for some hot spots, other more comprehensive models may be needed to explain other cases of hotspot volcanism.

CHAPTER 3

PROPERTIES OF OCEAN WATER

Case-in-Point
Driving Question
The Water Molecule
    Water as Ice, Liquid, and Vapor
    Temperature and Heat
    Changes in Phase of Water
    Specific Heat of Water
    Maritime Influence on Climate
Chemical Properties of Seawater
    Water as a Solvent
    Sea Salts
    Dissolved Gases
    Alkalinity of Seawater
Physical Properties of Seawater
    Water Density and Temperature
    Pressure
    Sea Ice
    Sound Transmission
Conclusions
Basic Understandings
ESSAY: Desalination
ESSAY: Sea Ice Terminology

An iceberg in the Ross Sea, Antarctica. [Courtesy of NOAA/NESDIS/ORA]

Case-in-Point

Since antiquity, people have used sea salt to meet their nutritional needs and, prior to the age of refrigeration, to preserve food for storage and transport. Jericho, an oasis at the northern end of the Dead Sea, was founded almost 10,000 years ago as a salt trading center. The Dead Sea, located on the Israeli-Jordanian border, has long been a source of salt. The lowest spot on Earth's surface at about 400 m (1300 ft) below sea level, the Dead Sea occupies a depression in the transform-fault complex that separates the African and Indian plates (Chapter 2). Salts were deposited when tectonic movements cut off the Mediterranean Sea about 6 million years ago. The isolated seawater evaporated forming massive salt deposits. Some nearby mountains are almost pure salt and persist because of the arid climate and meager runoff.

During the Middle Ages, salt cod was an important commodity for members of the Hanseatic League, a trading group of merchants in northern Germany and the Baltic, and later an important food source for American colonists. Salt was so highly valued that it was used as money in Tibet and in many parts of Africa. Today, salt is still used as a seasoning or preservative in meatpacking, fish curing, and food processing. It is also used in curing hides and for control of ice and snow on roads and walkways in winter in cold climates. In the chemical industry, salt is involved in the manufacture of baking soda, hydrochloric acid, and chlorine and as a flux in metallurgical processes. Salt is also used in water softeners to remove calcium and magnesium compounds from tap water.

At one time, most commercial salt was derived from seawater or natural brines in hot dry climates using solar evaporation. Saltwater flows into shallow ponds separated by dikes where suspended materials such as sand or clay gravitationally settle to the bottom. The water then moves on to crystallizing pans where water evaporates and salt precipitates and accumulates in layers on the bottom of the pan. Many different salts precipitate from saltwater in addition to common household table salt (sodium chloride); these include, for example, calcium carbonate, calcium sulfate, and magnesium sulfate. As water evaporates, different salts precipitate at different times because each salt has a different solubility in water. Salt is also mined from deposits of rock salt. In some cases, solution mining is used whereby water is pumped into underground salt beds to create brine that is recovered and then evaporated.

Driving Question:

How do the properties of water and dissolved salts affect the physical and chemical properties of ocean water?

Although water is a very common component of the Earth system, it has some very uncommon properties compared to other substances of similar molecular size or chemical composition. Water's unusually high freezing and boiling temperatures, coupled with the temperature range at Earth's surface, mean that water exists naturally in the Earth system in all three phases, as solid (ice), liquid, and gas (water vapor). In fact, the three phases of water can coexist in contact with each other such as at the edge of pack ice. Water frequently changes phase and during these transitions, unusually large amounts of heat energy are either absorbed from or released to the environment with no change in the temperature of the water. Furthermore, large amounts of heat are required to change the temperature of water. Water dissolves a wide variety of solids, liquids, and gases and is aptly described as the *universal solvent*. Salinity and temperature affect the density of ocean water, which in turn influences the circulation of the ocean.

In this chapter we examine water's unique properties, the fundamental reasons for those properties, and some of the implications for the functioning of the ocean in the Earth system. We begin by describing the structure of the water molecule and explaining how this structure is the principal reason for water's unique physical and chemical properties. We then explore the chemical properties of seawater, emphasizing the types, sources, and cycling of dissolved salts and gases. The closing discussion of the physical properties of seawater covers the effect of temperature and salinity on water density, pressure exerted by seawater, the formation of sea ice, and the use of sound transmission in the ocean to determine water depth and temperature.

The Water Molecule

Compared to other naturally occurring substances, water's thermal properties are unique. For example, based on water's molecular weight as well as the freezing and boiling temperatures of chemically related substances, fresh water should freeze at about −90 °C (−130 °F) and boil at about −70 °C (−94 °F). Actually, fresh water's freezing point is 0 °C (32 °F) and its boiling point is 100 °C (212 °F) at average sea level air pressure. Water's unusual properties arise from the physical structure of the water molecule (H_2O) and hydrogen bonding between molecules. Without hydrogen bonding, water would exist only as a gas within Earth's range of surface temperatures and pressures and Earth would have no water cycle, no ocean, no ice caps, and probably no life.

The water molecule consists of two hydrogen (H) atoms bonded to an oxygen (O) atom (Figure 3.1). Within the water molecule, bonding between hydrogen and oxygen atoms involves sharing of electrons, one from each hydrogen atom and two from the oxygen atom. (An *electron* is a negatively charged subatomic particle.) In this bonding, the electrons spend more time with the oxygen so that the oxygen acquires a small negative charge and the hydrogens are left with a small positive charge. This covalent bonding is very strong so that the water molecule resists dissociation into its constituent hydrogen and oxygen atoms. The 105-degree angle formed by the arrangement of the hydrogen-

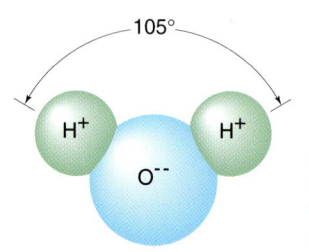

FIGURE 3.1
The water molecule consists of two hydrogen atoms and one oxygen atom.

oxygen-hydrogen atoms produces a charge separation in the water molecule. Molecules having a separation of positive and negative charges are described as *polar*.

Opposite electrical charges attract so that, like tiny magnets, neighboring water molecules link together. The positively charged (hydrogen) pole of a water molecule attracts the negatively charged (oxygen) pole of another water molecule; this attractive force is known as **hydrogen bonding**. Water molecules can form hydrogen bonds in three directions; that is, each molecule has three potential sites for hydrogen bonding, one on each of the hydrogen atoms, and one on the oxygen atom. Hydrogen bonding is roughly 10 to 50 times weaker than the bonds (covalent) between the hydrogen and oxygen atoms in individual water molecules. Nonetheless, they are strong enough to significantly influence the physical and chemical properties of water. Hydrogen bonding inhibits changes in water's internal energy so that it absorbs or releases unusually great quantities of heat when changing phase. Hydrogen bonding also means that greater additions or losses of heat are required to change water temperature as compared to other chemically related substances.

WATER AS ICE, LIQUID, AND VAPOR

Water is one of the very few substances that can occur naturally within the temperature and pressure ranges found at and near Earth's surface in all three phases, that is, as crystalline solid (ice), liquid (water), and gas (water vapor). In this section, we examine how water's molecular structure influences its properties in each of these phases.

Like all crystalline solids, ice has a regular internal three-dimensional framework consisting of a repeated pattern of molecules. A model of the *crystal lattice* of ice is shown in Figure 3.2. Each water molecule is bound tightly to its neighbors but intermolecular bonds are elastic (acting like springs) so that molecules vibrate about fixed locations in the lattice. For this reason, an ice cube retains its shape as long as temperatures are subfreezing. Hydrogen bonds are responsible for the ordered arrangement of water molecules in the crystal lattice that is responsible for the hexagonal (six-sided) structure of ice crystals. Because ice's internal framework is an open network of water molecules, the molecules in ice crystals are not as closely packed as a similar number of molecules in liquid water. At 0 °C (32 °F), ice has a density of about 0.92 g per cubic cm whereas pure liquid water at the same temperature has a density of nearly 1.0 g per cubic cm. This density difference explains why ice floats on liquid water. Most common solids would sink if placed in their liquid phase.

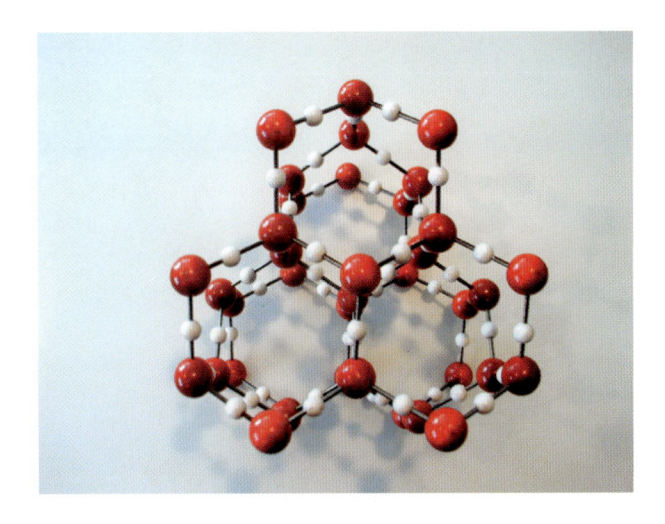

FIGURE 3.2
This is a model of the crystal lattice of ice. Each water molecule is bound tightly to its neighbors but intermolecular bonds are elastic so that molecules vibrate about fixed locations in the lattice. For this reason, an ice cube retains its shape as long as temperatures remain subfreezing. The ordered arrangement of water molecules in the crystal lattice is responsible for the hexagonal structure of ice crystals. Because ice's internal framework is an open network of water molecules, the molecules in ice crystals are not as closely packed as a similar number of molecules in liquid water so that ice is less dense than liquid water.

Despite the openness of ice's lattice, sea ice (formed when seawater freezes) contains less salt than does seawater. Salt ions are too large to fit into the empty spaces within the crystal lattice of ice and cannot substitute for water molecules. Hence, most impurities (salts and gases) are excluded from ice that forms when seawater freezes.

When ice melts, it becomes liquid water. After observing ice crystals disappearing during melting, we might expect that all hydrogen bonds between water molecules would break during the transition from ice to liquid water. This is not the case. Instead, many water molecules remain linked by hydrogen bonding as transient clusters of molecules surrounded by non-bonded (free) water molecules (Figure 3.3). Although molecular clusters persist into the liquid phase, water molecules exhibit much greater activity in the liquid than solid phase. In the liquid phase, water molecules undergo vibrational, rotational, and translational (straight-line) motions. This greater freedom of movement explains why liquid water takes the shape of its container.

When liquid water changes to vapor, essentially all hydrogen bonds are broken. Individual molecules move about with even greater freedom than in the liquid phase, diffusing rapidly to fill the entire volume of its container.

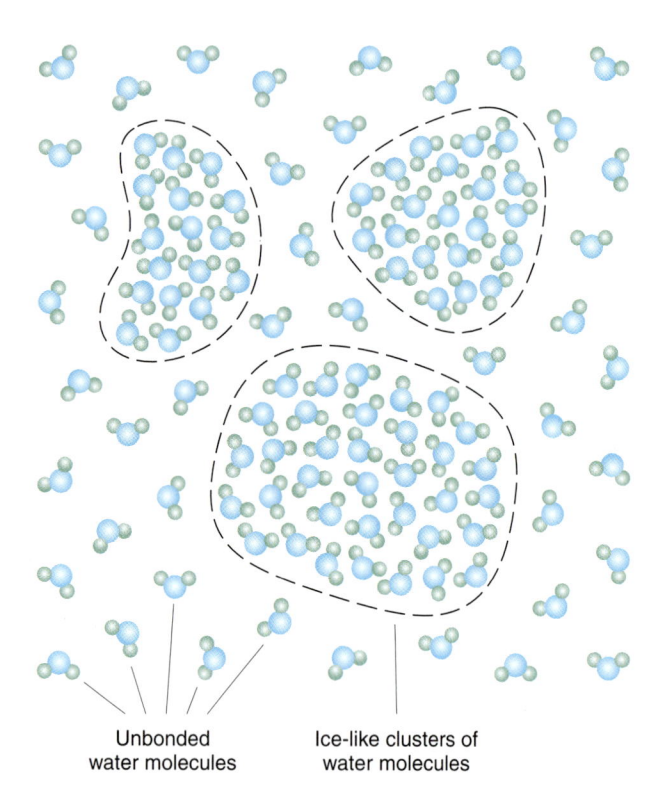

FIGURE 3.3
When water changes phase from ice to liquid, many water molecules remain linked by hydrogen bonding as transient clusters of molecules surrounded by non-bonded water molecules.

these phase changes of water in detail, we need to review the distinction between temperature and heat.

TEMPERATURE AND HEAT

From everyday experience, we know that temperature and heat are closely related. Heating a pan of soup on the stove raises the temperature of the soup whereas dropping an ice cube into a warm beverage lowers the temperature of the beverage. Although sometimes used interchangeably, temperature and heat are distinctly different concepts.

All matter is composed of atoms or molecules that are in continual vibrational, rotational, and/or translational motion. The energy represented by this motion is referred to as kinetic molecular energy or just *kinetic energy*, the energy of motion. In any substance, atoms or molecules actually exhibit a range of kinetic energy. **Temperature** is directly proportional to the *average* kinetic energy of atoms or molecules composing a substance. At the same temperature, one liter of water has the same average kinetic molecular energy per molecule as 50 liters of water.

Internal energy encompasses all the energy in a substance, that is, the kinetic energy of atoms and molecules plus the potential energy arising from forces between atoms or molecules. If two objects have different temperatures (different average kinetic molecular energies) and are brought into contact, energy will be transferred

Gas molecules exhibit vibrational, rotational, and translational motion and exert a force as they bombard a solid or liquid surface. Force per unit area is defined as *pressure*. Heating a closed rigid container of gas accelerates the activity of gas molecules so that they collide more frequently with the inside surfaces of the walls of the container and exert more pressure. Hence, if the volume is held constant, the pressure exerted by a gas increases as the temperature rises. The situation is more complicated for air (a mixture of gases) because the atmosphere is only bounded below (by continents and ocean) and a unit mass of air is free to change volume. When heated, air in the open atmosphere expands and its density and pressure decrease.

Water readily changes phase, contributing to the dynamic nature of the Earth system. In changing phase from ice to liquid to vapor, the energy state (i.e., molecular activity) of water increases. Hence, when water changes phase heat is either absorbed from the environment or released to the environment (Figure 3.4). Melting, evaporation, and sublimation are phase changes that absorb heat. Phase changes that release heat to the surroundings are freezing, condensation, and deposition. Before examining

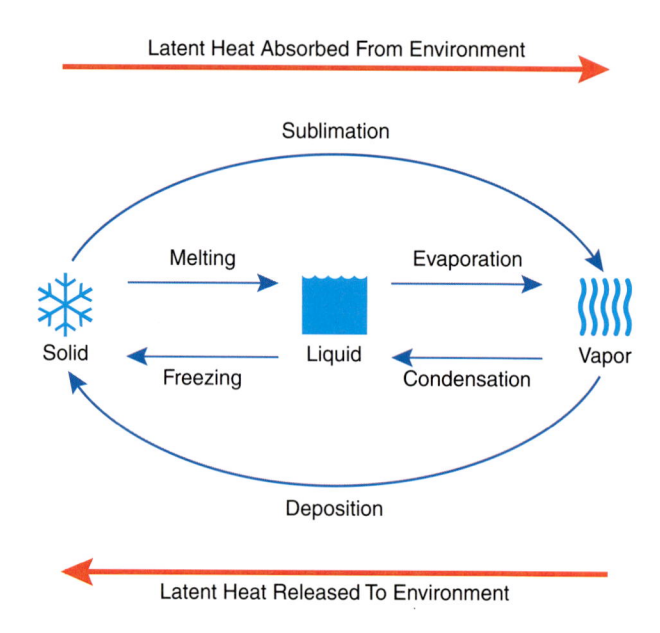

FIGURE 3.4
When water changes phase, heat energy is either absorbed from the environment (melting, evaporation, sublimation) or released to the environment (freezing, condensation, deposition).

between objects; we call this energy in transit, **heat**. Heat transferred from an object reduces the internal energy of that object whereas heat absorbed by an object increases its internal energy. Heat transferred to or from water brings about a change in temperature or a change in phase. An addition or loss of heat can also do work by bringing about a change in volume of a substance. For example, ice expands when heated and contracts when cooled.

Differences in temperature rather than differences in internal energy govern the direction of heat transfer. Heat energy is always transferred from a warmer object to a colder object. Heat is not necessarily transferred from an object having greater internal energy to an object with less internal energy. Consider, for example, a hot marble (at 40 °C) that is dropped into 5 liters of cold water (at 5 °C). The water has much more internal energy than the marble; nonetheless, heat is transferred from the warmer marble to the cooler water.

The following illustration makes clearer the distinction between temperature and heat. A cup of water at 60 °C (140 °F) is much hotter than bathtub water at 30 °C (86 °F); that is, the average kinetic molecular energy of water molecules is greater at 60 °C than at 30 °C. Although lower in temperature, the greater volume of water in the bathtub means that it contains more total kinetic molecular energy than does the cup of water. If in both cases the water is warmer than its environment, heat is transferred from water to its surroundings. However, much more heat energy must be removed from the bathtub water than from the cup of water for both to cool to the same temperature.

Through the years, scientists have devised various scales that express the temperature of an object by a number representing the degree of warmth. Temperature scales were originally derived using the freezing and boiling points of water as fixed points of reference. Eighteenth century scientists who developed temperature scales used water for this purpose because it was readily available and inexpensive. In this book we use the Celsius temperature scale primarily with the Fahrenheit equivalent in parentheses. The Celsius scale, devised by the Swedish astronomer Anders Celsius in 1742, has the numerical convenience of a 100-degree increment between the freezing point (0 °C) and boiling point (100 °C) of fresh water at average sea level air pressure. The United States is one of only a few nations still using the Fahrenheit temperature scale, introduced by the German physicist Gabriel Daniel Fahrenheit in 1714. The Fahrenheit temperature scale features a 180-degree increment between the freezing point (32 °F) and boiling point (212 °F) of fresh water at average sea level air pressure. Formulas for converting between the Celsius and Fahrenheit temperature scales are presented in Table 3.1.

A convenient unit of heat energy is the *calorie*, defined as the amount of heat needed to raise the temperature of one gram of water by one Celsius degree (technically, from 14.5 °C to 15.5 °C). (The calorie used to measure the energy content of food is actually 1000 heat calories or 1 kilocalorie.) Although the preferred unit of energy in any form, including heat, is the *joule (J)*, we generally use the *calorie* in this book because of its numerical convenience. Furthermore, thermal characteristics of water were used to define the calorie. For conversion purposes, one calorie equals 4.1868 J and one joule equals 0.239 calorie.

CHANGES IN PHASE OF WATER

A mixture of fresh water and ice has an equilibrium temperature of 0 °C (32 °F). Adding heat to the mixture causes ice to melt, whereas removing heat causes water to freeze. For that reason, 0 °C is called the freezing point of fresh water. Substances dissolved in water suppress the equilibrium temperature of a mixture of ice and water to temperatures below the freezing point of fresh water. The temperature at which seawater freezes varies with *salinity*, a measure of the mass (grams) of salts dissolved in a kilogram of seawater. We have more on the effect of salinity on the freezing point of seawater later in this chapter.

When water freezes, **latent heat** is released to the environment and for ice to melt an equivalent amount of latent heat is absorbed from the environment. The word *latent* means "hidden" and refers to the fact that this heat energy is used only to change the phase of water and not the temperature of the water. If heat is added to a mixture of ice and water at 0 °C (32 °F), the temperature of the ice-water mixture remains constant until all the ice melts. All available heat is used to bring about the phase change by breaking some of the hydrogen bonds that maintain water in the solid (crystalline) phase. Whether freezing or melting is taking place, the latent heat involved is commonly called the **latent heat of fusion**.

At the interface between liquid water and air (e.g., the sea surface), water molecules continually change phase: some crossing the interface from water to air and others

TABLE 3.1
Temperature Conversion Formulas

F = 9/5 C + 32
C = 5/9 (F − 32)

from air to water. If more water molecules enter the atmosphere as vapor than return as liquid, a net loss occurs in liquid water mass. This process is known as **evaporation**. Evaporation explains the disappearance of puddles following a rain shower. On the other hand, if more water molecules return to the water surface as a liquid than escape as vapor, net gain of liquid water mass results. This process is called **condensation**. Water vapor condenses on the cold surface of an aluminum beverage can on a humid summer day. Heat absorbed from the environment during evaporation and heat released to the environment during condensation are known as the **latent heat of vaporization** and the **latent heat of condensation** respectively.

All of us have experienced **evaporative cooling**. We are chilled upon stepping out of a shower or swimming pool. Water droplets evaporating from the skin absorb heat, lowering the skin's temperature. On a global scale, evaporative cooling is the most important process whereby heat at Earth's surface is transferred from the ocean to the atmosphere (Chapter 5). When water evaporates at Earth's surface, water vapor moves into the atmosphere where it may subsequently condense into clouds. Heat absorbed as water evaporates at the surface is later released to the atmosphere during condensation. This latent heat transfer mechanism is also important in powering storms, especially thunderstorms and tropical cyclones (e.g., hurricanes).

Water's latent heat of fusion is 80 calories per g at 0 °C. Considerably greater amounts of heat are required for water to vaporize because essentially all hydrogen bonds are broken during the phase change. (Recall that not all bonds are broken as water changes from solid to liquid and clusters of bonded water molecules persist into the liquid phase.) In fact, the magnitude of water's latent heat of vaporization is about seven times that of its latent heat of fusion and the highest of all common substances. The latent heat of vaporization varies with temperature from 597 calories per g at 0 °C (32 °F) to 540 calories per g at 100 °C (212 °F). More energy is needed to break the tighter and more numerous hydrogen bonds at lower temperature.

At the interface between ice and air (e.g., the surface of a snow cover), water molecules are also continually changing phase: directly from ice to vapor and from vapor to ice. If more water molecules enter the atmosphere as vapor than become ice, a net loss of ice mass occurs. **Sublimation** is the process whereby ice or snow becomes vapor without first becoming a liquid. Sublimation explains the gradual disappearance of a snow cover even while the air temperature remains well below freezing. On the other hand, if more atmospheric water molecules become ice than move from ice to vapor, a net gain of ice mass results. **Deposition** is the process whereby water vapor becomes ice without first becoming a liquid. During a cold winter night, the formation of frost on automobile windows is a common example of deposition.

Heat is absorbed from the environment during sublimation and heat is released to the environment during deposition. Latent heat involved in sublimation or deposition must equal the total amount of heat absorbed or released during the combined solid-liquid plus liquid-vapor phase changes. Sublimation requires the latent heats of fusion plus vaporization, known as the **latent heat of sublimation**. Deposition releases to the environment an equivalent amount of latent heat, that is, the **latent heat of deposition**. The magnitudes of the latent heats of sublimation and deposition are remarkably uniform, varying only from 677 calories per g at 0 °C to 678 calories per g at –30 °C.

SPECIFIC HEAT OF WATER

The temperature change associated with an input (or output) of a specified quantity of heat varies from one substance to another. The amount of heat that will raise the temperature of 1 gram of a substance by 1 Celsius degree is defined as the **specific heat** of that substance. Joseph Black, a Scottish chemist, first proposed the concept of specific heat in 1760. The specific heat of all substances is measured relative to that of liquid water, which is defined as 1 calorie per g per Celsius degree (at 15 °C). The specific heat of ice is about 0.5 calorie per g per Celsius degree (near 0 °C). Specific heats of other familiar substances are listed in Table 3.2. The variation in specific heat from one substance to another implies that different materials have different capacities for storing internal energy.

Upon absorbing the same amount of heat energy, a substance with a high specific heat experiences a smaller increase in temperature (warms less) than a substance having a low specific heat. Because of hydrogen bonding, water has an unusually high specific heat, in fact the highest specific heat of any naturally occurring liquid or solid. From Table 3.2, water's specific heat is about 5 times that of dry sand. Whereas one calorie of heat will raise the temperature of one gram of water by 1 Celsius degree, the same quantity of heat will raise the temperature of one gram of dry sand by slightly more than 5 Celsius degrees. This contrast in specific heat helps explain why at the beach in summer the sand feels considerably hotter to bare feet than does the water. This also largely explains why wet sand feels cooler than dry sand.

TABLE 3.2
Specific Heat[a] of Some Familiar Substances

Water	1.000
Wet mud	0.600
Ice (at 0 °C)	0.478
Wood	0.420
Aluminum	0.214
Brick	0.200
Granite	0.192
Sand	0.188
Dry air[b]	0.171
Copper	0.093
Silver	0.056
Gold	0.031

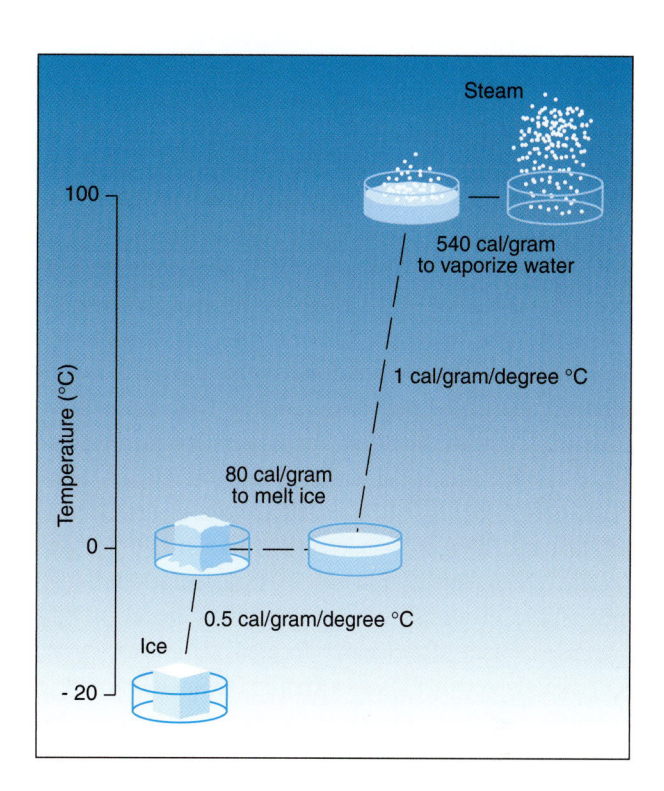

[a]Calories per gram per Celsius degree
[b]At constant volume

FIGURE 3.5
Heating a 1-gram ice cube causes a rise in temperature plus phase changes, initially to liquid and then to vapor.

A simple experiment conducted at sea level summarizes water's unusually high latent heats and specific heat (Figure 3.5). A one-gram ice cube initially at –20 °C (-4 °F) is heated to 0 °C (32 °F). Every one Celsius-degree (1.8 Fahrenheit-degree) rise in ice cube temperature requires an addition of 0.5 calorie of heat energy (the *specific heat* of ice). A total of 10 calories of heat (20 C° × 0.5 cal per C°) is required to warm the one-gram ice cube to 0 °C (32 °F). The temperature then remains constant while 80 calories of heat (*latent heat of fusion*) are added to melt the one-gram ice cube. Liquid water is then heated to 100 °C (212 °F). Every one-Celsius degree rise in temperature of the one gram of liquid water requires an addition of one calorie of heat energy (*specific heat of water*). If none of the water evaporates, a total of 100 calories of heat (100 C° × 1 calorie per C°) are required to warm the 1 gram of liquid water to 100 °C. At 100 °C, the water vaporizes requiring an input of 540 calories of heat (*latent heat of vaporization*).

MARITIME INFLUENCE ON CLIMATE

Water's exceptional capacity to store heat has important implications for weather and climate. A large body of water (such as the ocean or Great Lakes) can significantly influence the climate of downwind localities. The most persistent influence is on air temperature. Compared to an adjacent landmass, a body of water does not warm as much during the day (or in summer) and does not cool as much at night (or in winter). In other words, a large body of water exhibits a greater resistance to temperature change, called **thermal inertia**, than does a landmass. Whereas the higher specific heat of water versus land is the major reason for the contrast in thermal inertia, differences in heat transport also contribute. Sunlight penetrates water to some depth and is absorbed (converted to heat) through a significant volume of water. But sunlight cannot penetrate the opaque land surface and is therefore absorbed only at the surface. Furthermore, ocean and lake-waters circulate and transport heat through great volumes of water, whereas heat is conducted only very slowly into soil. The input (or output) of equal amounts of heat energy causes a land surface to warm (or cool) more than the equivalent surface area of a body of water.

Air temperature is regulated to a considerable extent by the temperature of the surface over which air resides or travels. Air over a large body of water tends to take on similar temperature characteristics as the surface water. Places immediately downwind of the ocean experience much less contrast between average winter and summer temperatures and the climate is described as *maritime*. Places at the same latitude but well inland experience a much greater temperature contrast between winter and summer and the climate is described as *continental*. However, in winter the climatic influence of ice-covered bodies of water (e.g., the Bering and Greenland Seas) is more like a landmass than an ocean.

Consider an example of the contrast in climate between continental and maritime locations. The latitude of San Francisco, CA (37.8 degrees N) is almost the same as that of St. Louis, MO (38.8 degrees N) so that the seasonal variation in the amount of solar radiation striking Earth's atmosphere (due to astronomical factors) is about the same at both places. St. Louis is situated far from the moderating influence of the ocean and its climate is continental. St. Louis' average summer (June, July, August) temperature is 25.6 °C (78.0 °F) and its average winter (December, January, February) temperature is 0.6 °C (33.0 °F), giving an average summer-to-winter seasonal temperature contrast of 25 Celsius degrees (45 Fahrenheit degrees). San Francisco, on the other hand, is located on the West Coast, immediately downwind of the Pacific Ocean; its climate is distinctly maritime. The average summer temperature at San Francisco is 17 °C (62.6 °F) and the average winter temperature is 10.2 °C (50.4 °F), giving an average seasonal temperature contrast of only 6.8 Celsius degrees (12.2 Fahrenheit degrees).

The moderating influence of the ocean is also evident in the contrast in climate between Western Europe and Eastern North America. At mid-latitudes, prevailing winds blow from west to east so that the maritime influence of the North Atlantic Ocean is much more apparent in Western Europe than Eastern North America. In the same latitude belt, winters are considerably milder in Western Europe than in Eastern North America. Consider, for example, the contrast in average January temperatures for Montreal, Quebec (45.5 degrees N) versus London, England (51.5 degrees N). In January, the average daily high temperature is –6.1 °C (21 °F) at Montreal and 6.7 °C (44 °F) in London. The January average daily low temperature is –14.4 °C (6 °F) at Montreal and 1.7 °C (35 °F) in London. Although London is considerably farther north than Montreal, its January temperatures are significantly milder.

Chemical Properties of Seawater

Pure water is unknown in the Earth system; that is, water free of all dissolved and suspended materials does not occur naturally. This is because water is an excellent solvent for a wide range of materials—in fact, water is sometimes referred to as the *universal solvent*. Water dissolves solids, liquids, and gases. The ocean, by far the largest reservoir of water in the Earth system, contains so much salt in solution that it cannot be used for most domestic, agricultural or industrial purposes. In the first Essay at the close of this chapter, we describe desalination techniques that are designed to remove salts from seawater and augment the supply of fresh water. In this section, we consider water as a solvent and then describe the types and sources of dissolved salts and gases in seawater.

WATER AS A SOLVENT

The polar nature of the water molecule favors the solution (i.e., dissolving) of both ionic and non-ionic substances. Many inorganic materials (primarily salts) are bonded ionically, whereas many organic chemicals have non-ionic bonds. River water, groundwater, and ocean water dissolve some of the rock or sediment (both organic and inorganic) that water contacts and some Earth materials dissolve in water more readily than other Earth materials. In Chapter 4, we consider the types and sources of sediments that dissolve or are suspended in ocean water.

Consider what happens when a pinch of common household table salt is added to water. Table salt is sodium chloride (NaCl), the mineral known as halite. In salt's cubic crystalline form, ionic bonds hold the positively charged sodium ions (Na$^+$) and the negatively charged chloride (Cl$^-$) ions together (Figure 3.6A). (An *ion* is an electrically charged atom.) Once salt enters the water, however, the hydrogen-bonded complexes of water molecules greatly reduce the force of attraction between oppositely charged sodium and chloride ions. That is, the strength of ionic bonding between sodium and chloride diminishes so that the compound readily dissociates into sodium and chloride ions (Figure 3.6B). Sodium ions are attracted to the negatively charged pole of the water molecule while chloride ions are attracted to the positively charged pole of the water molecule. In this way, salt dissolves in water.

SEA SALTS

Seawater is a salt solution of nearly uniform composition; only the relative amount of water in the solution varies. **Salinity** is a measure of the amount of salt dissolved in seawater. On average, seawater is 96.5% water and 3.5% dissolved salts. If all ocean water evaporated, the precipitated salts would cover the entire planet to a depth of about 45.5 m (150 ft).

In the 19th century, chemists began examining the composition of seawater in some detail, and one of them, William Dittmar, verified an important observation made by

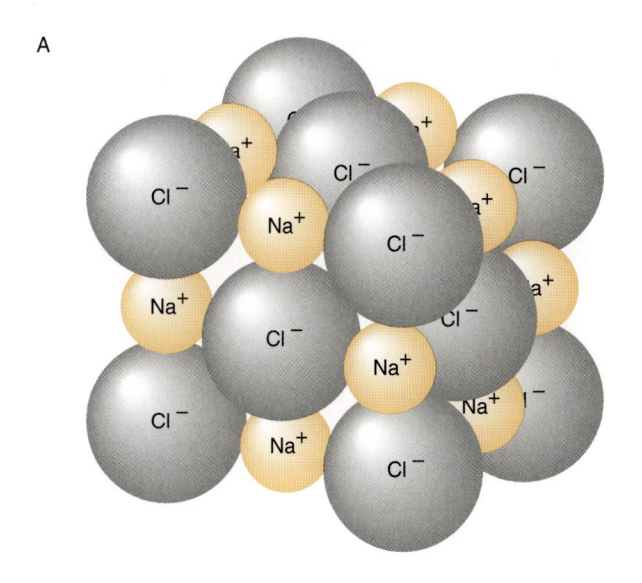

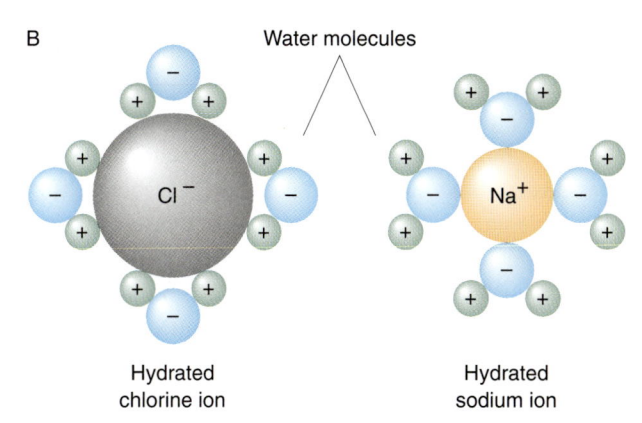

Chapter 3 PROPERTIES OF OCEAN WATER

The major dissolved constituents of ocean water, such as chloride (Cl⁻) and sodium (Na⁺) ions, are *conservative properties* of seawater; that is, they occur in constant proportions and change concentration very slowly by mixing or diffusion. Seawater constituents that participate in biogeochemical or seasonal cycles have variable concentrations and are described as *non-conservative properties*. Recall from Chapter 1 that the ocean is a reservoir in all biogeochemical cycles operating in the Earth system. Examples of non-conservative properties of seawater are silica and calcium compounds, nitrates, phosphates, and aluminum.

The principle of constant proportions implies that measurement of the concentration of one of the major conservative constituents of seawater is all that is needed to determine the concentration of any other conservative constituent (as well as the sum of all other conservative constituents). This principle made it possible for ocean scientists to define and measure salinity. Prior to World War II, salinity was defined as the total amount of solid materials in grams dissolved in one kilogram of seawater. The standard method for measuring salinity was chemical titration to determine the concentration of the chloride ion (Cl⁻), which was then substituted into a formula to calculate salinity. Salinity was expressed as grams of dissolved material per kilogram of seawater or parts per thousand (ppt). The accuracy of this method was about ±0.02 ppt.

After World War II, scientists found that salinity could be determined from measurements of the electrical conductivity of seawater with 10 times greater accuracy than the old titration method. In 1978 the international oceanographic community accepted a new definition of salinity based on conductivity measurements; that is, the salinity of a water sample is defined as the ratio of the conductivity of the sample to the conductivity of *standard seawater*. Samples of standard seawater, prepared by Ocean Scientific International in southern England, are supplied to laboratories around the world for use in calibrating salinity-measuring instruments. As a ratio, salinity has no units and may be written as a pure number, but for convenience some ocean scientists express salinity in *practical salinity units (psu)*. In this book, we present salinity values as a number without units. The conductivity-based definition of salinity was designed to retain the validity of measurements made the old way; that is, a salinity of 35 is essentially the same as 35.0 parts per thousand, or 3.5%. A seawater sample having a salinity of 34.82 contains about 34.82 grams of dissolved materials per kilogram.

Although more than 70 elements are dissolved in seawater, six make up more than 99% of all sea salts: chlo-

FIGURE 3.6
Common household table salt, sodium chloride (NaCl), dissolves in water. (A) In its cubic crystalline form, ionic bonds hold together the positively charged sodium ions (Na⁺) and the negatively charged chloride (Cl⁻) ions. (B) Once salt enters the water, however, the hydrogen-bonded complexes of water molecules greatly reduce the force of attraction between oppositely charged sodium and chloride ions. The compound readily dissociates into sodium and chloride ions with the sodium ions attracted to the negatively charged pole of the water molecule and chloride ions attracted to the positively charged pole of the water molecule.

earlier scientists. Dittmar analyzed the constituents of 77 water samples obtained at various depths and locations in the ocean during the worldwide voyage of *HMS Challenger* (from December 1872 to May 1876). He found that although the total amount of dissolved solids varied among water samples, the ratio of the concentrations of the major constituents of seawater was the same in all samples. That is, the major constituents of seawater occur in the same relative concentrations throughout the ocean system, a characteristic of seawater described as the **principle of constant proportions**.

62 Chapter 3 PROPERTIES OF OCEAN WATER

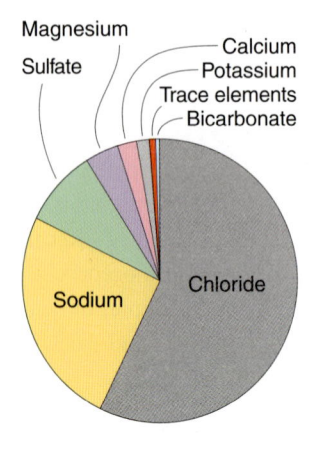

FIGURE 3.7 Although more than 70 ions are dissolved in seawater, six make up more than 99% of all sea salts: chloride (Cl⁻), sodium (Na⁺), sulfate (SO_4^{2-}), magnesium (Mg^{2+}), calcium (Ca^{2+}), and potassium (K⁺).

ride (Cl⁻), sodium (Na⁺), sulfate (SO_4^{2-}), magnesium (Mg^{2+}), calcium (Ca^{2+}), and potassium (K⁺) (Figure 3.7). Common table salt (NaCl) alone accounts for nearly 86%. Numerous trace components include aluminum, chromium, gold, lead, nickel, and zinc.

What is the origin of the salts dissolved in seawater? One source is weathering and erosion of rock on land and transport by rivers to the ocean (Chapter 4). As the largest reservoir in the global water cycle, the ocean receives most of its water from rivers. Although seawater is considerably more saline than river water, we might expect the ratios of the different chemical constituents to be essentially the same in river and ocean waters. But this is not the case. Whereas sodium and chloride account for 86% of solids dissolved in seawater, they typically constitute less than 16% of dissolved solids in river water. Calcium (Ca^{2+}) and bicarbonate (HCO_3^-) ions are minor ingredients of seawater (less than 2% of dissolved solids) but major components of river water (almost 50% of dissolved solids). On average, silica (SiO_2) accounts for about 14.5% of the dissolved components of river water but is a minor component of seawater.

What explains the difference in the chemical makeup of dissolved solids in river water versus seawater? For one, marine organisms extract calcium and silica from seawater to build their shells and skeletons. Also some forms of marine life concentrate, secrete, or excrete certain chemical elements. Differences in solubility and rates of physical-chemical reactions among ions also play a role by limiting the concentration of certain substances in seawater (e.g., calcium) or causing some chemicals to precipitate from solution (e.g., manganese). Furthermore, there are other sources of salts dissolved in seawater besides rivers.

Hydrothermal vents and chemical reactions between seawater and recently formed oceanic crust are important controls of ocean water salinity (Chapter 2). The entire volume of the world ocean cycles through fractures in new oceanic crust about every 10 million years. Suspended particles settle out of or are washed from the atmosphere by precipitation. For example, gases emitted during volcanic eruptions dissolve in rainwater and enter the ocean contributing chloride and sulfate ions.

The proportionality of the principal sea salts has remained nearly constant for about the past 1.5 billion years implying that the rate of addition of new salts to the ocean must balance the rate of removal. This is evident from the composition of salt strata (sedimentary rock) formed from the evaporation of seawater in the geologic past, which has changed little over that period of time. Processes that remove salt ions from the ocean include sea spray that is blown ashore and isolation of arms of the sea from the ocean followed by evaporation to form salt deposits. For example, about 6 million years ago, northward movement of the African plate cut off the Mediterranean Sea from the Atlantic Ocean at Gibraltar. In roughly 1000 years, the Mediterranean almost completely dried up precipitating thick layers of salt in the process. Certain ions chemically react with one another to form insoluble precipitates and some ions adsorb onto suspended sediments. In both cases, ions settle into the seafloor and mix with other sediments. Compounds used by marine organisms to build shells or skeletons, such as calcium and silica, are removed from seawater and upon death of the organism cycle to the ocean floor as shells, bones, or teeth (Chapter 4). Also, chemical reactions between seawater and newly formed oceanic crust remove some constituents of seawater such as magnesium ions (Mg^{2+}). Residence times of salts in seawater range from hundreds of years (e.g., aluminum, iron) to millions of years (e.g., sodium, potassium).

Processes operating at the atmosphere/ocean interface add or remove water molecules from seawater and largely explain spatial variations in sea-surface salinity. Most dissolved substances are left behind when seawater evaporates or freezes, increasing the surface salinity locally. On the other hand, precipitation, runoff from rivers, and melting ice add fresh water and decrease the local surface salinity. On an average annual basis, freezing of water and melting of ice have a negligible net effect on sea-surface salinity because the increase in salinity produced by ice formation in winter is offset by the decrease in salinity produced by melting ice during summer. Where large rivers enter the ocean, salinity is reduced as freshwater and seawater mix; salinity is also reduced where rainfall is heavy. Runoff is only important in coastal regions, so evaporation and precipitation are the principal processes governing surface salinity over most of the ocean. The salinity of

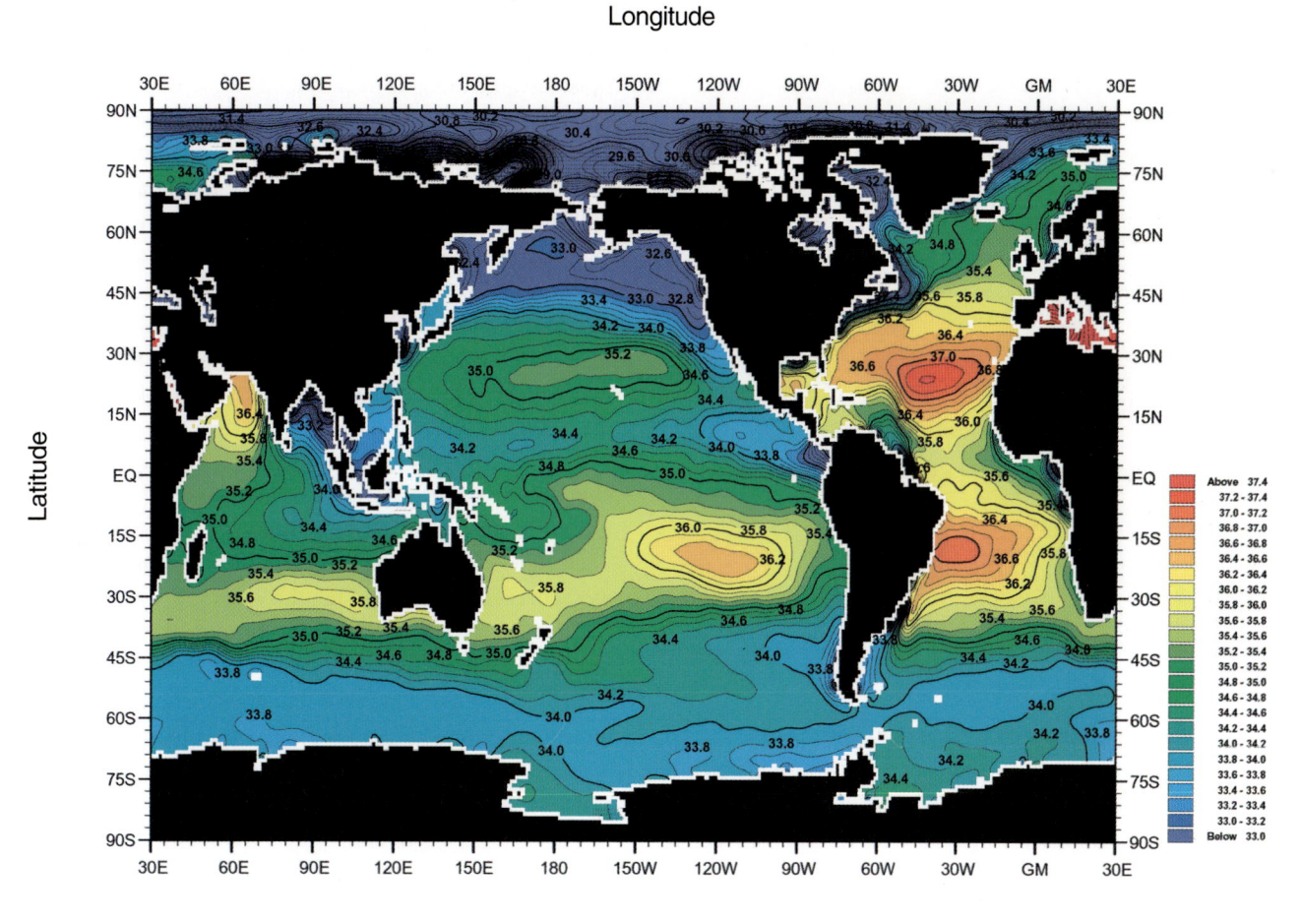

FIGURE 3.8
Global pattern of average annual sea-surface salinity shows the highest values in the subtropics of the Northern and Southern Hemispheres. Salinity values are in parts per thousand. [NOAA, National Oceanographic Data Center]

surface waters of the open ocean averages about 35 and rarely falls below 33 or rises above 38. The surface salinity of the open ocean peaks near the center of each ocean basin, that is, in the subtropics (between 20 and 30 degrees N and between 15 and 25 degrees S) where the annual evaporation rate exceeds precipitation (Figure 3.8). Average sea-surface salinity is generally lowest near the equator and in polar areas where annual precipitation is greater than the rate of evaporation.

DISSOLVED GASES

We are well aware from personal experience that solids such as salt and sugar dissolve in water. What may be less obvious is that gases also dissolve in water. For example, carbon dioxide (CO_2) dissolves under pressure in a carbonated beverage. If you open a can of cola, tiny CO_2 bubbles escape, giving the drink its fizz. Dissolved gases are also present in ocean water; these include carbon dioxide, nitrogen (N_2), and oxygen (O_2). (Note that dissolved oxygen is not the O in the H_2O.)

Gases are exchanged between the atmosphere and ocean at the ocean surface. If seawater is not saturated with a specific gas, the gas is transferred from air to ocean. If seawater is supersaturated with the gas, the transfer is in the opposite direction. When water is saturated with a gas, the rate at which the gas dissolves in water equals the rate at which the gas escapes to the atmosphere. The saturation concentration of a gas in water depends primarily on temperature in fresh water bodies such as lakes and rivers and a combination of temperature, salinity, and pressure in seawater. Almost all gases are more soluble in cold water than in warm water (Figure 3.9). For example, all other factors being equal, we would expect less dissolved oxygen in lakes in summer than in the cooler seasons. As the temperature or salinity of seawater increases, water holds less gas at saturation; of the two factors, temperature is much more significant.

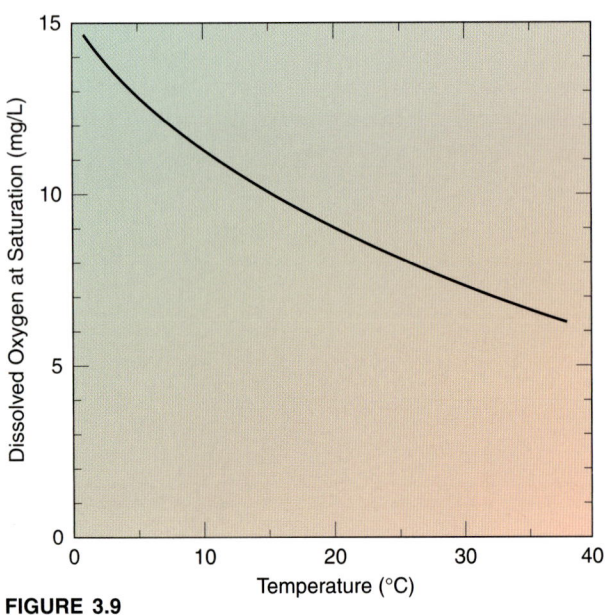

FIGURE 3.9
The saturation value of dissolved oxygen in fresh water decreases with rising temperature.

Waves on the ocean surface facilitate the transfer of gas between the atmosphere and ocean. Waves roughen the surface and increase the surface area of ocean water exposed to the atmosphere for gas exchange. On the other hand, in cold climates, sea-ice cover is a barrier to gas transfer at the ocean surface.

Below the ocean-atmosphere interface, biochemical processes play important roles in controlling the proportions of certain dissolved gases. Within the **photic zone**, the sunlit upper layer of the ocean where photosynthesis takes place, dissolved oxygen is enhanced relative to carbon dioxide. Recall from Chapter 1 that *photosynthesis* is the process whereby green plants (e.g., phytoplankton) use sunlight, water, and carbon dioxide to manufacture their food and generate oxygen as a byproduct. In surface ocean waters, the principal dissolved gases are nitrogen (48%), oxygen (36%), and carbon dioxide (15%). Beneath the photic zone, photosynthesis is absent and decomposer activity and cellular respiration are the most important biochemical processes. Recall from Chapter 1 that *cellular respiration* is the process whereby organisms break down food and release energy in a usable form. Through cellular respiration, marine organisms use dissolved oxygen and release carbon dioxide as a byproduct. For the ocean as a whole, CO_2 is the most abundant dissolved gas, accounting for 83% of the total.

Carbon dioxide is much more abundant in the ocean than the atmosphere. As noted in Chapter 1, CO_2 is only 0.037% by volume of the gases composing the atmosphere below an altitude of 80 km (50 mi). About seven times as much CO_2 is sequestered in the ocean. Carbon dioxide is highly soluble in water because it reacts with water to produce carbonate (CO_3^{2-}) and bicarbonate (HCO_3^-) ions that readily dissolve in water. At a temperature of 0 °C (32 °F) and sea level air pressure, one liter of water that is saturated with carbon dioxide has dissolved about 1.7 liters of CO_2. The ocean's ability to take up atmospheric carbon dioxide is an important consideration in recent discussions regarding the possible climatic implications of the rising levels of atmospheric CO_2. We have much more to say on this topic in Chapter 12.

If a dissolved gas does not participate in any biochemical process such as photosynthesis or cellular respiration, its concentration in a parcel of seawater remains unchanged except by the relatively slow movements of gas molecules (*diffusion*) through the water or by mixing with other water masses containing different amounts of dissolved gas. Generally, nitrogen and inert (chemically nonreactive) gases such as argon and neon behave in this way and their concentrations are described as *conservative properties*. As noted above, biochemical processes influence the concentration of some gases dissolved in seawater, primarily oxygen and carbon dioxide; the concentrations of these dissolved gases are examples of *nonconservative properties*.

ALKALINITY OF SEAWATER

Water in the various reservoirs of the Earth system can vary in acidity and alkalinity. An acid is a hydrogen-containing compound that releases hydrogen ions (H^+) when dissolved in water. Strong acids more readily release hydrogen ions than weak acids. An alkaline substance releases hydroxyl ions (OH^-) when dissolved in water and may be weak or strong. Pure water has properties of both acids and alkaline materials as water molecules continually break up (into hydrogen and hydroxyl ions) and form. That is,

$$H_2O \leftrightharpoons H^+ + OH^-$$

The acidity and alkalinity of water (or any other substance) is expressed as pH, a measure of the hydrogen ion concentration. On the **pH scale**, the pH increases from 0 to 14 as the hydrogen ion concentration decreases (Figure 3.10). Pure water has a pH of 7, which is considered neutral; a pH above 7 is increasingly alkaline whereas a pH below 7 is increasingly acidic. The pH scale is logarithmic; that is, each unit increment corresponds to a tenfold change

Chapter 3 PROPERTIES OF OCEAN WATER 65

all depths; that is, seawater is slightly alkaline. The relatively narrow variation in the pH of seawater is very important for marine organisms whose shells and skeletons are composed of calcium carbonate ($CaCO_3$). If ocean water were even slightly acidic, calcium carbonate would dissolve and so would be unavailable for these organisms.

Carbon dioxide plays a key role in controlling the pH of seawater. A substance that stabilizes a chemical system in this way is known as a **buffer**. Atmospheric CO_2 dissolves in ocean water producing carbonic acid (H_2CO_3) that dissociates into hydrogen (H^+), carbonate (CO_3^{2-}), and bicarbonate (HCO_3^-) ions. A chemical equilibrium develops in which carbon dioxide, carbonic acid, hydrogen ions, carbonate ions, and bicarbonate ions co-exist. Adding acid to the ocean shifts the equilibrium so that there are fewer carbonate ions, which in turn reduces the hydrogen ion concentration and raises the pH. Adding alkaline materials to ocean water raises the pH but the equilibrium shifts in such a way as to return the pH to the normal range.

Lake waters are generally not as well buffered as seawater and can exhibit greater variations in pH in response to influxes of acid precipitation or runoff. Rain and snow are normally slightly acidic as a result of dissolving atmospheric CO_2. Where air also contains oxides of sulfur (e.g., from coal burning) or oxides of nitrogen (e.g., byproducts of high-temperature industrial processes), precipitation can become much more acidic than normal. Rainwater that is saturated with CO_2 has a pH of 5.6 and any rain or snow having a pH below this value is designated **acid rain** (or *acid snow*). Lakes whose basins consist of carbonate rocks (i.e., limestone or dolostone) usually have sufficient carbonates to neutralize acid rain whereas lakes with non-carbonate basins (e.g., granite) cannot buffer the acid rain, snow, and runoff they receive. Lowering the pH of lake waters can endanger aquatic life.

With the steady rise in the concentration of atmospheric carbon dioxide, some ocean scientists are concerned about the potential effects on the ocean's natural buffering ability and ocean life. We consider this issue in Chapter 12 in the context of global climate change.

Physical Properties of Seawater

Some properties of water vary with salinity. Adding salts to water changes its temperature of initial freezing and the temperature at which it reaches maximum density. Because salt ions do not fit into the crystal lattice of ice, dissolved salts inhibit the formation of hexagonal ice crystals and this

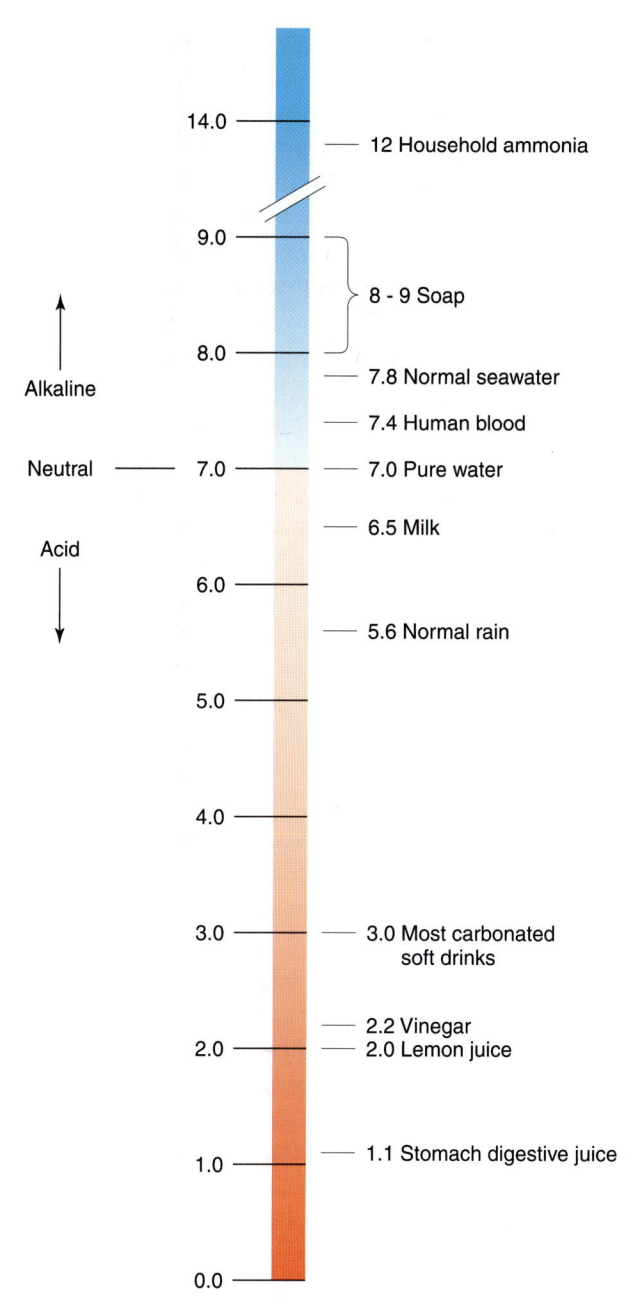

FIGURE 3.10
The acidity and alkalinity of water are expressed as pH, a measure of the hydrogen ion concentration. On the pH scale, the pH increases from 0 to 14 as the hydrogen ion concentration decreases. Pure water has a pH of 7, which is considered neutral; a pH above 7 is increasingly alkaline whereas a pH below 7 is increasingly acidic.

in acidity or alkalinity. Hence, a two-unit drop in pH (e.g., from 5.6 to 3.6) represents a hundred-fold (10 × 10) increase in acidity.

The pH of seawater ranges between 7.5 and 8.5 with an overall average of about 8.0 for the entire ocean at depresses the initial freezing point to temperatures below 0 °C. For seawater of average salinity (about 35), the temperature of initial freezing is −1.9 °C. Furthermore, seawater does not remain at the same temperature while freezing as does fresh water. Since salts are excluded from the ice structure as seawater freezes, the remaining unfrozen water becomes saltier and therefore freezes at still lower temperatures. Unless cooled to very low temperatures, some concentrated liquid brine remains trapped in cells of sea ice, although the brine does migrate downward over time toward the warmer water under the ice.

WATER DENSITY AND TEMPERATURE

Density is defined as mass per unit volume, which may be expressed as grams per cubic cm. When placed in water an object that is less dense than water will float to the surface whereas an object that is denser than water will sink. Fresh water density varies primarily with temperature whereas seawater density varies chiefly with temperature and salinity. Water is only slightly compressible (1-2%) so that pressure arising from the weight of overlying water does not significantly impact its density except in the deep ocean.

Most substances contract when cooled and expand when heated; their density increases with falling temperature and decreases with rising temperature. As the average kinetic molecular energy decreases (i.e., as the temperature falls), the same number of molecules occupies a progressively smaller volume. However, for fresh water, it is not quite that simple (Figure 3.11). As the temperature of fresh water falls steadily from say 25 °C (77 °F), the water contracts and its density increases. The density of fresh water reaches a maximum at about 4 °C (39.2 °F), but with additional cooling (below 4 °C), the water expands (its density decreases).

Ice-like molecular clusters continually form and break-up in liquid water and are responsible for the anomalous behavior of fresh water density at temperatures below 4 °C. As the temperature drops, the water molecules have less kinetic molecular energy and move closer together so that the number of hydrogen bonds increases resulting in more ice-like clusters. These clusters occupy more volume than the unorganized water molecules in the liquid phase. Recall that ice crystals are open hexagonal (six-sided) structures with widely spaced water molecules (Figure 3.2). As the water temperature falls below 4 °C, the decrease in water density caused by increasing numbers of ice-like molecular clusters more than offsets the increase in water density that would accompany a decline in kinetic molecular activity. Whereas most liquids contract when they solidify, as water freezes, its molecules bond into an open hexagonal structure so that the density of ice is about 92% of liquid water.

The unique temperature-density behavior of fresh water explains why the less dense ice floats in more dense water and lakes freeze from the top down rather than the bottom up. If ice were denser than liquid water, ice that forms at the air-water interface would sink and in winter lakes in cold climates would freeze solid from the bottom up, destroying all aquatic life. In autumn, lakes begin cooling at the air-water interface when the temperature of the overlying air falls below that of the lake surface. With continuing cooling in fall, lake-surface waters cool, contract, become denser and sink to the bottom. This process of surface cooling and sinking is repeated until the entire lake has a uniform temperature of 4 °C (and uniform density). Then, with additional cooling of surface waters below 4 °C, water density decreases and the coldest water remains at the surface. At 0 °C an ice cover begins to form. Once formed, ice contracts (becomes denser) as its temperature falls. But no matter how cold, ice remains less dense than liquid water.

With the addition of salt to water, the situation changes. At constant temperature, the density of seawater increases with increasing salinity because the atomic mass of dissolved salts is greater than that of water molecules. Hence, less dense fresh water floats on more dense seawater. The salinity of seawater also affects the temperature of maximum density and the freezing temperature for the same

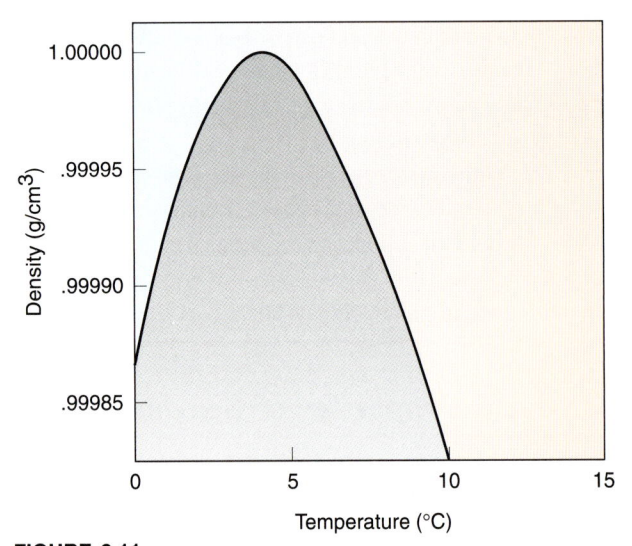

FIGURE 3.11
The density of fresh water reaches a maximum at about 4 °C (39.2 °F) and, with additional cooling (below 4 °C), liquid water expands until it freezes at 0 °C.

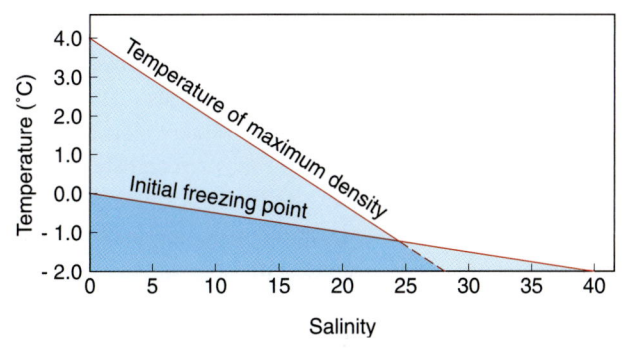

FIGURE 3.12
Water's temperature of maximum density and initial freezing point temperature decrease at different rates as the salinity increases. At a salinity of 24.7, the temperature of maximum density is the same as the freezing temperature (-1.33 ºC).

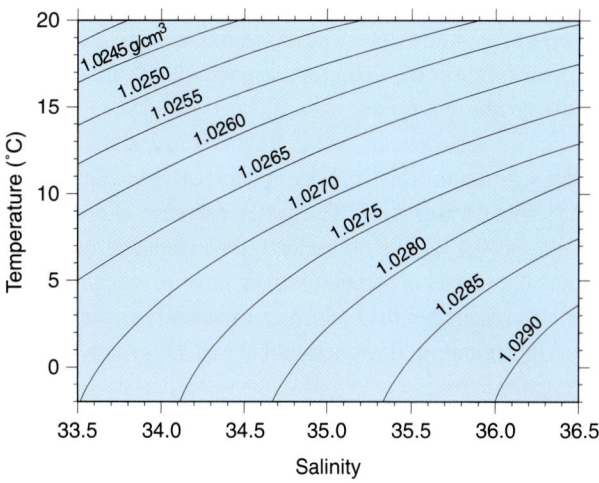

FIGURE 3.13
The density of seawater varies with temperature and salinity.

reason: adding dissolved materials such as salt apparently interferes with the formation of ice-like clusters. As shown in Figure 3.12, the temperature of maximum density decreases linearly with increasing salinity. At any salinity less than 24.7, the maximum density of water occurs at a temperature above the freezing point (which also decreases linearly with increasing salinity). At a salinity of 24.7, the temperature of maximum density is the same as the freezing temperature (−1.33 °C). For water with a salinity greater than 24.7, the temperature of maximum density is lower than its freezing point so that the water freezes before reaching maximum density. The density of seawater of average salinity (about 35) varies inversely with temperature; that is, seawater density always increases with falling temperature and decreases with rising temperature. Hence, like fresh water, seawater becomes denser with cooling and sinks. Except in a few important high latitude locations in the North Atlantic and around Antarctica (Chapter 6), relatively dense seawater usually does not sink to the deep ocean bottom but descends to the level where the density of the surrounding water is the same.

Figure 3.13 shows the relative effects of temperature and salinity on seawater density. At 15 °C, increasing the salinity from 33.7 to 34.9 raises the density from 1.025 g per cubic cm to about 1.026 g per cubic cm. (Note that ocean scientists usually write density in a shorthand notation. Because the density of seawater almost always starts with 1.0, they subtract 1 and then multiply by 1000. In this way, a seawater density of 1.02178 g per cubic cm becomes 21.78 with no units.) Density is changed an equal amount by cooling water with a constant salinity of 34.6 from 18 °C to 14 °C, a change of 4 Celsius degrees. Such variations in temperature are common at the ocean surface. As we will see in Chapter 6, differences in seawater density, caused by variations in temperature and salinity, are important controls of the vertical circulation of ocean water.

PRESSURE

Ocean water exerts pressure (i.e., force per unit area). It is useful to compare water pressure to air pressure. Atmospheric pressure can be thought of as the weight of a column of air acting over a unit area at the base of the column. By convention, *standard* atmospheric pressure is the average air pressure at sea level at 45 degrees latitude and an air temperature of 15 °C (59 °F). This pressure is equivalent to the weight exerted by a mass of 1.03 kg on a square cm (14.7 lbs per square in.) or 1013.25 millibars (mb). (A millibar is a unit of pressure where 1000 mb equals one bar.) Water is much denser than air so that a column of equivalent height produces much greater pressure.

A column of water having a height of 10.33 m (33.9 ft) exerts a pressure at its base equal to one standard atmosphere. Hence, the pressure at the bottom of the Marianas Deep, where the ocean depth is about 11,000 m (36,000 ft), is over 1000 times greater than standard atmospheric pressure. At that depth, the water pressure is 1097 kg per square cm (7.8 tons per square in.). For all practical purposes water is *incompressible*; that is, water density does not vary significantly with increasing pressure. To a good first approximation, the pressure at any point in a water column is directly related to depth and the relationship is linear; that is, doubling the depth doubles the pressure. At a depth of 10.33 m, the water pressure is 1.03 kg per square cm. At ten times this depth, 103.3 m (339 ft), the water pressure is ten times as great, that is, 10.3 kg per square cm. The tremendous pressures encountered in deep ocean waters prevent direct exploration except by special submersible vehicles designed to withstand potentially crushing stresses.

Ocean scientists, like their counterparts in meteorology, commonly use the *bar* and its derivatives as a standard unit of pressure. The water pressure expressed in *decibars* (0.1 bar) is numerically equivalent to the water depth expressed in meters (with an error of less than 2%). The interchangeability of the two measures (i.e., water pressure in decibars and water depth in meters) greatly simplifies data analysis. For example, the water pressure at 100 m (330 ft) depth is about 100 decibars.

SEA ICE

Each winter, seawater freezes at high latitudes and even in mid-latitude coastal areas. When seawater is chilled below its freezing point, microscopic ice crystals form, later growing into hexagonal needles 1 to 2 cm (0.4 to 0.8 in.) long. At this stage, the sea surface takes on a dull appearance. As freezing continues, individual ice crystals freeze together and cover the surface like a blanket of wet snow. Eventually ice crystals begin to grow downward and form a thin, flexible, plastic-like ice layer, honeycombed with small cells that fill with seawater. Ice crystals themselves contain no salt, but the brines trapped in the small cells can be saltier than seawater. Typically 1 kg of newly formed sea ice consists of about 800 g of ice (salinity 0) and 200 g of seawater (salinity 35). Thus, the average salinity of newly formed sea ice is about 7. As temperatures fall, more ice forms beneath the initial ice layer. The brines in the cells also partially freeze, making the remaining brines saltier. If temperatures continue to fall, the salt in the brine eventually crystallizes as the last bit of water freezes.

The salt content of newly formed sea ice depends on temperature. At temperatures near freezing, sea ice forms slowly, which allows brines to flow out leaving little seawater in the cells. Such ice therefore contains little salt. At lower temperatures, ice forms more rapidly and traps seawater; this sea ice contains more salt but is less salty than the seawater from which it formed. Ice forms in the upper colder region of a brine cell while ice melts in the lower warmer region of the brine cell, causing the brine cell to migrate toward higher temperatures.

An ice layer up to 1 m (3 ft) or so thick can form in one winter; this is called *first-year ice* (Figure 3.14). First-year ice dominates the Southern Ocean around Antarctica. But in the Arctic pack ice of the central basin, sea ice melts little during summer and *multi-year ice* dominates. Over several seasons, the maximum thickness of multi-year ice is usually 2 to 3.5 m (about 7 to 11 ft). Winds, however, can pile up floes of ice, forming pressure ridges offshore and ice-shove ridges along the coast. Pressure ridges extend many meters above and below the ice pack and are hazardous to submarines navigating under the Arctic ice and impede icebreakers traveling through the ice pack. Many different terms are used to describe the various forms of sea ice. For more on sea ice terminology, see this chapter's second Essay.

SOUND TRANSMISSION

Seawater is essentially transparent to sound just as the atmosphere is nearly transparent to sunlight. Whales use sound to communicate across ocean basins and to attract mates. Ocean scientists have developed techniques that use the sound propagating ability of seawater to determine ocean depth, locate underwater objects such as submarines and schools of fish, and to determine small changes in water temperature over great distances.

Sound propagates through some medium (e.g., air or water) as compression (push-pull) waves with the speed of propagation dependent on the properties of the medium. The speed of sound in ocean water averages about 1500 m per sec (5000 ft per sec)—more than four times the average speed of sound in air. Knowing the speed of sound in seawater is the principle behind an *echo sounder* used by most seagoing vessels to determine the depth of water beneath the ship. This instrument sends a narrow beam of sound vertically to the seafloor where it is reflected back to the ship. The time interval between emission and return of the sound signal (the *echo*) is calibrated in terms of water depth. *Fish finders* are echo sounders that send and receive sound waves that are reflected by schools of fish.

SONAR (*SOund NAvigation and Ranging*) is similar to an echo sounder except that the operator of the instrument can alter the direction of the sound signal. Pulses of sound are sent out to locate targets such as sub-marines and the return echoes are displayed electronically on a monitor. Complicating the use of SONAR for target location are changes in the speed of sound in water that can cause the sound wave to bend (refract). Sound waves bend toward regions where sound travels more slowly and away from regions where sound waves travel faster. The speed of sound in seawater is influenced by its temperature, pressure, and salinity.

Refraction of sound waves as they travel through ocean waters gives rise to the **SOFAR channel**, a zone centered at an ocean depth of about 1000 m (3300 ft) where the

Chapter 3 PROPERTIES OF OCEAN WATER 69

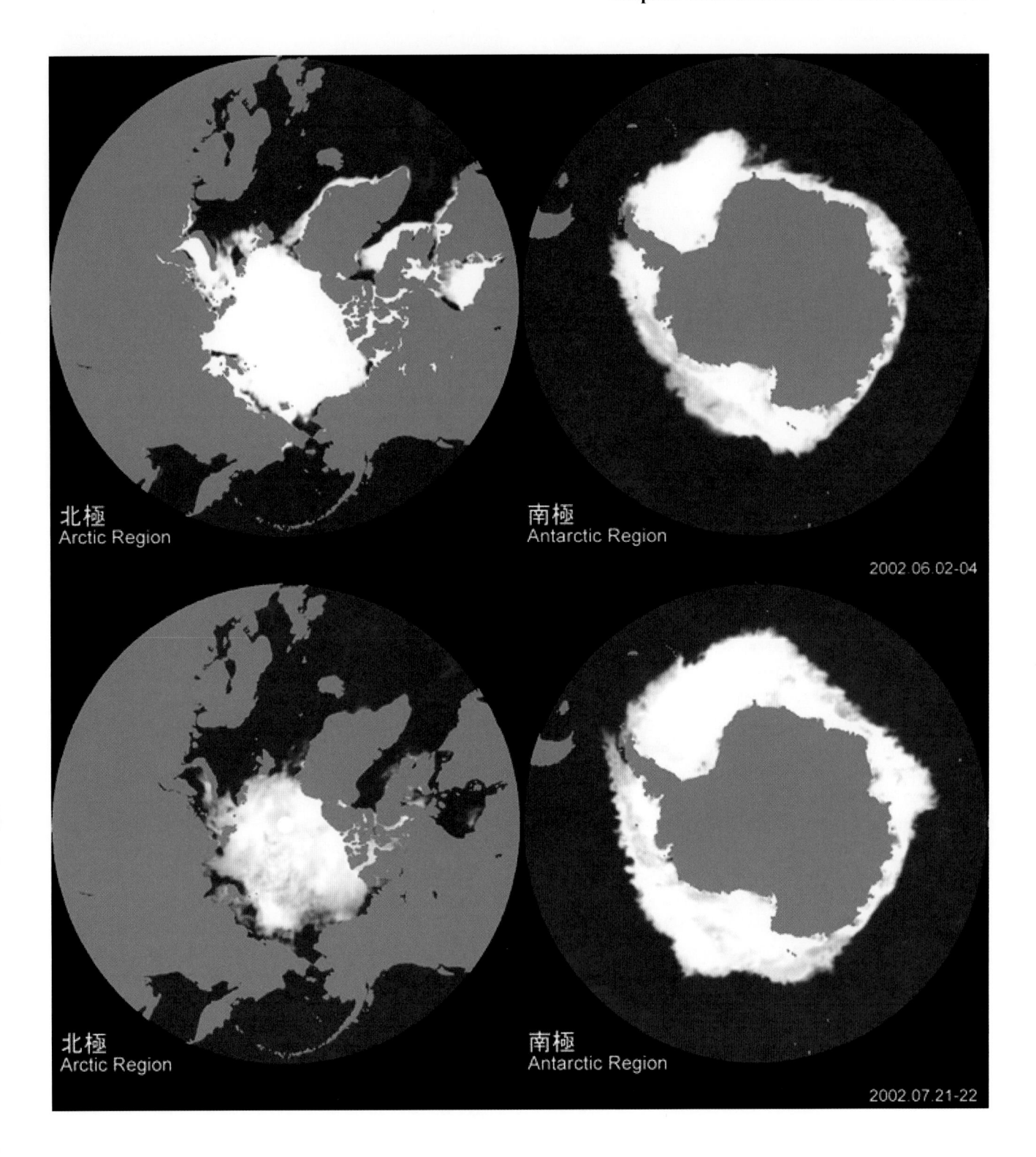

FIGURE 3.14
Differences in sea ice coverage between the Arctic region on the left and the Antarctic region on the right. The two top images are dated 2-4 June 2002 whereas the two bottom images are almost two months later on 21-22 July 2002. In the Arctic, the ice is melting leaving multi-year ice in the central basin while first-year ice is shown advancing around the Antarctic continent.

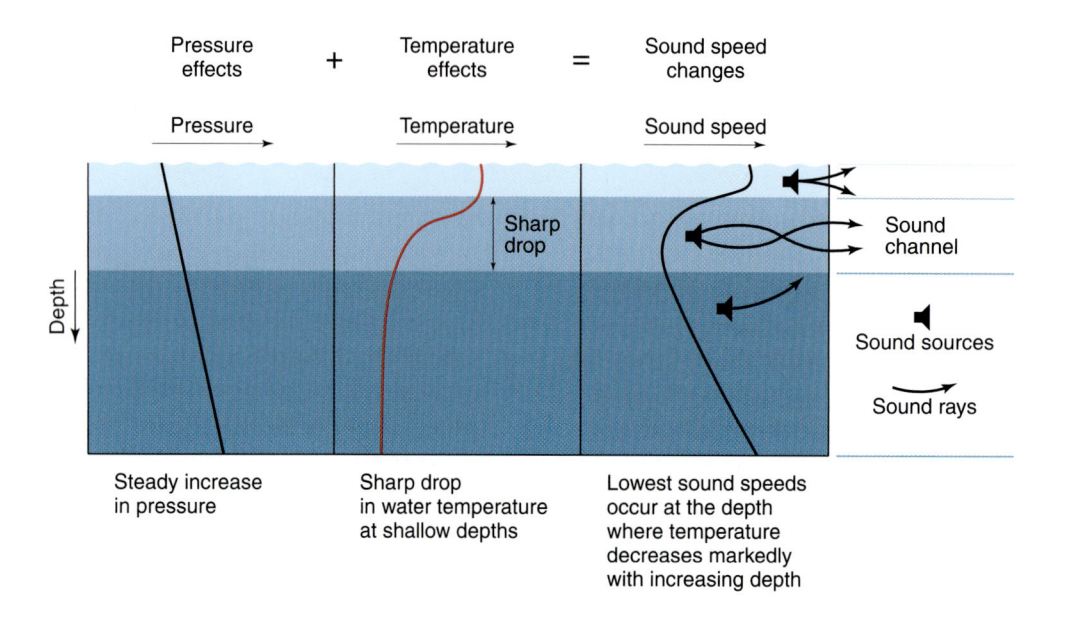

FIGURE 3.15
Refraction of sound waves as they travel through ocean waters gives rise to the SOFAR channel, a zone centered at an ocean depth of about 1000 m (3300 ft) where the speed of sound is at a minimum value.

speed of sound is at a minimum value (Figure 3.15). (*SOFAR* is the acronym for *SOund Fixing And Ranging*). Sound waves are essentially trapped in the SOFAR channel and can travel thousands of kilometers with little loss of energy. Why does the speed of sound reach a minimum and produce the SOFAR channel? The speed of sound in seawater increases with increasing temperature, salinity, and depth (i.e., pressure). From the sea surface down to a depth of about 1000 m, temperature usually decreases and is the primary control of sound speed. Hence, sound speed decreases with depth to 1000 m. At greater depths, the water temperature tends to be uniformly low and sound speed variation depends chiefly on pressure change with depth. With increasing depth and pressure, sound speed increases. As shown in Figure 3.15, sound waves are refracted upward (from below the SOFAR) or downward (from above the SOFAR) toward the region of minimum sound velocity within the SOFAR channel.

During the 1990s, with increased interest in global climate change, scientists at the Massachusetts Institute of Technology and Scripps Institution of Oceanography developed and tested a method of monitoring temperature changes within the SOFAR channel. The approach, known as **acoustic tomography**, is based on the dependency of the speed of sound on ocean water temperature and can measure very small temperature changes—to a few thousandths of a degree—over great distances in the ocean. An increase in sound speed between sound transmitters and receivers over a period of time would indicate warming of the ocean. Indeed, such warming has been observed in the Arctic basin using acoustic tomography.

Conclusions

Water has unusual physical and chemical properties resulting from the water molecule's unique polar structure that gives rise to hydrogen bonding. Water is a powerful solvent, dissolving both salts and gases, and is an effective heat-storage and transporting medium. Consequently, the ocean plays a major role in Earth's weather and climate. Temperature, salinity, and pressure affect most physical properties of seawater such as density and sound velocity. The density of seawater, in turn, influences ocean currents. In the next chapter, we continue our examination of ocean properties by focusing on the types and sources of marine sediments.

Basic Understandings

- The unusual physical and chemical properties of water (H_2O) arise from the unique structure of the water molecule, consisting of two hydrogen (H) atoms bonded at 105° to an oxygen (O) atom. The bonding angle of 105° allows a slight positive charge to reside on each of the hydrogen atoms and a slight negative charge to reside on the oxygen atom. This makes the water molecule a polar molecule. The positively charged (hydrogen) poles of a water molecule attract the negatively charged (oxygen) poles of other water molecules; this attractive force is known as hydrogen bonding which is the source of the unique properties of water.

- Water is one of the few substances occurring naturally in all three phases at the usual range of temperatures and pressures at Earth's surface. Ice has a three-dimensional internal framework consisting of a hexagonal ring of molecules that vibrate about fixed locations. Liquid water contains un-hydrogen bonded molecules and transient clusters of hydrogen bonded molecules. When liquid water changes to vapor, all hydrogen bonds are broken and individual molecules rapidly diffuse to fill the entire volume of its container.

- Temperature is directly proportional to the average kinetic energy of the atoms or molecules composing a substance. Heat is the name given to the energy transferred between objects at different temperatures and is always transferred from the warmer object to the colder object.

- When water changes phase heat is either absorbed from the environment to break hydrogen bonds (melting, evaporation, sublimation) or released to the environment when hydrogen bonds form (freezing, condensation, deposition). Heat energy that is involved in the phase change of water is known as latent heat.

- The thermal response of a unit mass of a substance to a given input of heat, identified as the specific heat of the substance, differs for various materials and is highest for water. Water's high specific heat has consequences for climate. Compared to an adjacent landmass, a body of water does not warm as much during the day (or in summer) and does not cool as much at night (or in winter). Places immediately downwind of the ocean have a maritime climate in which the average temperature contrast between summer and winter is less than it is at the same latitude but well inland.

- Water is an excellent solvent, readily dissolving solids, liquids, and gases. The polar nature of the water molecule favors the solution of both ionic and non-ionic substances. Seawater is a solution of nearly constant salt composition with a concentration of about 3.5%.

- Salinity is based on measurements of the electrical conductivity of seawater and is expressed as the ratio of the conductivity of the seawater sample to the conductivity of *standard seawater*. As a ratio, salinity has no units but numerically is essentially equivalent to grams of salt per kilogram of seawater (parts per thousand).

- Salts dissolved in seawater are derived from weathering and erosion of rock and sediment on land, chemical reactions between cold seawater and newly formed oceanic crust, and volcanic eruptions.

- Gases are exchanged between the atmosphere and ocean at the ocean surface. If seawater is not saturated with a specific gas, the gas is transferred from air to ocean. If seawater is supersaturated with the gas, the transfer is in the opposite direction. When water is saturated with a gas, the rate at which the gas dissolves in water equals the rate at which the gas escapes to the atmosphere. The saturation concentration of a gas in water depends primarily on temperature in fresh water bodies such as lakes and rivers and a combination of temperature, salinity, and pressure in seawater. In general, cold water can hold more dissolved gases than warm water. Also, solubility of a gas in water depends on the gas species so that, for example, oxygen and carbon dioxide are more soluble than nitrogen.

- Below the ocean-atmosphere interface, biochemical processes control the proportions of certain dissolved gases. Within the photic zone, the sunlit upper portion of the ocean where photosynthesis takes place, dissolved oxygen is enhanced relative to carbon dioxide. Beneath the photic zone, decomposer activity and cellular respiration are the most important biochemical processes. For the ocean as a whole, CO_2 is the most abundant dissolved gas, accounting for 83% of the total.

- The pH of seawater varies between 7.5 and 8.5 with an overall average of about 8 for the entire ocean at all depths; that is, seawater is slightly alkaline. The relatively narrow range in the pH of seawater is very important for marine organisms whose shells and skeletons are composed of calcium carbonate ($CaCO_3$).

- The density of fresh water reaches a maximum at a temperature of 4 °C (39.2 °F) and fresh water ice is less dense than fresh water. This explains why ice floats and why lakes freeze from the top down. At constant temperature, seawater density increases with increasing salinity so that less dense fresh water floats on more dense seawater. As salinity increases, the freezing temperature and temperature of maximum density decrease.

- Temperature and salinity are the principal controls of seawater density. Lower temperature and increasing salinity cause density to increase. In the open ocean, temperature dominates density in warmer waters whereas salinity dominates density in colder waters. Changes in salinity are important near shorelines (due to freshwater runoff), in polar regions (due to cold water and sea ice formation and melting), and in regions of high precipitation or evaporation.
- Pressure is the force exerted by a column of water per unit area at the base of the column. Water is essentially incompressible so that water pressure increases linearly with depth. Water pressure expressed in units of decibars is numerically equivalent to depth when expressed in meters.
- Sea ice forms in polar regions and mid-latitude coastal areas. As seawater freezes, salts are excluded from the sea ice and the resulting brines increase the salinity of surrounding waters. First-year ice dominates the Southern Ocean around Antarctica whereas multi-year ice dominates in the Arctic Ocean.
- Ocean scientists have developed techniques that use the sound propagating ability of seawater to determine ocean depth, locate underwater objects such as submarines and schools of fish, and to determine small changes in water temperature over great distances.
- Refraction of sound waves as they travel through ocean waters gives rise to the SOFAR channel, a zone centered at an ocean depth of about 1000 m (3300 ft) where the speed of sound is at a minimum value. Sound waves are essentially trapped in the SOFAR channel and can travel thousands of kilometers with little loss of energy.

ESSAY: Desalination

The Red Sea port of Jeddah, Saudi Arabia is a growing city of about 1.5 million. Rapid population growth is taking place in spite of the arid climate and no significant freshwater sources nearby. There are no rivers or lakes and very little groundwater. For their freshwater supply, residents of Jeddah depend mostly on water piped from the Red Sea and desalinated. *Desalination* is a process whereby dissolved solids (principally salts) are removed from saline water to make the water potable, that is, suitable for domestic and agricultural uses. Desalination utilizes either of two basic technologies: distillation or reverse osmosis.

With *distillation*, water is purified through phase changes (usually from liquid to vapor and back to liquid). The simplest (and least expensive) distillation device is a transparent plastic dome placed over a reservoir of seawater. Solar radiation penetrates the dome and provides the heat energy that evaporates the water. Water vapor then condenses as freshwater drops on the inside surface of the dome and the drops drip into a collection trough. This so-called *solar still* works best in sunny tropical climates where solar radiation is intense year-round. For much more rapid distillation of large quantities of salty water, the intake water is heated to steam that subsequently condenses into fresh water. At Jeddah and other locations in Saudi Arabia electricity and fresh water are co-generated; that is, desalination plants utilize the waste heat from electric power plants as the energy source for desalination. Jeddah's five desalination plants produce more than 380 million liters (100 million gal) of fresh water per day.

In *reverse osmosis*, pressure applied to water forces water molecules through a thin semi-permeable membrane. Up to 97% of the substances dissolved in seawater cannot pass through the membrane and are left behind. Pretreatment of the input water with substances such as chlorine, hydrogen peroxide, or sulfuric acid may be necessary to prevent biological fouling or calcium carbonate scaling of the membrane surface. The highly saline wastewater stream is usually discharged into the ocean. Reverse osmosis is used at desalination facilities on the Outer Banks of North Carolina where fresh water is in limited supply. At the Kill Devil Hills facility, operating since 1989, brackish (somewhat saline) water is pumped from 10 wells at a rate of up to 1900 liters (500 gal) per well per minute and then filtered and pretreated with sulfuric acid. Water is pumped under pressure to three reverse osmosis units, each of which produces about 3.8 million liters (1 million gal) of water per day. Chlorine, fluoride, and a corrosion inhibitor are then added to the desalinated water. About 25% of the water delivered to the facility becomes a highly saline waste stream that empties into the Atlantic.

The principal drawback to desalination of large quantities of water is the relatively high cost of energy-intensive technologies (distillation more so than reverse osmosis). For example, California's desalinated water costs more than twice that of water from watershed transfers or groundwater pumping. (The U.S. Navy constructed California's first desalination plant (using reverse osmosis) on San Nicolas Island in 1990.) According to the International Desalination Association, high cost is a major reason why the approximately 12,500 desalination plants operating worldwide produce less than 0.15% of all fresh water consumed by human activity. Desalination (via both distillation and reverse osmosis) is most common in the Middle East where energy is abundant and inexpensive. In that arid region of the world where fresh water resources are in short supply, desalination accounts for about 60% of the world's capacity.

ESSAY: Sea Ice Terminology

Over the years, a complex terminology has evolved concerning sea ice, its development and various forms. The World Meteorological Organization (WMO) standardized definitions for many of these terms more than three decades ago. The purpose of this Essay is to define some terms that you may encounter while exploring sea ice.

Sea ice refers to any form of ice floating in ocean waters that originated from the freezing of seawater. As noted elsewhere in this chapter, the salt dissolved in seawater depresses slightly the freezing point of water. For example, seawater having a salinity of 34 psu (about 3.4% salt) begins to freeze when its temperature drops to about –1.9 °C (28.6 °F) whereas the freezing point of fresh water is 0 °C (32 °F). Note that icebergs are not sea ice; they are massive chunks of ice that break away from the snout of a glacier advancing from land to sea (or a large lake) or calve off of the edge of an ice shelf. Icebergs consist of frozen fresh water and may be floating or grounded.

Sea ice is broadly classified based on age (and thickness) as *new ice* (less than 10 cm thick), *young ice* (10-30 cm thick), *first-year* ice (30 cm to as much as 120 cm thick), or *old ice* (second-year or multi-year ice). *First-year ice* is the product of no more than one winter's growth of sea ice. *Second-year ice* has survived one melt season (summer) whereas *multi-year ice* has persisted through at least two summers' melt. The latter is typically around 3 m (10 ft) thick and is more common in the Arctic Ocean than the Southern Ocean. Sea ice near Antarctica varies seasonally with summer ice cover usually ranging up to 150 km (93 mi) wide whereas the width of winter ice can vary from 450 km (280 mi) to more than 1700 km (1050 mi). Maximum sea-ice coverage occurs in the Weddell Sea and accounts for about 80% of the multi-year ice in the Antarctic.

In the initial stage of sea ice formation, frazil ice develops in the upper few centimeters of ocean waters. *Frazil ice* consists of millimeter-sized platelets or discs that give the sea surface an oily appearance. Wave action can stir frazil ice to a depth of several meters. With continued freezing, frazil ice crystals coagulate to form a soupy mix that reflects little light and gives the surface a matte appearance, sometimes referred to as *grease ice*. Frazil ice and grease ice are forms of new ice but do not form distinctive ice floes. An *ice floe* is any contiguous piece of ice that can vary greatly in size from a few meters to more than 10 km across. Other types of new ice include frozen *slush* (an accumulation of water-saturated snow on an ice surface or floating in water) and *nilas* (a thin crust of elastic gray-colored ice on a calm sea that readily bends into interlocking fingers.

Grease ice aggregates into small chunks of ice, which then become pancake-shaped ice floes. *Pancake ice* consists of nearly circular pieces of ice from 0.30 to 3.0 m (1 to 3 ft) in diameter and having a thickness of up to 10 cm (4 in.). Collisions with other ice floes are responsible for the upturned rim that is characteristic of pancake ice. With falling temperature, the ice cover thickens. Winds, waves, and currents cause ice floes to override one another (known as *rafting*) further thickening the ice cover generally to 40-60 cm (15-24 in.). However, in some fierce winter storms in the Bering Sea, scientists have reported 3-m (10-ft) thick ice floes rafting onto other ice floes and building the ice to a thickness of 6 m (20 ft), of which about 90% is below sea level. Rafting also produces a rough surface on the ice floe.

Fast ice forms along the coast and is attached to the shore or shallow sea bottoms (e.g., typical off the coast of northern Alaska) so that it cannot move laterally. It may extend from a few meters to several hundred kilometers offshore and may be more than one year old. *Pack ice* is any area of sea ice that is not anchored to land and moves with the wind and ocean currents. Pack ice is described as very open (1/10 to 3/10 ice cover), open (4/10 to 6/10 ice cover), close (7/10 to 8/10 ice cover), very close (9/10 to less than 10/10), and compact (10/10 ice cover with no water visible). If ice floes are frozen together, the compact pack ice is described as consolidated. The *extent of sea ice* is defined as the area in which ice covers at least 15% of the ocean surface.

A significant portion of Antarctic sea ice thickens through a process whereby seawater floods the snow that has accumulated on top of the ice turning it to slush; at low temperatures the slush then freezes into *snow ice*. If the snow cover is sufficiently massive, the ice/snow interface is suppressed below sea level allowing seawater to intrude and turn the snow to slush. Alternately, water may migrate upward through brine channels within the sea ice and enter the snow.

Stresses produce fractures in sea ice that vary from a few meters to many kilometers in length. Where fast ice is attached to the shore, tide-induced vertical motions of the ice may produce *tide cracks*. These fractures are important for some forms of marine life. They allow penguins and seals access to the ocean. A fracture that is wide enough to be navigable by surface vessels is known as a *lead* and may occur between the shore and the pack ice or between fast ice and pack ice or simply between large ice floes or in the ice pack at sea. At sea, leads provide breathing holes for whales. Leads also permit exchange of heat and moisture between the ocean and atmosphere. In addition, converging ice floes (perhaps wind- or current-driven) may be forced upward into a wall of fractured ice, known as a *pressure ridge*. Pressure ridges can extend several meters above the surrounding sea ice and ocean surface. The fractured ice forced downward meters below the ridge is known as an *ice keel*.

A non-linear shaped kilometer-scale opening in the sea ice cover that persists or reoccurs regularly (often annually), is called a *polynya*. These holes in the ice would not normally exist because of freezing temperatures were it not for some mechanism that keeps them open. In some cases, persistent gravity-driven cold winds (*katabatic winds*) blow ice away from land (e.g., Antarctica), islands, and grounded ice leaving surface waters ice-free. In a second mechanism, warm water welling up from below supplies sufficient heat to melt the ice cover. Also, polynyas are common at the mouths of large rivers. Major polynyas in the Northern Hemisphere include the St. Lawrence polynya in the Bering Sea and polynyas off the northeast and northwest coasts of Greenland. These are called *North East Water* and *North Water* respectively.

CHAPTER 4

MARINE SEDIMENTS

Case-in-Point
Driving Question
Sediment Size and Accumulation
    Size Classification
    Terminal Velocity
Classification of Marine Sediments
    Lithogenous Sediment
    Biogenous Sediment
    Hydrogenous Sediment
    Cosmogenous Sediment
Marine Sedimentary Deposits
    Continental-Margin Deposits
    Deep-Ocean Deposits
Marine Sedimentary Rock
Resources of the Seafloor
    Oil and Natural Gas
    Mineral Resources
    Exclusive Economic Zone
Conclusions
Basic Understandings
ESSAY: Heinrich Events
ESSAY: Burgess Shale: A Glimpse into Ancient Marine Life

Yellow iron oxide from hydrothermal vents in Pele's Vents, Hawaii. [Courtesy of OAR/National Undersea Research Program (NURP); Univ. of Hawaii - Manoa.]

Case-in-Point

Waste water enters rivers and streams via overland flow (e.g., agricultural runoff), groundwater, or direct input from drainage pipes (e.g., storm sewer systems). The net flow of water from land to sea in the global water cycle implies that all water-borne wastes eventually enter the ocean, either dissolved or suspended in water. Also as part of the global water cycle, rain and snow wash particles from the atmosphere into the ocean. Furthermore, in some cases, industries located along the coast discharge wastes directly into the ocean or barge them out to sea for dumping.

    Some wastes discharged at sea rapidly decompose physically, chemically, or biologically; others resist decomposition or, like mercury, occur in elemental form and cannot breakdown any further. These persistent chemicals enter food chains and move from one trophic level to the next higher trophic level increasing in concentration along the way. Problems caused by persistent chemicals are especially serious when they are toxic or hazardous to living organisms. In such cases, remediation requires either removal of the contaminants by dredging or capping them with layers of uncontaminated sediments to prevent their reentry into the ecosystem.

    Perhaps the most infamous example of a toxic discharge to the sea that had serious consequences for human health occurred in the small coastal village of Minamata,

Japan. For many years, the Chisso Chemical Plant discharged industrial waste containing mercury into Minamata Bay. Because elemental mercury is insoluble in water, it was expected to sink to the bottom sediments and remain inert, causing no harm.

Fish and shellfish taken from local waters were part of the staple diet of Minamata residents. But, in the early 1950s, they began noticing bizarre behavior in their cats, including twitching and stumbling, which we now recognize as signs of brain damage caused by mercury poisoning. In 1956, a five-year-old girl who lapsed into a convulsive delirium was diagnosed with neurological damage. A few weeks later, numerous people reported a variety of debilitating symptoms including numbness, headaches, loss of muscle control, and slurred speech. In some individuals, symptoms worsened and they developed violent trembling and paralysis. Children were born with physical deformities as well as severe mental retardation. Some victims died.

Upon investigation, fish and shellfish taken from Minamata Bay were found to contain high levels of methyl mercury in their tissues, a soluble and highly toxic form of mercury. Methyl mercury that enters the body attacks the central nervous system. In 1959, scientists demonstrated that bacteria in the bottom sediments converted mercury to methyl mercury. Methyl mercury was subsequently taken up by aquatic organisms, readily moved up the food chain, and was consumed by people. In eating fish and shellfish from Minamata Bay, people were exposed to dangerous levels of a highly toxic material. More than 3500 people were seriously affected and about 50 died from methyl mercury poisoning—now referred to as *Minamata Disease*. After many years of litigation, the Chisso Company finally accepted some responsibility for the Minamata tragedy and was required to pay reparations to victims.

Discharge of mercury into Minamata Bay ceased in the late 1960s and the cleanup began. This involved dredging and removal of mercury-contaminated sediment deposits. These sediments were placed in reclamation areas where they were surrounded by dikes with impervious fabric liners and covered by clean sands to prevent run-off and further contamination of bay waters. By July 1997, the waters of Minamata Bay and its marine organisms were found to meet Japan's environmental standards. Minamata Bay was declared safe and reopened for human use including fishing and recreational activities but monitoring of mercury levels in the water, sediments, and marine life continues.

Driving Question:

What are the types and sources of sediments that enter the ocean?

Particles (sediments) blanket much of the ocean floor. Most of these particles form at the interfaces among Earth's various subsystems and are transported by rivers, wind, ice, and gravity to the ocean. Some particles originate in the sea (e.g., the excretions and secretions of marine organisms). Marine sediment deposits consist of rock fragments, soil particles, and the shells, bones and teeth of marine organisms, and some material from outer space. Locked in the layers of ocean bottom sediments is a chronology of the history of Earth's subsystems over the past 200 million years, that is, since formation of the present ocean basins (Chapter 2).

The primary focus of this chapter is the types, sources, and distribution of marine sediments. We first describe how sediments are classified by size and the factors governing their rate of accumulation on the ocean floor. We then focus on the classification of marine sediments based on mode of origin, which largely determines their composition. There follows a description of the sedimentary deposits of the continental margin and the deep ocean floor and how geological processes modify these deposits. All this provides background for an overview of seafloor resources including oil, natural gas, and minerals.

Sediment Size and Accumulation

Particles that accumulate on the sea floor are known as **sediment**. Marine sediments differ in source, composition, size, and the rate at which they accumulate on the sea floor. In this section we describe the size range of sediments and the factors governing the settling rate of particles in ocean water.

SIZE CLASSIFICATION

Marine sediments are classified by size into three broad categories: mud, sand, and gravel. As shown in Table 4.1, these categories are further subdivided by size using common descriptive terms. The smallest particle of mud (clay) has a diameter less than 0.0039 mm, too small to be seen without a microscope. At the other end of the scale, a boulder is more than 256 mm (10 in.) in diameter. Accumula-

TABLE 4.1
Wentworth Classification of Sediments by Size[a]

Sediment	Type	Diameter (mm)
Gravel	Boulder	>256
	Cobble	64-256
	Pebble	4-64
	Granule	2-4
Sand	Very coarse	1-2
	Coarse	0.50-1.0
	Medium	0.25-0.5
	Fine	0.125-0.25
	Very fine	0.0625-0.125
Mud	Silt	0.0039-0.0625
	Clay	<0.0039

[a]Adapted from C.K. Wentworth, *Journal of Geology* 30 (1922): 377-392.

tions of sediment (*sediment deposits*) on the seafloor also vary in the range of grain size, known as **sorting**. A well-sorted sediment deposit has a narrow range of grain sizes whereas a poorly sorted deposit has a broad range of grain sizes. In general, the farther sediment is transported by running water, ocean currents, or wind, the better sorted it becomes. For example, marine sediments tend to be larger and more poorly sorted on the continental margin than on the deep ocean floor.

Deposits of marine sediment are thickest on the continental margins (and near islands) where accumulation rates are relatively high. Most of these sediments are transported to the sea in suspension by rivers. As rivers flow downhill under the influence of gravity, some of their *kinetic energy* (energy of motion) is used to erode their channel and transport sediment in suspension. Rivers also transport dissolved materials in solution and strong currents can push or roll larger sediment along the channel bottom. A river slows and diverges as it enters the ocean; dissipation of its kinetic energy reduces the river's ability to transport particles in suspension. More energy is required to transport larger particles than smaller ones. Hence, with diminishing kinetic energy, larger particles settle out of suspension almost immediately whereas finer particles are carried farther away from the coast before settling to the sea bottom.

Near the mouths of large sediment-transporting rivers, sediment accumulation rates can be enormous—perhaps as much as 8000 m (26,000 ft) per 1000 years. Waves and currents transport this material up or down the coast or offshore. Typical accumulation rates on the continental shelf and slope, on the other hand, range from 10 to 40 cm (4 to 16 in.) per 1000 years.

The rain of sediment in the deep-ocean has been likened to the barely perceptible fall of dust in a classroom. Accumulation rates generally average from 0.5 to 1.0 cm (0.2 to 0.4 in.) per 1000 years. The thickness of deposits depends on the supply of particles and the length of time that sediments have been settling onto the ocean floor. As noted in Chapter 2, ocean floor sediments generally are thickest where the ocean crust is oldest so that deposits become thicker with increasing distance from the mid-ocean ridges (divergent or spreading plate boundaries). We would expect very little sediment on top of undersea mountains at a spreading center, but on the order of 1000 to 2000 m (3000 to 6000 ft) of sediment farthest away from the spreading center at the edge of the ocean basins where the underlying oceanic crust approaches 200 million years old.

The length of time it takes for a particle to sink to the ocean bottom primarily depends on particle size. For example, a sand-sized particle may take a few days to sink to the bottom of the ocean whereas a clay-sized particle may take more than a century to cover the same distance. The longer it takes a particle to reach the ocean bottom, the greater the horizontal displacement of the particle by deep ocean currents and the more likely it is for soluble particles to dissolve in seawater. An important concept in understanding the ability of some fluid to transport a particle in suspension and the rate at which particles accumulate on the seafloor is terminal velocity.

TERMINAL VELOCITY

Terminal velocity is the constant speed attained by a particle falling through a motionless fluid such as water or air (Figure 4.1). The speed of a falling particle in calm water or air is regulated by (1) *gravity*, the force that accelerates the particle directly downward towards Earth's surface, and (2) the *fluid resistance* offered by the medium through which the particle falls. A downward accelerating particle meets increasing fluid resistance while gravity remains essentially constant. The magnitude of the resisting force eventually equals gravity; that is, the two opposing forces come into balance. When forces are balanced, the downward moving particle attains a constant speed. According to **Newton's first law of motion**, an object in con-

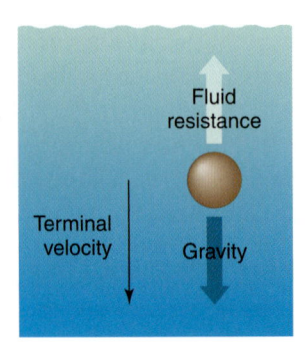

FIGURE 4.1 Terminal velocity is the constant downward-directed speed of a particle within a motionless fluid due to the balance between gravity (acting downward) and fluid resistance (directed upward).

stant straight-line motion or at rest remains that way unless acted upon by an unbalanced force. In this case, the fluid resistance (directed upward) balances gravity (acting downward) and the particle continues downward at a constant speed. That speed is the particle's terminal velocity.

For a given medium, the terminal velocity increases with increasing particle size (assuming that the density and shape of particles vary little). This is the reason why sand-size particles settle to the ocean bottom faster than clay-size particles. Furthermore, the terminal velocity of a given particle varies with the medium. Water is more viscous (offers more frictional resistance) than air. Hence, a particle of a given size has a greater terminal velocity in air than in water. For a particle to remain in suspension, turbulent motions in the medium must counter the terminal velocity. A fast-flowing river can be quite turbulent and can transport relatively large particles. The wind, on the other hand, ordinarily can transport only sand-sized particles within a meter or so of the Earth's surface but can transport clay and silt-sized sediment to altitudes of tens of thousands of meters and over horizontal distances of thousands of kilometers. Hence, as the wind or running water slows and becomes less turbulent, suspended particles settle out of suspension in an orderly sequence from larger to smaller particles.

Because of differences in terminal velocity, we would expect a deep ocean current to transport small slowly sinking particles a greater horizontal distance than a large rapidly sinking particle. For example, for a slow current of 1 km (0.6 mi) per day, the sand-sized particle mentioned above which takes a few days to sink to the bottom of the ocean will be displaced horizontally only a few kilometers, whereas the clay-sized particle which may take more than a century to reach the bottom can be displaced thousands of kilometers. For this reason, fine sediments on the seafloor may be quite different from the particles directly above them in the surface waters. This is often not the case, however, because in seawater small particles can aggregate together and form larger particles that rapidly settle to the sea bottom. The attractive force between particles having opposite electrical charges may be a reason why particles aggregate.

Classification of Marine Sediments

Ocean scientists classify marine sediments based on their source as *lithogenous* (from rock), *biogenous* (from living organisms or their remains), *hydrogenous* (precipitated from seawater), and *cosmogenous* (from outer space). Cosmogenous sediments are much less common than the other three. In this section, we describe each of these sediment types and the processes that form them.

LITHOGENOUS SEDIMENT

Lithogenous sediment accounts for about three-quarters of all marine sediments and owes its origin mostly to weathering and erosion of pre-existing rock. As noted in Chapter 2, **weathering** is the physical disintegration and chemical decomposition of rocks that are exposed to the atmosphere. **Erosion** is the transport of the products of weathering by running water, wind, glaciers, and gravity. Agents of erosion deliver particles to the ocean where they are further dispersed by waves and currents. Where highlands form the coastline, waves undercut cliffs and under the influence of gravity rock debris slides or slumps into the sea. Explosive volcanic eruptions also contribute lithogenous fragments of various sizes, shapes and composition that fall through the air and accumulate in the ocean; these particles are collectively known as **tephra**.

The composition of lithogenous sediments produced through weathering depends on their source rock. The most abundant elements in the Earth's crust are oxygen (O) accounting for 46.6% by weight and silicon (Si) accounting for 27.7% by weight (Table 4.2). The most common rocks in the Earth's crust are igneous and are made up of mostly silicate minerals. The primary chemical building block of silicate minerals is the silicon-oxygen tetrahedron, consisting of one silicon atom bonded to four oxygen atoms (Figure 4.2). In the three-dimensional crystal lattice of a silicate mineral, silicon-oxygen tetrahedra are linked together through a sharing of oxygen atoms. Hence, the actual ratio of silicon atoms to oxygen atoms varies among silicate minerals. Quartz, one of the most common silicate minerals, has the chemical formula SiO_2. Quartz makes up

TABLE 4.2
The Most Common Elements in the Earth's Crust

Element	Weight Percent
Oxygen (O)	46.6
Silicon (Si)	27.7
Aluminum (Al)	8.1
Iron (Fe)	5.0
Calcium (Ca)	3.6
Sodium (Na)	2.8
Potassium (K)	2.6
Magnesium (Mg)	2.1
All others	1.5

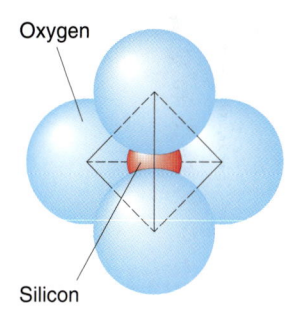

FIGURE 4.2
The silicon-oxygen tetrahedron is the building block of silicate minerals, the most common rock-forming minerals in the Earth's crust.

about 12% of Earth's crust and is the principal component of most beach sand.

The silicon-oxygen tetrahedron is combined chemically with ions such as calcium, magnesium, and aluminum. *Ferromagnesian silicate minerals* contain iron and magnesium ions and are dark in color and relatively dense. An igneous rock that is rich in ferromagnesian minerals is also dark and dense; basalt, the chief constituent of oceanic crust, is an example. *Nonferromagnesian silicate minerals* contain aluminum, calcium, sodium, or potassium ions and are relatively light in both appearance and density. An igneous rock that is rich in nonferromagnesian silicate minerals is similarly light in color and comparatively less dense; granite, the chief constituent of continental crust, is an example.

As a general rule, igneous rocks that are rich in ferromagnesian silicate minerals weather more rapidly than those that are composed of mostly non-ferromagnesian silicate minerals. But given sufficient time, all rocks and minerals eventually break down chemically and physically. Differences in rates of weathering determine the dominant composition of lithogenous sediments that are delivered to the sea—mainly quartz grains and clay particles. Quartz is the most resistant of the common silicate minerals and clay particles are common weathering products of silicate minerals such as feldspars.

Tropical and subtropical rivers account for the bulk of river-borne marine sediments at least in part because of the greater rate of chemical weathering in warm and humid climates (Figure 4.3). As noted earlier in this chapter, rivers transport the products of weathering in suspension, solution, or as part of the bed load. When a river reaches the coast, the denser and larger particles settle out first and are deposited in deltas, estuaries, or shallow bays. Currents flowing nearly parallel to the shoreline supply sediment to beaches (Chapter 8) while the finer sediment can be carried to the continental shelf and beyond.

Just about every part of the ocean receives wind-borne dust. In fact, this is the primary mechanism whereby lithogenous particles reach the far-from-land deep ocean. Such fine particles compose much of the deep-sea red and brown clays located near the centers of major ocean basins, especially around 30 degrees N and 30 degrees S. Because of the planetary-scale atmospheric circulation regime (Chapter 5), these latitudes are characterized by arid and semi-arid climates, resulting in large subtropical deserts such as the Sahara. Furthermore, these are regions of the ocean that are nutrient-poor and have low biological productivity. Hence the supply of biogenous sediments that would cover the clays on the bottom is very limited. High mountains and dry-lake beds also are sources of wind-blown dust because their barren surfaces are particularly susceptible to erosion during frequent episodes of strong winds.

Weathering removes carbon dioxide from the atmosphere and the potassium feldspar breaks down into a clay mineral ($KAl_3Si_3O_{10}(OH)_2$), plus bicarbonate (HCO_3^-) and potassium (K^+) ions and another silicate mineral (H_4SiO_4) that dissolve in water. As an example of the chemical weathering of a common silicate mineral, consider the reaction between potassium feldspar ($KAlSi_3O_8$), carbon dioxide (CO_2), and water (H_2O):

$$3KAlSi_3O_8 + 2CO_2 + 14H_2O = KAl_3Si_3O_{10}(OH)_2 + 2HCO_3^- + 2K^+ + 6H_4SiO_4$$

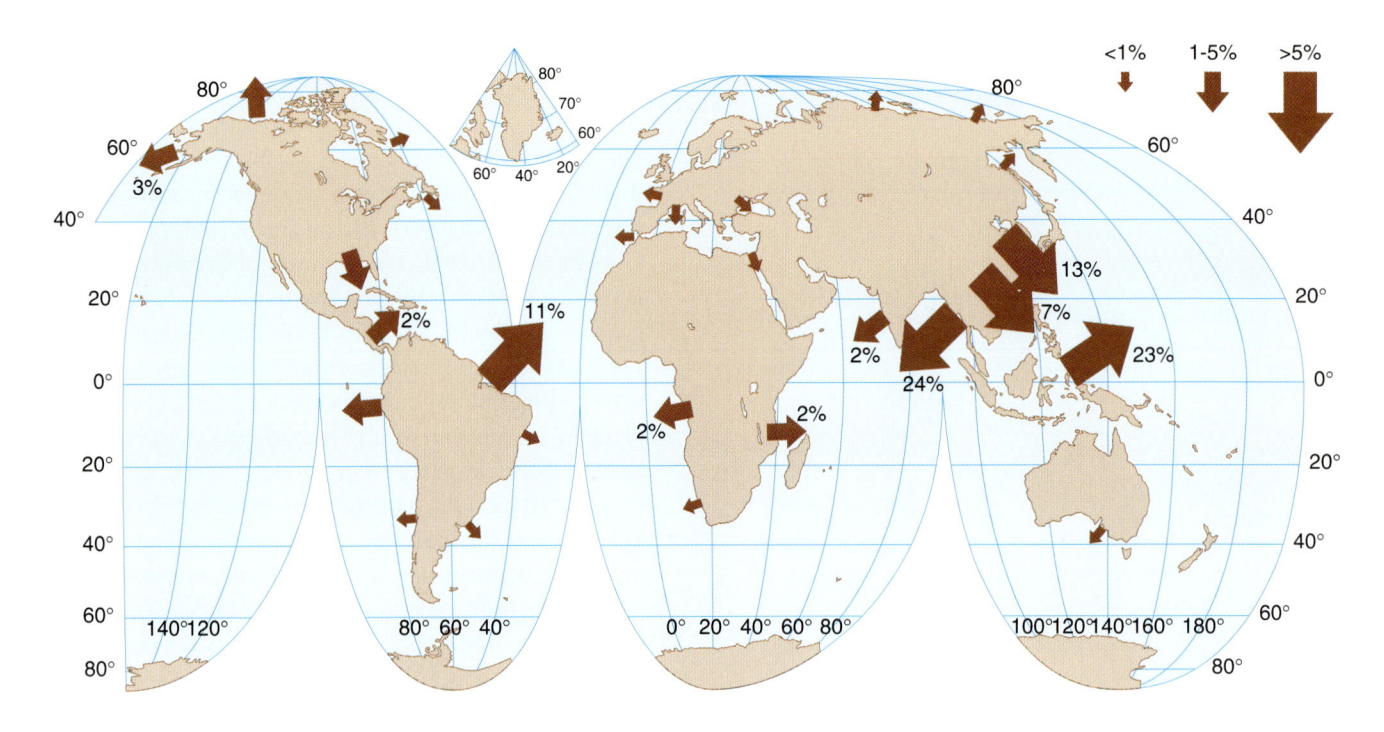

FIGURE 4.3
Sediment delivered to the ocean by Earth's major rivers expressed as a percentage of the total river-borne marine sediments. Tropical and subtropical rivers are the most important source of marine lithogenous sediments.

Volcanic ash is another source of windblown particles that settle into the ocean.

Remote sensing by satellite confirms wind transport of dust from the deserts of North Africa to the Atlantic basin (Figure 4.4). Strong winds associated with weather systems that cross North Africa pick up dust particles from the dry topsoil and carry them to altitudes of 3000 m (10,000 ft) or so. Trade winds blowing from the northeast toward the southwest transport plumes of the smallest dust particles over the Atlantic and Caribbean and into Central

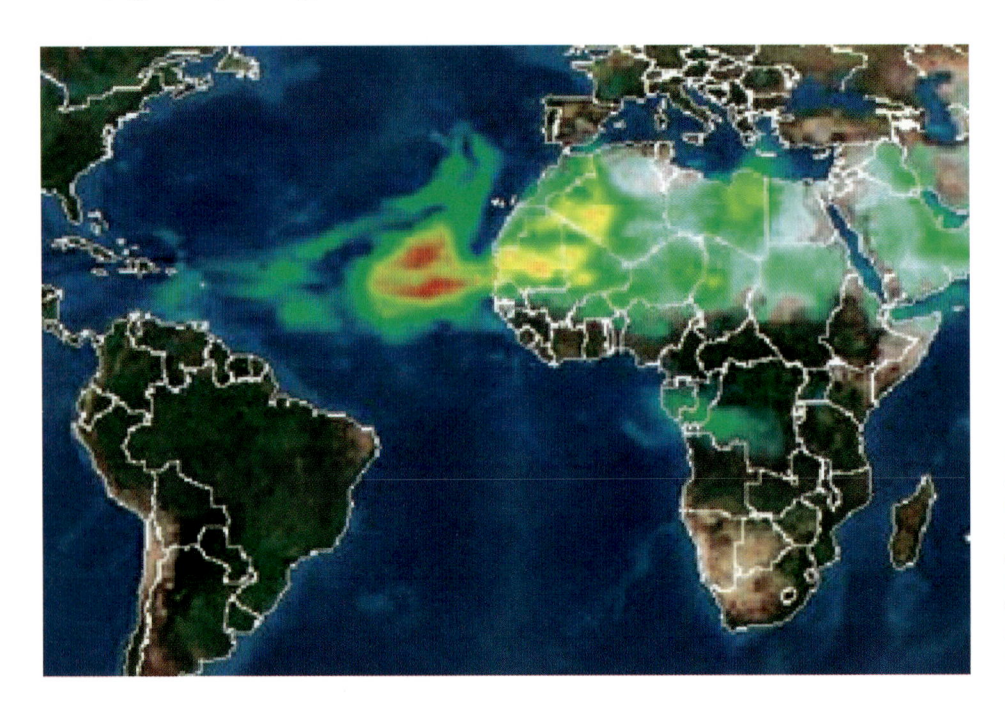

FIGURE 4.4
Satellite image obtained by NASA's Total Ozone Mapping Spectrometer (TOMS) showing dust coming off land sources in North Africa and moving westward across the Atlantic as a dust plume on 17 June 1999. [Courtesy of NASA]

America and the southeastern United States. This transoceanic journey takes about 1 to 2 weeks and occurs mostly from June to October, peaking in July.

Recently, scientists identified possible links between North African dust and red tides in the Gulf of Mexico and threatened coral reefs in the Caribbean. In the summer of 2001, scientists reported that iron in North African windborne dust particles fertilizes the Gulf of Mexico waters, increasing the frequency of toxic algae blooms, commonly known as **red tides**. Red tides have been implicated in the die-off of great numbers of fish, shellfish, marine mammals, and birds as well as respiratory problems and skin irritations in humans. Enhanced levels of iron enable specialized bacteria to convert nitrogen gas dissolved in seawater to a form that can be used by marine organisms. The greater availability of iron may trigger an explosive growth in populations of microscopic marine algae, including toxin-producing species that form red tides. North African dust may be harming coral reefs in the Caribbean through nutrient enhancement, spurring the growth of populations of algae and phytoplankton that colonize the same environment as coral and interfere with its growth. Furthermore, North African dust may also harbor a soil fungus that attacks coral reefs.

Glaciers erode bedrock and transport rock fragments of varying size and shape to the ocean. Sediments carried by glaciers are usually angular and can range in size from a fine powder to blocks of rock the size of an automobile or even larger. As a glacier flows into the ocean it floats on the denser seawater, giving rise to stresses that fracture the ice into floating bergs. The process whereby the leading edge of a glacier breaks up into icebergs upon entering the ocean (or large lake) is known as **calving**. Transported by ocean currents and wind, icebergs eventually melt in warmer environments, releasing a poorly sorted mix of sand and boulders, which quickly sinks to the bottom. Such ice-rafted *glaciomarine* sediments occur on about 20% of the sea floor; they cover the Antarctic continental shelf and are common in the Arctic Ocean and on the nearby deep-ocean bottom. Layers of such glacial debris, alternating with non-glacial sediments, have also been identified in cores extracted from deep-ocean sediments of the North Atlantic. These deep-sea sediment layers record sudden releases of icebergs during the last Ice Age, known as *Heinrich Events*. For more on this, see the first Essay at the end of this chapter.

BIOGENOUS SEDIMENT

Biogenous sediment includes the excretions, secretions, and remains of organisms. Examples include shells, fragments of coral, and skeletal parts. The chemical composition of most biogenous sediments is either calcium carbonate ($CaCO_3$) or silica (SiO_2)—substances secreted by organisms to form their shells. Organic carbon in these particles is largely consumed by organisms or decomposed by bacteria while sinking to the bottom or after deposition while the particle is exposed on the sea floor. Biogenous sediments dominate 30% to 70% of the ocean's mid-depths and skeletal remains alone account for 25% to 50% of all particles suspended in seawater.

Calcareous sediments are the most abundant of all biogenous sediment on the sea floor (Figure 4.5). They consist of calcium carbonate shells of *foraminifera* (single-celled organisms), shells of *pteropods* (small, floating snails), and *coccoliths* (platelets secreted by tiny, one-celled algae known as coccolithophores). Figure 4.6A is a scanning electron microscope photograph of a common species of coccolithophore with platelets that resemble tiny hubcaps. Robust shells such as many foraminifera make it to the deep-ocean bottom without entering into solution (Figure 4.6B). If covered over by later sediment deposits, they are preserved. This is the origin of calcareous muds that cover nearly half the deep-ocean bottom and are most abundant where ocean depths are shallower than about 4500 m (15,000 ft). They accumulate at rates of between 1 and 4 cm (0.4 and 1.6 in.) per 1000 years.

Many of the larger biogenous particles are **fecal pellets**, undigested organic matter that is concentrated and excreted by animals feeding in near-surface waters. Because of their large size, fecal pellets have relatively high terminal velocities and can sink hundreds of meters per day, reaching the ocean bottom perhaps within a few days. In this way, fecal pellets transport organic matter from surface waters and serve as a food source for bottom-dwelling organisms in the deep ocean. Smaller particles that are not ingested directly at the surface are often trapped by shredded pieces of large mucous webs produced by certain marine animals. These mucous fragments cause particles to clump together, making them larger so that they descend more rapidly to the sea floor (somewhat like fecal pellets although probably not as fast). The continual fall to the deep ocean floor of the remains of organisms from the upper, sunlit layer of the ocean, along with their fecal pellets and various forms of non-living matter, is sometimes called *marine snow*.

In the deep ocean, the smaller and more soluble sediments sink slowly and dissolve prior to reaching the bottom, thereby altering the chemical composition of deep-ocean waters. In addition to organic matter, these particles

84 Chapter 4 MARINE SEDIMENTS

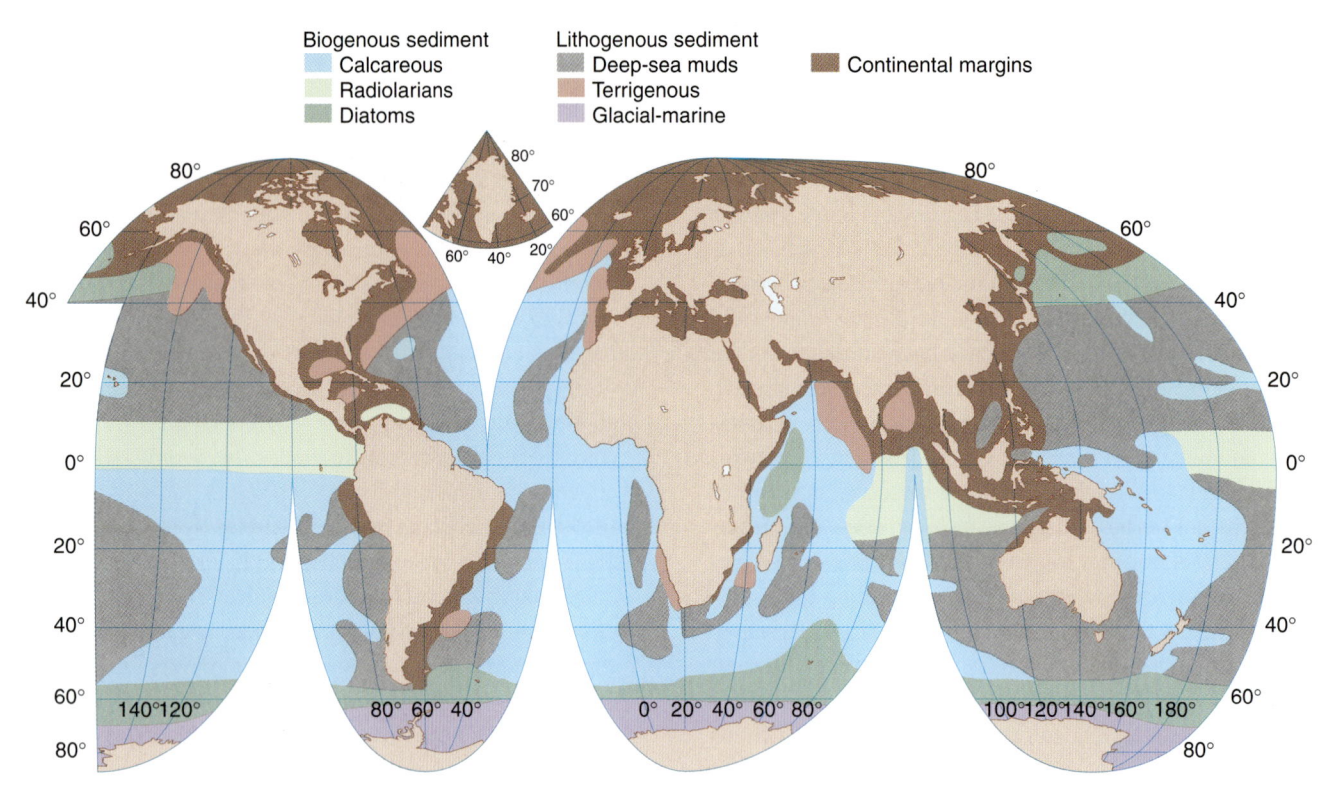

FIGURE 4.5
Distribution of lithogenous and biogenous sediments in the world ocean.

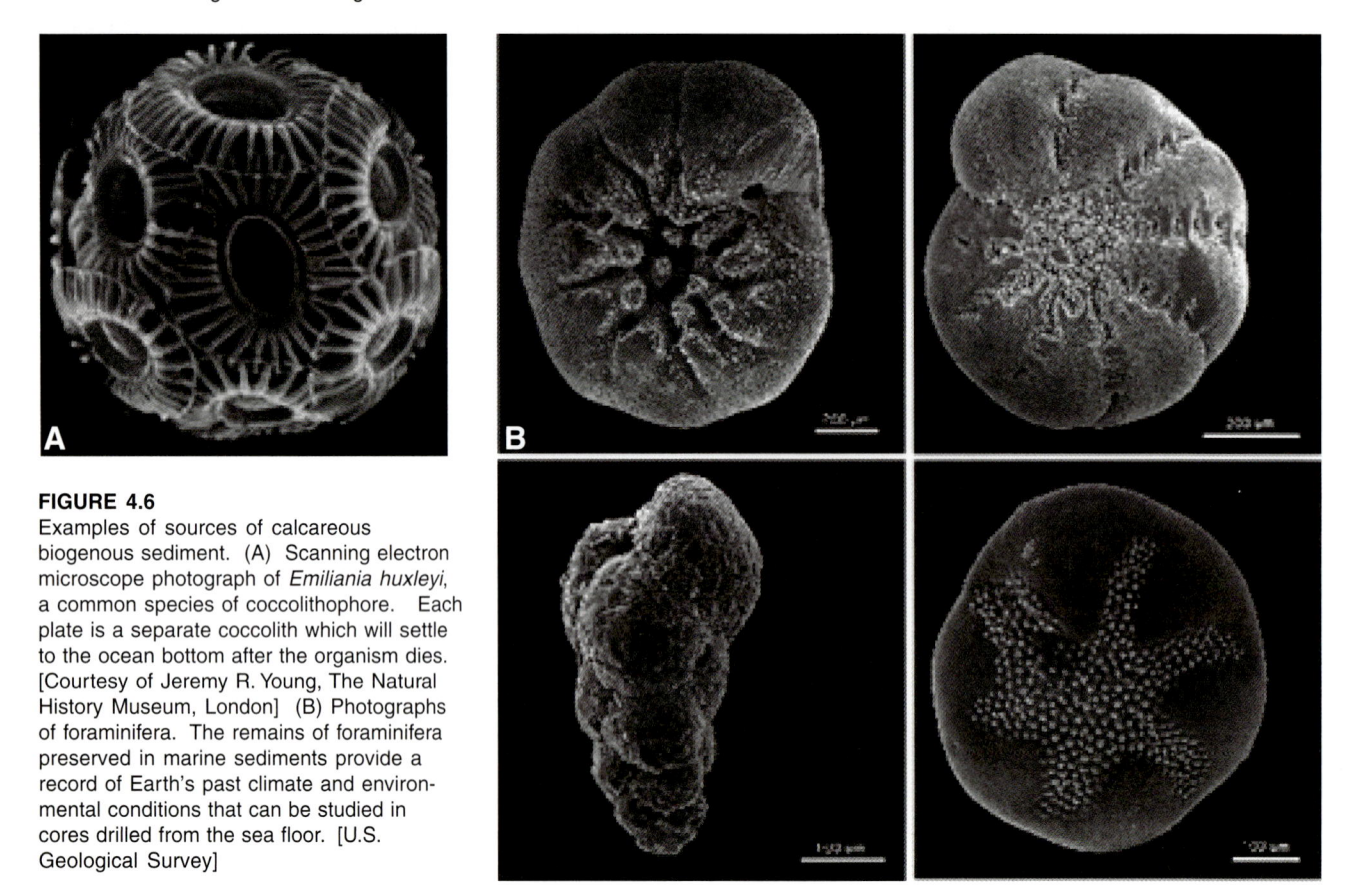

FIGURE 4.6
Examples of sources of calcareous biogenous sediment. (A) Scanning electron microscope photograph of *Emiliania huxleyi*, a common species of coccolithophore. Each plate is a separate coccolith which will settle to the ocean bottom after the organism dies. [Courtesy of Jeremy R. Young, The Natural History Museum, London] (B) Photographs of foraminifera. The remains of foraminifera preserved in marine sediments provide a record of Earth's past climate and environmental conditions that can be studied in cores drilled from the sea floor. [U.S. Geological Survey]

transport nutrients (nitrogen, phosphate, and silicate compounds) that are essential for photosynthetic organisms living in surface sunlit waters. When these particles dissolve in seawater, they release these nutrients to the deep-ocean waters. Nutrients may be cycled back to the surface layer through upwelling or other ocean circulation (Chapters 6 and 9).

The most soluble sediments dissolve rapidly and hence accumulate only on shallow ocean bottoms below regions where shell-forming organisms occur in abundance. For example, some organisms such as pteropods are abundant in near-surface waters, but their fragile shells are made of a particularly soluble form of calcium carbonate and readily dissolve in the more acidic deeper seawater. Consequently, pteropod shells are present only in sediment deposits on the tops of shallow seamounts; at greater depths, pteropod shells are not preserved.

Siliceous sediments are second in abundance to calcareous sediments on the ocean floor. These sediments consist of "shells" that are secreted by *diatoms*, single-celled algae (Figure 4.7A), and *radiolaria*, single-celled organisms (Figure 4.7B). Siliceous particles dissolve at all ocean depths (slightly more in shallow warm water) so their presence on the seafloor indicates areas where source organisms are particularly abundant. Siliceous muds are most common in the Pacific; diatom-rich muds nearly surround the Antarctic continent and occur in the northernmost Pacific. Radiolaria are abundant in warm waters so that radiolarian-rich sediments dominate the ocean floor in equatorial latitudes.

Phosphatic sediments (rich in phosphate) are rare in marine sediment deposits. Such deposits, consisting primarily of fish bones, teeth, and scales, occur primarily on shallow, isolated banks or near coastal areas where rates of biological productivity are especially high. Such near-shore phosphatic deposits may someday be exploited for their phosphate, which is used in manufacturing fertilizers.

HYDROGENOUS SEDIMENT

Hydrogenous sediment encompasses particles that are chemically precipitated from seawater, in some cases forming coatings on other seafloor sediment. In addition, some hydrogenous sediment is the product of chemical reactions taking place in hot seawater discharged by hydrothermal vents on the sea floor (Chapter 2). Examples of hydrogenous sediment include some carbonates, halite (NaCl), gypsum ($CaSO_4 \cdot 2H_2O$), and manganese nodules. A rise in the temperature of shallow water may cause dissolved carbonate to precipitate as tiny pellets known as

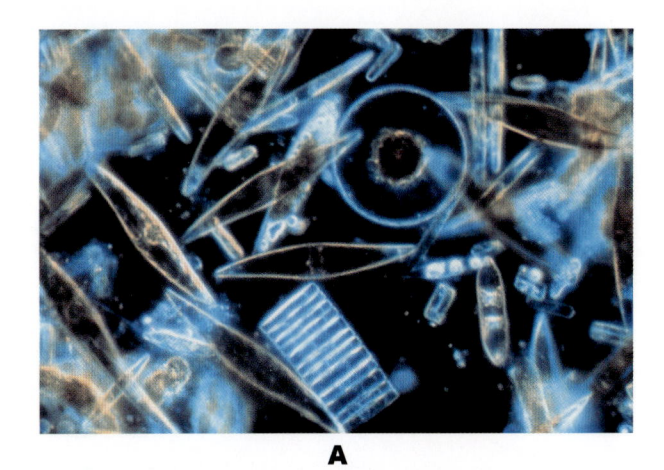

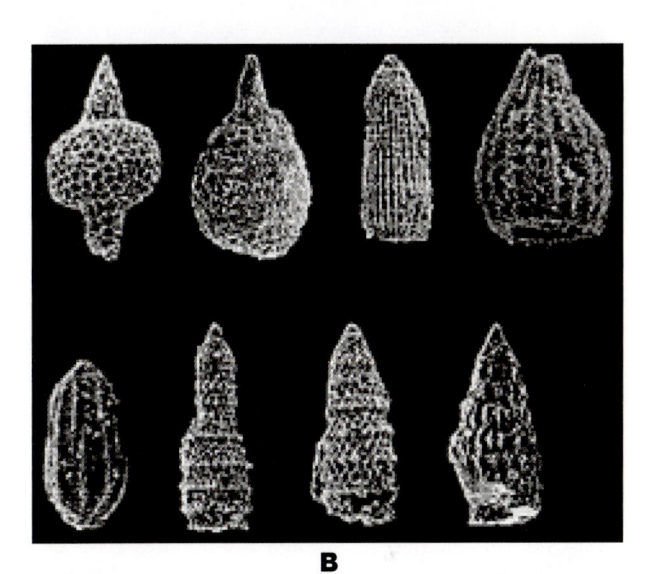

FIGURE 4.7
Examples of sources of siliceous biogenous sediment. (A) Diatoms as viewed through a microscope; they have silica exoskeletons. [Courtesy of NOAA and Dr. Neil Sullivan, University of Southern California] (B) These radiolaria have diameters of 0.5 to 1.5 mm. [Courtesy of National Park Service, U.S. Department of the Interior]

ooliths, perhaps 0.5 to 1.0 mm (0.02 to 0.04 in.) in diameter. As noted in Chapter 3, where evaporation rates are high and rainfall is low, salts precipitate from seawater in the sequence: carbonate salts, sulfate salts, halite.

Most conspicuous of all hydrogenous sediments are **manganese nodules**, irregularly shaped, sooty black or brown nodules on the sea floor (Figure 4.8). Manganese nodules on average contain about 18% manganese by weight, 17% iron, and more importantly, small amounts (generally less than 1%) of copper, cobalt, and nickel. They range in size from tiny grains to large slabs weighing hundreds of kilograms; most, however, are the size of potatoes. Manganese nodules occur on the floor of all oceans except

FIGURE 4.8
A box core from the floor of the tropical Pacific showing a relatively high density of manganese nodules. Box cores preserve the character of the undisturbed ocean bottom. [Courtesy of NOAA]

the Arctic and in a variety of marine environments from abyssal plains to mid-ocean ridges. They are most abundant in an east-west belt about 5000 km (3000 mi) long on the floor of the tropical Pacific southeast of Hawaii and north of 10 degrees N.

Manganese nodules begin as coatings on hard objects (e.g., rock fragments, whale ear bones, a shark's tooth) that are exposed on the ocean bottom for lengthy periods of times. Burrowing organisms turn the nodules over, exposing all sides to seawater so that a coating forms on all sides of the object. Manganese nodules grow extremely slowly, ranging from about 1 to 10 mm (0.004 to 0.04 in.) per million years. With such slow growth rates, manganese nodules must remain unburied on the ocean bottom. Otherwise, rapidly accumulating sediments would cover the nodules, isolating them from contact with seawater so that they could not grow substantially. In areas of the ocean receiving an abundant influx of sediment, only small (pea-sized or smaller) micronodules develop because they are buried before they can grow into larger nodules. Accordingly, rich manganese nodule deposits are found only in regions of the ocean far from shore, where input of lithogenous sediments is small and in unproductive areas where the rate of accumulation of biogenous particles is also low.

COSMOGENOUS SEDIMENT

Cosmogenous sediment comes from outer space, for example, as meteorite fragments. Perhaps 90% of the solid particles that enter Earth's atmosphere burn up due to frictional heat prior to reaching the surface. Most particles that survive the journey through the atmosphere and enter the ocean are so small that they dissolve before reaching the bottom. Nonetheless, cosmogenous particles are found mixed with other ocean bottom sediments. Cosmogenous particles also enter the ocean from the Greenland and Antarctic ice sheets by melting snow and ice.

Some cosmogenous sediments are remnants of the formation of planets in the solar system. These iron-rich particles likely have a chemical composition similar to Earth's core and mantle and their unique composition make them readily recognizable in deep ocean sediment deposits. Other particles, formed from silicate rocks blasted off other planets or the moon by meteorite impacts, are more difficult to identify as cosmogenous because they resemble lithogenous sediment.

A special type of sediment that is indirectly cosmogenous in origin consists of solidified droplets of rocks melted when huge meteorites struck the Earth. These small black fragments of silica-rich glass are known as **tektites**. Tektites are typically 2.5 to 5 cm (1 to 2 in.) across and have a teardrop or dumb-bell shape indicating that they were once fluid. They formed when droplets of molten rock produced by the meteorite impact were blasted into Earth's atmosphere and rapidly cooled and solidified. For example, the meteorite impact about 65 million years ago that is hypothesized to have killed off the dinosaurs generated great numbers of tektites that accumulated in and around the impact site, that is, Mexico's Yucatán Peninsula (Chapter 1). Tektites also have been identified in ocean bottom sediments off southern Australia and in the Indian Ocean.

Marine Sedimentary Deposits

The various types of marine sediments (i.e., lithogenous, biogenous, hydrogenous, and cosmogenous) occur in varying proportions on the ocean bottom as marine sedimentary deposits. In this section, we describe marine sediment deposits of the continental margin and the deep ocean basin.

CONTINENTAL-MARGIN DEPOSITS

Most marine sediment deposits in the continental margin (called **neritic deposits**) are lithogenous and occur in a wide range of sediment sizes. Most river-borne lithogenous particles do not travel very far seaward of the shoreline. About 95% of the largest sediments transported to the ocean by rivers are trapped and deposited in bays, wetlands, estuaries, beaches or deltas. Only about 5% of river-borne sediment brought to the shoreline reaches the continental shelf or slope. Very little terrestrial sediment is transported beyond the continental margin into the deep-ocean basins. A notable exception is seaward of the mouths of major sediment-transporting rivers such as the Mississippi, Ganges (Bangladesh), and Yangtse (China). Massive submarine avalanches and turbidity currents can transport sediments hundreds or thousands of kilometers out onto the continental rise and to the sea floor beyond.

Where a river enters the ocean, the water slows so that sediments begin settling out of suspension. As long as currents do not carry off sediments as fast as they are deposited, sediments accumulate at the mouth of the river as a **delta**, so-called because this deposit resembles a triangle or the Greek letter *delta* (Δ) when viewed from above. Water flowing in a branching series of channels, known as *distributaries*, spreads sediment over the delta. Unless artificially stabilized, the main current of the river may abruptly shift from one distributary to another.

Figure 4.9 shows the Nile River Delta looking toward the northwest, photographed from a NASA Space Shuttle. The delta stretches about 160 km (100 mi) north and south and 240 km (150 mi) west and east. Beginning at Cairo, Egypt, the delta disperses the Nile's flow through a fan-shaped network of channels into the Mediterranean Sea. At the edge of the delta, the Nile splits into two branches, the Rosetta to the west and the Damietta to the east, each about 240 km (150 mi) long, that cross the delta and empty into the Mediterranean Sea. For thousands of years, suspended sediment delivered and deposited by the Nile and its branches provided the Nile Delta with the most fertile soils on the African continent. But with completion of the Aswan High Dam in 1971, the delta's normal evolution was disrupted. Prior to the Aswan High Dam, perhaps 90% of the sediment transported by the Nile reached the delta; today, almost all the sediment is trapped in Lake Nasser, the huge reservoir behind the dam. With much less sediment input coupled with continued compaction and subsidence of the delta, erosive forces (e.g., waves) are dominating and the delta's shoreline is retreating rapidly—in some places more than 150 m (500 ft) per year. Saline water from the Mediterranean is also seeping into parts of the delta. And nutrients formerly transported downstream have been reduced, thereby depleting the populations of some fish species in the Nile and the eastern Mediterranean.

Wetlands are low-lying flat areas that are covered by water or have soils that are saturated with water for at least part of the year. They are common on deltas and coastal plains of the mid-Atlantic U.S. and Gulf of Mexico and in many other areas in passive continental margins worldwide. Wetlands such as salt marshes accumulate large amounts of brown or black organic matter and support numerous and diverse populations of marine plants, animals and birds. Wetlands help control coastal flooding by acting as a sponge taking up water during high water episodes.

As noted in Chapter 2, near the mouths of many major sediment-transporting rivers that empty into the ocean, intermittent avalanches of dense, mud-rich waters, called **turbidity currents**, flow down submarine canyons, carrying sediment onto the ocean floor (Figure 4.10A). Such flows are rarely directly observed on the ocean floor but are well known in lakes and reservoirs behind dams. Turbidity currents are powerful enough to break submarine cables, especially near mouths of major rivers, such as the Congo (Zaire) and Ganges. Sedimentary deposits produced by turbidity currents are known as **turbidites** (Figure 4.10B).

Although occurring infrequently, turbidity currents transport and deposit large amounts of sediment that form thick beds over broad areas of the deep ocean floor. For example, in 1929 an earthquake triggered a submarine avalanche that produced a turbidity current off the Grand Banks, south of Newfoundland. Sediments were deposited on the nearby deep-ocean floor over an area that measured 100 km (60 mi) by 300 km (190 mi). Based on the time elapsed between the earthquake and submarine-cable breaks, scientists calculated the speed of the turbidity current at about 20 km per hr (12 mi per hr), about the speed

FIGURE 4.9
The Nile River Delta. Photograph taken from NASA Space Shuttle in October 1984. [Courtesy of NASA]

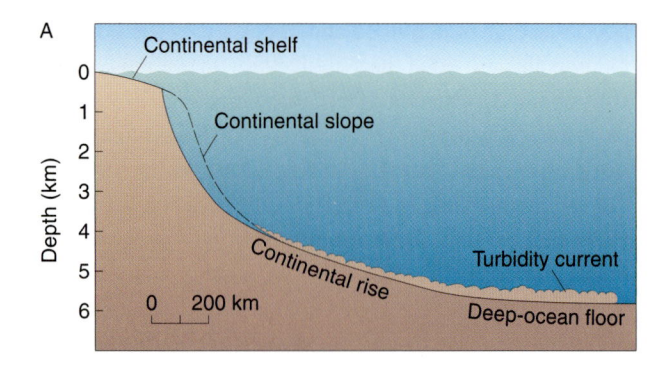

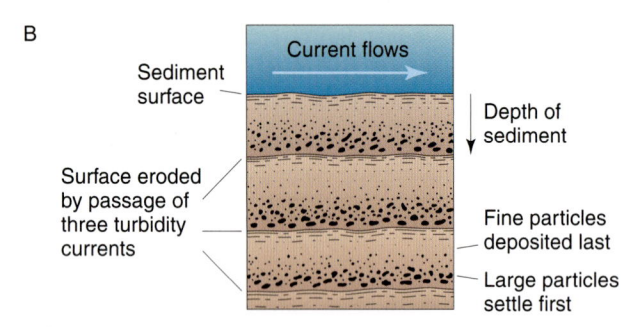

FIGURE 4.10
A turbidity current can transport lithogenous sediment into the deep-ocean bottom. (A) Vertical profile of a turbidity current flowing on the continental slope and rise. (B) Vertical cross-section of turbidites deposited by three successive turbidity currents.

of a slow freight train. Submarine avalanches not only cause turbidity currents but also may produce dangerous tsunamis (Chapter 7).

Turbidity currents are much denser than seawater because of the added suspended sediment load. Hence, they flow along the ocean bottom, often eroding channels just as rivers do on land. Such flow behavior explains why turbidites have unusual textures and contain abundant shells and other remains of shallow-water dwelling organisms. Submarine canyons near the mouths of major sediment-transporting rivers would soon fill with sediment deposits unless periodically scoured by turbidity currents.

Obstructions on the ocean bottom deflect turbidity currents and can prevent their flow onto the deep-ocean floor. This is especially the case in the Pacific basin, where submarine ridges associated with volcanic island arcs and trenches block near-bottom waters flowing toward the adjacent deep-ocean bottom. In the Atlantic, northern Indian, and Arctic Oceans, thick sediment deposits form the conspicuous continental rise (Chapter 2). The tops of seamounts and submarine ridges typically are unaffected by turbidity currents and are free of turbidites.

In coastal areas, especially near large rivers or in deltas, marine sediment accumulates rapidly, typically at rates of several meters per thousand years. Sediments are buried too quickly to react fully with seawater or water's dissolved oxygen. Furthermore, bottom-dwelling organisms cannot consume all of the food sources. These sediments exhibit a wide variety of colors, including greens and blues, due to the various oxidation states of iron on the particles.

DEEP-OCEAN DEPOSITS

Fine-grained sediments that gradually accumulate particle-by-particle on the deep-ocean floor form **pelagic deposits**. Most of these sediments are biogenous and their accumulation rates are considerably slower than neritic (near shore) sediments. On average, a 1 mm-thick layer forms in about 1000 years. (A 1-inch thick layer forms in about 25,000 to 250,000 years.) In spite of these very low accumulation rates, sufficient time has passed since formation of the ocean basins that the average thickness of pelagic deposits is 500 to 600 m (1600 to 2000 ft).

Very small particles have low terminal velocities and sink slowly through the ocean depths. Many of these particles are eaten by filter feeders or trapped in mucus nets secreted by marine organisms. Their relatively long suspension times in the water mean that they are transported and widely dispersed by currents. There is also ample time for chemical reactions to occur. For example, iron minerals in suspended clay particles react chemically with dissolved oxygen in seawater, forming a rusty (iron oxide) coating. The abundance of such red- or brown-stained particles in deep-sea muds accounts for the colors and common names of these deposits, that is, *red clay* and *brown mud*. Colors of deep-sea clays range from brick red (derived from the Saharan Desert) in the Atlantic Ocean to chocolate brown in the Pacific.

Pelagic deposits that are more than 30% biogenous by weight are called either calcareous ooze or siliceous ooze depending on composition. **Calcareous oozes**—made up of the tests (protective shells) of coccolithophores, pteropods, and foraminifera—are the most abundant of pelagic oozes. Calcareous oozes are generally confined to ocean waters shallower than the **carbonate compensation depth (CCD)**, the depth of the ocean below which material composed of calcium carbonate ($CaCO_3$) dissolves and does not accumulate (Figure 4.11). The CCD averages about 4500 m (14,800 ft). The rate at which calcium carbonate dissolves in water increases with falling temperature because cold water can hold more carbon dioxide (CO_2) which dissociates in water to form a weak acid that dissolves $CaCO_3$. Because

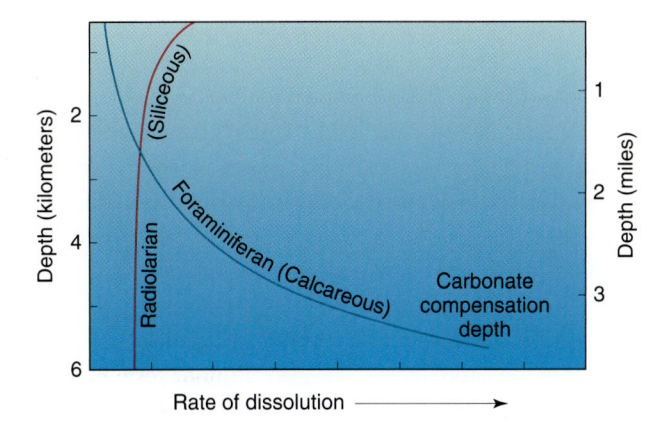

FIGURE 4.11
The rate at which radiolaria (siliceous) and foraminifera (calcareous) shells dissolve with increasing ocean depth in terms of percent weight loss. The carbonate compensation depth is defined as the depth at which the amount of calcareous particles declines to less than 20% of the mass of all particles present.

ocean temperatures usually drop with increasing depth below the surface waters, more and more calcareous particles dissolve with increasing water depth. Calcareous oozes are found on features that stick up from the ocean bottom to depths shallower than 4500 m, such as the slopes of the mid-ocean ridges. If the ocean basins were to suddenly dry up, the tops of volcanic peaks rising above the CCD would be light in color due to the accumulation of carbonate-rich deposits. Hence, the CCD is somewhat analogous to the snow line on terrestrial mountain peaks.

Siliceous oozes are composed of tests of diatoms and radiolaria (described earlier in this chapter). Ocean water is under-saturated with silica so that these tests dissolve at all ocean depths (Figure 4.11). They occur on the sea floor only below surface waters where source organisms are particularly abundant. Siliceous and calcareous oozes consist of clay size particles and dominate the deep-ocean bottom sediment (Figure 4.5). Sand-sized particles make up less than 10% of the deep-ocean sediments and the coarsest deep-ocean sediment deposits are the products of explosive volcanic eruptions and ice rafting.

Marine Sedimentary Rock

As noted in Chapter 1, sediment is generated at the interfaces between the lithosphere, atmosphere, hydrosphere, cryosphere, and biosphere. Physical and chemical weathering processes break down exposed bedrock forming rock fragments (sediments) that are transported by rivers, glaciers, wind, and gravity to the ocean. Weathering also releases soluble constituents, such as calcium and sodium that dissolve in water and are transported in solution to the ocean.

As part of the rock cycle (Figure 4.12), over the millions of years that constitute geologic time, sediment that is deposited on the ocean floor is gradually converted to solid marine sedimentary rock through lithification. **Lithification** usually involves both the compaction and cementing of sediments at relatively low temperatures (under 200 °C or 390 °F). Sediments are compacted by the increasing weight of sediments accumulating above that squeeze deeper sediments closer together. Siliceous and calcareous fluids migrating through the tiny openings between individual sediment grains precipitate minerals that fill the pore spaces, cementing grains to one another. The product of lithification is sedimentary rock such as shale, sandstone, or limestone depending on the composition of the constituent sediments. With deeper burial and further increases in temperature, pressure, and access to chemically active fluids, sedimentary rock may be converted to metamorphic rock such as slate, schist or gneiss (Chapter 2).

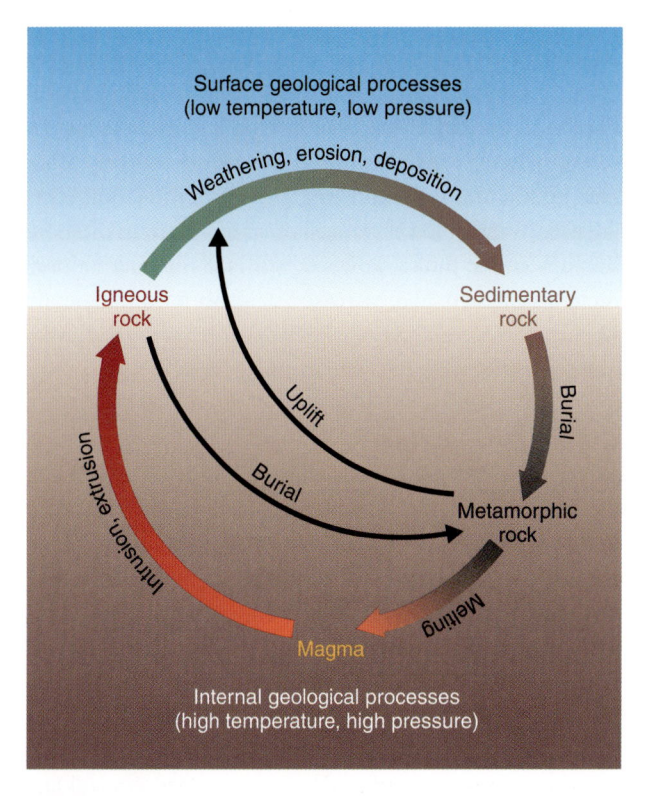

FIGURE 4.12
The conversion of marine sediments to sedimentary rock is part of the rock cycle.

Most sediment and sedimentary rock deposited on the ocean floor are transported with the underlying moving lithospheric plate, eventually entering a subduction zone, where they are drawn down into the mantle and metamorphosed or melted into magma. Some of this magma migrates to the ocean floor via seafloor spreading and volcanic eruptions, cools and solidifies. While most marine sedimentary strata on the continents originated in extensive shallow seas, some marine sedimentary deposits were scraped off subducting oceanic plates and physically attached to a continental plate. This plus other tectonic activity such as mountain building explains why marine sedimentary rocks are found high in continental mountains. For an example of this, see this chapter's second Essay.

Resources of the Seafloor

Resources extracted from the seafloor include oil, natural gas, sand, gravel, and minerals. While oil and natural gas account for more than 95% of the total monetary value of resources extracted from the seafloor, sand and gravel are the seafloor resources most commonly mined worldwide. Perhaps 25% of the oil and 20% of the natural gas consumed in the U.S. each year comes from offshore sources including the Gulf of Mexico, Persian Gulf, and North Sea.

Global economics plays a central role in governing the amount of ocean floor mining and is the principal reason why few resources are mined from the sea (other than oil, natural gas, sand, and gravel). Production costs are usually much lower for onshore mining versus offshore mining. In the future, however, shifts in the international minerals markets and increased demand for strategically important minerals may make it feasible to begin mining more seabed resources such as manganese nodules.

OIL AND NATURAL GAS

Oil and natural gas, the chief forms of petroleum, are derived from marine plant and animal remains and occur in the pore spaces of marine sedimentary rock. Both oil and natural gas consist of hydrocarbons, compounds whose molecules contain only hydrogen and carbon atoms. Oil is a mixture of thousands of different hydrocarbons. By volume, natural gas is up to 99% methane (CH_4), plus small quantities of ethane (C_2H_6), propane (C_3H_8), and butane (C_4H_{10}).

Today's major deposits of oil and natural gas developed under very restricted and unusual biological and geological conditions millions of years ago. Petroleum formation requires that substantial amounts of organic matter must accumulate on the bottom of a shallow quiet sea. Decomposer organisms reduced the dissolved oxygen to levels that could be tolerated by only anaerobic bacteria. Products of anaerobic decomposition include methane and other light hydrocarbons. With continued accumulation of organic and other sediment on the sea floor, the deeper organic-rich sediments were subject to increasing temperature and pressure that spurred their lithification and the conversion of organic matter to oil and natural gas.

Conditions favorable to oil and natural gas production occurred primarily during the geologic periods known as the Ordovician (505-438 million years ago), Permian (286-245 million years ago), Jurassic (208-144 million years ago) and Cretaceous (144-65 million years ago). At those times, sea level was unusually high and ocean waters spread over low-lying portions of the continents producing large shallow seas. Enhanced biological productivity in those seas provided the raw material for the eventual development of oil and natural gas deposits.

Oil and natural gas migrated upward from their source rock into more porous layers of sandstone or limestone, displacing the water that occupied the pore space in these strata. An overlying layer of less permeable rock trapped the petroleum in the so-called *reservoir rock*. Extraction wells must penetrate this reservoir rock to tap the oil or natural gas. Oil formed when sediments were buried to depths of at least 2 km (1.2 mi) and generally does not occur at depths greater than 3 km (1.9 mi). Natural gas is usually found in marine sedimentary rocks at depths of less than 7 km (4.3 mi).

MINERAL RESOURCES

Great quantities of sand, gravel, and shells are mined from the near-shore, shallow ocean bottom, especially near coastal cities. These resources are used primarily in road construction and the production of cement and concrete. Also, sand dredged from the shallow ocean bottom is used to replenish and restore nearby beaches eroded by storm waves (Chapter 8).

In some locales, valuable metallic and non-metallic minerals such as iron, tin, platinum, gold, and diamonds occur mixed with coastal sands. Most of these resources are products of weathering and erosion of continental rock and sediment and are transported to the sea in suspension by rivers along with other lithogenous particles. Ocean waves and currents sort and concentrate metals and gemstones in coastal or submarine deposits, known as *placer deposits*. Placer minerals typically are relatively dense and resistant and are left behind as a *lag concentrate* after waves and currents remove the less dense sand grains. Although hundreds of such deposits are known from around the world, very few are actually being exploited. Today, dredging of placer deposits usually takes place in shallow waters just offshore and yields tin (Thailand and Indonesia), gold (Alaska, New Zealand, and the Philippines), and diamonds (Namibia and South Africa). Rivers also deliver minerals in solution to the sea including phosphorite (a deposit rich in phosphate that precipitates from seawater and accumulates on the seafloor under upwelling or other biologically productive areas), and manganese (which forms manganese nodules via biogeochemical cycling).

As noted earlier in this chapter, manganese nodules on the deep ocean floor have been studied extensively as potential sources of copper, nickel, and cobalt. Cobalt-rich manganese crusts occur at intermediate depths (2 to 3 km, or 1 to 2 mi) on the slopes of extinct volcanoes that form many Pacific islands. But after decades of exploration and development of deep-ocean mining techniques, there is still no economic incentive for commercial production from the deep-ocean floor.

Over the past several decades, development of the plate tectonics theory has spurred interest in another source of marine mineral deposits: geological processes taking place at submarine plate boundaries (Chapter 2). Ocean basins include sites where mineral deposits form in place (a process called *mineralization*) rather than being delivered to the sea by rivers. Seawater, magma, and new oceanic crust interact at plate boundaries (including spreading centers and subduction zones), exchanging heat and chemicals and producing *hydrothermal mineral deposits*. Magma from Earth's interior heats the seawater that circulates through the fractures in the oceanic crust. Hot seawater dissolves metals from the magma and crust, and those metals react with sulfur in the seawater to precipitate as sulfide minerals on the ocean floor. These hydrothermal deposits are potential sources of copper, zinc, silver, gold, and other metals.

As in the case for seabed placer deposits, future development of hydrothermal mineral deposits depends on favorable economic conditions as well as consideration of potential environmental impacts. In any event, hydrothermal sulfide metals associated with subduction zones (e.g., in the western Pacific) are economically more promising for several reasons: a greater percentage of precious metals, occurrence at shallower ocean depths (1000-2000 m or 3300-6600 ft), and location within the 370-km (200-nautical mi) jurisdiction of coastal nations.

EXCLUSIVE ECONOMIC ZONE

Since 1958, the United Nations has worked to formulate international policies concerning the exploitation of seabed resources, including fuels and minerals. But progress has been slow because of conflicts among the more than 150 nations involved in negotiations. A fundamental philosophical division exists between less-developed and more-developed nations. Many less-developed nations view ocean resources as the common heritage of all people, but they also fear that the world's richest and most technologically advanced nations will reap the bulk of the harvest. While the U.S. and many other more-developed nations adhere to the common heritage concept, they fear that too much power would be vested in the governments of less-developed nations if they were granted significant say in shaping ocean resource policy.

The 1982 U.N. Convention on the Law of the Sea granted jurisdiction over an **exclusive economic zone (EEZ)** to each of 151 coastal nations. In March 1983, the U.S. (later joined by fifty other nations) defined its jurisdiction over ocean resources (including minerals, fuels, and fisheries) to extend 370 km (200 nautical mi) offshore, encompassing an area that is 1.7 times that of the total land of the U.S. and its territories. Within the EEZ, the federal government regulates all economic activity beyond the limits of the individual state's jurisdiction. The National Oceanic and Atmospheric Administration (NOAA) and the Interior Department share responsibility for managing seabed mineral resources.

The 370-km exclusive economic zone may or may not encompass the entire continental shelf associated with the landmass of a specific coastal nation (Figure 4.13). Hence, a provision of the 1994 U.N. Convention on the Law of the Sea allows a nation to expand its EEZ to the edge of the continental shelf, if it can establish that the new territory is a "natural prolongation" of its landmass. (Recall from Chapter 2 that the continental shelf is a submerged extension of a continent.) The challenge is to establish scientifically the outer boundary of the continental shelf, that is, where the continental shelf ends and the slope begins. Nations have until 2009 to prove their case. By one estimate, the U.S. stands to gain an additional 750,000 square km (290,000 square mi) of the Atlantic, Pacific and Arctic Oceans as the EEZ includes the shelf surrounding every island and possession, no matter how small or remote.

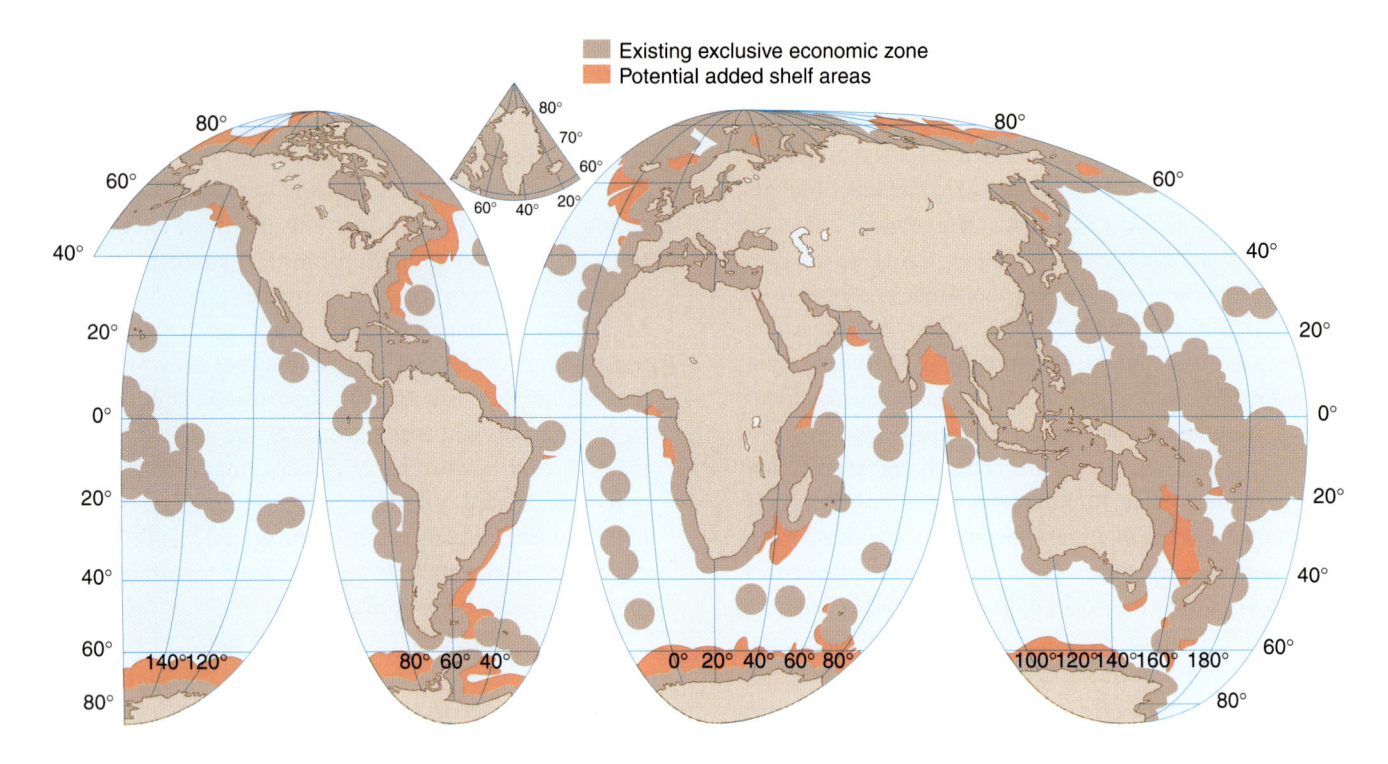

FIGURE 4.13
Current exclusive economic zones (brown) in the near future could be extended to the limits of the continental shelf (orange).

Conclusions

Sediments are produced at the interfaces between the lithosphere, atmosphere, hydrosphere, cryosphere, and biosphere. Rivers, winds, glaciers, and gravity transport sediment from land to sea while some sediment originates in the ocean. Sediment deposits on the ocean floor record changes in the various subsystems of the Earth system over millions of years, that is, since formation of the present ocean basins. In later chapters, we revisit deep-sea sediments as we seek to understand how the various subsystems have changed through Earth history and what they reveal about past variations in Earth's climate.

Sedimentary processes (e.g., weathering, erosion, lithification) are key players in the functioning of the Earth system. For example, sequestering of carbon in sediments and sedimentary rocks is responsible for our fossil fuel resources and accounts for the relatively small amount of carbon dioxide in the Earth's modern atmosphere. Atmospheric carbon dioxide is one of many gases that contribute to the planet's so-called greenhouse effect. We learn more about Earth's climate system in the next chapter where we consider the flux of heat energy and water between the ocean and atmosphere.

Basic Understandings

- Particles that accumulate on the sea floor are known as marine sediments. They differ in source, composition, size, and the rate at which they accumulate on the sea floor.
- Sediments are classified by size into three broad categories as mud, sand, or gravel. Accumulations of sediment on the seafloor also vary in the range of grain size, known as sorting. A well-sorted sediment deposit has a narrow range of grain sizes whereas a poorly sorted one has a broad range of grain sizes.
- Terminal velocity is the constant speed attained by a particle as it falls through a motionless fluid such as water or air. The speed of a falling particle in calm water or air is regulated by gravity and the resistance offered by the medium through which the particle falls. A downward accelerating particle meets increasing fluid resistance while gravity remains essentially constant. The magnitude of the resisting force eventually equals gravity; that is, the two forces come into balance. When forces are balanced, then according to Newton's first law of motion, the downward moving particle attains a constant speed (the terminal velocity). For a given

- medium, the terminal velocity generally increases with increasing particle size.
- Ocean scientists classify marine sediments based on their source as lithogenous (from rock), biogenous (from organisms or their remains), hydrogenous (materials precipitated from seawater), and cosmogenous (from outer space).
- Lithogenous particles (most of which originate from weathering of rock on land) are transported to the ocean by rivers, winds, glaciers, and gravity. Some lithogenous sediments are of volcanic origin. The largest lithogenous particles are deposited close to where they enter the ocean; smaller particles are transported further out to sea. Fine windblown sediment (i.e., clay and silt particles) can travel thousands of kilometers from their source.
- Biogenous sediment includes the excretions, secretions, and remains of organisms living in the ocean, usually in the sun-lit, near-surface waters. The chemical composition of most biogenous sediments is either calcium carbonate ($CaCO_3$) or silica (SiO_2).
- Hydrogenous sediment encompasses particles precipitated from seawater, in some cases forming coatings on other seafloor sediment. In addition, some hydrogenous sediment is the product of chemical reactions in seawater circulating in hydrothermal vents.
- Most marine sediment deposits on the continental margin, called neritic deposits, are lithogenous and occur in a wide range of sizes. About 95% of the largest sediments transported to the ocean by rivers are trapped and deposited in bays, wetlands, estuaries, beaches or deltas. Only about 5% of river-borne sediment brought to the shoreline reaches the continental shelf or slope.
- Fine-grained sediments that slowly accumulate on the deep-ocean floor form pelagic deposits. Most of these sediments are biogenous and their rate of accumulation is considerably slower than neritic (near shore) sediments.
- Calcareous oozes—made up of the tests (protective shells) of coccolithophores, pteropods, and foraminifera—are the most abundant of pelagic oozes. They occur where the deep ocean is shallower than the carbonate compensation depth (CCD). Siliceous oozes are composed of tests of diatoms and radiolaria. Ocean water is under-saturated with silica so that these tests dissolve at all ocean depths. Hence, they occur on the sea floor only in those portions of the ocean where source organisms in surface waters are abundant.
- Over the millions of years that constitute geologic time, sediment that is deposited on the ocean floor is gradually converted to solid marine sedimentary rock through lithification, which usually involves both compaction and cementing of sediments at relatively low temperatures (under 200 °C or 390 °F).
- Oil and natural gas, the chief forms of petroleum, are derived from the remains of marine organisms and occur in the pore spaces of marine sedimentary rock.
- Great quantities of sand, gravel, and shells are mined from the near-shore, shallow ocean bottom, especially near coastal cities primarily for use in road construction and the production of cement and concrete. In some locales, valuable metallic (e.g., tin) and non-metallic (e.g., diamonds) minerals occur mixed with coastal sands as lag concentrates.
- Since 1958, the United Nations has been working on international policies concerning the exploitation of seabed resources, including fuels and minerals. But progress has been slow because of long-standing conflicts among the more than 150 nations involved in negotiations. At present, the exclusive economic zone (EEZ) extends to 370 km (200 nautical mi) offshore with the possible extension to the edge of the continental shelf by the year 2009.

ESSAY: Heinrich Events

Cores extracted from sediment deposits on the deep ocean floor of the North Atlantic Ocean contain layers of pebbles and other coarse rock fragments, called *Heinrich layers*, named after the German researcher Hartmut Heinrich who first described them in 1988. These sediments apparently were released by unusually large numbers of melting icebergs in the North Atlantic and settled to the seafloor. Heinrich layers were associated with massive discharges of icebergs from northeastern Canada (Hudson Bay and the St. Lawrence River) during the last Ice Age. From analysis of North Atlantic sediment cores, Heinrich identified six layers dating from the past 100,000 years. The layers were deposited during episodes when sea surface temperatures were exceptionally low.

During the last Ice Age, Earth's mean surface temperature may have fluctuated as much as 2 to 5 Celsius degrees (5 to 10 Fahrenheit degrees) every 2000 to 3000 years with deposition of Heinrich layers coinciding with the close of a cold episode. In addition to producing Heinrich layers, the massive influx of icebergs chilled the surface waters and the addition of large quantities of fresh water from the melting ice may have altered the ocean's deep current system that exerts a major influence on the climate of northern Europe (Chapter 6). Another effect of injections of melting icebergs was a rapid rise in sea level that drowned Caribbean coral reefs.

The cause of Heinrich events is still disputed but Doug MacAyeal of the University of Chicago has proposed an intriguing explanation. Scientists have known for several decades that major fluctuations in the planet's glacial ice cover arise from regular variations in Earth's orbital parameters that control the seasonal and latitudinal distribution of incoming solar radiation (Chapter 12). During episodes when Earth's orbital parameters favored warmer winters and cooler summers in central and northern Canada, some of the winter snows persisted year-round. In time, these climatic conditions gave rise to the Laurentide ice sheet that thickened and eventually spread over much of Canada and the northern tier of the U.S. According to MacAyeal, over Hudson Bay the ice sheet initially was frozen to the bedrock but things changed as the ice sheet thickened. The growing ice sheet acted as an insulating blanket over Earth's surface and trapped enough geothermal heat conducted from Earth's interior to thaw the bottom layer of the ice sheet. Loaded with rock debris, this lubricated ice flowed rapidly into the North Atlantic releasing a massive surge of icebergs responsible for the Heinrich layers, changes in ocean circulation, and sea level rise. The now thinner ice sheet then re-froze to the bedrock and the cycle began anew as the ice sheet again thickened.

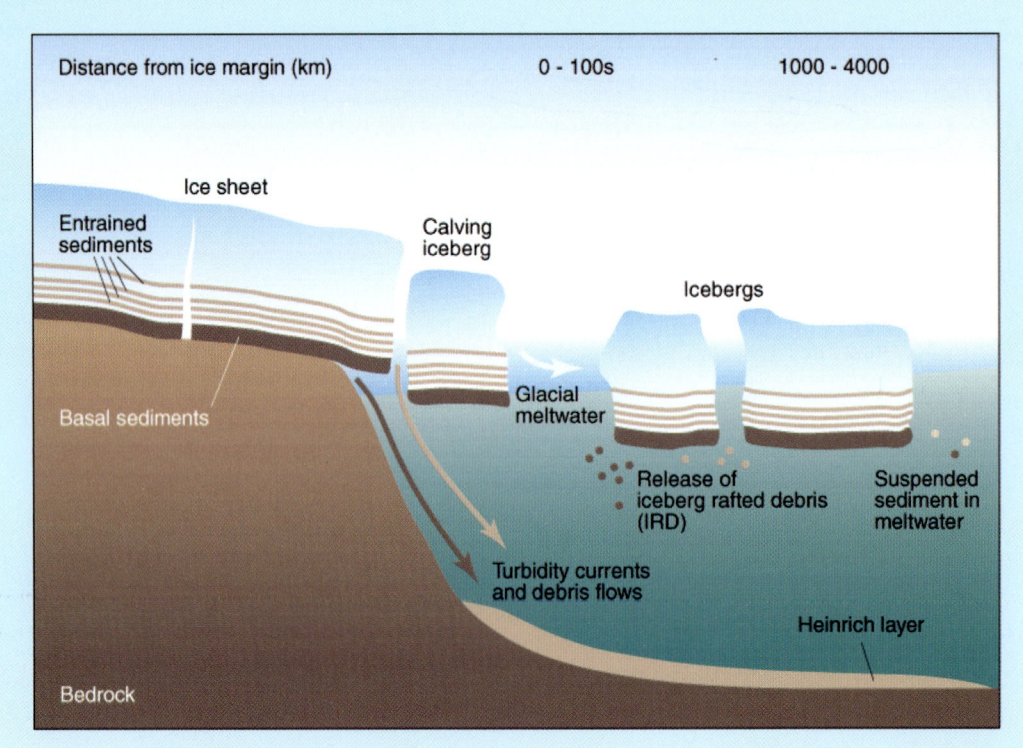

FIGURE
Ice rafting delivered relatively coarse sediment to the deep ocean floor. [From John T. Andrews and Thomas G. Andrews, NOAA Paleoclimatology Program and INSTAAR, University of Colorado, Boulder, CO]

ESSAY: Burgess Shale: A Glimpse into Ancient Marine Life

The earliest fossil record of life on Earth appears during the Precambrian Era around 3.5 billion years ago, about a billion years after Earth formed, and about half billion years after ocean waters first covered portions of the planet. Over the next 2 billion years or so, life on Earth consisted of simple soft-bodied unicellular marine organisms. The earliest hard-shelled animals appeared in the ocean about 650 million years ago, near the close of the Precambrian. Then, in an evolutionary big-bang known as the *Cambrian explosion*, a broad spectrum of complex animal forms appeared in the ocean in just 10 to 30 million years—a mere blink of an eye in the perspective of geological time. While these organisms were still primarily invertebrates, this event represented a huge expansion of marine biodiversity and included the ancestors of modern shellfish, corals, crustaceans, and other inhabitants of the sea. The Cambrian explosion also brought creatures that would eventually disappear. What caused the Cambrian explosion is still debated, although it is known that the transition between the Precambrian and Paleozoic Eras was a time of extreme fluctuations in climate.

A lack of fossils has hindered detailed study of the Cambrian explosion. Two conditions favor the preservation of the remains of organisms as fossils: Rapid burial and possession of hard parts (e.g., shells). Rapid burial slows or prevents the decay of soft body parts and prevents scavenging of both soft and hard body parts. Almost none of the early unicellular life forms from the Precambrian appear in the fossil record. In fact, in spite of nearly continuous sedimentation and burial in the ocean, even fossils from the diverse community of animals from the Cambrian explosion are almost nonexistent—except in a very few unique rock outcrops. One of these outcrops is that of the *Burgess Shale*, consisting of layers of shale and mudstone located high in the Canadian Rockies near the little town of Field in eastern British Columbia. The Burgess Shale contains an abundance of fossils of soft-bodied animals from the Cambrian Period (570 to 505 million years ago).

The Burgess Shale provides us with a rare window on marine life some 40 million years after the Cambrian explosion. So far, the Burgess Shale has yielded tens of thousands of unique specimens representing 170 species, the majority of which are benthic (sea-floor dwelling) organisms, many of which do not exist today. The most remarkable feature of the Burgess Shale fossils is the excellent preservation of soft-bodied organisms; 60-80% of the fossils found are preserved soft bodies. How were they preserved?

During the Cambrian Period, the paleo-North American continent was located in the tropics astride the equator. Life was restricted to the ocean while the land was barren and uninhabited. The continental margin of this paleo-continental coast was a warm, shallow tropical sea and offshore, a great reef had formed creating a steep underwater cliff or escarpment. This algal reef and cliff, now known as the Cathedral Escarpment, fell away hundreds of meters deeper into the ocean. Between the reef and the cliff, sediment accumulated forming submarine mud banks. A rich and diverse community of marine organisms lived in and on the mud banks around the productive reef. This community consisted of mostly large, soft-bodied benthic invertebrates including brachiopods, coelenterates, echinoderms, mollusks, worms, and sponges. Among them were soft, leaf-like animals 20 cm (8 in.) in length, and flat velvet-like worms with hook-like appendages. There were also a few pelagic (open-ocean) species such as Pikaia, an early chordate.

Periodically the accumulated mud became unstable and flowed down the escarpment as turbidity currents (described elsewhere in this chapter). Turbidity currents transported the animals within a slurry down slope to the base of the reef escarpment where most were buried fast and deep enough that they were preserved as fossils, sealed off from scavengers and decomposing bacteria. This process was repeated many times over a period of about 360 million years, building a 10,000-m (33,000-ft) sequence of fine-grained fossil-rich layers of sediment. In time, sediments were lithified into shale and mudstone.

Beginning about 175 million years ago (during the Jurassic Period), stresses associated with mountain building elevated and transported (via thrust faulting) the fossils beds from their ocean burial ground many kilometers eastward to their current position high on a mountain ridge in the Yoho Canadian National Park. Fortunately, thrust faulting carried the shale beds above and out in front of the region where tectonic forces were forming the Rocky Mountains. Otherwise, the shale would have been metamorphosed and the fossils destroyed. The Burgess Shale is located at the eastern edge of the Canadian Rockies just west of the Alberta plains. Since their close escape from tectonic metamorphism, these ancient shale and mudstones with their fabulous treasure of fossils have been gradually exposed by erosive forces including glaciers, wind, running water, landslides, avalanches, and people.

In 1886, R.G. McConnell of the Geological Survey of Canada was the first geologist to visit the area. He described one of the fossils he found as an "odd shrimp." It was almost 100 years later that scientists figured out that in fact "odd shrimp" was the molted claw of a giant predator. Charles D. Walcott (1850-1927) heard reports about "stone bugs" (trilobites) and first visited the site in 1907. He served as director of the U.S. Geological Survey (1894-1907) and director of the Smithsonian Institution (1907-1927). Walcott is credited with discovering the Burgess Shale (named after nearby Mt. Burgess), and found fossils unlike any he had ever seen before. He collected more than 65,000 specimens for the Smithsonian and attempted to classify them. Eventually, Walcott identified 100 of the 170 recognized species.

In the late 1960s, paleontologist Harry Whittington of Cambridge, England, along with his graduate students Derek Briggs and Simon Conway Morris, began a thorough study of the Burgess Shale fossils (including those in storage). They were unable to classify all the fossil animals using the modern classification system and described those they didn't know as "unknown phyla." The implication was that there had been a greater diversity of basic animal forms half a billion years ago than today. Stephen J. Gould, in his book *Wonderful Life: The Burgess Shale and the Nature of History* (1989), based on Whittington's work, concluded that "more than half the major animal groups are extinct." This was one of five major extinctions during Earth history.

Continuing work on the Burgess Shale fossils resulted in classification of some of the previously unclassified fossil species. However, just as many new unclassifiable forms have been found. The biodiversity represented by the Cambrian explosion is still greater than exists in the modern ocean. The fossil record reveals odd configurations of feeding devices, and body configurations and proportions, some of which are so odd and enigmatic that they do not belong to a known phyla and would have been unknown if not for the Burgess Shale. None of the body plans discovered seem to be any more viable than any other and perhaps luck resulted in survival of certain species to become ancestors of living floral and fauna. Debate, research, and further exploration continue to surround the possible implications of the Burgess Shale fossils for evolutionary processes.

While the fossils of the Burgess Shale are very rare, similar fossils have been found in Cambrian shale deposits near the town of Chengjiang in Yunnan Province of China. This mud/shale is about 15 million years older than Burgess Shale. Over 100,000 specimens have been found there, with some similar species found in both places so we know that the Cambrian explosion was not geographically limited but rather ocean wide.

The Burgess Shale is an exceptional site of fossil preservation and records a diversity of rare animals. Its beautifully preserved fossils constitute a valuable snapshot of Cambrian life, far more complete than deposits containing fossils with only hard parts. In 1981, the Burgess Shale was declared a UNESCO World Heritage Site.

CHAPTER 5

THE ATMOSPHERE AND OCEAN

Case-in-Point
Driving Question
Weather and Climate
Heating and Cooling Earth's Surface
    Solar Radiation
    Solar Radiation Budget
    Solar Radiation and the Ocean
    Infrared Radiation and the Greenhouse Effect
Heating Imbalances: Earth's Surface versus Atmosphere
    Latent Heating
    Sensible Heating
Heating Imbalances: Tropics versus High Latitudes
    Heat Transport by Air Mass Exchange
    Heat Transport by Storms
    Heat Transport by Ocean Circulation
Circulation of the Atmosphere: The Forces
    Pressure Gradient Force
    Coriolis Deflection
Circulation of the Atmosphere: Patterns of Motion
    Planetary-Scale Circulation
    Synoptic-Scale Weather Systems
Conclusions
Basic Understandings
ESSAY: Location at Sea, An Historical Perspective
ESSAY: The Stratospheric Ozone Shield and Marine Life

Cumulus clouds forming over water. Some clouds appear to be developing into a thunderstorm. [Courtesy of NOAA.]

Case-in-Point

At middle latitudes, prevailing winds blow from west to east so that the moderating influence of the North Atlantic Ocean on climate is much more evident over Western Europe than Eastern North America. It is interesting to speculate on what would happen to the climate of Western Europe if this moderating influence were to weaken as it has at times in the past.

    In Western Europe, the air temperature contrast between summer and winter is less than it is in North America. For reasons discussed in Chapter 3, sea surface temperatures (SST) change relatively little through the course of a year and this stable SST regime dampens the summer-to-winter temperature contrasts of air flowing over the ocean to downwind Western Europe. While summer average air temperatures are somewhat lower, winter average air temperatures in Western Europe are mild compared to upwind North America. The northward-moving warm Gulf Stream parallels the U.S. coastline from Florida to the Mid-Atlantic states and then the current turns east and northeastward across the North Atlantic. In winter, the relatively warm ocean surface moderates cold air masses as they surge from polar areas southeastward toward the British Isles and Western Europe.

Compare, for example, January and July temperatures at Cork, Ireland (51 degrees, 54 minutes, N) and Saskatoon, Saskatchewan, Canada (52 degrees, 8 minutes, N). At both places on average, January and July are the coldest and warmest months of the year respectively. Although located at about the same latitude, the two cities have markedly different climates. The average temperature contrast between July and January is about 36.5 Celsius degrees (65.7 Fahrenheit degrees) at continental Saskatoon but only about 11 Celsius degrees (20 Fahrenheit degrees) at maritime Cork. The reduced seasonal contrast at Cork is mostly due to much higher winter temperatures. January average temperature is 4.5 °C (40.1 °F) at Cork but −18.5 °C (-1.3 °F) at Saskatoon. July average temperatures are not much different at the two locations with 18 °C (64.4 °F) at Saskatoon and 15.5 °C (59.9 °F) at Cork.

Historical records indicate that during periods in the past, winters were much colder in the British Isles and other parts of Western Europe. The Little Ice Age, which lasted from about A.D. 1400 to 1850, was one such period when cold winters were more frequent than today. Sea ice cover expanded over the North Atlantic, mountain glaciers advanced, and growing seasons shortened over Western Europe bringing erratic harvests and much hardship for many people. Evidence of other cold episodes in the North Atlantic comes from climate signals unlocked from deep-sea sediment cores and annual ice layers in the Greenland ice sheet (Chapter 12). The past 10,000 years (since the end of the last Ice Age) were punctuated by multi-century cold periods that began abruptly (within decades) and occurred about every 1500 years. Across Northern and Western Europe, average winter temperatures during cold episodes were up to 7 Celsius degrees (13 Fahrenheit degrees) lower than today.

A possible cause of the Little Ice Age and prior cold episodes in Western Europe is periodic weakening of the North Atlantic circulation. The Gulf Stream is part of a planetary-scale conveyor-belt-like ocean circulation that transports enormous amounts of heat throughout the world ocean (Chapter 6). The flow in the Atlantic conveyor is about 100 times that of the Amazon River. Researchers at Lamont-Doherty Earth Observatory of Columbia University propose that runoff of unusually great amounts of freshwater into the North Atlantic alters the salinity (and density) of surface ocean waters and disrupts the conveyor-belt circulation. Shut down of the circulation would cause winter average temperatures in Western Europe to plunge abruptly by perhaps 5 Celsius degrees (9 Fahrenheit degrees). For Cork, this would mean winters more like those experienced at Spitsbergen, some 1000 km (600 mi) north of the Arctic Circle.

Driving Question:

What role does the ocean play in the long-term average state of the atmosphere?

To this point in our investigation of the ocean in the Earth system we have emphasized primarily the physical properties of ocean water and the ocean basin. In this chapter, our principal focus shifts to the flow of energy into and out of the Earth system especially as it involves processes operating at the interface between the ocean and atmosphere. In doing so, it quickly becomes apparent that the ocean plays a key role in the global radiation budget, the transport of heat between Earth's surface and atmosphere, the flow of heat from the tropics to higher latitudes, and the development of storm systems. For these reasons, the ocean is a major player in the state of the atmosphere (*weather*) and Earth's climate system.

In this chapter, we examine radiational heating and cooling of the **Earth-atmosphere system** (Earth's surface plus overlying atmosphere), the interaction of incoming solar radiation with the atmosphere, ocean, and continents, the flow of infrared radiation to space, and the greenhouse effect. In response to heating imbalances within the Earth-atmosphere system, temperature gradients develop and heat is transferred via phase changes of water, conduction and convection, exchange of air masses, and ocean currents. Atmospheric circulation operating at various spatial scales transports heat from where it is warmer to where it is colder. We begin our discussion by distinguishing between weather and climate.

Weather and Climate

Weather and climate are closely related concepts. We can think of **weather** as the state of the atmosphere at some place and time described in terms of such variables as temperature, precipitation, cloud cover, and wind speed. A place and time must be specified when describing weather because the atmosphere is dynamic; that is, its state is always changing from one place to another and with time. At the same hour the weather may be cold and snowy in Philadelphia, sunny and warm in Dallas, and cool and rainy in Seattle. *If you don't like the weather, wait a minute* is an old saying that is not far from the truth in many places. From personal experience, we know that tomorrow's weather may be markedly different from today's weather.

Ultimately, climate governs the supply of fresh water, the geographical distribution of plants and animals, and the type of crops that can be cultivated. **Climate** is popularly defined as weather at a particular place averaged over a specific interval of time. By international convention, average values of weather elements such as temperature or precipitation are computed over a 30-year period beginning with the first year of a decade. At the close of a decade the averaging period is shifted forward ten years. As of this writing, the official averaging period is 1971-2000. Thirty-year average monthly and annual temperatures and precipitation totals are commonly used to describe climate. Other useful climatic parameters include average seasonal snowfall, length of growing season, and frequency of thunderstorms.

Climate encompasses extremes in weather in addition to average values of weather elements. Tabulation of extreme values usually covers the entire period of record (or at least for the period when the weather station was at the same location). Specifying extremes in weather provides information on the variability of climate at a particular place and gives a more complete and useful description of climate. Climate researchers study not only trends in average temperature and precipitation, but also changes in the frequency of extreme events such as excess heat, cold, drought, or rainfall.

Heating and Cooling Earth's Surface

As Earth orbits the sun, its atmosphere and surface absorb a tiny fraction of the total radiation that the sun continually emits to space. Absorption of solar radiation heats the Earth-atmosphere system. At the same time the planet continually emits infrared radiation to space, which cools the Earth-atmosphere system. Overall, radiational heating of the planet is balanced by radiational cooling of the planet so that Earth remains in radiative equilibrium with surrounding space.

The sun emits a band of electromagnetic radiation having wavelengths mostly between 0.25 and 2.5 micrometers. (One *micrometer* is a millionth of a meter or about one-tenth the thickness of a human hair.) Solar radiation is most intense at a wavelength of about 0.5 micrometer, in the green of the visible portion of the electromagnetic spectrum (Figure 1.9). The Earth-atmosphere system, on the other hand, emits to space a broad band of electromagnetic radiation having wavelengths mostly between 4 and 24 micrometers, in the infrared portion of the electromagnetic spectrum. The intensity of infrared radiation emitted by Earth's surface peaks at a wavelength of about 10 micrometers. In this section, we take a closer look at radiational heating and cooling of the Earth system.

SOLAR RADIATION

Earth's motions in space govern daily and seasonal variations in the amount of solar radiation that strikes Earth's surface. Once every 24-hrs, Earth completes one rotation on its axis so that at any minute, half the planet is illuminated by solar radiation (day) while the other half is dark (night). The tilt of Earth's spin axis is responsible for the seasons. Earth's spin axis is tilted 23 degrees 27 minutes from the perpendicular to the plane defined by the planet's annual orbit about the sun (Figure 5.1). During its annual revolution about the sun, Earth's spin axis remains in the same alignment with the background stars (the North Pole always pointing toward *Polaris*, the North Star) while the planet's orientation to the sun changes continually. Simply put, the Northern Hemisphere tilts away from the sun in fall and winter and toward the sun in spring and summer. Astronomical winter begins on the winter solstice, about 21 December (in the Northern Hemisphere), and ends on the first day of spring (the vernal or spring equinox), on or about 21 March. Summer begins on the summer solstice, about 21 June, and continues until the first day of autumn (the autumnal equinox), on or about 23 September. Precise dates of solstices and equinoxes vary because Earth completes one orbit of the sun in 365.24 days, necessitating a leap year adjustment.

Regular changes in solar altitude and length of daylight accompany the annual periodic changes in the planet's orientation to the sun. These changes, in turn, affect the amount of solar radiation that strikes Earth's surface at a point. **Solar altitude** is the angle of the sun above the horizon and varies from 0 degrees (at sunrise or sunset) to as much as 90 degrees (where and when the sun is directly overhead), and *length of daylight* is the number of hours and minutes between sunrise and sunset. At middle and high latitudes, the

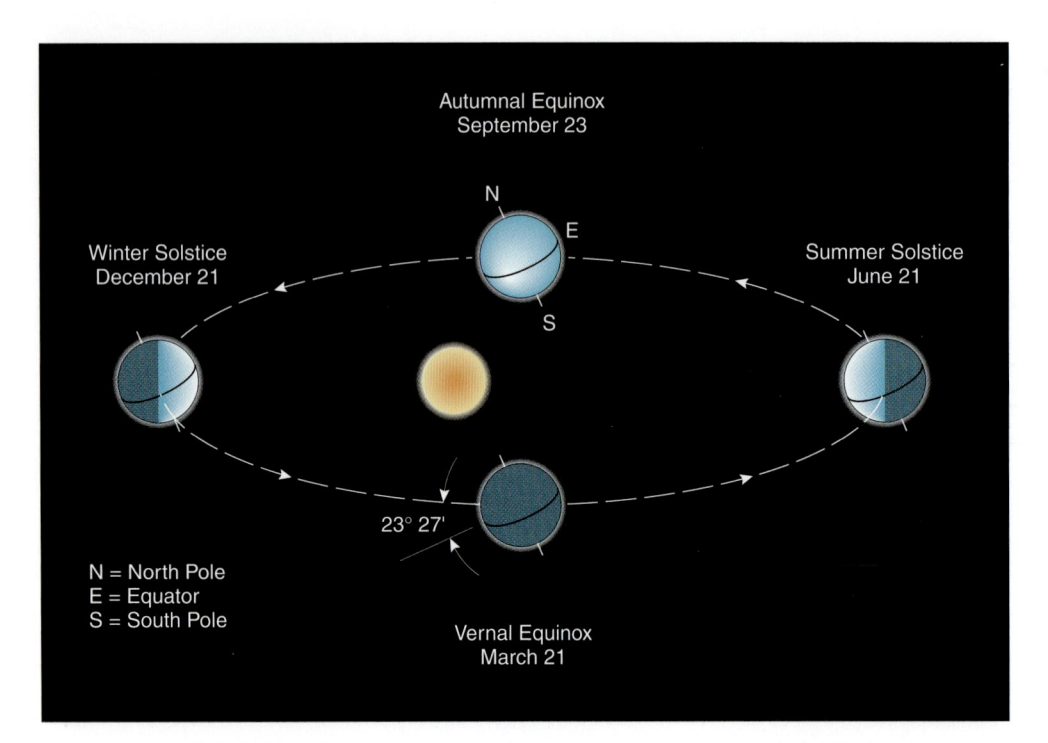

100 Chapter 5 THE ATMOSPHERE AND OCEAN

FIGURE 5.1
The seasons change because Earth's equatorial plane is inclined (at 23 degrees, 27 minutes) to the orbital plane. Seasons are given for the Northern Hemisphere. Note that the eccentricity of Earth's orbit is greatly exaggerated.

altitude of the noon sun is higher, daylight is longer, and solar radiation is more intense in summer than in winter.

With clear skies, the intensity of solar radiation striking Earth's surface at a point varies directly with solar altitude. With increasing solar altitude, more solar energy strikes a unit area of Earth's surface in a unit of time (Figure 5.2).

Earth is so far from the sun (a mean distance of 150 million km or 93 million mi) that solar radiation reaches the planet as nearly parallel beams of essentially uniform intensity. But the almost spherical Earth presents a curved surface to incoming solar radiation so that the noon solar altitude tends to be higher in the tropics than at higher latitudes. Greater solar

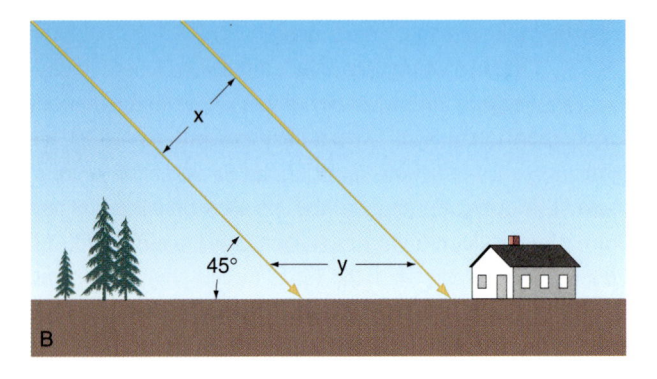

FIGURE 5.2
The intensity of solar radiation striking Earth's surface per unit area varies with solar altitude. (A) Incident solar radiation is most intense when the sun is directly overhead (solar altitude of 90 degrees). (B) With decreasing solar altitude, solar radiation received at Earth's surface spreads over an increasing area (y is greater than x) so that radiation is less concentrated (less radiational energy per unit area).

altitudes in the tropics translate into more intense radiation and higher temperatures at Earth's surface.

The daily path of the sun through the sky on the solstices and equinoxes is shown schematically in Figure 5.3 for the equator, a mid-latitude Northern Hemisphere location, and the North Pole. At the mid-latitude location and North Pole, the altitude of the noon sun is greatest on the summer solstice but at the equator the noon solar altitude is maximum on the equinoxes (when the sun is directly overhead). The only places on Earth where the solar altitude ever reaches 90 degrees during the course of a year are within the latitude belt bounded by the Tropic of Cancer (23.5 degrees N) and the Tropic of Capricorn (23.5 degrees S).

Daylight is shortest on the winter solstice and longest on the summer solstice. On the equinoxes, the length of daylight and night are about the same (12 hrs) everywhere on the planet except at the poles. Daylight is longer than night during spring and summer, and daylight is shorter than night during autumn and winter. The difference in length of daylight between the summer and winter solstices increases from zero at the equator to a maximum (24 hrs) at the Arctic and Antarctic Circles (Figure 5.4). Regular variations in maximum solar altitude and length of daylight through the year are ultimately responsible for changes in receipt of solar radiation and monthly average temperatures (Figure 5.5). Little annual variation in maximum solar altitude and length of daylight in the tropics translates into relatively uniform monthly mean temperatures through the year. In the tropics, the temperature difference between day and night often is greater than the summer-to-winter temperature contrast. At middle and high latitudes, however, maximum solar altitude and length of daylight vary considerably through the year and are responsible for marked contrasts between average summer and winter temperatures.

For reasons presented in Chapter 3, a large body of water such as the ocean influences the climate of downwind localities. Places downwind of the ocean experience a smaller contrast between average winter and summer temperatures and have a *maritime climate*. Places at the same latitude but well inland experience a greater temperature contrast between winter and summer and have a *continental climate*. Proximity to large bodies of water also affects the timing of the average warmest and coldest times of the year. Outside of the tropics, the annual temperature cycle lags the annual solar radiation cycle; that is, the warmest period of the year on average occurs after the summer solstice while the coldest period of the year occurs after the winter solstice. The Earth-atmosphere system takes time to adjust to seasonal changes in solar energy input. In the interior United States, the air temperature cycle lags the solar radiation cycle by an average of 27 days. But in coastal localities having a strong maritime influence (e.g., coastal California, Florida), the average lag time is up to 36 days.

Regular changes in the location of the sun and other celestial bodies in the sky were utilized by mariners to locate

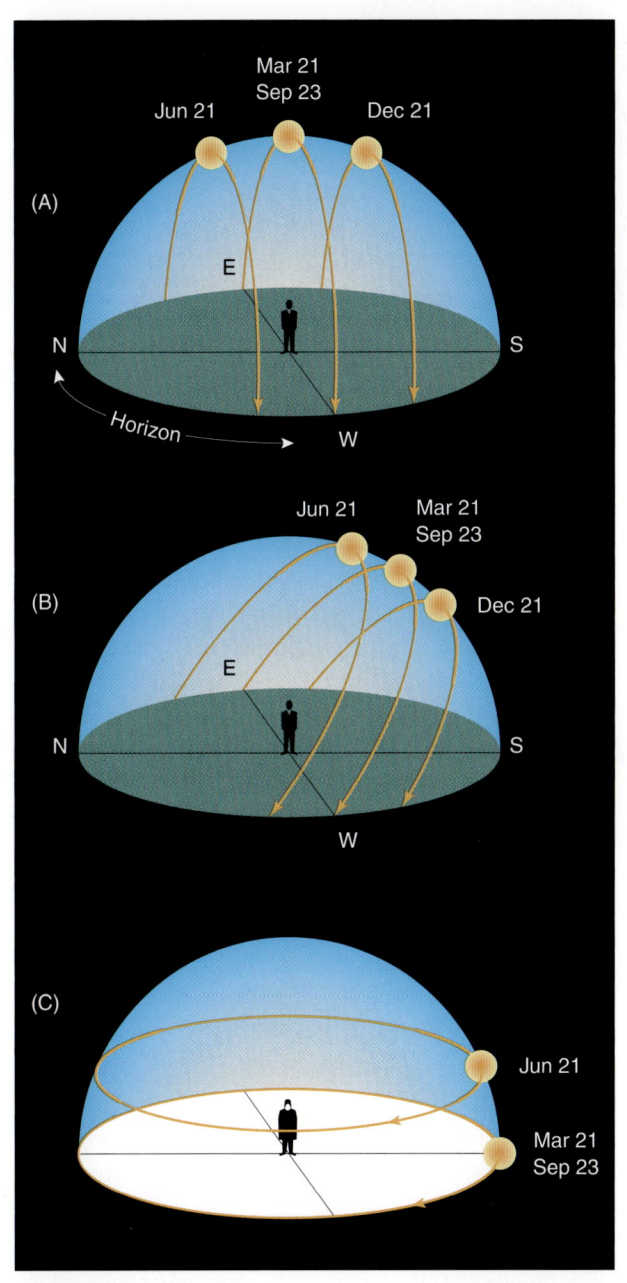

FIGURE 5.3
Path of the sun through the sky on the solstices and equinoxes at (A) the equator, (B) a middle latitude location in the Northern Hemisphere, and (C) the North Pole.

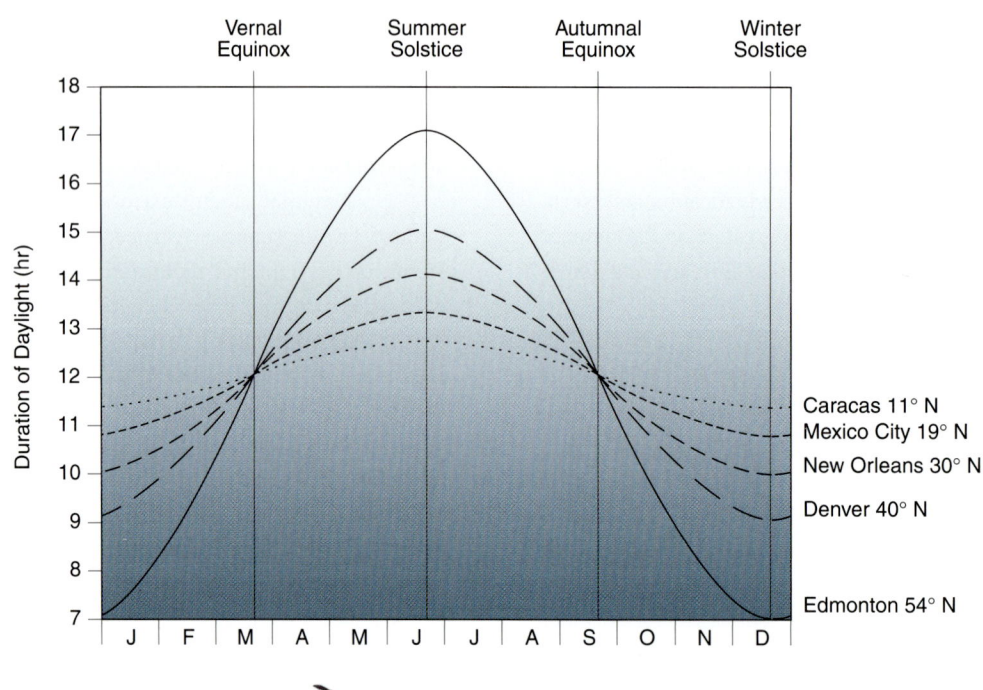

102 Chapter 5 THE ATMOSPHERE AND OCEAN

FIGURE 5.4
The variation in the number of hours of daylight through the year increases with increasing latitude.

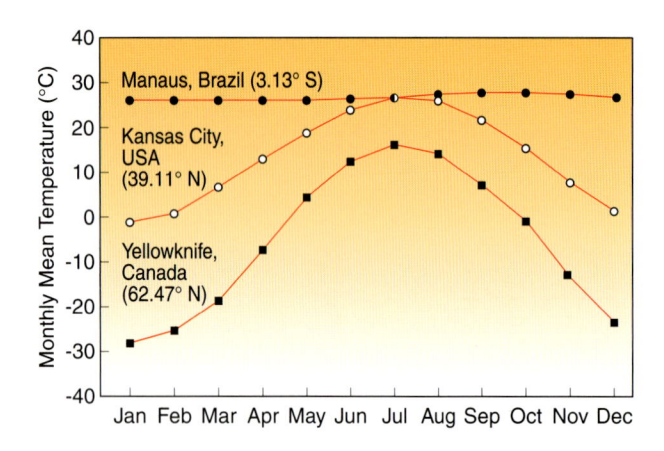

FIGURE 5.5
The seasonal contrast in monthly mean temperature generally increases with increasing latitude.

themselves at sea. For an historical perspective on the challenges of navigating at sea, refer to this chapter's first Essay.

SOLAR RADIATION BUDGET

Solar radiation intercepted by Earth travels through the atmosphere and interacts with its component gases and aerosols. These interactions consist of scattering, reflection, and absorption. Solar radiation that is not absorbed or scattered or reflected back to space reaches Earth's surface where additional interactions occur.

With **scattering**, a particle disperses radiation in all directions: up, down, and sideways. Within the atmosphere, both gas molecules and aerosols (including the tiny water droplets and ice crystals that compose clouds) scatter solar radiation. Scattering explains the blue color of the daytime sky. Visible sunlight is made up of all colors (from violet at the short wavelength end of the solar spectrum to red at the long wavelength end). Air molecules preferentially scatter short wavelength visible light (blue-violet).

Reflection is a special case of scattering in which a large surface area redirects radiation in a backward direction. The fraction of incident radiation that is reflected by a surface is known as the **albedo** of that surface, that is,

Albedo = [(reflected radiation)/(incident radiation)] × 100%.

Surfaces having a high albedo reflect a relatively large fraction of incident solar radiation and appear light in color. Surfaces having a low albedo reflect a relatively small fraction of incident solar radiation and appear dark in color.

Within the atmosphere, the tops of clouds are the most important reflectors of solar radiation. Cloud top albedo depends primarily on cloud thickness and varies from under 40% for thin clouds to 80% or more for thick clouds. The average albedo for all cloud types and thicknesses is about 55%, and at any point in time, clouds cover about 60% of the planet. All other factors being constant, solar radiation striking the Earth's surface is more intense and surface air temperatures are higher when the daytime sky is clear rather than cloudy.

Reflection and scattering within the atmosphere alter the direction of solar radiation without conversion to heat. **Absorption**, however, is a process in which some of the radiation that strikes an object is converted to heat energy. Oxygen, ozone, water vapor, and various aerosols (including cloud particles) absorb solar radiation. Absorption by atmospheric gases varies with wavelength; that is, each gas absorbs strongly in some wavelengths and weakly or not at all in other wavelengths. Essential for life on Earth is the strong absorption of ultraviolet (UV) radiation by oxygen and ozone (O_3) in the stratosphere, which shields organisms from exposure to potentially lethal intensities of UV. These absorption processes create the so-called *stratospheric ozone shield*. For more on the stratospheric ozone shield and marine life, see this chapter's second Essay.

Solar radiation that is not reflected or scattered to space or absorbed by atmospheric gases or aerosols strikes Earth's surface where it is either reflected or absorbed. The portion that is not reflected is absorbed (i.e., converted to heat) and the portion that is not absorbed is reflected. High-albedo surfaces, such as snow-covered ground or pack ice, reflect a considerable amount of incident solar radiation whereas low-albedo surfaces, such as an asphalt road or a conifer forest, reflect much less incident solar radiation. Albedos of some common surfaces are listed in Table 5.1.

In the visible satellite image in Figure 5.6, the ocean surface appears dark because of its strong absorption of solar radiation and low albedo. The albedo of the ocean surface varies with the angle of the sun above the horizon (*solar altitude*). Under clear skies, the albedo of a flat, tranquil water surface decreases with increasing solar altitude (Figure 5.7). The albedo is almost a mirror-like 100% near sunrise and sunset (when the solar altitude is near 0 degrees) but declines sharply as the solar altitude approaches 20 degrees. When the sun is low in the sky (small solar altitude), light rays reflect off the water surface with little penetration. However, when the sun is high in the sky, light rays penetrate the water to some depth and most of the sunlight is scattered below the surface with little scattered back to the atmosphere or space. With cloud-covered skies, only diffuse solar radiation strikes the water surface and the albedo varies little with solar altitude and is uniformly less than 10%.

TABLE 5.1

Average Albedo (Reflectivity) of some Common Surface Types for Visible Solar Radiation

Surface	Albedo (% reflected)
Deciduous forest	15-18
Coniferous forest	9-15
Tropical rainforest	7-15
Tundra	15-35
Grasslands	18-25
Desert	25-30
Sand	30-35
Soil	5-30
Green crops	15-25
Sea ice	30-40
Fresh snow	75-95
Old snow	40-60
Glacial ice	20-40
Water body (high solar altitude)	3-10
Water body (low solar altitude)	10-100
Asphalt road	5-10
Urban area	14-18
Cumulonimbus cloud	90
Cirrus cloud	40-50

FIGURE 5.6
In this visible satellite image, the ocean surface appears dark because of its low albedo for visible solar radiation. White areas are clouds.

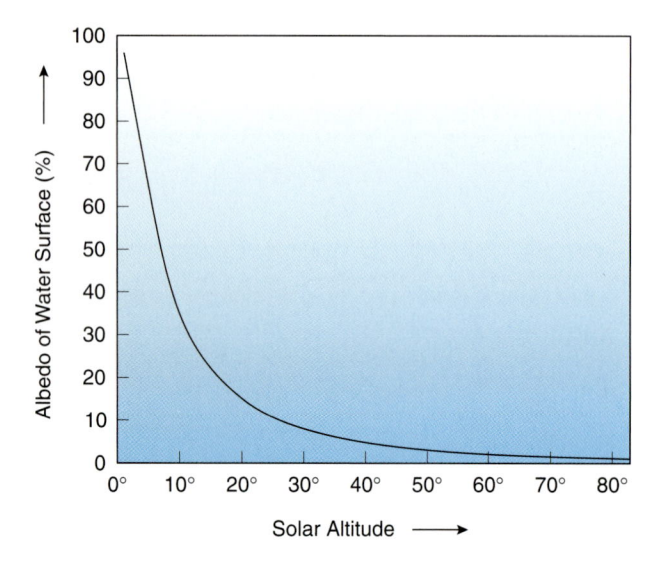

FIGURE 5.7
Under clear skies, the albedo of a flat and undisturbed water surface changes with solar altitude. A wave-covered water surface has a slightly higher albedo at high solar altitudes and a slightly lower albedo at low solar altitudes.

The roughness of a wave-covered water surface decreases its albedo. On a global basis, the albedo of the ocean surface averages only about 8%; that is, the ocean absorbs 92% of incident solar radiation. Considering that the ocean covers about 71% of the surface of the planet, the ocean is the major sink for solar radiation.

Measurements by sensors aboard satellites indicate that the Earth-atmosphere system reflects or scatters back to space on average about 31% of the solar radiation intercepted by the planet. This is Earth's **planetary albedo**. The atmosphere (i.e., gases, aerosols, clouds) absorbs only about 20% of the total solar radiation intercepted by the Earth-atmosphere system. In other words, the atmosphere is relatively transparent to solar radiation. The remaining 49% of solar radiation is absorbed by Earth's surface—mostly the ocean.

Earth's surface is the principal recipient of solar heating, and heat is transferred from Earth's surface to the atmosphere, which eventually radiates this energy to space. Hence, Earth's surface is the main source of heat for the atmosphere; that is, the atmosphere is heated from below. This is evident from the average vertical temperature profile of the troposphere (Figure 1.4). Normally, air is warmest close to the Earth's surface, and the temperature drops with increasing altitude, that is, away from the main source of heat.

SOLAR RADIATION AND THE OCEAN

Whereas the atmosphere is relatively transparent to solar radiation, the ocean absorbs most solar radiation within relatively shallow depths. As shown in Figure 5.8, the ocean's absorption of the visible portion of solar radiation is selective by wavelength. Water absorbs the longer wavelengths (i.e., reds and yellows) of visible light more efficiently than the shorter wavelengths (i.e., greens and blues) so that green and blue penetrate to greater depths. Within clear, clean water, red light is completely absorbed within about 15 m (50 ft) of the surface, whereas green and blue-violet light may penetrate to depths approaching 250 m (800 ft). More green and blue light is scattered to our eyes, explaining the blue/green color of the open ocean. Suspended particles significantly boost absorption so that sunlight often is completely absorbed at shallower depths. In fact, some near-shore waters are so turbid (cloudy) that little if any sunlight reaches much below 10 m (35 ft). Suspended particles preferentially scatter yellow and green light, giving these waters their characteristic color.

The **photic zone** is the sunlit surface layer of the ocean, down to the depth where light is just sufficient for photosynthesis. The base of the photic zone is generally where the light is just 1% of the radiation incident on the surface. In clear ocean waters, this depth is usually from 100 to 200 m (330 to 650 ft) but is much shallower in highly productive or turbid waters. Although a small amount of light penetrates below the photic zone (into the so-called *twilight zone*), light is not enough for plants to survive.

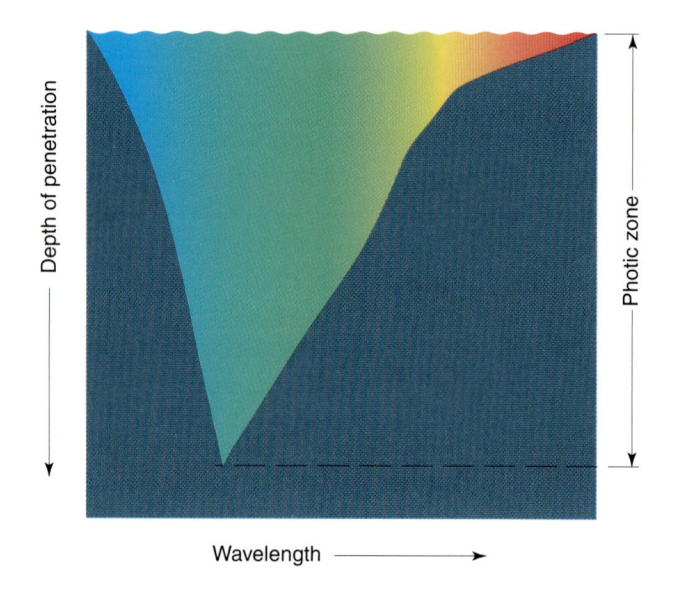

FIGURE 5.8
Visible solar radiation is selectively absorbed by wavelength as it penetrates the ocean's surface waters of the open ocean.

As the concentration of particles and dissolved organic matter in seawater increases, the color of light that penetrates deepest into the water shifts to yellow-green in coastal areas and to red in the most turbid estuarine waters. Hence, as light becomes dimmer with increasing depth, its color also changes. This color change affects plant production because each plant pigment is most efficient with a specific color of light. The combination of pigments in any type of phytoplankton determines its optimal depth distribution.

With some notable exceptions, marine life depends directly or indirectly on sunlight and organic productivity in the ocean's photic zone (Chapters 9 and 10). Even the diverse community of animals living at great depths on the ocean floor depends on organic particles produced within the photic zone that settle to the sea floor (Chapter 4); exceptions are organisms living near hydrothermal vents (Chapter 2).

INFRARED RADIATION AND THE GREENHOUSE EFFECT

If solar radiation were continually absorbed by the Earth-atmosphere system without any compensating flow of heat out of the system, Earth's surface temperature would rise steadily. Eventually, life would be extinguished and the ocean would boil away. Actually, the global air temperature changes very little from one year to the next. **Global radiative equilibrium** keeps the planet's temperature in check; that is, emission of heat to space in the form of infrared radiation balances solar radiational heating of the Earth-atmosphere system. Although solar radiation is supplied only to the illuminated half of the planet, infrared radiation is emitted to space ceaselessly, day and night, by the entire Earth-atmosphere system. This explains why nights are usually colder than days and why air temperatures drop throughout the night.

While the clear atmosphere is relatively transparent to solar radiation, certain gases in the atmosphere impede the escape of infrared radiation to space thereby elevating the temperature of the lower atmosphere. This important climate control, the so-called **greenhouse effect**, refers to the heating of Earth's surface and lower atmosphere caused by strong absorption and emission of infrared radiation (IR) by certain gaseous components of the atmosphere, known as **greenhouse gases**. Solar radiation and terrestrial infrared radiation peak in different portions of the electromagnetic spectrum, their properties differ, and they interact differently with the atmosphere. As noted earlier, the atmosphere absorbs only about 20% of the solar radiation intercepted by the planet. The atmosphere absorbs a greater percentage of the infrared radiation emitted by Earth's surface and the atmosphere, in turn, radiates some IR to space and some to Earth's surface. Hence, Earth's surface is heated by absorption of both solar radiation and Earth-emitted infrared radiation.

The similarity in radiational properties between infrared-absorbing atmospheric gases and the glass panes of a greenhouse is the origin of the term *greenhouse effect*. Window glass, like the atmosphere, is relatively transparent to visible solar radiation but strongly absorbs infrared radiation. A greenhouse (where plants are grown) takes advantage of the radiational properties of glass and is constructed almost entirely of glass panes. Sunlight readily penetrates greenhouse glass and much of it is absorbed (converted to heat) within the greenhouse. Objects in the greenhouse emit infrared radiation that is strongly absorbed by the glass panes. Glass, in turn, emits IR to both the atmosphere and to the greenhouse interior, thereby raising the temperature within the greenhouse. The analogy between the atmosphere and a greenhouse is not strictly correct. A greenhouse acts as a shelter from the wind and this is the principal reason for the elevated temperature within most greenhouses. Nonetheless, "greenhouse effect" is such a commonly used term (especially by the media) that we use it in this book.

The greenhouse effect is responsible for considerable warming of Earth's surface and lower atmosphere. Viewed from space, the planet (Earth-atmosphere system) radiates at about –18 °C (0 °F) whereas the average temperature at Earth's surface is about 15 °C (59 °F). The temperature difference is due to the greenhouse effect and amounts to

$$[15 °C - (-18 °C)] = 33 \text{ Celsius degrees}$$
or
$$[59 °F - (0 °F)] = 59 \text{ Fahrenheit degrees}$$

Without the greenhouse effect, Earth would be too cold to support most forms of plant and animal life.

Water vapor is the principal greenhouse gas. Other greenhouse gases include carbon dioxide, ozone, methane (CH_4), and nitrous oxide (N_2O). As shown in Figure 5.9, the percentage of infrared radiation absorbed by these gases varies with wavelength. An **atmospheric window** is a range of wavelengths over which little or no radiation is absorbed. A *visible window* extends from about 0.3 to 0.9 micrometers and the major *infrared window* is from about 8 to 13 micrometers. Significantly, this latter window includes the

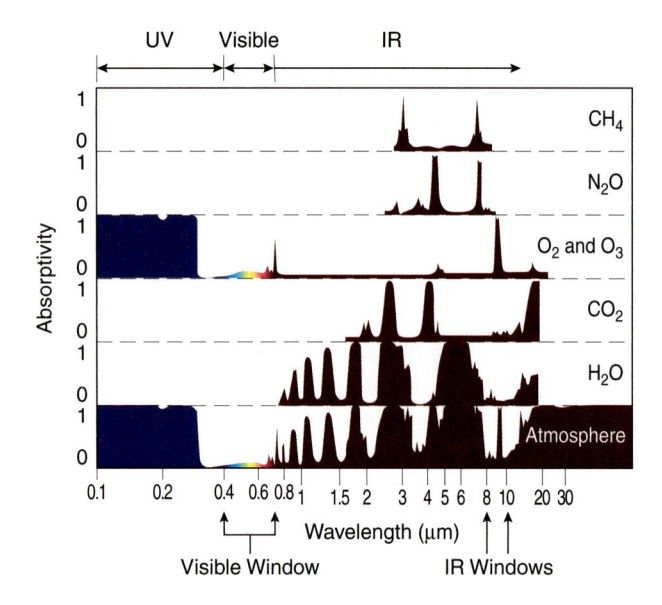

106 Chapter 5 THE ATMOSPHERE AND OCEAN

FIGURE 5.9
Absorption of radiation by selected gaseous components of the atmosphere as a function of wavelength. *Absorptivity* is the fraction of radiation absorbed and ranges from 0 to 1 (0% to 100% absorption). Absorptivity is very low or near zero in the *atmospheric windows*. Note the infrared (IR) windows near 8 and 10 micrometers.

wavelength of the planet's peak infrared emission (about 10 micrometers). Through this window, most heat from the Earth-atmosphere system escapes to space as infrared radiation. Also, IR sensors on Earth-orbiting satellites monitor this upwelling radiation (Chapter 1). As discussed in this chapter's second Essay, water vapor was not always Earth's principal greenhouse gas.

The warming effect of atmospheric water vapor is evident even at the local or regional scale. Consider an example. Locations in the desert Southwest and along the Gulf Coast are at about the same latitude and receive essentially the same input of solar radiation on a clear day. In both places, summer afternoon high temperatures commonly top 32 °C (90 °F). At night, however, air temperatures often differ markedly. Air is relatively dry (low humidity) in the Southwest so that infrared radiation readily escapes to space and air temperatures near Earth's surface may drop well under 15 °C (59 °F) by dawn. People who hike or camp in the desert are aware of the dramatic fluctuations in temperature between day and night. Infrared radiation does not escape to space as readily through the Gulf Coast atmosphere where the air is more humid. Water vapor strongly absorbs outgoing IR and emits IR back towards Earth's surface so that early morning low temperatures may dip no lower than the 20s Celsius (70s Fahrenheit). The smaller day-to-night temperature contrast along the Gulf Coast is due to more water vapor and a stronger greenhouse effect.

Clouds are composed of IR-absorbing ice crystals and/or water droplets and also contribute to the greenhouse effect. All other factors being equal, nights usually are warmer when the sky is cloud-covered than when the sky is clear. Even high thin cirrus clouds through which the moon is visible can reduce the nighttime temperature drop at Earth's surface by several Celsius degrees.

Natural biogeochemical cycles continually transport greenhouse gases into and out of the atmosphere and ocean. In Chapter 3, for example, we saw how carbon dioxide cycles out of the atmosphere and into the ocean. Human activities alter the rate of biogeochemical cycling so that, for example, the atmospheric concentration of certain greenhouse gases is increasing. All other factors being equal, higher concentrations of greenhouse gases are likely to *enhance* the natural greenhouse effect and may lead to warming on a global scale with implications for all sectors of society. Over the past 150 years, since the beginning of the Industrial Revolution, combustion of fossil fuels (i.e., coal at first, oil and natural gas later) and clearing of forests and other vegetation have altered the global carbon cycle so that the atmospheric concentration of CO_2 is now about 30% higher than it was in the pre-industrial era and continues to increase because of human activities. We have more to say about prospects for global warming and possible impacts in Chapter 12.

Heating Imbalances: Earth's Surface versus Atmosphere

Sensors onboard satellites that monitor the flux of incoming solar radiation and outgoing infrared radiation reveal imbalances in rates of radiational heating and radiational cooling. One important aspect of this heating imbalance involves Earth's surface versus the atmosphere.

Figure 5.10 shows how solar radiation intercepted by planet Earth interacts with the atmosphere and Earth's surface. Numbers are global and annual averages. For every 100 units of solar radiation that enters the upper atmosphere, the Earth-atmosphere system reflects or scatters 31 units (31%) to space, the atmosphere absorbs 20 units (20%) and Earth's surface (principally the ocean) absorbs 49 units (49%). In response to radiational heating, Earth's surface emits 114 units of infrared radiation. Atmospheric gases and clouds absorb 105 units of in-

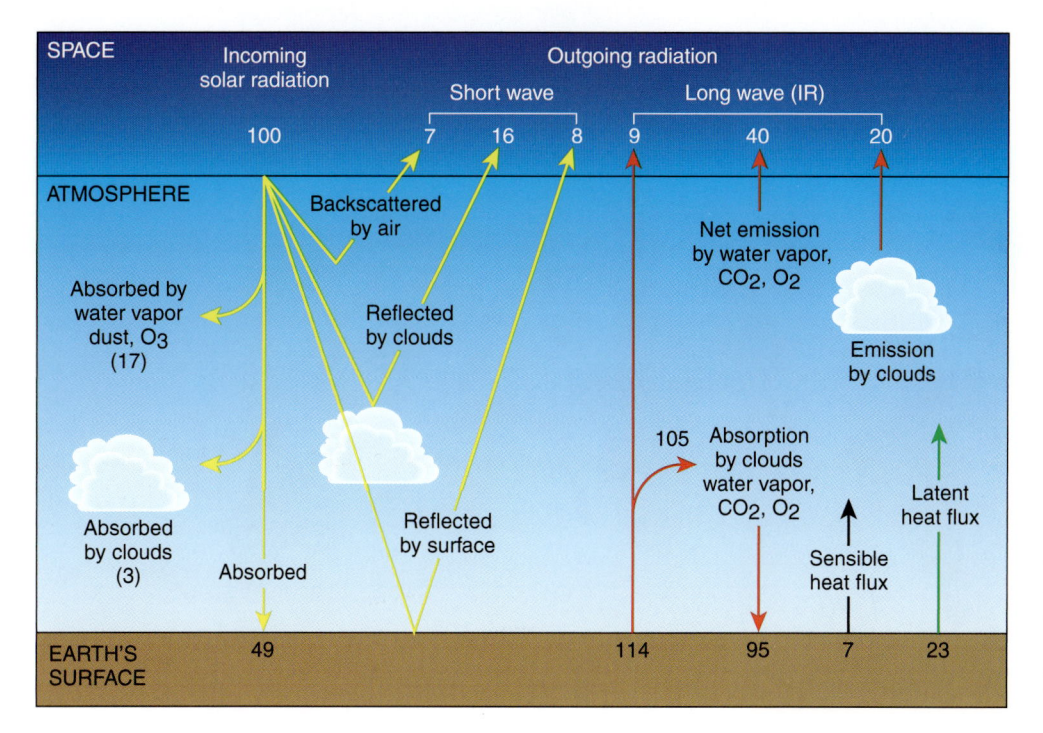

FIGURE 5.10
Globally and annually averaged distribution of 100 units of solar radiation entering the top of the atmosphere. Solar radiation fluxes are depicted at the left, infrared radiation fluxes in the middle, and latent and sensible heat fluxes at the right.

frared radiation and emit 95 units to Earth's surface (*greenhouse effect*). A total of 69 units of IR radiation are emitted out the top of the atmosphere and to space, equal to the amount of solar radiation absorbed by the Earth-atmosphere system.

The global annual distribution of incoming solar radiation and outgoing infrared radiation implies net warming of Earth's surface and net cooling of the atmosphere (Table 5.2). At Earth's surface, absorption of solar radiation is greater than emission of infrared radiation. In the atmosphere, on the other hand, emission of infrared radiation to space is greater than absorption of solar radiation. That is, on a global average annual basis, Earth's surface undergoes net radiational heating and the atmosphere undergoes net radiational cooling.

The atmosphere is not actually cooling relative to Earth's surface because radiation is not the only energy transfer mechanism at work. In response to the radiationally induced temperature gradient between Earth's surface and atmosphere, heat is transferred from Earth's surface to the atmosphere. A combination of latent heating (phase changes of water) and sensible heating (conduction and convection) is responsible for this transfer of heat. As shown in Figure 5.10, on a global annual average basis, 30 units of heat energy are transferred from Earth's surface to the atmosphere: 23 units (about 77% of the total) by latent heating and 7 units (about 23%) by sensible heating.

TABLE 5.2
Global Radiation Balance

Solar radiation intercepted by Earth	100 units
Solar radiation budget	
Scattered and reflected to space (7 + 16 + 8)	31
Absorbed by the atmosphere (17 + 3)	20
Absorbed at the Earth's surface	49
Total	100 units
Radiation budget at the Earth's surface	
Infrared cooling (95 – 114)	-19
Solar heating	+49
Net heating	+30 units
Radiation budget of the atmosphere	
Infrared cooling (- 40 – 20 + 105 – 95)	-50
Solar heating	+20
Net cooling	-30 units
Heat transfer: Earth's surface to atmosphere	
Sensible heating (conduction plus convection)	7
Latent heating (phase changes of water)	23
Net transfer	30 units

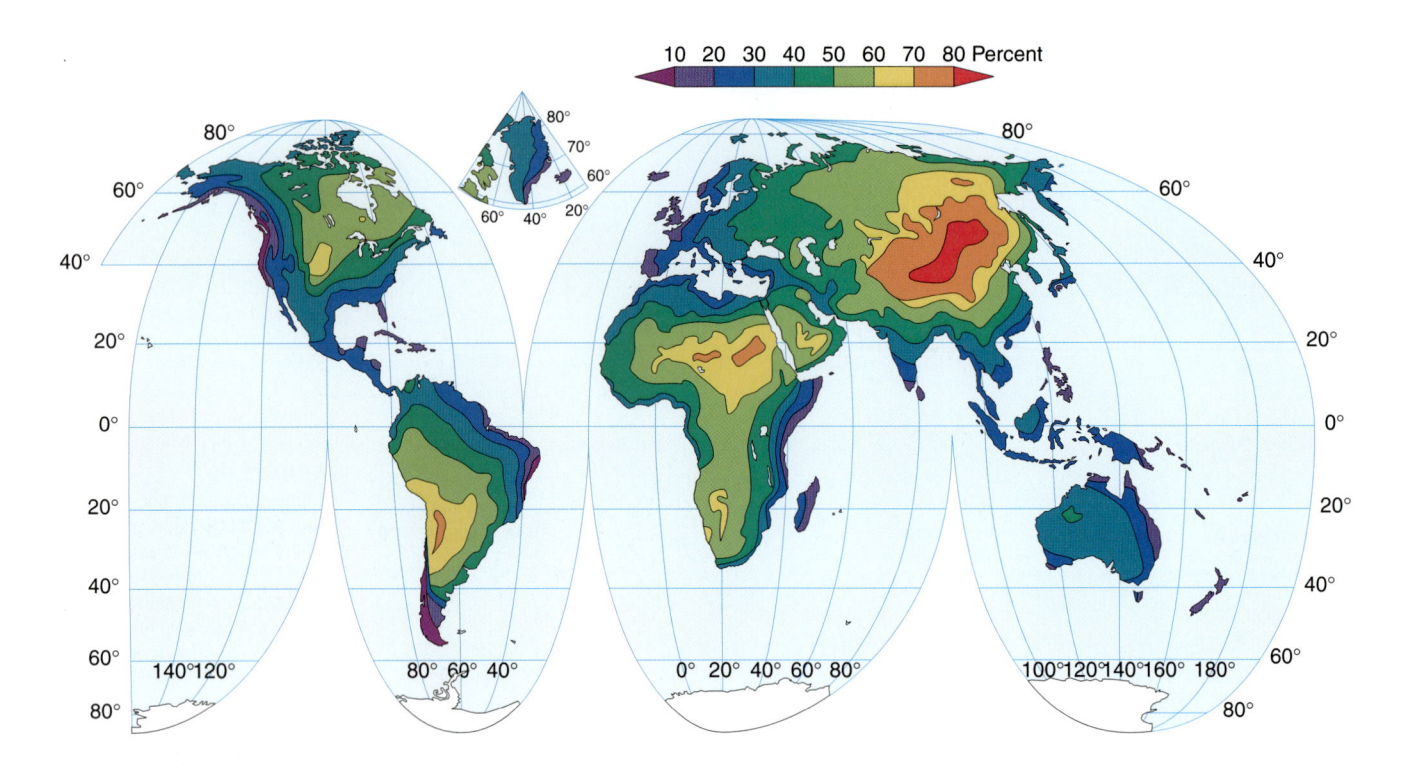

FIGURE 5.11
Percentage of precipitation over land that originated as evaporation on the continents, annually averaged over 15 years. In many land areas, the principal source of water for precipitation is evaporation from the ocean. [World Climate Research Programme, Global Energy and Water Cycle Experiment]

LATENT HEATING

Latent heating refers to the transfer of heat energy from one place to another as a consequence of phase changes of water. As discussed in Chapter 3, when water changes phase, heat energy is either absorbed from the environment (i.e., melting, evaporation, sublimation) or released to the environment (i.e., freezing, condensation, deposition). As part of the global water cycle, latent heat that is used to vaporize water at the Earth's surface is transferred to the atmosphere when clouds form. Significantly for Earth's climate, ocean water covers a large portion of Earth's surface and is the principal source of water vapor that eventually returns to Earth's surface as precipitation. In general, only well inland does most precipitation originate as evaporation on the continents (Figure 5.11). Also, the ocean is a major source of salt crystals that spur condensation and cloud development in the atmosphere. These *cloud condensation nuclei* have a special chemical affinity for water molecules and readily promote cloud development. When sea waves break, drops of salt water enter the atmosphere and evaporate leaving behind sea-salt crystals that function as nuclei.

As Earth's surface absorbs radiation (both solar and infrared), some of the heat energy is used to vaporize water from the ocean, glaciers, lakes, rivers, soil, and vegetation (*transpiration*). The latent heat required for vaporization (evaporation or sublimation) is supplied at the Earth's surface, and heat is subsequently released to the atmosphere during cloud development. Within the troposphere, clouds form as some of the water vapor condenses into liquid water droplets or deposits as ice crystals. During cloud formation, water changes phase and latent heat is released to the atmosphere. Through latent heating, then, heat is transferred from Earth's surface to the troposphere. In fact, latent heat transfer is more important than either radiational cooling or sensible heat transfer in cooling Earth's surface (Figure 5.12).

SENSIBLE HEATING

Heat transfer via conduction and convection can be monitored (*sensed*) by temperature changes; hence, **sensible heating** encompasses both of these processes. Heat is conducted from the relatively warm surface of the Earth to the cooler overlying air. Heating reduces the

FIGURE 5.12
Earth's surface is cooled through (A) vaporization of water, (B) net emission of infrared radiation, and (C) conduction plus convection. Numbers are global annual averages based on 100 units of solar radiation entering the top of the atmosphere.

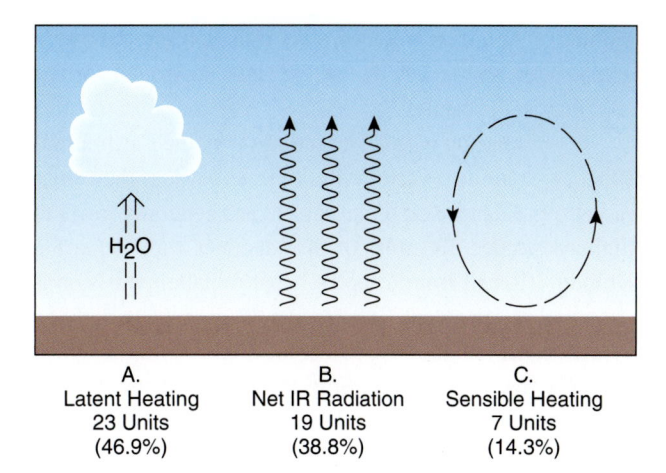

A. Latent Heating 23 Units (46.9%)
B. Net IR Radiation 19 Units (38.8%)
C. Sensible Heating 7 Units (14.3%)

Often sensible heating combines with latent heating to channel heat from Earth's surface into the troposphere. This happens during thunderstorm development. Updrafts (ascending branches) of vapor-laden air in convection currents often produce *cumulus clouds*, which resemble puffs of cotton floating in the sky (Figure 5.14A). These clouds are sometimes referred to as *fair-weather cumulus* because they seldom produce rain or snow. On the other hand, if atmospheric conditions are favorable, convective currents can surge to great altitudes, and cumulus clouds merge and billow upward to form

density of that air, which is forced to rise by cooler denser air replacing it at the surface (Figure 5.13). In this way, convection transports heat from Earth's surface to the troposphere. Because air is a relatively poor conductor of heat, heat convection is much more important than conduction as a transfer mechanism within the troposphere.

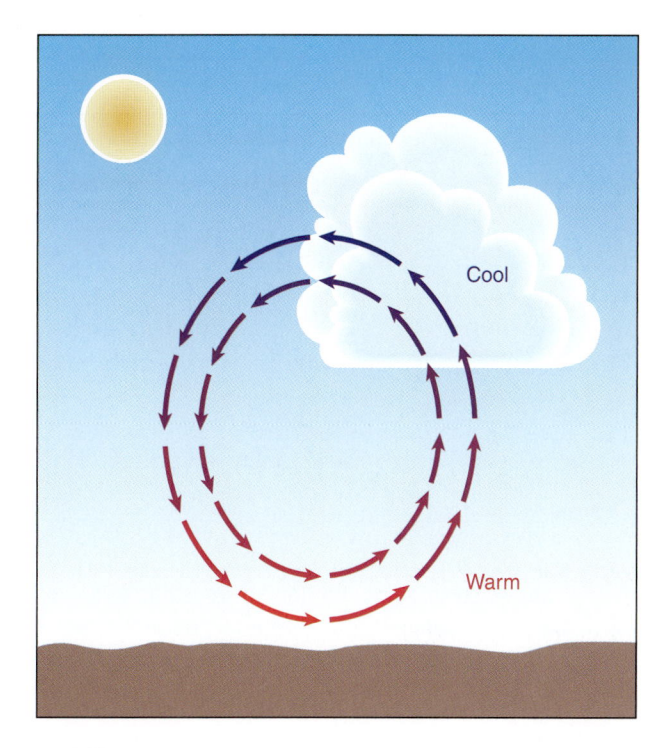

FIGURE 5.13
Convection currents transport heat from Earth's surface into the troposphere.

FIGURE 5.14
Latent heating and sensible heating are combined in the formation of (A) cumulus clouds and (B) cumulonimbus (thunderstorm) clouds.

towering *cumulonimbus clouds*, also known as thunderstorm clouds (Figure 5.14B). In retrospect, two important heat transfer processes (a combination of latent heating and sensible heating) took place last summer when that thunderstorm sent you scurrying for shelter at the beach.

At some times and places, heat transfer is directed from the troposphere to Earth's surface, the reverse of the global average annual situation. This reversal in direction of heat transport occurs, for example, when mild winds blow over cold, snow-covered ground or when warm air moves over a relatively cool ocean surface. Heat transport from the atmosphere to Earth's surface is the usual situation at night (especially when skies are clear) when radiational cooling causes Earth's land surface to become colder than the overlying air.

The **Bowen ratio** describes how the heat energy received at Earth's surface (by absorption of solar and infrared radiation) is partitioned between sensible heating and latent heating. That is,

Bowen ratio = [(sensible heating)/(latent heating)]

At the global scale,

Bowen ratio = [(7 units)/(23 units)] = 0.3.

The average Bowen ratio varies from one locality to another depending on the amount of surface moisture. The more moist the surface, the less important is sensible heating and the more important is latent heating. The Bowen ratio ranges from about 0.1 (one-tenth as much sensible as latent heating) for the ocean to about 5.0 (five times as much sensible as latent heating) in deserts. Ocean waters cover much of Earth's surface so it is not surprising that the global Bowen ratio is relatively low (0.3).

Heating Imbalances: Tropics versus High Latitudes

On a global scale, imbalances in radiational heating and radiational cooling occur not only between Earth's surface and atmosphere but also between the tropics and higher latitudes. Because the planet is nearly a sphere, parallel beams of incoming solar radiation strike the tropics more directly than higher latitudes. (That is, solar altitudes are higher in the tropics and lower at higher latitudes.) At higher latitudes, solar radiation spreads over a greater area and is less intense per unit horizontal surface area than in the tropics.

Emission of infrared radiation by the Earth-atmosphere system also varies with latitude but less than solar radiation. Because air temperatures are generally lower at higher latitudes, IR emission also declines with increasing latitude. (Recall from Chapter 1 that radiation emission is temperature dependent.) Consequently, over the period of a year at higher latitudes, the rate of infrared cooling to space exceeds the rate of warming caused by absorption of solar radiation. At lower latitudes the reverse is true; that is, over the course of a year, the rate of solar radiational heating is greater than the rate of infrared radiational cooling (Figure 5.15). Averaged over the globe, incoming energy (absorbed solar radiation) must equal outgoing energy (IR emitted to space). That is, the areas under the two curves in Figure 5.15 are equal. The balance between energy entering and leaving the Earth-atmosphere system (global radiative equilibrium) is the prevailing condition on Earth.

Measurements by sensors onboard Earth-orbiting satellites indicate that the division between regions of net radiational cooling and regions of net radiational warming is close to the 35-degree latitude circle in both hemispheres. By implication, latitudes poleward of about 35 degrees N

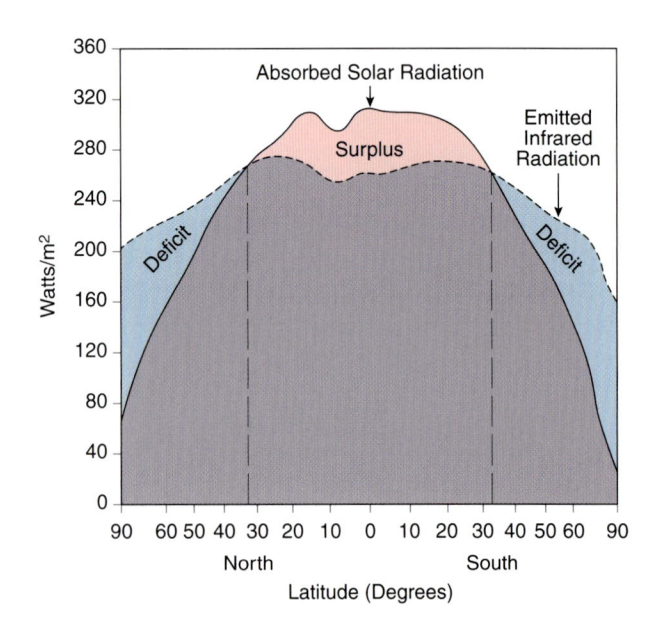

FIGURE 5.15
Variation by latitude of absorbed solar radiation and outgoing infrared radiation derived from satellite measurements. [From NOAA/NESDIS]

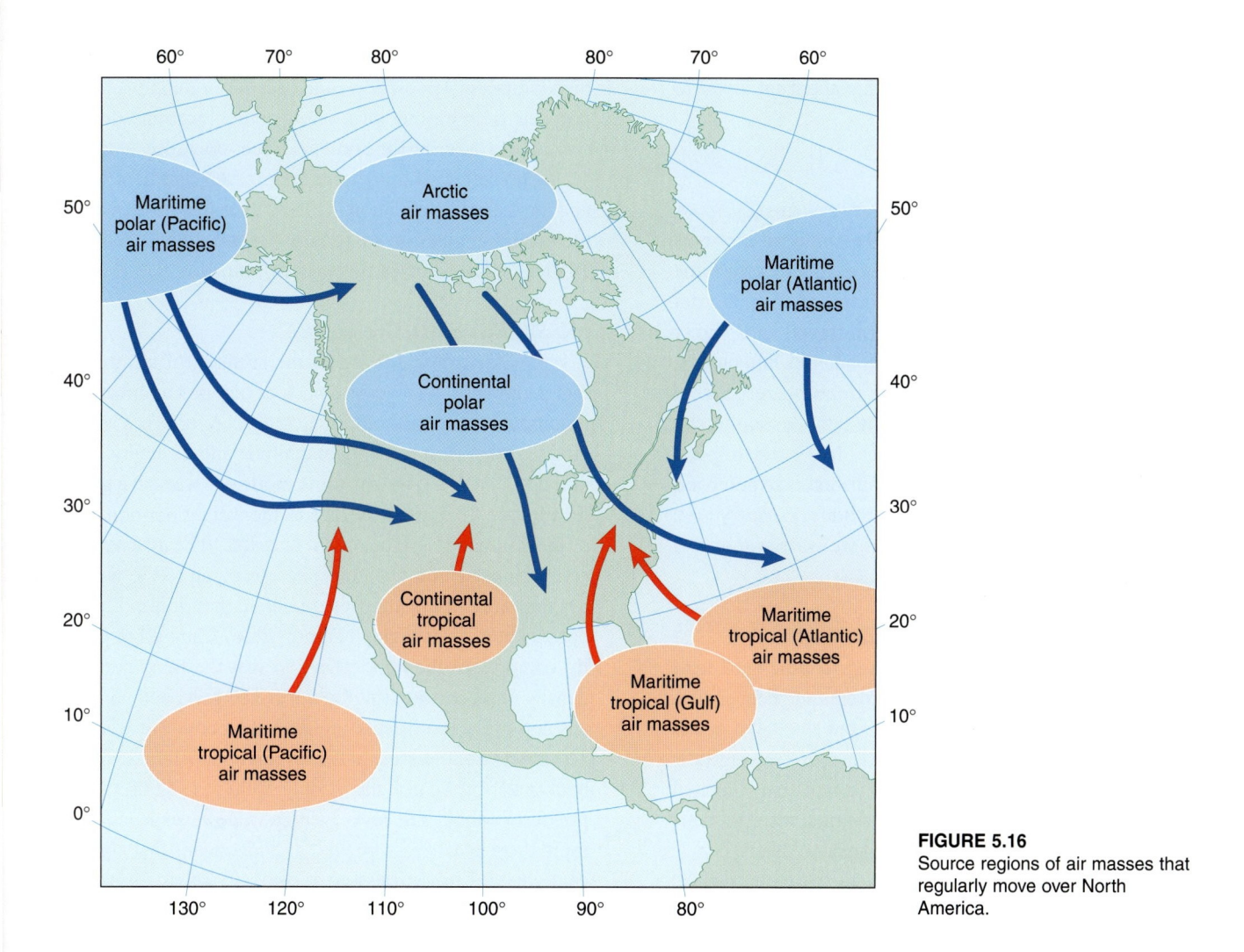

FIGURE 5.16
Source regions of air masses that regularly move over North America.

and 35 degrees S should experience net cooling over the course of a year, while tropical latitudes are sites of net warming. In fact, lower latitudes do not become progressively warmer nor do higher latitudes become cooler because heat is transported poleward from the tropics into middle and high latitudes. **Poleward heat transport** is brought about by (1) air mass exchange, (2) storm systems, and (3) ocean currents. Atmospheric processes account for 78% of total poleward heat transport in the Northern Hemisphere and 92% in the Southern Hemisphere. Oceanic processes account for the balance.

HEAT TRANSPORT BY AIR MASS EXCHANGE

North-south exchange of air masses transports sensible heat from the tropics into middle and high latitudes. An **air mass** is a huge volume of air covering thousands of square kilometers that is relatively uniform horizontally in temperature and humidity. The properties of an air mass largely depend on the characteristics of the surface over which the air mass forms (its *source region*) or travels (Figure 5.16). Air masses that form at high latitudes over cold, often snow- or ice-covered surfaces are relatively cold. Those air masses that form at low latitudes are relatively warm. Air masses that develop over the ocean are humid and those that form over land are relatively dry. Hence, there are four basic types of air masses: cold and humid, cold and dry, warm and humid, and warm and dry.

As shown in Figure 5.16, warm air masses that form in lower latitudes flow toward the pole while cold air masses flow toward the equator from source regions at high latitudes. Air masses modify (become cooler, warmer, drier, more humid) to some extent as they move away from their source region, gaining or losing heat energy in the process. In this north-south exchange of air masses, a net transport of heat takes place from lower to higher latitudes.

HEAT TRANSPORT BY STORMS

Release of latent heat in storm systems (*cyclones* or *lows*) plays an important role in the poleward transport of heat. At low latitudes, water that evaporates from the warm ocean surface is drawn into the circulation of a developing storm system. As the storm travels into higher latitudes, some of that water vapor condenses into clouds, thereby releasing latent heat to the troposphere. Latent heat of vaporization acquired at low latitudes is thereby delivered to middle and high latitudes. In general, tropical storms and hurricanes are greater contributors to poleward heat transport than ordinary middle latitude (*extratropical*) storms because they entrain much more water vapor and latent heat (Chapter 8).

HEAT TRANSPORT BY OCEAN CIRCULATION

The ocean also contributes to poleward heat transport via wind-driven surface currents and deeper conveyor-belt-like currents that traverse the lengths of ocean basins. Surface water that is cooler than the overlying air is a *heat sink* for the atmosphere; that is, heat is conducted from air to sea. Surface water that is warmer than the overlying air is a *heat source* for the atmosphere; that is, heat is conducted from sea to air. Cold surface currents, such as the California Current, flow from high to low latitudes, absorbing heat from the relatively warm tropical troposphere and greater solar radiation. Warm surface currents, such as the Gulf Stream, flow from the tropics into middle latitudes, supplying heat to the cooler middle latitude troposphere. We have more on wind-driven ocean currents in Chapter 6.

The ocean's conveyor-belt system transports heat and salt over great distances and to great depths in the ocean and plays an important role in Earth's climate system. As pointed out in this chapter's Case-in-Point, the ocean's circulation moderates the climate of Western Europe. We have more to say about the conveyor-belt system in Chapter 6.

Circulation of the Atmosphere: The Forces

As discussed earlier, the atmosphere circulates in response to temperature gradients within the Earth-atmosphere system. These temperature gradients are due to differences in rates of raditional heating and radiational cooling between (1) Earth surface and atmosphere, and (2) the tropics and high latitudes. Atmospheric and oceanic circulations transport heat from warmer locations to colder locations. In this section, we describe the principal forces operating in large-scale atmospheric circulation: the pressure gradient force and the Coriolis deflection. Other forces that influence atmospheric circulation are gravity (important within about 1000 m or 3300 ft of Earth's surface), and gravity.

PRESSURE GRADIENT FORCE

Air exerts a force on the surfaces of all objects that it contacts. (A *force* is a push or pull on an object and is computed as mass times acceleration. A force is a *vector* quantity; that is, it has both magnitude and direction.) As noted in Chapter 3, we can think of **air pressure** at a given location on the Earth's surface as the weight per unit area of the column of air above that location. The air pressure at any point within the atmosphere is equal to the weight per unit area of the atmosphere above that point. Unlike ocean water, air is highly compressible. The pull of gravity compresses the atmosphere so that the maximum air density and pressure is at the Earth's surface, and air density and pressure decrease rapidly with increasing altitude. The average air pressure at sea level is about 1013.25 millibars (mb). At an altitude of only 5500 m (18,000 ft), air pressure is about half of its average value at sea level. The rapid drop in air pressure with altitude means that significant changes in air pressure accompany relatively minor changes in land elevation. For example, the average air pressure at Denver, the *mile-high city*, is about 83% of the average air pressure at sea level.

Air pressure differs from one place to another, and variations are not always due to differences in the elevation of the land. In fact, atmospheric scientists are most interested in air pressure variations that arise from factors other than land elevation. Hence, weather observers determine an equivalent sea-level air pressure value; that is, for weather stations located above sea level, they adjust air pressure readings upward to approximately what the pressure would be if the station were actually located at sea level. When this adjustment to sea level is carried out everywhere, air pressure is observed to vary from one place to another and fluctuate from day to day and even from one hour to the next. Spatial and temporal changes in air pressure at Earth's surface arise from variations in air temperature (principally), humidity (concentration of water vapor in air), and atmospheric circulation.

In the free atmosphere, air density varies inversely with both temperature and humidity. That is, air density increases with falling temperature and decreasing humidity. Cold, dry air masses are denser and usually produce higher surface pressures than warm, humid air masses. Warm, dry air masses, in turn, often exert higher surface pressures than equally warm, but more humid air masses. As one air mass replaces another at a specific location, the air pressure at that location may change. Falling air pressure often signals a turn to stormy weather whereas rising air pressure indicates clearing skies or continued fair weather.

On a weather map, a *HIGH* or *H* symbol is used to designate places where sea-level air pressure is relatively high compared to the air pressure in surrounding areas. A *high* is also known as an *anticyclone* and is usually a fair weather system. A *LOW* or *L* symbol signifies regions where sea-level air pressure is relatively low compared to the air pressure in surrounding areas. A *low* is also known as a *cyclone* and, as we will see later, often brings stormy weather.

A change in air pressure from one place to another is known as an **air pressure gradient**. Air pressure gradients occur both vertically and horizontally within the atmosphere. A vertical air pressure gradient is a permanent feature of the atmosphere because air pressure always decreases with increasing altitude (at a rate dependent on the density of the air column). A horizontal air pressure gradient refers to pressure changes along a surface of constant altitude (e.g., mean sea level). Horizontal air pressure gradients can be determined on weather maps from patterns of *isobars*, lines joining points having the same air pressure (adjusted to sea level). Usually isobars are drawn on weather maps at 4-millibar intervals.

In response to an air pressure gradient, the wind blows from where the pressure is relatively high toward where the pressure is relatively low. The force that causes air to move as the consequence of an air pressure gradient is known as the **pressure gradient force** and is always directed across isobars and toward low pressure. The magnitude of the pressure gradient force is inversely related to the spacing of isobars. The wind is relatively strong where the pressure gradient is steep (closely spaced isobars), and light or calm where the pressure gradient is weak (widely spaced isobars).

CORIOLIS DEFLECTION

If Earth did not rotate, surface winds would blow directly from the cold poles (where surface air pressure is relatively high) to the hot equator (where surface air pressure is relatively low). And these winds would push ocean surface currents directly toward the equator. But because Earth rotates, anything moving freely (not frictionally coupled to the Earth), including air and water, is deflected to the right in the Northern Hemisphere and to the left in the Southern Hemisphere (Figure 5.17). This deflection is known as the **Coriolis effect**, named for Gaspard Gustav de Coriolis who first described the phenomenon quantitatively in 1835.

According to **Newton's first law of motion**, an object in constant, straight-line motion tends to remain that way unless acted upon by an unbalanced force. Winds in the atmosphere (and water moving in the ocean) exhibit this behavior. But these motions occur on a rotating Earth so that as air moves in a straight line, Earth rotates beneath the moving air. Even with no net force operating, the wind is displaced from a straight-line path when its motion is measured with respect to the rotating Earth. This inconsistency can be incorporated in Newton's first law of motion by explaining the deflection to be the result of an imaginary Coriolis force.

Reversal in the Coriolis deflection between the two hemispheres is related to the difference in an observer's sense of Earth's rotation. To an observer looking down from high above the North Pole, the planet rotates counterclockwise, whereas to an observer looking down from high above the South Pole, the planet rotates clockwise. For an Earth-bound observer, this reversal in the sense of

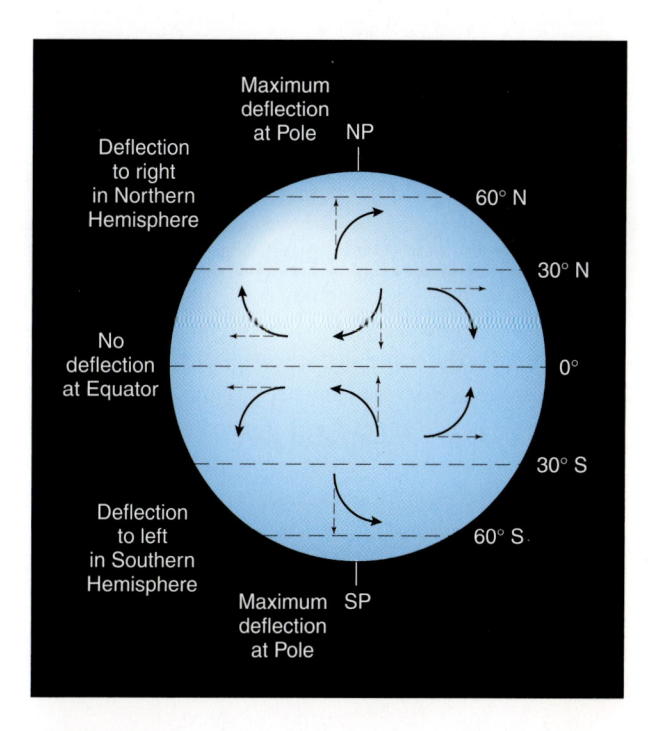

FIGURE 5.17
The Coriolis effect arises from the rotation of Earth on its axis and causes deflection of winds (and ocean currents) to the right in the Northern Hemisphere and to the left in the Southern Hemisphere.

Earth's rotation between the two hemispheres translates into a reversal in Coriolis deflection.

Although the Coriolis effect influences the wind regardless of its direction, the amount of deflection varies significantly with latitude; that is, the magnitude of Coriolis deflection varies from zero at the equator to a maximum value at the poles. This variation with latitude can be understood by visualizing the daily rotation of towers located at different latitudes. In a 24-hr day, Earth completes one rotation, as would a tower if located at the North or South Pole. In the same period, a tower at the equator would not rotate at all because of its orientation perpendicular to Earth's axis of rotation. At any latitude in between, some rotation of a tower occurs but not as much as at the poles.

The magnitude of the Coriolis effect also varies with wind speed and spatial scale of atmospheric circulation. The Coriolis deflection increases as the wind strengthens because, in the same period of time, faster moving air parcels cover greater distances than slower moving air parcels. The longer the trajectory, the greater is the underlying rotation of the Earth. For practical purposes, the Coriolis effect significantly influences the wind only in large-scale weather systems, that is, systems larger than ordinary thunderstorms. Large-scale weather systems also have longer life expectancies than small-scale systems so that air parcels cover greater distances over longer periods of time.

Circulation of the Atmosphere: Patterns of Motion

Air pressure gradients, the Coriolis effect, and the physical properties of Earth's surface shape the circulation of the atmosphere. For convenience of study, atmospheric scientists subdivide atmospheric circulation into discrete weather systems operating at various spatial and temporal scales (Table 5.3). The large-scale wind belts encircling the planet (e.g., westerlies of middle latitudes, trade winds) are **planetary-scale systems**. **Synoptic-scale systems** are continental or oceanic in scale; migrating storms and hurricanes are examples. **Mesoscale systems** include, for example, thunderstorms and sea breezes—circulation systems that are so small and short-lived that they may influence the weather in only a portion of a large city (Figure 5.18). A weather system covering only a very small area (e.g., a weak tornado) represents the smallest spatial subdivision of atmospheric circulation, **microscale systems**. In this section, we focus primarily on atmospheric circulation patterns at the planetary and synoptic scales.

PLANETARY-SCALE CIRCULATION

As shown in Figure 5.19, three broad wind belts encircle both the Northern and Southern Hemispheres. In the Northern Hemisphere, prevailing surface winds blow from the northeast *(trade winds)* between the equator and about 30 degrees N, from the southwest *(westerlies)* between about 30 and 60 degrees N, and from the northeast *(polar easterlies)* between about 60 degrees N and the North Pole. In the Southern Hemisphere, prevailing surface winds are southeasterly *(trade winds)* between the equator and about 30 degrees S, northwesterly *(westerlies)* between about 30 and 60 degrees S, and southeasterly *(polar easterlies)* between about 60 degrees S and the South Pole.

In both hemispheres, the trade winds blow out of the equatorward flank of the subtropical highs and the westerlies blow out of the poleward flank of the subtropical highs. The **subtropical highs** are massive semi-permanent systems that are centered over the ocean basins near 30 degrees N and S. They are *semi-permanent* features of the planetary-scale circulation in that they undergo seasonal changes in both location and strength. Viewed from above in the Northern

TABLE 5.3
Scales of Atmospheric Circulation

Circulation	Space scale	Time scale	Example
Planetary scale	10,000 to 40,000 km	weeks to months	Westerlies, Trade winds
Synoptic scale	100 to 10,000 km	days	Highs and Lows
Mesoscale	1 to 100 km	hours	Thunderstorms
Microscale	1 m to 1 km	seconds to minutes	Tornados

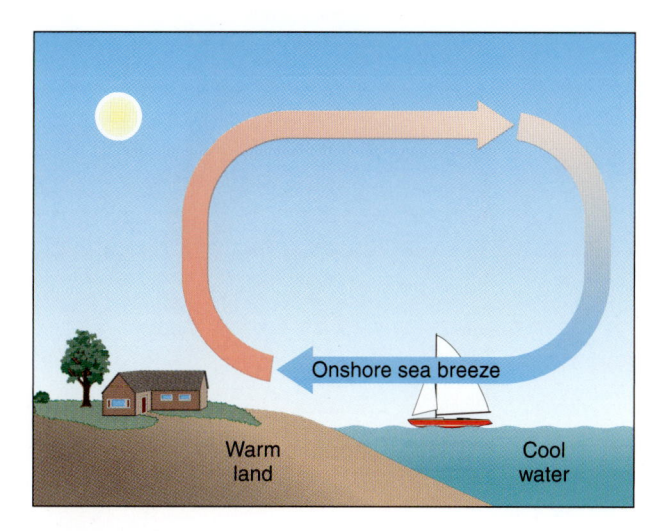

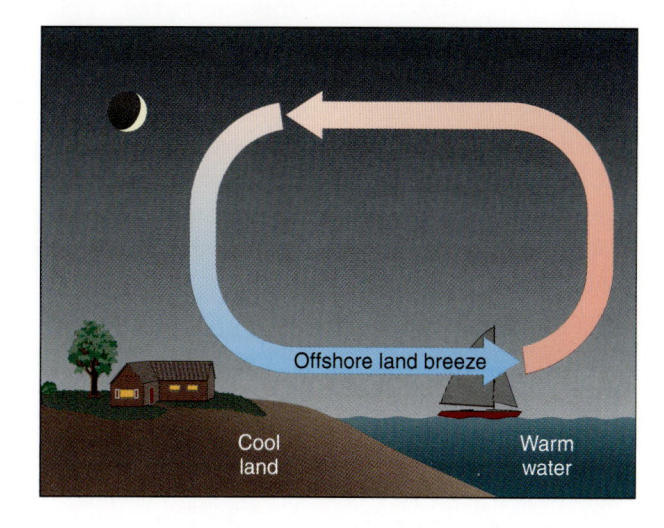

Chapter 5 THE ATMOSPHERE AND OCEAN 115

FIGURE 5.18
A sea breeze is a relatively cool mesoscale surface wind that develops during daylight hours and blows inland from the ocean in response to differential heating of land and sea. During the day, the land surface warms more than the sea surface inducing a horizontal air pressure gradient with high pressure over the ocean surface and low pressure over the land surface. At night, the circulation reverses and a land breeze blows offshore.

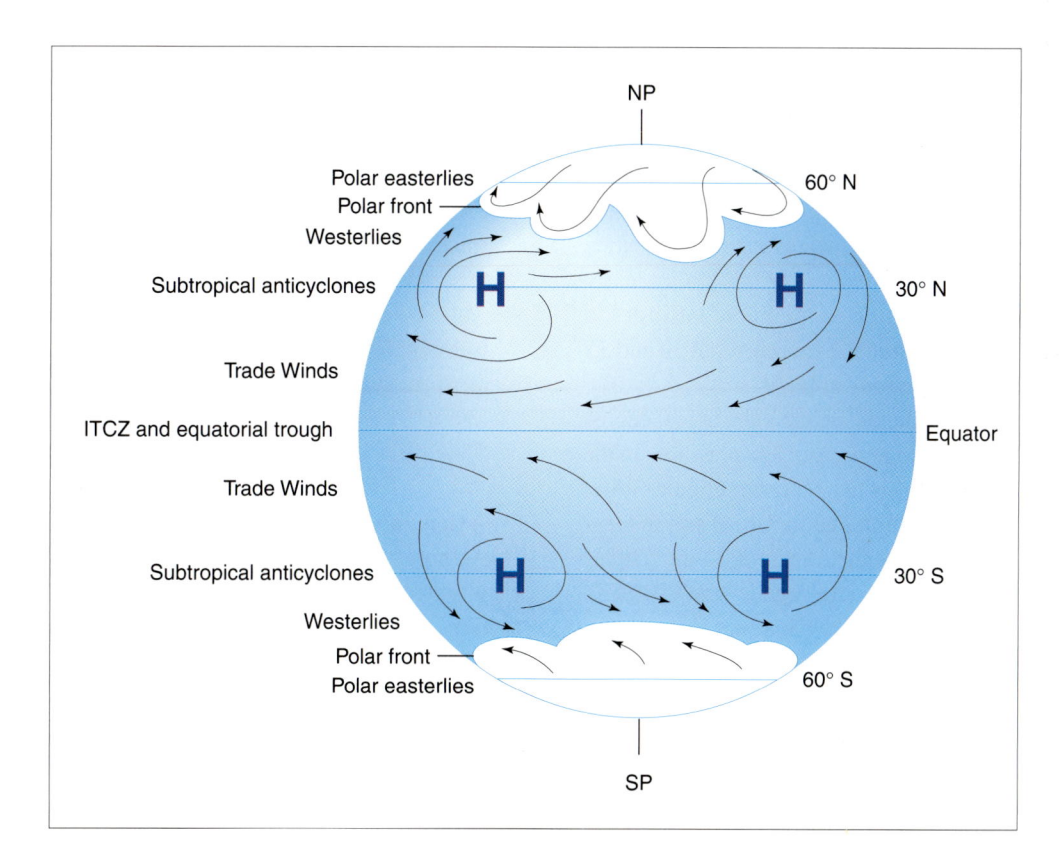

FIGURE 5.19
Schematic representation of the planetary-scale surface circulation of the atmosphere.

Hemisphere, surface winds blow clockwise and outward about the center of subtropical highs. Viewed from above in the Southern Hemisphere, surface winds blow counterclockwise and outward about the center of the subtropical highs. In both hemispheres, climates are relatively dry on the eastern flank of the subtropical highs and moist on the western flank. Broad areas of ocean basins under the subtropical highs experience persistent episodes of fair and warm weather. Low precipitation coupled with high temperatures cause high rates of evaporation and relatively high surface water salinity (refer back to Figure 3.8). Furthermore, the planetary-scale atmospheric circulation plays an important role in the horizontal transport of water vapor between ocean basins (as part of the global water cycle). For example, the trade winds transport water evaporated from the tropical Atlantic Ocean basin across Central America to the tropical Pacific Ocean where it condenses into clouds that produce rain. This transport of water increases the salinity of the Atlantic waters and freshens the Pacific waters.

Over a broad region surrounding the center of a subtropical high, the horizontal pressure gradient is weak so that surface winds are very light or the air is calm over extensive areas of the subtropical ocean. This situation played havoc with sailing ships, which were becalmed for days or even weeks at a time. Ships setting sail from Spain to the New World were often caught in this predicament and crews were forced to jettison their cargo of horses when supplies of water and food ran low. For this reason, early mariners referred to this region as the **horse latitudes**, a name now applied to all latitudes between about 30 and 35 degrees N and S under subtropical highs.

On the poleward side of the subtropical highs, surface westerlies flow into regions of low pressure. In the Northern Hemisphere, there are two separate **subpolar lows**: the *Aleutian low* over the North Pacific Ocean and the *Icelandic low* over the North Atlantic. These pressure systems mark the convergence of the middle latitude westerlies with the polar easterlies. By contrast, in the Southern Hemisphere, the middle latitude westerlies and the polar easterlies converge along a nearly continuous belt of low pressure surrounding the Antarctic continent.

At middle and high latitudes of the Northern Hemisphere, prevailing winds at altitudes from about 5500 to 12,000 m (18,000 to 40,000 ft) blow generally from west to east in wave-like pattern of ridges (clockwise turns) and troughs (counterclockwise turns) as illustrated in Figure 5.20. The temperature gradient between the relatively warm tropics and relatively cold polar areas induces a northward flow of air aloft that is deflected to the right (to the east) by

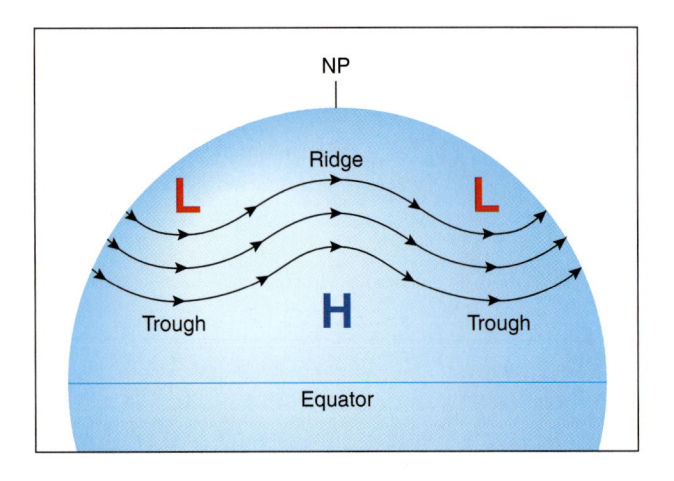

FIGURE 5.20
In the middle and upper troposphere, the Northern Hemisphere westerlies blow from west to east in a wave-like pattern of ridges (clockwise turns) and troughs (counterclockwise turns).

Earth's rotation (the Coriolis effect). A similar situation occurs in the Southern Hemisphere where the Coriolis effect deflects a southward-directed flow of air aloft to the left, resulting in a west wind in that hemisphere. The belts of westerlies encircle the planet and steer air masses, storms, and fair-weather systems generally from west to east.

Also part of the weaving westerlies are narrow corridors of exceptionally strong winds in the upper troposphere known as *jet streams*. In the Northern Hemisphere, the mid-latitude jet stream is situated over a boundary (called the *polar front*) between colder air to the north and warmer air to the south and contributes to the development of storm systems.

SYNOPTIC-SCALE WEATHER SYSTEMS

The two most important synoptic-scale weather systems are *highs* (or anticyclones) and *lows* (or cyclones). As noted earlier, **highs** usually are accompanied by fair weather whereas **lows** often bring clouds and precipitation. Whereas subtropical highs and subpolar lows are nearly stationary, synoptic-scale highs and lows move with the prevailing wind several kilometers above the surface, generally eastward across North America. Highs follow lows and lows follow highs. As a general rule, highs track toward the east and southeast whereas lows track toward the east and northeast. Important exceptions are tropical cyclones (e.g., hurricanes) that are imbedded in the trade wind flow and track generally from east to west over the tropical Atlantic and Pacific. Eventually these systems come under the influence of the mid-latitude westerlies and turn toward the north and northeast.

Highs originating over northwestern Canada bring cold, dry weather in winter and cool, dry weather in summer. Highs that develop farther south bring hot weather in summer and mild, dry weather in winter. Viewed from above in the Northern Hemisphere, surface winds in a high-pressure system blow in a clockwise and outward spiral as shown in Figure 5.21. Reversal of the Coriolis deflection in the Southern Hemisphere means that surface winds in a high blow in a counterclockwise and outward spiral south of the equator. Surface winds spiraling outward induce descending dry air near the center of a high—hence, fair weather.

Viewed from above in the Northern Hemisphere, surface winds in a low blow in a counterclockwise and inward spiral as shown in Figure 5.22A. Because the Coriolis deflection reverses direction in the Southern Hemisphere, surface winds in a low blow in a clockwise and inward spiral south of the equator. For middle and high latitude lows, surface winds spiraling inward bring together contrasting air masses and induce ascending air (Figure 5.22B). Air expands and cools as it ascends causing water vapor to condense into clouds—hence, stormy weather.

Lows that track across the northern United States or southern Canada are more distant from sources of moisture and usually produce less rain- or snowfall than lows that track farther south (such as lows that move along the Gulf Coast or up the Eastern Seaboard). Weather to the left side (west and north) of a storm's track (path) tends to be relatively cold, whereas weather to the right (east and south) of a storm's track tends to be relatively warm. For this reason, winter snows are most likely to the west and

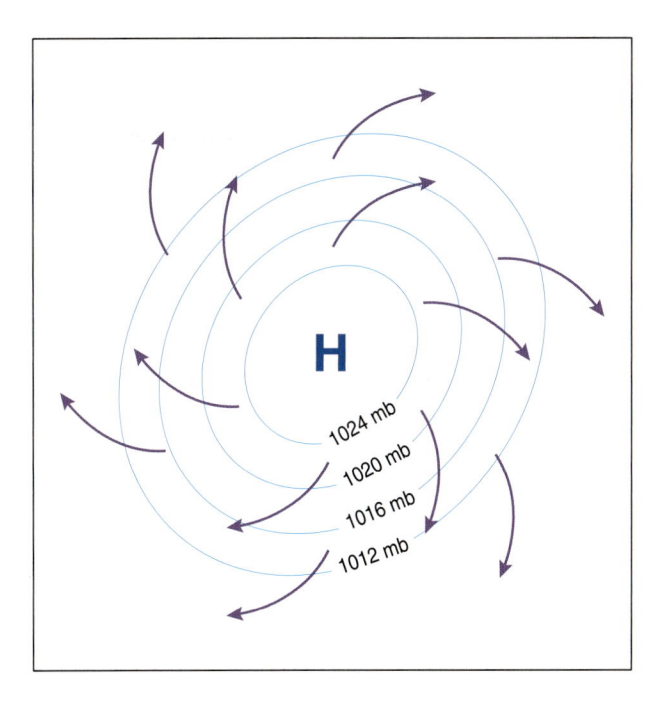

FIGURE 5.21
Viewed from above in the Northern Hemisphere, surface winds blow clockwise and outward in a high (anticyclone). Contour lines are isobars drawn at 4-millibar intervals.

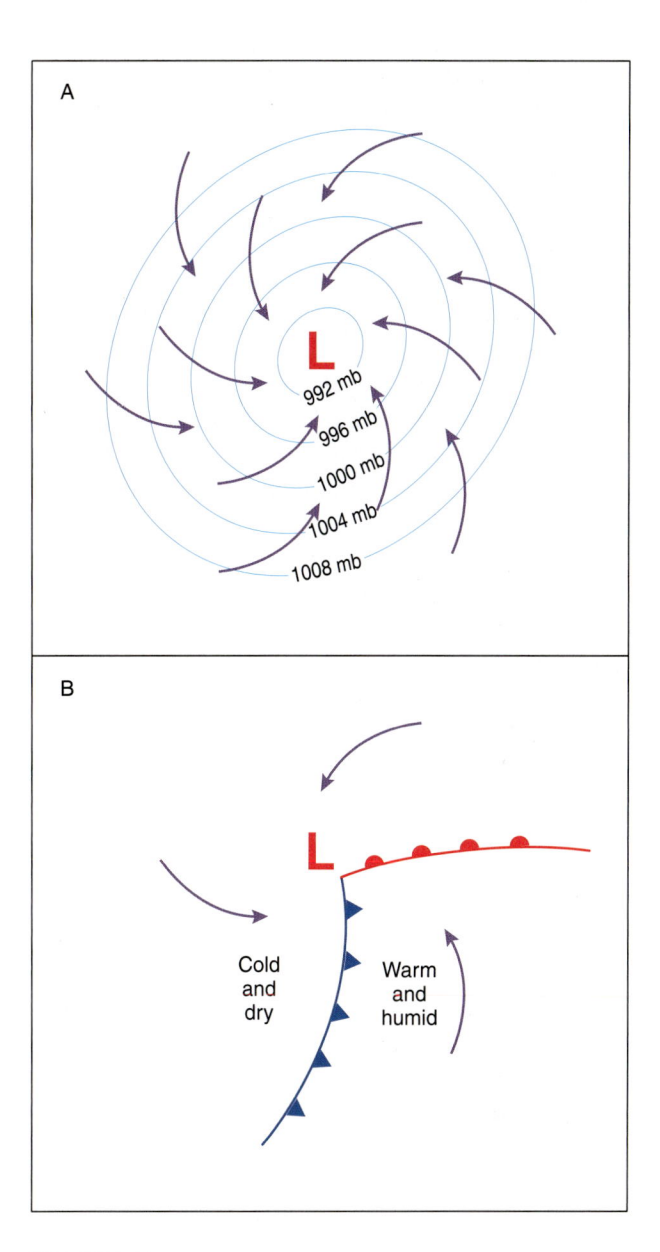

FIGURE 5.22
Viewed from above in the Northern Hemisphere, (A) surface winds blow counterclockwise and inward in a low (cyclone), (B) bringing together contrasting air masses to form fronts. Contour lines are isobars drawn at 4-millibar intervals. The curved blue line with spikes is the leading edge of relatively cold and dry air (a cold front) while the curved red line with half circles is the leading edge of relatively warm and humid air (a warm front).

north of the path of a low-pressure system. We have more to say about storms and their impact on the coastal zone in Chapter 8.

Conclusions

In this chapter we have seen how the ocean plays a key role in the global radiation balance. The ocean, covering about 71% of Earth's surface, has a very low average albedo for incident solar radiation and is the principal sink for incoming solar radiation. The ocean also is the chief source of water vapor and latent heat for the atmosphere. Latent heat is absorbed when water evaporates from the ocean surface and is released to the atmosphere when water vapor condenses during cloud formation. This latent heat transfer is the principal means whereby heat is channeled from the Earth's surface to the atmosphere. Furthermore, ocean currents contribute to heat transport from the tropics to higher latitudes and breaking ocean waves are an important source of cloud condensation nuclei.

The atmosphere is frictionally coupled to the ocean surface and winds supply the kinetic energy that drives surface ocean currents and generates ocean waves. Large-scale surface ocean currents mirror the prevailing planetary-scale atmospheric circulation. In the next chapter, we begin our examination of motion in the ocean with a look at wind-driven surface currents and the deeper thermohaline circulation.

Basic Understandings

- Weather is the state of the atmosphere at some time and place. Climate is popularly defined as weather at a particular location averaged over a specified interval of time (30 years, by international convention). Climate also encompasses extremes in weather.
- As Earth orbits the sun, its atmosphere and surface absorb a tiny fraction of the total radiation continually emitted by the sun. Absorption of solar radiation heats the Earth-atmosphere system. At the same time the planet continually emits infrared radiation to space, which cools the Earth-atmosphere system, resulting in a relatively constant planetary average temperature.
- As Earth orbits the sun, regular changes occur in the solar altitude and length of daylight, which affect the intensity of solar radiation striking Earth's surface at a point. Solar altitude is the angle of the sun above the horizon and varies from zero to a maximum of 90 degrees. The amount of solar energy striking a unit area of Earth surface in a unit of time increases with increasing solar altitude. The difference in length of daylight between the summer and winter solstices increases from zero at the equator to a maximum at the Arctic and Antarctic Circles. Variation of incoming solar radiation through the course of a year is reflected in the march of average monthly temperatures.
- The moderating influence of ocean water reduces the average temperature contrast between summer and winter. Locations immediately downwind of the ocean experience a maritime climate whereas locations well inland from the ocean have continental climates. Proximity to large bodies of water also affects the timing of the average warmest and coldest times of the year. In the United States, the air temperature cycle lags the solar radiation cycle by an average of 27 days. But in coastal localities having a strong maritime influence, the average lag time is up to 36 days.
- Solar radiation intercepted by Earth interacts with the atmosphere's component gases and aerosols. Interactions involve reflection, scattering, and absorption. Reflection and scattering only change the direction of the radiation, but with absorption radiation is converted to heat.
- Surfaces having a high albedo reflect a relatively large fraction of incident solar radiation and appear light in color. Surfaces having a low albedo reflect a relatively small fraction of incident solar radiation and appear dark in color. Within the atmosphere, cloud tops are the most important reflectors of solar radiation.
- Solar radiation that is not reflected or scattered to space or absorbed in the atmosphere strikes Earth's surface. There, some radiation is absorbed and some is reflected depending on the albedo (reflectivity) of the surface. On a global basis, the average albedo of ocean water is only about 8% so that the ocean, covering about 71% of Earth's surface, is the principal sink for solar radiation.
- On a global average annual basis, Earth's planetary albedo is about 31%. The atmosphere absorbs only about 20% of the total solar radiation intercepted by the planet, and Earth's surface (mostly ocean) absorbs 49%. Hence, the atmosphere is relatively transparent to solar radiation.
- Ocean waters absorb most incident solar radiation within relatively shallow depths. Water absorbs the

- longer wavelengths (i.e., reds and yellows) of visible light more efficiently than the shorter wavelengths (i.e., greens and blues) so that green and blue penetrate to greater depths. The upper portion of the ocean in which solar radiation is detectable is known as the photic zone.
- The greenhouse effect refers to the heating of Earth's surface and lower atmosphere caused by strong absorption and emission of infrared radiation by certain gaseous components of the atmosphere. The principal greenhouse gas is water vapor. Without the greenhouse effect, Earth would be much too cold to support most forms of plant and animal life.
- The global distribution of incoming solar radiation and outgoing infrared radiation implies net radiational warming of Earth's surface and net radiational cooling of Earth's atmosphere. In response to the resulting temperature gradient, heat is transferred from Earth's surface to the atmosphere via latent heating and sensible heating. Latent heating refers to the transport of heat from one place to another as a consequence of phase changes of water. Sensible heating involves heat transport via conduction and convection.
- The Bowen ratio describes how heat available at Earth's surface is partitioned between sensible heating and latent heating. The Bowen ratio is lowest for wet surfaces and highest for dry surfaces. The global average Bowen ratio is 0.3.
- On a global average annual basis, radiational heating exceeds radiational cooling between about 35 degrees N and S latitude. Radiational cooling exceeds radiational heating poleward of about 35 degrees N and 35 degrees S. Imbalances in radiational heating and cooling imply net heating in the tropics and subtropics and net cooling at middle and high latitudes. But lower latitudes are not warming relative to higher latitudes because heat is transferred from the tropics toward the poles (poleward heat transport).
- Poleward heat transport is brought about by north-south exchange of air masses, storm systems, and ocean circulation.
- When air pressure readings are adjusted to sea level everywhere (to remove the influence of land elevation), air pressure is observed to vary from one place to another and fluctuate from day to day and even from one hour to the next. Spatial and temporal changes in air pressure at Earth's surface arise from variations in air temperature (principally), humidity (concentration of water vapor in air), and atmospheric circulation.
- On a weather map, a *HIGH* or *H* symbol is used to designate places where sea-level air pressure is relatively high compared to the air pressure in surrounding areas. A *high* is also known as an *anticyclone* and is usually a fair weather system. A *LOW* or *L* symbol signifies regions where sea-level air pressure is relatively low compared to the air pressure in surrounding areas. A *low* is also known as a *cyclone* and often brings stormy weather.
- In response to an air pressure gradient, the wind blows from where the pressure is relatively high toward where the pressure is relatively low. The pressure gradient force is always directed across isobars and toward low pressure. The magnitude of the pressure gradient force is inversely related to the spacing of isobars.
- Because Earth rotates, anything moving freely over Earth's surface, including air and water, is deflected to the right in the Northern Hemisphere and to the left in the Southern Hemisphere. This deflection is known as the Coriolis effect. The magnitude of the Coriolis deflection increases from zero at the equator to a maximum at the poles.
- Viewed from above in the Northern Hemisphere, surface winds blow clockwise and outward in a high, and counterclockwise and inward in a low. Viewed from above in the Southern Hemisphere, surface winds blow counterclockwise and outward in a high, and clockwise and inward in a low.
- At the planetary scale, three broad wind belts encircle both the Northern and Southern Hemispheres. Trade winds characterize the tropics, the westerlies encircle middle latitudes, and the polar easterlies prevail at high latitudes. The trade winds and westerlies are linked to the subtropical highs, massive semi-permanent systems that are centered over the ocean basins near 30 degrees N and S.
- Whereas subtropical highs and subpolar lows are nearly stationary, synoptic-scale highs and lows move with the prevailing wind several kilometers above the surface. Highs follow lows and lows follow highs, shaping day-to-day weather.

ESSAY: Location at Sea, An Historical Perspective

When humans first left their caves they faced the challenge of determining their location, how to get to where they wanted to go, and how to get back to where they started. On land and inland waters, while not easy, the task was manageable. Using prominent landmarks such as shorelines or rock pinnacles as signposts, travelers developed some sense of direction and distance. Eventually common routes of travel for discovery, trade, and communication produced well-worn paths. Upon leaving land for the sea, the problem of location became many times more difficult. Once out of sight of the shore, with water in all directions as far as the eye could see, ancient mariners depended mostly on luck to bring them back to port. It is not surprising that early seafaring was mostly a coastal affair.

Eventually humans developed a rudimentary understanding of winds and ocean currents and became familiar with the local geography of coastlines and the seas they regularly plied. Knowledge of the true shape of Earth allowed sailors to comprehend the relationship between their latitudinal position and the position of the stars and sun in the sky. But they never knew with any certainty where their boat was located in terms of east and west. And when clouds obscured the sky or storm-tossed seas prevented an estimate of sun or star angle, even their north–south position was in doubt.

In about A.D. 150, the Egyptian astronomer and geographer Ptolemy (ca. A.D. 85-165) came up with the concept of latitude and longitude as imaginary reference lines on the globe by which location could be specified. His zero degree parallel (latitude) was set at the equator because he had learned that the sun, moon and planets pass almost directly over that location. His zero-degree longitude line (the *prime meridian*) was set in the Canary Islands. Perhaps for nationalistic reasons, others drew the prime meridian through Paris, Philadelphia, Moscow, and at least a half a dozen other places. Today, by international convention the prime meridian runs through Greenwich, England, a bit to the east of Ptolemy's mark. Longitude is measured in degrees to the east and west of that line to the common 180-degree meridian, essentially the International Date Line (the imaginary line on Earth that separates two consecutive calendar days).

The apparent lack of a connection between longitude and the regular motions of celestial bodies would challenge navigators and astronomers for more than 1000 years. By the early 18th century, many who searched for the connection were motivated by huge amounts of prize money offered by governments for the person who could solve the "longitude problem," that is, develop a reliable method of determining longitude at sea. Key to solving the problem was perfecting a way to determine precisely the time of the day. Earth is roughly a sphere that rotates through 360 degrees once every 24 hours, that is, 15 degrees every hour. If one knows the current time at the prime meridian and at one's location at sea (based on the sun's position in the sky), it is easy to calculate the number of degrees between the two to yield the longitude.

The most successful efforts to solve the longitude problem focused on the design and construction of a very accurate clock that could keep time over a lengthy sea voyage and withstand the harsh conditions at sea without suffering mechanical failure. The first in a series of such accurate and durable chronometers was invented by the English clockmaker John Harrison in 1733. Forty years later, he was awarded the 20,000 Pounds Sterling prize offered by Great Britain for a clock that could be used to determine longitude to within one-half a degree on a voyage from England to any port in the West Indies. It would be another 50 years or so for this technology to find its way onto virtually every ship. Meanwhile, most sea captains continued to rely on dead reckoning to determine their location. The clock-method for determining longitude was used for more than two centuries and then supplanted by satellite-based techniques including the Global Positioning System (GPS). (Refer to the Essay in Chapter 6 for more on this topic.)

ESSAY: The Stratospheric Ozone Shield and Marine Life.

Ozone (O_3) is a relatively unstable molecule made up of three atoms of oxygen and occurring naturally in the stratosphere mostly at altitudes below 48 km (30 mi). Stratospheric ozone shields marine and terrestrial organisms from exposure to potentially lethal intensities of solar ultraviolet (UV) radiation. Without this so-called *stratospheric ozone shield*, life as we know it could not exist on Earth. It was not until about 1 billion years ago that the atmosphere contained sufficient oxygen (from which ozone is generated), that an effective ozone shield was present. From that point on, for example, marine organisms could thrive in surface waters. However, in the latter portion of the 20th century, scientists became aware of a serious threat to the stratospheric ozone shield that poses a hazard for all forms of life including marine organisms.

Within the stratosphere, two sets of competing chemical reactions, both powered by solar ultraviolet radiation, continually generate and destroy ozone (see Figure). During ozone production, UV strikes an oxygen molecule (O_2) causing it to split into two free oxygen atoms (O). Free oxygen atoms then collide with molecules of oxygen to form ozone molecules (O_3). At the same time, ozone is destroyed. Ozone absorbs ultraviolet radiation, splitting the molecule into one free oxygen atom (O) and one molecule of oxygen (O_2). The free oxygen atom then collides with an ozone molecule to form two molecules of oxygen. The net effect of these opposing sets of chemical reactions is a minute reservoir of ozone that peaks at only about 10 parts per million (ppm) in the middle stratosphere. Ultraviolet radiation (at different wavelengths) powers both sets of chemical reactions so that much but not all UV radiation is prevented from reaching Earth's surface.

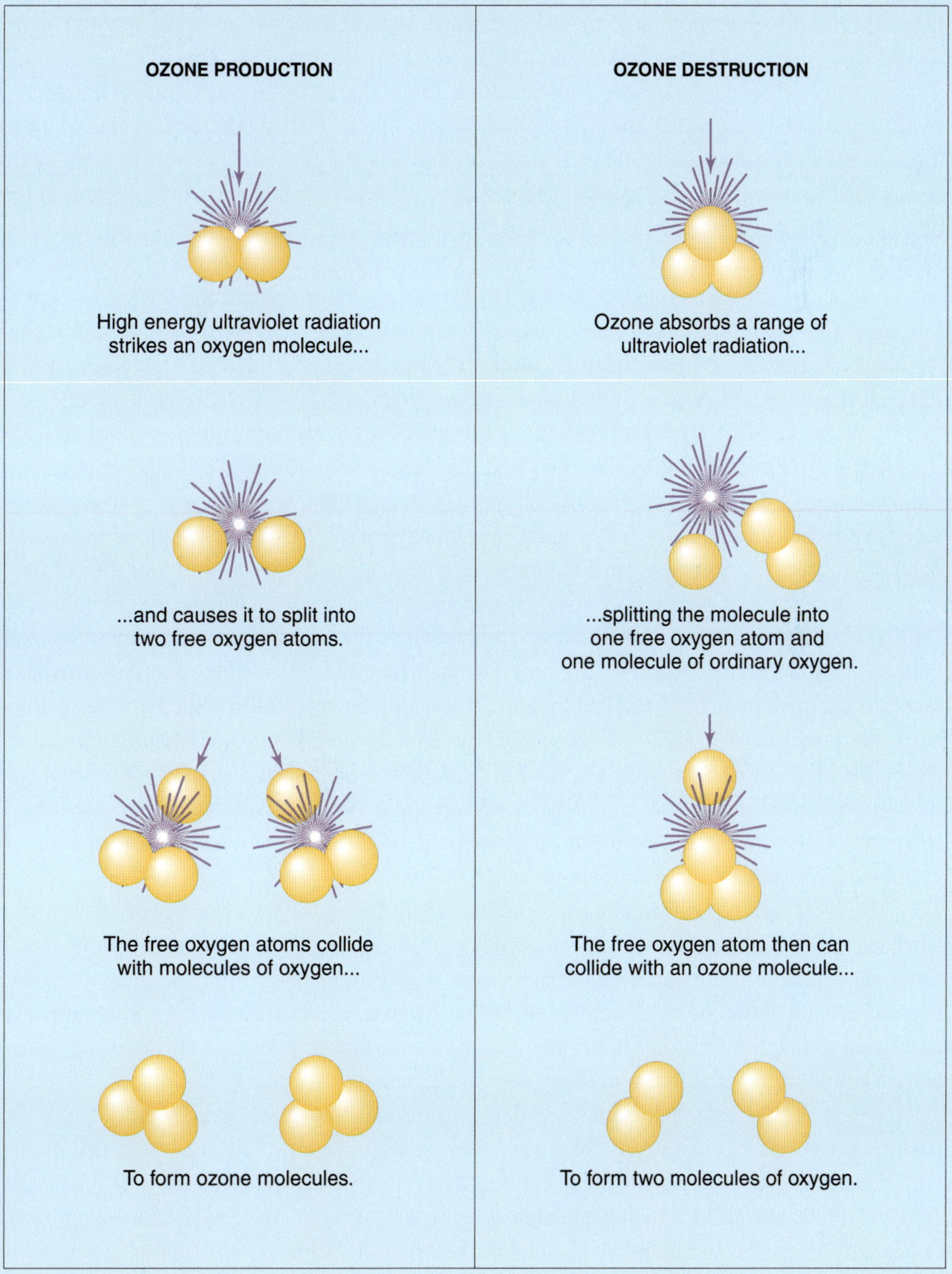

The most dangerous portion of UV radiation that reaches Earth's surface is designated *UVB* and spans the wavelength band from 0.28 to 0.32 micrometer. Prolonged exposure to

OZONE PRODUCTION	OZONE DESTRUCTION
High energy ultraviolet radiation strikes an oxygen molecule...	Ozone absorbs a range of ultraviolet radiation...
...and causes it to split into two free oxygen atoms.	...splitting the molecule into one free oxygen atom and one molecule of ordinary oxygen.
The free oxygen atoms collide with molecules of oxygen...	The free oxygen atom then can collide with an ozone molecule...
To form ozone molecules.	To form two molecules of oxygen.

UVB or an increase in the intensity of UVB reaching Earth's surface threatens the health and wellbeing of living organisms. As a general rule, every 1% decline in stratospheric ozone concentration translates into a 2% increase in the intensity of UV that passes through the ozone shield. The increase in UVB that actually reaches Earth's surface hinges on cloudiness and the dustiness of the atmosphere.

The effects on human health of exposure to too much UVB are well known and include increased incidence of skin cancer, premature aging of the skin, cataracts of the eye, and suppressed immune system. Perhaps less well known is the impact on marine ecosystems. Too much UV penetrating ocean surface waters can disrupt the orderly functioning of marine food webs, adversely affecting predator-prey relationships, competition, species diversity, and the dynamics of trophic (feeding) levels. Particularly important is the potential impact of elevated levels of solar ultraviolet radiation on autotrophs that form the base of marine food webs. Too much UV adversely affects the growth and reproduction of autotrophs thereby reducing the food (energy) supply for organisms occupying higher trophic levels (Chapter 9). Some phytoplankton are capable of moving out of harms way, but exposure to UVB impairs their orientation mechanisms and motility making even those species more susceptible to radiation damage. In addition, too much UV impairs larval development, harms young fish, shrimp, crabs, and amphibians, and threatens macro-algae and sea grasses.

More than three decades ago, scientists found that a group of chemicals known as CFCs (for *chlorofluorocarbons*) poses a serious threat to the stratospheric ozone shield. First synthesized in 1928, CFCs were widely used as chilling (heat-transfer) agents in refrigerators and air conditioners, for cleaning electronic circuit boards, and in the manufacture of foams used for insulation. F.S. Rowland and M.J. Molina of the University of California at Irvine first warned of the threat of CFCs to the stratospheric ozone shield in 1974. Use of CFCs as propellants in common household aerosol sprays such as deodorants and hairsprays was banned in the United States, Canada, Norway, and Sweden in 1979. Finally, in response to widespread acknowledgement of the CFC threat to the stratospheric ozone shield, by international agreement, worldwide production and use of CFCs for any purpose was phased out in 1996. For their pioneering studies of the depletion of stratospheric ozone, Rowland, Molina (now at the Massachusetts Institute of Technology), and P.J. Crutzen (of the Max Planck Institute for Chemistry, Germany) were awarded the 1995 Nobel Prize in chemistry.

Certain CFCs are inert (chemically non-reactive) in the troposphere, where they have accumulated for decades. Atmospheric circulation transports CFCs into the stratosphere where, at altitudes above about 25 km (16 mi), intense UV radiation breaks down CFCs, releasing chlorine (Cl), a gas that readily reacts with and destroys ozone. Products of this reaction are chlorine monoxide (ClO) and molecular oxygen (O_2). Chlorine (Cl) is a catalyst in chemical reactions that convert ozone to oxygen. In this way, each chlorine atom destroys perhaps tens of thousands of ozone molecules.

First signs of a thinner stratospheric ozone shield came from Antarctica. For about six weeks during the Southern Hemisphere spring (mainly in September and October), the ozone layer in the Antarctic stratosphere (mostly at altitudes between 14 and 19 km, or 9 and 12 mi) thins drastically with up to 70% of the ozone destroyed. Antarctic stratospheric ozone recovers during November. Satellite measurements indicate that this so-called *Antarctic ozone hole* has steadily deepened since the late 1970s. The 2000 ozone hole covered an area larger than Antarctica and exposed southern portions of Argentina and Chile to elevated levels of UV. Research conducted during the National Ozone Expeditions to the U.S. McMurdo Station in 1986-87 plus NASA aircraft flights into the Antarctic stratosphere in 1987 led to the discovery of relatively high concentrations of chlorine monoxide (ClO) in the Antarctic stratosphere. This discovery established a convincing link between ozone depletion and CFCs.

What causes the Antarctic ozone hole and why does it fill in by November? During the long, dark Antarctic winter, extreme radiational cooling causes temperatures in the stratosphere to plunge below –85 °C (-121 °F). At such frigid temperatures, what little water vapor that exists in the stratosphere forms clouds composed of ice crystals. Those ice crystals are key to ozone depletion by providing surfaces on which chlorine compounds that are inert toward ozone are converted to active forms that destroy ozone. Once the sun reappears in spring, solar radiation supplies the energy that causes active forms of chlorine to begin destroying ozone.

Ozone depletion takes place while the Antarctic atmosphere is essentially cut off from the rest of the planetary-scale atmospheric circulation by the *circumpolar vortex*, a belt of strong winds that encircles the outer margin of the Antarctic continent. A month or so into spring, however, the circumpolar vortex begins to weaken and allows warmer ozone-rich air from lower latitudes to invade the Antarctic stratosphere. Ice crystal clouds vaporize, and the stratospheric ozone concentration returns to normal levels; that is, the Antarctic ozone hole fills in.

Scientists investigating stratospheric chemistry in the Arctic in early 1989 discovered ozone-destroying chlorine compounds and a slight thinning of ozone. An Arctic ozone hole comparable in magnitude to the Antarctic ozone hole is unlikely for two reasons. For one, in winter the Arctic stratosphere averages about 10 Celsius degrees (18 Fahrenheit degrees) warmer than the Antarctic stratosphere, making formation of stratospheric ice crystal clouds improbable. Secondly, the circumpolar vortex that surrounds the Arctic weakens earlier than its Antarctic counterpart. However, an exceptionally cold Arctic winter coupled with an unusually persistent circumpolar vortex could translate into considerable ozone depletion in the Arctic.

Phase out of CFCs and other ozone-destroying chemicals slowed the release of chlorine into the atmosphere. However, CFCs have a long residence time in the atmosphere (up to 100 years or longer) so that the concentration of chlorine in the stratosphere is not expected to begin to decline until about 2010 and complete recovery of the Antarctic ozone layer is not likely to occur until at least 2050. Meanwhile, scientists continue to monitor stratospheric ozone levels worldwide.

CHAPTER 6

OCEAN CURRENTS

Case-in-Point
Driving Question
Ocean's Vertical Structure
Ocean in Motion: The Forces
    Wind-Driven Currents and Ekman Transport
    Geostrophic Flow
Wind-Driven Surface Currents
    Gyres
    Equatorial Currents
    Western Boundary Currents
    Rings
    Upwelling and Downwelling
Thermohaline Circulation
    Monitoring the Deep Ocean
    Water Masses
Oceanic Conveyer Belt
Conclusions
Basic Understandings
ESSAY: Predicting Oil-Spill Trajectories
ESSAY: The Global Positioning System
ESSAY: Profiling the Ocean Depths

Bottom mounted current meter being deployed from FERREL Project in the Gulf of Mexico to determine path of brine plume from salt dome solution and pumping. [Courtesy of NOAA.]

Case-in-Point

For much of human history, knowledge of the ocean and its currents was not recorded for many reasons. For one, most sailors could not write so their basic understandings of the sea were passed on orally. Even after the European nations (Portugal, Spain, Netherlands, England, France, and Russia) began systematic ocean exploration, their discoveries were closely guarded secrets, far too valuable to be made public. Hence, we know little of what sailors knew about ocean currents during the period of great ocean exploration from the 15th to 18th centuries.

Existence of the Gulf Stream, a major ocean surface current, was known as early as 1519. Bishop Resen of Copenhagen drew the first map of the Gulf Stream in 1605. He based his map on records from the trans-Atlantic voyages of the English explorer Martin Frobisher (1535-1594) and a crude sketch attributed to Icelanders. Subsequent charts showing the Gulf Stream were published in 1678 (by the Jesuit Athanasius Kircher) and 1685 (by a German named Happelius).

More widely known is Benjamin Franklin's study of the Gulf Stream. Franklin (1706-1790) served as colonial deputy postmaster general from 1753 to 1774. He found that British merchant ships arriving in the colonies from England took many days to weeks longer to make the voyage than American vessels. British postal authorities asked Franklin for an explanation. Franklin consulted with

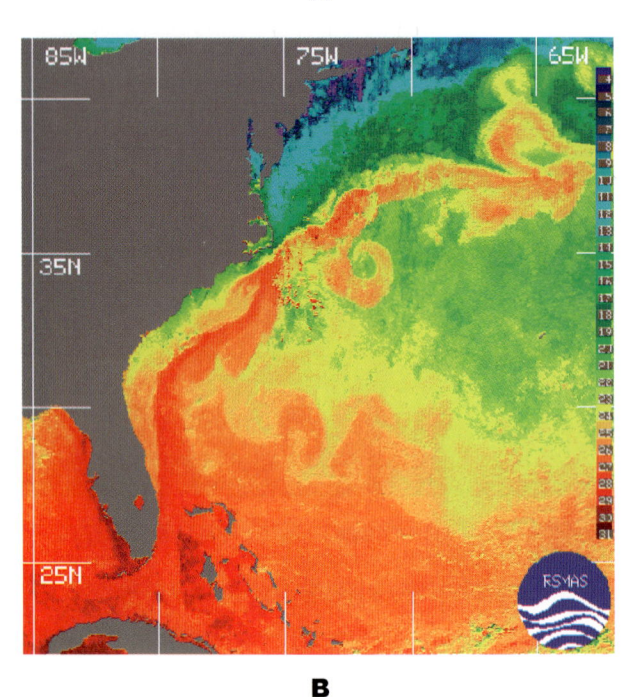

126 Chapter 6 OCEAN CURRENTS

sailed with the current but on the return voyage they would avoid it, cutting up to weeks off the journey. At the time, British captains were unaware of the Gulf Stream.

Franklin published his cousin's chart showing the location of the Gulf Stream and presented it to the British. Interestingly, the location of the Gulf Stream plotted on the 1769 Franklin-Folger map shown in Figure 6.1A is remarkably similar to the location of the Gulf Stream as revealed by satellite-derived sea-surface temperature patterns shown in Figure 6.1B. British authorities refused to take advantage of or even acknowledge the Franklin-Folger map. Franklin took measurements of sea-surface temperatures during his crossings of the Atlantic, thereby developing a navigation technique based on the location of the warm Gulf Stream waters.

In the mid 19th century, the American naval officer Matthew Fontaine Maury (1806-1873) conducted the first systematic study of the ocean's surface currents and winds (Figure 6.2). Maury compiled information on currents and winds from the logbooks of sailors' observations stored at the U.S. Navy's Depot of Charts and Instruments and published the first charts of the North Atlantic in 1847.

FIGURE 6.1
The location of the Gulf Stream (A) plotted on the 1769 Franklin-Folger map, and (B) on a much more recent infrared satellite depiction of sea-surface temperatures. Shades of red indicate the highest sea-surface temperatures. [Source: NOAA; University of Miami, Rosenstiel School of Marine & Atmospheric Science]

his cousin, Timothy Folger, a Nantucket, MA whaling ship captain, who told him about the Gulf Stream, the strongest surface current in the North Atlantic (often 10 to 12 km per hr or 6 to 7 mi per hr—roughly the speed of a modern sailing ship). On eastbound voyages, American captains

FIGURE 6.2
The American naval officer Matthew Fontaine Maury (1806-1873) compiled wind and ocean current data from thousands of ships' logs and produced the first reliable wind and current charts of the ocean. [Source: U.S. Navy]

He estimated current directions and speeds by analyzing deflections in ships' courses caused by surface ocean currents. Failure to correct a ship's course for current-induced deflections means that the ship's final position at the end of a run differs from its intended destination. Combining thousands of such observations, Maury constructed a map of average surface currents over much of the ocean. A skilled navigator with knowledge of currents can correct a ship's course to compensate for deflection by currents.

Studies of ocean currents continue today for many reasons, ranging from military operations to predicting oil spill trajectories (refer to this chapter's first Essay). Through the years, the instruments used to study currents have become more sophisticated. Satellite observations and data from instrumented platforms, ships and buoys have replaced analysis of ships' logs. Civilian Global Positioning Systems (GPS) now measure a ship's location to within 5-10 m (16-33 ft) so that the deflection caused by currents can be determined accurately and instantaneously. For more on GPS, see this chapter's second Essay.

Driving Question:

What causes the ocean to circulate and what are the patterns of ocean circulation?

Energy and matter (e.g., heat, water) are continually exchanged between the ocean and atmosphere, and these processes drive the ocean circulation. Evaporation, precipitation, plus heating and cooling bring about changes in the temperature and salinity of surface waters. Density changes that accompany changes in temperature and salinity can cause water to sink or rise in the ocean. *Kinetic energy* (energy of motion) is transferred from near-surface winds to the ocean's surface layer, driving the currents that dominate the motion of the upper few hundred meters of the ocean. Winds are responsible for not only horizontal currents but also vertical water motions within the surface layer.

Most ocean water (90%) is in the deep ocean, isolated from the atmosphere and its winds. Deep-ocean waters are cold and dark and come to the surface primarily at high latitudes where they interact with the atmosphere. Differences in water density drive the sluggish circulation of deep water. Typically, these waters flow at speeds of less than 1 cm per sec or 1 km per day—about 240 times slower than the Gulf Stream. Because of its volume and isolation from the atmosphere, the deep ocean is both a storehouse for heat (acting as a global buffer for temperature change) and a reservoir for gases such as carbon dioxide that remain there for centuries to millennia.

The ocean features two different circulation patterns: wind-driven surface-ocean currents and the deep ocean's slower density-driven thermohaline circulation. In this chapter, we examine the characteristics of these two circulation regimes, the governing forces, and the global-scale oceanic conveyor belt that links the two regimes. We begin by describing the vertical structure of the ocean.

Ocean's Vertical Structure

Except at high latitudes, the ocean is divided into three horizontal depth zones based on density: the mixed layer, pycnocline, and deep layer (Figure 6.3). At high latitudes, the pycnocline and mixed layer are absent.

Wind-driven surface currents are restricted to the ocean's uppermost 100 m (300 ft.) or less. The strongest currents occur in the ocean's surface layer, although some surface currents such as boundary currents like the Gulf Stream (discussed later) can be relatively strong to depths of several hundred meters. Surface currents are changeable, continually responding to variations in the wind, precipitation, and heating or cooling. Stirring of surface waters by the wind produces a well-mixed layer of uniform or nearly uniform density. For this reason, the surface ocean is called the **mixed layer**. We know most about the mixed layer because ships, aircraft, and Earth-orbiting satellites can readily monitor it.

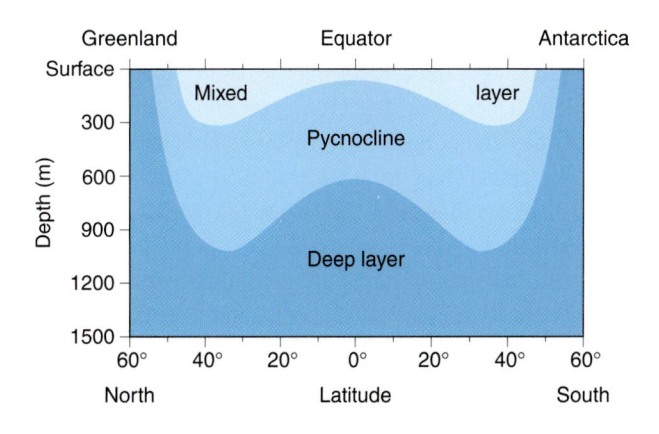

FIGURE 6.3
A cross-sectional longitudinal profile of the Atlantic Ocean from 60 degrees N to 60 degrees S showing the location of the mixed layer, pycnocline, and deep layer. Note that the ocean (and deep layer) extend to depths of 4000 to 6000 m.

The **pycnocline**, situated between the mixed layer and the deep layer, is where water density increases rapidly with depth because of changes in temperature and/or salinity. Recall that cold water is denser than warm water and salty water is denser than fresh water. Where a decline in temperature with depth is responsible for the increase in density with depth, the pycnocline is also a **thermocline**. On the other hand, if an increase in salinity is responsible for the increase in density with depth, the pycnocline is also a **halocline**. Typically, the pycnocline extends to a depth of 500 to 1000 m (1600 to 3300 ft). (However, in middle latitudes seasonal pycnoclines may develop within the mixed layer.) The dark, cold **deep layer** below the pycnocline accounts for most of the ocean's mass. Within the deep layer, density increases gradually with depth and water moves sluggishly; in only a few locations (usually near the bottom) are water movements fast enough to be considered currents.

The ocean's three-layer structure is an example of how gravity separates a fluid into layers such that the density of each layer is less than the density of the layer below it. More dense fluids sink and less dense fluids rise. The ocean's pycnocline is very stable thus suppressing mixing between the mixed layer and deep layer; that is, the pycnocline acts as a barrier to vertical motion within the ocean. The concept of stability is useful in understanding this property of the pycnocline.

Stability as used here refers to vertical motions of ocean water. A system is described as *stable* if it tends to persist in its original state without changing. Following a disturbance (i.e., vertical motion), a **stable system** returns to its initial state or condition. As noted above, the usual stable state of the ocean features a layer of water that is warmest near its interface with the atmosphere (the *mixed layer*) and the mixed layer overlies water that becomes denser with increasing depth (the *pycnocline*). Strong storm winds may temporarily disturb this stable stratification bringing colder than usual water to the surface. Once the wind slackens, however, the original layered structure is soon restored.

To demonstrate the effects of differences in water density on the circulation of ocean water, you can conduct a simple experiment. You will need two glasses or glass beakers, food coloring, salt, and a medicine dropper. Fill one glass with fresh water and the other with salt water. Add several drops of food coloring to the salt water and mix thoroughly. Let the glasses of water stand for a few minutes. Use a medicine dropper to put a drop of colored salt water into the glass of fresh water. You will observe that the denser salt water sinks to the bottom of the fresh water. Now empty and wash the two glasses. Again fill one glass with fresh water and the other glass with salt water. Add several drops of food coloring to the fresh water sample and mix thoroughly. Again, let the glasses of water stand for a few minutes. Use the medicine dropper to put a drop of the colored fresh water into the glass of salt water. You will observe that the fresh water remains at the surface because it is less dense than the salt water. Carefully add more fresh water with the dropper to form a two-layered system.

Suppose we add a drop of slightly salty water of intermediate density (using a different food coloring) to our two-layered system. The drop would come to rest at the boundary between the two layers due to the differences in density between the water layers and the drop. The density distribution is stable when the denser water (in this case the saltiest) is at the bottom and the less dense water is on top. Immediately after adding a drop of more dense salt water, we created an unstable density distribution. Since the drop is denser than the surrounding water, it sinks. However, if a drop of slightly salty water is carefully injected into the very salty water, it rises; this is an example of an *unstable* density distribution because the water drop is less dense than the surrounding waters and is buoyed upward. Hence, an **unstable system** will spontaneously shift toward a more stable density distribution.

Mixing the system so that water density is uniform throughout produces a *neutrally stable* density distribution. In fact, if the original two-layered system stands long enough, salt dissolved in water diffuses and water density eventually becomes uniform throughout the container. Following a disturbance, a **neutrally stable system** does not return to its initial state and is easily mixed.

Ocean in Motion: The Forces

Once the wind sets surface waters in motion as a current, the Coriolis effect, Ekman transport, and the configuration of the ocean basin modify the speed and direction of the current. In this section, we consider the forces involved in the coupling of wind and ocean surface waters.

WIND-DRIVEN CURRENTS AND EKMAN TRANSPORT
The wind blows across the ocean and moves its waters as a result of its frictional drag on the surface. Ripples or waves cause the surface roughness necessary

for the wind to couple with surface waters. A wind blowing steadily over deep water for 12 hrs at an average speed of about 100 cm per sec (2.2 mi per hr) would produce a 2 cm per sec current (about 2% of the wind speed).

If Earth did not rotate, frictional coupling between moving air and the ocean surface would push a thin layer of water in the same direction as the wind. This surface layer in turn would drag the layer beneath it, putting it into motion. This interaction would propagate downward through successive ocean layers, like cards in a deck, each moving forward at a slower speed than the layer above. However, because Earth rotates, the shallow layer of surface water set in motion by the wind is deflected to the right of the wind direction in the Northern Hemisphere and to the left of the wind direction in the Southern Hemisphere. We discussed this *Coriolis effect* in Chapter 5. Except at the equator, where the Coriolis effect is zero, each layer of water put into motion by the layer above shifts direction because of Earth's rotation.

Using vectors to plot the direction and speed of water layers at successive depths, we can show a simplified three-dimensional current pattern caused by a steady horizontal wind (Figure 6.4A). (A *vector* is an arrow representing a physical quantity so that length is directly proportional to magnitude and orientation represents direction.) This model is known as the **Ekman spiral**, named for the Swedish physicist V. Walfrid Ekman (1874-1954) who first described it mathematically in 1905. Ekman based his model on observations made by the Norwegian explorer Fridtjof Nansen (1861-1930). Nansen was interested in ocean currents in polar seas. In 1893, he froze his 39-m (128-ft) wooden ship, the *Fram*, into Arctic pack ice about 1100 km (680 mi) south of the North Pole. His goal was to drift with the ice and cross the North Pole thereby determining how ocean currents affect the movement of pack ice. The *Fram* remained locked in pack ice for 35 months but only came within 394 km (244 mi) of the North Pole. As the *Fram* slowly drifted with the ice, Nansen noticed that the direction of ice and ship movement was consistently 20 to 40 degrees to the right of the prevailing wind direction.

The Ekman spiral indicates that each moving layer is deflected to the right of the overlying layer's movement; hence, the direction of water movement changes with increasing depth. In an ideal case, a steady wind blowing across an ocean of unlimited depth and extent causes surface waters to move at an angle of 45 degrees to the right of the wind in the Northern Hemisphere (45 degrees to the left in the Southern Hemisphere). Each successive layer moves more toward the right and at a slower speed. At a depth of about 100 to 150 m (330 to 500 ft), the Ekman spiral has gone through less than half a turn. Yet water moves so slowly (about 4% of the surface current) in a direction opposite that of the wind that this depth is considered to be the lower limit of the wind's influence on ocean movement.

In the Northern Hemisphere, the Ekman spiral predicts net water movement through a depth of about 100 to 150 m (330 to 500 ft) at 90 degrees to the *right* of the wind direction (Figure 6.4B). That is, if one adds up all the vectors in Figure 6.4A, the resulting flow is at 90 degrees to the right of the wind direction. In the Southern Hemisphere, the net water movement is 90 degrees to the *left* of

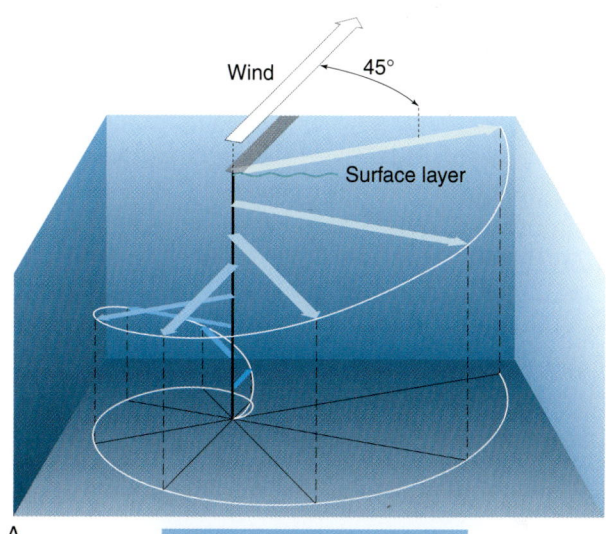

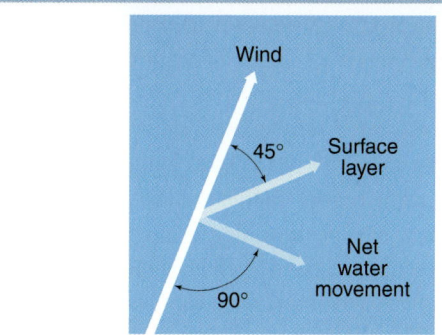

FIGURE 6.4
The Ekman spiral describes how the horizontal wind sets surface waters in motion. (A) As represented by horizontal vectors, the speed and direction of water motion change with increasing depth. (B) Viewed from above in the Northern Hemisphere, the surface layer of water moves at 45 degrees to the right of the wind. The net transport of water through the entire wind-driven column (Ekman transport) is 90 degrees to the right of the wind.

the wind direction. This net transport of water due to coupling between wind and surface waters is known as **Ekman transport**.

Because the real ocean does not match the idealized conditions of the Ekman spiral, wind-induced water movements often differ appreciably from theoretical predictions. In shallow water, for example, the water depth is insufficient for the full spiral to develop so that the angle between the horizontal wind direction and surface-water movements can be as little as 15 degrees. As waters deepen, the angle increases and approaches 45 degrees. The stable pycnocline inhibits the transfer of kinetic energy to deeper waters, restricting wind-driven currents to the mixed layer; that is, the pycnocline acts as a floor for Ekman transport and surface currents.

Ekman transport piles up surface water in some areas of the ocean and removes water from other areas, producing variations in the height of the sea surface, causing it to slope gradually. One consequence of a sloping ocean surface is the generation of horizontal differences (*gradients*) in water pressure. These pressure gradients, in turn, give rise to geostrophic flow.

GEOSTROPHIC FLOW

To a large extent, horizontal movement of ocean surface waters mirrors the long-term average planetary circulation of the atmosphere. As shown in Figure 5.19, three surface wind belts encircle each hemisphere: trade winds (equator to 30 degrees latitude), westerlies (30 to 60 degrees), and polar easterlies (60 to 90 degrees). The westerlies of middle latitudes and the trade winds of the tropics drive the most prominent features of ocean surface motion, large-scale nearly circular current systems known as **gyres**. Subtropical gyres are centered near 30 degrees latitude in the North and South Atlantic, the North and South Pacific, and the Indian Ocean. Gyres in the Northern and Southern Hemispheres are similar except that they rotate in opposite directions because the Coriolis deflection acts in opposite directions in the two hemispheres. Viewed from above, subtropical gyres rotate in a clockwise direction in the Northern Hemisphere but in a counterclockwise direction in the Southern Hemisphere.

Driven by the long-term average winds in the subtropical highs, Ekman transport causes surface waters to move from all sides toward the central region of a subtropical gyre. This transport produces a broad mounding of water as high as 1 m (3 ft) above mean sea level near the center of the gyre (Figure 6.5). As more water is transported toward the center of the gyre, the surface slope of

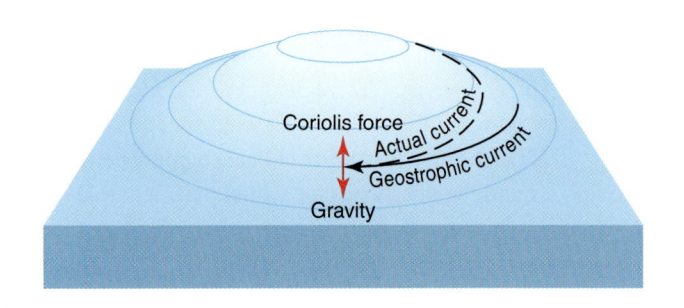

FIGURE 6.5
Ekman transport causes surface waters to move toward the central region of a subtropical gyre from all sides, producing a broad mound of water. Surface water begins flowing downhill. A balance develops between the Coriolis force and the force arising from the horizontal water pressure gradient such that surface currents flow parallel to the contours of elevation of sea level. This current is known as geostrophic flow.

the mound becomes steeper. At the same time, the horizontal water pressure gradient produced under the sloping sea surface increases. In response to the horizontal gradient in water pressure, water moves from where the pressure is higher toward where the pressure is lower, that is, downhill. Surface water parcels flow outward and down slope from the center of the gyre. The Coriolis effect causes these parcels to shift direction to the right in the Northern Hemisphere (to the left in the Southern Hemisphere). Eventually, the outward-directed pressure gradient force balances the Coriolis force and the water parcels flow around the gyre and parallel to contours of elevation of sea level.

The horizontal movement of surface water arising from a balance between the pressure gradient force and the Coriolis force is known as **geostrophic flow**. As noted earlier, viewed from above, geostrophic flow in a subtropical gyre is clockwise in the Northern Hemisphere and counterclockwise in the Southern Hemisphere.

Wind-Driven Surface Currents

In this section, we examine the characteristics of wind-driven surface-ocean currents, components of the huge gyres that dominate the central regions of the open ocean. Western boundary currents are the strongest segments of these gyres and often spawn warm- and cold-core eddies known as rings. In coastal and equatorial regions, coupling of wind and surface waters can cause upwelling and downwelling.

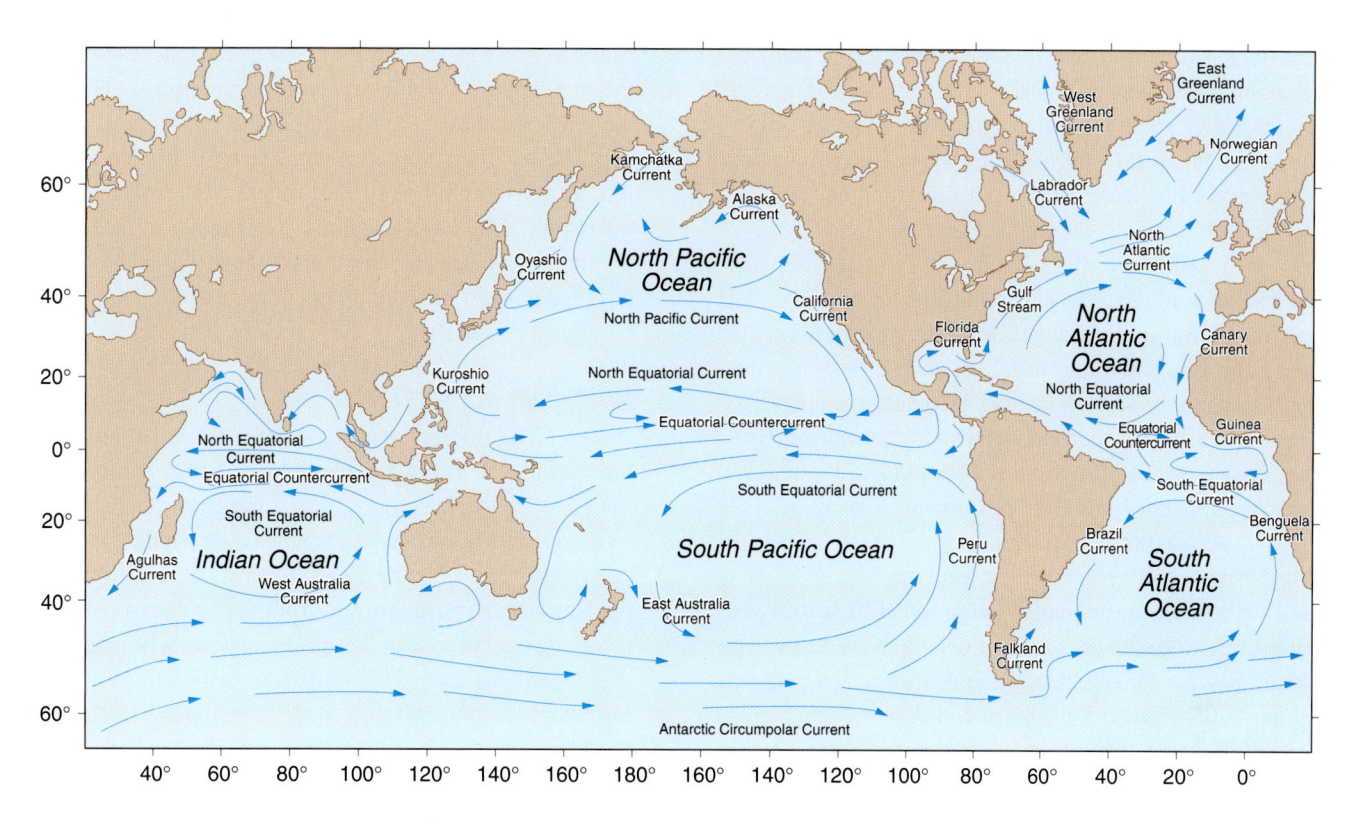

FIGURE 6.6
The long-term average pattern of ocean-surface currents.

GYRES

The long-term average pattern of ocean surface currents is plotted in Figure 6.6. Some currents are relatively warm whereas others are cold. Winds associated with a passing storm system can disturb the ocean surface and cause the actual flow of ocean currents locally to deviate temporarily from long-term average patterns.

Surface currents within gyres vary considerably in strength, width, and depth. The northeastward flowing Gulf Stream of the northwestern Atlantic and the Kuroshio Current of the northwestern Pacific are the swiftest surface currents with velocities averaging 3 to 4 km per hr (1.8 to 2.5 mi per hr). Those currents are also relatively deep and narrow, usually measuring no more than 50 to 75 km (30 to 45 mi) across. On the eastern arms of these gyres, the southward flowing Canary and California Currents, respectively, are hundreds of kilometers wide and rarely flow at more than 1 km (0.6 mi) per hr.

The westward flowing South Equatorial Current links the two subtropical gyres of the Atlantic Ocean. The eastward projection of Brazil splits the South Equatorial Current into two segments. The segment flowing southward forms the western arm of the South Atlantic gyre (the Brazil Current, a western boundary current). The segment flowing northward merges with the North Equatorial Current, which then splits into two currents that rejoin as they exit the Gulf of Mexico between Florida and Cuba to become the Florida Current. This current becomes the Gulf Stream that flows northeasterly and passes Cape Hatteras, NC. In that region, the current speed may be as great as 9 km (5.5 mi) per hr. Near Chesapeake Bay, the amount of water transported in the Gulf Stream exceeds 90 million cubic m per sec; the volume of water transported falls to about 40 million cubic m per sec by the time the current reaches southern Newfoundland. (For comparison purposes, 90 million cubic m per sec is equivalent to about 4500 times the discharge of the Mississippi River—enough to fill the Lake Superior basin in about 1.5 days.)

The North Atlantic subtropical gyre includes the Gulf Stream which becomes the North Atlantic Current at about 40 degrees N and 45 degrees W and flows easterly across the North Atlantic. The cold waters of the Labrador Current flow southeastward between Canada and Greenland while the East Greenland Current flows southwestward between Greenland and Iceland. Farther east, the North Atlantic Current splits into the Norwegian Current (which flows northeasterly between Iceland and Europe along the coast of Norway) and the Canary Current (which flows southward along the west coast of Spain, Portugal, and North Africa). The Canary Current merges with the North Atlantic Equatorial Current, thus completing the North Atlantic subtropical gyre.

Like their Northern Hemisphere counterparts, currents in the South Pacific and South Atlantic are narrowest and flow most rapidly along their western margins but are broad and sluggish along their eastern margins. The Indian Ocean gyre varies more than the others in response to seasonal reversal of monsoon winds. In the high-sun season, surface winds blow from sea to land and in the low-sun season, winds blow from land to sea.

Above the central regions of the ocean basins are broad expanses of generally light winds or calm air associated with the semi-permanent subtropical high-pressure systems centered near 30 degrees latitude (Chapter 5). These massive fair-weather systems are semi-permanent in that they persist throughout the year but undergo seasonal shifts in relative strength and location (following the sun). Persistent fair weather and high temperatures enhance the rate of evaporation in these regions of the subtropical ocean resulting in surface seawater with a salinity significantly higher than average (Figure 3.8). An example is the vast Sargasso Sea which lies under the Bermuda-Azores subtropical high in the North Atlantic and features an average surface water salinity of 36.5 to 37.0 (Chapter 10).

In addition to the South Pacific and South Atlantic Ocean gyres, prevailing winds generate the Antarctic Circumpolar Current. At about 60 degrees S, this current also makes up, at least in part, the southern edges of the gyres in the Atlantic, Pacific, and Indian Oceans. This easterly-flowing current encircles the Antarctic continent rather than rotating as a basin-centered gyre and features the ocean's greatest water flow. Such a globe-circling current is possible only in the Southern Ocean; elsewhere, continents interrupt east-west currents. The Drake Passage, the relatively narrow strait between Cape Horn (the southern tip of South America) and the Antarctic Peninsula, deflects some waters from the Antarctic Circumpolar Current to form a portion of the Peru Current that flows northward along the west coast of South America.

Sub-polar gyres, smaller than their subtropical counterparts, occur at high latitudes of the Northern Hemisphere; they are the Alaska gyre in the far North Pacific and the gyre south of Greenland in the far North Atlantic. The counterclockwise surface winds in the Aleutian and Icelandic sub-polar lows drive the sub-polar gyres. (The *Aleutian low* and *Icelandic low* are persistent features of the atmosphere's planetary-scale circulation.) Hence, viewed from above, the rotation in the sub-polar gyres is opposite that of the Northern Hemisphere subtropical gyres. Ekman transport moves surface waters away from the central region of the sub-polar gyres. The thinner surface layer permits more nutrient-rich waters from deeper in the ocean to move upward into the photic zone, thereby increasing biological productivity in these regions (Chapter 9).

EQUATORIAL CURRENTS

The tropical ocean encompasses broad areas of the Atlantic, Pacific, and Indian Ocean basins and is closely linked to the tropical atmosphere. The prevailing surface winds over the tropical ocean are the **trade winds** that blow persistently from the northeast (toward the southwest) in the Northern Hemisphere and from the southeast (toward the northwest) in the Southern Hemisphere. The name for these winds was coined by sea captains who sailed for trading companies and took advantage of their persistent speed and direction when crossing the ocean. Trade winds drive both North and South Equatorial Currents westward, thus transporting warm ocean-surface waters in that direction. Equatorial Counter Currents and Equatorial Under Currents return some warm waters eastward. Counter Currents flow along the surface whereas Under Currents flow at greater depths below the surface.

The trade winds of the two hemispheres converge in a narrow east-west zone located near the equator known as the *Intertropical Convergence Zone (ITCZ)*. The ITCZ is an important component of the planetary-scale atmospheric circulation that is particularly well defined over the tropical ocean. Warm and humid air ascending in the ITCZ gives rise to clusters of showers and thunderstorms that produce locally heavy rainfall. For mariners this region of the tropical ocean is known as the *doldrums*, feared by the captains of sailing ships because of light and variable winds. Seasonally, the ITCZ moves with the sun, shifting northward during the Northern Hemisphere spring and southward during the Northern Hemisphere autumn but generally remaining north of the equator, especially over the Atlantic Ocean. Consequently, the eastward-flowing Equatorial Counter Current, separating the surface current systems of the two hemispheres, also lies mostly just north of the equator.

The South Equatorial Current crosses the equator in the Atlantic and to a lesser extent in the Pacific. In this way, it transports surface waters and heat into the Northern Hemisphere. The return flow is through subsurface currents (discussed later in this chapter). The cape at the easternmost point of South America diverts part of the flow of the South Equatorial Current into the southward flowing Brazil Current. The remainder continues northwestward along South America's northeast coast into the Caribbean Sea.

The islands of Indonesia mark the boundary between the Indian and Pacific Oceans but do not completely block the flow of seawater between the two ocean basins. Although details are still lacking, it is clear that warm, low-salinity waters from the Pacific are transported into the Indian Ocean's South Equatorial Current. These waters flow through the many passages between the thousands of Indonesian islands and replenish the large amounts of water removed by evaporation from the northern Indian Ocean. The summer Asian monsoon circulation transports this water vapor over India and Southeast Asia where it falls as torrential rains. After flowing westward across the Indian Ocean, these waters enter the South Atlantic via the Agulhas Current flowing around southern Africa. The Indonesian islands partially block the inter-ocean flow, which leads to an accumulation of warm surface waters in the western equatorial Pacific Ocean that is linked to El Niño/La Niña (Chapter 11).

WESTERN BOUNDARY CURRENTS

As noted earlier in this chapter, surface currents on the western side of the subtropical gyres, so-called **western boundary currents**, are faster than their eastern counterparts (Figure 6.7). In fact, they are among the fastest surface currents in the ocean. One reason for the westward intensification of boundary currents has to do with the strengthening of the Coriolis effect with latitude. The Coriolis effect is stronger in the latitudes of the westerlies than in the latitudes of the trade winds. Transport of surface waters toward the western boundary of the ocean basins causes the ocean-surface slope to be steeper on the western side (versus eastern side) of a gyre (in either hemisphere). A steeper ocean-surface slope translates into a faster geostrophic flow on that side of the gyre.

Western boundary currents flow toward the poles, northward in the Northern Hemisphere and southward in the Southern Hemisphere. Waters moving in these currents transport large quantities of heat poleward from the tropics (Chapter 5). Among these boundary currents are the Gulf Stream and the Kuroshio (off Japan), which rank next in volume of water flow to the Antarctic Circumpolar Current. The Southern Hemisphere's western boundary currents include the Agulhas Current (off southeast Africa), the Brazil Current, and the East Australia Current. These currents are weaker than those in the northern ocean basins, in part because western boundary currents require extended land barriers that are generally absent in the Southern Hemisphere. Flows in the major western boundary currents are 50 to 100 times the total water discharged by all the world's rivers.

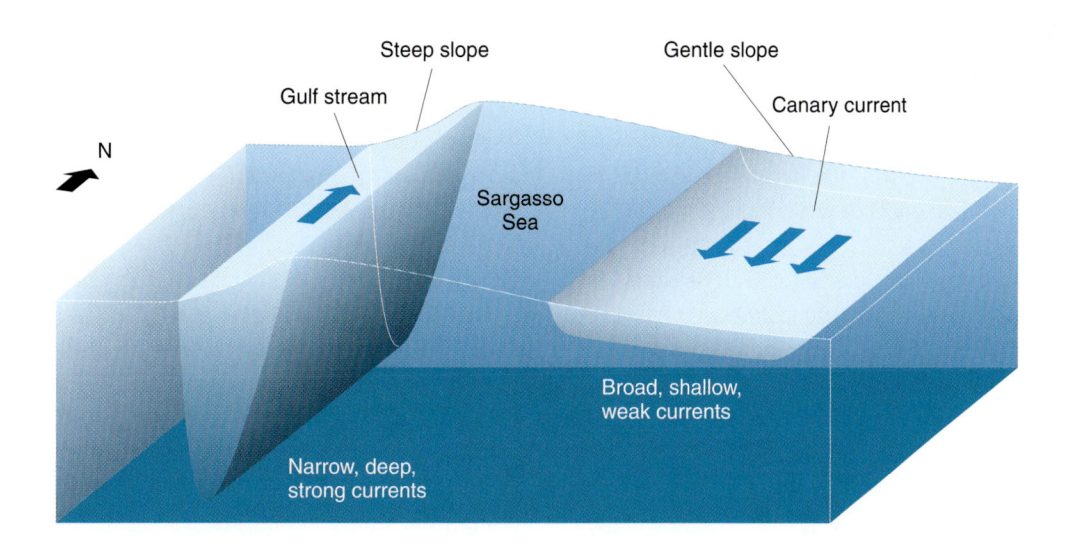

FIGURE 6.7
As shown here for the North Atlantic but also true for all ocean basins, the ocean surface slope is steeper on the western side of an ocean basin than on the eastern side. Consequently, surface ocean currents are stronger, narrower, and deeper in all western boundary regions. Note that the vertical scale is greatly exaggerated.

Waters in western boundary currents typically move 40 to 120 km (25 and 75 mi) per day. These currents also extend much deeper than most other surface currents, down to a depth of 1000 m (3300 ft) or more. Thus, the strong western boundary currents are so deep that they are deflected by the continental margins, which prevent these currents from flowing onto the shallow continental shelves. Western boundary currents therefore separate warm tropical open-ocean waters from cooler coastal waters, somewhat like a jet stream in the atmosphere that separates warm and cold air masses. Eastern boundary currents, such as the California Current and the Canary Current, are slower, shallower, and wider than the western boundary currents. Similar to the return flow in a household heating system, these currents transport colder waters into the tropics where they are heated and transported poleward in the western boundary currents.

RINGS

Instruments on ocean-observing satellites show the ocean surface to be much more dynamic than is indicated by maps and charts that portray ocean conditions averaged over decades. Satellite images show that currents change in response to variations in oceanic and atmospheric conditions. For example, relatively swift western boundary currents occasionally spawn large turbulent rotating warm-core and cold-core eddies, also known as **rings**. A ring forms when a meander in a boundary current (or the Antarctic Circumpolar Current) becomes a loop that pinches off (separates) from the main current and moves independently as an eddy. Currents bordering a ring can rotate at more than 1.0 knot (1.0 nautical mi per hr or 1.15 mi per hr or 1.86 km per hr), essentially isolating waters and organisms in rings from surrounding waters. Rings extend to some depth in the ocean and should be thought of as cylindrical pools of water rather than simply surface features.

Rings form on both sides of the Gulf Stream. On the north side, rings are typically 100 to 200 km (60 to 120 mi) across. These rings enclose warm waters from the Sargasso Sea located to the south and east of the Gulf Stream; thus, they are called **warm-core rings**. Viewed from above, these warm-core rings rotate in a clockwise direction. This circulation results in Ekman transport that piles up warm surface water toward the center of the ring. Because of the strong contrast in sea-surface temperatures, they are readily detected on infrared satellite images. Warm-core rings are also readily distinguished from the surrounding surface waters by their relatively low levels of biological production (Figure 6.8). The pool of water in a warm-core ring can extend to a depth of 1500 m (4900 ft) so that they cannot move onto continental shelves, which are shallow—typically 200 m (650 ft) deep or less. However, rings can come close enough to the shelf edge to modify coastal currents and bring unusual organisms onto the shelf. Occasionally boaters and fishers in normally cool coastal waters encounter organisms (e.g., sea turtles, tropical fish) that live in much warmer water having been transported in a warm core ring that spun off the Gulf Stream.

Rings that spin off the south side of the Gulf Stream entrain relatively cold and productive coastal waters and are called **cold-core rings**. These rings have diameters of about 300 km (185 mi) and, viewed from above, rotate in a counterclockwise direction. This circulation results in Ekman transport away from the center of the ring causing upwelling of nutrient-rich cold water from depth. Cold-core rings are more difficult for satellites to track because their originally cool surface water is warmed by absorption of solar radiation so that they become almost indistinguishable thermally from surrounding surface waters. However, cold water persists below the surface, sometimes extending down to the ocean floor at depths of more than 4000 m (13,000 ft), and can be detected in vertical profiles of temperature and salinity obtained by instrumented probes. Cold-core rings usually contain more nutrients and marine organisms than the biologically barren Sargasso Sea waters that surround them. Thus, they can also be identified in the subsurface by their unusually abundant marine life.

Rings move slowly (5 to 6 km or 3 to 4 mi per day), drifting southwestward in the weaker currents on either side of the northeast flowing Gulf Stream. The proximity of the Gulf Stream to the coast limits the southward movement of warm-core rings. Typically, after a few months to a year, a warm-core ring becomes caught between Cape Hatteras, NC and the Gulf Stream; the ring is then reabsorbed back into the Gulf Stream. Cold-core rings are not as restricted in their movements as warm-core rings and may persist for several years; on average, individual cold-core rings last for one and one-half years. Some of the ocean's largest warm-core rings, with diameters up to 400 km (250 mi), form in the Gulf of Mexico. These rings spin off the Loop Current at highly irregular intervals ranging from several months to 1.5 years. The Loop Current enters the Gulf from the Caribbean by flowing through the Yucatán Strait between Cuba and Mexico, heads northwestward in the general direction of Louisiana, then makes a clockwise turn, and exits the Gulf through the Straits of

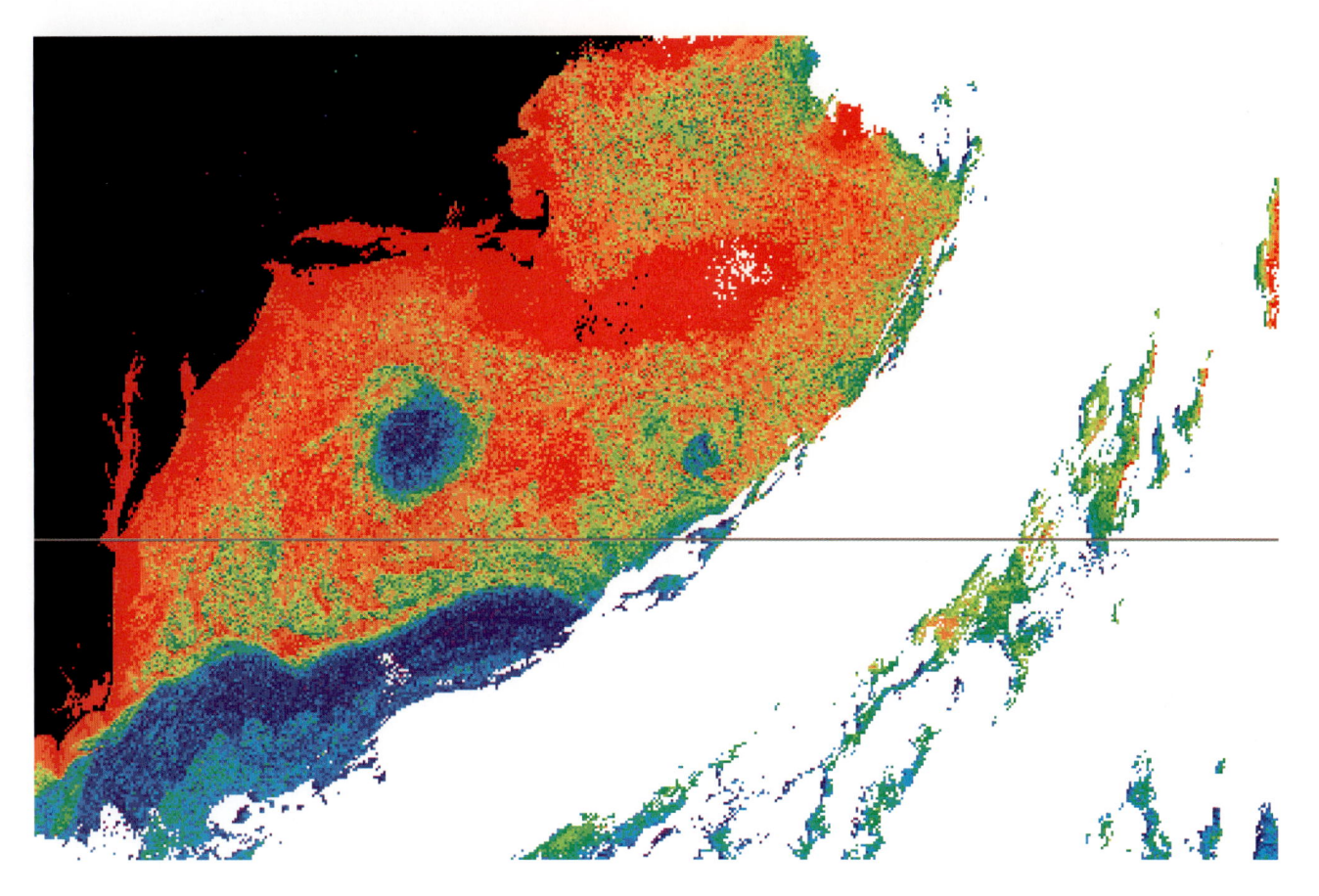

FIGURE 6.8
In this NASA Nimbus-7 CZCS (Coastal Zone Color Scanner) satellite image of ocean biological production, a prominent warm-core ring shows up as a nearly circular blue/violet region in ocean waters east of Delaware. Relatively low levels of phytoplankton pigment are indicated by blue/violet and relatively high levels are indicated by orange/red. [Source: NASA]

Florida (between Florida and Cuba). Rings drift westward across the Gulf at 2 to 5 km (1.2 to 3.1 mi) per day. Bordering currents of up to 4 knots (4.6 mi per hr) can play havoc with offshore oil platform operations, damaging equipment and increasing the risk of accidents.

The Agulhas Current, the Indian Ocean's western boundary current, also spawns rings. This relatively fast southward-flowing current averages about 7 km (4 mi) per hr. As it reaches Africa's southern tip, part of the current is caught up in the eastward flow around Antarctica and abruptly shifts direction back into the Indian Ocean. Some of the flow continues around South Africa as narrow (50 km or 30 mi wide) filaments that cool rapidly and mix with the surrounding waters in the large upwelling zone off Africa's Namibia coast. The Agulhas Current periodically sheds rings about 320 km (200 mi) across from its westernmost end. Rings in the Southern Hemisphere rotate in the opposite direction of those in the Northern Hemisphere. The Agulhas Current's warm-core rings rotate counterclockwise and contain Indian Ocean waters that are about 5 Celsius degrees (9 Fahrenheit degrees) warmer than nearby South Atlantic surface waters. They retain their identity as they move into the South Atlantic and transport heat, salt and organisms from the Indian Ocean into the South Atlantic. Over a two-year period in the mid-1990s, 14 of these rings formed.

UPWELLING AND DOWNWELLING

In some coastal areas of the ocean (and large lakes such as the North American Great Lakes), the combination of persistent winds, Earth's rotation (the Coriolis effect), and restrictions on lateral movements of water caused by shorelines and shallow bottoms induces upward and downward water movements. As explained above, the Coriolis effect plus the frictional coupling of wind and water (Ekman transport) cause net movement of surface water at about 90 degrees to the right of the wind direction in the Northern Hemisphere and to the left of the wind direction

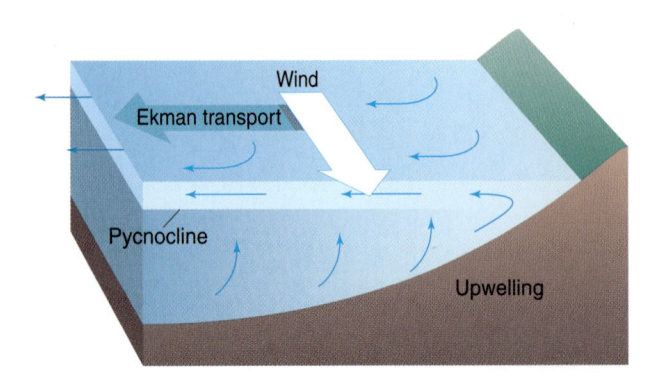

136 Chapter 6 OCEAN CURRENTS

FIGURE 6.9
Where Ekman transport moves surface waters away from the coast, surface waters are replaced by water that wells up from below in the process known as upwelling. This example is from the Northern Hemisphere.

in the Southern Hemisphere. Coastal **upwelling** occurs where Ekman transport moves surface waters away from the coast; surface waters are replaced by water that wells up from below (Figure 6.9). Where Ekman transport moves surface waters toward the coast, the water piles up and sinks in the process known as coastal **downwelling** (Figure 6.10). Upwelling and downwelling illustrate *mass continuity* in the ocean; that is, water is a continuous fluid so that a change in distribution of water in one area is accompanied by a compensating change in water distribution in another area.

Upwelling is most common along the west coast of continents (eastern sides of ocean basins). In the Northern Hemisphere, upwelling occurs along west coasts (e.g., coasts of California, Northwest Africa) when winds blow from the north (causing Ekman transport of surface water away from the shore). Winds blowing from the south cause upwelling along continents' eastern coasts in the Northern Hemisphere, although it is not as noticeable because of the western boundary currents. Upwelling also occurs along the west coasts in the Southern Hemisphere (e.g., coasts of Chile, Peru, and southwest Africa) when the wind direction is from the south because the net transport of surface water is westward away from the shoreline. Winds blowing from the north cause upwelling along the continents' eastern coasts in the Southern Hemisphere.

Upwelling and downwelling also occur in the open ocean where winds cause surface waters to diverge (move away) from a region (causing upwelling) or to converge toward some region (causing downwelling). For example, upwelling takes place along much of the equator (Figure 6.11). Recall that the deflection due to the Coriolis effect reverses direction on either side of the equator. Hence, westward-flowing, wind-driven surface currents near the equator turn northward on the north side of the equator and southward on the south side. Surface waters are moved away from the equator and replaced by upwelling waters.

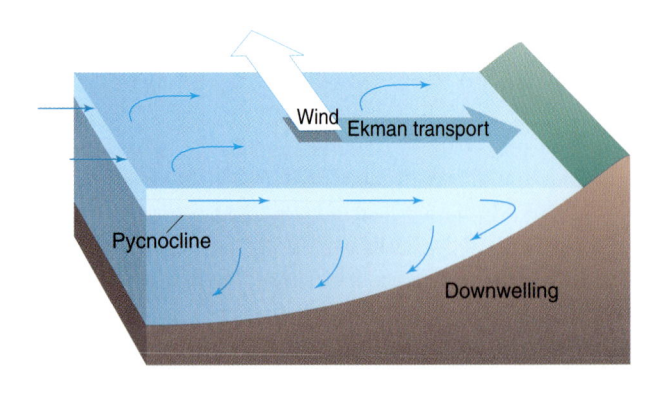

FIGURE 6.10
Where Ekman transport moves surface waters toward the coast, the water piles up and sinks in the process known as downwelling. This example is from the Northern Hemisphere.

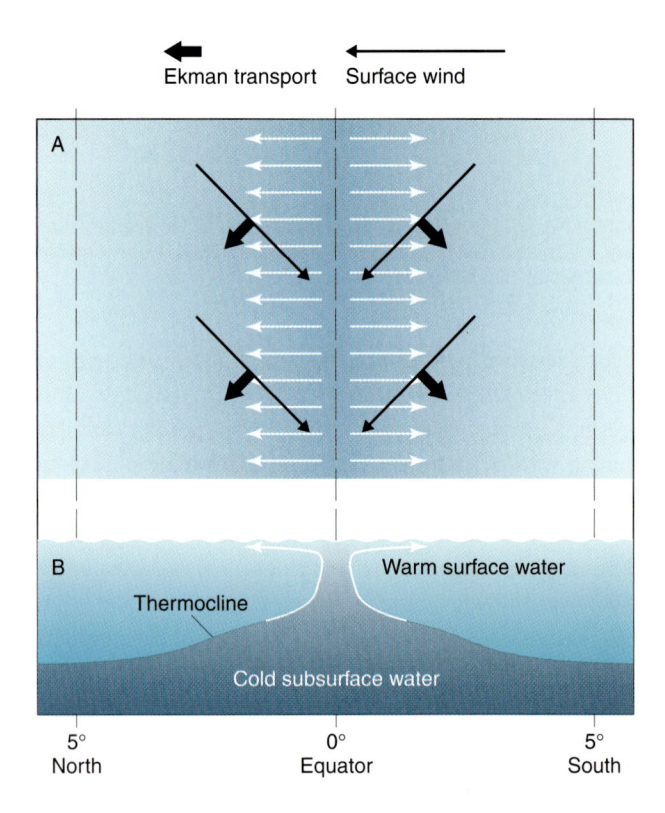

FIGURE 6.11
Equatorial upwelling. (A) In this plan view of the ocean from 5 degrees S to 5 degrees N, the trade winds of the two hemispheres are shown to converge near the equator. The consequent Ekman transport away from the equator gives rise to upwelling as shown in (B) a vertical cross section from 5 degrees S to 5 degrees N.

Upwelling and downwelling influence sea-surface temperature and biological productivity (Chapter 9). Upwelling waters originate below the pycnocline and are usually colder than the surface waters they replace. You may have experienced this phenomenon at the beach on a windy day when the warm surface water was blown offshore and replaced by chilly water from below. Coastal upwelling also transports waters rich in dissolved nutrients (nitrogen and phosphate compounds) from the ocean depths into the photic zone where sunlight penetrating the water supports phytoplankton growth. The world's most productive fisheries are located in areas of coastal upwelling (especially in the eastern boundary regions of the subtropical gyres); about half the world's total fish catch comes from upwelling zones (Chapter 10). On the other hand, in zones of coastal downwelling, the surface layer of warm, nutrient-deficient water thickens as water sinks. Downwelling reduces biological productivity and transports heat, dissolved materials, and surface waters rich in dissolved oxygen to greater depths. This occurs along the west coast of Alaska in the eastern boundary region of the Gulf of Alaska gyre (driven by winds in the Aleutian low).

Alternate weakening and strengthening of upwelling off the coast of Ecuador and Peru are associated with El Niño and La Niña episodes in the tropical Pacific (Chapter 11). During an El Niño event, upwelling wanes, and cold nutrient-rich water remains so deep that weak upwelling can bring only warm, nutrient-poor water into the photic zone. In extreme cases, nutrient-deficient waters coupled with over-fishing cause fisheries to collapse bringing about severe economic impacts.

Coastal upwelling and downwelling also influence weather and climate. Along the northern and central California coast, upwelling lowers sea surface temperatures and increases the frequency of summer fogs. Relatively cold surface waters chill the overlying humid marine air to saturation so that thick fog develops. Upwelling cold water inhibits formation of tropical cyclones (e.g., hurricanes). As we will see in Chapter 8, tropical cyclones derive their energy from warm surface waters. Also, seasonal upwelling and downwelling reduces the annual temperature range along the west coasts of the Americas. During El Niño and La Niña, changes in sea-surface temperature patterns associated with weakening and strengthening of upwelling off the northwest coast of South America and along the equator in the tropical Pacific affect the distribution of precipitation in the tropics and elsewhere.

Thermohaline Circulation

In the deep ocean, waters move sluggishly. Their slow, dominantly north-south movements are quite different from the well-defined gyres that characterize surface-ocean waters. The pycnocline isolates these deep currents from wind-driven surface currents. The deep-ocean circulation is driven primarily by slight differences in seawater density, caused by variations in water temperature and salinity (Chapter 3). Hence, the deep-ocean circulation is called the **thermohaline circulation**, the name coming from *thermo* meaning heat and *haline* referring to salt. Deep-ocean and near-bottom water movements are strongest on the western sides of ocean basins, analogous to western boundary currents in the surface-ocean gyres. The strongest flow in the deep-ocean circulation is in the western North Atlantic. These subsurface currents originate in the northern and southern polar regions.

Sea-floor topography influences the flow of near-bottom water. For example, low spots in the Mid-Atlantic Ridge channel water from the western Atlantic basin into the eastern Atlantic basin. The Greenland-Iceland-Faroe-Scotland Ridge separating the Arctic and Atlantic Ocean basins prevents the densest waters in the deep Arctic Ocean from flowing out over the ridge into the deep Atlantic. However, water masses that form near Greenland (south of the ridge) readily flow into the deep North Atlantic basin.

Waters slowly move upward from the deep ocean to the sea surface on time and space scales that are probably related to diffusion and to tidally induced mixing. The details and locations of this motion are not well understood but, for example, the bottom water in the Pacific Ocean is about 1500 years old, while the time required for circulation in the oceanic conveyor belt is about 1000 years. A rough estimate of the rate at which Pacific bottom water flows to the surface is about 4 m per year on average (4000 m depth divided by the 1000-year circulation period).

In this section, we examine thermohaline circulation and water masses. We begin with an overview of how ocean scientists monitor the deep ocean.

MONITORING THE DEEP OCEAN

While ocean scientists have measured near-surface temperature, salinity, and currents for more than a century, probing the ocean depths to determine the vertical structure and circulation of the ocean has been much more challenging. Accurate measurements of deep currents and water temperature and salinity at various depths not only require instruments that can withstand the stresses of the

138 Chapter 6 OCEAN CURRENTS

ocean environment, but also appropriate platforms that provide a means of delivering the instruments to the desired depth and retrieving the data once measurements are made. In the past few decades, however, systematic surveys of the deep ocean have greatly expanded our knowledge of the properties of seawater and water movements. Deep-sea casts from oceanographic ships have produced most of these data.

New promising ocean sensing technologies include submersible, instrumented *floats* that obtain vertical profiles of temperature, pressure (a measure of depth), and conductivity (a measure of salinity) (Figure 6.12). Research ships, commercial vessels, or low-flying aircraft drop floats into the sea. A float sinks to a prescribed depth (to 2000 m or 6600 ft), drifts with the current, and then returns to the surface monitoring ocean water properties along the way. At the surface, the float relays its collected data via satellites to computer databases. Tracking the position of the float over time also records water movements. In Chapter 3, we described another technology that uses changes in the speed of sound to measure ocean-water temperature over great distances within the ocean (*acoustic tomography*). In addition, the distribution of

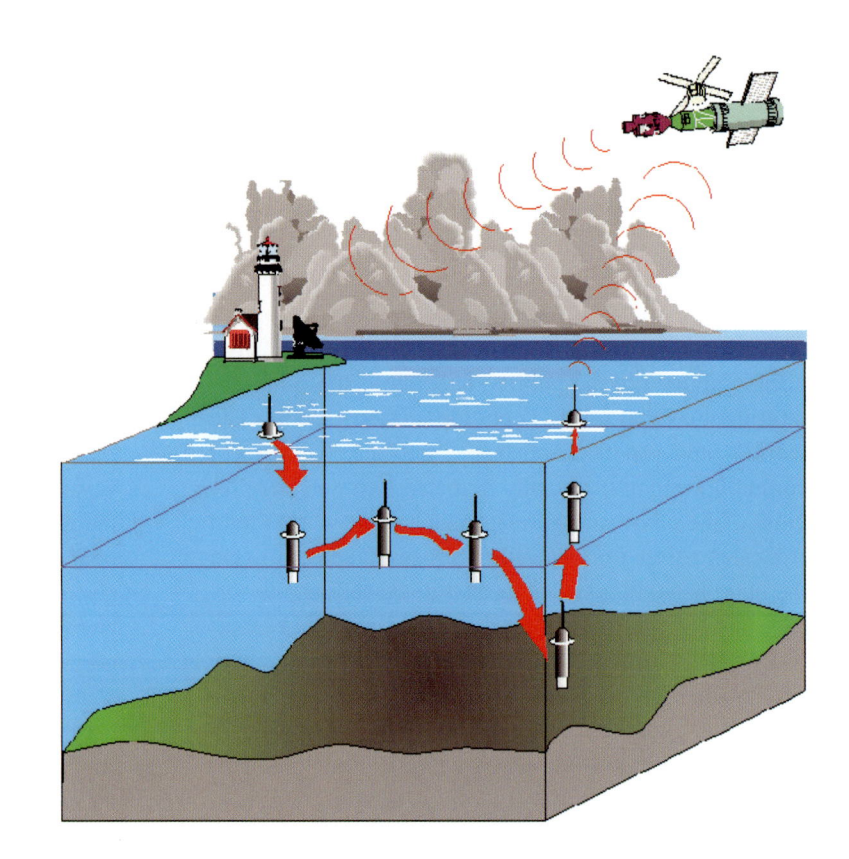

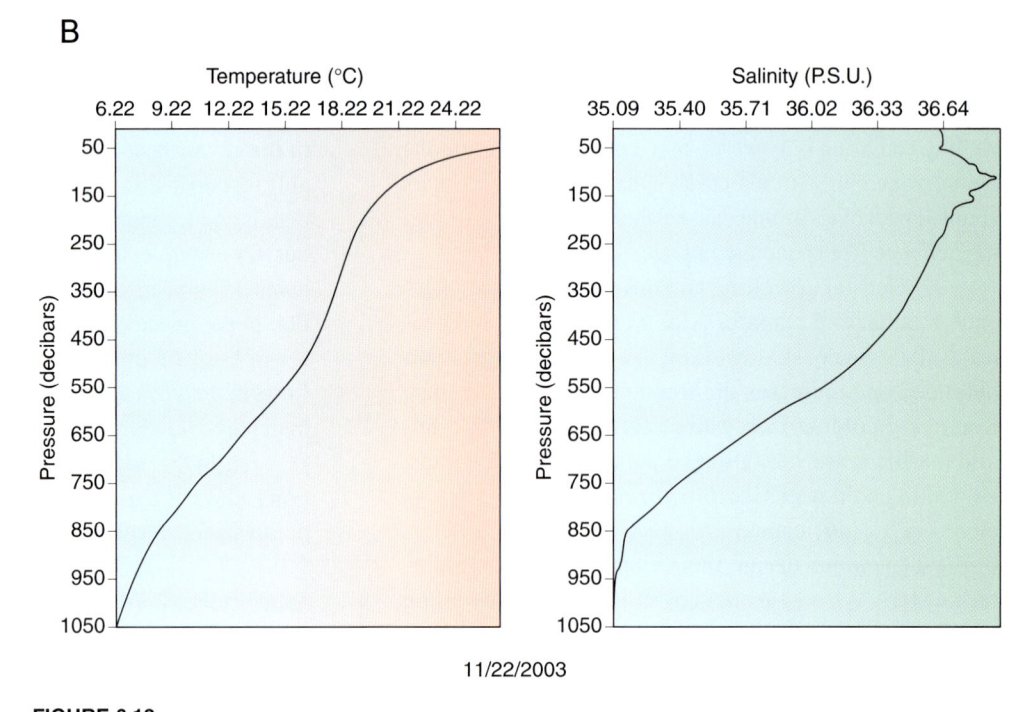

FIGURE 6.12

An ARGO float (A) obtains continuous profiles of ocean temperature and salinity to a maximum depth of about 2000 m (2000 decibars). The instrument then surfaces and sends data to a satellite for downloading to a laboratory. (B) This is a sample plot of float-derived temperature and salinity profiles to a depth of 1050 m obtained from a location off the coast of the Southeast U.S. [Courtesy of the University of Washington and Coriolis Data Centre.]

various tracer materials, such as dissolved oxygen, CFCs (chlorofluorocarbons), or radioactive materials from both atmospheric and oceanic testing of nuclear devices have been used to track movements of subsurface waters.

Atmospheric scientists use the term *sounding* to denote a sequence of measurements obtained in the vertical. Ocean scientists use the term *cast* (as in net cast or CTD cast) for measurements made at various depths in the ocean. This term likely originated with pilots (such as Mark Twain) on boats who cast soundings using a measured line to determine water depth. For more on oceanographic casting methods, refer to this chapter's third Essay.

WATER MASSES

A **water mass** is a large, homogeneous volume of water having a characteristic range of temperature and salinity. Most deep-ocean water masses form at high latitudes at the ocean surface where they acquire their characteristic low temperature and high salinity. Newly formed water masses then sink because they are denser (colder and more saline) than surrounding surface waters. Upon reaching a depth where its density is the same as that of the surrounding water, a water mass flows roughly horizontally along a constant-density surface like a single sheet inserted into a stack of paper. Let us examine how waters become sufficiently dense to sink into the deep-ocean.

At high latitudes, cold air chills surface waters, increasing their density. (Recall from Chapter 3 that lowering the temperature of seawater always increases its density.) The density of chilled surface waters further increases when seawater freezes because the salt that is excluded during ice formation is incorporated into nearby waters raising the salinity. At these high latitudes, for all practical purposes there is no pycnocline and very little density gradient down to the ocean bottom. As these relatively dense waters sink and move away from source regions, their temperature and salinity remain essentially constant, except near basin margins and rugged sea floors where mixing occurs. A few water masses also form in arid subtropical regions such as the Mediterranean and Red Seas where evaporation rates are relatively high. Surface waters become so saline and dense that they eventually sink below the pycnocline.

Oceanographers classify water masses based on their source region and the relative depth in the ocean where they reach equilibrium with their surroundings (Table 6.1). Source regions are the Atlantic, Pacific and Indian Oceans. In order of increasing depth, water masses are designated as *central waters* (the relatively warm, wind-driven ocean surface above the pycnocline down to about 1 km), *intermediate waters* (1 to 2 km), and *deep and bottom waters* (deeper than 2 km). As we will see later, heat transport by moving water masses, likened to a conveyor belt, is an important control of global climate. Surface currents transport heat poleward from the tropics whereas deep-ocean currents return cold water back to the tropics where they are warmed.

TABLE 6.1
Water Masses of the Atlantic, Pacific, and Indian Oceans

Atlantic Ocean
South Atlantic Central Water (SACW)
North Atlantic Central Water (NACW)
Antarctic Intermediate Water (AAIW)
North Atlantic Intermediate Water (AIW)
Mediterranean Intermediate Water (MIW)
Antarctic Deep Water (AADW)
North Atlantic Deep Water (NADW)
North Atlantic Bottom Water (NABW)
Antarctic Bottom Water (AABW)

Pacific Ocean
South Pacific Central Water (SPCW)
North Pacific Central Water (NPCW)
Antarctic Intermediate Water (AAIW)
North Pacific Intermediate Water (NPIW)
Pacific Subarctic Water (PSW)
Common Water (CoW)

Indian Ocean
Southern Indian Central Water (SICW)
Equatorial Central Water (ECW)
Antarctic Intermediate Water (AAIW)
Red Sea Intermediate Water (RSIW)
Common Water (CoW)

The Atlantic's subsurface water mass movements are the best understood of all deep ocean systems. They are also the most active. Deep and intermediate water masses form at both ends of the Atlantic basin (poleward of the subtropical gyres), primarily in the Greenland and Labrador Seas of the North Atlantic and in the Weddell Sea near Antarctica (Figure 6.13A). In addition, the Atlantic Ocean receives intermediate warm, salty waters from the Mediterranean Sea.

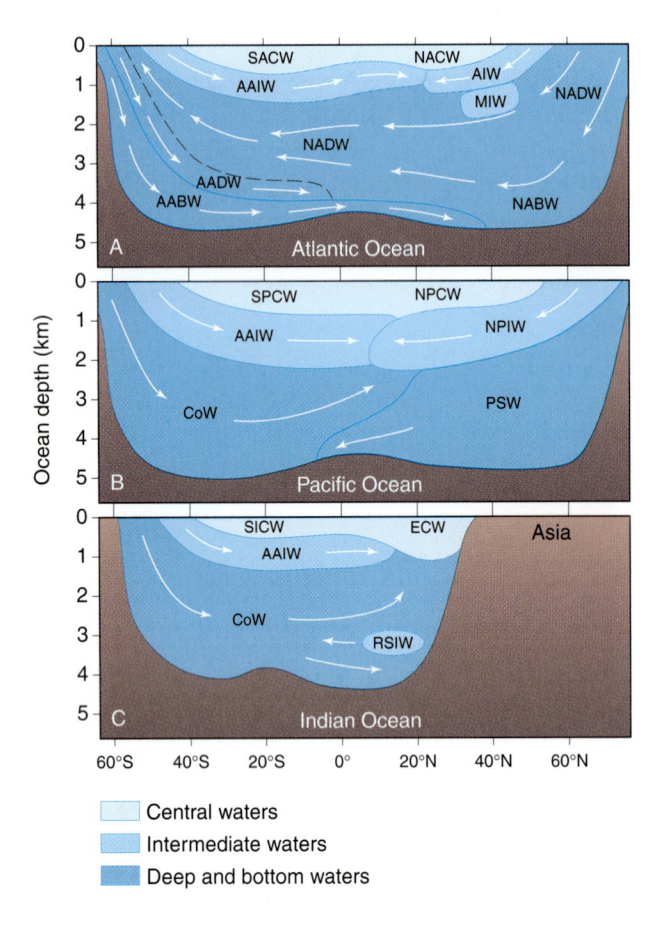

FIGURE 6.13
Distribution of water masses in the (A) Atlantic, (B) Pacific, and (C) Indian Oceans.

Newly formed Antarctic Bottom Water (AABW) sinks to the ocean floor and eventually flows north into the western North Atlantic basin. In fact, traces of this water mass are detected in the North Atlantic as far north as 45 degrees N. Another water mass, Antarctic Deep Water (AADW), which forms to the north of the AABW, is slightly warmer and more saline than Antarctic Bottom Water. Beneath the surface, AADW is sandwiched between the denser Antarctic Bottom Water below and the less-dense North Atlantic Deep Water (NADW) above. This zone of contrasting waters is rich in marine life primarily because upwelling NADW is nutrient-rich (Chapter 9). North Atlantic Deep Water forms near Greenland (through winter cooling, evaporation, and ice formation), sinks, and flows southward; it is the most abundant deep-water mass in the ocean.

Other water masses having densities between that of the surface waters and the deep-water masses occur in various parts of the Atlantic. The intermediate waters that form between the polar deep waters and central waters associated with the subtropical gyres include Antarctic Intermediate Water (AAIW) (about 40 to 50 degrees S), and North Atlantic Intermediate Water (AIW) (about 60 degrees N). An intermediate water mass of some importance but generated outside the boundaries of the Atlantic basin is Mediterranean Intermediate Water (MIW). It originates in the northwestern Mediterranean in winter as cold, dry *Mistral winds* cool the surface waters and enhance evaporation. Sea surface temperatures average about 15 °C (59 °F) and the salinity is near 39. Increased density of surface waters induces convection that reaches the sea floor at depths greater than 2000 m (6600 ft). MIW fills the bottom of the Mediterranean basin, flows westward along the North African coast at a depth averaging 400 m (1300 ft), and spills over the sill at the Strait of Gibraltar under the incoming (less dense) Atlantic waters. (Atlantic surface waters flow eastward into the Mediterranean to replace water lost to intense evaporation over the arid eastern portion of the Sea.) Although mixing occurs at the interface between MIW and the Atlantic waters, MIW is readily identified as it moves into the North Atlantic as a lens of relatively warm (about 13 °C or 55.4 °F), salty (37.3) water that flows down the continental slope until it ultimately reaches its neutrally buoyant depth of between 1000 and 2000 m (3300 and 6600 ft) and spreads outward in all directions.

Deep-water mass movements in the Pacific are the slowest and least understood among the three major ocean basins (Figure 6.13B). The shallow Bering Strait at a depth of only about 50 m (165 ft) allows little if any flow of Arctic Bottom Water into the North Pacific. Furthermore, low salinity surface waters (due to precipitation and runoff) prevent deep-water masses from forming in the North Pacific. Intermediate- and deep-water masses develop in the Pacific primarily near Antarctica. Even though these areas produce large volumes of cold, dense waters, they replenish the bottom waters very slowly because the Pacific basin is so large. Thus the age of the waters (time since the water was last in direct contact with the atmosphere) in the Pacific's deep- and intermediate- water masses is the oldest in the ocean. Radiocarbon dating indicates that bottom waters are approximately 1500 years old in the Pacific (compared to about 750 years old in the deep Atlantic).

As Antarctic Bottom Water flows around the Antarctic continent in the Southern Ocean, it mixes with Antarctic Deep Water and North Atlantic Deep Water producing Common Water (CoW), the principal water type in the deepest parts of the Pacific. The Antarctic Circumpolar Current, relatively strong to a depth of at least 2500 m (8200 ft) is responsible for the mixing. CoW sinks and flows northward to near the equator where it converges with Pacific Subarctic Water (PSW).

Relatively little Antarctic Intermediate Water is produced in the South Pacific. Consequently, the Pacific Ocean is only weakly layered at depths below about 2000 m (6600 ft) with remarkably uniform temperature and salinity. The deep circulation is particularly sluggish because of minimal density contrasts between water masses. Antarctic Intermediate Water (AAIW) and North Pacific Intermediate Water (NPIW) are the most common water masses in the Pacific's mid-depths (about 2000 m). They are formed by poorly understood mixing processes.

Most of the Indian Ocean, occupying the smallest of the three major ocean basins, is located in the Southern Hemisphere (Figure 6.13C). Deep-ocean circulation in the Indian Ocean is similar to the Pacific. No cold-water masses form along its northern boundaries, where the climate is tropical and subtropical. Bottom flows in the southern region connect with those in the South Atlantic south of about 40 degrees S. Deep water in the Indian Ocean is a Common Water (CoW), a mixture of Antarctic Bottom Water and North Atlantic Deep Water.

The warm, salty waters of the Red Sea Intermediate Water (RSIW) flow over the submarine sill at the entrance (Strait of Bab el Mandeb) and southward into the Indian Ocean at depths of about 3000 m (9800 ft). The salinity of RSIW is exceptionally high, typically greater than 40, because of high temperatures and evaporation rates in the Red Sea. Highly saline brines produced by the ancient salt deposits underlying the region also flow into the Red Sea. RSIW provides the only significant modifying effect in the entire deep Indian Ocean. Equatorial shallow layers in the northern Indian Ocean are not well defined, partially because of changes in winds and surface currents associated with the monsoon atmospheric circulation. Prevailing winds blow onshore during the wet monsoon (summer) and offshore during the dry monsoon (winter).

Oceanic Conveyor Belt

The **global oceanic conveyer belt** is a unifying concept that connects the ocean's surface and thermohaline circulation regimes, transporting heat and salt on a planetary scale.

The conveyor belt system can be thought of as beginning near Greenland and Iceland in the North Atlantic where dry, cold winds blowing from northern Canada chill surface waters (Figure 6.14). The combined chilling of surface waters, evaporation, and sea-ice formation pro-

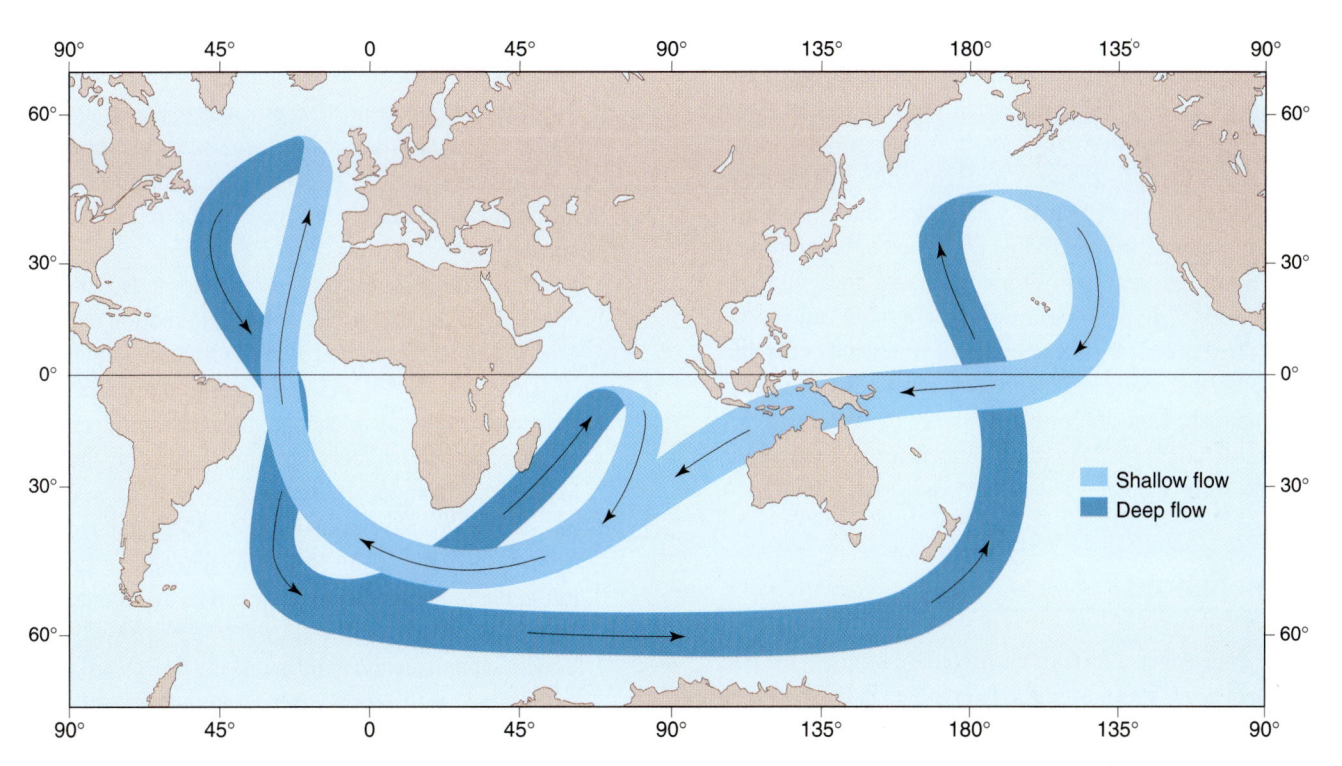

FIGURE 6.14
The global oceanic-conveyor belt transports heat and salt and is an important control of climate.

duces cold, salty North Atlantic Deep Water (NADW). The newly formed NADW sinks and flows southward along the continental slope of North and South America toward Antarctica where the water mass then flows eastward around the Antarctic continent (in the Antarctic Circumpolar Current). There the NADW mixes with Antarctic waters (i.e., AABW and AADW). The resulting Common Water, also called Antarctic Circumpolar water, flows northward at depth into the three ocean basins (primarily the Pacific and Indian Oceans).

These bottom waters gradually warm and mix with overlying waters as they flow northward. They move to the surface at a rate of only a few meters per year. After rising to the surface in the Pacific, the surface waters flow through the many passages between the Indonesian islands into the Indian Ocean. Eventually they flow into the Agulhas Current, the Indian Ocean boundary current that flows around southern Africa. After entering the Atlantic Ocean, the surface waters join the wind-driven currents in the Atlantic, becoming saltier by evaporation under the intense tropical sun. Trade winds transport some of this water vapor out of the Atlantic Ocean basin, across the Isthmus of Panama, and into the Pacific Ocean basin. Atlantic surface waters eventually return northward to the Labrador and Greenland seas in the North Atlantic.

Continued operation of the oceanic conveyor belt is important to northern Europe's moderate climate because of northward transport of heat in the Gulf Stream and North Atlantic Current. The system can weaken or shut down entirely if the North Atlantic surface-water salinity somehow drops too low to allow the formation of deep-ocean water masses. This apparently happened during the Little Ice Age (about 1400 to 1850 AD). The conveyer system shut down and northern Europe's climate became markedly colder. Old paintings from this era show Dutch skaters on frozen canals—something that would not occur during today's climatic regime. Cores extracted from deep-sea sediment deposits contain evidence of earlier cold periods, including the Heinrich events that we discussed in the Essay in Chapter 4.

Conclusions

The surface ocean, its currents driven by winds, is the part of the ocean that is most directly involved in Earth-system processes. It transports heat globally, supplies water vapor to the atmosphere, dissolves and transports salts, nutrients, and gases, supports fisheries, and plays a major role in day-to-day weather and in short-term climate variability. The deep-ocean is isolated from the atmosphere by the surface zone except at high latitudes; hence, it is mostly cold and dark. The relatively sluggish thermohaline circulation is driven by density contrasts between water masses and involves about 90% of the ocean's waters. The deep circulation is important in long-term climate variability, may play an important role in sequestering or buffering greenhouse gases such as carbon dioxide, and is important in transporting dissolved nutrients. We continue our investigation of the dynamic nature of the ocean in the next chapter where we examine waves and tides.

Basic Understandings

- The ocean is divided into three depth zones based on density: the surface mixed layer of uniform or near uniform density, the intermediate pycnocline where density increases markedly with depth because of changes in temperature and/or salinity, and the deep layer. The pycnocline is very stable and inhibits mixing of ocean waters between the mixed layer and the deep layer.
- Two circulation regimes dominate the ocean: wind-driven currents in the surface layer and the sluggish thermohaline circulation in the deep ocean.
- Wind-driven currents are maintained by kinetic energy transferred from the horizontal winds to ocean surface water. Once the wind sets surface waters in motion as a current, the Coriolis effect and the configuration of the ocean basin modify the speed and direction of the current.
- Frictional coupling of the horizontal winds with the ocean surface puts surface waters into motion and the Coriolis effect, combined with the frictional coupling of successive layers of water, cause the horizontal movement of water to change direction and decrease in magnitude with increasing depth, producing the Ekman spiral. In an ideal case, a steady wind causes surface waters to move at an angle of 45 degrees to the right of the wind in the Northern Hemisphere (to the left in the Southern Hemisphere). At a depth of about 100 to 150 m (330 to 500 ft), water moves slowly in a direction opposite the wind and is the downward limit of the wind's influence on ocean water movement. The Ekman spiral predicts a net water movement through a depth of about 100 m (330 ft) at 90 degrees to the right of the wind direction in the Northern Hemisphere and

90 degrees to the left of the wind direction in the Southern Hemisphere.
- Because of the wind circulation in subtropical highs, Ekman transport causes surface waters to move toward the central region of subtropical gyres from all sides, producing a mounding of surface water near the center of the gyre. In response to this horizontal gradient in water pressure, water moves from where the pressure is higher toward where the pressure is lower, that is, downhill and directly opposite Ekman transport. Surface water parcels flow outward and down slope from the center of the gyre and the Coriolis effect causes parcels to shift direction to the right in the Northern Hemisphere (to the left in the Southern Hemisphere). Eventually, the outward-directed pressure gradient force balances the Coriolis deflection and the water parcels flow around the gyre and parallel to contours of elevation of sea level. This horizontal movement of surface water is known as geostrophic flow.
- Ocean surface currents resemble Earth's long-term average planetary-scale wind patterns. Surface currents form gyres roughly centered in each ocean basin near 30 degrees latitude. Viewed from above, currents in these subtropical gyres flow in a clockwise direction in the Northern Hemisphere and a counterclockwise direction in the Southern Hemisphere. In the sub-polar gyres of the Northern Hemisphere, current directions are reverse that of the Northern Hemisphere subtropical gyres.
- Winds associated with a passing storm system can disturb the ocean surface and cause the flow of ocean surface currents locally to deviate temporarily from long-term average patterns.
- Trade winds of the Northern and Southern Hemispheres drive equatorial currents and warm surface water westward across the tropical ocean.
- The ocean-surface slope is steeper on the western sides of subtropical gyres in both hemispheres so that geostrophic currents are stronger on that side of each gyre. These western boundary currents are the fastest surface currents in the open ocean and flow northward in the Northern Hemisphere and southward in the Southern Hemisphere. The western boundary currents are in part a result of the Coriolis effect becoming stronger poleward, the same direction as the flow of the western boundary currents.
- Western boundary currents spawn large warm- and cold-core rings (eddies). Rings develop when a meandering current forms a large loop that pinches off and separates from the main current. In the Northern Hemisphere, warm-core rings form on the current's landward side and rotate clockwise whereas cold-core rings form on the current's ocean side and rotate counterclockwise.
- In coastal areas, the combination of persistent winds blowing parallel to the coast, the Coriolis effect, and restrictions on lateral movements of water caused by shorelines and shallow bottoms induces upward and downward movement of water. Where winds generate Ekman transport of surface waters away from the coast, colder nutrient-rich water wells up from below, a process called upwelling. Where winds generate Ekman transport of surface waters toward a coast, the water piles up and sinks; this is called downwelling. Regional scale coastal upwelling is characteristic of the eastern boundary regions off the west coasts of continents. Upwelling also occurs along the equator in response to Ekman transport associated with the trade winds of the two hemispheres. Upwelling supplies nutrient-rich waters to the sunlit, surface zone of the ocean spurring biological productivity.
- Over the past few decades, systematic surveys of the deep ocean have greatly expanded our knowledge of the properties of seawater and water movements. Casts denote a sequence of measurements (e.g., temperature, conductivity) obtained vertically through the ocean depths.
- Deep-ocean (thermohaline) circulation is sluggish and is driven primarily by slight differences in water density caused by variations in water temperature and salinity. Deep waters acquire their higher density through chilling, evaporation, and increased salinity because of sea ice formation at high latitudes.
- A water mass is a large, homogeneous volume of ocean water having a characteristic range of temperature and salinity. Oceanographers classify water masses by their source region and the depth in the ocean where they reach equilibrium with their surroundings.
- Deep- and intermediate-water masses form at the north and south ends of the Atlantic basin, primarily in the Greenland and Labrador Seas of the North Atlantic and in the Weddell Sea near Antarctica. In addition, the Atlantic Ocean receives warm, salty waters from the Mediterranean Sea. Deep-water mass movements in the Pacific are the slowest among the three major ocean basins in part because the Pacific is cut off from the Arctic Ocean plus the Pacific is less saline than the Atlantic.

- The global oceanic conveyer belt is a unifying concept that connects the surface and thermohaline circulation regimes and transports heat and salt on a planetary scale. It is an important climate control especially for Western Europe.

ESSAY: Predicting Oil-Spill Trajectories

Accidents involving spilled oil are all too frequent occurrences in coastal regions. Many factors (e.g., winds, surface currents, tides, air and water temperatures, and salinity) control the movement of spilled oil. Furthermore, the type and amount of spilled oil, and local shoreline and bottom features also influence movements of an oil slick. An effective response to an oil spill requires the input of scientists representing many different specialties and information on the chemical composition of the spilled oil, ocean currents, and weather. All these data are needed for mathematical models that predict movements (known as *trajectory analysis*) of the oil. When combined with biological resource information, trajectory analyses can be used to identify those areas that are most vulnerable to the oil so that equipment to contain the oil spill can be dispatched to where it will be most effective. Sensitive areas requiring protection include *marine sanctuaries* and unique habitats, especially those that are home to endangered species. Protection is key because oil not only causes immediate contamination but also has long-term effects on coastal ecosystems.

The National Oceanic and Atmospheric Administration (NOAA) Coastal Change Analysis Program uses remote sensing data, primarily aerial photography, to identify and classify sensitive habitats of bottom-dwelling organisms, such as sea grasses. Oil spills can affect sea grasses in many ways. Toxic materials in the spilled oil are introduced into sediments, the water, and ultimately living organisms. Oil can also block sunlight from reaching the plants. Heavy oils sink and mix with sediment and can coat sea grasses. When these grasses are damaged or killed, organisms that depend upon them as a food source or for habitat are also adversely affected. This has a ripple effect on the local ecosystem and ultimately damages economically valuable fish and shellfish, thus impacting local economies for many years.

Major sectors of Maine's economy, especially tourism and fishing, require clean coastal waters, aesthetically appealing coastlines, and functioning coastal ecosystems. Thus, the state requires timely and appropriate responses to oil spills. To demonstrate the important role of accurate oil-spill trajectory analysis and prompt response to mediate the effects of oil spills, we examine two cases separated by nearly 25 years.

In 1972, the tanker *Tamano* spilled 380,000 liters (100,000 gal) of oil into Casco Bay, near Portland in southern Maine, then the largest oil spill in the state's history. Plans for dealing with spills were antiquated and slow to be implemented. During the several days required to mobilize response teams, the oil spread along about 75 km (47 mi) of coastline, including beaches on 18 small islands, damaging commercially valuable fish and shellfish stocks. The cleanup that followed took 11 years and cost nearly $4 billion.

On 27 September 1996, the tanker *Julie N.* struck a bridge entering Casco Bay, spilling 680,000 liters (180,000 gal) of light fuel oil and much heavier bunker-C, a mix of oils. The light fuel oil evaporated quickly but left behind toxic components. The less toxic but heavier bunker-C sank to the bottom, covering vegetation and wildlife with a thick, sticky coating. The response to this incident was rapid and effective. The damaged tanker quickly docked and was immediately surrounded by floating barriers, called booms, to contain floating oil. Other booms were deployed quickly to prevent oil from reaching vulnerable biological resources or economically valuable locations. In addition, special clean-up vessels skimmed and collected oil from the water surface. Timely response combined with successful predictions of oil movements meant that much of the damage that had occurred 25 years earlier in the same area was not repeated.

ESSAY: The Global Positioning System

Satellite navigation was the brainchild of the U.S. Department of Defense. Like the 18th century governments who offered prize money for anyone who could develop accurate methods of navigation at sea, the U.S. government spent over $12 billion to develop a satellite-based system that would reliably determine location precisely, at all times and in all kinds of weather. Originally intended for military use, the *Global Positioning System (GPS)* dates to the 1978 launch of the first Navstar satellite. GPS can provide near pinpoint location accuracy and is jam and interference proof. (Not to be outdone by its then Cold War adversary, the former Soviet Union developed a similar system called GLONASS. However, recent funding problems have essentially disabled GLONASS.)

Civilian use of GPS began in the 1980s. A minimum of 24 Navstar satellites is needed to provide continuous service. Once this was achieved in the early 1990s, relatively inexpensive, small portable receivers were developed and marketed, and civilian use of GPS soared. Today, the civilian sector accounts for about 92% of GPS equipment sales. (Worldwide, sales of GPS equipment reached $3.5 billion in 2003 and are projected to climb to $10 billion in the next decade.) GPS is widely used for navigation by ships, planes, delivery trucks, and automobiles.

The principle behind GPS is trilateration (related to triangulation) from satellites that are in view of the receiver (see Figure). At least four satellites are needed for maximum precision. The time it takes for a ranging signal to travel the distance from a satellite to a receiver is determined using accurate clocks which record when the signal was emitted by the satellite and when the signal arrived at the receiver. The fact that the radio signal travels at a finite speed (300,000 km per sec, or 186,000 mi per sec, the speed of light), allows these times to be converted to distances if the position of the satellite in its orbit is known. This is no problem because all Navstar satellites are identifiable and their orbits are regular. Using an inexpensive civilian receiver, a person can locate his/her position within 5 to 10 m (16 to 33 ft). More sophisticated military receivers can achieve a location accuracy of 50 cm (20 in.).

Using one satellite, our location must be some point on a sphere centered on the satellite with the radius of the sphere equal to the distance to the satellite. Knowing our distance from a second satellite, our location must be some point on a sphere having a radius equal to the distance to that satellite. The two spheres have different radii and geometry indicates that the intersection of two spheres is a circle. Our location must be somewhere on that circle. A third satellite describes yet another sphere that intersects with the other two spheres, narrowing our location to one of two possible points, that is, where all three spheres intersect. Usually one of the two points is either in space or in Earth's interior. Hence, the other point is determined to be our location on the surface of Earth. A computer in the GPS receiver has algorithms that can distinguish between locations likely to be correct and locations that are spurious.

Use of a fourth (or additional) satellites is intended to compensate for the imprecision caused by slight differences in timing between the precise atomic clocks carried on the GPS satellite and the less accurate quartz clocks in the GPS receiver. Measurements made by the fourth satellite are used to synchronize the two clocks. Actually, most GPS receivers will choose the best signals from many satellites and use four of them to fix the position so that assumptions about which of two points of intersection is correct never have to be made. A mariner at sea level does not need to know altitude and this further simplifies the procedure.

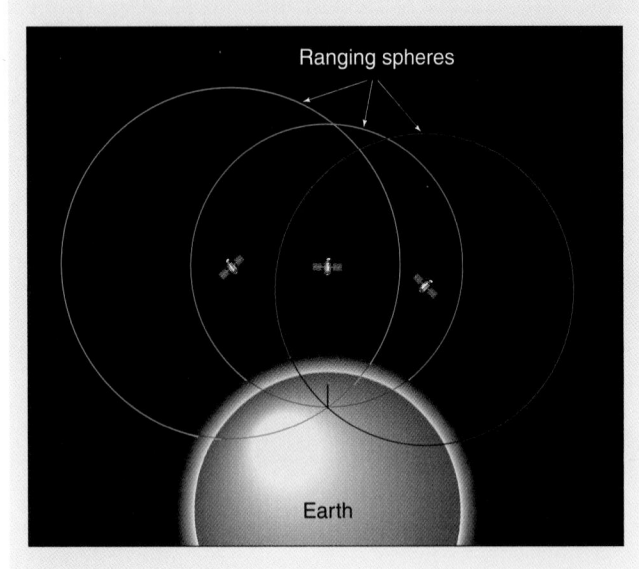

FIGURE
To determine its location on Earth's surface, a GPS receiver measures the travel times of radio signals sent by three satellites of known orbital location. Travel times are converted to distances. This trilateration technique is made more precise when a fourth satellite is used to synchronize the clocks on the GPS satellites and receiver.

Plans are currently underway to increase the location accuracy of GPS by broadcasting new signals from the navigation satellites. One of the anticipated benefits of these new signals will be a reduction in the interference caused by Earth's ionosphere (a region of the high atmosphere containing a relatively high concentration of electrically charged particles). These improvements to GPS should be in place by about 2008.

ESSAY: Profiling the Ocean Depths

Sampling at various depths in the ocean provides profiles of temperature, salinity (from conductivity), and density. Some of the first temperature and salinity casts made in the early 20th century employed *reversing thermometers* that measure both ocean temperature and sample depth and *collection bottles* for obtaining water samples at depth for measuring salinity (in the laboratory). Two liquid-in-glass thermometers go to depth on a wire upside down so they can adjust to temperature at depth—one accounting for pressure at depth and the other protected from the effects of pressure at depth. A messenger is sent down the wire to trip the bottle so that it flips over and captures a water sample at the same time reversing the two thermometers and recording the reading on the protected and unprotected thermometers. After bringing the thermometers (and collection bottle) to the surface, the depth of collection and the temperature at that depth are computed. Combined with the salinity of the seawater at that depth, the density of seawater is determined. Although this casting method is time consuming, reversing thermometers and collection bottles are still used today for calibrating other instruments.

In the mid 20th century, a mechanical instrument called a *bathythermograph (BT)* was developed that could obtain a nearly continuous temperature profile with depth from a slowly moving ship. As this torpedo-shaped device sank into the water, a temperature-sensitive element (typically a deformation or bimetallic thermometer) inside the BT moved a stylus across a metal- or smoke-coated slide recording temperature as a function of depth (obtained from pressure bellows). The resulting temperature trace was interpreted after the device was retrieved from the water.

Since the 1970s, newer electronic technologies have permitted more rapid soundings that produce an almost continuous profile of temperature. An *expendable bathythermograph (XBT)* is a non-recoverable device that is deployed from fast moving ships or airplanes and measures a nearly continuous ocean temperature profile to a depth of approximately 1800 m (5900 ft). This torpedo-shaped device consists of a thermistor (electronic thermometer) housed in an expendable casing. A thin conducting wire connects the XBT to the ship or plane and transmits the temperature signal to an onboard recorder. The electrical conductivity of seawater is used to complete the electrical circuit. As the device sinks into the ocean at a known rate, data are transmitted electronically at regular intervals, permitting the recording of ocean temperature as it changes with depth. A slightly different profiling instrument is an *expendable conductivity-temperature-depth profiler (XCTD)* that provides essentially continuous profiles of ocean conductivity (salinity) in addition to temperature and pressure.

NOAA's Atlantic Oceanographic and Meteorological Laboratory in Miami, FL oversees the XBT program. Volunteer commercial ships deploy as many as four XBTs daily along selected shipping lanes. The data are compiled by a computer onboard ship and then transmitted via satellite relay to the Laboratory for global distribution. More than 70 ships voluntarily produce 26 monthly transects across the three major ocean basins.

In 1985, the ten-year international Tropical Ocean Global Atmosphere (TOGA) program commenced. One of TOGA's projects was deployment of TAO (Tropical Atmosphere/Ocean), an array of moored buoys (small, non-piloted, instrumented platforms) in the tropical Pacific Ocean. Data from this array have been extremely valuable in detecting and predicting such atmospheric/oceanic episodes as El Niño and La Niña (Chapter 11). This instrument array, renamed TAO/TRITON in 2000, presently consists of approximately 70 deep-sea moorings that measure several atmospheric variables (air temperature, wind, relative humidity) as well as oceanic parameters (sea surface and subsurface temperatures at 10 depths in the upper 500 m or 1650 ft). Several newer moorings also have salinity sensors, along with additional meteorological sensors. Five moorings along the equator also measure ocean velocity using a Subsurface Acoustic Doppler Current Profiler. The data are collected and relayed in near real-time to shore via satellites. Real time data displays from the TAO/TRITON array are available from NOAA's Pacific Marine Environmental Laboratory (PMEL) in Seattle, WA.

In the last decade, a variety of autonomous (free-drifting) instrumented profilers have been developed to measure large-scale subsurface currents and make repeated vertical measurements of ocean variables. Early versions of these free-drifting profilers were identified with the term *PALACE (Profiling Autonomous Lagrangian Circulation Explorer)* whereas a subsequent version is called *APEX (Autonomous Profiling Explorer)*. These subsurface floats, approximately 1 m (3.3 ft) long and less than 20 cm (8 in.) in diameter, can be deployed from either ships or aircraft.

Once deployed, a PALACE/APEX profiler is designed to sink to a depth where it is neutrally buoyant, drift for approximately 10 days at that depth, and then rise back to the surface collecting data en route. The maximum drift depth to which the profiler can sink is 2000 m (6600 ft). Typically, the profiler sinks to 2000 m on one of every four monitoring cycles and to 1000 m (3300 ft) during the other three. The profiler moves vertically by pumping hydraulic oil between an internal reservoir and an external bladder. The profiler ascends when the oil flows into (and expands) the bladder (decreasing the density) and descends when oil flows in the opposite direction. During the float's ascent, onboard sensors record the temperature, pressure (depth), and conductivity (salinity) nearly continuously. Upon returning to the ocean surface, the float telemeters these data to a satellite for subsequent relay to data collection stations (Figure 6.12). The float's position is determined by satellite and ocean current information is inferred from the horizontal displacement of the float from one surfacing to the next. The geostrophic assumption is used to compute ocean currents at depth. Following a programmed interval at the surface, hydraulic oil is pumped back to the internal reservoir, and the float returns to depth for the next 10 to 14 day cycle. The anticipated lifetime of one PALACE/APEX float is 100 cycles.

To monitor the climate (long-term average conditions) of the upper ocean on denser spatial and temporal scales, a widely spaced array of 3000 instrumented floating profilers are to be deployed across the ocean at 3-degree latitude/longitude intervals within the next several years. The long-term observations provided by the float array known as *ARGO (Array for Real-time Geostrophic Oceanography)* is expected to be the basis for mapping the large-scale average oceanic flow. While this proposed global-scale array is to be an international effort, various institutions in the U.S. will deploy about half of the floats. Other participating nations include Australia, Canada, Japan, France, and the United Kingdom. As of March 2004, 1122 floats were operational. The U.S. portion of the ARGO program includes floats that are under the auspices of the University of Washington, Scripps Institution of Oceanography, and the Woods Hole Oceanographic Institution. NOAA's Atlantic Oceanographic and Meteorological Laboratory in Miami, FL is archiving ARGO data.

CHAPTER 7

OCEAN WAVES AND TIDES

Case-in-Point
Driving Question
Wind-Driven Waves
    Wind-Wave Generation
    Deep-Water and Shallow-Water Waves
    Seiche
    Atmosphere-Ocean Transfer
Ocean Tides
    Tide-Generating Forces
    Types of Tides
    Tides in Ocean Basins
    Tidal Currents
    Observing and Predicting Tides
    Open-Ocean Tides
Internal Waves
Tsunamis
Conclusions
Basic Understandings
ESSAY: The State of the Sea
ESSAY: Monitoring Sea Level from Space
ESSAY: Tidal Power

North Pacific storm waves as seen from the MV Noble Star. [Courtesy of NOAA.]

Case-in-Point

Gigantic waves, known as *rogue waves*, occur without warning in the open ocean and can damage or sink even the largest ships afloat. In September 1965, while crossing the Atlantic, one of the world's largest ocean liners, the *Queen Elizabeth 2 (QE2)*, was hit by a giant wave. She had altered course to avoid a hurricane and appeared to be out of serious danger, although she was encountering rough seas. It is extremely difficult to estimate the height of waves from onboard a ship, but the wave that struck the *QE2* was level with the captain's line of sight from the ship's bridge, some 29 m (95 ft) above the sea surface. An instrumented buoy in the same area recorded extreme waves of similar height. The *QE2* survived with minor damage, but many ships including modern supertankers are not so lucky and are extensively damaged or sunk. Rogue waves can also be destructive in coastal areas where they tear apart piers, breakwaters, and jetties.

    Rogue waves sometimes appear even when sea/air conditions are tranquil and have been reported from all ocean basins. These waves are especially common in the Agulhas Current, a western boundary current in the Indian Ocean off South Africa's east coast (Chapter 6). Rogue waves have a distinctive form unlike other sea waves—a steep forward face preceded by a deep trough. Some mariners have described encounters with rogue waves as "sailing into a hole in the sea" or "hitting a wall of water."

Mariners have long attributed rogue waves to an interaction, called *constructive interference*, between large storm-driven swell and a strong current. This would explain the relatively high frequency of rogue waves off the coast of South Africa where large storm swell from the Southern Ocean encounter the swift Agulhas Current. Another, more recent explanation is that eddies (rings) formed by the Agulhas Current may focus wave energy, much like an optical lens focuses light. Such focusing could also occur near the Gulf Stream. It might even partially account for the ominous reputation of the region near the Gulf Stream known as the "graveyard of ships" or even more colorfully as the "Bermuda Triangle" where ships have been known to vanish mysteriously.

Driving question:

What are sea waves and tides and what causes them?

In thinking about the ocean, most people visualize a vast water surface disturbed by continually changing and sometimes seemingly chaotic patterns of waves. Ceaselessly, waves break against the shore; waves and sea spray buffet a small fishing boat bobbing in the water just offshore; and waves the height of a three-story building batter a supertanker plying through storm-tossed waters a thousand kilometers at sea. On 7 February 1933, the *U.S.S. Ramapo* encountered a 34-m (112-ft) high wave during a storm in the central South Pacific. Destructive waves are not limited to the ocean. A huge wave probably caused the sinking of the *Edmund Fitzgerald* on Lake Superior in November 1976.

What are sea waves and what causes them? Most sea waves are the product of an interaction between the ocean and atmosphere in which *kinetic energy* (the energy of motion) of the wind is transferred to surface waters. Hence, these waves are often called **wind-waves**. Wind-waves are part of the current-generating processes described in the previous chapter. In this chapter, we discuss the formation and life cycle of these waves, their generation as wind-waves or "sea," plus their existence at sea as free waves and swell before expending their energy as breakers on a distant reef or shoreline. We also describe internal waves and tsunamis.

Coastal residents are very familiar with not only sea waves but also the periodic rise and fall of sea level known as tides. Tides are also a type of wave but with very long wavelengths approaching the dimensions of an ocean basin. Because of their importance to marine interests, tides were among the earliest ocean characteristics to be monitored on a regular basis. In fact, elaborate mechanical computers were developed early on to predict tides. In this chapter, we examine tide generation, types of tides and tide prediction.

Wind-Driven Waves

In a general sense, a **wave** is a regular oscillation that occurs in a solid, liquid, or gaseous medium as energy is transmitted through that medium. A pebble tossed into a still pond disturbs the water surface creating waves that move outward in all directions away from the point where the pebble entered the water. The principles of water waves apply regardless of the size of the body of water. A **sea wave** is an oscillation on the ocean surface that propagates along the interface between the atmosphere and ocean (Figure 7.1). A schematic drawing of an idealized sea wave is shown in Figure 7.2. The highest point reached by the oscillating water surface is called the **wave crest** and the lowest point is the **wave trough**. The **wave height** is the vertical distance between trough and crest. The horizontal distance between successive wave crests (or equivalently between any other two corresponding points of two consecutive waves) is the **wavelength**. The time needed for two successive wave crests (one

FIGURE 7.1
A water wave is an oscillation of the water surface that propagates along the interface between the water and atmosphere. [Photo by J.M. Moran]

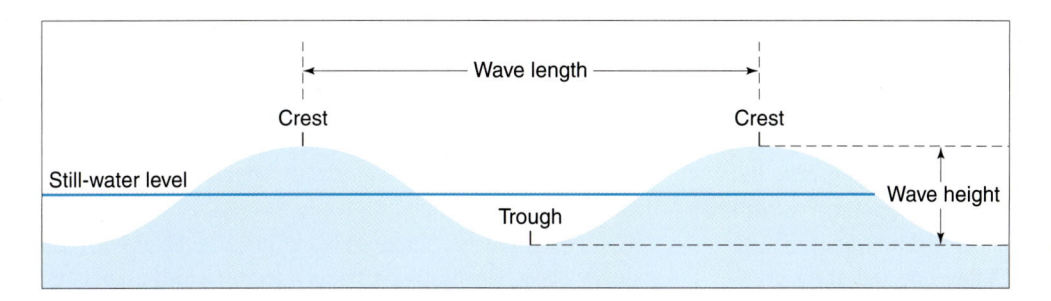

FIGURE 7.2
Cross-section of an idealized ocean wave.

wavelength) to pass a fixed point is the **wave period**. The number of waves passing a fixed point over an interval of time is the **wave frequency**, which is the inverse of the wave period.

WIND-WAVE GENERATION

If the air over the ocean were always calm, the sea surface would be smooth and motionless. (Sailors refer to this condition as a *flat calm*.) Wind disturbs this equilibrium state as some of its *kinetic energy* is transferred to surface ocean waters as the wave-generating force. Restoring forces return the ocean surface to its original horizontal equilibrium state. Waves begin as small ripples, called **capillary waves**, with wavelengths of less than 1.7 cm (0.7 in.). At these short wavelengths, water's surface tension is the restoring force that smoothes out and flattens these relatively small waves. At greater wavelengths, gravity provides the restoring force.

Surface tension, the attraction between molecules at or near the surface of a liquid, is a property of water arising from hydrogen bonding (Chapter 3). The density of water molecules in the air above the water surface is relatively low. Hence, H_2O molecules in the surface water layer are more strongly attracted by hydrogen bonding to each other and to the layer of water molecules immediately below. Surface tension explains why we can pour water into a glass to a level that is slightly above the rim of the glass. Among liquids, water is second only to mercury in the strength of its surface tension.

Strengthening winds increasingly disturb the water surface, producing larger waves having longer wavelengths and greater wave heights. For these larger waves, gravity is the restoring force working to level the wave crests and fill in the wave troughs. Gravity pulls the crested water downward, but momentum causes the water to continue downward and a trough is formed. The lowered water is buoyed upward and the next crest rises. Wind and gravity are responsible for most waves observed on the ocean surface.

In a propagating wave, water particles oscillate in circular orbits, thereby affecting the motion of neighboring water particles (Figure 7.3). In this way, waves (i.e., the changing shape of the water surface) move away from the disturbance that caused them. Water particles oscillate in approximately circular orbits as long as the frictional drag from the ocean bottom is negligible. The orbital diameter of a water particle at the ocean surface equals the wave height. As a complete wave (from crest to trough to crest) passes a fixed point, a water particle completes one orbit and returns to approximately its original position. Oscillations of water particles become smaller (orbits decrease in diameter) with depth away from the water surface and eventually dissipate completely. This depth of no wave motion, called **wave-base**, corresponds to approximately one-half the wavelength.

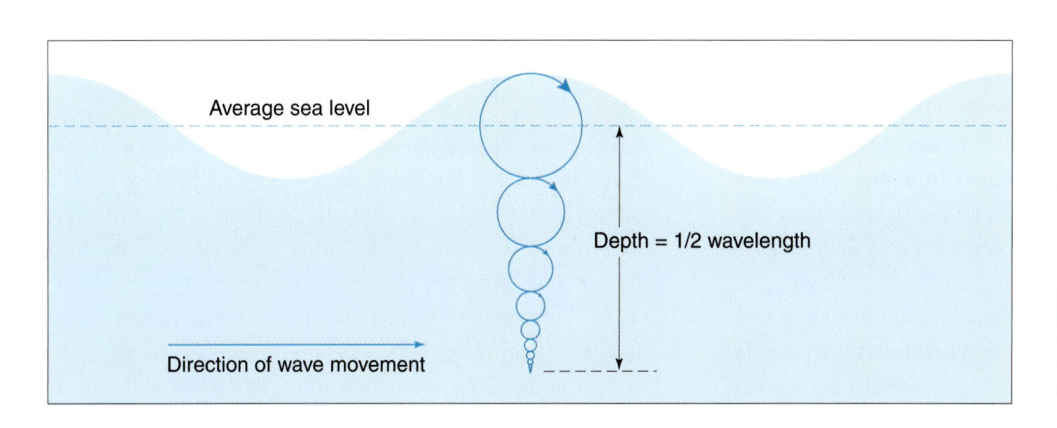

FIGURE 7.3
For waves passing through relatively deep water, the nearly circular orbits of water particles weaken with increasing depth and essentially disappear at a water depth of one-half the wavelength.

Only a very small net horizontal transport of water mass occurs with each wave cycle. Much more significant is the propagation of the changing shape of the water surface (the wave) and its energy. A wave crest travels twice as fast as the wave energy. If you carefully observe the movement of wind-generated waves on a pond, you will notice that they travel as "packets" of several waves. These are actually packets of energy. After a period of observation, you will be able to single out a particular wave crest and follow its movements. You will note that a wave arises in the back of the packet, grows in height, moves through the packet twice as fast as the packet, and disappears at the front of the packet. This same behavior is exhibited by ocean waves as they also propagate at twice the speed of the wave energy.

Water wave formation and evolution depend primarily on the wind's speed, turbulence, and duration as well as fetch. With increasing speed, the wind becomes more turbulent, that is, it consists of increasingly energetic irregular whirls of air motion known as *eddies*. We experience eddies as wind gusts and lulls on a windy day. Turbulent eddies have both horizontal and vertical components of motion so that the wind exerts forces that act parallel and perpendicular to a water surface. Forces acting parallel to a water surface drag surface water particles laterally whereas forces acting perpendicular to a water surface push the water up and down. At low wind speeds (less than a few km per hr), weak eddies produce capillary waves on water surfaces and, as noted earlier, the primary restoring force for these tiny waves is water's surface tension. However, these ripples roughen the water surface, thereby increasing the wind's drag on the ocean surface. Furthermore, growing waves present an increasingly sloped surface area against which the wind blows, creating even more wind resistance and wave growth. Increasing wind speed translates into more kinetic energy available for transfer to the ocean surface, therefore causing higher waves.

For the same wind speed, higher waves are produced with increasing **fetch**, the distance the wind blows over a continuous water surface. *Duration* refers to the length of time the wind blows from the same direction. For the same wind speed and fetch, waves become higher and longer the longer the wind blows from the same direction. Together, wind speed, fetch, and duration largely determine the amount of available kinetic energy that is transferred to the water surface.

In general, waves continue building in height and length as long as the energy supplied to the waves by winds exceeds the amount of energy dissipated in breaking waves. When the amount of dissipated energy equals the amount of energy supplied by the wind, no further wave build up (growth in wave height) occurs. Conversely, when the energy supplied by wind is less than that dissipated by the waves in breaking, waves decay (become weaker and smaller).

Wave interference also influences the growth and decay of sea waves. Waves generated by winds associated with two or more storm systems at sea may interfere constructively or destructively. In **constructive wave interference**, two or more wave crests coincide to form composite waves having heights greater than any of the original wave components (e.g. rogue waves). On the other hand, sets of waves can also interact such that the crests in one set of waves coincide with the troughs in another set, partially canceling both sets. The product is a composite wave whose height is smaller than that of the original wave components. This process is known as **destructive wave interference**. Both types of wave interference are almost always happening at sea and contribute to the continually changing patterns of waves on the open ocean.

DEEP-WATER AND SHALLOW-WATER WAVES

An ocean surface disturbed by storm winds becomes a confused mass of sharp-crested waves of various heights and lengths, moving in many different directions. This condition is known as **sea**. These are *forced waves*, that is, waves that are forced by the wind (while gravity is the restoring force). Waves continue to propagate well beyond the area of strong storm winds. Such waves, called **swell**, are lower and more rounded than waves forming directly under the storm winds. Swell consists of *free waves* where the only force acting on them is gravity.

Waves in water that is deeper than their wave-base (one-half the wavelength) are known as **deep-water waves**. **Celerity** is the speed of the wave relative to the water. The celerity of deep-water waves depends on wavelength and gravity and can be computed using the formula:

$$C = \sqrt{1.56 \times \text{wavelength}},$$

where C is the celerity in m per sec and the wavelength is given in meters. For example, a 10-m (33-ft) long wave's energy travels at 3.9 m (13 ft) per sec (9 mi per hr) whereas that of a 100-m (330-ft) long wave travels at 12.5 m (41.6 ft) per sec (28 mi per hr).

In relatively deep water, waves with longer wavelengths travel faster than waves with shorter wavelengths so that in the swell, waves sort themselves out with longer

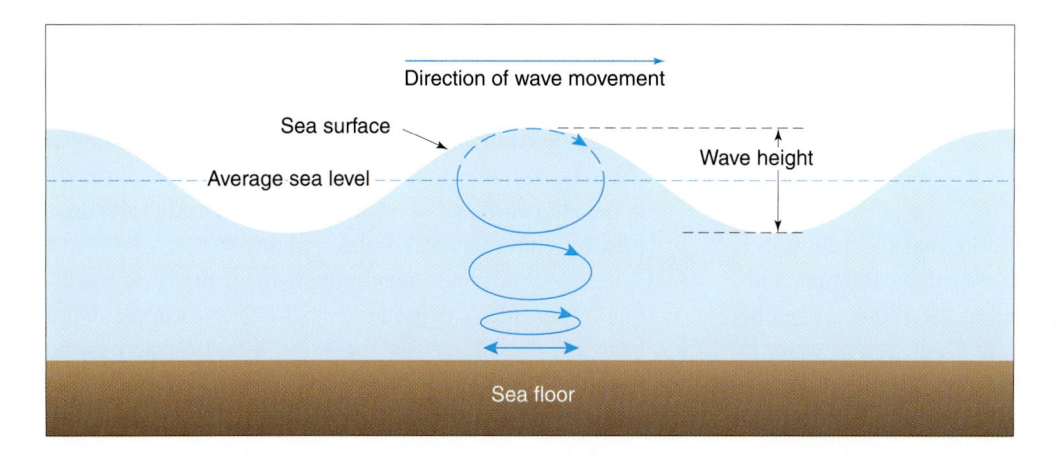

FIGURE 7.4
In shallow-water waves, the orbits of water particles gradually flatten with increasing depth, changing from circular to elliptical and ultimately to a back-and-forth motion near the ocean bottom. Figure not drawn to scale.

waves outdistancing shorter waves. With little dissipation due to friction, swell can travel thousands of kilometers from its source. This is why you sometimes observe swell from a distant storm striking a coastline even while the local weather is tranquil.

Recall from above that with deep-water waves, there is no interaction between the orbital motions of the wave and sea bottom. However, as water waves approach the coastline, they encounter increasingly shallow water. Eventually waves enter waters shallower than their wavebase and they begin to encounter frictional resistance as they start to "feel" the ocean bottom. At this point, these waves are no longer the free waves of the open ocean but are controlled by frictional contact with the ocean bottom. These are now **shallow-water waves** where the orbits of water particles gradually flatten with increasing depth, changing from circular to elliptical and ultimately to a back-and-forth motion near the bottom (Figure 7.4). Wave period is unchanged so that a wave interacting with the ocean bottom slows down, its wavelength shortens, and its wave height increases. (With constant period, the water and energy have no place to go but up!) A shallow-water wave occurs in water shallower than one-twentieth the wavelength, whereas **transitional wave** is the name given to a wave entering water having a depth that is between one-twentieth and one-half of the wavelength.

When waves enter waters having a depth of less than one-half the wavelength, the wave celerity depends on water depth and gravity but not wavelength. Shallow-water wave celerity can be computed using the formula:

$$C = \sqrt{g \times depth}$$

where C is the celerity in m per sec, g is gravity, and depth is given in meters. For example, the energy of waves in 10 m (33 ft) of water travels at 9.9 m per sec (22 mi per hr) whereas the energy of waves in 2 m (7 ft) of water travels at 4.4 m per sec (9.8 mi per hr). Hence, shallow-water waves slow as they enter shoaling water (water that becomes shallower).

As the water shoals, particles of water in the building wave crest move forward faster than the wave propagates toward shore. The wave becomes steeper and eventually unstable. The crest plunges forward, forming a *breaker*, dissipating wave energy in the process (Figure 7.5). Waves break when the ratio of wave height to wavelength approaches 1 to 7; at this point, the wave-crest angle (a measure of the steepness of the wave) is close to 120 degrees. Surfers take advantage of the fact that breakers are often much higher than deep-water wave crests. Breakers 2 m (7 ft) high exert a pressure of about 15,000 kg per square m, significantly impacting both natural and artificial features along the coast. A nearly continuous train of waves breaking along a shore is called **surf**.

FIGURE 7.5
A wave propagating through shallow water becomes steeper and unstable. When the ratio of wave height to wavelength approaches 1 to 7, the crest plunges forward as a breaker. [Photo by J.M. Moran]

156 Chapter 7 OCEAN WAVES AND TIDES

Over the years, mariners have developed methods of describing the state of the sea and relating that state to near-surface winds. For more on this, see this chapter's first Essay.

SEICHE

If you've ever sloshed water back and forth in a bathtub, you produced a seiche, a phenomenon first studied in Lake Geneva, Switzerland in the 1700s. A **seiche** (pronounced *say-sh*) is a rhythmic oscillation of water in an enclosed basin (e.g., bathtub, lake, or reservoir) or a partially enclosed coastal inlet (e.g., bay, harbor, or estuary). During a seiche, the water level in a basin rises at one end while simultaneously falling at the other end. A seiche episode may last from only a few minutes to a few days.

A seiche is a *standing wave*. This is in contrast to the wind-driven waves (described above) that are *progressive waves* in that they move through a body of water. With wind-driven waves, crests and troughs travel along the water surface but with standing waves, crests alternate vertically with troughs but at fixed locations. Gravity is the restoring force for both progressive and standing waves.

With a typical seiche in a simple enclosed basin, the water level near the center does not change at all (Figure 7.6). This location, called a *node*, is where water moves fastest horizontally. At either end of an enclosed basin, vertical motion of the water surface is greatest and horizontal movement of water is minimal; these are locations of *antinodes*. The motion of the water surface during a seiche is somewhat like that of a seesaw. The pivot point of the seesaw does not move up or down (analogous to a node) while people seated at either end of the seesaw move up and down (analogous to an antinode). Partially enclosed basins usually have a node at the mouth (rather than near the center) and an antinode at the landward end. Furthermore, some basins are complex and have several nodes and antinodes.

Wind, regional contrasts in air pressure, earthquakes, or tides can induce a seiche. For example, wind blowing persistently in the same direction across the broad expanse of a bay (or lake) causes water to pile up at the downwind shore. When the wind slackens, the piled up water is released and the water surface oscillates back-and-forth from one end of the bay (or lake) to the other as a seiche until eventually the water calms to a horizontal surface. The natural period of a seiche generally ranges from minutes to hours and is directly proportional to basin length. Hence, the natural period of a seiche in a large coastal inlet is considerably longer than in a small pond. Also, for the same basin, the natural period is inversely proportional to water depth; that is, the period shortens as water deepens.

Usually a seiche in a harbor or lake causes little concern because typically the vertical movements of water level are small—often only a few centimeters. Under certain conditions, however, a seiche may grow to great heights with serious consequences including coastal flooding and damage to moored vessels. A seiche grows as a consequence of **resonance**, that is, when the period of the disturbance (e.g., earthquake, wind) matches the natural period

FIGURE 7.6
A seiche in an enclosed basin with a single node. The maximum vertical motion of the water takes place at the antinodes and the maximum horizontal motion occurs at the node.

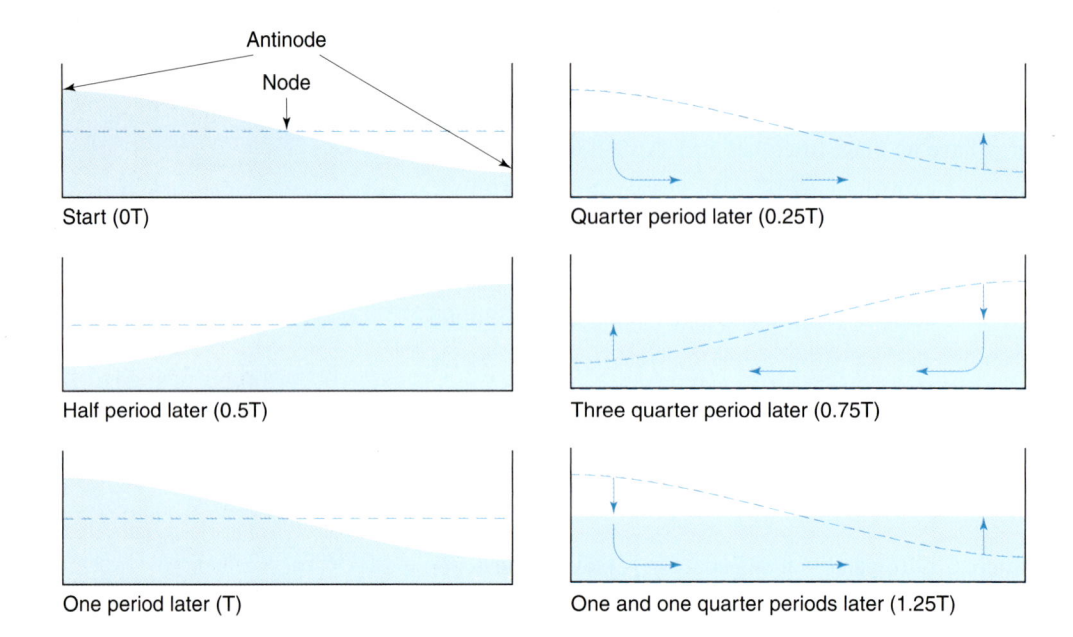

Start (0T)

Quarter period later (0.25T)

Half period later (0.5T)

Three quarter period later (0.75T)

One period later (T)

One and one quarter periods later (1.25T)

of oscillation of a basin. By timing your rhythmic disturbance of the water in a tub to match the natural period of the tub (about 1 second), you can cause the seiche to build until water splashes out of the tub and onto the floor. Through resonance, vibrations from the January 1994 Northridge, CA earthquake caused swimming pools to overflow throughout Southern California. In bays open to the ocean, if the period of tidal forcing equals the natural period of the bay, resonance can greatly increase the tidal range (difference in water level between high and low tide).

ATMOSPHERE-OCEAN TRANSFER

Sea waves help bring about the transfer of energy and matter between the atmosphere and ocean, a major interaction among subsystems of the Earth system. The largest wind-generated waves on the ocean surface are important in driving ocean currents by transferring momentum from the winds to ocean surface waters. Waves with shorter wavelengths play a major role in heat transfer from the ocean surface to the atmosphere through latent and sensible heating (Chapter 5). Latent heat derived from ocean's surface waters powers storms such as thunderstorms and tropical cyclones (e.g., hurricanes) (Chapter 8).

Waves with shorter wavelengths (especially breaking waves) also deliver salt particles to the atmosphere where they function as cloud condensation nuclei spurring development of clouds (Chapter 5). Droplets of ocean spray can also transfer microscopic marine algae and viruses into the atmosphere where winds can carry them long distances. Breaking waves also capture myriads of air bubbles that are carried tens of meters below the ocean surface. Gases in these bubbles dissolve in seawater; this process is an important source of dissolved oxygen and carbon dioxide for surface waters.

Ocean Tides

Astronomical tides are the regular rise and fall of the sea surface caused by the gravitational attraction between the rotating Earth and the moon and sun. Tides can be thought of as progressive planetary-scale waves that propagate across ocean basins; wave crests are high tides and wave troughs are low tides. A tide wave's length is considerably greater than the depth of the ocean so it behaves as a shallow-water wave.

In the theoretical case of an ocean-covered Earth of infinite depth with no continents, tides can be visualized as waves having lengths of one half of the circumference of the planet. Astronomical tides are *forced waves* in that they always follow the driving force of the moon and sun. Tide crests would be located directly below the the celestial body (moon or sun) that is responsible for the tide-generating force. On our theoretical Earth, the speed of propagation of the tide crest depends, at least in part, on the rotation of the planet relative to the sun or moon. On the equator where the planet's circumference is about 38,700 km (24,000 mi), the tide crest would travel at about 1600 km (1000 mi) per hr.

On the real Earth, continents break up the ocean into separate basins and the ocean has a finite depth. Tides are shallow-water waves so that wave celerity depends on water depth. For an average ocean depth of 4000 m (13,000 ft), the tidal celerity is about 200 m per sec (444 mi per hr). Tides speed up where the ocean is relatively deep and slow over ridges where the ocean is shallower. At sea, the amplitude of the tide wave is well under a meter but increases as it enters shallower coastal waters. Tides are measured mostly at coastal locations as local changes in sea level through time. The vertical difference in height between water levels at high and low tides is called the **tidal range** and generally varies between less than a meter to several meters (Figure 7.7). The time between successive high tides is the **tidal period**.

FIGURE 7.7
Low tide at a mooring in a harbor along the coast of Maine. Note the watermark on the dock indicating the level of the water at high tide. [Photo by J.M. Moran]

Tides are important because of their effects on ecosystems, local navigation, moorings, coastal structures, legal boundaries, fisheries, and recreation. Today, considerable scientific research focuses on the global nature of tides, including their influence on other physical processes in the ocean such as circulation, mixing, and wave generation. Furthermore, ocean and atmospheric scientists are interested in how storm-driven waves and surges combine with tides to affect the potential for coastal flooding and the rate of coastal erosion (Chapter 8). In this section, we focus on some fundamentals of ocean tides and their prediction.

TIDE-GENERATING FORCES

The tide-generating force is produced by the combination of (1) the gravitational attraction between Earth and the moon and sun, and (2) the rotation of the Earth-moon and Earth-sun systems. Forces combine to deform Earth's ocean surface into a roughly egg-shape with two bulges. One ocean bulge faces towards the moon and the other is on the opposing side of the planet, facing away from the moon (Figure 7.8). A similar interaction between Earth and sun produces two other ocean bulges that line up towards and away from the sun.

According to Isaac Newton (1642 - 1727), the gravitational attraction between two bodies is directly proportional to the product of the masses of the two bodies and inversely proportional to the square of the distance between them. Simply put, the greater the mass, the greater is the force of attraction whereas the greater the distance, the smaller the force of attraction. Although the sun is 10^7 times more massive than the moon, the moon is much closer to Earth and for that reason has a greater gravitational pull on Earth. In fact, the tide-generating force of the moon on the Earth is more than twice that of the sun on the Earth. For now, we will ignore the influence of the sun and focus on the moon.

The gravitational pull of the moon on Earth is primarily responsible for the bulge in the ocean surface that is directly under the moon. On the opposing side of the planet, the gravitational pull is weaker and the rotation of the Earth-moon system is primarily responsible for the tidal bulge. Earth and moon revolve around a common center of mass. Because Earth is much more massive than the moon, the center of mass of the system is within 4700 km (2900 mi) of Earth's center, that is, 1700 km (1060 mi) below the Earth's surface. This has been likened to a seesaw with an adult seated at one end and a child at the other end. The pivot point (center of mass of the adult-child system) must be moved toward the adult for the two individuals to balance the seesaw.

Newton's first law of motion predicts that a net force must operate in any rotating system and this net force in the rotating Earth-moon system gives rise to the tidal bulge on the side of the planet opposite the moon. Recall from Chapter 4 that according to *Newton's first law of motion*, an object in constant straight-line motion remains that way unless acted upon by an unbalanced (net) force. In a rotating system, the net force confines an object to a curved (rather than straight) path. Consider an analogy. Suppose that you are a passenger in an auto that rounds a curve at high rate of speed. You feel a force that pushes you outward from the turning auto. Actually, you experience the tendency for your body to continue moving in a straight path while the auto follows a curved path. In the same way, the rotation of the Earth-moon system causes the ocean to bulge outward on the side of the planet opposite the moon.

So far in our discussion of tide-generating forces, we have used the *equilibrium model of tides*, which assumes a frictionless Earth entirely covered by water. With this model, ocean bulges would always align with the celestial body that caused them. Furthermore, any location on the planet that is moved by Earth's rotation through the bulges would experience rising and falling sea level (i.e., tides). If only one celestial body (moon or sun) were present, each day a low-latitude locality would experience two high tides (when bulges pass) and two low tides (when halfway between the bulges). If the positions of the Earth and the other celestial body remained the same in space,

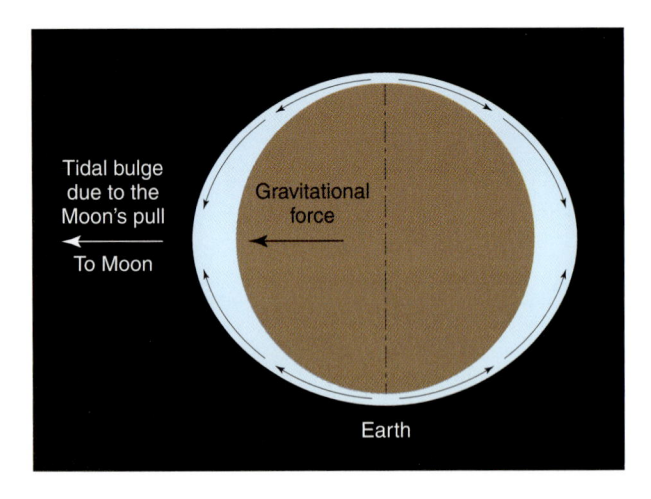

FIGURE 7.8
Two ocean tidal bulges produced by the gravitational attraction of the moon combined with the rotation of the Earth-moon system on an idealized water-covered Earth.

the period of these waves would be the time it takes for one half a rotation of Earth, about 12 hrs.

While Earth is rotating, however, the moon is revolving around Earth. The moon revolves around Earth once each lunar month (averaging 29.5 days, between new moons) and in the same direction as Earth's daily rotation. Hence, Earth must make more than a full rotation in order for a specific location on the planet to line up again with the advancing moon. Catching up with the advancing moon requires 24 hrs plus 1/29.5 of a day, which is approximately 24 hrs and 50 min. This moon-based day (24 hours, 50 minutes) is also called the **tidal day**. Because the tidal day is longer than the solar day, the times of high and low tide change by about 50 minutes from one solar day to the next.

The ocean's tidal bulges produced by the moon remain in the same alignment relative to the moon, but change their latitudinal (north-south) positions on Earth from day to day as they follow the moon during its monthly revolution about Earth. The plane of the moon's orbit is inclined by 5 degrees to Earth's equatorial plane so that during one lunar month, the moon's latitudinal position moves from directly over the equator northward to 28.5 degrees N (5 degrees beyond the Tropic of Cancer), back to the equator, on southward to 28.5 degrees S (5 degrees beyond the Tropic of Capricorn), and then back to the equator where another cycle begins.

When the moon is at its maximum latitudinal position, the center of one tidal bulge is just north of the Tropic of Cancer and the center of the other tidal bulge is on the other side of the planet just south of the Tropic of Capricorn. Consequently, a location along or near the Tropic of Cancer or the Tropic of Capricorn experiences only one significant tidal bulge in 24 hrs.

Sun-related ocean tidal bulges are produced in the same way as those caused by Earth-moon interactions. That is, the gravitational attraction between Earth and sun plus Earth's annual revolution around the mutual center of mass of sun and Earth generate a second set of similar but smaller tidal bulges that are aligned with the sun. As noted earlier, because of the sun's much greater distance from Earth, the sun's tidal pull on the ocean is less than half (about 46%) of the moon's tidal pull. These tidal bulges follow the sun (just as moon-related bulges track the moon), their latitudinal positions changing as Earth follows its yearly orbit about the sun (Figure 5.1).

The tide-generating force diminishes rapidly with increasing distance so that celestial bodies at greater distances from Earth than the sun are too far away to exert a significant tidal pull on Earth's ocean. The tide-generating force (arising from a combination of gravitational and rotational forces) is inversely proportional to the cube of the distance between Earth and any other celestial body. Hence, doubling the distance between two bodies reduces the tide-generating force by a factor of 2^3 or 8 times.

TYPES OF TIDES

Based on the number of high and low tides and their relative heights each tidal day, tides are described as semi-diurnal, mixed, or diurnal (Figure 7.9). When the moon is directly over Earth's equator, its associated tidal

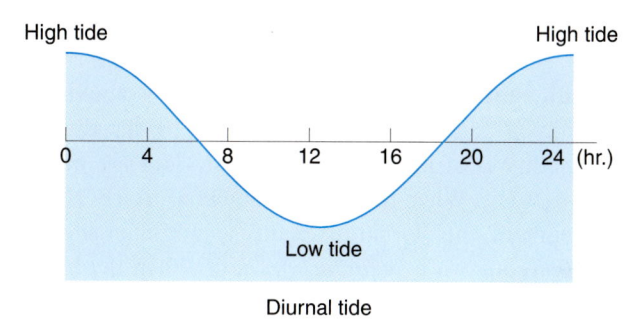

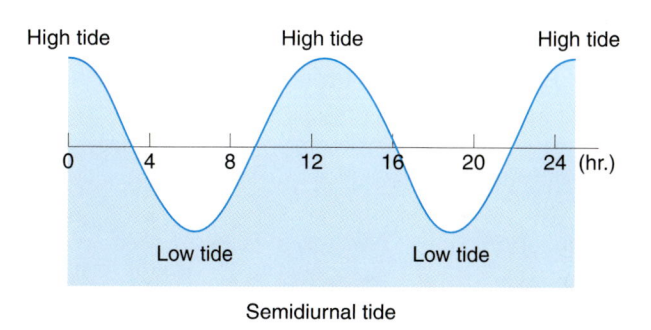

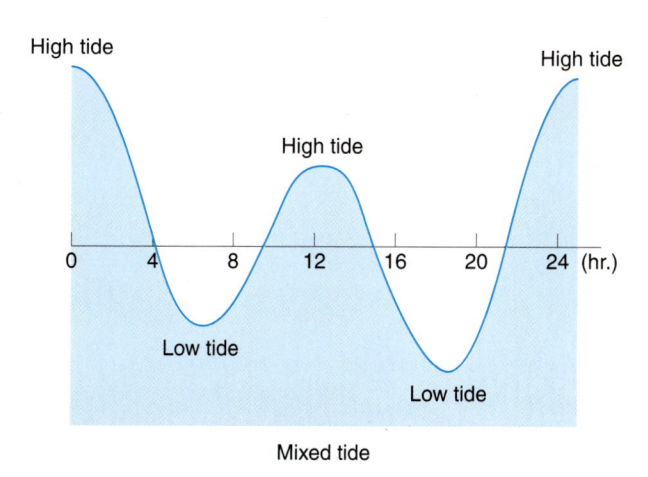

FIGURE 7.9
Types of astronomical tides observed in coastal locations.

bulges are centered on the equator. In theory, all locations on the planet except at the highest latitudes would rotate through the two tidal bulges and experience two equal high tides and two equal low tides per tidal day; this is known as a **semi-diurnal tide**. Semi-diurnal tides have a period of 12 hrs and 25 min, and theoretically have a wavelength of more than half the circumference of Earth.

Different types of tides occur when the moon is either north or south of the equator. Whereas semidiurnal tides are observed at the equator at all times, most locations north or south of the equator experience two unequal high tides and two unequal low tides per tidal day; this is called a **mixed tide** and the difference in height between successive high (or low) tides is called the **diurnal inequality**. When the moon is above the Tropic of Cancer or Tropic of Capricorn, the diurnal inequality is at its maximum and the tides are called *tropic tides*. When the moon is above or nearly above the equator, the diurnal inequality is minimum and the tides are known as *equatorial tides*. When the moon and its associated tidal bulges are either north or south of the equator, most points at high latitudes in theory would be impacted by one tidal bulge and would experience one high tide and one low tide per tidal day. This so-called **diurnal tide** has a period of 24 hrs and 50 min.

The separate sets of ocean bulges related to the moon and sun act at times together and at other times in opposition. About every two weeks, the positions of the sun, moon, and Earth form a straight line (Figure 7.10A). At these times of new and full moon phases as viewed from Earth, the lunar- and solar-related ocean bulges also line up (and add up) to produce tides having the greatest monthly tidal range (that is, the highest high tide and lowest low tide); these are called **spring tides**. Between spring tides, at the first and third quarter phases of the moon, the sun's pull on Earth is at right angles to the pull of the moon (Figure 7.10B). At this time, tides have their minimum monthly tidal range (that is, unusually low high tide and unusually high low tide); these are called **neap tides** or *fortnightly tides*. Furthermore, the moon orbits Earth in an ellipse (rather than a circle) so that the moon is closest to Earth (stronger tide-generating force) at *perigee* and farthest from Earth (weaker tide-generating force) at *apogee*. The moon completes one perigee-apogee-perigee cycle once every 25.5 days.

TIDES IN OCEAN BASINS

To this point in our discussion of tides, we have assumed a water-covered Earth that rotates on its spin axis through the tidal bulges. In a more realistic situation, many non-astronomical factors modify ocean tides, including the

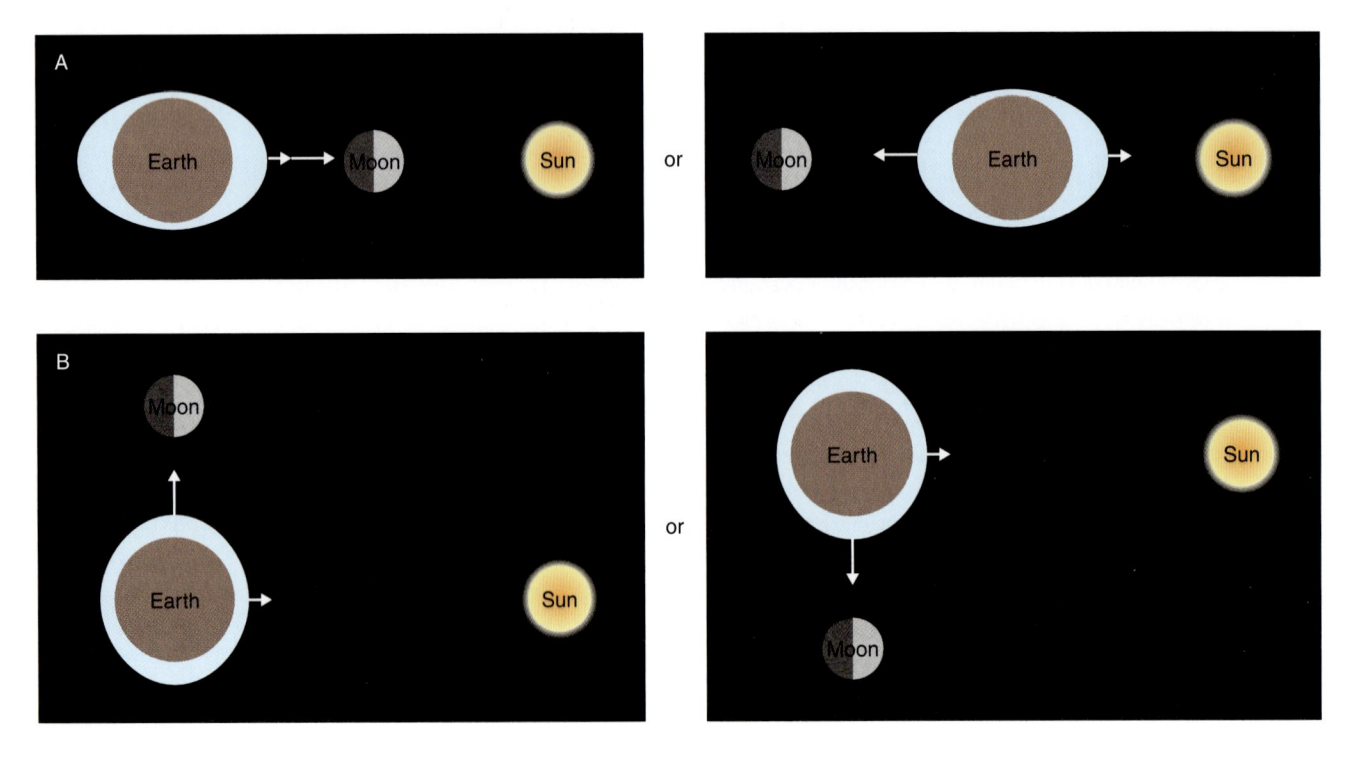

FIGURE 7.10
The configuration of Earth, moon, and sun that is responsible for (A) spring tides, and (B) neap tides.

presence of continents, the Coriolis effect, winds, and variations in coastline configuration, water depth, and bottom topography. Tidal bulges move relatively unimpeded around the globe only in the Southern Ocean near Antarctica. Consider, for example, the idealized case of tides in a Northern Hemisphere ocean basin of uniform depth that is completely surrounded by land. Assume also that the moon is providing the only tide-generating force and initially is situated directly above the ocean basin.

As Earth rotates from west to east, the tidal bulge shifts toward the western boundary of the ocean basin and the water surface slopes gently downward toward the east. The western boundary of the basin experiences high tide while the eastern boundary experiences low tide. Tide waves are shallow-water waves. Hence, as noted earlier in this chapter, the orbits of water particles flatten with increasing depth, changing from circular to elliptical and ultimately to a back-and-forth motion near the ocean bottom. The horizontal motion of water particles in a tide persists for long periods (because the tide wave is constantly being forced) so that the tide wave is subject to the Coriolis effect. As the tidal bulge at the western boundary begins to move down slope toward the east, the Coriolis deflects water particles to the right (in the Northern Hemisphere) so that the tidal crest (high tide) rotates into the southern portion of the basin. Now, the water surface slopes downward toward the north. The tidal crest continues to rotate around the basin in a counterclockwise direction (viewed from above). When the tide is high on one side of the basin, it is low on the opposite side. In Southern Hemisphere basins, the reversal of the Coriolis deflection causes the tide wave to rotate in a clockwise direction (viewed from above).

This so-called *dynamic model of tides* applies reasonably well to seas and large embayments as well as the open ocean. Ocean scientists graphically represent the rotary motion of the tide wave in a basin by a series of cotidal lines radiating outward from a central node like the spokes in a bicycle wheel (Figure 7.11). A **cotidal line** joins points where high tide occurs at the same time of day; they are usually drawn at one-hour intervals. The tidal range varies from zero at the node to a maximum at the antinode along the coast.

Diurnal tides make one complete circuit per tidal day whereas semidiurnal or mixed tides complete two circuits per tidal day. The period of the rotary tide waves is 12 hrs 25 min for semidiurnal tides and 24 hrs 50 min for diurnal tides. For an ocean of average depth (about 4000 m or 13,000 ft), a tide wave progresses as a shallow-water wave at about 645 km (400 mi) per hr. (As with wind-generated water waves, this is the speed of the wave energy, not the water.) Actual wave celerity along the coast varies greatly due to topography. Along the west coast of the U.S., tide waves travel at about 565 km (350 mi) per hr, whereas along the west coast of Africa in the Northern Hemisphere, speeds are about 360 km (225 mi) per hr. For coastal residents, however, these high speeds are not apparent because for a diurnal tide, a 2-m (6.5-ft) rise in water may take slightly more than 6 hrs.

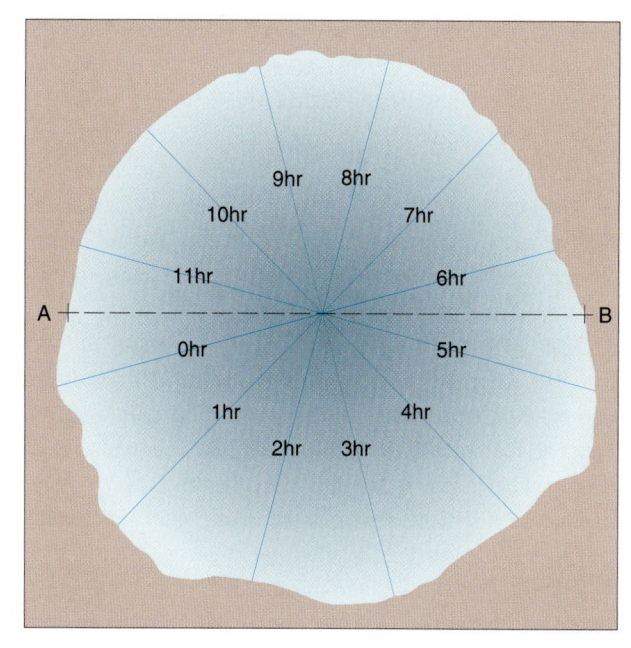

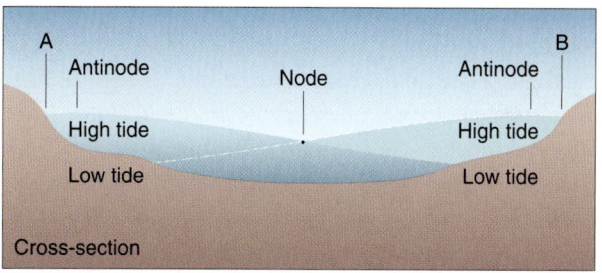

Shallow basins with just the right length may have a natural period of oscillation that matches the period of the tide-generating force. This *resonance* explains the extraordinary tidal range (as great as 16 m or 53 ft during a

FIGURE 7.11
In this idealized Northern Hemisphere ocean basin bordered on all sides by land (top), a tide wave rotates in a counterclockwise direction (viewed from above). Lines radiating outward from the central node are cotidal lines that join points where high tide occurs at the same time of day. This is a semi-diurnal tide. Also shown (bottom) is a vertical cross-section from point A to point B. Note that the tidal range varies from zero at the node to a maximum at the antinodes (along the coast).

spring tide) observed in the Bay of Fundy, Nova Scotia. The natural period of oscillation of the Bay of Fundy (about 12 hrs) is very close to the period of the moon's tidal forcing (12 hrs, 25 min). Non-astronomical factors help explain why locations on the U.S. Atlantic coast have predominantly semidiurnal tides whereas many places on the Gulf Coast have predominantly diurnal tides, and localities on the Pacific Coast have mostly mixed tides.

TIDAL CURRENTS

Alternating horizontal movements of water accompanying the rise and fall of astronomical tides in coastal areas are called **tidal currents**. Along the boundaries of an ocean basin (the location of the *antinodes) tidal ranges* and hence, tidal currents are at their maximum. Irregularities along the coast modify the rotary motion of tide waves so that tidal currents move more directly into and out of rivers and harbors (Figure 7.12). Tidal currents flow in one direction during part of the tidal cycle and in the opposite direction during the remainder of the tidal cycle. When tidal currents are directed toward the land, water levels rise in harbors and rivers; these are called **flood tides**. Tidal currents flowing seaward with falling sea levels are called **ebb tides**. Between flood and ebb tides are **slack water** periods (little or no horizontal movement).

In some coastal areas where the tidal range is relatively large and the flood tide enters a narrow bay or channel, a **tidal bore** forms and moves upstream in a river or shallow estuary. A tidal bore is a wall of turbulent water, usually less than a meter in height. Tidal bores are well known at the mouth of the Amazon River in Brazil, on the Severn River in England, and in Turnagain Arm off Cook Inlet, Alaska.

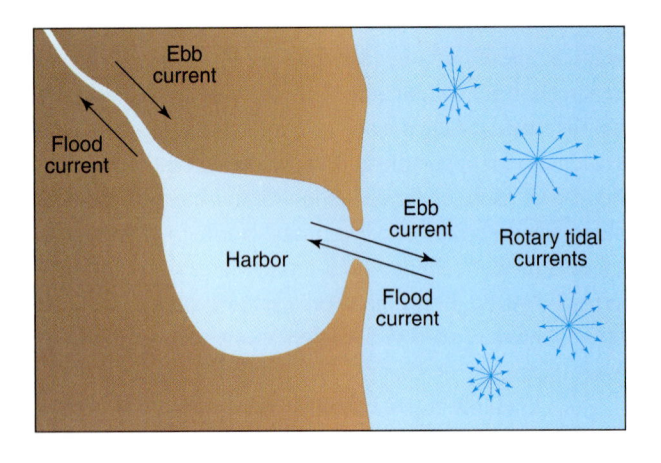

FIGURE 7.12
Ebb tide and flood tide in a harbor.

OBSERVING AND PREDICTING TIDES

Observing the changing water levels caused by astronomical tides is relatively simple and has long been important for major ports. Knowing tide levels helps pilots and ship captains avoid running aground in shallow stretches of harbor channels. In the 1850s, the port of New York began operating a real-time tide gauge that indicated to ship operators the tide level and whether it was rising or falling. Today government agencies (e.g., NOAA, Canadian Hydrographic Services, the British Admiralty) maintain tide observing and prediction systems to advise mariners based primarily on the output of numerical models and computers. On the other hand, fewer sources of information exist on tidal currents in harbors because they are much more difficult and expensive to observe than tide levels. Whereas tide levels are nearly uniform over broad areas, tidal currents change quickly with variations in winds and river discharge. Furthermore they are affected by complex shorelines and bottom topography. Nonetheless, NOAA's Physical Oceanographic Real Time Services (PORTS) provides information on tidal currents for 10 major U.S. ports.

Predictions and real-time observations of tides are distributed electronically to ship operators. In the United States such predictions are made by NOAA's National Ocean Service CO-OPS (Center for Operational Oceanographic Products and Services) for more than 3000 locations and are available to the public online. Consider how tide predictions are made.

Periods of motions of the Earth, sun, and moon in space (i.e., orbits and rotation) are fixed and known precisely. The predictability of the movements of tide-generating celestial bodies means that astronomical tides can also be predicted with great accuracy. Tides are waves so that local tides can be resolved mathematically into their various components, called **partial tides**. Partial tides are forecasted individually and added together to predict the height and timing of future local tides. Although as few as four partial tides can account for 70% of the total tidal range, some 60 components are commonly used (to account for both astronomical and non-astronomical factors). More than 100 components must be considered to predict accurately the tides along a complex coastline, such as that of Alaska. Local astronomical tides are best predicted when based on data collected for at least 18.6 years, a period that encompasses most of the astronomical configurations of the Earth-moon-sun system that generate the tides. However, only a single year of tide gauging data usually suffices for very reasonable tide predictions. In most cases,

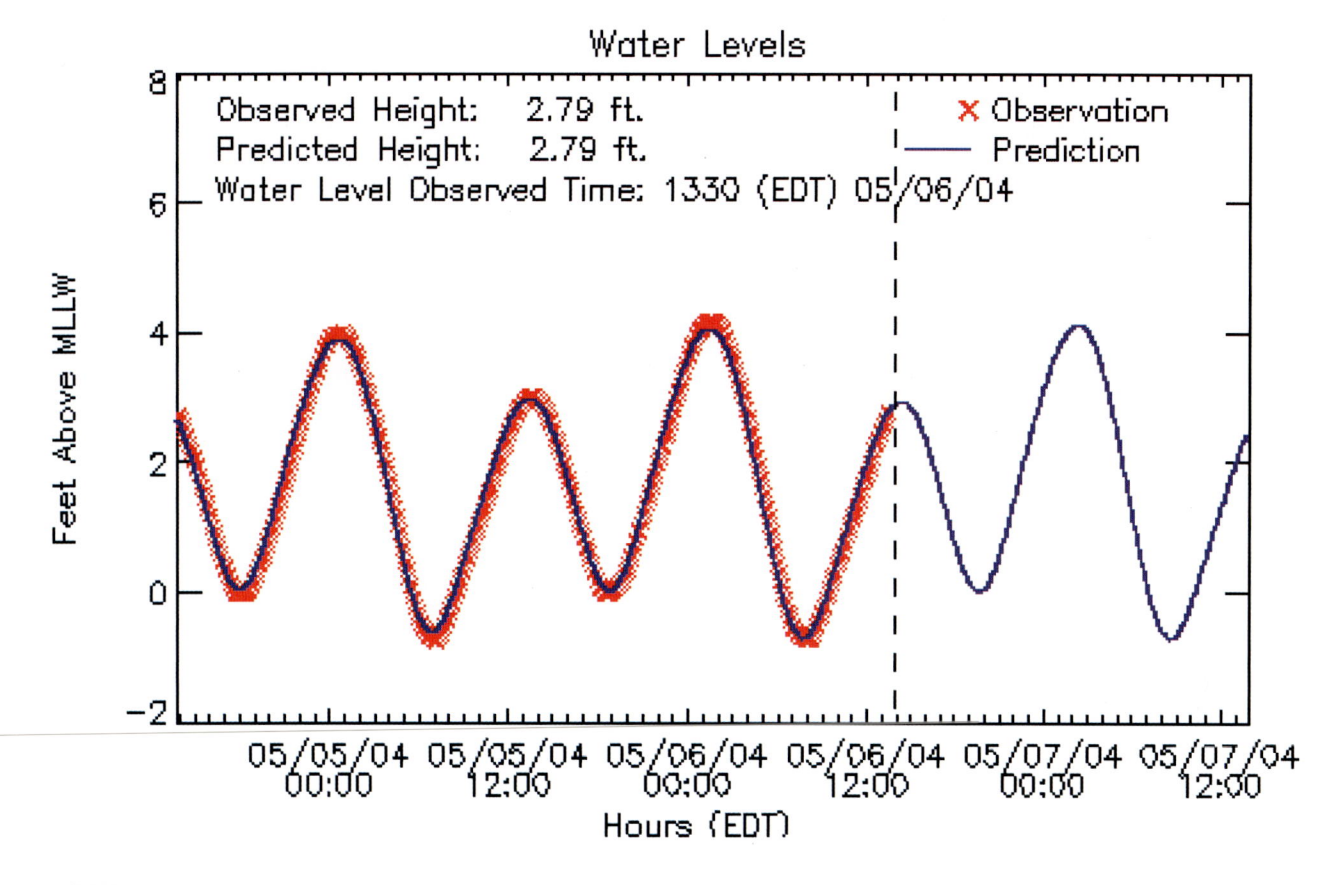

FIGURE 7.13
Observed (red) and predicted (blue) water level (in feet above Mean Low Low Water) for the tide gauge located at Nantucket Island, MA from 5 to 7 May 2004. Vertical dashed line marks the "present" time. [From NOAA, National Ocean Service, Center for Operational Oceanographic Products and Services.]

local winds and atmospheric pressure variations are the primary causes of the difference between the actual and predicted tide (Figure 7.13).

OPEN-OCEAN TIDES

Earth-orbiting satellites routinely measure tides over the deep-ocean. The U.S.-French TOPEX/POSEIDON satellite is equipped with a radar-altimeter that bounces microwave signals off the sea surface and precisely measures sea level. With such data, it is possible to determine what happens to tide waves and their energy as they travel across deep ocean basins. For more on measuring sea level from space, refer to this chapter's second Essay.

As shallow-water waves, tides lose energy through frictional drag with the ocean floor, especially in the ocean's shallow seas and along continental margins. Satellite measurements show that about three quarters of the global tidal energy dissipates in shallow seas bordering northern Europe, in the Yellow Sea off Asia, in the shallow seas around Australia, near Argentina, and in Canada's Hudson Bay.

Open-ocean tides are important in mixing deep-ocean water. Ocean scientists long assumed that wind was the principal mixing agent of the open ocean, but satellite altimeter data now show that tidal mixing in the deep ocean is about as important as the wind. Perhaps as much as half of the tidal energy in the ocean is dissipated in mixing processes when tidal currents in the deep ocean flow over seamounts, ridges, and other rugged features on the ocean floor or weave through passages between islands.

Tidal currents flowing over topographic irregularities on the ocean floor generate internal waves that propagate away from their source. These internal waves arise from the fact that water density increases gradually with increasing depth. As tidal currents encounter a seamount or submarine ridge, relatively dense water is forced upward into slightly less dense water. Then to the lee of the obstacle gravity pulls the denser water downward. However, the descending water gains momentum and overshoots its equilibrium level and descends into denser water. The water then ascends thereby forming an oscillating wave that propagates horizontally. Because these waves are generated by tides, they occur at tidal frequencies and are called **internal tides**. Internal tide waves can travel thousands of kilometers beyond the obstruction that formed them and can have very large wave heights. They also break, like surf on a beach but under water, locally mixing waters above and below the internal wave. Internal tides are important in mixing cold bottom waters with warmer surface waters as part of the global oceanic conveyer belt circulation (Chapter 6).

Recently ocean scientists gathered evidence that internal tides influence the gradient of the continental slope (Chapter 2). The inclination of the continental slope varies from very gentle (as small as one degree) to precipitous (up to 25 degrees where submarine canyons cut into the slope). About 80% of the continental slope is inclined at less than 8 degrees and the average inclination is about 4 degrees. According to geological studies, however, the sediments supplied to the continental slope (mostly by rivers) would support a stable average slope of perhaps 15 degrees or greater. Data acquired from model studies, dives in piloted submersible vessels, and moored instruments show that internal tides produce strong currents that prevent accumulation of sediment that would make the continental slope steeper. In fact, the internal waves ascending the continental slope apparently behave very much like ordinary sea waves entering the shoaling waters of a coastal area (with changes in amplitude, wavelength, and water velocity). Whereas the influence of internal tides is widespread along the continental slope, turbidity currents and tectonic forces can be important locally and regionally in shaping the slope.

A vast amount of energy is involved in ocean tides and waves so it is not surprising that considerable interest has focused on developing technologies to tap this energy to generate electricity. Although the potential is enormous, very little of this energy resource has been developed to date. For more on tidal power, see this chapter's third Essay.

Internal Waves

In the previous section, we saw that internal tides are waves that form in the deep ocean where density gradually increases with depth. **Internal waves** also form within the ocean along interfaces where the change in density with depth is more abrupt. Fridtjof Nansen is credited with discovering internal waves in the Arctic Basin in the late 1800s.

Fundamentally, waves are pulses of energy that travel along surfaces separating fluids having different densities. Hence, wind-driven waves and seiches occur at the interface between air and water. Similar interfaces also occur beneath the ocean surface. As described in Chapter 6, the ocean is not homogeneous; water density changes with depth principally in response to variations in temperature and salinity. Favorable sites for development of internal waves include the base of the *mixed layer* and interfaces (*pycnoclines*) between water masses having different densities. Internal waves also form in estuaries along the pycnocline between fresh river water and salty ocean water (Chapter 8).

The smaller the density contrast between two fluids in contact, the slower the internal wave propagates and the greater the wave height. Density contrasts between different water masses are about 1000 times less than that between air and water so that internal waves propagate more slowly than surface waves but wave heights are much greater. Simply put, with internal waves there is a much smaller density difference for gravity to act upon. Typically, internal waves have lengths in the hundreds of meters and heights of several meters but sometimes are considerably higher (to 100 m, 330 ft, or more). Besides astronomical tides, slumping on the ocean floor, turbidity currents, and water masses slipping over one another also can generate internal waves. Even ships moving across the sea surface can generate internal waves on shallow pycnoclines.

Tsunamis

On 1 April 1946 in a span of 27 minutes, two earthquakes shook the Scotch Cap Lighthouse on Unimak Island in Alaska's Aleutian chain. Shortly afterward, a huge sea wave appeared and obliterated the lighthouse, killing the five-man crew and washing debris 35 m (115 ft) above sea level. All told, this destructive sea wave, known as a **tsunami** (from *tsu-nami*, the Japanese word for harbor wave and pronounced *sue-nah-mee*), was responsible for more than 165 fatalities and $26 million (1946 dollars) in property damage. During the 1990s, 82 tsunamis were reported worldwide; 10 of them together claimed more than 4000 lives. The highest reported tsunami wave reached 31 m (102 ft), one of a series of huge waves that struck Okushiri, Japan on 12 July 1993, taking 239 lives. In the 20th century, the 141 most damaging tsunamis killed more than 70,000 people. Over the past 2000 years, tsunamis may have claimed the lives of more than 460,000 people living in the Pacific region.

Although the media sometimes refers to tsunamis as *tidal waves*, they have nothing to do with astronomical tides. Most tsunamis originate in the Pacific Ocean and are generated by submarine earthquakes occurring along convergent plate boundaries and associated subduction zones (Chapter 2). They are less of a hazard in the Atlantic except perhaps around the Caribbean plate boundary. Tectonic movements of the ocean floor disturb the overlying water column, generating a long-wavelength sea wave. In addition, volcanic eruptions, meteorite impacts, submarine landslides, and even calving glaciers can produce tsunamis. For example, the violent eruption of the Indonesian volcano Krakatoa on 26-27 August 1883 spawned huge tsunami waves (one reported to be the height of a 12-story building) that killed at least 36,000 people.

At sea, a tsunami travels as a series of waves with wavelengths typically 100 to 200 km (60 to 120 mi) and periods of 10 to 30 minutes. Tsunamis are shallow-water waves because their wavelengths are much greater than the depth of the ocean. For the average depth of the ocean of about 4000 m (13,000 ft), wave celerity is about 700 km (435 mi) per hr—approaching the speed of a commercial jetliner. But with a wave height of only a few meters or less and taking 10 to 30 minutes to pass, a tsunami would be unnoticed moving under a ship at sea. In deep water, tsunami wave energy is distributed through a huge volume of water. But like any other wave, when a tsunami enters shoaling coastal waters, it slows and its energy is forced into a decreasing volume of water. Wavelength shortens and wave height increases, sometimes building to extraordinary heights as the tsunami approaches the coast. The initial arrival of a tsunami is often a trough that draws water off the beach. This unusual event sometimes entices curious people to head for the beach where they are caught by the tsunami crest coming ashore. The first wave is often followed by a succession of other powerful tsunami waves. Tsunami waves crash ashore and take lives and destroy property.

In the United States, coastal residents of Hawaii and Alaska are at greatest risk from a tsunami although tsunamis have struck the West Coast. Hawaii has reported a dozen damaging tsunamis since 1895, the most destructive of which occurred on 1 April 1946. A tsunami destroyed most of the waterfront of Hilo and caused 159 deaths (the same tsunami that destroyed the Scotch Cap Lighthouse). A tsunami that struck the Seward, Alaska area on 28 March 1964 took more than 100 lives. Although tsunamis are rare along the West Coast of the U.S. outside of Alaska, recently discovered geological evidence suggests that the Cascadia subduction zone located off the Pacific Northwest Coast may generate tsunami-producing earthquakes about every 300 to 700 years. This finding has generated scientific interest in the tsunami hazard all along the West Coast of North America.

The Hawaiian tsunami of 1946 led to the establishment of the Pacific Tsunami Warning Center in Hawaii in 1948. The tsunami that struck Prince William Sound during the 1964 Alaskan earthquake led to formation of the West Coast and Alaska Tsunami Warning Center in 1967 in Palmer, Alaska. Scientists at these centers attempt to forewarn the public of an approaching tsunami based on earthquake occurrences detected by seismic networks around the Pacific rim plus tide gauge readings in coastal areas. However, these methods have not been very successful—the false alarm rate has hovered around 75% since the 1950s. Submarine earthquakes do not always generate a tsunami and warnings provided by coastal tide gauges generally arrive too late for adequate public warning.

Improved early warning of an approaching tsunami is the goal of NOAA's Deep-Ocean Assessment and Reporting of Tsunamis (DART) system. The centerpiece of DART is a collection of sensitive instruments placed on the ocean floor at depths to 500 m (1600 ft) that measures slight variations in pressure exerted by the overlying water column in response to a passing tsunami wave. The sensitive pressure sensor is capable of detecting water pressure changes associated with a tsunami wave as small as one centimeter. Data from the DART instrument is transmitted to a nearby moored buoy floating on the ocean surface, and then to a satellite for downloading at NOAA's two Tsunami Warning Centers in order that more timely warnings could be issued.

As of early 2004, Dart pressure sensors had been deployed at six locations on the Pacific Ocean floor: one along the equator between Hawaii and Chile, two off the U.S. West Coast, and three south of Alaska's Aleutian Islands. More sensors are planned. Data from DART sensors are essential components of the U.S. National Tsunami Hazard Mitigation Program, a partnership that includes representatives of various federal agencies (i.e., NOAA, FEMA, and the U.S. Geological Survey) and the Pacific Coast states. An important outcome of the program is the construction of maps that identify coastal areas most vulnerable to inundation by a

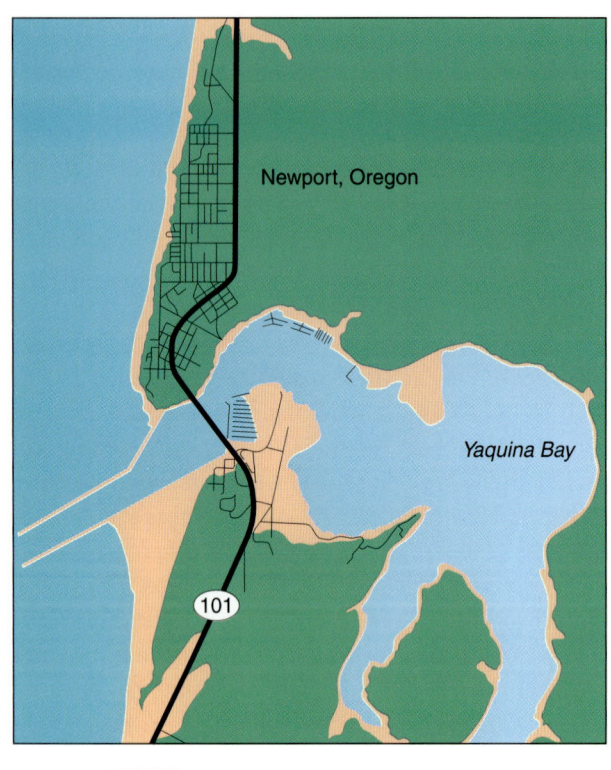

FIGURE 7.14
Tsunami-risk map for Newport, OR, identifying areas where tsunami flooding is likely. Such maps inform local planning for tsunami hazard mitigation. [Produced by the Oregon Department of Geology in collaboration with the Oregon Graduate Institute of Science and Technology]

tsunami (Figure 7.14). As of this writing, tsunami-risk maps have been drawn for 125 communities where a total of about 1.3 million people are at risk.

Conclusions

Winds supply kinetic energy to form waves on the ocean surface, which then propagate horizontally, driven by gravity, until they expend their energy by breaking on a distant shore. As noted in Chapter 6, this is an important aspect of ocean-atmosphere interaction that provides energy to drive surface currents and mix surface waters. Wave behavior depends on wind speed, fetch, and duration as well as water depth. In terms of interaction with the ocean bottom, a distinction is made between deep-water waves and shallow-water waves.

The gravitational attraction of the moon and sun combined with the rotation of the Earth-moon and Earth-sun coupled systems generate planetary-scale shallow-water tide waves in the ocean basins. The consequence is a periodic rise and fall of sea level that is so familiar to coastal dwellers as ocean tides. Tides also cause currents that are among the strongest of all currents in coastal regions and are responsible for some water movements in the deep ocean. In Chapter 8, we take a closer look at waves and tides in the coastal zone, where the majority of the human population lives.

Basic Understandings

- Kinetic energy of air in motion (the wind) is transferred to the ocean surface causing surface waves. Wind-driven ocean waves are oscillations of the sea surface that propagate horizontally along the ocean/atmosphere interface away from where they were generated. Energy is propagated with the wave—not the water (except where the wave breaks).
- The life cycle of a sea wave begins with an input of kinetic energy that disturbs the ocean surface. Water's surface tension is the restoring force for capillary waves—small waves having wavelengths less than 1.7 cm (0.7 in.). For sea waves longer than capillary waves, the restoring force is gravity. Open ocean deep-water waves propagate away from their energy source driven by gravity with little energy loss through friction. Waves expend their energy as they encounter the ocean bottom and shore, finally dissipating their energy in breaking.
- Wave height, the vertical distance between wave trough and wave crest, depends on wind speed (and turbulence), duration of the wind in the same direction, and the distance (fetch) over which the wind blows. Wave interference also affects the growth and decay of sea waves. Wind-driven waves associated with two or more storm systems at sea may interfere constructively or destructively. In active wave-forming areas, waves (called a sea) have a mixture of wavelengths and periods, plus sharp crests, giving the ocean surface a chaotic appearance. As waves move away from the area where they formed, they become more rounded and sort themselves by wavelength; these are called swell.
- Waves are either shallow-water waves or deep-water waves depending on the ratio of wavelength to ocean depth. The depth of no-wave action is approximately one-half the wavelength of a surface wave. Shallow-water waves are affected by frictional interaction with

- the ocean bottom. In deep-water (depths greater than one-half the wavelength), waves are unaffected by the ocean bottom.
- As a wave moves into water depths of less than one-half its wavelength, the wave shortens and slows but with no change in period. Wave energy has no place to go but up, so the wave builds and becomes steeper as the top of the wave moves faster than the bottom. The wave becomes unstable and breaks expending its energy in the surf zone.
- A seiche is a rhythmic oscillation of water in an enclosed basin or a partially enclosed coastal inlet. During a seiche, the water level in a basin rises at one end while simultaneously falling at the other. Under certain conditions, a seiche grows to great heights with serious consequences including coastal flooding and damage to moored vessels. A seiche grows through resonance, that is, amplification that occurs when the period of the disturbance (e.g., earthquake, wind) matches the natural period of oscillation of the basin.
- Astronomical tides are planetary-scale shallow-water waves responsible for the regular rise and fall of the ocean surface. Tides typically are measured at coastal locations as local changes in sea level as a function of time. The difference in height between water levels at high and low tides is called the tidal range and the time between successive high tides is the tidal period.
- Tides are caused by the gravitational attraction between the Earth and the moon and sun combined with the rotation of the Earth-moon and Earth-sun coupled systems. Although its mass is considerably smaller than that of the sun, the moon has a greater influence on ocean tides because it is much closer to Earth. The tide-generating force produces two bulges of water on Earth's ocean surface, one facing the moon (or sun) and one on the opposite side of the planet. Earth's rotation through these two tidal bulges causes the regular rise and fall of sea level.
- Depending on factors such as specific location and phases of the solar and lunar cycles, a coastal locality may experience two equal high tides and two equal low tides per tidal day (semi-diurnal tide), two unequal high tides and two unequal low tides per tidal day (mixed tide), or one high tide and one low tide per tidal day (diurnal tide). Many non-astronomical factors modify ocean tides, including the presence of continents, the Coriolis effect, winds, and variations in coastline configuration, water depth, and bottom topography.
- In a Northern Hemisphere basin, the Coriolis effect causes the shallow-water tide wave to rotate around the basin in a counterclockwise direction (viewed from above). When the tide is high on one side of the basin, it is low on the opposite side. In Southern Hemisphere basins, the reversal of the Coriolis deflection causes the tide wave to rotate in a clockwise direction (viewed from above).
- Ocean scientists graphically represent the rotary motion of the tide wave in an ocean basin by a series of cotidal lines radiating outward from a central node. A cotidal line joins points where high (or low) tide occurs at the same time of day.
- Alternating horizontal movements of water accompanying the rise and fall of tides in coastal areas are called tidal currents. Tidal range and hence, tidal currents are at a maximum along the boundaries of an ocean basin (coasts). In coastal embayments and harbors, tidal currents are constrained to flow back and forth as flood tides (flowing toward land) or ebb tides (flowing away from land).
- Treated as a wave, the tide at a particular location can be resolved into many component partial tides. Partial tides are predicted individually and added together to forecast the local astronomical tide.
- Internal waves form and propagate below the ocean surface on interfaces between water masses of different densities. Density contrasts between different water masses are about 1000 times less than that between air and water so that internal waves propagate more slowly than surface waves but wave heights are much greater.
- A tsunami is a shallow-water ocean wave that develops when a submarine earthquake, landslide, or volcanic eruption disturbs the ocean water. At sea, tsunami wave height is only a few meters with a period of 10 to 30 minutes. Its wavelength is typically 100 to 200 km (60 to 120 mi) with speeds of hundreds of kilometers per hour. With small heights, long wavelengths, and long periods, Tsunamis are imperceptible in the open ocean. However, when they reach shore, the wave can build to tremendous heights and take lives and cause considerable property damage.

ESSAY: The State of the Sea

Experienced sailors are usually adept at describing the state of the sea. The term *sea state* refers to the overall appearance of the ocean surface, including the dimensions of wind-generated waves and the presence of surface phenomena such as white caps, foam, or spray. Because the wind is largely responsible for the appearance of the sea surface, approximate wind speed can be inferred from the sea state. The sea state also provides an indication of wind direction because wave crests tend to move in approximately the same direction as the near-surface wind.

In the 19th century, prior to the invention of accurate instruments for measuring wind speed (*anemometers*), mariners developed a method to estimate the wind speed from the observed sea state. This method evolved from a wind force scale proposed in 1805 by Francis Beaufort (1774-1857), later an Admiral in the British Royal Navy. Initially, Beaufort devised a method for estimating the force of the wind using the amount of sail needed by a fully rigged sailing vessel, a British frigate of the day. He divided his scale into 13 increments, ranging from 0 (calm) to 12 (hurricane). Hence, for example, a wind of force 1 (light air) was described as "just sufficient to give steerage way" for a frigate, whereas a force 12 (hurricane) was "that which no canvas could withstand." By 1838, the Beaufort scale was modified to specify the sea state and was adopted by the British Navy. On this scale, force 1 winds produce "ripples with appearance of scales; no foam crests" (wave heights of about 7.5 cm or 3 in.) whereas a force 12 wind produces a chaotic sea state described as "air filled with foam; sea completely white with driving spray; visibility greatly reduced."

In the 20th century, the first reliable anemometers allowed mariners to quantify the Beaufort scale in terms of wind speed. For example, a force 1 wind is 1.6 to 4.8 km (1 to 3 mi) per hr whereas a force 12 wind is greater than 119 km (74 mi) per hr. In 1926, the Beaufort scale was extended to include the effects of wind on land. In addition to wind speed, the state of the sea depends on swell and the fetch and duration of the wind. Other factors such as water depth, heavy rain, or ice can also affect wave height. Furthermore, the Beaufort scale applies to waves on the open sea; waves in enclosed waters tend to be smaller and steeper.

Experience tells us that waves do not have a single wave height or wavelength, but occur in a spectrum of sizes. This variation is partly due to the inherent variability of wind speed and direction over a continuum of space and time scales. One of the most useful statistics for describing wave characteristics is *significant wave height*, the average height of the highest one-third of waves observed. Because smaller waves are usually not visible against the background of larger waves, we can assume that significant wave height approximates the visually observed mean wave height. Statistical analysis of waves indicates that the largest individual wave that one might encounter in a storm would be roughly twice as high as the significant wave height. The *dominant wave period* represents the period of waves exhibiting the maximum wave energy.

Spurred by the need for better weather and climate forecasts, the meteorological and oceanographic communities have expanded their monitoring of wind and wave conditions over the open ocean. These observations utilize more sophisticated techniques than visual observations from ships of opportunity traversing limited regions of the ocean. Today, sensors on moored automated buoys and orbiting satellites as well as ships at sea gather near-surface wind and wave data.

Moored buoys, deployed by various nations in their coastal waters, serve as instrumented platforms for making automated weather and oceanographic observations. The National Data Buoy Center (NDBC), part of the NOAA National Weather Service, operates approximately 70 moored buoys in the coastal and offshore waters of the western Atlantic Ocean, Gulf of Mexico, and the Pacific Ocean from the Bering Sea to southern California, around the Hawaiian Islands, and in the South Pacific, as well as the Great Lakes. Buoys are equipped with accelerometers or inclinometers that measure the heave acceleration or the vertical displacement provided to the buoy by waves passing during a specified time interval. An onboard computer uses statistical wave models to process these measurements and generate wind-sea and swell data that are then transmitted to shore stations. These data include significant wave height, average wave period, and dominant wave period during each 20-minute sampling interval. Selected buoys also measure directional wave data, such as mean wave direction.

Satellites equipped with radar scatterometers use the sea state to estimate near-surface wind speed and direction. A *scatterometer* is a sensor that measures the return reflection or scattering of a microwave (radar) signal sent to Earth's surface. A rough sea surface reflects back (backscatters) to the antenna on the satellite a stronger signal than does a smooth sea surface. Backscatter is proportional to roughness because more water surface faces directly at the radar when the water is rougher (with waves) than when the water is smooth. Computer algorithms estimate wave height and then the wind speed from differences measured in the strength of the return signals. If two emitted beams are spaced with a precise angular distance, slight variations in the return signals from the roughened surface permit determination of wind direction. As a feasibility study, the first space-borne scatterometer was onboard the U.S. Skylab Mission in 1973-74. From June to October 1978, the U.S. SEASAT-A Satellite Scatterometer (SASS) demonstrated that accurate winds could be obtained remotely from space. The NSCAT (NASA Scatterometer) was flown onboard the Japanese satellite ADEOS, launched into orbit in 1996 and operational until mid-1997. SeaWinds sensors accompanied the U.S. QuikSCAT (Quick Scatterometer) launched in 1999 and the Japanese Midori 2 satellite in December 2002.

ESSAY: Monitoring Sea Level from Space

Positioned on a polar-orbiting satellite, a *radar altimeter* measures the apparent distance to the sea surface based on the return time (round trip travel time) of a pulse of microwave energy directed downward by the instrument. When the microwave pulse strikes the sea surface, part of the signal is reflected back to the radar instrument onboard the satellite. The time between transmission of the original pulse and receipt of the reflected signal, coupled with the known speed of the microwave signal allows calculation of the distance between the satellite and the ocean surface. Using the latest satellite altimeters, oceanographers can determine the height of the ocean surface with an accuracy of a few centimeters. Remote sensing of the height of the sea surface takes into account different factors including those arising from the satellites' orbit, the atmosphere's influence on propagation of the microwave pulse, variations in the strength of gravity, and astronomical tides. Variations in sea-surface elevation are then compared to a mathematical model of Earth's shape. This model, flattened at the poles and bulging at the equator, is the *reference ellipsoid*, and is used as a benchmark to describe relative heights of the actual ocean surface.

Many variables govern fluctuations in sea level, with changes in local gravitational attraction being one of the most important. The local gravitational attraction between the solid Earth and overlying ocean water depends on both distance and mass of Earth's lithosphere. The greater the mass, the greater is the gravitational attraction, whereas the greater the distance, the smaller is the attraction. Ocean water is attracted horizontally toward areas of stronger gravitational attraction, piling up and increasing sea-surface height, and away from areas of weaker gravitational attraction, decreasing sea-surface height. Earth's lithosphere is not uniformly dense or level. Density differences in Earth's interior can cause sea-surface height variations as great as about 200 m (650 ft), roughly the height of a 60-story building. Sea level drops about 4 m (13 ft) for every 1000 m (3300 ft) increase in ocean depth, owing to local gravitational effects of Earth's crust on the overlying ocean water. Ocean bottom features such as ridges and trenches can cause sea height to vary by tens of meters. While the mounds and depressions in the sea surface are impressive, they are not apparent when sailing or flying over them because the horizontal dimensions are so great that the slopes are small.

Although efforts at remote sensing of sea surface topography by satellite were conducted in the 1970s and 1980s, the technique was not fully developed until the early 1990s. Sensors on the TOPEX/Poseidon satellite, launched in 1992, could measure sea heights with an accuracy of a few centimeters. Today, the Jason series of satellites continues to provide ocean altimetry data that has many valuable applications. The satellite radar altimeter has proven to be a versatile and powerful tool for remote sensing of the ocean. Analysis of satellite altimeter derived sea-surface elevations provides the most complete data set on the topography of the ocean floor. Physical oceanographers also use satellite altimetry to measure changes in the height of the ocean surface due to wind and density difference. This enables them to calculate patterns of surface currents (*geostrophic flow*) from observed sea-surface slopes (Chapter 6). Altimeter data are also used to determine wave heights, estimate surface wind speed, and track surface water masses including movement of surface water across the tropical Pacific basin during El Niño and La Niña (Chapter 11).

ESSAY: Tidal Power

Ocean tides and waves have long interested scientists and engineers as potential energy sources for generating electrical power. Tidal power has received the most attention—both the regular rise and fall of the sea surface as well as strong tidal currents. Indeed, tidal power was used for milling grain as early as the 11th century in Britain and France and the 17th century in Boston.

In the last half of the 20th century, large tidal-power installations were built in France, Canada, Russia, and China. These plants operate much like hydroelectric facilities except that water flows in both directions. Suitable coastal sites must have tidal ranges greater than 5 m (17 ft) and tidal currents that flow through a constricted inlet into a large basin. A dam (called a barrage) is constructed across the inlet with turbines submerged at the base of the dam. The ocean is on one side of the barrage and the basin is on the other side. With the incoming flood tide, sluice gates in the barrage are opened and the basin fills with water. During the slack water period the gates are closed. Then, with the outgoing ebb tide, the gates are opened and the exiting water drives the turbines that generate electricity.

The most successful of the tidal-power installations is La Rance Plant on the Brittany coast of France just south of Saint-Malo where the tidal range varies from 9 m (30 ft) to more than 13 m (43 ft). This 240-megawatt plant has 24 turbines that generate about 90% of Brittany's electricity. The 18-megawatt Canadian facility at Annapolis Royal, Nova Scotia, has been less successful and never operated commercially. No new tidal-power plants are planned for coastal locations because of high costs and lack of appropriate sites.

A small, tidal power plant was built on the sea floor near the Arctic tip of Norway to generate power from the area's strong tidal currents and provide electricity for Hammerfest, one of the world's most northerly towns. The power plant uses large windmill-like turbines mounted on the sea floor and transmits electricity to the local power grid via submarine cables. Initially the plant will generate enough electricity for about 300 homes but is planned for expansion to power perhaps 1000 homes.

Many engineering schemes exist for generating electrical power from ocean waves. A few have been built and some operated as demonstration projects. One demonstration plant was built in the mid-1990s on the remote island of Islay in the Western Islands of Scotland, an area known for its high waves. The plant operated for a short time before being destroyed by high waves in a storm. Small-scale wave-powered generators are often used to power aids to navigation on offshore buoys.

CHAPTER 8

THE DYNAMIC COAST

Case-in-Point
Driving Question
Coastline Formation
Coastal Features
    Beaches
    Barrier Islands
    Deltas and Salt Marshes
    Human Alterations
Estuaries
Coastal Storms and Storm Surge
Tropical Cyclones
    Where and When
    Hurricane Life Cycle
    Hurricane Hazards
    Evacuation
Extratropical Cyclones
Coastal Zone Management
Conclusions
Basic Understandings
ESSAY: Moving the Cape Hatteras Lighthouse
ESSAY: Restoring Salt Marshes
ESSAY: The Great Lakes and the Ocean

Pinnacle rocks and weather-beaten logs - trademarks of the Oregon coast. [Courtesy of NOAA.]

Case-in-Point

Many of the world's most populous cities are located along the coast, and efforts to protect them against flooding from storm-generated surges are becoming more challenging as global sea level rises and people continue to build in vulnerable locations. In many instances, the problem is not new as demonstrated by what has happened to Venice, a city in northern Italy situated on a cluster of 120 salt-marsh islands in a large coastal lagoon at the head of the Adriatic Sea (Figure 8.1). Bridges connect the islands and a causeway and ferries link Venice to the mainland. A long narrow barrier island separates the lagoon from the sea (Gulf of Venice) except for three major tidal inlets. At one time, the sea and the lagoon protected the city from foreign invaders. In fact, tradition has it that Venice was founded in 452 AD by people from northern Italian cities who fled to the islands seeking protection from invading Teutonic tribes. Today, however, the sea and lagoon threaten the city's continued existence.

    Slow subsidence of Venice combined with rising sea level has increased the frequency of flooding. Venice has been sinking slowly under its own weight for centuries. From the 1930s to 1970s, large withdrawals of groundwater for industrial use further exacerbated the problem. Groundwater occupies the tiny pore spaces in sediment and

FIGURE 8.1
ASTER satellite image of Venice, Italy. Note the saltwater lagoon between the mainland and barrier island. The image covers an area that measures 39 km by 35 km (24 mi by 22 mi) and was acquired on 9 December 2001. ASTER (Advanced Spaceborne Thermal Emission and Reflection Radiometer) is on NASA's Terra satellite orbiting Earth at an altitude of 705 km (430 mi). [Courtesy of NASA/GSFC/METI/ERSDAC/JAROS, and U.S./Japan ASTER Science Team]

as the water is pumped out, the sediment compacts and the ground subsides. Ten major floods inundated Venice over the past 67 years with smaller floods becoming much more frequent. During the often stormy months of October through January, some portions of the city are under water almost every day. The historic Piazza San Marco is flooded about 100 days of the year; a century ago, the flood frequency was about 9 days per year. Flooding forces businesses to close and inconveniences residents and tourists alike but more significantly, the salt water corrodes the brick underpinnings of historic churches and other buildings along the Grand Canal and elsewhere in the city.

To protect Venice against future flooding, the Italian government proposed a multi-billion dollar project that would construct a total of 78 floodgates at the three tidal inlets. Each gate will measure up to 5 m (16 ft) thick, 20 m (65 ft) wide, and 27 m (90 ft) long. Most of the time, the hollow gates will be filled with seawater and lie horizontally within a foundation (caisson) on the sea floor so as not to impede tidal currents or navigation. But when a storm surge is predicted to elevate the tide to a height of 110 cm (43 in.) or more, compressed air is pumped into the gates expelling the water. The gates are hinged at one end so that the free end can swing to the surface and block the flow of floodwaters through the inlets. Similar flood-control structures have been built in the Thames River in England to protect London and at the mouths of estuaries to protect the low-lying Dutch coastal plain.

Although the idea for the Venice floodgates dates back to at least the mid 1980s, it was not until May 2003 that a stone-laying ceremony signaled the beginning of construction on the so-called Moses Project. The project is slated for completion in 10 years at a cost of $2.7 billion. However, environmental concerns may delay or even halt construction. Opponents of the Venice floodgate scheme argue that closing the inlets will cause substantial environmental damage, especially to the lagoon ecosystem. Most of Venice's sewage empties untreated into the many canals dissecting the city or directly into the lagoon. Normally tidal currents dilute and transport these wastes out to sea but if the gates are closed for extended periods, wastes will accumulate in the city's canals or in the lagoon. Some scientists speculate that the gates could be closed for 100 to 120 days each year.

Driving Question:

Why and in what ways is the coastal zone a particularly dynamic and vulnerable portion of the Earth system?

The **coast**, the primary focus of this chapter, encompasses the relatively narrow region transitional between land and ocean. Although on a given day the coast can appear stable and benign, the many components of the Earth system (ocean, atmosphere, geosphere, cryosphere, and biosphere) continually interact in the coastal zone making it an extremely dynamic environment. Forces operating at the land/sea interface include tides, breaking waves, near-shore currents, discharge of sediments by rivers, changing sea level, tectonic processes, and the impact of living organisms including humans. These dynamic interactions are responsible for the wide variety of coastline types worldwide including, for example, rocky beaches, cliffed headlands, barrier islands, mangrove swamps, estuaries, sand dunes, and coral reefs.

Rapid human population growth is stressing the coastal zone. Statistics provided by the U.S. Census Bureau indicate that as of 2002, 96 million people (about one-third of the total U.S. population) reside in the 330 counties (or equivalent geographic units) bordering the Atlantic and Pacific Oceans, Gulf of Mexico, and Great Lakes.

Between 1990 and 2002, the total population of these coastline counties increased by more than 13%. Population density is now more than triple that of non-coastal counties. Furthermore, another 56 million people live in 343 near-coastal counties where much of the runoff from rain and snowmelt (along with a variety of dissolved and suspended contaminants) drains directly into the ocean. From 1980 to 2002, the population density of these counties plus the coastal counties (representing 17% of the total U.S. land area) increased at about 4.5 times the growth rate of other parts of the nation. Coastal population is expected to increase by 11 million by 2008 with population density projected to rise by 9% in the coastal Southeast Atlantic, 8% along the Gulf of Mexico, 6.5% along the Pacific coast, and 3.6% on the Atlantic coast between Virginia and Maine.

Increasing human population density in the coastal zone means greater demands for electrical power, fresh water, and roads. Environmental stressors that accompany population growth include waste disposal, recreational activities, tourism, expansion of industry, homes, and power plants, and the exploitation of fisheries, minerals, oil, and natural gas deposits. Adverse impacts on coastal ecosystems include shrinking plant and animal habitats, surface and groundwater pollution, and more frequent algal blooms due to enhanced runoff of fertilizers and other nutrients. Furthermore, population growth along the Atlantic and Gulf coasts exposes more people to the hazards of hurricanes and powerful extratropical storms.

In this chapter, we examine the various types of coasts and the processes responsible for their evolution and distinguishing features, and how human activity is altering the coastline. Special attention focuses on beaches, barrier islands, deltas, salt marshes, and estuaries. We describe hazards posed to coastal residents by tropical and extratropical storm systems and the key role played by air-sea interactions in powering these weather systems. We close with a brief description of coastal zone management. We begin with a description of coastline forming processes.

Coastline Formation

As shown schematically in Figure 8.2, the farthest inland extent of storm waves defines the **coastline**, in some cases marked by sand dunes or wave-cut cliffs. Land exposed at low tide up to the coastline is the **shore** and the average low tide line is the **shoreline**. The **intertidal zone**, the shore area between high and low tides, is alternately under water and exposed to the atmosphere as the tide rises and

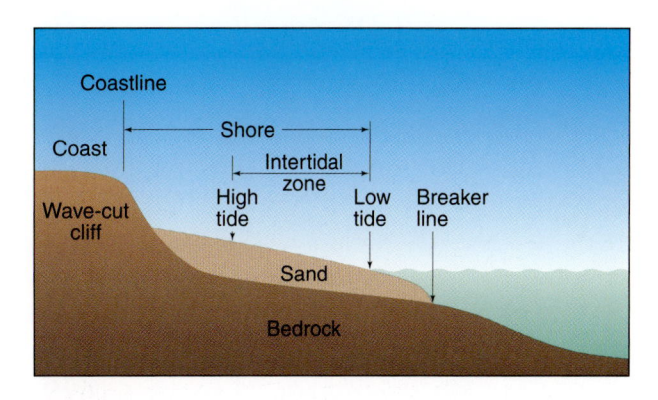

FIGURE 8.2
Idealized cross-section of a coastal zone.

falls. Moisture, temperature, and salinity fluctuate dramatically so that organisms living in the intertidal zone are equipped with special adaptations enabling them to survive drastic changes in their environment (Chapter 10).

A fundamental distinction among coastlines is whether they occur in tectonically active or inactive (passive) continental margins (Chapter 2). Mountain building and volcanic activity are dominant geological processes in tectonically active coastlines such as those that rim much of the Pacific Ocean basin. These processes can elevate coasts so rapidly that waves and currents cannot erode or deposit sediment fast enough to significantly modify long stretches of the shoreline. The few beaches that exist are small, isolated and usually restricted to near wave-cut cliffs or the mouths of rivers, which supply sand and gravel. For example, subduction is causing uplift along the Pacific Northwest coast. The Juan du Fuca plate is created at the submerged Juan de Fuca Ridge and is moving eastward, subducting beneath Washington, Oregon, and Northern California. Along the southern Oregon coast, the rate of tectonic uplift is greater than the global rise in sea level. Wave-cut benches are visible in the cliffs well above present high-water level. Meanwhile, along the northern Oregon coast, the rate of sea level rise is keeping pace with tectonic uplift.

Sediment erosion and deposition are dominant geological processes in passive shorelines such those that border the Atlantic Ocean and Gulf of Mexico. Along passive continental margins, sediments (mostly sand and silt) are deposited by rivers and streams entering the ocean and supplied by breaking waves undercutting shoreline cliffs that subsequently slump or slide into the ocean. Waves and currents transport these sediments along the shore, forming beaches, spits, and offshore barrier islands.

Rocky shorelines that are common in New England and Atlantic Canada formed during the Pleis-

176 Chapter 8 THE DYNAMIC COAST

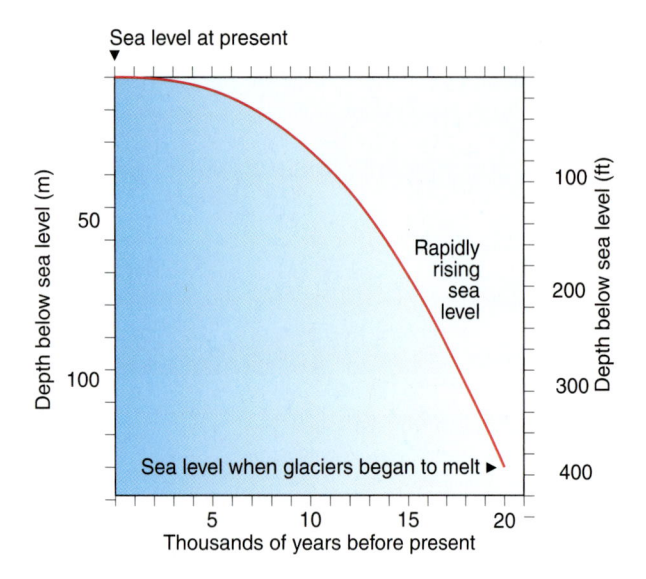

FIGURE 8.3
The post-glacial rise in global sea level.

tocene Ice Age when huge lobes of glacial ice scoured the land, eroding soils and loose sediment. As the ice retreated, sea level rose (Figure 8.3). The time since the ice retreated has been too short for accumulation of significant quantities of sands and gravels on most rocky shorelines. Furthermore, the glaciers left behind many lakes and bogs that trap the little sediment produced, preventing the remainder from reaching the shore. Hence, not enough sediment is available for waves and tides to smooth the complex coastlines. Such recently de-glaciated coastlines feature many small harbors and inlets, with rocky islands offshore.

As waves enter coastal waters of varying depth, they usually undergo **refraction**, bending of a wave in response to changing wave speed arising from frictional interaction with the bottom. Along the coast, wave refraction influences patterns of erosion, sediment transport, and deposition. Consider some examples.

Suppose that the shoreline is straight and coastal waters shoal uniformly toward the coastline. If wave crests approach the shoreline at an angle other than 90 degrees, the wave bends. The segment of the wave moving through deeper water progresses toward shore at its original speed while the segment of the wave entering shoaling water begins to "feel" (frictionally interact with) the bottom and slows (Chapter 7). Consequently, the wave crest is refracted toward the shallower water. Refraction causes approaching wave crests to come into near alignment with the depth contours of the ocean bottom and to closely conform to the shape of the shoreline.

On the other hand, suppose that the coastline is irregular and coastal waters do not uniformly shoal toward the shore. This situation is shown in Figure 8.4 viewed from above. Well offshore in deep water, approaching wave crests are essentially straight and parallel. Waves eventually enter water that is not uniformly deep and are refracted toward shallower water. Hence, waves approaching an irregular coastline are refracted (converge) toward headlands (rocky cliffs that project seaward) and away from coves or bays. Convergence concentrates wave energy causing undercutting and erosion of headlands whereas spreading out (divergence) of wave energy in bays weakens wave action and favors deposition of sediment and expansion of a beach. Eventually, the forces of erosion cut back the headlands

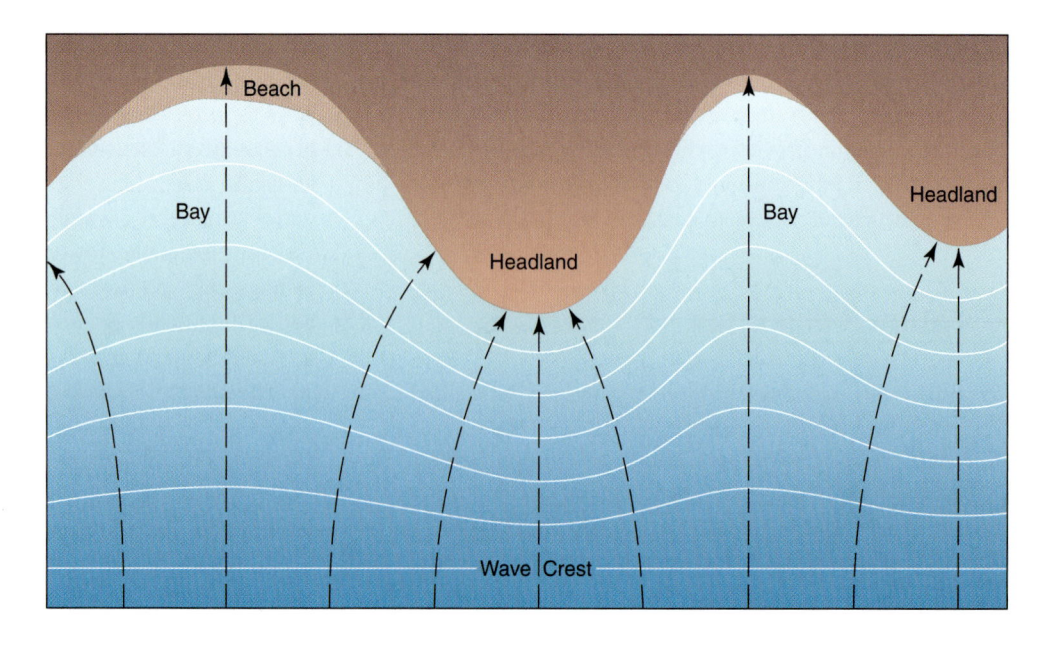

FIGURE 8.4
Viewed from above, waves slow in shoaling water and undergo refraction as they approach an irregular coastline.

FIGURE 8.5
This rocky knob along the Washington coast is what remains of a headland that was eroded by wave action. [Photo by J.M. Moran]

and deposition of sediment fills in coves and bays so that the coastline straightens. Figure 8.5 shows the remnants of a headland that was extensively eroded by wave action.

Coastal Features

Erosive and depositional processes impact both tectonically active and passive coastlines to some degree. The immediate source of energy for these processes is breaking waves. As we saw earlier in this book, winds transfer kinetic energy to the ocean surface forming waves (Chapter 7) and the sun drives atmospheric circulation (Chapter 5). Hence, solar energy ultimately powers erosive and depositional processes operating along the coast. In this section, we focus on selected features of the coast, that is, beaches, barrier islands, deltas, and wetlands. We also consider some of the ways human activity modifies the coastal environment.

BEACHES

A **beach** is an accumulation of wave-washed sediment (usually sand and gravel) at the landward margin of the ocean (or a lake). At the mean high-water mark, most beaches feature a prominent **berm**, a platform of sand that is nearly flat-topped and slopes steeply seaward. A beach is a dynamic system in which erosive and depositional forces continually do battle. Waves, tides, and currents erode cliffs, wash away dunes, and transport sand to and from the beach. The balance between sediment reaching a beach and sediment leaving a beach is always changing. In winter, erosive forces tend to prevail over depositional forces as powerful storm waves cut back the beach, moving sands onto shallow submerged bars near the shoreline. Hence winter beaches have a steeper slope. Strong storm winds blow sands from dry areas of the beach to dunes building behind the beach or into a lagoon. In summer, the weather is often more tranquil and depositional forces tend to prevail over erosive forces. Waves move sands from offshore bars back onto the beach and the beach widens. Summer beaches have a more gentle slope.

Sediment is delivered to the shore by rivers and eroded from seaside cliffs by ocean waves. As rivers and streams enter the sea, their current diverges and slows, depositing much of their suspended sediment load near their mouths (Chapter 4). Breaking waves attack the base of seaside cliffs, carving out a notch. The cliff slope becomes too steep and rock debris slumps or slides into the ocean (Figure 8.6). A number of factors govern the vulnerability of a cliff to slumping or sliding including composition (e.g., consolidated versus unconsolidated earth materials), geological structure (e.g., presence or absence of fractures), and rainfall. Furthermore, the relentless pounding of waves breaking against the coast fragments exposed rock. Coastal currents then transport these sediments to and from the beach. Even the rock debris that accumulates at the base of a seaside cliff is eventually removed, exposing the cliff to further erosion.

Earlier in this chapter, we noted that waves approaching the shore bend (are refracted) toward shallower water causing their crests to closely conform to the shape of the shoreline. Nonetheless, waves often approach the shore at an oblique angle up to 10 degrees or so (the angle between wave crests and the shoreline). Breaking waves striking the shore at an oblique angle give rise to sediment-transporting currents along the shore. If a breaking wave approaches the shore directly, a thin sheet of water rushes straight up the sloping beach face as an *upwash*, stops briefly, and then flows straight down slope as a *backwash*. Together, the upwash and backwash constitute the **swash**, so familiar to beachcombers (Figure 8.7). If, on the other hand, a breaking wave approaches the shore at an oblique angle, the upwash follows a diagonal up the beach face followed by the backwash flowing directly down slope (Figure 8.8). This zigzag or saw tooth motion of the swash rolls particles along the beach face and is known as *beach drift*. The oblique approach of waves to the shore produces a component of water motion that flows parallel to the shore known as a **longshore current** (Figure 8.9). A longshore current transports sediments as **littoral drift** (a river of sand) along the coast, either nourishing or cutting

178 Chapter 8 THE DYNAMIC COAST

FIGURE 8.6
Wave-eroded cliffs along the Oregon coast. [Photo by J.M. Moran]

FIGURE 8.7
Swash on a sloping beach face. [Photo by J.M. Moran]

FIGURE 8.8
Viewed from above, wave crests approaching the beach at an oblique angle. [Photo by J.M. Moran]

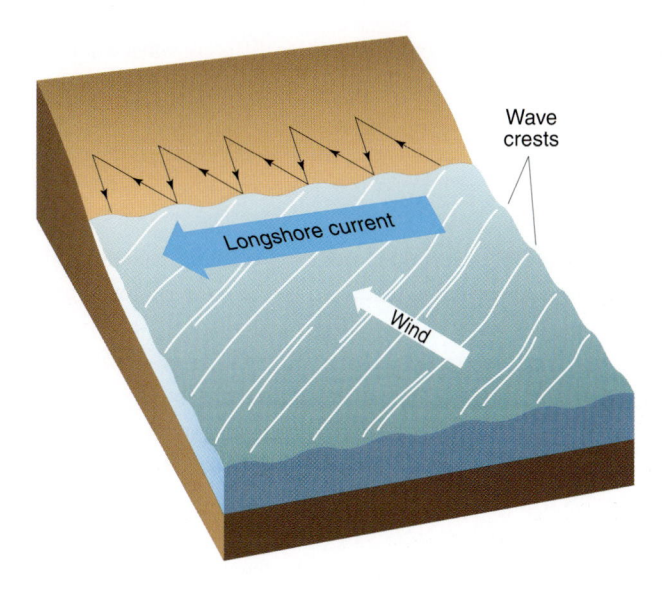

FIGURE 8.9
Viewed from above, waves approaching the coastline at an oblique angle produce a component of water motion roughly parallel to the shore (longshore current) that transports sediment (littoral drift). Zig-zag arrows represent the flow of water up and down the beach face.

back beaches. In some cases, littoral drift transports sediment offshore into the head of a submarine canyon where it is permanently lost to the beach.

In the *surf zone* (where shoreward moving waves break in shoaling coastal waters), water moves toward and usually along the shore. Whereas shoreward transport of water occurs over broad areas of the surf zone, the return seaward flow of water concentrates in narrow widely spaced belts often corresponding to depressions in the seafloor or breaks in sand bars. The narrow seaward flow of water must balance the broad shoreward flow of water so that the offshore flow occurs as a relatively swift surface or near-surface current with speeds up to 2 knots (1.7 mi per hr) or so. This is known as a **rip current**. Usually the strength of a rip current increases as wave height and wave period increase.

A rip current flows at nearly right angles to the shoreline and spreads seaward a few meters to hundreds of meters beyond the line of breaking waves. Viewed from above, a rip current is marked by choppy water made turbid by suspended sand and floating debris being transported seaward. Conventional thinking was that rip currents were transient phenomena lasting from a few minutes to a few hours. However, recent field studies conducted by University of Florida scientists indicate that a rip current may persist for weeks or even months at essentially the same location along the shore, varying in strength during that period.

A rip current is hazardous for people swimming in the surf zone and can develop off any surf beach including those bordering the Great Lakes. Most people are unable to swim against the strong current and are swept into deeper offshore waters. According to the U.S. Lifesaving Association, rip currents claim an estimated 100 lives on average each year along the nation's beaches. Perhaps as much as 80% of all rescues performed by surf beach lifeguards involve rip currents. In Florida, during the 1989-99 period, drowning deaths due to rip currents averaged 19 annually—more than the combined yearly death toll of hurricanes, tornadoes, and lightning. As noted above, however, a rip current is narrow. If caught in the current, swimmers are advised not to fight the current but to swim parallel to the shore until they escape into calmer waters. In coastal areas, the National Weather Service issues a daily rip current outlook that rates the risk of rip currents as low, moderate, or high.

Sometimes, a rip current is referred to as a "rip tide," but this is a misnomer. Rip currents have nothing to do with astronomical tides.

The **beach sediment budget** underscores the dynamic nature of the beach environment and is summarized in the following word equation:

Sediment Budget = Sediment Input - Sediment Output

In summary, sediment input processes include delivery by rivers and streams, erosion of cliffs, longshore current supply, and onshore movement. Sediment output processes encompass removal by longshore currents, wind erosion, and offshore movement. If the sediment balance is positive, (e.g., in summer) the beach grows; with a negative balance, (e.g., in winter) the beach is cut back. If input and output are equal, the beach is in a steady state.

BARRIER ISLANDS

Nearly 300 barrier islands fringe portions of the Atlantic and Gulf Coasts from Maine to Texas, the longest and best developed of any barrier island chain in the world. A **barrier island** is an elongated, narrow accumulation of sand oriented parallel to the coast and separated from the mainland by a lagoon, estuary, or bay. They occur where the supply of sand is abundant, the sea floor is gently sloping, and the continental margin is passive. Barrier islands vary in length from a few hundred meters to more than 100 km (62 mi). Padre Island is the longest of the nation's barrier islands extending more than 180 km (112 mi) along the Texas Gulf Coast. Tidal inlets segment barrier islands. Recall from this chapter's Case-in-Point that three major tidal inlets link the Gulf of Venice to the lagoon where the island city of Venice, Italy is located.

Beginning at the ocean side of a typical undeveloped barrier island and working our way toward the mainland, we would encounter a beach, dunes, flats, marsh, and lagoon. The near-shore and beach zones are high-energy environments with the potential for considerable impact from waves, longshore currents, and onshore winds. Just inland from the beach is a dune field consisting of one or more parallel ridges composed of wind-blown sand from the beach (Figure 8.10). The next zone is the back-island flat, an extensive plain consisting of grassland or perhaps woodland, depending on the age and size of the island and its exposure to storms and high winds. This area is prone to wash over by high waves generated by the winds of powerful coastal storms. The flat grades into a salt marsh that borders a lagoon, estuary, or bay; it is a low-energy environment where fine sediments accumulate. A **lagoon** is a partially enclosed body of water separating a barrier island from the mainland. Normally, lagoons receive essentially no input of river water and tidal fluxes maintain salinities near that of seawater. (In this way, lagoons differ from estuaries—discussed later in this chapter.) Where evaporation rates are relatively high, the salinity of water in some isolated lagoons may exceed the usual seawater salinity. In an extreme example, Laguna Madre between Padre Island and the Gulf coast has salinities as high as 60 psu.

FIGURE 8.10
Sand dunes located just inland from a beach on a barrier island on North Carolina's Outer Banks. [Photo by J.M. Moran]

Scientists recognize three different ways whereby barrier islands form. Many (especially along the southeast Atlantic and Gulf Coasts) date from the mid Holocene epoch (about 5000 years ago) when sea level was somewhat higher than it is now. (The Holocene epoch covers the past 10,500 years.) Higher sea level submerged beaches isolating the tops of dune ridges as offshore islands while flooding lower areas landward of the dune ridges, which became lagoons. Some barrier islands (e.g., small Gulf Coast islands) are the product of the vertical growth of a submarine sand bar that emerged from the sea as an island. Along an irregular coastline, sand spits can lengthen into barrier islands through longshore sand transport. A **sand spit** is a finger-like ridge of sand or gravel that projects from the shore into a body of water. During intense storms, the spit many be breached, forming barrier islands. This is the origin of barrier islands along the New York and New England coasts.

A barrier island is a constantly changing system. Waves breaking on the shores of a barrier island dissipate their energy by shifting sands and modifying the shape of the island. As sea level gradually rises, a barrier island migrates toward the mainland as episodes of storm-induced wash-over become more frequent. Sand is removed from the ocean-side of the island and transported to the lagoon-side.

Barrier islands absorb the brunt of powerful storm-driven sea waves thereby providing some protection for coastal beaches, estuaries, wetlands, and shoreline structures. But many barrier islands have been developed, especially for cottages and resorts, and their sands are now temporarily stabilized under a layer of asphalt or concrete. Some coastal cities, including Atlantic City, NJ, Miami Beach, FL, and Virginia Beach, VA are built entirely on barrier islands. Such exposed locations are particularly vulnerable to the ravages of the high winds and floodwaters associated with tropical cyclones (discussed later in this chapter). On developed barrier islands, much of the energy of storm waves would be expended in demolishing buildings and roads instead of shifting sands. Conflict is inevitable between rigid structures such as roads and buildings and the inherently dynamic (changeable) platforms (i.e., barrier island) upon which they are built.

DELTAS AND SALT MARSHES

A **delta** is a feature of some coastlines formed where a sediment-bearing river or stream enters a body of water and sediment settles out of suspension as the current diverges and slows (Chapter 4). Many factors govern the shape of a delta; these include river discharge, sediment load and compaction, subsidence of the sea (or lake) bottom, tides, and waves. Depending on which factor prevails, deltas are classified as river-dominated (e.g., Mississippi), wave-dominated (e.g., Niger), or tide-dominated (e.g., Ganges). In a river-dominated delta, the rate of input of sediments (from the river or stream) exceeds the rate of removal of sediments (by waves and currents) and the delta has the classic triangular shape (as viewed from above). In a wave-dominated delta, strong wave action and longshore currents produce only a slight bulging of an overall straight coastline. In a tide-dominated delta, tidal currents rework river-supplied sediments into long, narrow islands and submarine ridges.

Human activities have impacted natural delta-forming processes with important consequences for the coast. Dams built on rivers for flood control and navigation trap much of the sediment in reservoirs so that little reaches the delta. Furthermore, dredging of channels and construction of levees on the lower Mississippi delta for oil and gas exploration has greatly reduced the supply of sediments to that delta. With reduced sediment input, compaction of sediments, wave action, and rising sea level many deltas are eroding (Chapter 4). The impact on the southern Louisiana shore is particularly severe; there, the coastline in some places is retreating at more than 20 m (66 ft) annually.

A **salt marsh** is composed of salt-tolerant plants that colonize shores that are protected from wave action. Along the New England coast, the dominant plant species in undisturbed salt marshes consist of cordgrass (along the shore), marsh hay, black rush, and shrubs (inland often bordering trees). Tidal creeks meander through most salt marshes allowing seawater to flood the marshes on rising tides and drain on ebbing tides (Figure 8.11). Vegetation slows the movement of water through a salt marsh thereby inducing deposition of suspended fine sediment and the plants anchor the sediment in place. A salt marsh offers many benefits including protecting the coastline from storm wave erosion, filtering runoff thereby reducing the input of pollutants and excess nutrients into estuaries, and providing habitat for organisms such as crabs and oysters.

In tropical areas, such as South Florida and Puerto Rico, marine wetlands support thick growths of salt-tolerant trees called **mangroves**. Mangroves have extensive root systems that trap sediments and also provide habitat for both marine and land animals. Indeed, some organisms are especially adapted for survival in mangrove thickets, such as mangrove oysters, which attach themselves to roots and branches. Many wetlands are extremely produc-

FIGURE 8.11
A tidal creek meanders through a salt marsh. [Photo by J.M. Moran]

tive, in some cases equivalent to that of a cornfield. Furthermore, wetlands provide food-rich nursery grounds for many coastal and estuarine organisms, including many commercially important fish.

Roots of wetland plants—both grasses and trees—trap sediment and organic matter. Where sea level is rising, wetlands can build up fast enough that their surface remains at sea level, assuming an adequate supply of sediment. If sea level is rising too rapidly, marshes are eroded as is presently the case on the Mississippi River delta and around the Chesapeake and Delaware Bays. Where sea level is stable, wetland plants can trap enough sediment to build the marsh surface above normal sea level. Where this happens, the shore advances and grasslands or tropical forests eventually replace the wetland.

HUMAN ALTERATIONS

Humans alter the coastal environment to preserve beaches, control flooding, maintain harbors and navigation channels, and protect houses, roads, and other structures built on the beach, in the dunes behind them, or on the top of seaside cliffs. When erosion threatens to damage or destroy shoreline structures, three protective strategies are available: armor, artificial beach nourishment, and strategic retreat. Armor includes dams, breakwaters, jetties, groins, and sea walls. Artificial beach nourishment refers to the addition of sand to a beach, and strategic retreat involves inland displacement of a structure threatened by rising sea level and storm waves.

Coastal armor has both benefits and costs, but research suggests that the overall costs outweigh the benefits. Armor can disrupt littoral drift and alter the natural flow of sediments to and from a beach. Sediments settle out of suspension in the reservoir behind a dam, reducing the river's sediment load below the dam and the sand supply for the beach. Longshore currents transport existing sands away from the beach, sometimes to the extent that all that remains is rocky rubble (cobbles and boulders).

A **breakwater** is a long, narrow offshore structure, usually constructed of large blocks of rock or concrete and oriented parallel to the shoreline. Its purpose is to provide calm waters for docking boats or to protect beaches from erosion by absorbing the energy of breaking waves or reflecting waves back to sea. Some breakwaters are built on open coastlines where no natural harbors exist. However, sediments settle to the bottom of the relatively calm waters on the landward side of the breakwater and costly dredging may be required to keep the harbor open for navigation.

Breakwaters also disrupt littoral drift. Most sediment accumulates in the harbor and is unavailable for deposition on down-current beaches. Just such a situation occurred at Santa Barbara, CA after a breakwater was built in 1928. To keep the harbor open and maintain longshore sand transport, sand is continually dredged from the harbor and piped as a slurry to the down-current side of the harbor. A pipeline now transports sand formerly conveyed by natural longshore currents.

A **jetty** is similar to a breakwater except that it is oriented perpendicular to the shoreline and extends seaward up to a kilometer or more. In some cases, a jetty has the same function as a breakwater, that is, protecting harbors from storm waves. Often jetties are also used to keep a barrier island inlet open for navigation into a harbor on the mainland side of the island. Jetties constructed on either side of the ocean-side entrance of a tidal inlet constrict the flow of tidal currents. The constricted flow of water accelerates, reduces sediment deposition, and may excavate the channel bottom. On the other hand, jetties disrupt littoral drift across the inlet. Up current from a jetty, beaches widen due to enhanced sediment deposition (beach nourishment) whereas beaches disappear down current from a jetty due to lack of sand input and wave erosion. Consider an example.

Ocean City, MD is a beach resort that has undergone considerable development since the 1950s. The resort is located at the south end of Fenwick Island, a barrier island. Although steps were taken in the late 1970s to curb further development by limiting dredging and filling of wetlands (to create more land for dwellings), the resort remains a popular destination. Most residents and visitors are probably unaware that in 1933 a powerful hurricane lashed the island and opened an inlet at its southern end linking the backwater bay with the Atlantic Ocean. The U.S. Army Corps of Engineers erected two rock jetties to maintain the inlet for navigation. However, the jetties also altered the southward flowing littoral drift causing a wide beach to form at Ocean City north of the jetty while the beach at Assateague Island, south of the inlet, was deprived of sand and eroded back.

Other structures called **groins** resemble jetties but are smaller and usually more closely spaced along a beach. Their purpose is to widen a beach by trapping sand moving in the littoral drift. But once again, sand trapped by groins is not available to beaches down current so that local beach widening is at the expense of beach erosion elsewhere.

A **seawall** is a concrete or rock embankment intended to protect beaches, roads, buildings, and shoreline cliffs from erosion by storm waves. However, seawalls also cut off a source of sediment for the beach by preventing erosion of bluffs. Furthermore, evidence suggests that wave energy, rather than being mostly absorbed, is actually reflected by a seawall, stirring up the water and sand at the base of the structure, accelerating erosion, and causing beaches to narrow. Deeper water in the near shore area means that rather than breaking well offshore, storm waves break against the seawall, further exacerbating erosion and ultimately destroying the seawall.

Beaches are the economic life-blood of many coastal communities, attracting visitors for a variety of recreational activities. Perhaps 70% of all Americans on vacation will spend time at a beach. Breakwaters, jetties, groins, and seawalls are often constructed to preserve beaches, protect seaside property, and support the local economy. But for reasons discussed above, these strategies to *armor* or *harden* the coastline have serious drawbacks. In fact, these drawbacks convinced many coastal states to prohibit armoring of the coast to prevent beach erosion.

An alternate and widely used strategy to maintain existing beaches or restore badly eroded beaches is **artificial beach nourishment**. Sand is usually dredged from deep waters just offshore and then spread on the beach to be redistributed by waves, longshore currents, and winds. On the plus side, artificial nourishment attempts to re-establish the natural balance between sediment inputs and outputs on the beach. On the negative side, beach nourishment is expensive and short-lived (typically lasting 5 years or less) and must be repeated regularly to maintain the beach.

By way of an example of the transient nature of artificial beach nourishment, consider what happened at Folly Beach, SC, a barrier island southeast of Charleston, SC. In September 1989, Hurricane Hugo's storm surge severely eroded the beach. In 1992, the U.S. Army Corps of Engineers spent $15 million to restore the beachfront through artificial beach nourishment. But only two months after the project was complete, a coastal storm washed away 80% of the new sand. Similar experiences are common along U.S. coastlines.

When erosion threatens shoreline buildings, strategic retreat is an alternative to artificial beach nourishment or armoring the coast with seawalls, groins or other structures. *Strategic retreat* refers to the physical relocation of a building inland. In 1999, this strategy was used to preserve the historic Cape Hatteras Lighthouse. (For more on this, refer to this chapter's first Essay.) In some cases, entire settlements are moved inland. For example, the town of Port Valdez in Alaska's Port Valdez fjord (terminus of an oil pipeline) was relocated after being heavily damaged by a tsunami triggered by the 1964 Alaska earthquake (Chapter 7). Originally, the town was situated at the head of the fjord and was moved to one side for greater protection.

Salt marshes have also been the object of modification by human activity. For centuries most people considered wetlands to be wastelands and reclaimed them by ditching, draining, or filling. In this way, former wetlands were developed for airports, agriculture, housing, and shopping malls. Over the past 250 years, more than half of the original wetland acreage in the United States was converted to other purposes. But we now know that wetlands provide habitats that are necessary to support coastal marine ecosystems and efforts are underway to reclaim damaged wetlands. For example, restoration efforts are underway to return the area of saltpans in southern San Francisco Bay to its original wetland condition by removing dikes, grading, and replanting grasses. Comparable projects are underway in the Netherlands where polders (reclaimed continental shelf areas) are also being restored to their original condition. Wetlands are complex ecosystems, however, and it is difficult to re-establish a wetland that will function as well as an undisturbed one. For more on wetland restoration, see this chapter's second Essay.

Estuaries

An **estuary** is a partially isolated body of water where fresh water (from rivers and streams) mixes with seawater—from the lower extent of the photic zone to the most inland extent of tidal action. As pointed out in Chapter 1, estuaries are among the most productive ecosystems on Earth because of their special combination of physical and biological characteristics. They receive a continual supply of nutrients and organic matter delivered by river water and ocean tides. In Chapter 10, we focus on biological productivity in estuaries; in this chapter, we consider the origins of estuaries and their circulation patterns.

Most estuaries developed during the post-glacial (Holocene) rise in sea level. Rising seas inundated the land drowning the mouths of rivers. Other estuaries developed through glacial erosion of mountain valleys, tectonic processes, or through the action of longshore sediment transport. In high latitude mountainous regions during the Pleistocene Ice Age, tongues of glacial ice occupied former river valleys and eroded those valleys below present sea level. With deglaciation and the concurrent rise in sea level, ocean water flooded the valleys. These steep-walled enclosures of water are known as **fjords** (Figure 8.12). Along some tectonically active coasts, the sea has flooded down-faulted blocks of crust. And in other coastal locations, sand spits grew via littoral drift and partially isolated a body of water from the ocean.

Although some mixing of fresh river water and seawater occurs in all estuaries, the characteristic circulation of an estuary depends on the amount of river inflow versus the strength of tidal currents. Based on circulation type, ocean scientists distinguish among salt-wedge estu-

FIGURE 8.12
A fjord along the mountainous coast of Norway. [Photo by J.M. Moran]

aries, partially mixed estuaries, and well-mixed estuaries. At one extreme (salt-wedge estuaries), river water is the dominant factor and mixing is minimal and at the other extreme (well-mixed estuaries), tides are the dominant factor and mixing is extensive.

In estuaries where river inflow is relatively large and tidal currents are weak, the low-salinity river water is quite distinct from the high salinity ocean water. That is, the two layers of water remain stratified. Where they meet, the wedge of less dense fresh water overlies the wedge of denser seawater; the thin vertical transition zone between the two layers is a *halocline*. This is known as a **salt-wedge estuary** (Figure 8.13). Seaward flow of the fresh water is much stronger than the landward movement of the seawater. The only mixing that takes place is due to internal waves that develop and propagate along the halocline (Chapter 7). Breaking of these internal waves mixes a relatively small amount of salt water (from below) with the fresh water (above) so that strong stratification persists. During peak flow, a salt-wedge estuary forms near the mouth of the Columbia River but when river discharge is low, the salt-water wedge migrates upriver. Other examples of salt-wedge estuaries occur at the mouth of the Mississippi and Hudson Rivers. The Amazon River in Brazil is an extreme case of mixing in that the river's discharge is so great that seawater does not penetrate into the river's mouth and all mixing of seawater and river water occurs on the continental shelf.

With declining river discharge and strengthening tidal currents, vertical mixing within an estuary becomes more vigorous so that stratification and the halocline weaken. In these so-called **partially mixed estuaries**, salinity typically varies by less than 10 psu from the bottom water to the surface water (Figure 8.14). That is, the flow of fresh water seaward (above the halocline) and the flow of seawater landward (below the halocline) are both stronger than in a salt-wedge estuary. The Chesapeake Bay is a classic example of a partially mixed estuary and is the largest estuary in the United States. It is almost 300 km (185 mi) long, 65 km (40 mi) at its broadest, and averages about 20 m (65 ft) deep. The estuary was formed by the postglacial rise in sea level that flooded the ancient Susquehanna River Valley. The Bay receives about half its water from the Atlantic Ocean and the other half from the more than 150 rivers and streams draining a 166,000 square km (64,000 square mi) land area encompassing portions of New York, Pennsylvania, West Virginia, Delaware, Maryland, Virginia, and the District of Columbia. Other examples include San Francisco Bay and Puget Sound.

In a **well-mixed estuary**, strong tidal currents dominate the inflow from rivers and thoroughly mix the fresh water and saltwater. There is no vertical stratification and no halocline so that the only salinity gradient is a horizontal gradient from fresh water up estuary to seawater in the ocean. Delaware Bay is an example of a well-mixed estuary.

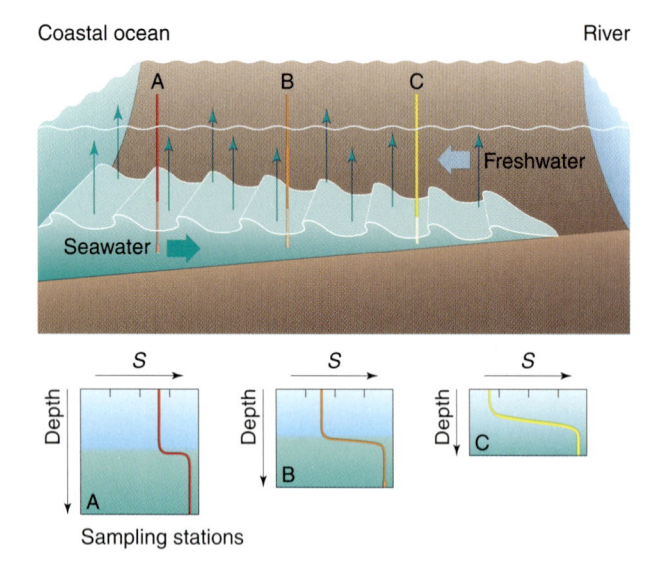

FIGURE 8.13
Circulation in a salt-wedge estuary. The charts below show the variation in salinity (S) with depth at three water-sampling stations.

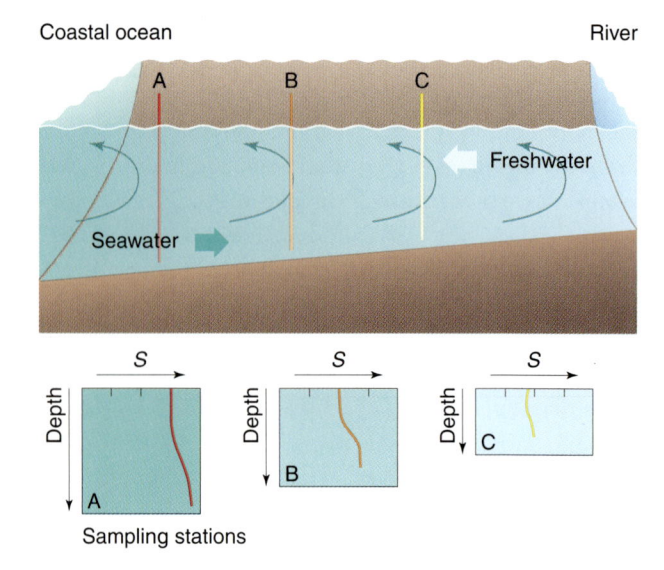

FIGURE 8.14
Circulation in a partially mixed estuary. The charts below show the variation in salinity (S) with depth at three water-sampling stations.

In broad estuaries, the Coriolis effect influences the circulation and mixing. In the Northern Hemisphere, both seaward and landward flowing currents are deflected to the right of their initial direction. That is the two currents are deflected toward opposite banks of the broad estuary, such as Chesapeake Bay, resulting in fresh river water flowing to the sea on one side and saltwater flowing up river on the other side. This results in a tilted halocline and lateral mixing in contrast to the vertical mixing in estuaries that are too narrow to be significantly influenced by the Coriolis effect.

Sediment distribution within an estuary depends on the relative strength of river discharge compared to tidal currents. River-borne sediments dominate salt-wedge estuaries whereas the sediment in well-mixed estuaries comes mostly from offshore sources. Partially mixed estuaries have a mixture of both types of sediments. The two-layered circulation pattern in salt-wedge and partially mixed estuaries trap fine river-borne sediment. River-transported sediments eventually encounter the weaker current associated with the halocline. Coarser sediments settle to the bottom of the estuary while finer sediments (clays) remain suspended and concentrate near the halocline, forming a **turbidity maximum**.

In arid coastal regions that receive less fresh water input than is lost through evaporation, vertical circulation patterns are opposite to those in most estuaries (Figure 8.15). Such systems are sometimes called *anti-estuarine circulations* or *reverse estuaries*. In these systems, landward-flowing surface waters replace waters lost through evaporation and seaward-flowing bottom waters transport high-salinity waters caused by intense evaporation. Such circulation systems characterize the Mediterranean and Red Seas.

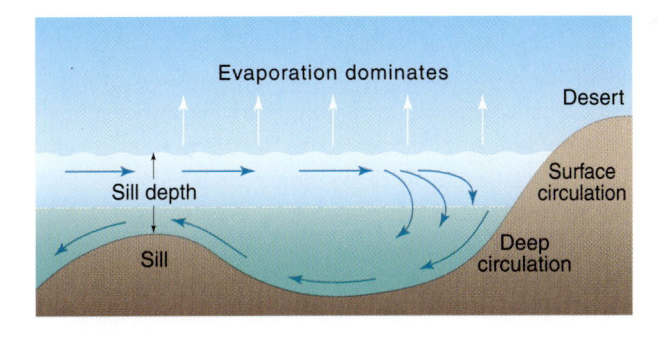

FIGURE 8.15
Anti-estuarine circulation.

Coastal Storms and Storm Surge

People who live in the coastal zone are vulnerable to a variety of natural hazards and those hazards can be exacerbated by human activity. Among these coastal hazards are sea level rise, tsunamis, land subsidence, salt-water intrusion into aquifers, and storm surge. In this section we describe the storm surge that can be produced by coastal storms. In Chapter 12, we consider the threat of coastal flooding caused by climate change and rising sea level.

Atmospheric scientists distinguish between two types of storm systems that pose a threat to the coast: tropical cyclones and extratropical cyclones. These are both low-pressure weather systems that are characterized by surface winds that blow in a counterclockwise and inward direction in the Northern Hemisphere (Chapter 5). A **tropical cyclone** originates in a uniform mass of warm humid air over the tropical ocean but may track into higher latitudes; it includes hurricanes, tropical storms, and their precursors. An **extratropical cyclone** forms in middle latitudes and spends its entire life cycle in middle and high latitudes; its circulation brings together contrasting warm and cold air masses to form fronts.

Potentially the most devastating impact of tropical and extratropical storms on the coastal zone is ocean water driven ashore by strong onshore winds (blowing from sea to land) associated with an intense storm system centered over or near the ocean. Strong winds (coupled with low air pressure) pile up a dome of seawater 80 to 160 km (50 to 100 mi) wide that sweeps over the coastline bringing floodwaters inland that can take lives and cause considerable property damage. This so-called **storm surge** can erode beaches, wash over barrier islands, cut new tidal inlets, wash out roads and railway beds, and demolish marinas, piers, cottages, and other shoreline structures.

Many factors influence the potential for a storm surge, including storm intensity, timing of astronomical tides, topography of the ocean bottom, and shoreline configuration. In general, a storm surge of 1 to 2 m (3 to 6.5 ft) can be expected with a weak hurricane, whereas the storm surge accompanying a violent hurricane may top 5 m (16 ft). Lower air pressure contributes to the wind-driven mounding of water, with sea level rising about 0.5 m (1.6 ft) for every 50-millibar drop in air pressure. A storm surge is superimposed on normal ocean tides so that the impact is greatest when the storm surge coincides with high tide and especially spring tide (Chapter 7). In addition, wind-driven waves having heights of at least 1.5 to 10 m (5 to 33 ft) top the storm surge, further exacerbating the impact on coastal areas. All other factors being equal, the more gradual the slope of the ocean bottom in coastal areas, the greater is the storm surge. Low-lying coastal plains are most vulnerable to a storm surge especially if no barrier islands are present to dissipate much of the energy of the surge.

A special numerical model developed in 1979 by NOAA accurately predicts the location and height of a storm surge. Weather forecasters report considerable success with the **SLOSH (Sea, Lake, and Overland Surges from Hurricanes)** model. Using the output from SLOSH and consulting local topographic maps, forecasters can identify areas most likely to be inundated by floodwaters when a hurricane threatens. Hypothetical hurricanes are simulated with various combinations of storm intensity, forward speed, and expected location of landfall to predict the maximum high water level for various SLOSH basins along the coast. Public safety officials use this information in developing evacuation plans. SLOSH model coverage includes the entire East and Gulf Coasts, as well as parts of Hawaii, Guam, Puerto Rico, and the Virgin Islands.

Understanding storm surges and why some coastal areas are more prone to storm surges than others requires a closer look at the basic characteristics of tropical and extratropical cyclones. We do this in the remainder of this chapter.

Tropical Cyclones

The most catastrophic storm surges are associated with hurricanes that make landfall, that is, hurricanes that track from sea to land. In the most deadly natural disaster in U.S. history, more than 8000 people perished, mostly by drowning, when a hurricane storm surge devastated Galveston, TX on 8 September 1900. Ocean waters flooded the entire city located on a barrier island. Some 3600 homes were destroyed; a dam of wreckage that built up some six blocks inland from the beach probably spared the business district total destruction. To protect the city from future storm surges, the U.S. Army Corps of Engineers elevated the city an average of some 3 m (10 ft) and erected a massive concrete seawall along the city's Gulf shore.

Hurricane Camille, an exceptionally intense system with winds to 300 km (186 mi) per hr, produced a maximum storm surge of 7.3 m (24 ft) at Pass Christian, MS, on 17 August 1969 (Figure 8.16). The death toll from the hurricane was 256, about half from coastal flooding. In 1970, one of the most disastrous storm surges ever flooded the Bay of Bengal coast of East Pakistan (now Bangladesh). A storm surge of nearly 7 m (23 ft) spread over the vast low-lying coastal plain of the Ganges Delta, claiming an estimated 300,000 lives mostly by drowning because there was no high ground for people to seek refuge.

FIGURE 8.16
Damage produced at Gulfport, MS by the storm surge of Hurricane Camille in August 1969. [Photo Courtesy of NOAA]

Hurricane is likely derived from *Haracan*, the name of the storm god of the Taino people who inhabited Caribbean islands at the time of Spanish exploration of the New World. A **hurricane** is an intense cyclone that originates over tropical ocean waters, usually in late summer or early fall (when tropical waters are warmest), and has a minimum sustained wind speed of 119 km (74 mi) per hr. By convention in the United States, *sustained wind speed* is a one-minute average measured at the standard anemometer height of 10 m (33 ft). A storm surge is most likely on the side of the hurricane where the system's counterclockwise surface winds blow toward the shoreline.

Hurricanes develop in a uniform mass of very warm and humid air. Typically, the central pressure at sea level is lower and the horizontal air pressure gradient is much steeper in a hurricane than in an extratropical cyclone. A hurricane is usually a much smaller system, averaging a third the diameter of a typical extratropical cyclone. Rarely do hurricane-force winds (119 km per hr or higher) extend much more than 120 km (75 mi) beyond the system's center. The circulation in a hurricane weakens rapidly with altitude and usually becomes anticyclonic (clockwise in the Northern Hemisphere) in the upper troposphere at altitudes above about 12,000 m (40,000 ft).

At the center of a hurricane is an area of almost cloudless skies, subsiding air, and light winds (less than 25 km or 16 mi per hr), called the **eye** of the storm (Figure 8.17). The eye generally ranges from 10 to 65 km (6 to 40 mi) across, shrinking in diameter as the hurricane intensifies and winds

FIGURE 8.17
NOAA geostationary satellite image of Hurricane Floyd as it approached the East Coast of the U.S. on 15 September 1999. Note the eye at the center of the system

strengthen. At a hurricane's typical rate of forward motion, the eye may take up to an hour to pass over a given locality. People are sometimes lulled into thinking the storm has ended when skies clear and winds abruptly slacken following a hurricane's initial blow. They may be experiencing passage of the hurricane's eye; heavy rains and ferocious winds will soon resume but blow from the opposite direction.

Bordering the eye of a mature hurricane is the **eyewall**, a ring of thunderstorm (cumulonimbus) clouds that produce heavy rains and very strong winds. The most dangerous and potentially most destructive part of the hurricane is the portion of the eyewall on the side of the advancing system where the wind blows in the same direction as the storm's forward motion. On that side, hurricane winds combine with the storm's forward motion producing the system's strongest surface winds. In the Northern Hemisphere, this dangerous semicircle of high winds and waves occurs on the right side of the hurricane when facing in the direction of the system's forward movement. Cloud bands, producing heavy convective showers and hurricane-force winds, spiral inward toward the eyewall.

WHERE AND WHEN

Three conditions are required for a tropical cyclone to form: (1) high sea-surface temperature, (2) adequate Coriolis deflection, and (3) weak winds aloft. To a large extent, these requirements dictate where and when tropical cyclones develop.

Tropical cyclone formation requires a sea-surface temperature (SST) of at least 26.5 °C (80 °F) through an ocean depth of 60 m (200 ft) or more. Such exceptionally warm ocean water sustains the system's circulation by the latent heat released when water vapor, evaporated from ocean surface, is conveyed upward and condenses within the storm system. Temperature largely governs the rate of evaporation of water, so the higher the SST, the greater is the supply of latent heat for the storm system. Furthermore, the spray from breaking ocean waves readily evaporates and adds to the supply of latent heat.

As a tropical cyclone makes landfall or moves over colder water, however, it loses its warm-water energy source and weakens. The strong winds of a tropical cyclone stir up surface ocean waters. Cyclonic winds induce Ekman transport and divergence of surface waters under the storm system, bringing cold water to the surface. Until the normal thermal structure of the ocean is restored, lower than usual sea-surface temperatures can inhibit development of subsequent tropical cyclones over the same region of the ocean. In recent years, atmospheric scientists discovered that changes in SST associated with warm- and cold-core ocean rings influence tropical cyclone development. Recall from Chapter 6 that rings are eddies that break off western boundary currents such as the Gulf Stream. Tropical cyclones may intensify over warm-core rings and weaken over cold-core rings.

The second requirement for tropical cyclone development is a significant Coriolis deflection; that is, the influence of Earth's rotation must be sufficiently strong to initiate a cyclonic circulation. As noted in Chapter 5, the Coriolis effect weakens toward lower latitudes and is zero at the equator. The minimum latitude where the Coriolis force is adequate for formation of tropical cyclones is about 4 degrees.

The first two conditions that favor tropical cyclone formation (i.e., high SST and sufficient Coriolis effect) occur only over certain portions of the ocean. The main ocean breeding grounds for tropical cyclones along with average storm trajectories are plotted in Figure 8.18. Most hurricanes form in the 8- to 20-degree latitude belts. Major breeding grounds are: (1) the tropical North Atlantic west of Africa (including the Caribbean Sea and Gulf of Mexico), (2) the North Pacific Ocean west of Mexico, (3) the western tropical North Pacific and China Sea, where a hurricane is called a *typhoon* (from the Cantonese *tai-fung*, meaning great wind), (4) the South Indian Ocean east of Madagascar, (5) the North Indian Ocean (including the Arabian Sea and Bay of Bengal), and (6) the South Pacific Ocean from the east coast of Australia eastward to about 140 degrees W. In the Indian Ocean and near Australia, hurricanes are simply called *cyclones*.

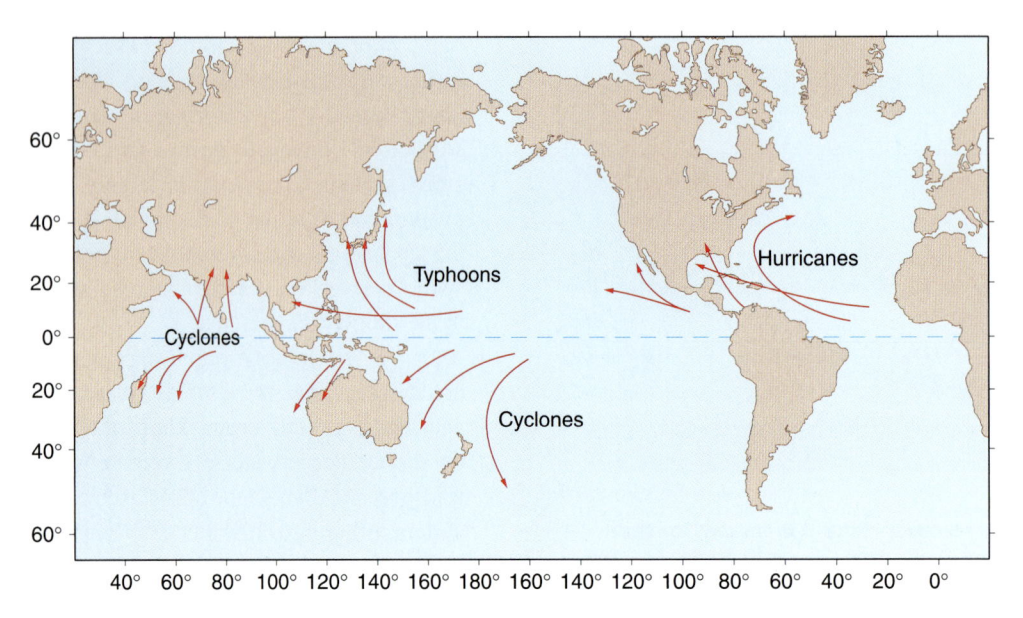

FIGURE 8.18
Tropical cyclone breeding grounds are located only over certain regions of the world ocean. Arrows indicate average hurricane trajectories.

The requirement of high sea-surface temperatures makes the occurrence of tropical cyclones distinctly seasonal. For reasons presented in Chapter 3, the temperature of surface ocean waters lags the regular seasonal variations in incoming solar radiation. Sea-surface temperatures reach a seasonal maximum roughly 6 to 8 weeks after the date of most intense solar radiation. Most Atlantic hurricanes develop when surface waters are warmest, that is, in late summer and early autumn; the official hurricane season runs from 1 June to 30 November, with the peak hurricane threat for the U.S. coastline between mid-August and late October.

The third requirement for tropical cyclone development is relatively weak winds aloft over oceanic breeding grounds. Weak winds aloft allow a cluster of cumulonimbus clouds to organize over tropical seas, the first step in the evolution of a hurricane. By contrast, strong west-to-east winds in the middle and upper troposphere shear off the tops of westward tracking thunderstorms, preventing the systems from building vertically and organizing. *Wind shear* is the principal reason hurricanes rarely form off the east or west coasts of South America (although Caribbean hurricanes occasionally impact the north coast of Venezuela).

A worldwide average of about 80 tropical cyclones develops each year and approximately one-half reach hurricane strength. The western Pacific Ocean, with its vast expanse of warm surface waters, is the most active area for tropical cyclones, with roughly 32 systems each season, of which about 18 intensify into typhoons.

Only hurricanes spawned over the tropical Atlantic, Caribbean Sea, and Gulf of Mexico pose a serious threat to coastal North America. A seasonal average 10 named tropical storms (precursors to hurricanes) form over these waters. Of these systems, on average about 6 intensify into hurricanes and 2 of these hurricanes strike the U.S. coast. The 1995 Atlantic hurricane season was the second most active in recorded history with 19 tropical storms, 11 of which attained hurricane strength. However, the annual number of hurricanes may have little bearing on the number of landfalling hurricanes and their impact. Whereas only 4 hurricanes formed during the 1992 hurricane season, one of them, Hurricane Andrew, was the most costly in terms of property damage in U.S. history, amounting to $34.3 billion (in year 2000 dollars). Hurricanes have hit every Atlantic and Gulf Coast state from Texas to Maine. Florida is the most hurricane-prone of all the states, with 57 hurricanes crossing its coastline between 1900 and 1996. In the same period, Texas was second with 36 and Louisiana and North Carolina had 25 each.

The Pacific Ocean off Mexico and Central America ranks second to the western North Pacific in the average annual number of tropical cyclones (13), the majority of which develop into hurricanes. And yet, the Pacific Coast of North America is rarely a target of hurricanes. Prevailing winds (northeast trades) are directed offshore and usually steer tropical cyclones that form west of Central America away from the coast. Also, the southward flowing cold California Current plus upwelling just off the Southern California (and Baja California) coast produce sea-surface temperatures that normally are too low to sustain tropical cyclones that travel toward the northeast. However, during unusual atmospheric/oceanic circulation regimes, hurricanes have struck coastal Southern California and even tracked over the Desert Southwest.

The Hawaiian Islands are sometimes threatened by tropical cyclones that develop over the central tropical Pacific or track into that region from the Pacific hurricane breeding grounds west of Mexico. Fortunately, in an average year, only 3 tropical storms affect the central Pacific and since 1957 only 4 hurricanes have struck the islands. However, the most recent one, Iniki in September 1992, devastated the island of Kauai. Seven people lost their lives. Total property damage was estimated at $2.3 billion, making this the most costly natural disaster in the history of the State of Hawaii.

The primary breeding ground for hurricanes in the Atlantic shifts east and west with the seasons. Early in the hurricane season (May and June), hurricanes form mostly over the Gulf of Mexico and the western Caribbean. By July, the main area of hurricane development begins to shift eastward across the tropical North Atlantic. By mid-September, most hurricanes form in a belt stretching from the Lesser Antilles (in the eastern Caribbean Sea) eastward to south of the Cape Verde Islands (off Africa's West Coast). After mid-September, hurricanes again originate mostly over the Gulf of Mexico and the western Caribbean.

HURRICANE LIFE CYCLE

The first sign that a hurricane may be forming is the appearance of an organized cluster of thunderstorm clouds over tropical seas. This area of convective activity is labeled a **tropical disturbance** if a center of low pressure is detected at the surface. If atmospheric/oceanic conditions favor hurricane development and if those conditions persist, the surface air pressure falls and a cyclonic circulation develops. Water vapor condenses within the storm, releasing latent heat of vaporization, and the heated air rises. Expansional cooling of the rising air triggers more condensation, release of even more latent heat, and a further increase in buoyancy. Rising temperatures in the core of the storm, coupled with an anticyclonic outflow of air aloft, cause a sharp drop in surface air pressure, which in turn, induces more rapid convergence of humid air at the surface. The consequent uplift surrounding the developing eye leads to additional condensation and release of latent heat.

Through these processes, a tropical disturbance intensifies and its winds strengthen. When sustained wind speeds top 37 km (23 mi) per hr, the developing system is called a **tropical depression**. When sustained wind speeds reach 63 km (39 mi) per hr, the system is classified as a **tropical storm** and assigned a name, such as Alberto or Beryl. Once sustained winds reach 119 km (74 mi) per hr, the storm is officially designated a hurricane. As a hurricane weakens and decays, the system is downgraded by reversing this classification scheme.

Hurricanes that form over the Atlantic near the Cape Verde Islands usually drift slowly westwards with the trade winds (along the southern flank of the Bermuda-Azores subtropical high) across the tropical North Atlantic and into the Caribbean. At this stage in the storm's trajectory, it is not unusual for the system to travel at a mere 10 to 20 km (6 to 12 mi) per hour and take a week to cross the Atlantic. Once over the western Atlantic, however, the storm usually picks up speed and begins curving northward along the western flank of the Bermuda-Azores high, and northeastward as the system is caught in the westerlies of middle latitudes. Precisely where this curvature takes place determines whether the hurricane enters the Gulf of Mexico (perhaps then tracking up the lower Mississippi River Valley or over the Southeastern States), moves up the Eastern Seaboard, or curves back out to sea.

When an Atlantic hurricane reaches about 30 degrees N, it may begin to acquire extratropical characteristics as colder air circulates into the system and fronts develop. From then on the storm resembles an extratropical cyclone and completes its life cycle usually over the North Atlantic. Many hurricanes, however, depart significantly from the track just described. Some of the hurricane tracks plotted in Figure 8.19, for example, are erratic. A hurricane can describe a complete circle or reverse direction. In addition, some hurricanes, fueled by warm Gulf Stream waters, maintain their tropical characteristics far north along the Atlantic Coast. Coastal New England and Atlantic Canada, for example, have been the targets of hurricanes.

HURRICANE HAZARDS

In addition to storm surge (discussed earlier in this chapter), hazards of hurricanes are flooding rains, strong winds, and tornadoes. Hurricanes and tropical storms produce very heavy rainfall with amounts typically in the range of 13 to 25 cm (5 to 10 in.). Even if the system tracks well inland, heavy rains often persist and may trigger costly flooding. According to statistics released by the Tropical Prediction Center/National Hurricane Center, from 1970 to 1999, fresh-

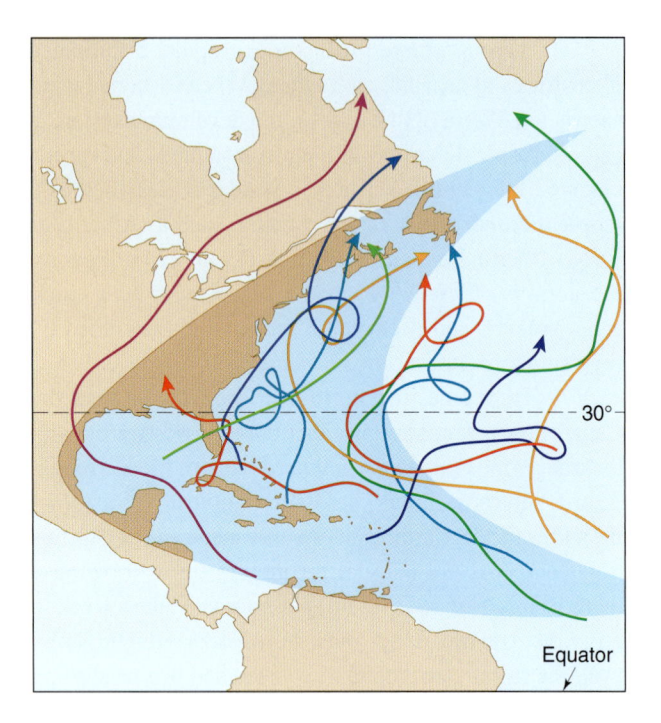

FIGURE 8.19
Tropical cyclone trajectories sometimes are erratic as shown by these samples. As indicated by the shaded area, however, most Atlantic tropical cyclones initially drift westward and curve toward the north and northeast when they reach the western Atlantic. [From NOAA, *Hurricane*. Washington, DC: Superintendent of Documents, 1977.]

water floods were responsible for almost 60% of the 600 U.S. deaths attributed to tropical cyclones or their remnants. In those three decades, more people (351) died from inland flooding than from coastal storm surge flooding (only 6 fatalities).

Winds pushing on the outside walls of buildings exert a pressure that increases dramatically as winds strengthen. (Wind pressure is a function of the square of the wind speed so that, for example, doubling the wind speed increases the pressure exerted by the wind on an external wall by a factor of four.) Furthermore, debris transported by the wind and hurled against structures exacerbates the damage potential of strong winds. Hurricane winds diminish rapidly once the system makes landfall, so that most wind damage is confined to within about 200 km (125 mi) of the coastline. Two factors account for the abrupt drop in wind speed once a hurricane makes landfall. A hurricane over land is no longer in contact with its energy source, warm ocean water. In addition, the frictional resistance offered by the rougher land surface slows the wind and shifts the wind direction toward the low-pressure center of the system. This wind shift causes the storm to begin to fill; that is, its central pressure rises, the horizontal air pressure gradient weakens, and winds slacken.

Although winds weaken once a hurricane makes landfall, the system may produce tornadoes. (A *tornado* is a small column of air that whirls rapidly about a nearly vertical axis and is made visible by clouds, dust and debris). Usually only a few tornadoes occur with a hurricane but in 1967, Hurricane Beulah reportedly spawned as many as 115 tornadoes across southern Texas. Tornadoes are most probable after the hurricane enters the westerly steering current and curves towards the north and northeast; they form mostly to the northeast of the storm center, often outside the region of hurricane-force winds.

In the early 1970s, H.S. Saffir, a consulting engineer, and R.H. Simpson, former director of the National Hurricane Center, designed a rating system for hurricanes known as the **Saffir-Simpson Hurricane Intensity Scale**. First included in hurricane advisories in 1975, the scale rates hurricanes from 1 to 5 corresponding to increasing intensity (Table 8.1). The scale provides an estimate of potential coastal

TABLE 8.1
Saffir-Simpson Hurricane Intensity Scale

Category	Central pressure mb (in.)	Wind speed km/hr (mi/hr)	Storm surge m (ft)	Damage potential
1	>980 (>28.94)	119-154 (74-95)	1-2 (4-5)	Minimal
2	965-979 (28.50-28.91)	155-178 (96-110)	2-3 (6-8)	Moderate
3	945-964 (27.91-28.47)	179-210 (111-130)	3-4 (9-12)	Extensive
4	920-944 (27.17-27.88)	211-250 (131-155)	4-6 (13-18)	Extreme
5	<920 (<27.17)	>250 (>155)	>6 (>18)	Catastrophic

flooding and property damage from a hurricane landfall. Wind speed is the primary determining factor for a hurricane's rating on the Saffir-Simpson scale as storm surge heights are highly dependent on underwater topography and other factors in the region of landfall. Each intensity category specifies (1) a range of central air pressure, (2) a range of maximum sustained wind speed, (3) storm surge potential, and (4) the potential for property damage.

Of the 162 hurricanes that struck the U.S. Atlantic or Gulf Coasts between 1900 and 1998, 65 (about 40%) were classified as major; that is, they rated 3 or higher on the Saffir-Simpson scale. Property damage potential rises rapidly with ranking on the Saffir-Simpson scale. In fact, destruction from a category 4 or 5 hurricane can be 100 to 300 times greater than that caused by a category 1 hurricane. The 25 major hurricanes that made landfall along the Gulf or Atlantic Coasts between 1949 and 1990 accounted for three-quarters of all property damage from all landfalling tropical storms and hurricanes during the same period.

The National Hurricane Center (a component of the Tropical Prediction Center in Miami, FL) watches for the development of tropical cyclones over the Atlantic, Caribbean, Gulf of Mexico, and the Eastern Pacific. The Center operates the SLOSH model for prediction of storm surges, prepares and distributes hurricane watches and warnings for the public, conducts research on hurricane forecasting techniques, and sponsors public awareness programs.

EVACUATION

From 1970 through 1989, 10 major hurricanes (category 3 or higher) struck the U.S. mainland—down from 15 during 1950-69 and 16 in 1930-49. Between 10 August 1980 and 17 August 1983, no hurricanes struck the United States. The infrequency of major hurricanes during the 1970s and 1980s lulled many coastal residents of the Southeast U.S. into a false sense of security and encouraged development and population growth in areas that could be devastated by a major hurricane. More and more resort hotels, high-rise condominiums, and expensive homes were constructed perilously close to the shoreline and even among coastal sand dunes (Figure 8.20).

About 45 million permanent residents inhabit hurricane-prone portions of the nation's coastline, and the population continues to climb. Today, perhaps 80% to 90% of Atlantic and Gulf Coast residents have never experienced the full impact of a major hurricane. Some of them may have weathered with relative ease a weak

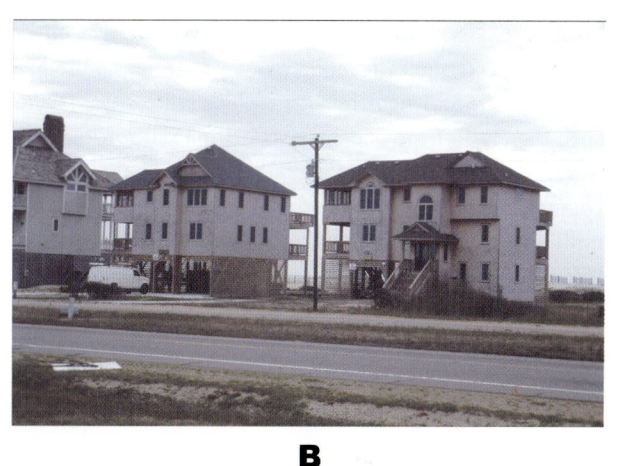

FIGURE 8.20
The relative infrequency of hurricanes in the 1970s and 1980s lulled some residents of coastal areas into a false sense of security and inspired construction of resort hotels (A) and vacation homes (B) within meters of high-tide level. [Photos by J.M. Moran]

hurricane or the fringes of a strong system. But such an experience may lull them into complacency so that they are less likely to prepare adequately should a major hurricane threaten. Compounding the problem of the growth of the resident population in the Southeast is the arrival of holiday, weekend, and seasonal visitors to seaside resorts. During vacation periods, the population of some of these locales swells ten- to one-hundred-fold. Many of these resorts are in low-lying coastal areas or on barrier island beaches that are subject to rapid inundation by a storm surge.

Evacuation of people from barrier islands, as well as other low-lying coastal areas, is the traditional strategy

FIGURE 8.21
In hurricane-prone areas, special road signs mark evacuation routes. [Photo by J.M. Moran]

in the event of a major storm threat (Figure 8.21). The effectiveness of coastal evacuation plans was tested in the late summer of 1985 when Hurricane Elena (category 3) menaced the Gulf of Mexico coast. For four days, Elena followed an erratic path over the Gulf of Mexico before finally coming ashore near Biloxi, MS on 2 September. Nearly a million people from Sarasota, FL to New Orleans, LA were forced to leave coastal communities and flee to inland shelters. Some returned home only to evacuate again as Elena changed course. Although property damage was considerable ($2 billion in year 2000 dollars) because of extensive flooding and winds that exceeded 160 km (100 mi) per hr, only 4 fatalities were attributed to Elena and none in the area of the hurricane landfall. Timely evacuation of residents of low-lying coastal areas also saved many lives when Hurricanes Hugo (September 1989, category 4) and Andrew (August 1992, category 5) threatened. Although both were major hurricanes and caused considerable property damage when they made landfall, the death toll was only 21 in Hugo and 23 in Andrew.

The potential downside of evacuation was illustrated in September 1999 when Floyd, an unusually massive hurricane, threatened much of the Eastern Seaboard. In the greatest evacuation in U.S. history, more than 2 million residents of the coastal area from South Florida to South Carolina took to the roads and fled inland. In many areas the result was gridlock—too many vehicles on too few highways. As it turned out, Floyd spared most of the evacuated area and made landfall near Cape Fear, NC on 16 September 1999 as a category 2 system. Torrential rainfall, totaling as much as 38 to 50 cm (15 to 20 in.) over portions of eastern North Carolina and Virginia caused extensive inland flooding, 56 deaths, and property damage of at least $6 billion.

Successful evacuation of coastal communities hinges on sufficient advance warning of a hurricane's approach but hurricanes are notorious for sudden changes in direction, forward speed, and intensity. (In retrospect, perhaps 75% of all coastal storm evacuations prove to be unnecessary.) A hurricane's erratic behavior is especially troublesome for people living in isolated localities (e.g., a barrier island linked to the mainland by a single bridge) and congested cities where highway systems have not kept pace with population growth. In such places the time required for evacuation may be lengthy. The Federal Emergency Management Agency (FEMA) estimates evacuation times during the peak tourist season as 50-60 hrs for New Orleans, LA and Fort Myers, FL, 40-49 hrs for Ocean City, MD, 30-39 hrs for the Florida Keys and Cape May County, NJ, and 20-29 hrs for Long Island, NY, Atlantic City, NJ, the Outer Banks of NC, and Galveston, TX.

As coastal communities continue to grow, the time required for evacuation of their population lengthens. Evacuations must begin earlier when a storm is farther away and greater uncertainty surrounds its likely track. Such uncertainty necessitates a broader zone of evacuation that translates into greater economic losses associated with evacuation (e.g., closed businesses). Cognizant of the problem of lengthy evacuation times, some coastal communities are looking into alternative strategies such as establishing shelters in public and commercial buildings and growth control ordinances.

Extratropical Cyclones

Of all the extratropical cyclones to affect the U.S. coastal zone, the so-called nor'easter has received the most attention because of its severity and the number of people impacted. A **nor'easter** is an intense extratropical cyclone that tracks along the East Coast of North America and is named for the direction from which the most destructive winds blow. Unlike hurricanes, nor'easters may originate over land, forming along the boundary between air masses that contrast in temperature and humidity. Like hurricanes, they often intensify over warm ocean waters. Nor'easters are most frequent from October through April, when the seasonal contrasts in air mass characteristics (temperature and humidity) are greatest. The system moves toward the northeast and, if centered just offshore, strong onshore winds on the northern and northwestern flanks of the system (blowing from the east and northeast) can cause a storm surge resulting in considerable coastal erosion, flooding, and property damage.

Although rarely attaining the strength of even a category 1 hurricane, a nor'easter can pack a powerful punch. In an intense nor'easter, winds can produce wave heights of 1.5 to 10 m (5 to 33 ft) superimposed on a storm surge of up to 5 m (16 ft). A nor'easter often impacts a much greater swath of coastline than a typical East Coast hurricane. As noted above, the diameter of an average extratropical cyclone is three times that of an average hurricane. Whereas a hurricane tends to track directly from sea to land, impacting perhaps only about 100 to 150 km (60 to 90 mi) of coastline, a nor'easter often moves parallel to the coast so that its onshore winds may sweep over more than 1500 km (900 mi) of coastline. A less intense but slower-moving nor'easter can actually cause more damage than a more intense but faster-moving nor'easter. Winds in the slow-moving system blow in the same onshore direction for a longer period than is the case with the fast-moving system. Persistent onshore winds increase the likelihood of a destructive storm surge and high waves.

Powered by latent heat acquired from evaporation of warm ocean water, some nor'easters develop very rapidly. By convention, a rapidly intensifying extratropical cyclone is labeled a *bomb* if its central pressure drops at least 24 millibars in 24 hrs. Few cyclones actually meet this criterion, and most of those that do, develop over warm ocean currents such as the Gulf Stream or the Kuroshio Current. An extreme example of an East Coast bomb was observed during early January 1989. An incipient cyclone was first identified off of Cape Hatteras, NC at about 7:00 p.m. (EST) on 4 January; its central air pressure was 996 mb. Twenty-four hours later, the storm was centered about 700 km (430 mi) south of Newfoundland with a central pressure of 936 mb. The storm had intensified by 60 mb in 24 hrs, 2.5 times the criterion for a bomb.

One of the most noteworthy nor'easters of the 20th century in terms of its widespread impact on human activities occurred on 12-15 March 1993. A major storm system tracked from the Gulf of Mexico up along the Eastern Seaboard. Although the system's size and central pressure were not particularly unusual, its human impact was enormous primarily because of its track along the East Coast. This track enabled the storm to draw into its circulation a tremendous amount of moisture from the warm ocean waters just off the coast, while the cold air inland helped the storm produce heavy snow over the highly populated areas to the north and west of its track from Alabama to Maine. In addition to property damage caused by heavy snow and high winds, a *squall line* (an elongated group of strong to severe thunderstorms) associated with the storm produced 27 tornadoes in Florida, and a 3-m (10-ft) storm surge in the Gulf of Mexico flooded the Apalachicola area. All told, this nor'easter claimed about 208 lives, more than three times the combined death toll of Hurricanes Hugo and Andrew, and caused damage estimated at $6 billion, the costliest extratropical cyclone in U.S. history.

Many residents and visitors to Ocean City on Fenwick Island, MD are probably unaware of the impact of the great nor'easter of 5-8 March 1962. This slow moving and intense storm system produced waves as high as 9 m (30 ft) superimposed on a 2-m (7-ft) storm surge. Coastal erosion and flooding were extensive. During five successive high-tide cycles over two days, storm waves washed over almost all of Fenwick Island. Property damage at Ocean City was estimated at $7.5 million. In view of the considerable development that has occurred on the island since 1962, can you imagine the impact of a similar storm today?

R.E. Davis and R. Dolan of the University of Virginia proposed an intensity scale for nor'easters that is analogous to the Saffir-Simpson Scale for hurricanes. Their scale was based on research conducted along the Outer Banks of North Carolina. Davis and Dolan recognized that sea waves generated by a nor'easter largely govern the storm's coastal impact and based their classification scheme on a *wave power index*. The Davis-Dolan wave power index is computed by multiplying a storm's duration by the square of the maximum significant wave height. Storms are classified into 5 categories, ranging from Class I (very common weak systems) to Class V (very rare extreme systems). A Class I storm might last for 4 hrs with an average peak wave height of 2 m (6.6 ft) and produce minor beach erosion and little or no property damage. At the other extreme, Class V storm might last for 4 days or longer with an average peak wave height of 7 m (23 ft) and produce extreme beach erosion, destroy dunes, and cause region-wide property damage.

Coastal Zone Management

The need to balance multiple use and preservation of the coastal zone spurred the federal government to partner with states having coastlines to develop and implement plans for sustainable coastal management. The federal government assists coastal states in managing and protecting their coastal resources through the **Coastal Zone Management Program (CZMP)** authorized by the 1972 Coastal Zone Management Act. The goals of this program are "to preserve, protect, develop, and where possible, to restore or enhance the resources of the Nation's coastal zone" and to encourage states to develop and implement plans "to achieve wise use of the land and water resources of the coastal zone." At the federal level, NOAA's Office of Ocean and Coastal Resource Management administers the CZMP.

The CZMP promotes comprehensive management of coastal resources. This program seeks to protect those resources for future generations while balancing competing economic, cultural, and environmental issues. The federal government supports state efforts at coastal zone management by providing financial assistance (matched by state funds), mediation of conflicts, technical services, and information. Coastal states that show satisfactory progress in implementing their management plan are eligible for special federal grants (matched by state funds) to help states preserve or restore coastal areas, redevelop urban waterfronts and ports, and provide access to public beaches. Actual day-to-day coastal zone management decisions rest with the individual states.

As of this writing, 33 of the 34 eligible coastal states and U.S. territories have federally approved coastal zone management plans that apply to much of the more than 160,000 km (100,000 mi) of national coastline. All approved state coastal management plans must address a number of issues in balancing use with preservation. These issues include (1) an inventory and designation of areas of particular concern in the coastal zone, (2) a definition of permitted land and water uses that directly impact coastal waters, (3) identification of how those uses will be controlled, and (4) an outline of broad guidelines to determine priority uses in coastal areas.

Another provision of the 1972 Coastal Zone Management Act established the National Estuarine Research Reserve System (NERRS). The U.S. Secretary of Commerce can authorize grants to coastal states to acquire, develop, and operate estuarine research reserves. Since 1972, the NERRS has expanded to encompass 26 sites in 19 states and Puerto Rico.

The nation's coastal zone also borders the Great Lakes. For more on the Great Lakes and how their properties and processes compare with those of the ocean, refer to this chapter's third Essay.

Conclusions

Essentially all components of the Earth system come together and interact in the coastal zone: ocean, atmosphere, cryosphere, geosphere, and biosphere (including humans). The land/sea interface is a particularly dynamic portion of the Earth system where numerous forces shape the coast. Beaches, barrier islands, cliffed headlands, and estuaries are among the many features of the coastline. People are increasing their presence in the coastal zone and their activities sometimes conflict with the forces of nature including rising sea level, storm surges, and littoral drift.

Now that we have examined the physical properties of the ocean, we are ready to focus on the living portion of the ocean. The next chapter covers marine ecosystems and the processes that support life in the ocean. Chapter 10 considers marine organisms and how they have adapted to the varying physical conditions of their environments.

Basic Understandings

- The coast encompasses the relatively narrow region that is transitional between land and the ocean. Although the coast may appear stable and unchanging, the many components of the Earth system (ocean, atmosphere, biosphere, geosphere, and cryosphere) interact in the coastal zone making it an extremely dynamic environment.
- The farthest inland extent of storm waves defines the coastline, in some cases marked by sand dunes or wave-cut cliffs. Land exposed at low tide up to the coastline is the shore and the average low tide line is the shoreline. The intertidal zone, the area of the shore between high and low tides, is alternately under water and exposed to the atmosphere as the tide rises and falls.
- Mountain building and volcanic activity are dominant geological processes in tectonically active coastlines such as those that rim the Pacific Ocean basin. Sediment erosion and deposition are dominant geological processes in passive shorelines such as those that surround the Atlantic Ocean and the Gulf of Mexico.
- As sea waves move into coastal waters of varying depth, they undergo refraction, that is, bending of a wave in response to changing wave speed caused by changing depth. Wave crests bend toward the shallower water. In this way, waves approaching an ir-

regular coastline are refracted (converge) toward headlands (rocky cliffs that project seaward) and away from coves or bays. Convergence concentrates wave energy causing erosion of headlands whereas spreading out (divergence) of wave energy in bays reduces wave action and favors deposition of sediment and expansion of a beach.
- A beach is an accumulation of wave-washed sediment (usually sand and gravel) at the landward margin of the ocean (or lake); it is a dynamic system in which erosive and depositional forces continually do battle.
- When breaking waves approach the shore at an oblique angle, a zigzag motion of the swash moves particles along the beach face; this is known as beach drift. Waves approaching the shore at an oblique angle also produce a component of water motion that is directed parallel to the shore known as a longshore current. A longshore current transports sediments as littoral drift along the coast, either nourishing or cutting back beaches.
- A barrier island is an elongated, narrow accumulation of sand oriented parallel to the coast and separated from the mainland by a lagoon, estuary, or bay. They occur where the supply of sand is abundant and the sea floor is gently sloping. Beginning at the ocean side of a typical undeveloped barrier island and working our way toward the mainland, we would encounter a beach, dunes, flats, marsh, and lagoon.
- A delta is a feature of some coastlines that is also formed by transport and deposition of river sediment. Salt marshes, are nearly flat, low-lying, protected coastal areas where fine-grained sediment accumulates and salt-tolerant plants grow. Tidal creeks meander through most salt marshes allowing seawater to flood the marshes on a rising tide and to drain on the ebb tide.
- Artificial structures, including dams, breakwaters, jetties, groins, and sea walls, disrupt littoral drift and alter the natural flow of sediments to and along a beach. An alternate strategy to maintain existing beaches or restore badly eroded beaches is artificial beach nourishment. With this approach, sand is usually dredged from deep waters and then spread on the beach to be redistributed by waves, longshore currents, and winds. However, artificial beach nourishment is an expensive and temporary solution.
- An estuary is a partially isolated body of water where fresh water (from rivers and streams) mixes with seawater. Most estuaries developed during the post-glacial (Holocene) rise in sea level. Other estuaries developed through glacial erosion of mountain valleys, tectonic processes, or longshore sediment transport. Based on differences in circulation regime, a distinction is made among salt-wedge estuaries, partially mixed estuaries, and well-mixed estuaries.
- A tropical cyclone originates in a uniform mass of warm humid air over the tropical ocean but may track into higher latitudes; it includes tropical storms and hurricanes. An extratropical cyclone forms in middle latitudes and spends its entire life cycle in middle and high latitudes; its circulation brings together contrasting warm and cold air masses to form fronts.
- Potentially the most devastating impact of tropical and extratropical storms in the coastal zone is a storm surge, a dome of ocean water driven ashore by strong onshore winds associated with an intense storm system centered over or near the ocean. Factors that influence the potential for a storm surge include storm intensity, timing of tides, topography of the ocean bottom, and shoreline configuration.
- A hurricane is an intense tropical cyclone that originates over tropical ocean waters, usually in late summer or early fall, and has a minimum sustained wind speed of 119 km (74 mi) per hr. In addition to storm surge, hazards of hurricanes are flooding rains, strong winds, and tornadoes.
- The three conditions required for a tropical cyclone to form are (1) relatively high sea-surface temperature, (2) adequate Coriolis deflection, and (3) weak winds aloft. To a large extent, these requirements dictate the geographical and seasonal distribution of tropical cyclones. The energy source for tropical cyclones is latent heat released when water vapor evaporated from the ocean condenses in the system. As a tropical cyclone intensifies, the successive stages in its life cycle are designated tropical disturbance, tropical depression, tropical storm, and finally hurricane.
- The Saffir-Simpson Hurricane Intensity Scale rates hurricanes from 1 to 5 corresponding to increasing intensity and provides an estimate of potential coastal flooding and property damage from a hurricane landfall.
- A nor'easter is an intense extratropical cyclone that tracks along the East Coast of North America and is named for the direction from which its most destructive winds blow. Unlike hurricanes, nor'easters may originate over land, forming along the boundary between air masses that contrast in temperature and humidity. They are most frequent from October through April.

ESSAY: Moving the Cape Hatteras Lighthouse

When beach erosion threatens to damage or destroy shoreline structures, three protective strategies are available: armor (e.g., groins, seawalls), artificial beach nourishment, and strategic retreat (relocation of structures inland). At various times, all three strategies were used to keep the encroaching sea from claiming the historic Cape Hatteras Lighthouse.

In 1870, a 63-m (208-ft) brick lighthouse was built at Cape Hatteras on one of the barrier islands that form North Carolina's Outer Banks. At the time, the lighthouse with its distinctive white and black diagonal stripes was about 460 m (1500 ft) from the shoreline. But the beacon's location on a barrier island and the steady rise in sea level (presently about 2 mm or 0.08 in. per year) made it increasingly vulnerable to attack by powerful sea waves generated by the winds of tropical and extratropical cyclones that often track near Cape Hatteras. In less than 50 years, about 335 m (1100 ft) of the east-facing beach in front of the lighthouse had washed away. In the 1930s and early 1940s attempts to stabilize the beach with groins failed and after World War II, the lighthouse was taken out of service (replaced by a beacon mounted on a metal tower located well inland) and assigned to the National Park Service (NPS).

The Cape Hatteras Lighthouse soon became one of North Carolina's most popular tourist attractions and public support for preserving the lighthouse grew. Artificial beach nourishment was tried many times in the 1960s and 1970s but with short-lived success. The growing threat to the lighthouse was made clear in 1982 when an intense winter storm brought storm waves to the base of the lighthouse. During the height of the storm, quick acting NPS personnel tore up slabs of asphalt from a newly constructed parking lot and piled them around the base of the lighthouse to protect the foundation. Subsequently, sand bags were piled on top of the asphalt slabs.

FIGURE
This photo of the Cape Hatteras Lighthouse was taken in April 1995. Notice the sandbags at the base of the tower. [Photo by J.M. Moran]

Coastal engineers advised against a plan to build a seawall around the lighthouse. They pointed out that the structure's brick and stone foundation is less than 2 m (7 ft) deep and rests on yellow pine timbers that must remain submerged in freshwater to keep from decomposing. Surrounded by a seawall, the lighthouse eventually would become an island, saltwater would replace freshwater in the subsurface and rot the pine timbers, the foundation would collapse, and the lighthouse would fall into the sea.

By 1987, the Cape Hatteras Lighthouse was only 49 m (160 ft) from the sea. The only option to save the lighthouse was a strategic retreat inland, a plan endorsed by the National Park Service in 1989. In effect, retreat is a preservation/ protection strategy that avoids rather than confronts the forces of nature. After years of public debate, the lighthouse was physically moved on rollers and iron rails 884 m (2900 ft) to a new location (and a new foundation) some 488 m (1600 ft) from the edge of the Atlantic. The lighthouse began its journey on 17 June 1999 and arrived at its destination on 9 July 1999. The

Cape Hatteras Lighthouse again opened to the public on 26 May 2000. Is this the end of the Cape Hatteras Lighthouse story? Not likely. Sea level will probably continue to rise, tropical and extratropical cyclones will generate storm surges, and barrier islands will erode and move inland. In fact, on 18 September 2003, Hurricane Isabel made landfall on the Outer Banks as a category 2 system causing considerable coastal erosion and flooding. The storm surge cut a new inlet and came precariously close to the Cape Hatteras Lighthouse, damaging roads and parking lots.

ESSAY: Restoring Salt Marshes

For centuries, people have attempted to reclaim wetlands, including salt marshes, near cities and in densely populated coastal regions, such as Japan or the Netherlands. Much of Lower Manhattan's shoreline is built on such reclaimed lands, as is New York City's John F. Kennedy International Airport. The Dutch pioneered reclamation techniques, converting shallow sea bottoms to agricultural fields surrounded by dikes; such reclaimed lands are called polders. Other former coastal marshes were filled and are now used as landfills for solid waste.

In more recent years, studies have demonstrated the many valuable functions these wetlands perform, including protecting coastal lands against flooding from storm surges and from excessive rainfall. Wetlands filter water flowing through as well as acting as sponges, soaking up water at high tide and during flooding. Again, the Dutch are leading the way in restoring wetlands by breaking the dikes surrounding certain polders, permitting them to flood again and return to their original condition as salt marshes or shallow sea beds. On smaller scales, individual farmers have removed drains to permit low-lying fields to flood seasonally, thereby permitting marsh plants to invade and provide habitats. These areas are used by water birds and by fish as nursery grounds for their larval and juvenile stages.

Some marshes were damaged by introduced species that destroyed the native grasses, thereby accelerating erosion as sea level rises. That happened at the Blackwater Natural Wildlife Refuge on Chesapeake Bay's Eastern Shore, which is the largest unbroken stretch of marshland on the Chesapeake. It was damaged by the deliberate introduction in the 1930s of nutria, a rodent native to South America, in order to provide local trappers with a source of furs. Unlike its cousin the native muskrat, which eats only the top of the marsh grasses, nutria eats the plant roots. Its voracious appetite destroyed the native grasses, leaving large marshland areas unprotected against erosion by waves and tidal currents. In 2001, the original 9700-hectare (24,000-acre) refuge was losing 60 hectares (150 acres) of salt marsh each year, leaving large open-water areas and accelerating erosion of adjoining salt marshes.

In the 1990s, restoration projects began, such as the Chesapeake Marsh Restoration/Nutria Control Project, to reclaim marshes such as the Blackwater National Wildlife Refuge. The restoration project for the marshes at the Blackwater Natural Wildlife Refuge also helps alleviate another major environmental problem in the Chesapeake Bay, that is, the disposal of sediment dredged from shipping channels. A special dredge removes mud from channel bottoms and discharges it over the marshes. The newly deposited sediment layers are sufficiently thin that they do not kill the grasses, which grow over them, thus stabilizing the restored marsh. Each year, many thousands of migrating waterfowl as well as large numbers of bald eagles use the restored marshlands. These areas are also home to endangered fox squirrels. The shallow waters are nursery grounds for crabs, the most important commercial fishery in the Chesapeake. Finally sediment trapping and nutrient retention by the marsh helps improve water quality in the nearby Chesapeake Bay.

ESSAY: The Great Lakes and the Ocean

The Great Lakes of North America are components of the Great Lakes of the World (GLOW). GLOW also includes Great Bear and Great Slave Lakes of Canada, Lake Baikal of Russia, Lakes Victoria and Malawi of Africa, Lake Titicaca in South America and others. GLOW and particularly the Great Lakes are often referred to as "models of the ocean." How true is this statement?

Combined, the Great Lakes constitute one of the largest reservoirs of surface fresh water on the planet, accounting for about 18% of the total. (Only the polar ice sheets and Lake Baikal contain more fresh water.) The Great Lakes are a chain of lakes (see Figure). Lake Superior, the largest and deepest of the five lakes, drains into Lake Huron via the St. Marys River. Lake Michigan waters also empty into Lake Huron through the Straits of Mackinac. From Lake Huron, water flows via the St. Clair and Detroit Rivers into Lake Erie, the smallest and shallowest of the five lakes. Water then flows from Lake Erie through the Niagara River and Welland Canal into Lake Ontario and empties into the St. Lawrence River, the Gulf of St. Lawrence, and ultimately the North Atlantic Ocean. Water draining into Lake Superior along its western shore travels some 3600 km (2230 mi) to reach the Gulf of St. Lawrence. *Residence time* of water in the Great Lakes ranges from 2.6 years in Lake Erie to 191 years in Lake Superior.

The Great Lakes are a legacy of the Pleistocene Ice Age that began about 1.7 million years ago and ended about 10,500 years ago. Prior to the Pleistocene, rivers likely flowed through valleys where lake basins are now located. Huge lobes of glacial ice advancing southward from what is now Canada widened and deepened the river valleys, and the weight

FIGURE
The Great Lakes of North America with the drainage basin outlined. Weather and climate over the lakes and the drainage basin govern lake levels.

of the ice further depressed the evolving lake basins. Ocean basins, on the other hand, are much older (hundreds of millions of years) and are the products of tectonic processes (Chapter 2).

The Great Lakes are small compared to ocean basins; their dimensions are similar to those of the continental shelf, that is, 10s to 100s of kilometers (10s to 100s of miles) in the horizontal, and 10s to 100s of meters (10s to 100s of yards) in the vertical. The Great Lakes are sufficiently large that the Coriolis effect plays a role in the circulation of their waters (Chapter 5). The Coriolis effect along with winds and stratification generate circulation phenomena that are the same as those observed in the ocean (e.g., wind-driven Ekman transport, coastal upwelling and downwelling, and geostrophic flow).

Unlike small lakes but like the ocean, the Great Lakes are sufficiently large and deep that much of their water mass is unaffected by frictional interaction with the coast and bottom. If wind sets the surface water in motion at the nearly frictionless center of the lake basin, inertia keeps the water in motion. (Recall from Chapter 4 that according to *Newton's first law of motion*, an object in constant straight-line motion or at rest remains that way unless acted upon by an unbalanced force. *Inertia* is the name given to this tendency for an object to continue moving or to stay at rest.) With no other forces operating, only the Coriolis deflection affects the flow. In the Great Lakes, Coriolis deflection is to the right, so that a surface current describes a loop (called an *inertial circle*) having a period of about one pendulum day (about 17 hrs at 45 degrees North—near the middle of Lake Michigan). (Actually, the current follows a straight path while the basin rotates with the spinning Earth). For a current speed of 10 cm per sec (0.2 mi per hr) at 45 degrees N, the radius of an inertial circle is 1 km (0.6 mi), whereas for a current speed of 100 cm per sec (about 2 mi per hr), the radius is 10 km (6 mi).

Ekman transport operates in the Great Lakes. If the wind blows parallel to the coast, depending upon the direction of the wind, Ekman transport causes either coastal upwelling or downwelling (Chapter 6). In response to upwelling or downwelling, the thermocline tilts near the coast creating a horizontal gradient in density (or pressure). The balance that develops between this horizontal density gradient and the Coriolis effect produces *geostrophic flow* (Chapter 6). In the Great Lakes, geostrophic flow is observed as a coastal current or *jet*.

When the wind slackens, the release of the tilted thermocline (caused by upwelling or downwelling) generates two types of free internal waves involving the Coriolis effect: inertial waves or oscillations and Kelvin waves. Inertial oscillations occur in the thermocline across the entire lake but with maximum currents at the center of the lake. A Kelvin wave is an *edge wave*, meaning that maximum wave height is at the coast but decreases to lake level within 5 to 10 km (3 to 6 mi) offshore. Kelvin waves depend upon the Coriolis effect and progress cyclonically around the lake with the "edge" to the right of the direction of motion. Crests and troughs of a Kelvin wave are not readily observed except by temperature readings at municipal water intakes. A Kelvin wave crest brings deeper cooler water to the intake whereas a Kelvin wave trough delivers shallower warmer water. The period of a Kelvin wave is typically much longer than the period of an inertial oscillation. Large scale eddies are also observed in the Great Lakes coastal waters that resemble those on the shelf and in the open ocean.

Both inertial oscillations and Kelvin waves also occur in the ocean. Kelvin waves develop in the equatorial Pacific when the trade winds weaken during El Niño and the water piled up in the western tropical Pacific sloshes back toward South America (Chapter 11). The sloshing consists of Kelvin waves with maximum height on the equator where the Coriolis deflection changes sign. When these Kelvin waves strike the South American coast, they split into two waves, one progressing northward and the other southward along the coast. The northward moving Kelvin wave eventually reaches the U.S. west coast.

An obvious difference between the Great Lakes and the ocean is salinity: Great Lakes waters are fresh whereas seawater is saline. In the Great Lakes, water density depends on temperature and maximum density is at 4 °C (39.2 °F). On the other hand, seawater density depends on both temperature and salinity and its temperature of maximum density is below its initial freezing point (Chapter 3). An important implication of this contrast in physical properties is the occurrence of seasonal turnover in lakes but not the ocean. Spring and fall turnover of lakes in temperate latitudes is important for organisms living in lakes because turnover replenishes the dissolved oxygen supply of those water bodies. Lake turnover also recycles nutrients, especially nitrogen and phosphorus compounds, from bottom sediments to the overlying water, where they become available to aquatic plants, especially algae.

Bright summer sun penetrates a lake to shallow depths, warming the surface layer of water. Meanwhile, the dark deep water does not benefit from solar heating and remains cold, with an average temperature as low as 4 °C. This stable stratification (layering) of warm, less dense surface water and colder, denser water at depth persists through most of the summer, with little mixing between the layers. The upper layer of the stratified lake is known as the *epilimnion* and the lower layer is known as the *hypolimnion*, with the transition zone between the two layers being the *thermocline*.

In a stratified lake, oxygen that is supplied by the atmosphere and photosynthetic organisms replenishes the dissolved oxygen supply of only the epilimnion. In the hypolimnion, cellular respiration by decomposers and other organisms remove dissolved oxygen. With very little transfer of oxygen from the epilimnion, dissolved oxygen levels in the hypolimnion steadily diminish. If lake stratification were to persist, the dissolved oxygen concentration could decline to levels that would severely stress cold-water fish species (e.g., trout, whitefish) living in the hypolimnion. Fortunately, these fish and other inhabitants of the hypolimnion usually do not perish because lake stratification eventually breaks down, lake waters turn over, and the dissolved oxygen content of the hypolimnion is replenished.

As summer gives way to autumn, the sun is lower in the sky, daylight becomes shorter, and air temperatures drop. Heat is lost from the warm surface waters to the overlying cool air and ultimately to space. Eventually the temperature of the epilimnion cools to that of the hypolimnion, the thermocline disappears, and the lake has a uniform temperature and density from top to bottom. Winds blowing across the lake transport surface water from the upwind shore toward the downwind shore. Oxygen-depleted water wells up from below on the upwind shore and oxygen-rich surface water sinks at the downwind shore. This *fall turnover* of lake-waters brings oxygen-depleted water from the lake bottom to the surface where it is exposed to the atmosphere and replenished in dissolved oxygen.

Cooling during late autumn and early winter eventually drops the temperature of surface waters to 4 °C and lower. The lake begins to stratify with the coldest and least dense water at the surface. With continued cooling, the surface water temperature eventually drops to 0 °C and a skim of ice forms on the lake surface. While large areas of the Great Lakes (especially sheltered inlets such as Green Bay) may develop an ice cover, stirring of waters by strong winds brings slightly warmer water to the surface and keeps much of the lake surface ice-free through the winter. For the Great Lakes, the entire water column must overturn to reach maximum density (at 4 °C) before the lake can cool below 4 °C, stratify, and freeze. For Lake Superior, the deepest of the Great Lakes, this is a maximum depth of 400 m (1300 ft). Lake-water temperatures typically vary from 0 °C just under the ice to 4 °C at the lake bottom. Ice forms a barrier that prevents exchange of oxygen between the lake-water and atmosphere, and cellular respiration within the lake causes a gradual decline in dissolved oxygen levels. When ice finally melts in spring, the lake waters may be seriously depleted in dissolved oxygen.

With the arrival of spring, daylight lengthens, air temperatures rise, and surface waters warm. Eventually, the temperature of the surface water reaches 4 °C and once again the temperature and density of the lake become uniform from top to bottom. Just as in autumn, strong winds trigger turnover of lake waters and the mixing that accompanies *spring turnover* brings oxygen-depleted water to the surface where oxygen is replenished directly from the atmosphere and via photosynthesis. As spring gives way to summer, lake stratification is reestablished. What is intriguing and different from ocean water is that in spring the Great Lakes typically begin to stratify in shallower water near shore and stratification progresses toward the center of the Lakes at about 1 km (0.6 mi) per day.

Wind waves are similar on the ocean and Great Lakes although the scales are different. Swell and larger-size waves are more common on the ocean because of more persistent winds and longer fetch (Chapter 7). Through the years, many powerful storms have swept over the Great Lakes, producing huge waves that sank thousands of vessels with a considerable loss of life. Even very large ships may be unable to withstand the fury of storm waves on the lakes. For example, on 10 November 1975, a fierce storm tracking over Lake Superior produced winds gusting to 145 km (90 mi) per hr. Waves of 3.5 to 5 m (12 to 16 ft) sank the *Edmund Fitzgerald*, at the time the largest ore carrier on the Great Lakes at 222 m (729 ft) long and carrying a load of 26,116 tons of taconite pellets. The *Edmund Fitzgerald* sank in 160 m (5200 ft) of water at the eastern end of Lake Superior about 27 km (17 mi) from Whitefish Bay, MI with the loss of its crew of 29. Speculation is that huge waves lifted the bow and stern enough to cause the ship to break amidships and sink before the crew could react.

The ocean and the Great Lakes also share some similarities in long-term changes in water level but there are also fundamental differences. Astronomical tides are evident in the ocean, but are not very noticeable on the Great Lakes. On the other hand, seiches are common on the Great Lakes but are not prominent in the open ocean or on the continental shelves, although they are observed in harbors and embayments. For both tides and seiches the time scale is hours and the height scale is centimeters to meters.

Over the past 3000 to 3500 years, Great Lakes levels have fluctuated in a quasi-periodic manner about an established datum (based on mean lake levels during the modern record of 1918-1998). Departures from this so-called International Great Lakes Datum (IGLD) were never much more than a meter during the 3000 to 3500 year period. As of this writing, Great Lake levels are rising from near historic lows of the past decade. Great Lakes levels are governed by precipitation, runoff from spring snowmelt, and air temperature. Temperature affects the rate of evaporation of water and controls ice cover that also influences evaporation. Should the present large-scale warming trend continue, lake levels are expected to fall whereas sea level is predicted to rise. We have more to say about these prospects and their implications in Chapter 12.

CHAPTER 9

MARINE ECOSYSTEMS

Case-in-Point
Driving Question
Requirements for Marine Life
Structure of Marine Ecosystems
    Producers
    Consumers
    Decomposers
    Trophic Structure of Ecosystems
    Bioaccumulation
Ecosystem Processes
    Energy for Growth and Reproduction
    Production in the Photic Zone
    Nutrients and Trace Elements as Limiting Factors
    Microbial Marine Ecosystems
Ocean's Role in the Global Carbon Cycle
    Physical Pump
    Biological Pump
Ecosystem Observations and Models
Conclusions
Basic Understandings
ESSAY: Iron Fertilization and Climate Change
ESSAY: Gas Hydrates, A Future Energy Source
ESSAY: Ocean Color and Marine Productivity

Sea anemones clustered on a rocky slope off Hawaii. [Courtesy National Undersea Research Program (NURP)/NOAA.]

Case-in-Point

In December 1987, many people in Montreal, Canada became seriously ill and some died after eating mussels later traced to one area of the eastern Canadian province of Prince Edward Island. Their symptoms included abdominal cramps, nausea, neurological problems, and memory loss. The disease was later called amnesic shellfish poisoning.

    Scientists worked around the clock for days at a Canadian government laboratory in Halifax, Nova Scotia to identify the toxin (poison) responsible for the illness and its source. They found that the mussels had consumed a species of phytoplankton that produced domoic acid, a neurotoxin, which accumulated in their tissues. Tests for this toxin enabled fishery managers to halt immediately the harvesting and sale of mussels from contaminated areas thereby preventing additional illness and death.

    Amnesic shellfish poisoning is one of the many possible consequences of a phenomenon referred to as *harmful algal blooms* (*HAB*) or *red tides*. Not all harmful algal blooms are red or toxic and they have nothing to do with astronomical tides. Furthermore, they are not all caused by algae. Many toxic blooms are caused by dinoflagellates, microscopic one-celled organisms that behave like both plants and animals. Some red-tide producing organisms are in the ballast-water discharged by ships far from their port of origin and may contribute to the increasingly widespread occurrence of such events. For example, poisonous algal blooms increased 10-fold between 1975 and 1995 in coastal waters of China. However, the principal reason for more frequent red tides is probably increased runoff of nutrients into coastal waters due to human activities.

Blooms of phytoplankton, microscopic algae, occur widely in surface ocean waters. Under normal conditions, the phytoplankton community consists of many species. Sometimes, however, a bloom of a single species occurs and can cause serious harm to the environment, the demise of marine organisms, economic losses, and human health problems, including death. Blooms of toxic algae have been blamed for fish kills and the deaths of marine mammals around the world. In late February 2002, the deaths of 20 dolphins that washed ashore in Southern California were attributed to the domoic acid secreted by a species of dinoflagellates. A toxin that enters a marine food chain and is taken up by shellfish may threaten human health. Several neurological conditions, some of them fatal, can result from ingesting contaminated shellfish. Fortunately, commercial supplies of mussels, clams, oysters and scallops from most areas are now rigorously tested for toxins.

In addition to toxin-producing blooms, a second type of harmful algal bloom, known as a *high biomass bloom*, depletes the supply of dissolved oxygen in seawater (Chapter 3). If excess amounts of nutrients (mostly nitrogen and phosphorus compounds) enter coastal waters, phytoplankton blooms can become extremely dense. Runoff from agricultural fields or discharge of sewage water into the ocean is usually responsible for this condition. When these algae die, their remains sink to the ocean bottom where they are acted upon by aerobic decomposers, organisms that use dissolved oxygen. Eventually the water may become severely depleted of dissolved oxygen, making it impossible for other organisms (except anaerobic organisms) to survive. Such conditions can cause massive fish kills.

Between 1987 and 1992, harmful algal blooms caused direct economic losses of more than $50 million annually. Besides losses to commercial fisheries, algal blooms cost the tourism industry millions of dollars each year because the decomposing residue of algal blooms that washes up on beaches produces a foul odor and unpleasant appearance that drive away tourists.

Driving Question:

What are the basic components and structure of marine ecosystems and what is their source of energy?

With this chapter, our focus shifts to life in the ocean. The biosphere, atmosphere, and ocean interface over nearly 71% of Earth's surface. The physical and chemical processes examined in earlier chapters influence all marine organisms and their interactions. In this chapter, we describe how interactions among the biosphere, atmosphere, and ocean govern the distribution and abundance of life in the ocean.

A fundamental subdivision of the Earth system is the *ecosystem*. This chapter examines the components of marine ecosystems including producers, consumers, and decomposers. We emphasize the source of energy (photosynthesis, chemosynthesis) and the inefficient transfer of energy among organisms occupying the various feeding (trophic) levels within food webs. We then take a closer look at the various processes operating within marine ecosystems including energy supply for growth and reproduction, factors influencing biological productivity in the ocean, the role of nutrients and trace elements as limiting factors, and the importance of microbes in marine ecosystems. In view of the contemporary interest in the carbon cycle and prospects for global climate change, we also examine the physical and biological processes governing the flux of carbon into and out of the oceanic reservoir over a range of time scales. We begin by identifying the essential requirements for marine life.

Requirements for Marine Life

Many of the fundamental requirements for life in the ocean are the same as those needed by terrestrial organisms. Essential for all life on Earth are a source of energy (e.g., sunlight), liquid water, the appropriate mix of chemical constituents (e.g., nutrients), and the right combination of environmental conditions (e.g., range of temperature).

One major difference between life in the ocean and life on land is the much greater space and variety of marine habitats. On land, the tops of the tallest plants are only a few tens of meters above the ground and only a few species of birds can fly higher than 100 m (330 ft). In the ocean, microscopic plants live at depths as great as about

200 m (650 ft) and animals inhabit all depths. Even the deepest trenches (about 10,000 m or almost 33,000 ft deep) on the ocean floor support life. The ocean is home to a huge number of different types or species of organisms—perhaps as many as 10 million—and many more are yet to be discovered and studied. The wide variety of marine habitats makes possible this great biodiversity. Some marine ecosystems, such as coral reefs (Chapter 10), are structurally complex and provide niches for many types of organisms that have evolved different adaptations to their environment. It has been only two decades since discovery of deep-sea hydrothermal vents and their associated thousands of species of marine organisms (Chapter 2).

One of the most basic needs for all forms of life and the chief constituent of all living organisms is liquid water. All organisms require energy, which they obtain either as producers (via photosynthesis or chemosynthesis) or as consumers eating other organisms. Photosynthetic organisms use light energy to convert simple inorganic compounds (carbon dioxide and water) into complex energy-rich organic substances, which provide food. Chemosynthetic organisms live in the absence of light and derive energy from substances such as hydrogen sulfide (H_2S).

In addition to water and energy, marine life forms require nutrients and trace elements in sufficient quantities—just as humans require certain minerals and vitamins in small quantities in their diet. Nutrients that are required by marine organisms in relatively large quantities include phosphorus and nitrogen compounds, calcium, and magnesium—all of which are essential for plant growth. Iron and several other elements, so-called *trace elements* or *micronutrients*, are also required, but in very small quantities. Iron is especially important because it plays a role in the formation of *chlorophyll*, the most common photosynthetic pigment in plant cells, and it is essential for certain plant enzymes. When any one of these essential factors is not available in sufficient quantities, plant growth is limited.

According to the **law of the minimum**, the growth and well being of an organism is limited by the essential resource that is in lowest supply relative to what is required. This most deficient resource is known as the **limiting factor**. For more on this topic, see this chapter's first Essay.

Structure of Marine Ecosystems

An **ecosystem** is a fundamental subdivision of the Earth system in which communities of organisms interact with one another and with the physical conditions and chemical substances of their habitats. All ecosystems have both living (*biotic*) and non-living (*abiotic*) components and energy sources. Biotic components include producers, consumers, and decomposers whereas abiotic components constitute their physical and chemical environment. As we shall see, ecosystems vary enormously in structure and complexity.

The basis of all ecosystems is a group of organisms collectively known as **producers**, photosynthesizing organisms. On land, they are primarily green plants. In the ocean, producers include familiar seaweeds and sea grasses in shallow waters, algae, and some bacteria. Producers are also called *autotrophs* (meaning *self-nourishing*) because they manufacture the food they need from inorganic constituents of their physical environment. That manufactured food constitutes primary production, the base of most common ecosystems. More precisely, **primary production** is the amount of organic matter synthesized by organisms from inorganic substances and is usually measured in units of grams of carbon per square m per unit time. Autotrophs store energy-rich organic matter in their cells for use in all life processes, such as growth, maintenance of cells, and reproduction.

Organisms (usually animals) that feed on autotrophs are called **consumers** or *heterotrophs* (meaning *other-feeders*). The organic material produced in the growth of consumers is known as **secondary production**. Heterotrophs may be herbivores, carnivores, or omnivores. *Herbivores* feed exclusively on plants, whereas *carnivores* eat herbivores or other carnivores. *Omnivores* eat both plants and animals. In this section we take a closer look at marine producers, consumers, and decomposers (organisms that break down dead plants and animals).

PRODUCERS

In the ocean, the producers are plants, algae, and some bacteria. Multi-celled plants are relatively rare in the ocean. Seaweeds (large algae) and sea grasses (true vascular plants) grow attached to some substrate on the ocean bottom. They can survive only in shallow, relatively clear water where light levels are high enough to support photosynthesis. Floating, microscopic unicellular algae, photosynthetic bacteria, and other groups of organisms capable of photosynthesis are responsible for more than 99% of biological production in the photic zone. Collectively these floating organisms are referred to as **phytoplankton**, from the Greek words *phyto* for plant and *plankton* for wanderer.

Phytoplanktonic organisms are very small and make up only 1% of the total mass of plants on Earth, but they are responsible for considerable global primary production. As is typical of very small organisms, phytoplankton reproduce rapidly during favorable environmental conditions (as often as eight times a day) so that populations of phytoplankton species can fluctuate dramatically. Some types have very complex life cycles with numerous stages and different, poorly known forms. This is especially true of certain toxic and otherwise harmful algal species, discussed in this chapter's Case-in-Point.

Here we describe four major groups of phytoplanktonic organisms: diatoms, coccolithophores, dinoflagellates, and bacteria. Although all are single-celled, some may form colonies or chains of individual cells. Some types of phytoplankton (i.e., dinoflagellates and bacteria) can be confusing because they have characteristics of both plants and animals. They may carry out photosynthesis and they may also act as heterotrophs. Organisms that can be both autotrophic and heterotrophic are called **mixotrophs**. The various types of phytoplanktonic organisms range in size from less than 2 micrometers to about 2 mm (visible to the naked eye).

Diatoms are a diverse group of *protista* (or *protoctista*) having properties of both plants and animals. They are encased in shell-like cases (called *frustules*) made of silica (Figure 4.7A). These cases can be very ornately decorated. Diatoms contain photosynthetic pigments (primarily *chlorophyll*) and contribute significantly to primary production. When diatoms die, they sink to the bottom where their frustules form a major component of marine biogenous sediment deposits in some areas of the ocean bottom (Chapter 4).

Coccolithophores are also single-cell photosynthesizing organisms; they are covered with tiny calcium carbonate plates (Figure 4.6A). Coccolithophores are unusual in that they appear to thrive in nutrient-poor waters, forming large blooms that turn surface waters greenish blue as viewed from space (Figure 9.1). The plates of coccolithophores are major contributors to marine calcareous sediment deposits including, for example, the White Cliffs of Dover on the southeast coast of England. About 70 million years ago a white mud composed of mostly coccolithophore fragments accumulated at the bottom of a shallow sea and gradually converted to chalk, a fine-grained calcareous rock. Although the chalk was deposited at a very slow rate (about 0.5 mm per year), up to 500 m (1600 ft) of chalk eventually accumulated in some places. Tectonic forces elevated the chalk layers above sea level and later, toward the end of the Pleistocene Ice Age, formation of the English Channel made Britain an island and exposed the White Cliffs, rising up to 90 m (300 ft) or more above sea level.

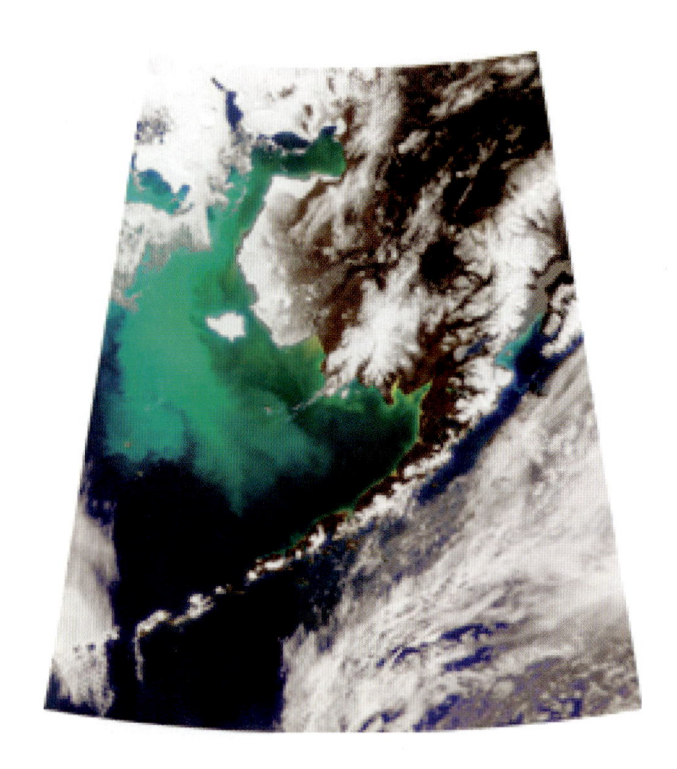

FIGURE 9.1
A coccolithophore bloom in the Bering Sea off Alaska, north of the Aleutian Island chain. Bright white areas are clouds or snow covered land. [NASA SeaWiFS image]

Dinoflagellates have two flagella (thread-like structures) and no solid covering so that they are not preserved in marine sediment deposits. Many species of dinoflagellates are mixotrophs, many are bioluminescent (Chapter 10), and some species form red tides (Figure 9.2). Species of dinoflagellates that do not contain photosynthetic pigments are exclusively heterotrophs and are considered to be zooplanktonic (described below).

The smallest but probably also the most numerous of phytoplanktonic organisms are certain groups of bacteria. The many types of bacteria play important roles in the functioning of marine ecosystems. Some are autotrophs; others are heterotrophs; and still others are decomposers, breaking down organic material from dead organisms (known as *detritus*) and recycling the nutrients back into the ecosystem. Cyanobacteria have a blue-green pigment and are mixotrophs (Figure 9.3). Some bacteria play an important role in the ocean's nitrogen cycle.

Whereas photosynthesis drives the most familiar marine ecosystems, scientists have discovered extraordinary ecosystems in the deep ocean in which chemical energy, rather than sunlight, drives the basic biological processes. This method of primary production is referred to as **chemosynthesis**. For example, bacteria oxidize sulfides,

FIGURE 9.2
A large red tide formed by the dinoflagellate *Noctiluca*. [Courtesy of Peter Franks, Scripps Institution of Oceanography, University of California, San Diego.]

FIGURE 9.3
Cyanobacteria are the most abundant photosynthetic organisms in the ocean. [NASA]

such as hydrogen sulfide (H_2S), to obtain energy to fix carbon from carbon dioxide into simple sugars. Regardless of whether an ecosystem is driven by photosynthesis or by chemosynthesis, the total amount of carbon converted from carbon dioxide into organic matter is known as primary production. The process is also referred to as *carbon fixing* because carbon is changed, or fixed, from a simple (inorganic) to a more complex (organic) form, usually energy rich sugars and carbohydrates.

CONSUMERS

Small herbivores are the chief consumers of phytoplankton; small and large carnivores, in turn, prey on herbivores. Single-cell and multi-cellular consumers that drift passively with ocean currents or are weak swimmers are referred to collectively as **zooplankton** (animal plankton). They include bacteria and dinoflagellates as small as 2 micrometers (called *microplankton*). Very large organisms, such as jellyfish and other gelatinous animals, are also zooplankton; these animals may measure as much as a meter (3 ft) across. Active swimmers can pursue and capture their prey, but most zooplanktonic organisms are suspension or **filter feeders**. They use tiny hairs (called *cilia*) or mucus-covered surfaces to capture particles suspended in the water.

The most ubiquitous members of the zooplankton are **copepods**, a diverse group of tiny crustaceans covered with an exoskeleton made of chitin, a colorless and amorphous substance (Figure 9.4). Copepods include both herbivorous and carnivorous species. In some areas of the ocean, the feeding activities of herbivorous copepods can limit the growth of phytoplankton populations. Copepods play an important intermediate role in marine food chains, feeding on phytoplankton and, in turn, being eaten by larval and juvenile fish.

Larger members of the zooplankton community include **euphausiids** and other shrimp-like crustaceans called **krill** (Figure 9.5). Krill, up to 6 cm (2.5 in.) long, are so numerous in the ocean that their biomass equals the combined weight of all humans on Earth! A major reason for their abundance in the Southern Ocean is the upward

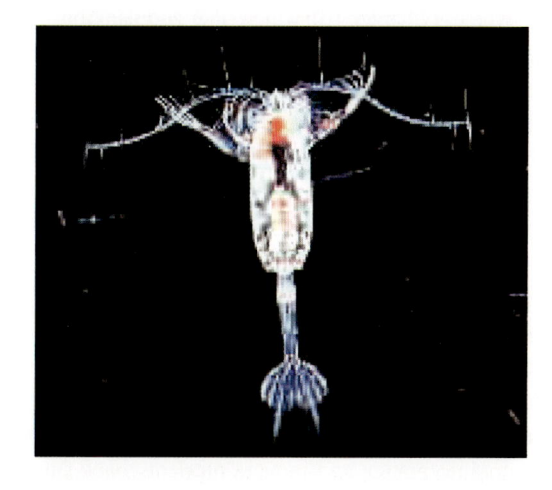

FIGURE 9.4
Copepods are tiny crustaceans that are an important food source for juvenile fish and shellfish. [NOAA]

FIGURE 9.5
Euphausiids (krill) occur in large swarms in the Southern Ocean and feed on algae on the under surface of sea ice. [NOAA, National Marine Fisheries Service, La Jolla, CA]

transport of nutrients in the North Atlantic Deep Water around Antarctica. Krill is a direct or indirect food source for a variety of marine organisms including whales, fish, squid, seals, penguins, and seabirds. In fact, krill is regarded as the keystone species in the Southern Ocean. A *keystone species* is an essential member of an ecosystem that contributes to biodiversity; its loss would likely lead to the extinction of other forms of life in the ecosystem. Larval forms of fish, crabs, lobsters, starfish and other larger animals are also members of the ocean's zooplankton community.

Zooplankton is the main food source for larger consumers in the sea. These include larval and juvenile fish and even the much larger basking sharks (e.g. whale sharks) and baleen whales (Chapter 10), which filter zooplankton out of the water by passing seawater through comb-like structures in their mouths. It is important to remember, however, that larger animals are rare in the sea compared to the enormous number of smaller zooplanktonic organisms, protozoans, and bacteria. In fact, the size of the average animal in the open ocean is no larger than a mosquito!

DECOMPOSERS

Decomposers, or detritivores (detritus eaters), are consumers that feed on dead organic matter, either on the ocean bottom or in the water column. Decomposers are essential components of all ecosystems because they break down organic matter and recycle nutrients back into the ecosystem. Whereas most decomposers are bacteria size, some decomposers such as worms may be complex organisms living on the sea floor and in ocean bottom sediments.

TROPHIC STRUCTURE OF ECOSYSTEMS

A **food chain** is a sequence of feeding relationships among organisms (e.g., producers, consumers) that provides pathways along which energy and materials move within an ecosystem. For example, in a marine food chain, phytoplankton is eaten by a herbivore, which in turn is eaten by a carnivore, which is then eaten by another carnivore and so on. In reality, such simple food chains rarely exist in the ocean and marine heterotrophs are more likely to feed on a wide variety of prey organisms. Most ecosystems consist of several levels of consumers including fish, marine mammals, and sea birds. Most humans feed at the highest trophic levels in many marine ecosystems. This complex of feeding relationships (food chains) is more realistically described as a **food web** and each feeding position occupied by a given organism within a food web is called a **trophic level** (Figure 9.6).

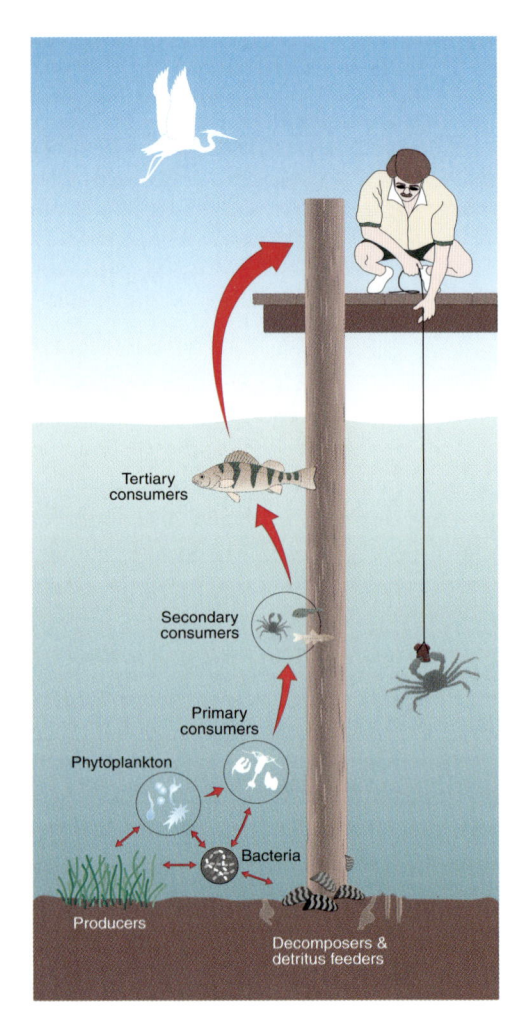

FIGURE 9.6
This simple food web is typical of an estuary. [NOAA]

Within food webs, energy is transferred in only one direction, from lower trophic levels to higher trophic levels (e.g., from producers to consumers). But this energy transfer is inefficient. As an organism occupying one trophic level feeds on its prey, only a small portion of the energy stored in the tissues of the prey organism is passed to the next higher trophic level. We can visualize the energy distribution within a food web as a pyramid with producers forming a broad base containing most of the ecosystem's energy. Each successively higher trophic level contains considerably less energy than the one below it. That is, energy transfer within food webs is much less than 100% efficient. **Ecological efficiency** is defined as the fraction of the total energy available at one trophic level that is transformed into work or some other usable form of energy at the next higher trophic level.

Worldwide, ecological efficiency varies depending on types of organisms and ecosystems but is never very high. In some ecosystems, no more than 5% of the energy at one trophic level becomes incorporated in the tissues of organisms that occupy the next higher trophic level. An ecological efficiency of 10% is typical for the open ocean. In some highly productive marine ecosystems having few trophic levels and where strong upwelling brings an abundant supply of nutrients to the surface, ecological efficiency can be as high as 20%.

Biomass is easier to measure than energy so that scientists usually follow and describe the flow of energy within food webs in terms of biomass. (*Biomass* is the total mass or weight of organisms.) Applying the *10% rule of ecological efficiency*, some 10,000 grams of primary producers will supply food for 1000 grams of herbivores (e.g., copepods). They, in turn, will feed 100 grams of primary consumers (e.g., fish larvae) and only 10 grams of secondary consumers—salmon, for example. If humans consume the salmon, then only 1 gram of human biomass will result from the original primary production of 10,000 grams of phytoplankton. As we have noted, however, few marine ecosystems are as simple as this example; most marine food webs consist of many different organisms feeding on multiple food sources. Many of the fishes listed on restaurant menus (e.g., tuna, shark, salmon) are carnivores that feed at higher trophic levels and require considerable food energy from the ecosystem. Furthermore, a complex food web with diverse food sources tends to be more stable and less vulnerable to environmental change than a simple ecosystem. The loss of one species as a food source may not destroy an entire complex food web as it might a simple food chain. On the other hand, a short, simple food chain is more efficient at transferring energy.

Why is ecological efficiency so low? There are many reasons. For one, not all the biomass at each trophic level is harvested. Many organisms have characteristics (adaptations) that enable them to avoid predation thereby reducing the harvest. Not all the harvested biomass is ingested. Parts of the prey organism (e.g., skeleton) may not be eaten even though they contain food energy. Not all the ingested biomass is digested (assimilated). If consumed, indigestible materials are excreted and subsequently broken down by decomposers. Furthermore, not all the assimilated biomass is converted to usable energy. Energy losses occur when assimilated food enters the organism's cells and is processed to liberate energy for maintenance, growth, and reproduction. This process of energy liberation, known as **cellular respiration**, occurs in all living cells. However, cellular respiration is inefficient. Less than half the energy in sugars (the direct source of energy for cellular respiration) is converted to a usable form. The rest is transformed to heat that cannot be used to perform work (biological activities) and eventually is released to the environment. Carbon dioxide and water, the other products of cellular respiration, also escape to the environment.

Low ecological efficiency has implications for aquaculture such as rearing salmon in *fish farms*. Farm salmon are generally fed fishmeal, made by grinding up marine fish caught in the wild. Given an ecological efficiency of 10%, it takes at least 10 grams of fishmeal (actually, more than 10 grams of whole fish since it is dried prior to grinding) to produce one gram of salmon.

BIOACCUMULATION

Food webs are pathways not only for energy but also for toxins, especially those that persist in the environment. They persist because they do not break down physically, chemically, or biologically. Examples include polychlorinated biphenyls (PCBs) and heavy metals such as mercury. Such substances tend not to dissolve in water but to concentrate in fatty tissue. These toxins may interfere with the metabolism and reproduction of sensitive organisms and in some cases pose health threats for humans. We described a tragic example of bioaccumulation in the Case-in-Point of Chapter 4 when mercury contamination of sediments in Minamata Bay, Japan entered the marine food web and caused human illness and death.

Some persistent toxins move from one trophic level to the next, building in concentration as one organism consumes another. This process of continually increasing concentration within a food web is called **bioaccumulation**. Bioaccumulation is a consequence of the ecological inefficiencies that prevail between trophic levels—the biomass declines at each successively higher level but the toxins remain. Furthermore, organisms do not readily break down or excrete persistent toxic substances. Hence, from one trophic level to the next higher trophic level, the amount of toxin lost is small compared to the amount that is transferred. Because most of an ingested toxin is retained and most of the biomass is lost, a toxin becomes more concentrated at successively higher trophic levels (Figure 9.7).

Bioaccumulation is especially pronounced in aquatic (marine and fresh water) food webs, which usually consist of four to six trophic levels rather than the two or three levels that are common in terrestrial ecosystems. In addition to contamination via bioaccumulation, some fish and shellfish can absorb certain toxins directly through their gills from the surrounding water. The combination of absorption plus bioaccumulation causes a toxin to become anywhere from 1000 to more than a million times more concentrated in upper-trophic level organisms (compared to the toxin's concentration in the environment). Consequently, people and fish-eating birds (e.g., eagles, osprey) who feed on fish at these high trophic levels run the greatest risk of consuming harmful levels of toxins with their meal. Especially susceptible individuals (e.g., children, pregnant women, nursing mothers) are advised to restrict their intake of fish that may be exposed to toxins.

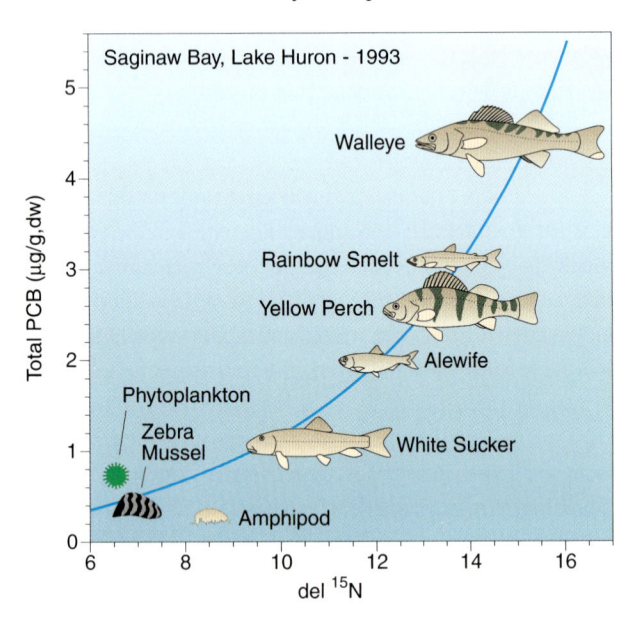

FIGURE 9.7
Field studies in the fresh water ecosystem of Saginaw Bay, Lake Huron, MI show bioaccumulation of PCBs in successively higher trophic levels. Scientists use "del ^{15}N" as a measure of the distance between trophic levels where ^{15}N is an isotope of nitrogen. [Adapted from NOAA Great Lakes Environmental Research Laboratory]

Persistent toxic substances can enter the ocean in the discharge of rivers or can settle out of the atmosphere. As we saw in this chapter's Case-in-Point, a toxin may even be a substance that is naturally produced by a marine organism.

Ecosystem Processes

Sunlight provides energy for photosynthetic organisms, which inhabit sunlit surface ocean waters and provide the food for other marine organisms. Chemosynthesizing organisms, which live in dark, deep-ocean waters and within the rock and sediment on the seafloor, obtain their energy through complex chemical reactions involving a mix of chemicals in the hydrothermal fluids circulating through rock fractures. In this section, we examine in more detail the sources of energy for growth and reproduction of marine organisms, marine productivity, how nutrients and trace elements can be limiting factors, and the important role of microbes in marine ecosystems.

ENERGY FOR GROWTH AND REPRODUCTION

Photosynthesis is represented by the following chemical equation:

$$6H_2O + 6CO_2 + ENERGY \rightarrow C_6H_{12}O_6 + 6O_2$$

Photosynthetic pigments in marine plants and bacteria absorb light energy and use it to combine carbon dioxide and water to form simple carbohydrates ($C_6H_{12}O_6$) and release oxygen gas. That is, photosynthesis converts solar energy into chemical energy. However, less than 0.5% of the available solar energy becomes chemical energy; much of the rest of the solar energy is converted to heat. Chlorophyll is the most common photosynthetic pigment and is the ingredient that makes the leaves on most terrestrial plants green. (Chlorophyll reflects radiation in the green portion of the visible spectrum and absorbs radiation at all other wavelengths.) Other pigments dominate in some plants and are responsible for the brown and red colors of some algae. The complex chemical structure of these pigments enables them to trap the energy of light photons and transfer it to electrons in the sugar or carbohydrate molecules formed during photosynthesis.

Because solar radiation is essential for photosynthesis, this process occurs only during daylight and primarily in relatively shallow waters (Chapter 5). Most sunlight is absorbed very near the ocean's surface; typically 80% is absorbed in the upper 10 m (33 ft) of water. The blue and green portions of the visible spectrum penetrate to the greatest depth in the water column (Figure 5.8) and photosynthesis is most efficient within this range (peaking at wavelengths of about 0.44 and 0.65 micrometers). The depth to which sunlight penetrates ocean water depends on the clarity of the water. Where waters are made turbid by abundant phytoplankton or suspended particles, or discolored by dissolved materials, the depth of the photic zone is reduced. In muddy rivers the photic zone may be only a few centimeters deep, whereas in very clear areas of the open ocean, it may be more than 200 m (660 ft) deep. Coastal waters, with their abundant dissolved and suspended sediments, tend to be more turbid and absorb light at much shallower depths.

Energy-releasing chemical reactions can occur in the ocean without sunlight. Many microbes synthesize organic matter, obtaining energy from chemical reactions fueled by energy-rich substances, such as hydrocarbons (petroleum or methane gas), hydrogen sulfide, and various metals. These energy-rich substances are discharged in the waters of hydrothermal vents (Chapter 2), from sea-floor petroleum seeps, decomposing whale carcasses, or decaying salt-marsh vegetation. For example, hydrogen sulfide (H_2S) takes the place of the sun in supplying energy for chemosynthesis:

$$H_2S + 2O_2 \rightarrow H_2SO_4 + Energy$$

Hydrogen sulfide (H_2S) combines with dissolved oxygen (O_2) to produce sulfuric acid (H_2SO_4) and releases energy. Chemosynthetic organisms can use this energy to produce carbohydrates (just as in photosynthesis):

$$24H_2S + 6O_2 + 6CO_2 \rightarrow C_6H_{12}O_6 + 24S + 18H_2O$$

Hydrogen sulfide (H_2S) combines with dissolved oxygen (O_2) and carbon dioxide (CO_2) to produce carbohydrates ($C_6H_{12}O_6$), sulfur (S), and water. Hence, chemosynthesis supports ecosystems on the ocean bottom just as photosynthesis in the surface ocean supports ecosystems in much of the rest of the ocean. The biomass of such ecosystems on and within the ocean floor is unknown but may be many times that of ecosystems in the open ocean.

PRODUCTION IN THE PHOTIC ZONE

As noted earlier, cellular respiration makes available energy in a form that organisms use for growth, maintenance, and reproduction. In plants photosynthesis is the dominant process during daylight hours but both plants and animals respire continually. Also, whereas photosynthesis is confined to the photic zone, respiration takes place at all depths in the ocean.

Net production is the amount of organic matter produced during photosynthesis that exceeds the amount consumed in the process of cellular respiration. Because respiration occurs at all depths and at all times in the ocean while photosynthesis occurs only in light, net primary production is confined to the surface sun-lit waters. The total amount of carbon fixed into organic matter through photosynthesis in a given unit of time is the **gross primary production** (expressed in units of grams of carbon per square m per day or year). But respiration by producers releases some of the energy stored in these carbon-containing compounds for their own metabolic processes such as growth and reproduction. The amount of carbon remaining is called **net primary production** and this is the total food and energy supply available for the rest of the ecosystem.

No net primary production occurs below the **compensation depth**—usually the ocean depth where the light level diminishes to about 1% of what it is at the surface. The compensation depth varies with location and time of day. At that depth, the amount of carbohydrates and oxygen produced by photosynthesis exactly equals the amount consumed in respiration. Photosynthetic organisms cannot survive long below the compensation depth because they cannot fix enough carbon or produce enough organic carbon based food to meet their own respiratory needs.

When organisms die, their tissues decay and nutrients are remineralized and released back to the water. If decomposition occurs within the surface zone, living organisms quickly recycle nutrients. Primary production using recycled nutrients is called *regenerated production*. Where decomposition occurs below the pycnocline, nutrients cannot readily mix back up into the photic zone. Therefore, recycling is much slower because little new growth occurs below the photic zone. Primary production based on nutrients brought into the ecosystem by processes such as upwelling or winter mixing is called *new production*. Pulses of phytoplankton growth may occur when storm winds and waves transport nutrients from below the pycnocline upward into the sunlit surface waters.

In summary,

Gross Primary Production = New Production + Regenerated Production

and

Net Primary Production = Gross Primary Production − Respiration

The currently accepted estimate of the total gross primary production in the world ocean is 50 gigatons (Gt) of carbon per year, where one Gt equals one billion tons. Given the area of the ocean, this figure translates into about 145 grams of carbon per square m per year—about the same production as terrestrial forests, grasslands, and cropland. To put these numbers into some perspective, a standard paper clip weighs about 0.5 gram so that each square meter of the ocean (summed to the depth where the light level drops to 1% of its surface value) fixes the equivalent of about 290 paper clips of carbon. Actual values of primary production (in grams of carbon per square m per year) range from less than 50 in the open ocean, up to 500 in coastal upwelling zones, and to as high as 1250 in shallow estuaries.

Primary production requires sunlight, nutrients, and phytoplankton (above the compensation depth) and varies with both location and season. Primary production is very low in much of the tropical ocean. The chief reason for this desert-like condition is lack of nutrients. In the tropics, incoming solar radiation is intense throughout the year so that strong stratification persists year-round inhibiting new nutrients from moving into the photic zone from below. Brief exceptions are caused by the strong winds of a tropical cyclone that mix water to great depths and transport nutrients into sun-lit surface waters. In addition, tropical coral reefs are highly productive. Overall, the tropical ocean has considerable biological diversity but relatively little biomass.

Low productivity in the central subtropical ocean is also due to lack of nutrients. Recall from Chapter 6 that Ekman transport in the subtropical gyres produces a broad mound of warm surface waters near the center of the gyres. This mound of water is also nutrient-poor and depresses nutrient-rich water to depths well below the photic zone.

At temperate latitudes, productivity varies with the season. In winter, sunlight is weak and the winds of winter storms mix water to the extent that the shallow seasonal pycnocline disappears and most autotrophs are carried to depths below the photic zone. In spring, more intense solar heating reestablishes the pycnocline and stratification, halting the deep mixing of autotrophs. With abundant nutrients, more sunlight, and a more stable water column, conditions are favorable for a dramatic increase in phytoplankton populations, an event known as the **spring bloom** (Figure 9.8). Once the spring bloom is well underway, there is an explosive growth of zooplankton populations that graze on the phytoplankton. (This is sometimes referred to as a *zooplankton bloom*). In some areas of the ocean, such as the North Atlantic, the spring bloom ends when the zooplankton have reduced the phytoplankton to levels that are not sustainable. In other areas of the ocean, the spring bloom is limited by nutrient supply. Often, a secondary bloom occurs in early fall while sunlight is sufficiently intense and the first storms of the season transport new nutrients upward into the sunlit surface layer.

Phytoplankton grows abundantly where upwelling transports nutrient-rich waters into the photic zone. Areas of persistent wind-induced upwelling occur along the equator and on eastern sides of ocean basins, due to prevailing winds and Ekman transport (Chapter 6). Such areas of upwelling support about one quarter of all fish production in the ocean. For example, upwelling along the west coast of South America supports the rich sardine and anchovy fisheries off Chile, Ecuador, and Peru. These fisheries collapse when El Niño suppresses upwelling and nutrients are not available to fuel phytoplankton growth (Chapter 11).

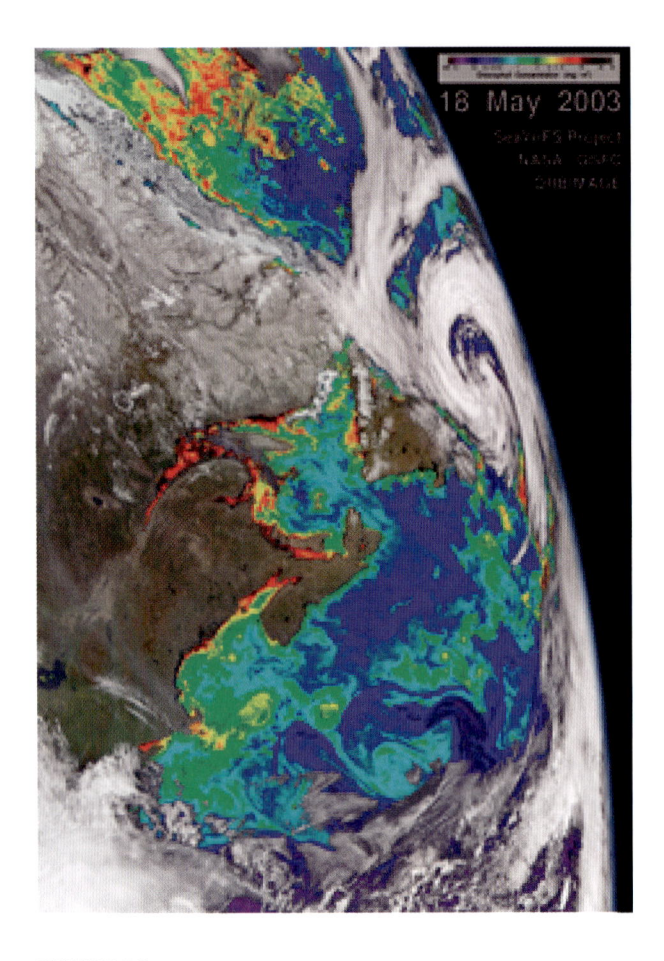

FIGURE 9.8
This satellite image shows the spring bloom around Arctic Canada in May 2003. Areas of high chlorophyll concentration are shown in red, orange, and yellow whereas areas of lower production are in shades of green and blue. [NASA]

In polar regions sunlight limits productivity. Stratification is generally absent and nutrients are abundant. (High latitude seas such as the Bering Sea are seasonally stratified.) In summer after some of the ice cover melts, the long polar day triggers high primary production.

NUTRIENTS AND TRACE ELEMENTS AS LIMITING FACTORS

Over large areas of the ocean, waters contain relatively high concentrations of nitrogen and phosphorus compounds, the usual limiting nutrients, yet biological production is low. These areas are often in the middle of ocean basins, far from land, and are known as **HNLC regions** (for high nutrients, low chlorophyll). Note the areas of very low productivity in the central ocean basins shown in Figure 9.9.

HNLC regions have long puzzled biological oceanographers. If the nutrient supply is sufficient, why is primary production so low? They suspected that this condition might arise from the lack of some element that usually occurs in trace quantities. Research conducted in the 1980s and 1990s demonstrated that iron is often a limiting trace metal in marine ecosystems. Iron is very insoluble in oxygen-rich ocean waters and usually attaches to particles and quickly sinks to the ocean bottom. However, winds transport iron-rich dust particles from land sources to the ocean where they settle into surface waters and often promote phytoplankton blooms (Chapter 4). HNLC regions of the ocean may be so distant from terrestrial sources of wind-borne iron particles (e.g., Sahara Desert) that input of iron-rich dust is insufficient to make these areas productive. For more on this topic, refer to this chapter's first Essay.

MICROBIAL MARINE ECOSYSTEMS

The traditional view of marine ecosystems is that large organisms such as fish dominate food webs. However, oceanographers now recognize that such large organisms are actually very rare in the ocean. In fact, microbes, single-celled organisms including bacteria, apparently dominate life in the ocean along with viruses and a group of primitive organisms called *archaea*. A single drop (1 milliliter) of seawater contains as many as one million bacteria and 1 billion viruses! The biomass of bacteria and other microbes in the ocean is so large that they are probably responsible for processing and recycling more carbon and nutrients than any other component of the ecosystem.

Now scientists are unraveling the role of these microbes in the ocean. In the 1970s the American oceanographer, John Steele, then at Woods Hole (MA) Oceano-

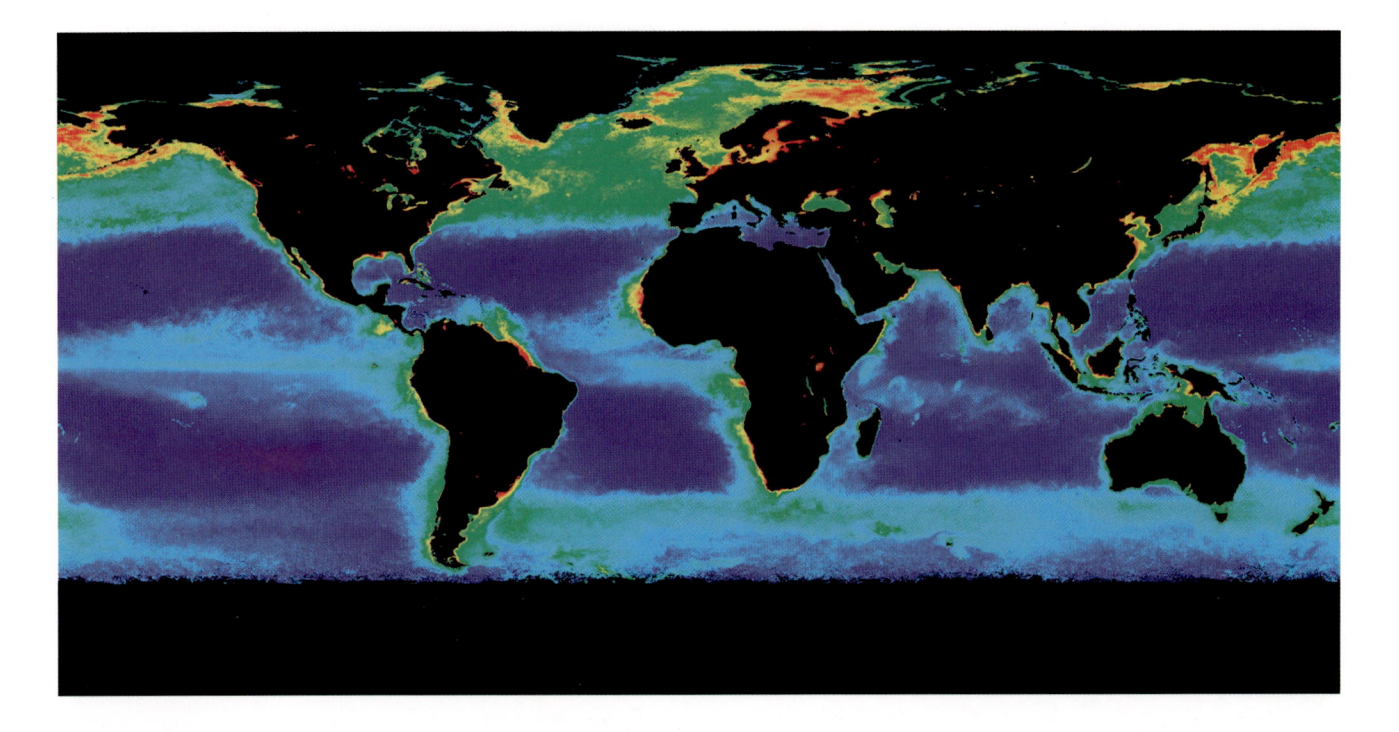

FIGURE 9.9
SeaWiFS image of marine biological production during Northern Hemisphere spring, compiled from data for the years 1998-2003. Areas of relatively high production are shown in red, orange, and yellow whereas areas of relatively low production are in shades of blue. [NASA]

graphic Institution, found that he could not account for the production of all the fish catches reported in government statistics from around the world. His estimate of total global primary production simply did not provide enough food to support the reported commercial fish catches, let alone all the other organisms in the sea. Where was the extra primary production coming from? Until that time, bacteria and protozoa had been considered to be important in the ocean only for their role in decomposing and recycling materials in ecosystems. Using improved microscopes and microbiological techniques, scientists discovered that microbes act as both producers and consumers. Furthermore, they concluded that microbes contribute enormously to global oceanic primary production and are especially important in open-ocean waters. Consequently, oceanographers are revising their traditional view of the marine ecosystem in which plants are the major producers.

To observe these one-celled organisms, scientists must use light microscopes or electron scanning microscopes (for the smaller forms). The Dutch scientist Antony van Leeuwenhoek (1632-1723), who ground his own magnifying lenses, was the first to observe microbes in fresh water. Some types of bacteria can be grown in the laboratory by placing a few drops of water on a special jelly (agar, made from marine algae). Hence, a few types of microbes, such as bacteria, are relatively well known.

The ocean is home to three major groups of microbes: bacteria, archaea, and viruses. Archaea and bacteria together form the class of organisms called *prokaryotes*. All prokaryotes are unicellular and primitive. They differ from all other organisms (*eukaryotes*) in that their DNA is not enclosed within a cell nucleus. In fact, prokaryotes have essentially no internal cellular structure.

The genetic structure of archaea differs from that of bacteria and is more like that of much more highly evolved organisms. They have some different cell structural properties and unusual physiological adaptations. These are likely related to the fact that archaea often inhabit extreme environments such as hydrothermal vents, Antarctic sea ice, or extremely salty pools. In the 1980s, Farooq Azam of the Scripps Institution of Oceanography in La Jolla, CA discovered archaea living in ocean water. Subsequent research has shown that archaea play important roles in decomposing simple organic compounds and dissolving organic carbon in the ocean.

Archaea living within the pore spaces of marine sediments appear to be involved in the generation of gas hydrate deposits. Gas hydrates are a potential source of methane (the chief constituent of natural gas), may contribute to global climate change, and may be a factor in the potential for submarine landslides. For more on gas hydrates, refer to this chapter's second Essay. Eleven different types of microbes (nine bacteria and two archaea) dominate life in the ocean in numbers and probably in mass. For example, one photosynthetic bacterium, *Prochlorococus*, accounts for up to half the primary production in the tropical ocean, making it the most abundant and most important primary producer on Earth. This particular organism was discovered in 1988.

Viruses are even stranger organisms. Some question whether they are even living in the traditional sense of the word because they cannot reproduce without a host organism. Their structure is even simpler than that of prokaryotes. They are studied using molecular techniques and may be many times more abundant in ocean waters than bacteria.

There are also many different types of unicellular *eukaryotes* of very small sizes. These have complex internal cellular structure, similar to the cells of more complex organisms, and are collectively known as *protozoa*. The major groups of protozoans in the ocean include ciliates and dinoflagellates. Many of these single-celled organisms are capable of weak swimming, propelled by hair-like structures.

Scientists now know that bacteria are most abundant in surface ocean waters to depths of about 150 m (500 ft). As we have seen, in addition to their importance as primary producers, bacteria and archaea are significant consumers in marine food webs. They can directly absorb and digest *dissolved organic carbon (DOC)* that cannot be used as food by larger organisms. DOC comes from many sources, including the contents of ruptured phytoplankton cells and microbial cells (as they die or are partially eaten by zooplankton), and from the liquid excretions of zooplankton.

Bacteria and archaea are used as food by the smallest heterotrophs, including dinoflagellates and ciliates. All of these microbes and protozoa may be too small to be a significant food source for most other zooplankton. Thus, the organic carbon they consume seldom reaches the rest of the marine food web, but is recycled within a semi-separate food web, the so-called **microbial loop** that operates parallel to the larger food web (Figure 9.10). The microbial loop is important in regulating the flow of carbon and energy in the marine ecosystem, particularly in recycling within the photic zone. The microbial loop is a self-sustaining system that scientists are just beginning to understand. It is clear, however, that the ocean is home to large numbers of bacteria, though of only a few types compared to terrestrial ecosystems. Viruses apparently infect bacte-

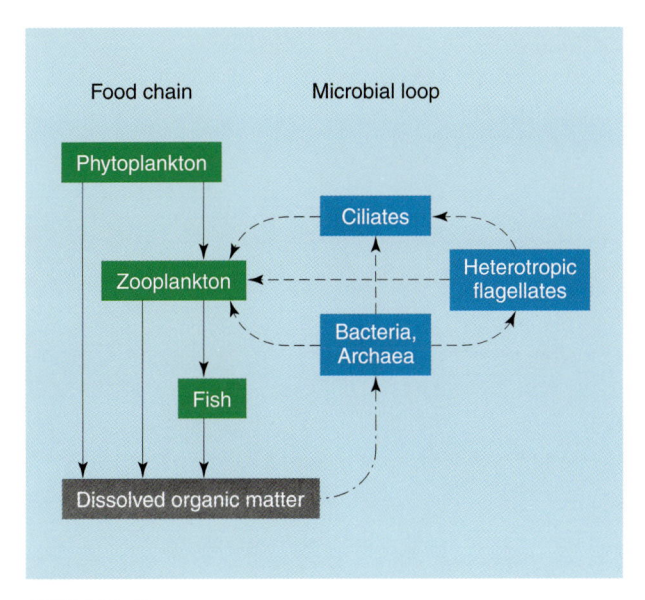

FIGURE 9.10
A simple microbial loop linked to a schematic marine food chain.

ria and other microbes, thereby holding their populations in check. Otherwise bacteria alone could consume all the available nutrients in seawater.

Ocean's Role in the Global Carbon Cycle

One of today's most widely debated environmental concerns is global climate change and the possible role played by the carbon cycle, one of the chief biogeochemical cycles that operate in the Earth system. The ocean is a major reservoir in the carbon cycle and an important player in Earth's climate system. Carbon dioxide, a greenhouse gas, is transferred between the atmosphere and ocean as part of the carbon cycle.

In terms of changes taking place over a time frame of a few thousand years, the ocean—primarily the deep ocean—is the most significant reservoir of carbon (mostly as dissolved CO_2) in the Earth system. The ocean holds about 50 times the amount of carbon dioxide contained in the atmosphere and about 20 times the amount of carbon stored in the biosphere (primarily as organic matter in soils). Over time frames of millions to billions of years the largest carbon reservoir consists of sedimentary rock (e.g., limestones, shales) and fossil fuels (i.e., coal, oil, and natural gas). Together these large reservoirs contain almost 1400 times more carbon than in ocean water. In this section, we focus on the cycling of carbon among the ocean, atmosphere, and biosphere.

On time scales of years to millennia, two different sets of processes, biological and physical, govern the cycling of carbon into and out of the ocean. Both sets of processes maintain relatively high concentrations of dissolved CO_2 in the ocean's depths. Some is cycled in the global conveyor belt circulation and is isolated from the atmosphere for centuries whereas some cycles at a faster rate as in primary production.

PHYSICAL PUMP

An important process controlling the ocean's carbon cycle is primarily physical in nature. This is called the **physical pump** or the *solubility pump*. The latter name refers to the fact that CO_2 dissolves more readily in cold water and hence more CO_2 sinks with the deep water as it forms at high latitudes of the North Atlantic as well as around Antarctica (Chapter 6). That is, dissolved CO_2 is pumped deep into the ocean below the photic zone.

The ocean's surface waters take up carbon dioxide primarily at middle and high latitudes. The major Northern Hemisphere western boundary currents (Gulf Stream and Kuroshio) that originate in the tropics are chilled during their flow northward. (Similar processes occur in the Southern Hemisphere.) Because of lower temperatures and high biological production in spring and summer, surface ocean waters take up large amounts of CO_2. These dense CO_2-rich waters sink as NADW in the Greenland and Labrador Seas and as AADW and AABW around Antarctica.

As noted in our discussion of the global oceanic conveyor belt in Chapter 6, NADW mixes with AADW and AABW to form Common Water (CoW). This deep water (CoW) circulates slowly throughout the ocean on a time scale of 1000 years. It gradually warms and diffuses upward into the surface layer where CO_2 is released (*outgassed*) to the atmosphere because warm water cannot dissolve as much carbon dioxide. The largest natural releases of CO_2 from the ocean's depths take place in the narrow strips of coastal and equatorial upwelling. As cold waters from below move up into the surface zone, they warm and release some of their dissolved CO_2. During El Niño, the presence of a thick layer of warm water in the central and eastern tropical Pacific inhibits upwelling of cold, CO_2-rich waters, thus suppressing the outgassing of CO_2 from deep ocean water (Chapter 11).

BIOLOGICAL PUMP

The second set of processes by which carbon cycles through the ocean involves biological activity and is known as the **biological pump** (Figure 9.11). Eighty percent of the primary production in the ocean takes place in

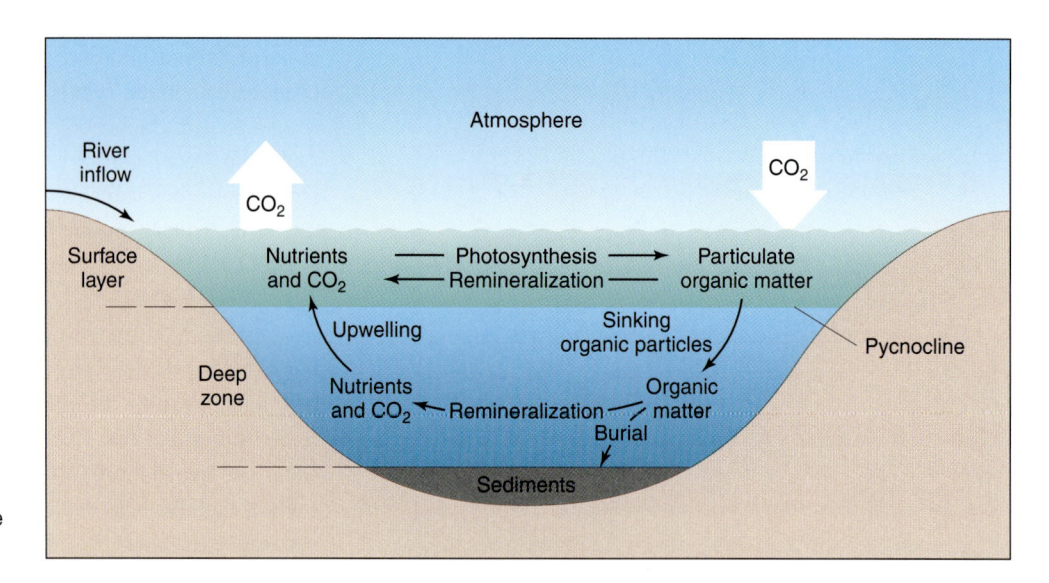

FIGURE 9.11 Biological processes that control the distribution of carbon in the ocean constitute the biological pump.

the photic zone of open ocean and roughly equals the amount of organic matter produced by all land plants on an annual basis. Unfortunately the ocean's carbon cycle is not well known and poorly measured, especially in the intermediate- and deep-ocean zones. Consequently we must examine how carbon is fixed as organic matter and then decomposed by bacteria and recycled through the ocean's ecosystems. This set of processes constitutes the biological pump. Consider first the open ocean.

As we have seen, in the photic zone, phytoplankton and other photosynthetic organisms convert inorganic carbon into organic carbon and organisms grow and reproduce rapidly. Having very brief life spans, their cells divide several times a day and individuals die or are eaten by herbivores. In either case, the organic particles they produce sink out of the surface layer as *marine snow*, fecal pellets of zooplankton, or as aggregates of dead phytoplankton cells. The carbon contained in this material plus the carbon in the sinking bodies of zooplankton and other marine animals, is called **particulate organic carbon (POC)**. Up to 50% of the carbon fixed in the photic zone settles out of the surface layer in this form. POC may be eaten by zooplankton or decomposed by bacteria or other microbes. Some of it is quickly converted back to dissolved inorganic carbon in this process. The POC that is eaten by zooplankton is repackaged into even larger fecal pellets that sink more quickly through the water column. Some organic matter reaches the ocean bottom and is buried by the subsequent rain of particles (Chapter 4).

Only about 10% of the POC that sinks out of the photic zone reaches depths greater than 1000 m (3300 ft). This intermediate zone of the ocean, below the photic zone, but above the greater depths where there is less biological activity, is known as the **twilight zone**. The twilight zone has too little light for photosynthesis, but sufficient light for visual predators to locate prey. This is also the depth of the oxygen minimum zone caused by the accumulation of POC in the main pycnocline, which results in higher demand for dissolved oxygen for decomposition and respiration. This decline in dissolved oxygen begins just below the photic zone and reaches a minimum just beneath the pycnocline.

Particles of organic matter that reach the sea floor are buried in sediment deposits and the carbon they contain can be removed from the water and from the biosphere, probably for millions of years. However, only about 3% of the carbon fixed in the photic zone is removed from the ocean in this way for longer than 1000 years; most organic carbon finds its way back to the surface waters and atmosphere in less time via the global oceanic conveyor belt.

Another way in which organic carbon is recycled in the ocean is by *remineralization*, that is, by being converted back to soluble inorganic forms. Respiration by all organisms releases CO_2 into the water. If this happens at shallow depths, the CO_2 enters the surface layer and potentially escapes to the atmosphere relatively quickly (within a few years). Dissolved carbon dioxide in the deep ocean remains there for a very long time—centuries or longer. It may eventually return to the surface via the physical pump described earlier. When a single-celled organism dies, its cell may rupture, spilling the contents into the seawater. Also, zooplankton and other organisms excrete liquid in addition to fecal pellets. These organic carbon-containing substances are ideal food for microbes, which ingest them, returning CO_2 in solution to the water. This process begins in the surface layer and continues through the water column as bacteria and other microbes feed on and decompose organic particles of all sizes. The ocean depth at which decomposition occurs determines the length of time that carbon is retained. Larger and heavier particles are more likely to pass through the mid-ocean depths without significant alteration by microbes until they reach the ocean floor.

All processes involved in the ocean's carbon cycle operate over a wide range of temporal and spatial scales. Temporal scales are hours for photosynthesis, days for particles to sink to the bottom, millennia for circulation of water in the oceanic conveyor belt, and hundreds of millions of years for tectonic recycling of marine sedimentary rocks. Spatial scales encompass the atmosphere, hydrosphere, biosphere, and lithosphere.

Cycling of carbon into and out of the ocean is a key player in Earth's climate system (Chapter 12). Human activities, especially the burning of fossil fuels and deforestation are adding carbon dioxide to the atmosphere. The pre-industrial atmospheric CO_2 concentration was about 280 parts per million by volume (ppmv). In 2002 it was about 372 ppmv and could top 550 ppmv by the close of this century if present trends continue. If there were no photosynthesis or biological pump operating in the ocean, the present atmospheric CO_2 concentrations would be 1000 ppm and Earth's climate would be much warmer because of a greatly enhanced greenhouse effect (Chapter 5). If the efficiency of the biological pump were perfect, on the other hand, only 100 ppm of CO_2 would be in the atmosphere.

CO_2 fluxes at the interface between the ocean and atmosphere vary greatly from one region of the ocean to another. For example, across much of the central and eastern equatorial Pacific Ocean, upwelling transports cold deep-ocean waters rich in CO_2 to the sea surface where it is released to the atmosphere as surface waters warm. In the North Atlantic, on the other hand, chilling high salinity surface waters forms dense waters that sink, carrying CO_2 from the surface down into the deep ocean. In addition, nutrient-rich waters ensure high primary production in the North Atlantic. Hence, in this region of the ocean, physical and biological processes combine to transport relatively large amounts of CO_2 from the atmosphere into the deep ocean. Similar conditions exist seasonally in the Southern Ocean. Decades-long studies of the ocean carbon cycle have shown how the ocean "breathes," that is, where it absorbs CO_2 from the atmosphere and where it returns CO_2 to the atmosphere.

To this point, we have been discussing the operation of the carbon cycle in the open ocean. Consider now the carbon cycle in the coastal ocean. The coastal ocean is highly productive, receiving organic matter from many sources including tidal currents that flow through wetlands, the discharge of rivers and streams, winds, and erosion of sediment deposits on continental margins. The total amount of organic carbon of terrestrial origin in the coastal ocean is estimated to be equivalent to about one-fifth of the primary production of the open-ocean surface waters.

Ecosystem Observations and Models

Studies of the oceanic carbon cycle on various temporal and spatial scales rely on data obtained remotely by satellite-borne instruments. One of these satellite sensors is NASA's Sea-viewing Wide Field-of-View Sensor (SeaWiFS). SeaWiFS maps *ocean color*, that is, the distribution of photosynthetic pigments and particularly chlorophyll-a, the most significant photosynthetic pigment in marine algae. With frequent global observations, we can follow seasonal changes in the distribution of chlorophyll-a concentrations (Figure 9.9).

Global images obtained from SeaWiFS show the narrow bands of highly productive coastal ocean waters, especially along the western sides of the continents, and the moderately productive waters of the central and eastern equatorial Pacific and the margins of the Southern Ocean. Using such data, the ocean can be divided into similar biological provinces. In addition, the Advanced Very High-Resolution Radiometer (AVHRR) sensor on NOAA's polar-orbiting satellites measures sea surface temperatures on a global scale.

By combining remote sensing data from satellite sensors such as SeaWiFS and the AVHRR, one biological and the other physical, scientists can estimate the global (or regional) annual primary production (exclusive of that due to autotrophic microbes or chemosynthesis). In this way, they can deduce the amount of carbon fixed through photosynthesis. This information adds to our understanding of the role of the oceanic carbon cycle in global climate change. For more on SeaWiFS, refer to this chapter's third Essay.

For information on ocean properties at depths below the range of satellite sensors, ocean scientists rely on observations at specific locations taken over a lengthy period, spanning years, decades, or longer. A major impetus for developing such *time series* of data is the Joint Global Ocean Flux Study (JGOFS). This multi-disciplinary international program (with participants from 20 nations) was established in 1987 under the auspices of the Scientific Committee of Oceanic Research (SCOR). The principal goal of JGOFS is to "assess more accurately, and understand better the processes controlling, regional to global and seasonal to inter-annual fluxes of carbon between the atmosphere, surface ocean and ocean interior, and their sensitivity to climate changes." In late 1988, JGOFS established long-term ocean time series projects at sites near Bermuda and Hawaii. At these stations, observations are routinely collected on more than 40 physical, chemical, optical, biological, and atmospheric variables from the surface through the ocean depths.

These time series data permit scientists to construct numerical models ranging in complexity from those applicable to a single location to three-dimensional models of the global ocean. A major goal of ocean modelers is to produce coupled models that integrate biogeochemical and physical processes. However, physical data that drive the models (e.g., solar radiation, wind, cloud cover) currently are more readily available than biogeochemical data. Biogeochemical aspects of ocean modeling are essential and data are being collected, processes are being studied, and the gap between physical modeling and biogeochemical modeling is closing.

When the model results match observational data, researchers can be reasonably confident that they have included most of the major processes affecting ecosystems and biogeochemical cycles and their model accurately reflects what is happening in the real ocean. In cases where the models cannot reproduce the observed data, important processes are still not adequately understood or represented. In some cases, more research on the processes is required and in others, more environmental data must be collected to permit more realistic simulations of ocean processes in time and space.

Conclusions

Although many of the fundamental ingredients of life in the ocean are the same as those required by terrestrial life, the ocean has a larger life zone and a greater variety of habitats. Marine ecosystems are composed of producers, consumers, and decomposers and their energy is supplied via photosynthesis or chemosynthesis. For several reasons, energy transfer between trophic levels in food webs is inefficient. Persistent toxic chemicals may enter marine food webs and bioaccumulate, threatening the well being of organisms (including humans) feeding at higher trophic levels.

We have seen that primary production in the ocean's photic zone varies spatially and temporally because of seasonal changes in sunlight, weather, fluctuations in nutrient supply, the availability of trace elements, stratification of ocean waters, and upwelling. We have also examined the physical and biological processes that govern the role of the ocean in the global carbon cycle. We return to this topic in Chapter 12 where we discuss prospects for an enhanced greenhouse effect and global climate change. Our examination of ocean life continues in the next chapter.

Basic Understandings

- Many of the fundamental requirements for life in the ocean are the same as those needed by terrestrial organisms. Essential for all life on Earth are energy sources, liquid water, the appropriate mix of chemical constituents (e.g., nutrients), and a favorable combination of environmental conditions (e.g., range of temperature).

- Major differences between the ocean and land include the much greater space and variety of marine habitats. Also, terrestrial plants require some sort of substrate to support them, usually soil. In the ocean for phytoplankton, stratification plays this role and is an important part of their ecosystem, to keep them in the photic zone so that they can produce more than they respire.

- When any one of the essential factors (e.g., nutrients) is not available in sufficient quantities, plant growth is limited. According to the law of the minimum, the growth and well being of an organism is limited by the essential resource that is in lowest supply relative to what is required. This most deficient resource is known as a limiting factor.

- An ecosystem is a fundamental subdivision of the Earth system in which communities of organisms interact with one another and with the physical conditions and chemical constituents of their surroundings. Ecosystems vary in complexity but all have biotic (living) and abiotic (non-living) components.

- The simplest organisms in marine ecosystems consist of producers (or autotrophs), most commonly photosynthesizing plants, algae and bacteria. Autotrophs manufacture their food from the physical and chemical environment. That food constitutes primary production, the base of most common ecosystems. More precisely, primary production is the amount of organic matter synthesized by organisms from inorganic substances and is usually measured in units of grams of carbon per square m per unit time.

- Organisms (usually animals) that feed on autotrophs are called consumers or *heterotrophs*. The organic material produced in the growth of consumers is known as secondary production. Heterotrophs may be herbivores, carnivores, or omnivores. Organisms that function as both autotrophs and heterotrophs are called mixotrophs.
- Marine producers are plants, algae and some bacteria, which use sunlight to produce food through photosynthesis. Floating, microscopic unicellular algae, photosynthetic bacteria, and other groups of organisms capable of photosynthesis are responsible for much of biological primary production in the ocean. Collectively these organisms are referred to as phytoplankton. Four principal groups of phytoplanktonic organisms are diatoms, coccolithophores, dinoflagellates, and bacteria.
- While photosynthesis drives most marine ecosystems, scientists have discovered ecosystems in the deep ocean in which chemical energy, rather than sunlight, drives the basic biological processes. This method of primary production is referred to as chemosynthesis.
- Uni-cellular and multi-cellular consumers that drift passively with ocean currents or are weak swimmers are referred to collectively as zooplankton. They include bacteria and dinoflagellates as small as 2 micrometers as well as very large organisms, such as jellyfish and other gelatinous animals.
- Decomposers or detritivores (detritus eaters) are consumers that feed on dead organic matter, either on the ocean bottom or in the water column. They are essential components of all ecosystems because they break down organic matter and recycle nutrients back into the ecosystem.
- Food chains and food webs are sequences of feeding relationships among organisms (e.g., producers, consumers) through which energy and materials move within an ecosystem. Within food chains and food webs, energy transfer occurs in one direction, from lower trophic (feeding) levels to higher trophic levels (e.g., from producers to consumers). But, for a variety of reasons, this energy transfer is inefficient. As an organism occupying one trophic level feeds on its prey, a small portion of the energy stored in the tissues of the prey organism is passed on to the next higher trophic level. In the open ocean, ecological efficiency is about 10% where ecological efficiency is defined as the fraction of the total energy at one trophic level that is transformed into work or some other usable form of energy at the next higher trophic level.
- Food webs are pathways not only for energy but also for toxins, that is, poisons that persist in the environment because they do not break down physically, chemically, or biologically. Some persistent toxins move from one trophic level to the next, building in concentration as one organism consumes another. This process of continually increasing concentration within a food web is called bioaccumulation and is a consequence of ecological inefficiencies.
- Sunlight supplies the energy needed by photosynthetic organisms to convert organic matter from water and CO_2, substances that are readily available from the atmosphere and ocean. Because solar radiation is essential for photosynthesis, this process occurs only during daylight and primarily in relatively shallow waters. Other energy-releasing chemical reactions can occur in the deep ocean in the absence of sunlight. Many microbes synthesize organic matter, obtaining energy from chemical reactions fueled by energy-rich substances, such as hydrocarbons (petroleum or methane gas), hydrogen sulfide, and various metals. This process is known as chemosynthesis.
- The total amount of carbon fixed into organic matter through photosynthesis in a given unit of time is the gross primary production (expressed in units of grams of carbon per square m per day or year). But respiration by producers releases some of the energy stored in these carbon-containing compounds for their own metabolic processes. The amount of carbon remaining is the net primary production and this is the total food and energy supply available for the rest of the ecosystem. No net primary production occurs below the compensation depth—usually the depth where the light level diminishes to about 1% of what it is at the surface.
- Photosynthetic primary production requires sunlight, nutrients, and phytoplankton above the compensation depth and varies with both location and season. Primary production is very low in much of the tropical and subtropical ocean chiefly because of a lack of nutrients or trace elements. At temperate and polar latitudes, productivity varies seasonally.
- Phytoplankton grows abundantly where upwelling brings nutrient-rich waters into the photic zone. Areas of persistent wind-induced upwelling occur along the equator and on eastern sides of ocean basins, due to prevailing winds and Ekman transport.
- Tiny microbes far outnumber large organisms in the ocean. The ocean is home to three major groups of microbes: bacteria, archaea, and viruses.

- An important process controlling the ocean's role in the global carbon cycle is primarily physical in nature. This is called the physical pump or the solubility pump. The latter name refers to the fact that CO_2 dissolves more readily in cold water and hence more CO_2 sinks with the deep water as it forms at high latitudes of the North Atlantic as well as around Antarctica. The second set of processes by which carbon cycles through the ocean involves biological activity and is known as the biological pump. That is, the biological pump describes how carbon is fixed as organic matter in the photic zone and then decomposed by bacteria and recycled through marine ecosystems.
- Studies of the oceanic carbon cycle on large temporal and spatial scales depend on data obtained by remote sensing from satellite-borne instruments. (These instruments must be calibrated by field data.) One of these satellite sensors is NASA's Sea-viewing Wide Field-of-View Sensor (SeaWiFS). SeaWiFS maps surface ocean color, that is, the distribution of photosynthetic pigments and particularly chlorophyll-a, the most significant photosynthetic pigment in marine algae.

ESSAY: Iron Fertilization and Climate Change

As described elsewhere in this chapter, large areas of the open ocean are nearly devoid of one-celled plants (algae) that form the base of marine food webs. In some areas the reason is low concentrations of dissolved phosphorus and nitrogen compounds, nutrients that are essential for most plant growth. Other areas, however, have high concentrations of these nutrients but low concentrations of chlorophyll (plant pigments). Such waters occur in the Antarctic, subarctic, and equatorial Pacific Ocean. Apparently, something is keeping the algae from growing.

The mystery was solved by John H. Martin (1935-1991), an oceanographer based at Moss Landing (CA) Marine Laboratories. Initially, Martin studied how the classic plant nutrients (nitrogen, phosphorus) limit algal growth. Later his research focused on the role of metals that occur in trace concentrations in seawater and algae. Martin's study of trace metals was made difficult by potential contamination, which usually occurred while sampling seawater aboard ship and in laboratory chemical analyses. To avoid contamination, Martin used special clean rooms with filtered air and ultra-clean plastics. He even made his own ultra-pure chemical reagents so that he could accurately measure concentrations of trace metals in uncontaminated seawater. This work showed that most trace-metal concentrations reported in the scientific literature were too high because of contamination. His fellow scientists were not pleased to have their research criticized and many arguments ensued over Martin's analytical methods and results.

Martin discovered that algal growth normally is inhibited by the presence of copper and zinc in the water, but low levels of these metals were necessary for their growth. Martin then proposed that the high nutrient-low chlorophyll (HNLC) areas of the ocean were deficient in iron and other trace elements. To test his hypothesis, Martin proposed a field experiment in which he would add large quantities of iron to high nutrient waters to see if this stimulated algal growth.

Building on his iron limitation hypothesis, Martin also turned his attention to the possible role of marine algae in Earth's climate system. For one, he proposed that algae might have played a role in the large-scale climatic fluctuations of the Pleistocene Ice Age. Algae not only give ocean waters a greenish color but also they take up CO_2 through photosynthesis. Martin proposed that during cold, dry glacial climatic episodes, strong planetary-scale winds transported iron and other trace metals in dust blown off mountains and deserts. Dust plumes moved over the ocean and particles settled into the ocean stimulating the growth of phytoplankton. More phytoplankton meant more photosynthesis, less atmospheric CO_2, a weaker greenhouse effect, and a cooler planet. During mild, wet interglacial climatic episodes, less dust reached the ocean and phytoplankton populations declined. Less phytoplankton meant reduced photosynthesis, more CO_2 in the atmosphere, an enhanced greenhouse effect, and a warmer planet.

Lecturing at the Woods Hole (MA) Oceanographic Institution in July 1988, Martin said, "Give me a half-tanker of iron and I'll give you an ice age." Martin's hypothesis came at a time of growing scientific interest in the prospects for global warming due to a CO_2-enhanced greenhouse effect. Martin argued that the build up of atmospheric CO_2 (due to fossil fuel combustion and deforestation) could be offset to some extent by increasing the ocean's uptake of CO_2. He proposed an experiment in which large amounts of dissolved iron would be released into open-ocean waters where nutrient concentrations are high and chlorophyll levels are low. However, Martin died before he could carry out his experiment. In 1993, Kenneth Coale, a researcher at Moss Landing (CA) Marine Laboratories, and his colleagues carried out Martin's experiment in a patch of water near the Galapagos Islands in the equatorial eastern Pacific. From their ship, the *R.V. Melville*, they released half a ton of iron over an ocean area of 100 square km (38 square mi). Within a day the clear water turned soupy green indicating an explosive growth of phytoplankton and confirming Martin's hypothesis that at least some species of phytoplankton benefited from the addition of iron. However, the hoped for sequestering of carbon dioxide in the deep ocean did not materialize because zooplankton consumed much of the phytoplankton releasing CO_2 to surface waters. In 1995, Coale and colleagues conducted another iron-fertilization experiment in the eastern Pacific that increased phytoplankton biomass and reduced the carbon dioxide concentration of surface waters. The different result was attributed at least in part to changes in experimental procedures. Subsequently, three open-ocean iron fertilization experiments were conducted in the Southern Ocean. All three experiments increased phytoplankton biomass and reduced the amount of carbon dioxide dissolved in surface waters. However, evidence that particulate organic carbon sunk to the deep ocean was limited.

Some scientists and policymakers vigorously oppose geoengineering schemes such as iron fertilization of the ocean aimed at combating human contributions to global climate change. They argue that such quick fixes deflect attention away from the root cause of this global environmental issue, that is, humankind's increasing dependence on fossil fuels. Besides, there are ethical considerations and many uncertainties in tinkering with the Earth system on a grand scale. Not enough is understood about the workings of the Earth system and its numerous feedbacks loops. How might the explosive growth in the populations of certain phytoplankton species impact the marine ecology? Others counter by pointing out that large-scale geoengineering projects such as iron fertilization of the ocean may be the only feasible way to stem global warming.

ESSAY: Gas Hydrates, A Future Energy Source

The basic chemical structure of a *gas hydrate* is a single gas molecule, usually methane (CH_4), enclosed in a rigid cage of six water molecules. Cages are linked together forming an ice-like crystalline solid. Gas hydrates are also called *methane ice* or *methane hydrate*. Hydrates are similar to water ice except that the crystal structure is stabilized by the caged gas molecule. Many gases can form hydrates such as carbon dioxide, hydrogen sulfide, as well as light hydrocarbon molecules; "light" means that they have a small number of carbon atoms (e.g., methane).

Marine gas hydrates form not in water or on the ocean floor, but rather in pore spaces within ocean sediments. Hydrates are not stable at typical surface temperatures and pressures but are found in sediments in water deeper than 400 to 500 m (1300 to 1600 ft) where temperatures are sufficiently low and/or pressures are high enough to squeeze water and methane into hydrates. When sediment cores containing gas hydrate, or solid chunks of gas hydrate, are brought up from the ocean floor, the reduction in pressure and rise in temperature cause hydrates to become unstable and they decompose, much like ice melting when warmed. Without precautions, gas hydrate will melt and fizz away before reaching the ocean surface. On the other hand, because of its volatility, chunks of methane hydrate will burn with the touch of a match.

The precise origin of gas hydrates is unknown, although scientists suspect that microbes known as archaea living within marine sediments generate methane from carbon and hydrogen extracted from rich organic materials. That is, biogenic methane is concentrated where organic detritus (from which archaea generate methane) and sediments (which protect detritus from oxidation) accumulate rapidly. In the upper 500-m (1600-ft) of sediments beneath the seafloor, where the pressure is high and temperatures are lower than about 20 °C (68 °F), the gas molecules are captured within the frozen hydrate structure. By contrast, conventional deposits of methane (the chief ingredient of natural gas) form through a different process. In that case, seafloor sediments are buried far deeper and are exposed to much higher temperatures, so that the organic material in the sediments simmers until it transforms into petroleum and eventually methane. Free methane then migrates upward through sediments and is trapped below the hydrate layer.

Although chemists first discovered gas hydrates in the early part of the 19th century, geoscientists have only recently started documenting their presence in submarine and underground deposits and exploring their potential as an energy source. Ocean scientists first drilled through methane hydrates unintentionally on an expedition in 1970. Although that encounter was uneventful, research-drilling cruises avoided suspected gas hydrate deposits for two decades afterward, fearing they might hit an over-pressurized pocket of methane gas, which could blast away the drilling equipment. Concerns over pressurized gas gradually diminished, and mounting scientific curiosity emboldened researchers to try boring through gas hydrate fields. Starting in 1992, researchers with the international Ocean Drilling Program (ODP) intentionally breached hydrate deposits several times without incident. Through continued exploration and research, gas hydrates are now known to occur typically in the upper 500 m of sediments in polar regions (shallow water) and in continental slope sediments (deep water) where pressure and temperature conditions combine to make gas hydrates stable. Marine geologists have identified deposits at hundreds of sites around the globe along most continental margins of the world ocean (see figure on the next page). Terrestrial deposits of gas hydrates occur at shallow depths at high latitudes where low temperatures (rather than high pressure) keep them stable. For example, petroleum companies have encountered hydrates while drilling through permafrost in Alaska, Canada, and Siberia.

Gas hydrates are important because of their potential as a future energy resource—both in the immense amounts of methane bound up in hydrates in sea-floor sediments as well as in the free methane trapped beneath hydrate ice layers. Hydrate is an efficient gas concentrator. That is, gas migrating upward through the sediments first forms a gas hydrate layer which caps and concentrates methane gas beneath. Gas hydrate itself is a concentration of methane in that the breakdown of one unit of gas hydrate ice at sea level pressure produces about 160 units of methane gas. Methane in marine hydrates is conservatively estimated to contain at least 10,000 gigatons of carbon—about twice the amount of all other fossil fuel reserves on Earth. The United States Geological Survey (USGS) estimates that gas hydrates on the continental margin of North Carolina alone contain about 350 times the energy consumed by the U.S. in one year.

The difficulty is bringing gas hydrates to market at competitive prices. In 2004, geologists reported that most deposits are too thin to be recovered economically. Another major challenge is getting it up from the ocean bottom without losing the methane to hydrate decomposition. The first successful extraction of methane from a gas hydrate deposit took place in 2003 at a site in the Mackenzie delta in northwest Canada. Research continues on methods to mine methane in gas

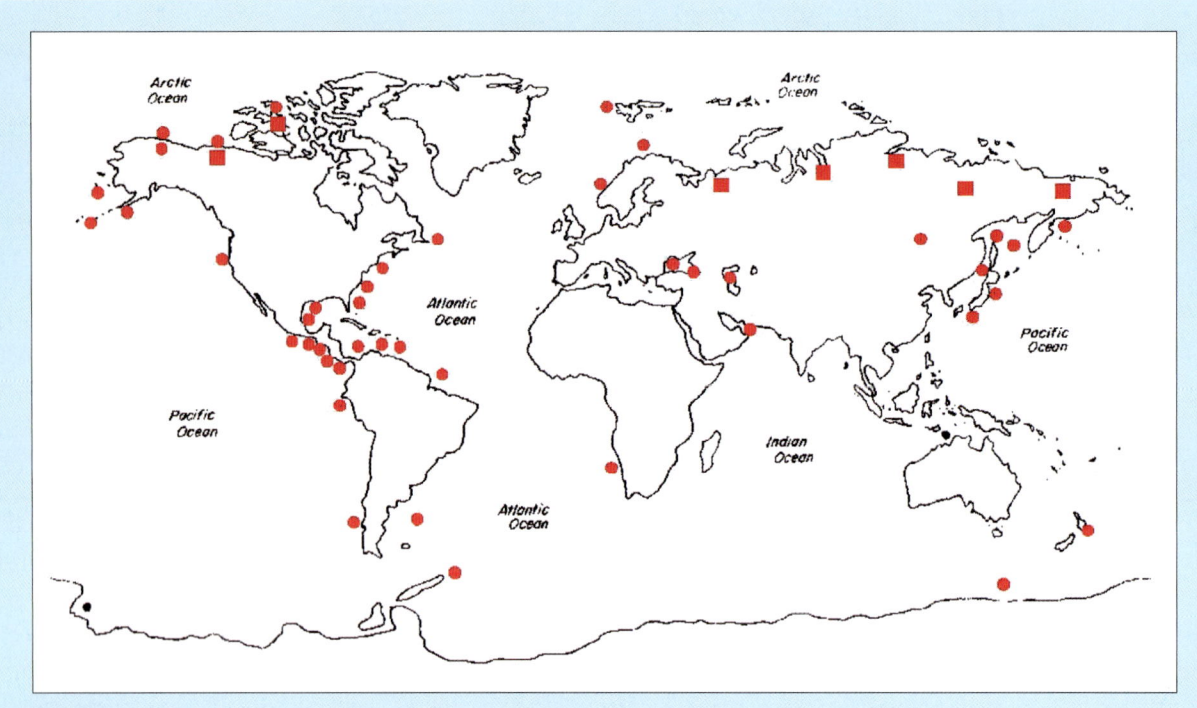

FIGURE
Worldwide distribution of gas (methane) hydrate deposits. [Lawrence Livermore National Laboratory]

hydrates in continental margin deposits worldwide, but it may be many decades before gas hydrates contribute significantly to the world's energy supply. Meanwhile, Japan is drilling 30 exploratory wells in the Nankai Trough subduction zone in search of gas hydrates.

Gas hydrates may exert their greatest impact on the Earth system by contributing to global climate change. Hydrates may function as a major source or sink for atmospheric methane, a greenhouse gas (Chapter 5). At present, gas hydrates are estimated to sequester three times more methane than exists in the atmosphere. Massive melting of hydrates and the ensuing release of methane gas could raise Earth's surface temperature. On the other hand, cooling could bind up more methane in hydrates, which could further lower Earth's surface temperature by reducing the greenhouse effect. As both temperature and pressure play a role in the stability of gas hydrates, multiple feedback loops are likely.

Consider the following possible feedback scenario: Global warming melts methane hydrate, releasing methane to the atmosphere and enhancing the greenhouse effect. More warming leads to melting of glacial ice and sea level rise. However, higher sea level increases pressure on ocean sediments, which could increase the formation of methane hydrates. Gas hydrates form at the expense of atmospheric methane, reducing the greenhouse effect and the Earth cools. Glaciers form and expand and sea level drops. While this scenario is pure speculation, evidence exists for dramatic shifts in methane concentration in the ocean over the past 70,000 years that coincide with episodes when Earth's climate suddenly warmed. For example, release of methane from gas hydrates may have contributed to global warming at the close of the Paleocene Epoch about 55 million years ago.

Gas hydrates may play an important role in submarine landslides. Gas hydrates cement sediments, so their formation and breakdown influence the strength of sedimentary deposits and possibly the potential for submarine landslides on the continental slope and rise. Evidence of gas hydrate instability pockmarks the ocean floor along the Blake Ridge off the coast of South Carolina and Georgia. Numerous craters and depressions, 500 to 700 m (1600 to 2300 ft) wide and 20 to 30 m (65 to 100 ft) deep, apparently formed when gas hydrates melted, releasing methane. In other cases, melting at the base of the hydrate layer destabilized seafloor slopes. Hydrate weakness and failure is thought to be a factor in landslides off Alaska, the U.S. Atlantic coast, British Columbia, Norway, and Africa. Such inherent instability could spell problems for future drilling platforms resting on top of hydrate-rich deposits, for example in the Gulf of Mexico. If the gas hydrate collapses are large enough, they could also trigger destructive tsunamis (Chapter 7).

Scientists have discovered ice worms living in gas hydrates. As noted earlier, gas hydrate is usually buried deep in marine sediments. The Gulf of Mexico is one of the few places where gas hydrates are found exposed on the ocean bottom, where occasionally solid methane bursts through in mounds, often 1.8 to 2.4 m (6 to 8 ft) across. In 1997, a team of scientists using a mini research submarine on a NOAA-funded research cruise discovered, photographed, and sampled what appears to be a new species of centipede-like worms living on and within mounds of methane ice at a depth of 550 m (1,800 ft) about 240 km (150 mi) south of New Orleans. Although it had been hypothesized that bacteria might colonize methane ice mounds, this is the first time that animals have been found living in the mounds (see Figure).

FIGURE
Ice worms living in a methane hydrate deposit at a location on the floor of the Gulf of Mexico about 240 km (150 mi) south of New Orleans, LA. [NOAA, Pennsylvania State University]

Ice worms, which are 2.5 to 5 cm (1 to 2 in.) in length, flat and pinkish in color, were observed in dense colonies burrowing into the methane hydrate. The worms were observed using their two rows of oar-like appendages to move about the honeycombed, yellow and white surface of the icy mound. Researchers speculate that the worms may be grazing off chemosynthetic bacteria that grow on the methane or are otherwise living symbiotically with them. Scientists have also managed to keep a number of the exotic worms alive in shore side laboratories for further study.

ESSAY: Ocean Color and Marine Productivity

Satellite remote sensing of ocean-surface color is contributing to our understanding of the global carbon cycle, other biogeochemical cycles, and global climate change. Ocean-surface color is an index of the global distribution and concentration of photosynthetic organisms living in the upper layers of the ocean. The *Coastal Zone Color Scanner (CZCS)* flown onboard NASA's Nimbus-7 satellite pioneered the technique between 1978 and 1986. SeaWiFS (Sea-viewing Wide Field-of-view Sensor), the successor to CZCS, is on the SeaStar satellite, launched on 1 August 1997 into a sun-synchronous orbit 705 km (437 mi) above Earth's surface. SeaWiFS is an important component of NASA's Earth Science Enterprise, designed to investigate the Earth system from space. These satellite sensors eliminate many of the sampling problems associated with ship-based measurements of marine productivity—greatly increasing spatial and temporal coverage. (However, remote sensing instruments must be periodically calibrated by surface data.) A satellite sensor can view every square kilometer of cloud-free ocean surface once every 24 hours.

Ocean color sensors indirectly measure variations in the distribution and concentration of phytoplankton within the upper tens of meters of the open ocean (at somewhat shallower depths in more turbid coastal waters). As noted elsewhere in this chapter, phytoplankton are microscopic, single-celled organisms that form the base of most marine food webs. Chlorophyll-a (and other pigments) in phytoplankton absorbs selected wavelengths of visible solar radiation that penetrate seawater and that energy is used for photosynthesis. (The spectrum of *visible light* is composed of all colors, from violet at the short wavelength end to red at the long wavelength end.) Chlorophyll, the commonest of the phytoplankton pigments, absorbs primarily in red and blue portions of the visible spectrum and reflects green light. Some of that reflected light is scattered to the ocean surface and then into the atmosphere. The more phytoplankton present, the greater the concentration of plant pigments, and the greener the water.

Of the scattered radiation intercepted by an ocean-color sensor, only about 10% to 20% actually originates in ocean waters; the rest is sunlight scattered back by atmospheric aerosols and air molecules (primarily nitrogen and oxygen). Ocean waters scatter very little solar radiation having wavelengths in the red (0.67 micrometer) and near infrared (0.75 micrometer). Using these two spectral channels, the atmosphere's contribution to scattered sunlight can be subtracted from the total radiation scattered by the ocean-atmosphere system. The product is a color-coded image of phytoplankton concentration (see Figure on the next page).

SeaWiFS estimates near-surface phytoplankton concentration from measurements of scattered sunlight over a swath width of 2800 km (1740 mi) with a pixel resolution on the order of 1.1 km by 1.1 km. The sensor measures energy in eight spectral bands (six visible, one near-infrared, and one far-infrared) chosen because of their sensitivity to chlorophyll in phytoplankton. From data acquired over a succession of orbits, SeaWiFS provides a composite global view of variations in marine productivity. SeaWiFS imagery clearly shows upwelling zones off the coasts of Peru, northwest Africa, and the U.S. West Coast, among the most productive regions of the world ocean. As described in Chapter 6, in these upwelling zones, coupling of surface winds with the sea surface transports water away from the shore and nutrient-rich water wells up from below, spurring marine productivity.

Marine phytoplankton plays an important role in regulating biogeochemical cycles. Perhaps as much as half of the carbon dioxide released to the atmosphere by human activities (e.g., fossil fuel burning) ultimately dissolves in ocean water. Through photosynthesis, phytoplankton removes carbon dioxide dissolved in water and releases oxygen as a by-product. A better understanding of the distribution and concentration of photosynthetic organisms in the sea will improve our understanding of the flux of carbon dioxide into and out of the ocean and the possible implications for global climate change. Furthermore, ocean-color remote sensing is now combined with data from other passive and active satellite-borne sensors for a wide variety of practical applications from locating productive fishing grounds to helping plan the lowest fuel consumption routes for ocean shipping.

Chapter 9 MARINE ECOSYSTEMS 227

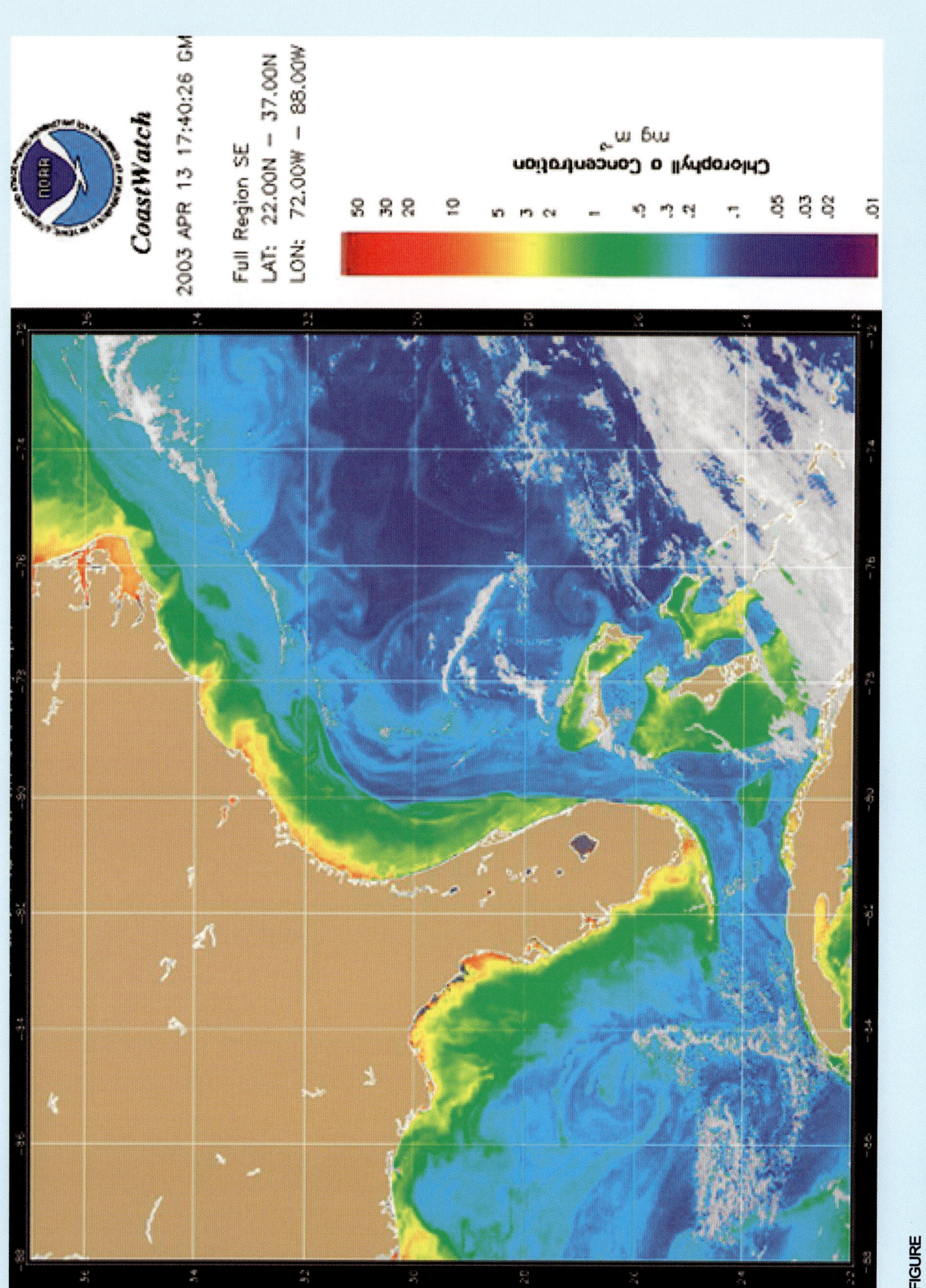

FIGURE
SeaWiFS image of biological production for the Gulf Coast, Cuba, Bahamas, and the Eastern Seaboard on 13 April 2003. Red and orange in coastal waters indicate relatively high chlorophyll concentrations; light grey patches are clouds. [Courtesy of NOAA CoastWatch and ORBIMAGE]

CHAPTER 10

LIFE IN THE OCEAN

Case-In-Point:
Driving Question
Marine Habitats
    Oceanic Life Zones
    Plankton in the Pelagic Zone
    Nekton in the Pelagic Zone
Life Strategies and Adaptations
    Vertical Migration
    Light and Vision
    Sound
    Feeding Strategies
Life at the Ocean's Edge
    Intertidal Zone
    Sea Grass Beds and Salt Marshes
    Kelp Forests
    Coral Reefs
    Benthic Feeding Habits
    Life on the Deep-Sea Floor
Marine Animals
    Fishes
    Marine Mammals
    Marine Reptiles
    Sea Birds
Conclusions
Basic Understandings
ESSAY: Marine Sanctuaries and Reserves
ESSAY: Ecosystem Approach to Fisheries Management

A Sea Turtle in Jobos Bay, Puerto Rico. [Courtesy of NOAA]

Case-in-Point

Codfish were once so plentiful in the western North Atlantic, especially on the Grand Banks southeast of Newfoundland, that early explorers reported that their ships were slowed by the cod and they could catch them easily using hand nets from small boats. For nearly 500 years the cod fishery was the basis of the economy of Newfoundland and other parts of Atlantic Canada. Some experts argue that the availability of easily transported salt cod provided the food that sustained western European sailors as they explored the North Atlantic and colonized its shores.

    Natural fluctuations in cod populations meant periods of low catches but fish stocks always recovered. All that changed after the middle of the 20th century when huge factory trawlers began fishing intensively for cod on the offshore banks, in waters beyond the limits of Canada's national jurisdiction. In 1977, Canada declared a 323-km

(200-mi) exclusive fisheries zone. At that time, Canadian factory trawlers replaced foreign vessels. Although strict regulation ensued, major damage had been done to the cod fishery.

Despite years of regulation, by 1992 cod stocks were so depleted that the Grand Banks fishery was closed. Almost immediately, some 40,000 people lost their jobs in Newfoundland where many small coastal villages depended entirely on the cod fishery for employment. Now, a decade later, the cod population has shown no sign of recovery; in fact, their numbers continue to decrease. Declines in the cod fishery elsewhere prompted the Canadian government in April 2003 to close much of the remaining cod fishery in other areas of the western North Atlantic off Newfoundland and Atlantic Canada. Apparently, the ecosystem that formerly supported cod stocks was altered in such a way as to favor other fish and even lobsters in place of cod. Given our present limited understanding of the way this ecosystem functions, it is unclear if the fish stock and the ecosystem that sustained it for centuries can be restored.

In the early years of the 21st century, roughly the same scenario played out as the fisheries authorities of the European Union struggled to find a way to protect the dwindling cod stocks of the North Sea. Heavily subsidized fishing fleets, modern fish finding equipment, and better nets permitted fishers to catch so many fish that stocks fell below the levels needed to maintain the fishery. Efforts to regulate the fisheries met with strong opposition from the fishers and the fishing industry throughout most of the European Union. Although fish catches were reduced, scientists argue that these cuts were too little and too late to restore stocks sufficiently to permit a sustainable fishery.

The impact of the decline in cod stocks on the fishing industry in northwestern Europe is expected to be serious with the loss of many thousands of jobs and devastation of fishing communities—comparable to what happened earlier in Newfoundland. The effects on the fish consumer will be significant including, for example, loss of the traditional fish and chips from the menus in British pubs and salt cod from dinner tables in Portugal. Furthermore, extensive disruption of the ecosystem may also lead to the demise of other species, including fish such as haddock that replaced cod on restaurant menus.

The problem of the declining cod fishery also extends into New England where federal courts and NOAA's National Marine Fisheries Service have reduced the length of time that fishers can spend at sea catching cod and other fish. Similar problems also occur off the Pacific coast of the United States where more stringent restrictions on catching groundfish (fish species that live near the sea bottom) have been enacted.

Driving Question:

How have the large and diverse populations of marine organisms adapted to environmental conditions in the ocean?

In the previous chapter we examined the components, structure, functioning, and energy flow of marine ecosystems. In this chapter we continue our look at life in the ocean by describing the various flora and fauna, marine habitats, and some of the ways these marine organisms have adapted to the diverse environmental conditions in the ocean.

Marine plants and animals are distributed unevenly throughout the ocean, often occurring in scattered patches. We open this chapter with a description of marine habitats, oceanic life zones, and the characteristics of plankton and nekton in the open ocean. Central to this discussion is the variety of adaptations evolved by marine organisms including specialized means of floating and moving in the ocean, defense mechanisms to protect against predators, plus feeding and reproductive strategies. We then examine marine plant and animal life at the boundaries of the ocean including organisms living in the dynamic intertidal zone, wetlands and estuaries, the deep-sea floor, and coral reefs. Our discussion closes with brief descriptions of marine fishes, mammals, and reptiles plus sea birds.

Marine Habitats

The ocean contains about 99.5% of Earth's potentially inhabited living space; the remaining 0.5% is on land. Living space in the ocean not only vastly exceeds that on land but also consists of many habitats unfamiliar to land dwellers. Ocean scientists know most about life in the ocean's relatively shallow surface zone (especially the cosatal zone and over the shelf), and less about life in the mid-depths, the deepest ocean waters, and the deep-sea floor.

Living organisms inhabit all parts of the ocean, even extreme environments such as Arctic and Antarctic sea ice and hot hydrothermal vents on the sea floor.

The diverse marine habitats include estuaries, tropical coral reefs, polar oceans, and deep-sea trenches. Even sediments and bedrock on the sea floor harbor abundant life. We begin our survey of marine organisms and their habitats by identifying the ocean's major life zones and then examining the plankton and nekton of the open ocean.

OCEANIC LIFE ZONES

There are several ways to define *oceanic life zones*. The most commonly used system defines marine habitats in terms of distance from shore and depth (Figure 10.1). In general, open ocean waters constitute the **pelagic zone**, from the ancient Greek *pelagos* meaning the deep or open ocean. Pelagic organisms include **plankton** (passive floaters or weak swimmers such as copepods, larval fish, and jellyfish) and **nekton** (strong swimmers including most fish, squid, turtles, and marine mammals). The environment of the sea floor at all depths is called the **benthic zone**. Benthic organisms (*benthos*) live either on the ocean bottom or within sediment deposits. They include attached (or sessile), burrowing, and mobile organisms, such as sea stars, crabs, worms, clams, sea cucumbers, sea anemones, urchins, snails, and barnacles.

The pelagic and benthic zones can also be subdivided according to water depth. The upper part of the pelagic zone, up to 200 m (650 ft) deep, is the *epipelagic zone* and roughly corresponds to the photic zone. Below that, from 200 to 1000 m (650 to 3300 ft), is the *mesopelagic zone*, and the deep waters of the open ocean from 1000 to 4000 m (3300 to 13,000 ft) constitute the *bathypelagic zone*. (*Bathy* is a Greek prefix meaning deep). Even greater depths are referred to as the *abyss* or *abyssopelagic zone*.

The area along the shore between high- and low-tide lines is the **intertidal zone** (also known as the *littoral zone*). Two familiar intertidal habitats are rocky shores and sandy beaches. Intertidal zones are home to ecosystems such as salt marshes and mangrove swamps. The area seaward from the shore, across the continental shelf, to the shelf break at a depth of 120-200 m (390-650 ft) forms the **neritic zone**. The neritic zone includes the *intertidal zone* and is commonly referred to as the *coastal zone* (Chapter 8).

Oceanic life zones are also defined in terms of nutrient supply (e.g., phosphorus and nitrogen compounds) and productivity. Marine organisms are relatively sparse in large, open-ocean areas far from land. Apparently, many of these areas are lacking in one or more nutrients essential for photosynthetic plants and bacteria. Such nutrient-poor waters having low primary production are described as *oligotrophic*. Waters are exceptionally clear and appear luminous blue in sunlight due to the lack of organisms and suspended particles. The Mediterranean Sea and large areas of open-ocean within the subtropical gyres (e.g., the Sargasso Sea) are examples of *oligotrophic waters*. On the other hand, ocean waters over continental shelves are especially rich in marine plants and animals. These relatively shallow waters are much more productive than most open-ocean waters because of coastal upwelling and river discharge of nutrients. Nutrient-rich waters having high primary productivity are described as *eutrophic*. In coastal areas where the nutrient supply is enhanced by runoff from agricultural lands or sewage discharge, waters may become excessively eutrophic. Excessive amounts of nutrients stimulate an unusually abundant growth of phytoplankton (an algae bloom), which die, sink to the bottom, and deplete waters of dissolved oxygen as they decompose.

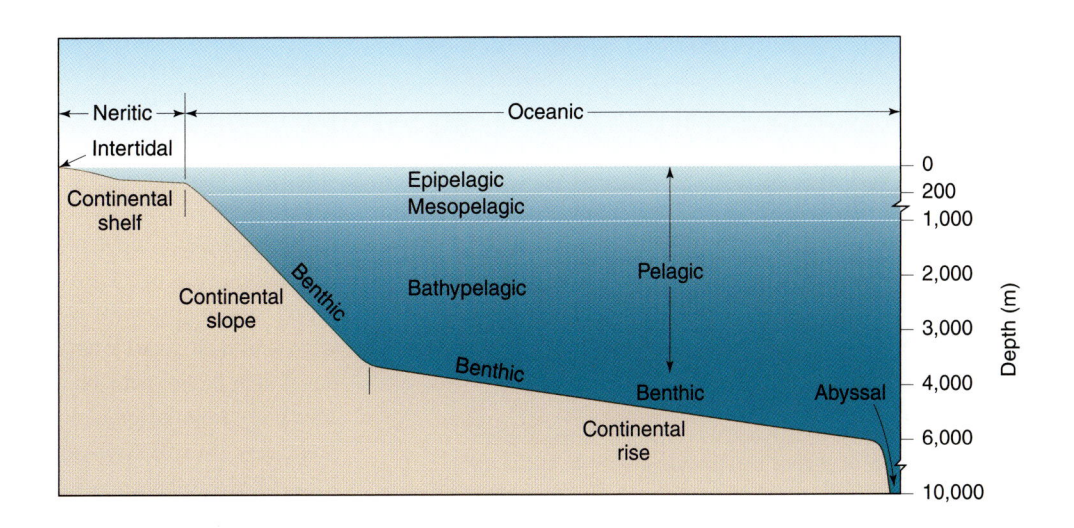

FIGURE 10.1
Life zones in the ocean.

PLANKTON IN THE PELAGIC ZONE

Most plankton are very small and are members of phytoplankton, zooplankton or microbial communities. In Chapter 9, we described the roles of these organisms in the functioning of marine ecosystems. We also saw earlier in this book how their remains are incorporated into marine sediments and sedimentary rocks, thereby linking the biosphere and geosphere through the rock cycle (Chapter 4). In this section we describe strategies and adaptations plankton have developed for life in the pelagic zone. Here we define **adaptation** as a genetically controlled trait or characteristic that enhances an organism's chances for survival and reproduction in its environment.

Phytoplankton are slightly denser than seawater and would gradually sink below the sunlit photic zone were it not for characteristics that counteract this tendency to sink. Some species have flattened shapes; others are spiny or occur as long chain-like colonies. These shapes give the cells a relatively large surface area compared to their volume. This adaptation significantly slows sinking, especially in less dense, warm waters, and allows them to remain in the photic zone above the pycnocline. Not surprisingly, the most ornate plankton species inhabit warm tropical waters. Diatoms have a similar adaptation (Figure 10.2). The delicately ornamented outer shell (*frustule*) of diatoms consists of porous silica (SiO_2) that allows for direct uptake of nutrients and gases from seawater as well as rapid expulsion of wastes. One of the most common types of diatoms is shaped like a pillbox; others are elongated or form chains of individual cells. Diatoms are most abundant in cool, nutrient-rich waters, especially at high and middle latitudes and where silica concentrations are relatively high.

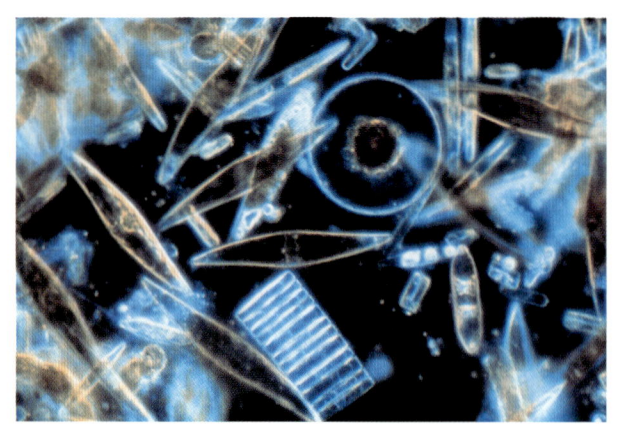

FIGURE 10.2
Although microscopic in size, diatoms can be beautiful and occur in a variety of shapes that provide them with buoyancy. [NOAA Photo Library]

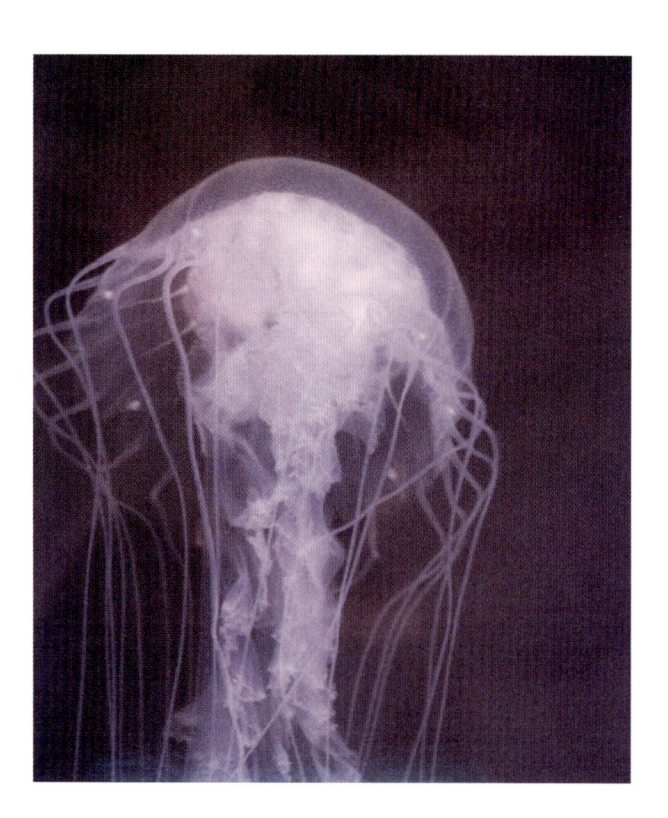

FIGURE 10.3
The bane of swimmers in the Chesapeake Bay, the stinging sea nettle, *Chrysaora quiquecirrha*. [NOAA Photo Library, photo by Mary Hollinger, NODC]

The complex shapes and spiny structures of zooplanktonic organisms also increase their surface area to volume ratio, thereby adding to their buoyancy. This is especially obvious in the larvae of many organisms, which have beautifully complex shapes. Another adaptation to increase the buoyancy of planktonic animals is the storage of fats in large, oily globules within their body cavities. Fat is less dense than the surrounding seawater.

Some large planktonic animals have gelatinous bodies, consisting of 95% to 98% water so that they are almost neutrally buoyant (having nearly the same density as the surrounding waters). Jellyfish, for example, are gelatinous zooplankton having contractile tissues (primitive muscles) that enable them to pulsate their bell-shaped bodies to maintain their position and move slowly in the water (Figure 10.3). Along the edges of their bodies are sensory organs that respond to light or gravity, allowing them to swim weakly upward and then sink slowly downward, trapping organisms under their bodies with their tentacles. The Portuguese man-of-war is a colonial jellyfish-type organism (a siphonophore). It consists of a collection of specialized polyps of four types: a gas-filled float that holds the colony at the surface oriented with the wind, stinging tentacles that trap food, bag-like polyps that secrete enzymes to digest the trapped prey, and reproductive polyps. The specialized stinging cells (nematocysts) in the tentacles produce a powerful toxin, which can be dangerous to humans.

NEKTON IN THE PELAGIC ZONE

Larger, free-swimming pelagic animals (collectively called nekton) include fish of all sizes, squid, sea turtles, and marine mammals. Some migrate long distances. For example, California gray whales annually migrate from their breeding grounds in warm, shallow lagoons off the Pacific coast of Baja California, Mexico, to feeding grounds on the Arctic Ocean's continental shelf between Alaska and Siberia, a distance of about 13,000 km (8250 mi). On the other hand, the distribution of many pelagic animals is limited by their tolerance for water temperature, salinity, or food availability.

Except for the smallest zooplankton and jellyfish, marine animals are much denser than seawater. Hence, these animals must swim actively (requiring considerable energy) or utilize other mechanisms to maintain themselves at suitable depths in the ocean. Many adaptations for buoyancy are found in nektonic organisms. The simplest of these are gas bladders. The beautiful chambered nautilus (a primitive relative of squid and octopus) has an external, spiral shell with internal chambers (Figure 10.4). These animals cannot control the air pressure in the chambers—it is always the same as at sea level—so they cannot swim too deep or the shell will collapse under the increasing water pressure. However, they can adjust the amount of fluid, and hence the air volume, in the shell chambers, thereby having some control over buoyancy. In fact, it appears that the nautilus may have been the first large swimming predator to appear in Earth's ancient ocean perhaps 500 million years ago. They are referred to as *living fossils*, that is, organisms that have remained essentially unchanged since their appearance in the geologic past.

Squids are invertebrates; they are mollusks and related to mussels, clams, and snails. They are extremely fast and highly maneuverable swimmers, living at mid-depths of the ocean and migrating to surface waters at night. These fast-growing animals are very successful predators. Most are from 10 to 100 cm (4 to 40 in.) in length and have acute vision. The mysterious giant squid, up to 23 m (75 ft) long, lives in deep waters where it is prey for sperm whales. A few specimens have been caught in fishing nets or have drifted ashore. Their movements are abrupt and powerful and they use their ten long tentacles, covered with suckers, to trap prey and bring it to the mouth, which has a sharp beak-like structure for tearing food apart.

The elusive giant squids are neutrally buoyant and use a special mechanism to prevent sinking. Muscles and cavities in their bodies contain abundant ammonia ions. Since ammonium chloride is less dense than the sodium chloride in seawater, they gain a buoyancy advantage. There is a disadvantage as well. Being neutrally buoyant means that if they are trapped in warm water, they gradually float upward. Their hemocyanin (a form of invertebrate blood) carries oxygen less efficiently in warm water and the squid suffocates and dies. Much of what is known about these mysterious creatures has been learned from dying animals recovered from warm surface waters. Other types of squid must swim continuously in order to maintain their preferred depth in the water column. They are equipped with an internal jet propulsion system. By contracting muscles, the squid forces water into its body through a pipe-like structure called a siphon and then pumps the water out through the siphon as a jet.

Many fishes have gas-filled swim bladders that control buoyancy and regulate the amount of gas in the bladder through a connection to the gut. This makes them sensitive to changes in water pressure with depth, however. Deep-sea fishes with swim bladders are rarely caught alive. If they are brought to the surface too fast, their bladders expand and burst because they cannot equilibrate to the decreasing pressure fast enough. Fat is less dense than seawater so that another means of increasing buoyancy is to increase the proportion of fat in the body. To achieve neutral buoyancy, about 33% of a fish's tissue must be fat.

FIGURE 10.4
The nautilus is the only living cephalopod with a shell. They are ancient animals related to squid and octipi. [OceanLink and Bamfield Marine Sciences Center]

Some sharks come close to this—their large liver is rich in a fatty substance that is much less dense than seawater. Still, many sharks must swim all their life to maintain lift in the water or they will sink.

Most pelagic fish rely on active swimming to maintain their level in the ocean and obtain food. Some fish lie on the bottom, waiting for their prey and suddenly lunge to catch it. Others swim all the time, seeking prey over large areas of the ocean perhaps covering hundreds of kilometers each day. The streamlined shape of their body reduces frictional resistance and increases their swimming efficiency. Fish with this feeding strategy require a great deal of energy, efficient muscles, and a rich oxygen supply. Tuna is one such predator. These fishes usually feed at a trophic level that is high in a marine food web and are referred to as *top predators*.

Life Strategies and Adaptations

Marine organisms have evolved many different adaptations to obtain food and avoid being eaten by predators. These include vertical migration, special coloration, eyes sensitive to low light levels, bioluminescence, use of sound, and specialized feeding strategies.

VERTICAL MIGRATION

The photic zone is where photosynthetic primary producers live and food is most plentiful. However, it is also a dangerous place for marine organisms because predators can easily see them. Many types of zooplankton get around this problem by daily **vertical migration**. Each day at dusk, they come to the surface zone to feed on phytoplankton. As daylight comes, they return to the relative safety of darker, deep waters (the *twilight zone*). These migrations, typically over a vertical distance of about 200 m (650 ft), require expenditure of enormous amounts of energy. In a single daily migration, most small zooplanktonic animals will travel a distance equal to tens of thousands of times their body length. Some fish follow the zooplankton in their daily migration and prey on them. Vertical migration also plays a role in the carbon cycle in that carbon consumed by zooplankton feeding on phytoplankton near the surface at night is transported to deeper water as the animals respire during their return to depth at dawn.

World War II submarine sonars first detected vertical migration as a sound-scattering layer (known as the *deep scattering layer* or *DSL*) that moved up through the water after sunset and down again at sunrise. Sound waves were reflected by millions of zooplanktonic organisms moving together up and down in the water column to and from the surface. Submarine commanders were able to use this knowledge to hide their vessels beneath the sound-scattering layer.

LIGHT AND VISION

Near the ocean surface, light is abundant and predators have no problem locating prey. Therefore, species of smaller fish that are prey for larger ones have adaptations that make them less visible and therefore less likely to be eaten. The most common adaptation is **adaptive coloration** or camouflage where the animal's color pattern closely matches its background substrate. That is, they blend in with their surroundings thereby avoiding detection. Many fish have **countershading**, that is, their dorsal side (or back) is a dark color, making it difficult for predators above to see them against the dark, deep water. Their ventral side (or underbelly) is lighter colored, making the fish more difficult for predators to see them from below against the more brightly lit surface waters. Squids and octopi have yet another defense against predators. They can eject a cloud of black or brown "ink" into the water and escape without being seen by a predator.

Marine animals that rely on sight to locate their prey in the dim light of the twilight zone have large, sensitive eyes. In this vast area of the ocean, from 200 to 1000 m (660 to 3300 ft) deep, there is just enough light for predators to locate their prey. In general, the deeper in the twilight zone, the dimmer is the light, and the larger are the eyes of predators living there. Many of these deeper-living fish tend to be reddish in color; red light is absorbed near the surface of the ocean so that red objects are virtually invisible at greater depths.

Below the twilight zone, in the greatest depths of the ocean, light is absent and vision is not useful for locating prey. There is a major exception to this rule, however. Many marine animals (including for example, squid, dinoflagellates, and some species of fish) emit light, a phenomenon known as **bioluminescence** (Figure 10.5). The light usually is a product of a chemical reaction that takes place in specialized cells (*photocytes*) or organs (*photophores*). When a substance known as luciferin reacts with oxygen in the presence of the enzyme luciferase, the chemical product gives off blue-green light. In some squid and fish, bacteria living within certain organs are responsible for bioluminescence. Note that bioluminescence differs from *phosphorescence* or *fluorescence* in which light is received from other sources and re-emitted. Bioluminescence is mainly a marine phenomenon and is thought to

FIGURE 10.5
Bioluminescence is a property of a wide variety of marine organisms such as the jellyfish *Aequorea aequorea* pictured here. [Courtesy of NSF and Osamu Shimomura, Marine Biological Laboratory, Woods Hole, MA]

have evolved a number of different times because it is a characteristic of a broad array of organisms.

Marine animals use light emission to attract mates or prey, frighten or confuse predators, or to disguise themselves. Various deep-sea animals exhibit characteristic patterns of light flashes or luminescent shapes to help them attract a mate. Some fish use bioluminescence to distinguish between male and female. Short bursts of bioluminescent light apparently are especially effective at disorienting prey. Some bathypelagic fishes have strange shapes and many use bioluminescence in their feeding strategies. The female anglerfish has an unusual growth protruding from its head that resembles a fishing rod with a light-producing organ located at the tip of the rod. The structure is a modified dorsal fin that slides along a canal in the back of the fish. When an unwary prey is attracted to the light, the anglerfish consumes it with a sudden snap of its huge mouth.

Light can also be a defensive adaptation for animals in the pelagic environment. The ink ejected by some squid species is luminescent, thereby further confusing a potential predator. Some shrimp emit a luminescent cloudy substance in the direction of an approaching predator while moving rapidly out of sight into the darkness. Also, some marine animals (e.g., squid) are able to camouflage themselves by using bioluminescence to match their body's coloration with that of their background. Even phytoplankton use bioluminescence defensively. Some organisms, such as certain species of dinoflagellates emit light when disturbed. For example, copepods feeding on dinoflagellates stimulate their light emission. But sometimes this attracts the attention of nearby fish to the copepods, allowing the dinoflagellates to escape predation.

SOUND

Marine organisms have other ways of locating prey including sound. Analogous to hearing, many marine animals have evolved ways of sensing the vibrations produced by other organisms moving through the water. A variety of structures and specialized organs have evolved for this purpose. As pointed out in Chapter 3, seawater is essentially transparent to sound. Many marine mammals, including whales, routinely communicate great distances, even across an entire ocean basin. Scientists are beginning to understand some of their vocabulary. Whales and dolphins possess highly developed auditory senses. In addition to communication and locating prey, they may use sound for navigating over long distances. Fish too are capable of producing sound by using their swim bladders like drums. Predatory seals, dolphins, and whales take advantage of this sound to search for prey.

Concern is growing over increasing noise levels in the ocean caused by human activities. Sounds such as those produced by large ships, military weapons testing, scientific research, and undersea oil and gas exploration significantly disturb many types of whales. Noise in the ocean may be the reason for some whale strandings in shallow waters or on beaches. Noise need not be very loud to cause problems; low frequency sound waves have higher energy and travel much farther than high frequency sound waves, especially in the ocean's SOFAR channel (Chapter 3).

FEEDING STRATEGIES

Tiny marine animals and the smaller one-cell plants and bacteria mostly move with the waters around them because they cannot swim against ocean currents. Some marine animals have evolved special mechanisms for capturing drifting organisms such as appendages covered with fine bristles that act like strainers to trap small prey. Such structures are very common among copepods and are often very ornate, especially in warm water where they also add buoyancy and act as parachutes to slow sinking out of the photic zone. The general name for marine animals equipped with features that strain food particles out of large volumes of water is **filter feeder**. Even blue whales, the largest animal ever known to have lived on Earth, are filter feeders. It and eleven other species of baleen whales use plates of fibrous baleen (a substance similar to human finger nails) hanging in a row from their upper

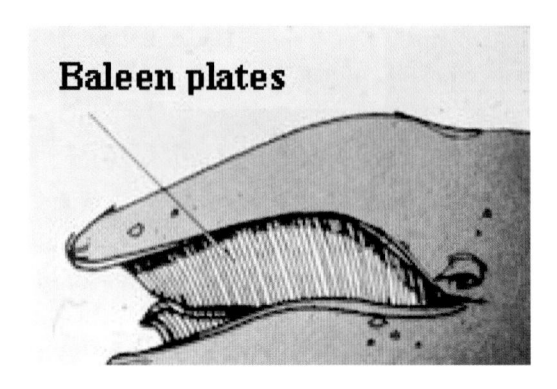

FIGURE 10.6
Diagram showing the baleen plates in the mouth of a baleen whale. In the 19th and early 20th centuries, this material was highly sought after for use in corsets, umbrella ribs, and buggy whips. [Courtesy of NOAA]

jaw to filter plankton from seawater pumped through their mouths by contractions of their massive tongue muscles (Figure 10.6). By feeding directly on primary production phytoplankton, baleen whales shorten the food chain to only two trophic levels.

Filter feeders eat most of the food items trapped in their feeding structures; that is, they are not selective. Other organisms are selective feeders and eat only specific prey organisms. Selective feeding may require special adaptations. For example, some starfish eat mussels, using strong suction to pry open their shells.

Some planktonic organisms, such as the planktonic snails called pteropods (also known as sea butterflies), produce a large sticky net of mucus, which functions much like spider webs to trap small water-borne organisms and other food particles. The entire net is then eaten and digested along with the trapped organisms.

Life at the Ocean's Edge

After the pelagic zone, the next largest environment for marine life is the benthic zone, the sea floor at all depths, from the intertidal zone to the deep ocean. The benthic environment is essentially two-dimensional (limited vertically), unlike the vast three-dimensional pelagic zone. Marine organisms that live in the benthic zone collectively are called **benthos**.

There are three basic life strategies in the benthic zone. Organisms may live attached to a firm surface, they may construct burrows or tunnels or simply dig into sediment deposits, or they may move freely on the sea floor. In general, these habitats dictate their feeding strategies. Attached animals are filter feeders, mobile animals are active predators, and burrowing organisms ingest sediment, digesting the organic material it contains and excreting the rest. Some also dig into the bottom sediments for protection from predators.

Attached organisms include seaweed (technically known as *macroalgae*) and sea grasses, the only truly marine form of the more complex plants (angiosperms). Macro-algae have root-like structures called *holdfasts* that attach to the bottom. They absorb nutrients and water directly from the surrounding seawater. Some familiar attached animals are barnacles, mussels, clams, oysters, corals, and anemones. In general, benthic animals that are not attached move relatively slowly. Starfishes, urchins, and snails are examples of mobile benthic predators. Others are crabs and lobsters, which can move quickly in short spurts to catch their prey or to avoid predation.

Burrowing animals are less familiar to us because they largely disappear beneath the surface of sediment deposits, even in shallow tide pools. They live on sandy and muddy bottoms and include various forms of mollusks and worms. Their presence can often be detected by small holes in the sediment surface and piles of mud or sand ejected from their tunnels during burrowing.

INTERTIDAL ZONE

As mentioned earlier in this chapter, the intertidal (or littoral) zone is the area along the shore between low- and high-tide levels. Most intertidal zones are readily accessible so that properties and components are well known to scientists and amateur naturalists. Because the intertidal zone is where the atmosphere, hydrosphere, biosphere, lithosphere, and cryosphere interface and interact, this zone is among the most dynamic portions of the Earth system (Chapter 8). Waves, winds, and tidal currents continually disturb the intertidal zone. With the rise and fall of the tides, the intertidal zone is alternately exposed to seawater and the atmosphere with accompanying drastic changes in temperature and salinity. In winter, winds, tides, and currents may drive floating pans of sea ice ashore where they abrade and erode the intertidal zone. Special adaptations enable animals and plants to survive in this continually changing environment.

Although all intertidal zones share some common characteristics, a distinction is made among intertidal habitats based on the type of substrate (e.g., rocky, muddy, sandy). The most energetic of these habitats are rocky intertidal shorelines where organisms are frequently subject to strong wave action, especially in winter. Rocky shorelines, especially in middle latitudes, are home to complex marine ecosystems with organisms adapted for life in particularly harsh conditions. Where the distance between low and high tide levels is relatively great, organisms may be exposed to dry air and sunshine for many hours at a stretch. Depending on the season, they may be forced to withstand extremely high or low air temperatures when they are not covered by seawater.

Some intertidal zones feature extensive mud flats. A **mud flat** is a nearly level area of fine silt along the shore. Streams and creeks deliver the sediment and wave action is typically weak. Mud flats are often ideal habitats for submerged aquatic vegetation, salt marshes, and many forms of benthic animals. In addition, sandy beaches are intertidal habitats (Chapter 8). Seaweeds are important components of the intertidal zone and provide habitat for animals. For example, many snails, sea urchins, and other marine herbivores graze on seaweed and may control its population in some areas. Seaweeds are distributed according to their light requirements and their ability to resist wave action as well as drying and/or freezing conditions during low tide.

All seaweeds are algae and are green, brown, or red, depending on the color of the dominant photosynthetic pigment. In general, brown and red algae are more robust and often have a rubbery feel, whereas green algae are more delicate. Bright green sea lettuce, with thin leaves up to one meter (3.3 ft) long, lives from just above the low tide mark (so it is not exposed to air for very long periods) to depths of about 10 m (33 ft). It requires abundant light. Red algae photosynthesize most efficiently at lower light levels; hence, they are more likely to be found in deeper water or in tide pools that are shaded by rock outcrops or cliffs. They also prefer the higher temperatures of tropical waters.

Brown algae are the most common seaweeds found along the rocky coasts of the northeastern and western United States. They flourish in cold water and usually have a heavier structure to resist strong waves. They often have many small, gas-filled bladders that help keep their blades (not true leaves) floating near the water surface. Another adaptation of brown algae for life in the intertidal zone is the presence of gelatinous substances that helps them retain seawater at low tide when they are exposed to drying air. These substances have many commercial uses as stabilizers in ice cream and as emulsifiers in many food products. Algin extracted from brown algae is also used to thicken textile dyes, in paints, and even by dentists in taking impressions of teeth.

Open-ocean waters are too deep for plants to attach to the bottom and survive. Some plants do survive, however, in an area in the western Atlantic Ocean where Gulf Stream waters flow around the western side of the North Atlantic subtropical gyre (Chapter 6). This area is known as the Sargasso Sea, after the abundant, brown seaweed (*Sargassum*) that floats on the surface. *Sargassum* weed originates in the Caribbean Sea, where it grows attached to hard surfaces in the intertidal zone. It is broken off by storm waves and becomes trapped in the current system of the North Atlantic subtropical gyre where it accumulates as large, floating rafts.

Sargassum weed is supported by bubble-like gas filled structures that keep the plants floating near the surface where light for photosynthesis is abundant. Surface waters of the subtropical gyre are nutrient poor and cannot support large phytoplankton and fish communities. However, rafts of *Sargassum* weed provide a unique habitat for animal life. The small animals that can survive in the nutrient-poor waters maintain a close relationship with the seaweed, either attached to it or swimming within the floating mats. Some small fishes and crabs have color patterns that camouflage them against the background of *Sargassum* weed. Many invertebrates such as snails, crabs, and anemones live attached to or moving about the weed. All of these animals are feeding generalists (omnivores), eating whatever comes their way in the water. Sea turtles, especially young ones, also find a protective environment within the *Sargassum* weed of the Sargasso Sea.

As with seaweed, animals are distributed in the intertidal zone according to their tolerance for drying and changes in temperature and salinity. Highly mobile animals, such as crabs, move up and down with the changing water levels as the tide floods and ebbs. Animals such as barnacles and mussels attach themselves to hard surfaces and can shut their shells completely, thereby retaining enough seawater inside to survive until they are covered again by the flood tide. For this reason, barnacles are the dominant animals in the upper, more exposed portion of the intertidal zone. Although they resemble shellfish, barnacles are actually crustaceans. They feed only when covered by water, opening their shells and extending their feathery legs through their shell top to filter food from the water. Barnacles are so firmly cemented to rock surfaces that waves cannot loosen them. Mussels are mollusks that usually live slightly lower in intertidal zones and are firmly attached to rocks by sticky threads.

Other intertidal mollusks include limpets and chitons that attach to rocks with a strong suction created by a muscular foot-like appendage. Both move so slowly that they are unable to keep pace with tidal oscillations. They are fairly resistant to drying out and are only active when they are covered with water. Their food is plant material that they scrape off rock surfaces with a radula, a tongue-like structure consisting of rows of chitinous teeth. Still lower in intertidal zones is a greater diversity of animals including starfishes, anemones, urchins, and crabs.

While walking along a rocky shore at low tide you may have encountered tide pools sheltering a diverse animal community. A **tide pool** is a volume of water left behind in a rock basin or other intertidal depression by an ebbing tide (Figure 10.7). If a tide pool is exposed to direct sunshine, organisms inhabiting the pool must contend with great fluctuations in water temperature. Rainfall can also alter the salinity of tide-pool waters. Where tide pools are shaded or in caves, temperature and salinity variations are much less so that a greater variety of more delicate animals can survive. Typically, tide pools harbor species of organisms that have evolved to deal with wide fluctuations in environmental conditions; these organisms include certain algae, sea anemones, starfish, snails, small crustaceans, and fish.

Just below low-tide level on rocky shores live the most diverse communities of plants and animals. The complex substrate and the seaweed provide protection for many small animals. Here soft bodied and delicate animals like anemones and sea cucumbers can thrive. Predators, such as lobsters, crayfishes, small fishes, octopi, and sea otters are important in this ecosystem.

SEA GRASS BEDS AND SALT MARSHES

On mud flats and other soft-bottomed habitats (such as along sandy and estuarine shorelines), the most important plants are *sea grasses*, also known as submerged aquatic vegetation (Figure 10.8). These are *angiosperms*, flowering and seed bearing vascular plants with true roots. They are limited to water depths where light is sufficient for photosynthesis and prefer clear water. Sea grass beds are highly productive with some rivaling the primary production of intensively developed agricultural land in the amount of carbon fixed (500 to 1000 gm per square m per year). Sea grass beds export large amounts of organic matter to nearby coastal waters. However, water pollution, sediment runoff from land, and dredging are among the many human activities that threaten sea grasses. Where there is excessive nutrient input, such as in runoff from agricultural areas, small algae may grow on the blades of sea grasses, shading and eventually killing them.

Perhaps the most extensively studied sea grass ecosystem is the Chesapeake Bay. When Europeans ar-

FIGURE 10.7
A tide pool on the coast near Carmel, CA. [NOAA, Monterey Bay National Marine Sanctuary. [Photo by Kip Evens]

FIGURE 10.8
A sea grass meadow in the Florida Keys National Marine Sanctuary. [NOAA photo by Heather Dine]

rived in the Bay area in the early 1600s, sea grasses covered an estimated 240,000 hectares (600,000 acres) of the Bay. By 1978, the sea grass area had shrunk to only 16,000 hectares (41,000 acres). Destruction of sea grass beds is a concern for several reasons. Sea grass anchors sediment and dampens wave action thereby controlling erosion and turbidity. It is a food source for many organisms including waterfowl and small mammals and serves as a primary nursery for crabs and many species of fish.

Reduction of the Chesapeake Bay sea grass habitat along with over-fishing has been implicated in the decline of populations of blue crabs, a mainstay of the Bay fishery for more than a century. Over the past decade or so, the number of adult female crabs plunged by about 80%. Without adequate protection by sea grass, the blue crab is more vulnerable to predation by striped bass (i.e., rockfish). Striped bass turned to blue crabs as a food source when fishing reduced the numbers of menhaden, their preferred food. Menhaden is a marine fish in the herring family and the Bay's top fishery by weight. Fortunately, restoration efforts have had some success and by 2001, the area of sea grasses in the Chesapeake had increased to more than 34,000 hectares (85,000 acres).

Salt marshes commonly occur along sheltered shorelines and are ecologically similar to sea grass beds in estuaries (Figure 10.9). Salt marsh grasses and bushes differ from sea grasses by being salt-tolerant true land plants, pollinating in the air. Salt marshes are flooded at high tide, but the grasses are never completely covered by salt water. Salt marshes are also home to abundant marine life, and are refuges for waterfowl and other wildlife. They are true transition zones between marine and terrestrial ecosystems.

In estuaries, the species of plants and animals vary from truly marine species near the mouth of the estuary to brackish water types where the salinity is lower, to those that prefer a nearly fresh water environment at the head of the estuary (Chapter 8). The salinity of estuarine waters occasionally changes drastically as when the discharge of fresh water entering the estuary increases abruptly following a period of heavy rainfall. This change in salinity can impact the types of species and their distribution in the estuary depending on how fast they can move or their tolerance.

In the tropics, between about 30 degrees N and 30 degrees S, mangrove swamps are common along muddy, low-lying coastlines and in estuaries. A **mangrove swamp** consists of tropical plant species including trees that grow in low marshy areas and can tolerate salt water flooding of their roots and lower stems (Figure 10.10). They generally compete successfully with local marsh grasses. Mangroves form dense growths and their aerial roots provide a complex and protective habitat for many organisms. For example, man-

FIGURE 10.9
A salt marsh near Galveston, TX. [NOAA, National Marine Fisheries Service]

FIGURE 10.10
Red mangroves are common in Florida. [South Florida Water Management District]

grove oysters and other organisms attach to the roots, which trap sediment and organic material. Mangrove swamps are also important nursery grounds for many marine species. However, many hectares of mangrove swamps are lost each year, destroyed to create farmland or for coastal development projects such as housing, shopping malls, resorts, airports, and industrial parks. Furthermore, some coastal residents have removed mangrove trees because they obstructed their view of the ocean. Ironically, mangrove swamps may be the final line of defense against a storm surge (Chapter 8).

KELP FORESTS

Seaward of the intertidal zone, kelp forests grow where waters are cool and nutrient-rich (Figure 10.11). **Kelp** includes various species of brown algae that grow to enormous size. In clear waters, individual plants grow at depths of about 30 to 40 m (100 to 130 ft) and their stipes (stem-like structures) and leaf-like fronds reach all the way to the surface. Found in cool waters worldwide, they are especially abundant in coastal upwelling zones off California and the Pacific Northwest.

Kelp plants grow attached to rocky bottoms by root-like structures (*holdfasts*). Holdfasts are quite small so that in many species, the weight of the plant is supported at the surface by a bulbous gas filled float below the fronds. Hence, strong waves easily destroy kelp beds and huge amounts of detached kelp wash onto beaches following storms. Fortunately, kelp can grow rapidly—as much as tens of centimeters per day. Because of this, it can be easily harvested and will regrow if only the tops of the blades are removed. Kelp's primary productivity is high, rivaling that of some of the richest farmlands.

FIGURE 10.11
A forest of giant kelp in Baja California. This large brown algae can grow as much as 0.6 m (2 ft) per day in water depths up to 45 m (150 ft). [NOAA Restoration Center, photo by M. Golden]

A dense stand of kelp is like a submarine forest, and it supports a rich community of animals that lives below its canopy. Small fishes, urchins, crabs, and lobsters are common on the sea floor around the holdfasts of kelp plants. Smaller algae, sea anemones, mollusks, and other invertebrates may attach to these structures. Further up the stipe where light is a little brighter, red and brown algae may live on the surface of the kelp. Sea urchins climb up the kelp plant eating these algae, or even the kelp itself. The canopy protects large schools of young fishes such as herring and sardines. Also, many organisms such as snails graze upon kelp fronds.

Along the Pacific coast of North America, kelp forests are home to large numbers of sea otters. Sea urchins are the favorite food of otters and urchins abound in the kelp forest habitat. In fact, urchins can destroy a kelp forest through grazing if they become too numerous. Sea otters feeding on urchins keep their populations in check. However, when sea otters were hunted (for their fur) to near extinction in the 18th and 19th centuries, sea urchin populations exploded and many kelp beds were lost. A similar relationship exists among kelp, urchins, and lobsters in North Atlantic kelp habitats, and over-fishing of lobsters has had similar results.

CORAL REEFS

A **coral reef** is among the ocean's most spectacular biological features. These slow growing structures can be centuries old and may be so large that they are visible from space. In many parts of the tropical ocean, reefs stand hundreds of meters high above the sea bottom and can extend hundreds of kilometers along the shoreline. Carbonate-secreting colonial animals are the primary builders of coral-reef frameworks. These open coral structures are bound together by layers of calcareous algae. Coral reefs grow along coastlines or cap extinct undersea volcanoes. They consist of thin veneers of living organisms growing on older layers of dead coral (limestone) or volcanic rock. Whereas corals are found in all oceans of the world, large reefs occur only in tropical waters, between about 30 degrees N and 30 degrees S. A few slow growing, small coral reefs occur along shelf breaks in many ocean areas.

The Great Barrier Reef stretches more than 2,000 km (1,200 mi) along the Queensland coast of eastern Australia (Figure 10.12). It is the world's most extensive reef system and also the largest structure on Earth made by living organisms. At its northern end, the reef is nearly continuous and is only about 50 km (30 mi.) offshore. In the south, individual reefs are more common with some located

FIGURE 10.12
Stretching more than 2000 km (1200 mi) along eastern Australia's Queensland coast, the Great Barrier Reef is considered one of the world's natural wonders. It is the most extensive coral reef system and the largest structure made by living organisms on Earth. In the north, the reef is essentially continuous and is located only about 50 km (30 mi) from shore. In the south, individual reefs are more common, and in some places are up to 300 km (190 mi) offshore. This Landsat-7 image depicts the southern portion of the reef system. [NASA Goddard Space Flight Center]

up to 300 km (190 mi) offshore. Hundreds of low-lying carbonate-sand islands mark the tops of the reefs.

Most coral reefs are in the tropical Pacific and Indian Oceans. Reef-building corals prefer waters having an average annual temperature of 23 °C to 25 °C (73 °F to 77 °F) and most corals cannot tolerate prolonged exposure to either very low or high water temperatures or to large fluctuations in temperature. For this reason, even small changes in sea surface temperature—perhaps associated with large-scale climate change—threaten coral reefs. Corals also require clear water and are endangered by sediment runoff from land, oil spills, and other forms of water pollution. Eutrophication caused by nutrient input from the land can stimulate the growth of algae on the surface of the coral reefs, smothering the coral polyps.

Each type of coral animal builds a characteristic structure that is conspicuous on reef surfaces. Some corals (e.g., brain corals) form robust compact structures; others build delicate, complex branching forms. Many corals have growth rings, much like trees, that can be used to reconstruct past variations in ocean conditions and climate. Recently coral growth rings were used to study El Niño events in the tropical Pacific Ocean. Coral reefs provide shelter for many species of fish, invertebrates, and plants and are among Earth's most productive habitats.

Individual coral animals, called *polyps* (figure 10.13), capture tiny plants and animals floating in the waters flowing over the reef. Reef-building corals also obtain large amounts of energy from microscopic dinoflagellates, called *zooxanthellae*, living within the tissues of polyps. The photosynthetic pigments in zooxanthellae are responsible for the bright color of corals. More importantly photosynthesis by zooxanthellae provides corals with food, permitting them to grow more rapidly than if they depended only on the food they captured from the water flowing past. Zooxanthellae, in turn, use the metabolic wastes produced by coral polyps as a source of nutrients. Hence, the corals and their zooxanthellae have a symbiotic relationship. A **symbiotic relationship** is a mutually beneficial association in which each organism benefits from the other. The sunlight requirements of zooxanthellae mean that the actively growing parts of a coral reef must be near enough to the ocean surface to receive abundant sunlight.

FIGURE 10.13
Coral polyps. [NOAA, National Ocean Service, Office of Response and Restoration]

FIGURE 10.14
Satellite image of an atoll in the western Pacific Ocean. [NOAA Photo Library, photo by Richard B. Mieremet]

Without zooxanthellae, corals cannot flourish. As noted above, coral are very sensitive to changes in water temperature. In response to unusually high sea surface temperatures, coral polyps expel zooxanthallae. Without zooxanthallae, coral polyps have little pigmentation and appear nearly transparent on the coral's white skeleton, a condition known as **coral bleaching**. A sea surface temperature rise of one to two Celsius degrees (2 to 4 Fahrenheit degrees) is sufficient to cause temporary bleaching. However, if bleaching episodes persist or are unusually severe such as may occur during El Niño, coral polyps and the reef die (Chapter 11).

Another potential threat to coral reefs is rising sea level. If sea level rises slowly, healthy coral reefs can grow upward fast enough to maintain themselves near the ocean surface and sunlight. However, if sea level rises too rapidly, reefs may not be able to grow fast enough. If they are too deeply submerged to receive sufficient sunlight for photosynthesis, they die. Many deeply submerged Pacific seamounts are capped by reef limestone. In these cases, the coral could not grow fast enough to keep up with the rapid rise in sea level or sinking platforms due to plate movement down slope and away from the ocean ridge spreading centers.

Coral reefs growing on volcanic islands typically exhibit a characteristic sequence of forms first described by Charles Darwin (Figure 2.17). A coral reef grows along the shore of an active volcano as a relatively narrow *fringing reef*, similar to those now found on the "Big Island" of Hawaii. After the volcano becomes inactive and eventually extinct, it sinks as the plate under it subsides while moving away from the hot spot that originally formed the volcano. At first the coral reef grows upward from its original position to form a *barrier reef*, separated from the main island by a shallow lagoon. Barrier reefs occur on the older Hawaiian Islands to the northwest of the "Big Island" such as off the coast of Oahu. As the volcano erodes and continues to sink into the ocean, the volcanic island becomes smaller. Eventually the volcanic island sinks beneath the waves or erodes away, leaving only the ring of coral reefs surrounding a shallow lagoon where the island once stood. At this stage, the reef is known as an **atoll** (Figure 10.14). Atolls may be ring- or horseshoe- shaped and they vary from 1 to 100 km (0.6 to 60 mi) across. Midway and Kure Islands, at the northwestern end of the Hawaiian-Island chain, are examples of atolls. Atolls are most common in the Pacific Ocean (Figure 10.15).

Corals described thus far are *hermatypic*, that is, they live in warm, shallow water and build large reefs, and possess zooxanthellae. Solitary corals and small coral reefs also live in cold, deep water along continental shelf breaks in some parts of the ocean (Figure 10.16). Without light, these so-called *ahermatypic corals* cannot rely on zooxanthellae, so they depend exclusively on trapping food directly from the water using their stinging cells or nematocysts. Some ahermatypic corals are hard structures whereas others are softer and more delicate, branching forms, which do not build limestone structures. They provide excellent habitats for deep-water fishes. But therein lies a threat to their survival. Trawlers, fishing boats that drag huge nets along the sea bottom, are destroying these slow growing deep-sea corals. Although little is known about the distribution and biology of deep-sea corals, some nations such as Norway are beginning to ban harmful fishing practices in areas where they occur.

In an effort to preserve coral reefs and other underwater marine resources for present and future genera-

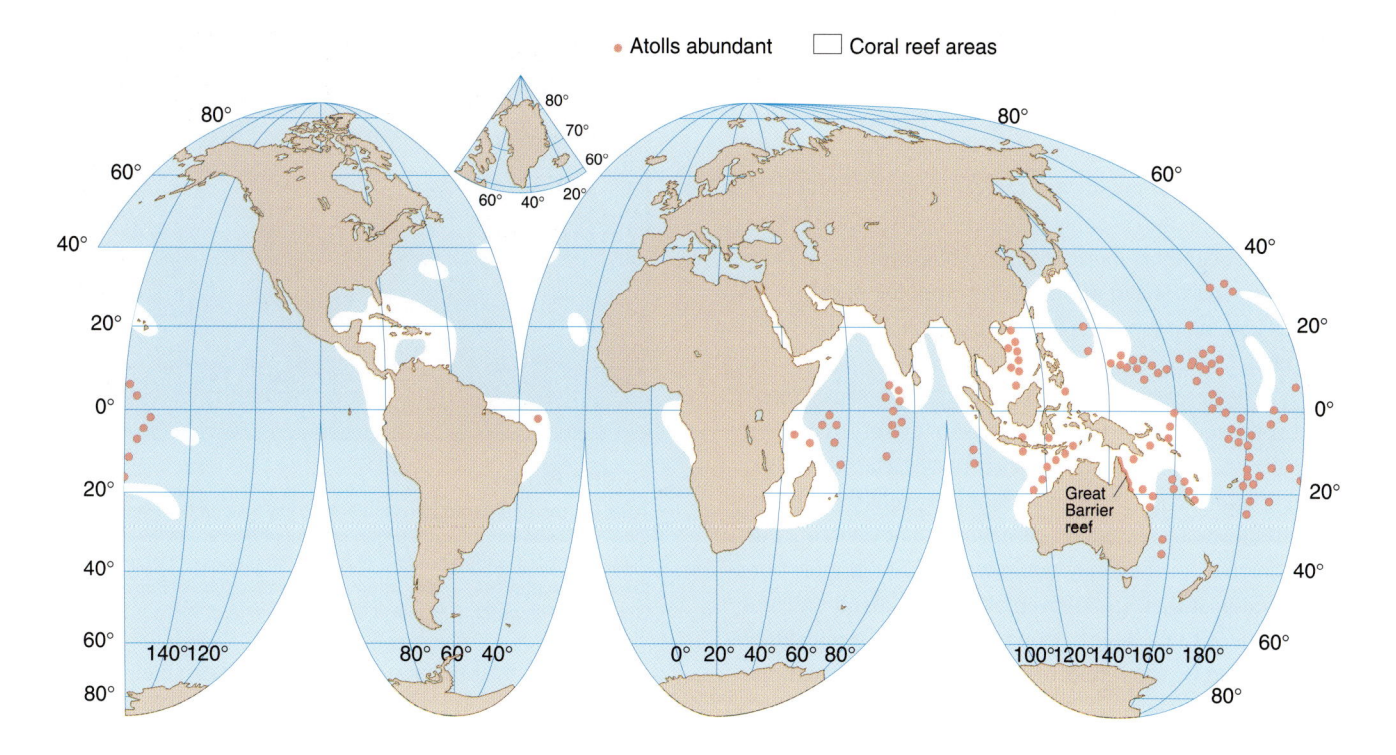

FIGURE 10.15
Location of atolls and coral reefs in the world ocean.

tions, the U.S. federal government has designated certain areas of its jurisdictional waters as *marine sanctuaries* or *marine reserves*. For more on this topic, refer to this chapter's first Essay. In managing marine resources, authorities are advocating an ecosystem approach. For an overview of this effort at sustainable development, see this chapter's second Essay.

FIGURE 10.16
Deep sea corals on a mud-covered rock outcrop at a depth of 865 m (2855 ft) in Oceanographer Canyon in the Gulf of Maine. Photo taken from a camera sled in 1978. [NOAA photo by Barbara Hecker]

BENTHIC FEEDING HABITS

Animals living on and in soft-bottomed habitats (whether in the intertidal zone, shallow coastal areas, or in the deep sea) are divided into two categories: infauna and epifauna. **Infauna** live within sediment deposits whereas **epifauna** live on the sea floor. Some of these benthic animals are *unselective* in their feeding, moving through or on the mud and simply eating everything. They ingest bottom deposits and digest the organic detritus they contain whereas the remaining materials (mud or very fine sand) are excreted. Other benthic animals are more *selective* in their feeding habits, ingesting food particles from the mud and sand particles, rather than consuming the sediment itself. As benthic animals feed on or in sediment, they disturb the deposits, a process called **bioturbation**. Their feces bind sediments into harder, more durable aggregates that are not readily re-suspended by currents. Close observation of the surface of a water-covered mud flat reveals a variety of feeding structures of infaunal worms and clams extending above the entrances to their burrows.

Soft bottoms also harbor many filter feeders that separate phytoplankton, small zooplankton, and other edible materials from large volumes of water. The distribution of animals in muddy and sandy habitats is controlled primarily by the grain size of the sediment deposits. Filter feeders cannot live in fine-grained mud, because it clogs their feeding apparatus. Hence, they usually inhabit coarser grained sands. Filter feeders are generally rare where deposit feeders are abundant because bioturbation makes the near-bottom water muddy and therefore unsuitable for filter feeding. On the other hand, mud is ideal for unselective feeders. It contains more organic matter brought in from terrestrial sources and raining down from the photic zone plus an abundant supply of decomposing bacteria and other microbes.

Whereas some animals living in the rocky intertidal zone, such as starfishes, can absorb dissolved oxygen from the water directly through their body surfaces, benthic animals living in muddy habitats require specialized breathing structures. A respiratory structure may extend outside their burrows or they may pump water through their burrows and past their gills for uptake of dissolved oxygen. Some animals line their burrows with mucus or with shell fragments to prevent mud from clogging these breathing structures.

LIFE ON THE DEEP-SEA FLOOR

In the deep ocean, detritus that reaches the sea floor is mostly decomposed and has little nutritional value. For this reason, far fewer organisms can survive in this habitat but it is not the dead zone envisioned by most 19th century scientists. Many unusual creatures live on the deep-sea floor. For example, a starfish-like animal, called a sea lily, lives attached to the bottom by a long, stem-like structure (Figure 10.17). Its arms, covered with a mucous substance, are suspended in the water above the sea floor and trap particles in the near-bottom waters. The particles are moved along grooves to the mouth. Predatory starfishes

FIGURE 10.17
This delicate sea lily (crinoid), a member of the phylum that includes starfish, can orient toward the current to increase food capture. [NOAA Undersea Research Program]

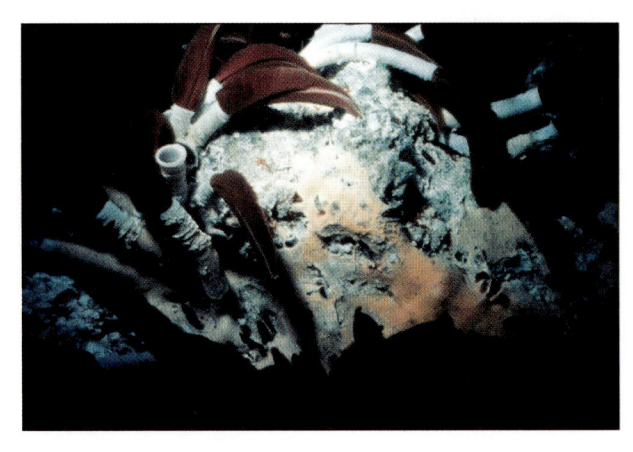

FIGURE 10.18
Tube worms at a hydrothermal vent on the floor of the Pacific Ocean. Discovery of such organisms launched a new avenue of inquiry into our understanding of biological processes operating in the deep ocean (especially chemosynthesis). [OAR/National Undersea Research Program (NURP); College of William & Mary]

move across the bottom on long legs that keep their bodies above the sediment surface. Other filter feeders in the deep ocean include sponges, crustaceans, and some kinds of worms and mollusks. Deposit feeders are similar to those living in shallow benthic habitats, although animals in the deep, cold ocean are generally smaller, an adaptation to the scarcity of food. In these dark waters color has no strategic value so that most animals are very drab.

Communities of specialized animals have evolved to live on the deep ocean bottom near hydrothermal vents (Chapter 2). Large colonies of tubeworms have a symbiotic relationship with bacteria that live within their bodies (Figure 10.18). The bacteria use hydrogen sulfide (H_2S) as an energy supply and make food for the tubeworms through chemosynthesis (Chapter 9). These worms can grow to a length of 3 m (10 ft). Also living in this extreme environment are long legged crabs and specialized mollusks and clams.

Marine Animals

In this section, we briefly describe larger marine animals and their adaptations for life in the sea. These organisms include fishes, mammals, reptiles, and birds.

FISHES

Fishes are the most familiar type of nekton. Unlike mammals, which regulate their body temperature, fish are cold blooded and hence are as warm or cold as the surrounding water. (Some fish that live in water at temperatures slightly below 0 °C have a natural antifreeze.) Fishes range in size from the smallest of all vertebrates, a marine goby only 8 to 10 mm long, to the huge whale shark up to 15 m (50 ft) long. Fish may be herbivorous, carnivorous, or omnivorous. They inhabit all parts of the ocean and possess special adaptations for buoyancy, swimming, and life in the dimly lit twilight zone as well as the greatest depths of the ocean. Many open-ocean pelagic fishes are streamlined for fast swimming in search of prey and to flee from predators. Fishes living among plants or rocks often have body shapes and colors that allow them to blend unnoticed with their background. Because the blood and body fluids of marine fish are only about one third as salty as seawater, they need special physiological adaptations in order to survive in the ocean.

There are two major groups of fishes: elasmobranchs (or cartilaginous fishes) and teleosts (or bony fishes). **Elasmobranchs** or **cartilaginous fishes** include sharks, skates, and rays and are considered to be more primitive. Their skeletons lack true bones and consist entirely of cartilage. Their skin feels much like fine sandpaper due to small, embedded tooth-like structures called denticles. Sharks' teeth are large, specialized denticles that occur in rows in the mouth and are continually growing forward, being lost and subsequently replaced from behind.

A popular misconception portrays sharks (figure 10.19) as ferocious predators who menace swimmers. In fact, a few shark species are ferocious but their threat to humans is not significant. Nonetheless, shark attacks—no matter how infrequent—are shocking and frightening. Some are quite selective predators, feeding mostly on preferred types of prey. Other species, such as the tiger shark, are sometimes referred to as "garbage cans of the sea" because they eat almost anything, living or dead. Some types of sharks and rays are bottom feeders, with an upper jaw designed to pick up food from the sea floor. Basking sharks, whale sharks, and many rays are passive filter feeders; that is, they swim continuously, straining plankton from the water passing over their gills. Whale sharks are thus very docile even though they weigh more than 40,000 kg (about 45 tons) and grow to a length of 15 m (50 ft).

Sharks have some unique problems in that they do not have a swim bladder for buoyancy, their blood pressure is low, and most cannot pump water over their gills to obtain oxygen. To help compensate for all three of these problems, most sharks constantly swim their entire life. With swimming motion, their pectoral fins provide lift much like an airplane wing. Movement causes a flow of water over their gills to oxygenate the blood (known as *ram ventilation*), and continuous swimming produces muscle action that helps the heart pump blood. Without active swimming, a shark would sink to the bottom and most would asphyxiate. Some sharks, however, possess spiracles, and others have special muscles that force water over the gills, helping them to obtain oxygen even while stationary.

There are over 800 species of cartilaginous fishes. They have reproductive strategies that favor survival of their young. Unlike most bony fishes, cartilaginous fishes have internal fertilization. Although they bear very small numbers of young (whereas bony fishes release thousands of eggs), the chance of reproductive success is increased. Some sharks and rays lay their eggs in leathery cases that harden in a few hours. Beachcombers sometimes find these leathery pouches, called "mermaid's purses," washed up on the beach. The young are hatched fully formed; there is no planktonic larval stage among cartilaginous fishes.

Some species of sharks are *viviparous*; that is, embryonic development occurs entirely within the female shark's body and they bear live young. Hammerhead sharks are a viviparous species. Cartilaginous fishes usually lay their eggs or bear their young in coastal nursery areas where there is less danger of predation by larger fishes.

Cartilaginous fishes are *isosmotic* with their environment; that is, the salt concentration in their body is approximately the same as the surrounding seawater. Hence, there can be an equal amount of water leaving the body as enters it. However, the types of salts are very different. They have a very high concentration of organic solutes, mostly urea, in their tissues. The concentration of sodium chloride in their body fluids is only about half that of seawater so that a special rectal gland contains a cell that selectively excretes chloride ions into their urine.

FIGURE 10.19
This large white shark is cruising offshore from the Farallon Islands off San Francisco Bay. [NOAA photo by Scott Anderson]

Teleosts (or *bony fishes*) have bony skeletons, scales, and a flap covering the gills. This is a large, diverse, and familiar group of fishes that live in a variety of marine and fresh water environments. Most bony fish have a swim bladder, a gas-filled structure that can be inflated or deflated enabling the fish to adjust its buoyancy with changes in water depth. Normally, the salt concentration in bony marine fish is only about one quarter to one third that of seawater. Hence, they must also have a mechanism to regulate the salt content of their bodies. Otherwise they would continually lose water via diffusion through their gills and body surface to the environment. (Ions or molecules tend to migrate from regions of higher concentration to regions of lower concentration.) To prevent this from happening, they must drink a lot of water. Their gills have special chloride excretory cells that transport the excess salt out of their bodies. Their kidneys excrete the larger ions, such as calcium and magnesium, retaining water in the body. Marine teleosts produce only minute amounts of concentrated urine compared to freshwater fish.

Fishes living on or near the ocean bottom are called *demersal fishes* and many of them are commercially important, such as cod, halibut, haddock, and sole. *Flatfishes* are demersal fishes that have a special coloration that camouflages them against the ocean bottom. They include halibut, flounder, and sole. Because these fish spend most of their lives lying on the sea floor, an eye on the lower side would be useless to them (Figure 10.20). When larval flatfishes hatch as free-swimming zooplankton, they have eyes on either side of their head. As larvae mature, however, one eye moves over the top of its head, so that when a young fish finally takes up life on the bottom, both eyes are looking up on the same side of the head. Also its body thins and is laterally compressed. In addition, they have the ability to quickly change the color and pattern of their skin so that they closely resemble the background, another example of adaptive coloration.

As noted above, many pelagic fishes are fast swimmers. Tuna and mackerel are examples of fishes well adapted for strong swimming over long distances in open waters. For example, blue fin and yellow fin tuna cruise at 16 km (10 mi) per hr and are capable of brief bursts of 95-130 km (60-80 mi) per hr. Such high speeds require large amounts of energy and these fast swimmers usually occupy a high trophic level in marine food webs. Their dense muscles make their meat very desirable and they are much in demand by humans.

Smaller pelagic fishes such as herring and sardines are abundant in coastal waters and in upwelling areas where nutrients are plentiful. Some of the world's largest fisheries are based on these species. Smaller fishes are converted to fishmeal and used as high protein feed for cattle, poultry, and fish farms and also as fertilizer.

Most marine teleosts spawn; that is, the females release their eggs into the water where they are fertilized after males release sperm. A single female herring can release between 50,000 and 700,000 eggs per breeding season. In general, larger and older fishes are the most productive. The eggs hatch, releasing larvae, which exist for varying periods of time in the plankton as they mature and become stronger swimming juveniles. Only a tiny fraction of the larvae become juvenile fish and far fewer survive to adulthood. There are many reasons for this relatively low reproductive success. Currents can keep the eggs and sperm apart or they may carry the planktonic larvae away from the protective nursery grounds in estuaries. Furthermore, many animals prey upon the eggs, larvae, and young fishes.

Most teleosts broadcast their eggs and sperm into the water and rely upon chance fertilization. By producing thousands of eggs and sperm, they increase the odds of producing a few successful young fishes. Some teleosts have evolved to increase their reproductive success through strategies such as internal fertilization, extended periods of parental care of the young, brooding of the young in the mouth of one of the parents or in a brood pouch (such as in the male seahorse). As mentioned earlier, elasmobranches increase their reproductive success through internal fertilization and protection of their young, but they can produce only a very small number of offspring.

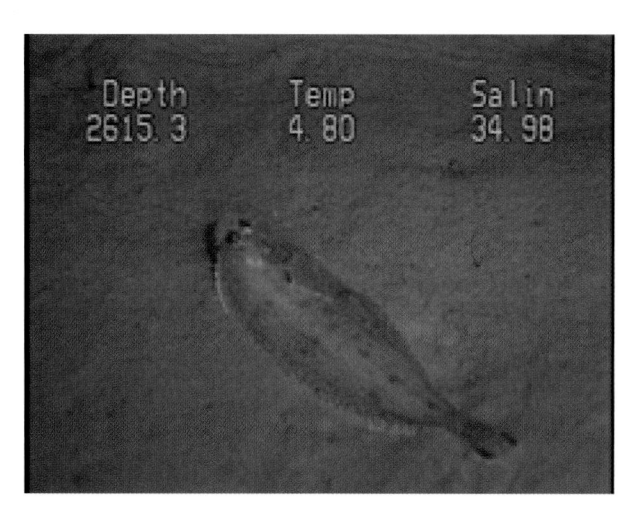

FIGURE 10.20
Note the two eyes on one side of the head of this flounder photographed at an ocean depth of more than 2600 m (8600 ft) off the North Carolina coast. [NOAA, Office of Ocean Exploration]

Many environmental factors influence the reproductive success of bony fishes. Some fishes undertake long migrations. **Anadromous fishes** such as salmon, shad, sturgeon, and striped bass are born in rivers and streams, but spend most of their lives in cold regions of the ocean, returning to the same river to breed when they are sexually mature. Pacific salmon may migrate as far as 1300 km (800 mi) to complete this cycle. They reproduce only once, expending considerable energy to reach the headwaters of their particular river. Along the way, they use much of their body mass in this effort or convert it to eggs or sperm. They die after spawning. On the other hand, Atlantic salmon can reproduce many times.

Dams can be formidable obstacles to salmon migration. In the Pacific Northwest, for example, only about 5% of the juvenile salmon survive passage through dams and reservoirs on the Columbia and Snake Rivers. Largely ineffective are fish ladders intended to help the salmon move upstream and other structures that guide them downstream around hydroelectric turbines. These obstacles have been likened to giant food processors for smolt (young salmon) attempting to swim through them. Furthermore, on their downstream passage smolt are held up in reservoirs where they are exposed to predators, pathogens (disease-producing organisms), and water that is too warm. Atlantic salmon suffer a similar fate. More than 900 dams on New England and European rivers prevent most Atlantic salmon from reaching their freshwater spawning grounds. Consequently, their population has declined to less than 1% of historical levels. Today, almost all "Atlantic" salmon sold at fish markets come from fish farms.

Less well known than anadromous fishes are **catadromous fishes** that breed in the open ocean, but spend their adult lives in fresh water. In North America, the American eel is the only catadromous species. For many years the location of their breeding grounds was a mystery. Scientists now know that eels from both European and North American rivers travel thousands of kilometers to congregate in the Sargasso Sea where they spawn. Planktonic eel larvae remain in that location for about a year before migrating to the coastal waters where their parents matured. Young eels mature into adults in the coastal ocean before swimming up river. After eight to twelve years, they return to the Sargasso Sea where they reproduce once and die. Many questions remain unanswered about how these migratory fishes navigate and identify their specific home rivers and how the specifics of navigation are transferred from generation to generation.

FIGURE 10.21
A school of northern anchovies (*Engraulis mordax*). Schooling is a behavioral adaptation of teleosts that may offer protection from predators. [NOAA National Undersea Research Program]

Many species of fish, such as herring and mackerel, reduce predation by swimming together in organized groups, keeping a specific distance between one another (Figure 10.21). This so-called **schooling** behavior is particularly advantageous in the open ocean where hiding places are few and far between. Schooling confuses potential predators and functions as an alarm system when a would-be intruder approaches. Typically, the individuals in a school disperse in all directions when the alarm is sounded and then reassemble after the danger has past. In addition, fish that form schools swim more efficiently over long distances using less energy than if each individual were swimming alone. This greater efficiency is accomplished by drafting, similar to the strategy employed by racing cars. Fish in a given school are of one species, similar in size, and may be a mixture of males and females or a single sex (depending on the species). There is no evidence of a hierarchical structure (leaders and followers) among individuals in the school. Wide-angle vision or sound vibrations enable each fish in the school to sense the location and movement of its neighbors thereby maintaining order.

MARINE MAMMALS

Marine mammals are warm blooded, air breathing animals that bear live young, which they nurse. They include whales, dolphins, seals, walrus, sea lions, and polar bears (a terrestrial mammal that has adapted to marine habitat). The largest marine mammals, baleen whales, live primarily in the open ocean where they filter zooplankton from the water as they swim slowly with their mouths open. These filter-feeding whales include the largest animal to have ever lived on Earth, the blue whale. Grey whales are also very large baleen whales. They forage for crustaceans by sucking up large amounts of sediment from the sea floor that they expel through their baleen, which retains the food. Smaller whales, such as sperm whales, killer whales, porpoises and dolphins are faster swimming, toothed carnivores that actively hunt their prey. Sperm whales feed on squid captured in the mid-depths of the ocean. Cetaceans (whales, porpoises, and dolphins) spend their entire lives at sea.

Polar bears are unique to the Arctic and are the region's largest land predator growing to a height of 2.5 to 3 m (8 to 10 ft) and weight of 250 to 800 kg (550 to 1700 lb). Polar bears inhabit the waters and coastal areas of Alaska, northern Canada, northern Russia, Greenland, and Norway. These animals depend on sea ice as the platform from which they hunt seals, their principal food source. They will also take walrus and in some cases small whales (e.g., Beluga whales) that have been trapped in small open areas in the ice. They travel across the sea ice and open water to reach their dens on land. About half of all female polar bears bear their cubs in dens on land; the rest build dens on ice floes. Their dense fur and blubber allow them to comfortably swim from one ice floe to another in the Arctic Ocean for distances as great as 100 km (62 mi). In short, polar bears are uniquely adapted to living on Arctic sea ice and are often considered to be marine mammals. However, because of their dependence on sea ice, they face possible extinction as the sea ice cover in the Arctic thins and shrinks (Chapter 12).

Pinnipeds are named for their distinctive swimming flippers. They include seals, walruses, and sea lions; many of them live in coastal waters (Figure 10.22). Unlike cetaceans, pinnipeds come ashore to breed, give birth, and rear their young. All pinnipeds are fur bearing and many species have been hunted nearly to extinction, although some populations are recovering due to protective legislation.

Manatees and dugongs are large herbivorous mammals that feed on vegetation in shallow waters along the coast and in the estuaries of Florida and many parts of Asia and South America. They are now seriously endangered in many areas because of increasing coastal development and encounters with motorboats whose propellers can severely injure these gentle, slow-moving animals. Loss of sea grass beds, a primary feeding habitat for some manatees and dugongs, is another threat to their survival.

FIGURE 10.22
The Steller sea lion (*Eumetopias jubatus*) breeds in large colonies such as this one on an island off the Alaskan coast. [NOAA photo by Captain Budd Christman, NOAA Corps]

The National Marine Fisheries Service of NOAA, through its Office of Protected Resources, is responsible for enforcing the Marine Mammal Protection Act in the United States. This legislation, enacted in 1972 and re-authorized in 1994, established a moratorium on the hunting of marine mammals in U.S. waters and by U.S. citizens on the high seas.

MARINE REPTILES

The few reptiles that live in the ocean fall into three groups, the best known being sea turtles (Figure 10.23). Marine lizards, such as the iguana of the Galapagos Islands, and sea snakes are examples from the other two groups of marine reptiles.

FIGURE 10.23
Leatherback sea turtles occasionally nest on the beach at Canaveral National Seashore in Florida. The leatherback, an endangered species, is one of the largest sea turtles in the world. It can grow to a length of over 2 m (7 ft). [NOAA and the National Park Service, Canaveral National Seashore]

All sea turtles live in the ocean, but come ashore to lay their eggs. Female sea turtles dig their nests on sandy beaches, lay up to one hundred leathery eggs, and then return to the sea. Newly hatched turtles must fend for themselves and find their way back down the beach to the sea. Along the way, sea birds and other animals prey upon them. If they make it to the sea, predatory fishes reduce their numbers still further. Only a few hatchlings survive to adulthood. Sea turtles have been hunted for their meat as well as their shells, which are used to make combs and jewelry. Plastic litter in the ocean is a serious threat to sea turtles (as well as other marine animals) that frequently mistake drifting plastic bags for jellyfish, a favorite prey. These bags, when swallowed, can block the digestive tract, eventually killing them. Turtles also get caught in illegal drift nets and on long-lines of hooks intended to catch swordfish.

Sea snakes are rare, occurring only in the tropical Pacific and Indian Oceans, usually around reefs. Unlike moray eels, which are actually fishes with gills, sea snakes are reptiles and must come to the surface to breathe; they have special valves that close their nostrils while submerged. Some species of sea snakes go ashore to reproduce and lay eggs; others mate and bear live young without leaving the sea. They are among the most venomous animals on Earth. A single snake can produce about 10 to 15 milligrams of venom at a time, but the fatal dose for humans is only one-tenth this amount. Fortunately sea snakes are extremely shy animals and rarely attack humans unless provoked.

Marine iguanas are herbivores, feeding mostly on seaweeds in the intertidal zone. They occur only in the Galapagos Islands in the equatorial Pacific Ocean. Having evolved from land lizards, they need a special adaptation to deal with the large amount of salt water they swallow while feeding. A special gland connected to the nostrils excretes excess salt, which the iguana expels by sneezing. Being cold blooded, iguanas regulate their body temperature by spending much time sunning themselves on the rocky shore.

The Galapagos Islands are the product of hot spot volcanism in the eastern tropical Pacific (Chapter 2). Prevailing trade winds cause equatorial upwelling of cold, nutrient-rich waters that provide sufficient productivity to support marine iguanas. Apparently, the abundant food supply was a factor in the iguana's adaptation to cold, salty ocean waters. During El Niño episodes, however, upwelling is suppressed and populations of iguanas and other inhabitants of the Galapagos Islands are threatened (Chapter 11).

SEABIRDS

Unlike most birds on land, seabirds are carnivorous predators that occupy a high trophic level in marine food webs. They have high metabolic rates and require energy-rich, fatty foods. Wading birds, such as herons, have long legs and inhabit wetlands and shallow soft-bottom habitats where they find prey in the shallow waters and soft sediment. Some, such as flamingos and spoonbills, have specialized bills that allow them to strain bottom deposits to feed on small plants and animals. Herons have a sharp bill enabling them to snatch small fishes from the shallow water. Other seabirds dive for food; these include sea ducks, cormorants, pelicans, terns, and loons (Figure 10.24). Pelicans are among the largest and heaviest birds on Earth, yet they make spectacular dives as they hunt for fish.

No seabird can stay at sea for its entire life. All of them must come ashore to breed and lay eggs on a solid surface. Also, their body heat is necessary to hatch the eggs. The chicks require feeding by their parents until they are ready to fly and hunt for themselves. However, some seabirds spend very long periods flying over the open ocean. The albatross is a classic example, spending years at a time far from land.

FIGURE 10.24
Brown pelican in breeding colors. [NOAA National Marine Fisheries Service, photo by William B. Folsom]

Where fishing grounds are exceptionally rich, seabirds establish huge colonies on rocky shores and cliffs, especially during their breeding season. Some seabirds migrate long distances searching for food. Terns migrate from the Canadian Arctic, where they spend the Northern Hemisphere summer, all the way to Patagonia at the southern tip of South America, where they spend the Southern Hemisphere summer.

One type of seabird, however, has lost its ability to fly. Penguins are adapted to live at sea and their streamlined bodies enable them to swim with considerable speed and agility (Figure 10.25). They can dive as deep as 250 m (820 ft) and can stay submerged for up to twenty minutes while searching for fish and small squid. They lay eggs and rear their young on land ice. Nearly all species of penguins (17) live in the Southern Hemisphere, mostly around

FIGURE 10.25
Although ungainly on land, the flightless penguin is an expert swimmer. Shown here is a Gentoo Penguin (*Pygoscelis papua*), largest of the "bush-tail" penguins. They live on Antarctic islands as well as the Antarctic Peninsula. [U.S. Geological Survey photo by John Chardine]

Marine plants and animals are distributed unevenly in the ocean; their distribution is controlled primarily by environmental factors such as water temperature, salinity, light, and availability of food. Most marine organisms live near the ocean's surface, edges, or on the bottom. In the open ocean, organisms are most abundant in the surface waters, less so at mid-depths, and are least abundant in the deep ocean.

This chapter completes our survey of marine organisms and how they cope with the physical, chemical, and biological properties of the ocean in the Earth system. In the next chapter, we examine some of the large-scale interactions between the ocean and atmosphere and their impacts on the rest of the Earth system including marine life.

Basic Understandings

- Living space in the ocean not only vastly exceeds that on land but also features very different types of habitats. Organisms live in all parts of the ocean, even extreme environments such as Arctic or Antarctic sea ice and hydrothermal vents on the sea floor. Other marine habitats include rocky shores, tide pools, estuaries, tropical coral reefs, polar oceans, deep-sea trenches, and sediments and bedrock on the sea floor.
- Oceanic life zones are commonly defined in terms of distance from shore and water depth. Open ocean waters constitute the pelagic zone, home to plankton (passive floaters or weak swimmers) and nekton (strong swimmers). The environment of the sea floor at all depths is called the benthic zone, home to organisms that live either on the ocean bottom or within sediment deposits.
- The area along the shore between high- and low-tide lines is the intertidal (or littoral) zone, home to ecosystems such as salt marshes and mangrove swamps. The area seaward from the shore to the shelf break at a depth of about 200 m (650 ft), including the intertidal zone, forms the neritic zone—commonly referred to as the coastal zone.
- Oceanic life zones are also defined in terms of nutrient supply and productivity. Nutrient-poor waters having low primary production are described as oligotrophic. Broad areas of open-ocean within the subtropical gyres are examples of oligotrophic waters. Nutrient-rich waters having high primary production are described as eutrophic. Coastal upwelling zones are examples of eutrophic waters.

Antarctica and on the southern extremities of Africa, South America, and Australia. One species lives near the equator on the Galapagos Islands. All these areas are nutrient-rich with cold currents. Some penguin species spend up to 75% of their lives at sea, away from land for months at a time. For birds, penguins have unusually dense bones, an adaptation that counteracts their buoyancy problem.

Conclusions

Most of Earth's living space is in the ocean where habitats are many and diverse, ranging from coastal intertidal zones to soft sediments and hydrothermal vents on the deep-ocean floor. Organisms exploit all these habitats. But survival in the ocean presents many challenges that differ from those faced by terrestrial organisms. Marine plants and animals have evolved many specialized adaptations involving buoyancy, feeding, reproduction, and protection from predators.

- Plankton, members of phytoplankton, zooplankton or microbial communities, evolved adaptations and strategies for life in the pelagic zone. An adaptation is defined as a genetically controlled trait or characteristic that enhances an organism's chances for survival and reproduction in its environment. For example, the shape of some plankton produces a relatively high surface area to volume ratio that provides buoyancy so that they do not sink below the sunlit photic zone.
- Larger, free-swimming pelagic animals, collectively called nekton, include fish of all sizes, squid, sea turtles, and marine mammals. Some migrate over long distances; the distribution of many others is limited by their tolerance for water temperature, salinity, or food availability. Nektonic organisms also have adaptations for buoyancy—the simplest of these are gas bladders.
- Most pelagic fish rely on active swimming to obtain food. Some fish lie on the bottom, waiting for their prey and suddenly lunge to catch it. Others swim all the time, seeking prey over great areas of the ocean perhaps covering hundreds of kilometers each day.
- The photic zone is where food is most plentiful but it is also the most dangerous for marine organisms because predators can easily see them. Many types of zooplankton avoid this problem by daily vertical migration. Each day at dusk, they come to the surface zone to feed on phytoplankton. Then, as daylight returns, they move back to the relative safety of darker, deep waters.
- Species of smaller fish that are prey for larger ones have adaptations that make them less visible and therefore less likely to be eaten. The most common adaptation is adaptive coloration where the animal's color pattern closely matches its background substrate. Many fish have countershading, that is, their dorsal side is a dark color, making it difficult for predators above to see them against the dark, deep water. Their ventral side is lighter colored, making the fish more difficult for predators to see them from below against the more brightly lit surface waters.
- Some marine animals emit light, a phenomenon known as bioluminescence. The light usually is a product of a chemical reaction that takes place in specialized cells or organs. Marine animals use light emission to attract mates or prey, frighten or confuse predators, or to disguise themselves.
- The general name for animals that are equipped with features that strain food particles out of large volumes of water is filter feeder. Some planktonic organisms produce large mucous nets, which function much like spider webs to trap small water-borne organisms and other food particles.
- Benthic organisms may live attached to a firm surface, they may construct burrows or tunnels or simply dig into sediment deposits, or they may move freely on the sea floor. Attached animals are filter feeders; mobile animals are active predators; and burrowing organisms ingest sediment, digesting the organic material it contains and excreting the rest.
- A distinction is made among intertidal habitats based on the type of substrate (e.g., rocky, muddy, sandy). The most energetic of these habitats are rocky intertidal shorelines where organisms are frequently subject to strong wave action, especially in winter. Where the distance between low and high tide levels is relatively great, organisms may be exposed to dry air and sunshine for many hours each day. Depending on the season, they may be forced to withstand extremely high or low air temperatures when they are not covered by seawater.
- Seaweeds are important components of the intertidal zone. All are algae and are green, brown, or red, depending on the color of the dominant photosynthetic pigment. They are distributed according to their light requirements and their ability to resist drying and freezing during low tide. Seaweeds provide habitat for many marine organisms living in the intertidal zone.
- On mud flats and other soft-bottomed habitats, the most important plants are sea grasses, also known as submerged aquatic vegetation. These are angiosperms, flowering and seed bearing vascular plants with true roots. They are limited to water depths where light is sufficient for photosynthesis. Sea grass beds are highly productive.
- Salt marshes commonly occur along sheltered shorelines and are ecologically similar to sea grass beds in estuaries. The salt marsh grasses and bushes differ from sea grasses by being salt-tolerant true land plants, pollinating in the air. Salt marshes are flooded at high tide, but the grasses are never completely covered by salt water. They are home to abundant marine life and are refuges for waterfowl and other wildlife.
- In estuaries, plants and animals vary from truly marine species near the mouth of the estuary to brackish water types where the salinity is lower, to those that prefer a nearly fresh water environment at the head of the estuary.

- In the tropics, mangrove swamps are common along muddy, low-lying coastlines and in estuaries. They consist of tropical tree species that are very salt-tolerant and can withstand tidal flooding of their roots and lower stems. Mangroves form dense growths and their aerial roots provide a complex and protective habitat for many organisms such as mangrove oysters.
- Seaward of the intertidal zone, kelp forests grow where waters are cool and nutrient-rich. Kelp includes various species of brown algae that grow to enormous size. They are especially abundant in coastal upwelling zones off California and the Pacific Northwest. Kelp grows attached to rocky bottoms by root-like structures known as holdfasts and much of the plant is supported at the surface by a bulbous gas filled float below the fronds.
- Coral reefs are among the ocean's most spectacular biological features. They grow along coastlines or cap extinct undersea volcanoes. Coral reefs are only thin veneers of living organisms growing on older layers of dead coral (limestone) or volcanic-rock.
- Hermatypic corals live in warm, shallow water and build large reefs. Solitary corals and small coral reefs also live in cold, deep water along continental shelf breaks in some parts of the ocean. These are known as ahermatypic corals. Carbonate-secreting colonial coral animals are the primary builders of coral-reef frameworks. Coral reefs provide shelter for many species of fish, invertebrates, and plants and are among the ocean's most productive habitats.
- Coral reefs growing on volcanic islands typically exhibit a characteristic sequence of forms caused by either sea level rise or sinking of the island: fringing reef, barrier reef, and atoll.
- Animals living on and in soft-bottomed habitats are either infauna or epifauna. Infauna live within sediment deposits whereas epifauna live on the sea floor. Some of these benthic animals are unselective in their feeding, moving through or on the mud and simply ingesting everything. Other benthic animals are more selective in their feeding habits, ingesting only food particles from among the mud and sand particles.
- Soft ocean bottoms harbor many filter feeders that separate phytoplankton, small zooplankton, and other edible materials from large volumes of water. The distribution of animals in muddy and sandy habitats is controlled primarily by the grain size of the sediment deposits.
- Many unusual creatures live on the deep-sea floor. Deposit feeders are similar to those living in shallow benthic habitats, although animals in the deep, cold ocean are generally smaller, an adaptation to the scarcity of food. Communities of specialized animals have evolved to live on the deep ocean bottom near hydrothermal vents utilizing chemosynthesis.
- Unlike mammals, which regulate their body temperature, fish are cold blooded and hence are as warm or cold as the surrounding water. Fish may be herbivorous, carnivorous, or omnivorous. They inhabit all parts of the ocean and possess special adaptations for buoyancy, swimming, and life at the surface, in the dimly lit twilight zone, as well as the greatest depths of the ocean. The two major groups of fishes are elasmobranchs (or cartilaginous fishes) and teleosts (or bony fishes). Elasmobranchs include sharks, skates, and rays and are considered to be more primitive. Their skeletons lack true bones and consist entirely of cartilage. Teleosts have bony skeletons and scales. This is a large, diverse, and familiar group of fishes that live in a variety of marine and freshwater environments.
- Anadromous fishes are born in freshwater rivers and streams, but spend most of their lives in the ocean, returning to the same river to breed when they are sexually mature. Less well known are catadromous fishes that breed in the open ocean, but spend their adult lives in fresh water.
- Many species of fish reduce predation by swimming together in organized groups called schools, keeping a specific distance between one another. This strategy confuses potential predators and functions as an alarm system when a would-be intruder approaches.
- Marine mammals are warm blooded, air breathing animals that bear live young, which they nurse. The largest marine mammals, baleen whales, live primarily in the open ocean where they filter zooplankton from the water as they swim slowly with their mouths open. Pinnipeds are named for their distinctive swimming flippers and include seals, walruses, and sea lions; many of them live in coastal waters.
- The few reptiles that live in the ocean fall into three groups, the best known being sea turtles. Marine lizards and sea snakes are examples from the other two groups of marine reptiles.
- Unlike most birds on land, seabirds are carnivorous predators that feed at high levels in marine food webs.
- Penguins are flightless marine birds that are primarily confined to the Southern Hemisphere.

ESSAY: Marine Sanctuaries And Reserves

In the United States, a major part of the effort to preserve the natural resources of coastal waters is the *marine sanctuaries* program. At the state level, California has led the way. Point Lobos, CA (near Monterey) was founded in 1960 as the nation's first underwater sanctuary. Since then California has set aside 11 other sites as sanctuaries and has been studying several more potential sites. Although the federal government has moved more slowly than California, federal legislation passed more than 30 years ago holds much promise for the future.

Provisions of the 1972 Marine Protection, Research, and Sanctuaries Act authorize the President to designate national marine sanctuaries in coastal waters of the continental shelf and in the Great Lakes. Any of several characteristics may qualify a locality as a marine sanctuary, but basically, potential sites must have special biological, aesthetic, archaeological, cultural, or historical significance. The objective of federal legislation is to preserve and protect those areas by managing the multiple demands placed on them. Marine sanctuaries are not places of refuge for marine life although activities that threaten marine resources are prohibited.

As of this writing, 13 national marine sanctuaries have been designated. They include near-shore coral reefs, whale migration corridors, deep-sea canyons, and historical sites (see Figure). All are administered by the National Marine Sanctuary Program, a division of NOAA. The first national marine sanctuary, established in 1975, is the wreck site of the Civil War ship, the *USS Monitor*, off the North Carolina coast. The most recent sanctuary, the first in the Great Lakes, is the Thunder Bay National Marine Sanctuary and Underwater Preserve, resting place of about 160 shipwrecks. It is located in

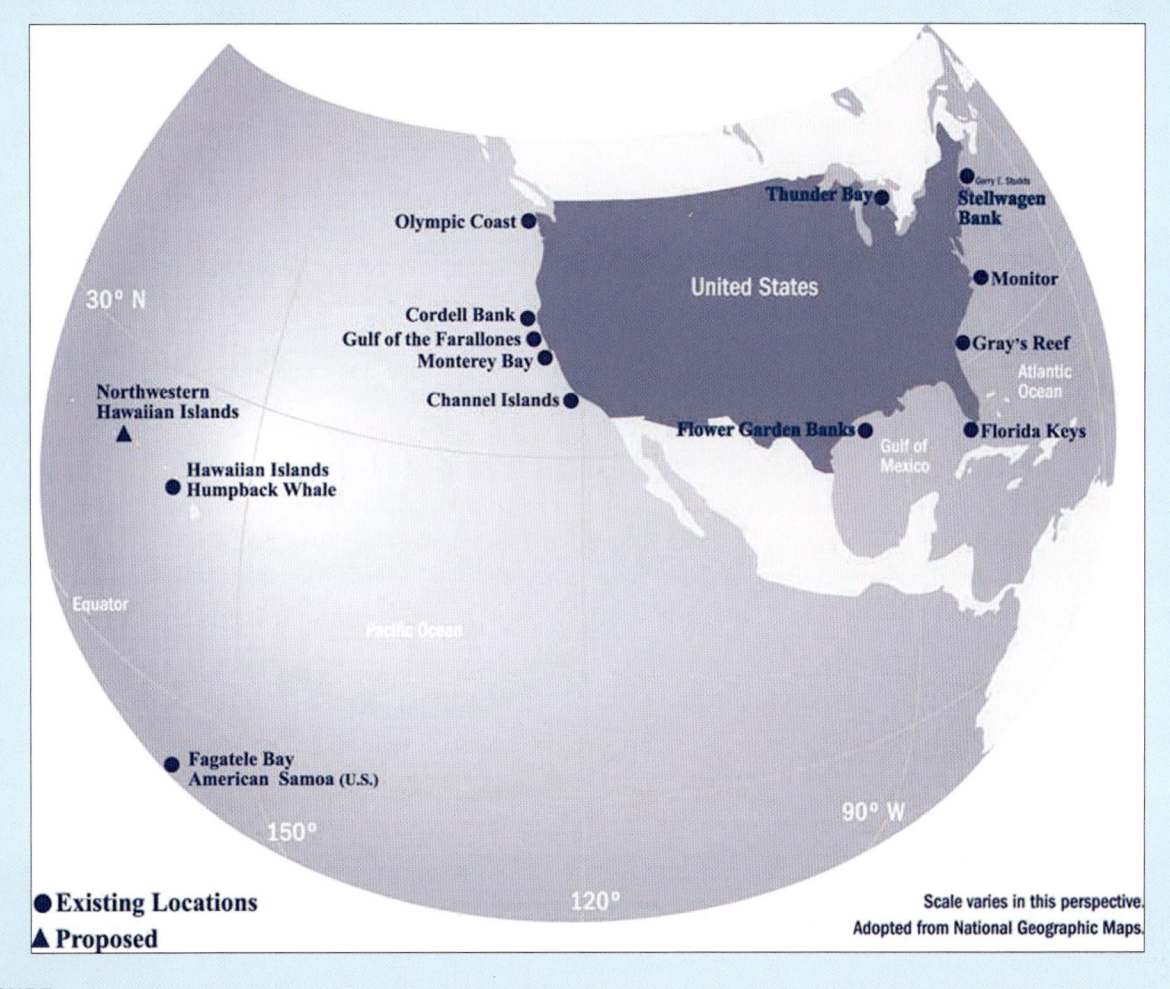

FIGURE
Map showing the location of marine sanctuaries in U.S. waters. [NOAA]

northern Lake Huron near Alpena, MI. The smallest sanctuary is Fagatele Bay, American Samoa, home to a tropical coral reef covering an area of 0.65 square km (0.19 square nautical mi) whereas one of the largest is Monterey Bay, CA with over 13,700 square km (4000 square nautical mi.) Together, the thirteen national marine sanctuaries cover an area along the U.S. coast that is nearly equivalent to the combined areas of the states of New Hampshire and Vermont (more than 18,000 square nautical mi).

Consider some other examples of national marine sanctuaries. The Channel Islands National Marine Sanctuary is located about 40 km (25 mi) off the coast of Santa Barbara, CA and was designated a sanctuary in 1980. Waters surrounding the five islands of the sanctuary are a rich breeding ground for numerous species of plants and animals and home to a kelp forest. Annually, more than twenty-four species of whales and dolphins visit the sanctuary. The islands are also home to seabird colonies and pinniped rookeries. The Florida Keys National Marine Sanctuary encompasses the waters immediately surrounding most of the 1700 islands of the Florida Keys. Designated in 1990, the sanctuary forms a 355-km (220-mi) arc extending from the southern tip of Key Biscayne (south of Miami) southwest to, but not including the Dry Tortugas Islands. This complex of subtropical marine ecosystems includes offshore coral reefs, fringing mangroves, and sea grass meadows home to more than 6000 species of plants, fishes, and invertebrates. The sanctuary covers 9500 square km (2800 square nautical mi).

In November 2000, President Clinton reauthorized the 1972 Marine Protection, Research, and Sanctuaries Act reaffirming the nation's commitment to conserve special areas of the marine environment for the appreciation and enjoyment of present and future generations. However, some conservationists argue that not enough is being done. They point out that at the time of the 1972 legislation, the principal threats to the marine environment were oil spills and the plundering of sunken ships. Since then, over-fishing has emerged as the chief threat to marine resources. Nearly half the U.S. marine fisheries are either depleted or over-fished and, according to the U.S. Department of Commerce, depletion of the fisheries costs the nation's economy billions of dollars in lost revenue each year. Yet, marine sanctuaries permit fishing (except for a few small areas) and most allow recreational boating and mining of some resources—activities that are potentially disruptive of marine habitats.

Conservationists argue for more "no-take" refuges or *marine reserves* where all fishing and activities that threaten marine habitats are prohibited. The goal of marine reserves is to allow recovery of over-fished stocks, preserve seafloor communities or rebuild ones destroyed by trawling. Studies of existing marine reserves show that compared to the surrounding waters, biomass and species diversity are greater and fish are larger within the boundaries of a typical refuge. These ecological benefits develop rapidly and occur regardless of the size of the marine reserve. At present, only about 430 square km (125 square nautical mi) are marine reserves. The most recent and largest is the 253 square km (74 square nautical mi) Tortugas reserve within the Florida Keys National Marine Sanctuary. Individual states or local municipalities protect other small no-take areas. All told, these refuges account for only about 0.01% of U.S. coastal waters.

The fishing industry has opposed any new or expansion of existing marine reserves. Already beleaguered by dwindling fish stocks (some on the brink of extinction), fishers are adamantly against any action that would further reduce the area of open fishing grounds. However, research reported in 2001 demonstrates that marine reserves actually boost populations of fish in the waters outside their borders. Some fishers are now beginning to realize the potential benefits of marine reserves for the fishing industry.

ESSAY: Ecosystem Approach To Fisheries Management[*]

In 1993, the 24th Annual Report of the President's Council on Environmental Quality (CEQ) recommended that the President issue a directive establishing a national policy to encourage sustainable development through ecosystem management. More than ten years later natural resource managers still struggle with this concept. Managers accept that humans do not manage ecosystems but rather it is the human activities that use and impact ecosystems which are managed. Recognizing that, ecosystem management is better referred to as *ecosystem-based management* or an *ecosystem approach to management*. Ecosystem approaches integrate ecological principles, human systems, and goals of sustainability for use in the management decision-making process.

Since the release of the CEQ report, much has been written about the ecosystem approach to management. For marine ecosystems, the concept is arguably best developed for fisheries, with publications in the scientific literature (e.g., Sissenwine and Mace, 2002), a report by the National Research Council (1999), a Congressionally mandated report (Ecosystems Principles Advisory Panel, 1999), reports by non-governmental international "think tanks" (e.g., World Humanity Action Trust, 2000), and a report by United Nations organizations (FAO, 2002). While every author and organization characterizes an ecosystem approach to management somewhat differently, there are several common themes that are captured by the following definition:

> *An ecosystem approach to management is geographically specified adaptive management, which takes account of knowledge and uncertainties about ecosystems (including humankind), and strives to balance multiple societal objectives.* [Modified from Sissenwine and Mace, 2002]

An important point that all of the reports cited above agree on is that an ecosystem approach to management needs to evolve incrementally from existing approaches. Applications are becoming increasingly sophisticated in response to more intense and diverse public interest, with application of new knowledge, and by giving uncertainty explicit consideration. Some key challenges in advancing ecosystem approaches are to:
1. Improve and create processes for public participation in objective setting and in prioritizing objectives;
2. Develop operational protocols for taking into account food web complexity (recognizing that only a crude level of predictability is likely, at best) and climate change; and
3. Determine the functional value of habitat relative ecosystem goods (e.g., food) and services (e.g., absorption of atmospheric carbon dioxide) that benefit humanity.

Through an incremental and collaborative transition to an ecosystem approach, managers can expect to achieve healthy ecosystems with increased social and economic value of its resources as well as an informed public by which the benefits of ecosystem management can be maintained.

---

[*]Becky Allee, NOAA, National Marine Fisheries Service

CHAPTER 11

THE OCEAN, ATMOSPHERE, AND CLIMATE VARIABILITY

Case-in-Point
Driving Question
Earth's Climate System
    Climate Controls
    Role of the Ocean
The Tropical Pacific Ocean/Atmosphere
    Historical Perspective
    Neutral Conditions in the Tropical Pacific
    El Niño, The Warm Phase
    The 1997-98 El Niño
    La Niña, The Cold Phase
    Predicting and Monitoring El Niño and La Niña
    Frequency of El Niño and La Niña
North Atlantic Oscillation
Arctic Oscillation
Pacific Decadal Oscillation
Conclusions
Basic Understandings
ESSAY: Sea Surface Temperature and Drought in Sub-
    Saharan Africa
ESSAY: El Niño in the Past

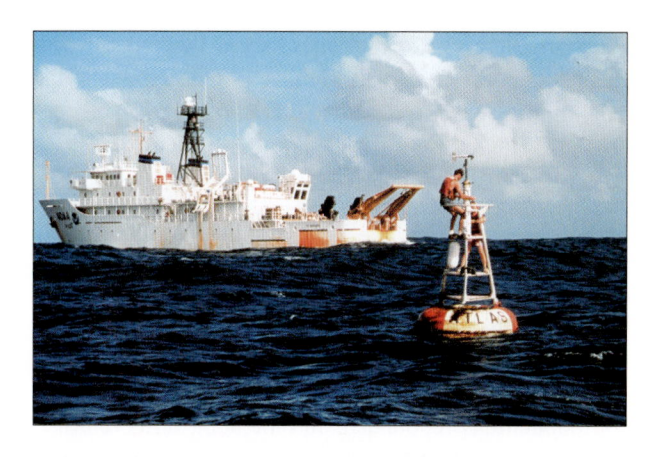

Personnel off NOAA Ship KA'IMIMOANA servicing Atlas Buoy on equatorial Pacific array. These buoys are instrumented to measure ocean temperature at varying depths and give forewarning of El Niño or La Niña events. [Courtesy of NOAA.]

Case-in-Point

In 1982-83, the weather seemed to go wild in many parts of the world and after it was over the total worldwide impacts included thousands of deaths and an estimated $13 billion in property damage. Excessive rains between mid-November 1982 and late January 1983 caused the worst flooding of the century in usually arid Ecuador. Strong winds and torrential rains produced by six tropical cyclones lashed the islands of French Polynesia in a span of only three months. (By comparison, on average one tropical cyclone strikes this region of the eastern South Pacific about every 5 years.) At the other extreme, drought parched eastern Australia, Indonesia, and southern Africa. Huge drought-related wildfires broke out in Australia and Borneo. Australia's drought was that nation's worst in 200 years, causing $2 billion in crop losses and the deaths of millions of sheep and cattle. Meanwhile, drought in sub-Saharan Africa grew worse. Over North America, the winter storm track shifted hundreds of kilometers south of its usual location, bringing episodes of destructive high winds and heavy rains to portions of California. Flooding rains also caused havoc across the southeastern United States. Meanwhile ski resorts in the northern U.S. experienced a snow drought and suffered considerable economic loss.

Just prior to these worldwide weather extremes, the ocean circulation off the northwest coast of South America changed drastically with dire implications for marine productivity. Along the coast of Ecuador and Peru, plankton populations plunged to about 5% of their normal level. The decline in plankton reduced the numbers of anchovy, which feed on plankton, to a record low. Other fishes dependent on plankton, such as jack mackerel, suffered a similar fate. Commercial fisheries off the coast of Ecuador and Peru collapsed. With the decline in fish populations, marine birds (e.g., frigate birds and terns) and marine mammals (e.g., fur seals and sea lions) also experienced major population declines as scarce food supplies caused breeding failures as well as migration or starvation of adults.

The ecological impact of these 1982-83 environmental changes was particularly severe on the remote island of Kiritimati (Christmas Island) at 2 degrees N, 157 degrees W, in the central tropical Pacific. An estimated 17 million seabirds, which feed on fish and squid, normally nest on the island. But scientists arriving on Kiritimati in November 1982 found that nearly all the adult birds had abandoned the island, leaving their young to starve. Apparently, sharp declines in food sources forced fish and squid to search for better feeding grounds, and the adult birds followed their prey. By July 1983, the usual ocean circulation returned, plankton populations rebounded, and the adult birds returned to Kiritimati. Nesting began and seabird populations started to recover.

What caused these drastic changes in atmospheric and oceanic conditions? Early on, some scientists attributed the weather extremes to the violent eruption of the Mexican volcano El Chichón in March-April 1982. But it quickly became apparent that the worldwide weather extremes were linked to large-scale ocean/atmosphere circulation changes in the tropical Pacific and soon a new scientific term was added to the public's vocabulary: El Niño. The El Niño of 1982-83 spurred further research on ocean-atmosphere interactions and the deployment of an array of direct and remote sensing instruments in the tropical Pacific to provide early warning of the development of El Niño. In the late 1990s, when El Niño returned in its full fury, the global community was able to better prepare thereby lessening the impact.

Driving Question:

How do interactions between the ocean and atmosphere impact worldwide weather and short-term climate variability?

Some weather extremes such as drought and episodes of unusually heavy rains are linked to coupled changes in atmospheric and oceanic circulation. The principal focus of this chapter is short-term climatic fluctuations that involve the interaction between the ocean and atmosphere. One of the most intensively studied of these interactions occurs in the tropical Pacific and is known as El Niño/La Niña. During El Niño, trade winds weaken, upwelling diminishes off the South American coast and in the equatorial Pacific, sea-surface temperatures (SST) rise well above long-term averages over the central and eastern tropical Pacific, and areas of heavy rainfall shift from the western into the central tropical Pacific. La Niña sometimes (but not always) follows El Niño and is a period of exceptionally strong trade winds in the tropical Pacific, vigorous coastal and equatorial upwelling in the Pacific, and lower than usual SST in the central and eastern tropical Pacific. Based on changes in SST in the eastern tropical Pacific, some scientists refer to El Niño as the *warm phase* and La Niña as the *cold phase* of this air/sea interaction.

Broad-scale changes in SST patterns over the tropical Pacific that accompany El Niño and La Niña influence the prevailing circulation of the atmosphere in middle latitudes, especially in winter. Weather extremes that most often accompany El Niño are essentially opposite those that usually occur during La Niña. Although we devote much of this chapter to El Niño and La Niña, we also consider other examples of short-term climate variability stemming from air/sea interaction including the North Atlantic Oscillation, the Arctic Oscillation, and the Pacific Decadal Oscillation. We set the stage for this discussion by first describing controls operating within Earth's climate system.

Earth's Climate System

In Chapter 5, we defined **climate** as weather at a particular location averaged over a specific interval of time (by convention, thirty years). A complete description of climate also includes extremes in weather (e.g., highest and lowest temperatures on record). In a more general sense, climate is the state of the Earth-atmosphere system that gives a locality its characteristic weather patterns.

Climate varies both spatially and temporally. The globe is a mosaic of many different climate types including, for example, the hot wet tropics, warm and cold deserts, temperate regions, and polar ice caps. Climate changes over a broad range of temporal scales, from years to decades to centuries to millennia. In this section, we describe the various controls operating in Earth's climate system with emphasis on the role of the ocean.

CLIMATE CONTROLS

Many factors working together shape the climate of any locality. Controls of climate consist of (1) latitude, (2) elevation, (3) topography, (4) proximity to large bodies of water, (5) Earth's surface characteristics, (6) long-term average atmospheric circulation, and (7) prevailing ocean circulation. Over a period of at least several million years, all climate controls are variable. On time scales that extend to hundreds of millions of years, continents have drifted to different latitudes, ocean basins have opened and closed, and mountain ranges have risen and eroded away—all with implications for climate. On shorter time scales (e.g., the range of human existence), for all practical purposes, the first four climate controls are essentially fixed and exert regular and predictable influences on climate.

Seasonal changes in incoming solar radiation, as well as length of daylight, vary with latitude and air and sea surface temperatures respond to those regular variations (Chapter 5). Air temperature drops with increasing elevation and determines whether precipitation falls in the form of rain or snow. Topography can affect the distribution of clouds and precipitation. For example, the windward slopes of high mountain barriers (facing the oncoming wind) usually are wetter than the leeward slopes (facing downwind). The relatively great thermal inertia of large bodies of water (especially the ocean) moderates the temperature of downwind localities, reducing the temperature contrast between summer and winter and lengthening the growing season (Chapter 3). Earth's surface characteristics (e.g., ocean versus land, type of vegetative cover, semi-permanent snow and ice cover) influence the amount of incident solar radiation that is converted to heat and how that heat is used (e.g., raising air temperature, evaporating water).

Atmospheric circulation encompasses the combined influence of all weather systems operating at all spatial and temporal scales ranging from sea breezes to the prevailing winds that encircle the planet. Although strongly influenced by the other climate controls, atmospheric circulation is considerably less regular and less predictable than the others. This variability is especially evident in weather systems such as thunderstorms and hurricanes that are smaller than the planetary scale. Planetary-scale circulation systems (e.g., the prevailing wind belts, subtropical anticyclones), exert a more systematic influence on climate, determining for example, where precipitation is seasonal and the location of the major subtropical deserts.

ROLE OF THE OCEAN

The ocean is a major player in Earth's climate system operating on time scales of days to millennia and spatial scales from local to global. The ocean influences radiational heating and cooling of the planet. Covering about 71% of Earth's surface, the ocean is a primary control of how much solar radiation is absorbed (converted to heat) at the Earth's surface. Also, the ocean is the main source of the most important greenhouse gas (water vapor) and is a major regulator of the concentration of atmospheric carbon dioxide (CO_2), another greenhouse gas.

On an annual average, the ocean absorbs about 92% of the solar radiation striking its surface; the balance is reflected to space. Most of this absorption takes place within about 200 m (650 ft) of the ocean surface with the depth of penetration of sunlight limited by the amount of suspended particles and discoloration caused by dissolved substances. On the other hand, at high latitudes highly reflective multi-year pack ice greatly reduces the amount of solar radiation absorbed by the ocean. The snow-covered surface of sea ice absorbs only about 15% of incident solar radiation and reflects away the rest. At present, multi-year pack ice covers about 7% of the ocean surface with greater coverage in the Arctic Ocean than the Southern Ocean (mostly in Antarctica's Weddell Sea). (The Arctic is an ocean surrounded by continents whereas the Antarctic is a continent roughly centered on the pole and surrounded by ocean. Without an Antarctic continent, much more of the Southern Ocean would be covered by sea ice.)

Recall from Chapter 5 that the atmosphere is nearly transparent to incoming solar radiation but much less transparent to outgoing infrared (heat) radiation. This is the basis of the *greenhouse effect*. Most water vapor, the principal greenhouse gas, enters the atmosphere via evaporation of seawater (Figure 11.1). Carbon dioxide, a lesser greenhouse gas, cycles into and out of the ocean depending on the sea surface temperature, circulation patterns, and biological activity in surface waters (Chapter 9). Cold water can dissolve more carbon dioxide than warm water so that carbon dioxide is absorbed from the atmosphere where surface waters are chilled (at high latitudes) and released to the atmosphere where upwelling brings cool water to the surface and is heated (at low latitudes). Photosynthetic organisms take up carbon dioxide and all organisms release carbon dioxide via cellular respiration.

FIGURE 11.1
The ocean interacts with the atmosphere exchanging heat, water, and gases. The ocean is the primary source of atmospheric water vapor, the principal greenhouse gas. [Photo by J.M. Moran]

The ocean influences the planetary energy budget not only by affecting the radiational heating and cooling of the entire planet, but also by contributing to the non-radiative latent and sensible heat fluxes at the air-sea interface. Recall from Chapter 5 that heat is transferred from Earth's surface to the atmosphere via latent heating (vaporization of water at the surface followed by cloud formation in the atmosphere) and sensible heating (conduction plus convection). On a global average annual basis, about ten times more heat is transferred form the ocean surface to the atmosphere via latent heating than sensible heating.

The ocean and atmosphere are closely coupled. This coupling is most apparent in the ocean's surface waters where temperatures and wind-driven currents respond to variations in atmospheric conditions within hours to days. On the other hand, the deeper basin-scale circulation responds more sluggishly to changes in atmospheric conditions, taking decades to centuries or longer to fully adjust. In turn, ocean currents strongly influence climate. Cold surface currents, such as the California Current, are heat sinks; they chill and stabilize the overlying air, thereby increasing the frequency of sea fogs and reducing the likelihood of thunderstorms. Relatively warm surface currents, such as the Gulf Stream, are heat sources; they supply heat and moisture to the overlying air, destabilizing the air, thereby energizing storm systems. As noted in Chapter 5, ocean surface currents and the global-scale oceanic conveyor belt transport heat from the tropics to higher latitudes.

Although the importance of the ocean in Earth's climate system is apparent, we need to remember that the ocean and atmosphere work together in governing climate. Some recent research underscores the ocean-atmosphere climate connection with somewhat surprising findings. Winters are significantly milder in Western Europe than in Eastern North America at the same latitude. Since at least the mid-1800s, scientists have attributed this climate contrast primarily to the moderating influence of the warm Gulf Stream and North Atlantic Current on Western Europe. However, research results published in late 2002 call into question this assumption. Some scientists argue that the Gulf Stream is not as important as atmospheric circulation in explaining Western Europe's relatively mild winters.

Richard Seager of Columbia University's Lamont-Doherty Earth Observatory and David Battisti of the University of Washington in Seattle, WA reported that the key is the pattern exhibited by the prevailing westerlies. Recall from Chapter 5 that aloft the westerlies blow from west to east in a wave-like pattern of ridges and troughs (Figure 5.20). In winter, a cold pool of air (a *trough*) is anchored over eastern North America resulting in a cold northwesterly flow on the western flank of the trough, but an intrusion of warm air over the North Atlantic (a *ridge*) is associated with a milder southwest flow over Western Europe. That is, in winter, the westerlies tend to blow from the colder northwest over Eastern North America but from the milder southwest over Western Europe. In addition, the relatively great thermal inertia of ocean water means that summer heat persists in the North Atlantic surface waters long after the North American continent has cooled in fall. Hence, southwest winds blowing toward Western Europe cross relatively warm waters and are heated from below. The direction of the prevailing winds over the North Atlantic and Western Europe is primarily responsible for delivering relatively mild air masses in winter. This finding also implies that changes in the direction of the prevailing winds could alter the winter climate of Western Europe.

Patterns of sea surface temperatures exert a strong influence on the location of major features of the atmosphere's planetary scale circulation. When sea surface temperature patterns change so too do the location of those circulation features. For example, changes in the location of the highest sea surface temperatures in the tropical Atlantic affect the north-south shifts of the **intertropical convergence zone (ITCZ)**. The ITCZ is a discontinuous belt of thunderstorms paralleling the equator and marking the convergence of the trade winds of the Northern and Southern Hemispheres. The ITCZ encircles the globe and shifts north and south with the seasonal excursions of the sun—more so over land and less so over the ocean—reaching its most northerly location in July and its most southerly location in January. Displacement of the ITCZ over the tropical Atlantic Ocean affects the timing and amount of rainfall along the east coast of South America from Brazil northward into the Caribbean from March to May and in the western part of sub-Saharan Africa in August and September. With north-south shifts in the location of the highest SST, the latitude of the ITCZ changes, and regional rainfall patterns also vary.

Changes in sea surface temperature (SST) patterns in the eastern tropical Atlantic have been implicated in the disastrous multi-decade droughts that have afflicted sub-Saharan Africa. For more on this, see this chapter's first Essay.

The Tropical Pacific Ocean/Atmosphere

Scientists have been aware of short-term (inter-annual) variations in climate at many locations for more than a century. One of the regions where these inter-annual variations in the Earth-atmosphere system are readily apparent is the tropical Pacific. There ocean/atmosphere conditions varying on a quasi-periodic basis are identified as El Niño and La Niña. Within the last several decades, the scientific community has realized that these episodes can cause weather extremes not only in the tropical Pacific but also in many other parts of the world.

HISTORICAL PERSPECTIVE

Originally, El Niño was the name given by fishermen to a seasonal period of an unusually warm southward flowing ocean current and poor fishing off the coast of Peru and Ecuador that often coincided with the Christmas season. (*El Niño* is the Spanish reference to the Christ child). Typically, these warm water episodes are relatively brief, lasting perhaps a month or two, before sea-surface temperatures and the fisheries return to normal levels. Every three to seven years, however, El Niño persists for 12 to 18 months or even longer and is accompanied by significant changes in SST over vast stretches of the tropical Pacific, major shifts in planetary-scale oceanic and atmospheric circulations, and collapse of important South American fisheries. Today, oceanic and atmospheric scientists reserve the term **El Niño** for these long-lasting ocean/atmosphere anomalies.

An important step in understanding El Niño came in 1924 with the discovery of the so-called southern oscillation by the Englishman Sir Gilbert Walker (1868-1958). Monsoon failure in 1899-1900 caused terrible famines in India with the loss of a million lives. In 1904, Walker was appointed director general of observatories in India and charged with developing a method to predict the Indian monsoon. He set out on an extensive search for any possible relationship between monsoon rains and weather conditions in various parts of the world. One of his discoveries was the **southern oscillation**, a seesaw variation in air pressure across the tropical Indian and Pacific Oceans. When air pressure was low over the Indian Ocean and the western tropical Pacific, it was high east of the international dateline in the eastern tropical Pacific. During these conditions, monsoon rains were plentiful over India. With the opposite pressure pattern (high pressure west of the dateline and low pressure east of the dateline), monsoon rains were lighter than usual. Today, the *southern oscillation index (SOI)* is based on the difference in air pressure between Darwin (on the north coast of Australia at 12 degrees S, 130 degrees E) and Tahiti (an island in the central south Pacific at about 18 degrees S, 149 degrees W). When air pressure is anomalously low at Darwin, air pressure is anomalously high at Tahiti. Conversely, when air pressure is high at Darwin, it is low at Tahiti (Figure 11.2).

The broader significance of Walker's discovery of the southern oscillation was not fully recognized for more than four decades. In 1966, the Norwegian meteorologist Jacob Bjerknes, while at the University of California at Los Angeles, demonstrated a relationship between El Niño and the southern oscillation. The air pressure difference (*air pressure gradient*) across the tropical Pacific changes as air pressure to the west rises and air pressure to the east falls (and vice versa). Using oceanic/atmospheric observations gathered from the tropical Pacific during the International Geophysical Year of 1957-58, Bjerknes found that an El Niño episode begins when the air pressure gradient across the tropical Pacific begins to weaken, heralding the slackening of the trade winds. Scientists refer to this link between El Niño and the Southern Oscillation by the acronym **ENSO**.

Not until the El Niño of 1982-83, one of the two most intense of the 20th century, did the scientific community and general public fully realized the potential worldwide impact of ENSO. Some of the effects of the 1982-83 El Niño are described in this chapter's Case-in-Point. That event spurred development of numerical models to simulate ENSO as well as deployment of a network of instrumented buoys and satellites to provide early warning of a developing El Niño. Also, the last two decades have seen

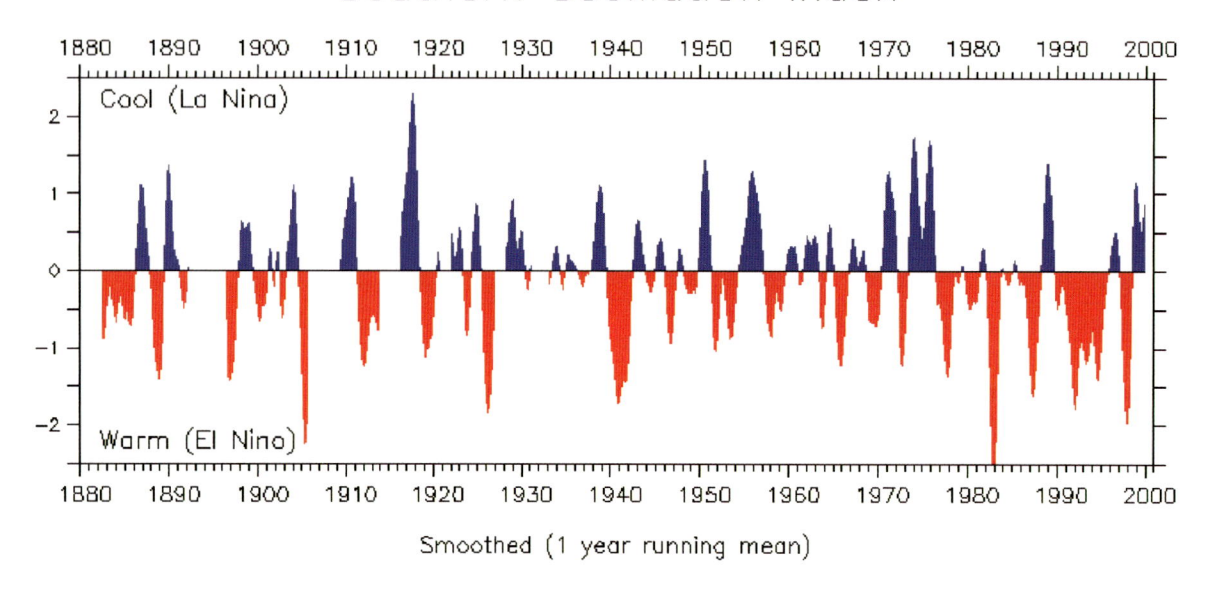

Southern Oscillation Index

FIGURE 11.2
The Southern Oscillation Index (1880s to 1990s) based on the sea level pressure at Tahiti minus the sea level pressure at Darwin, Australia, divided by its standard deviation. Strongly positive values indicate La Niña conditions and strongly negative values indicate El Niño conditions. [NOAA, Pacific Marine Environmental Laboratory, Seattle, WA]

increasing interest in **La Niña** (*the girl*), the name coined in the mid-1980s for an ocean/atmosphere interaction that is essentially the opposite of El Niño. Some scientists refer to the warm El Niño and cold La Niña as opposite extremes of the *ENSO cycle*.

So important are El Niño and La Niña in year-to-year climate variability that signs of a developing El Niño or La Niña are now routinely incorporated into long-range seasonal weather outlooks worldwide. Such outlooks identify areas of expected anomalies in temperature and precipitation and make possible the development of regional agricultural and water management strategies. Adoption of these strategies helps lessen the impact of attendant weather extremes on water supply and food production in various parts of the world.

NEUTRAL CONDITIONS IN THE TROPICAL PACIFIC

El Niño and La Niña represent departures from the long-term average or *neutral* ocean/atmosphere conditions in the tropical Pacific. Understanding El Niño and La Niña first requires a look at neutral conditions. The prevailing trade winds impact ocean currents, sea-surface temperatures, and rainfall across the tropical Pacific.

Most of the time, southerly or southwesterly winds blowing along the west coast of South America drive warm surface waters to the left (westward), via Ekman transport, away from the coast (Chapter 6). This causes cold, nutrient-rich waters to well up from below the pycnocline, which is normally only 50 to 100 m (160 to 325 ft) deep along the coast, and replace the warm, nutrient poor surface waters that are transported offshore. Although this zone of coastal upwelling is narrow (typically less than 15 km or 10 mi wide), the abundance of nutrients conveyed into the photic zone spurs an explosive growth of phytoplankton populations for an additional 15 km (10 mi) offshore. Those populations in turn, support a diverse marine ecosystem and highly productive fishery. Peru's fishing industry boomed in the 1950s and 1960s and by 1970 was one of the largest in the world. At that time Peru was responsible for about 20% of the total global catch of anchovies. Most of the catch was exported for poultry feed supplement and accounted for almost one-third of the nation's foreign exchange income. However, a combination of over-fishing plus the effects of the 1972-73 El Niño sent the Peruvian fishery into a tailspin from which it has not recovered.

Upwelling is also responsible for the tongue of relatively cool surface waters along the equator in the eastern tropical Pacific. Near the equator, the northeast trade winds of the Northern Hemisphere converge with the southeast trade winds of the Southern Hemisphere. The associated Ekman transport (although weak because of minimal Coriolis effect) causes surface waters to diverge away from the equator and colder water wells up from below replacing the departing surface waters. *Equatorial upwelling* produces a strip of relatively low sea surface temperatures along the equator from the coast of South America westward to near the international dateline (180 degrees longitude).

Meanwhile, these same trade winds drive a pool of relatively warm surface waters westward toward Indonesia and northern Australia (both the north and south equatorial current). The wedge of warm water increases the depth of the pycnocline and raises sea level in the western tropical Pacific. The pycnocline is at a depth of about 150 m (490 ft) in the western tropical Pacific and shoals to about 50 m (165 ft) depth in the eastern tropical Pacific. Water expands when heated (as well as being piled up by trans-Pacific trade winds) so that sea level is some 60 cm (2 ft) higher in the west than in the east. The contrast in sea surface temperature between the western and eastern tropical Pacific (averaging about 8 Celsius degrees or 14.5 Fahrenheit degrees) has important implications for precipitation across the tropical Pacific (Figure 11.3). Relatively cool surface waters in the central and eastern tropical Pacific chill the overlying air and suppress convection so that rainfall is light in that region and along the adjacent western coastal plain of South America. Over the western tropical Pacific, warm surface waters heat the overlying air, strengthening convection that gives rise to heavy rainfall.

The contrast between relatively high air pressure over the central and eastern tropical Pacific and relatively low air pressure over the western tropical Pacific ultimately drives the trade winds. Winds initially blow from regions where air pressure is relatively high toward regions where air pressure is relatively low. The greater the air pressure contrast (i.e., horizontal pressure gradient), the stronger are the winds. Once in motion, the Coriolis effect deflects winds to the right in the Northern Hemisphere and to the left in the Southern Hemisphere. Trade winds blow from the northeast in the Northern Hemisphere and from the southeast in the Southern Hemisphere. As long as high pressure and low SST persist over the eastern Pacific basin and low pressure and high SST persist over the western Pacific basin, the trade winds are strong and upwelling remains vigorous along the coast of Ecuador and Peru and the equator east of the international dateline.

High SST in the western tropical Pacific lower the surface air pressure whereas low SST in the eastern

Tropical Rainfall Measuring Mission (TRMM)

Five - Year TRMM Climatology

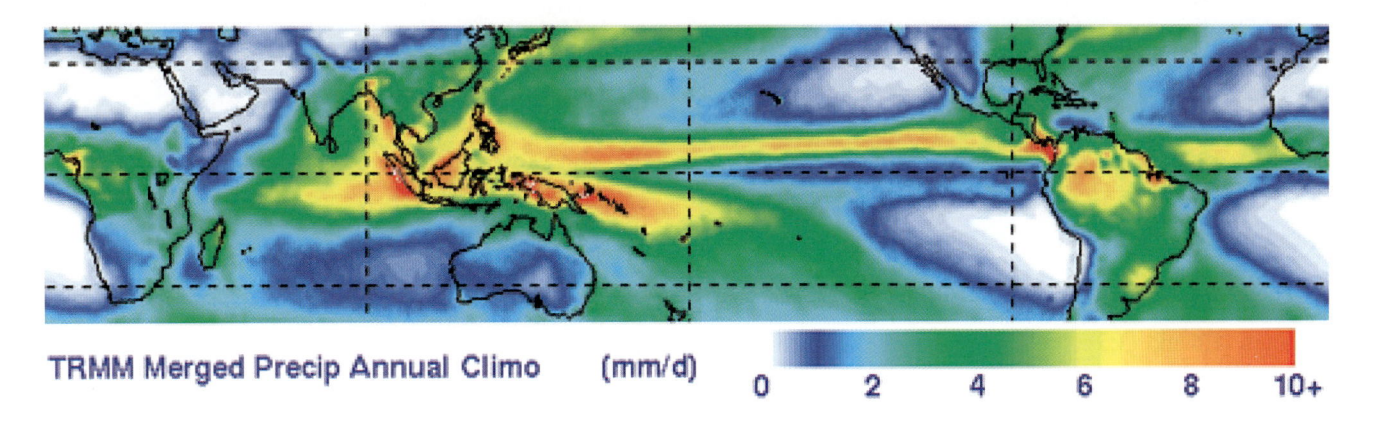

January 1998 - December 2002

FIGURE 11.3
Average rainfall in millimeters per day across the tropical Pacific Ocean for the five-year period January 1998 through December 2002. The heaviest rainfall is in the western tropical Pacific. These data were obtained from the TRMM Microwave Imager and IR sensors onboard geosynchronous satellites supplemented by conventional rain gauge measurements.
[Source: Tropical Rainfall Measuring Mission (TRMM), NASA]

tropical Pacific raise the surface air pressure. Hence, during neutral conditions the east-west SST gradient reinforces the trade winds by strengthening the east-west pressure gradient. In flowing over the ocean surface, the trade winds become warmer and more humid. In the western tropical Pacific warm humid air rises, expands, and cools. Water vapor condenses into towering cumulonimbus (thunderstorm) clouds that produce heavy rainfall. Aloft this air flows back eastward and sinks over the cooler waters of the eastern tropical Pacific. Sinking air is compressed and warmed so that clouds vaporize or fail to develop. This completes the large convective-type circulation known as the *Walker Circulation*, named for Sir Gilbert Walker.

EL NIÑO, THE WARM PHASE

With the onset of El Niño, air pressure falls over the eastern tropical Pacific and rises over the western tropical Pacific as part of the southern oscillation. The air pressure gradient across the tropical Pacific weakens and the trade winds slacken in the western and central equatorial Pacific. During a particularly intense El Niño, trade winds west of the international dateline may reverse direction and blow toward the east.

In response to these shifts in atmospheric circulation over the tropical Pacific, changes occur in surface ocean currents, SST, sea level, depth of the pycnocline, and upwelling (Figure 11.4). With relaxation of the trade winds, the westward flow of the equatorial currents weakens and at times reverses direction. Hence, the thick layer of warm surface water normally in the west drifts slowly eastward across the tropical Pacific until deflected toward the north and south by the continental landmasses, sometimes reaching as far north as the coast of western Canada and as far south as central Chile. Eastward drift of warm water is so slow that it may take several months to reach the west coasts of North and South America. In the western tropical Pacific, SSTs drop, sea-level falls, and the pycnocline rises whereas in the eastern tropical Pacific, SSTs rise, sea-level climbs, and the pycnocline deepens (Figures 11.4 and 11.5).

The impacts of these environmental changes can be severe on marine ecosystems. Arrival of warm surface waters in the eastern tropical Pacific effectively blocks upwelling along the coast of Ecuador and Peru. Deprived of nutrients, phytoplankton populations decline and the commercial fish harvest plummets. Warmer surface waters can also severely stress coral reefs living in shallow tropical waters. In response to unusually high sea surface temperatures, coral expels zooxanthallae, the symbiotic

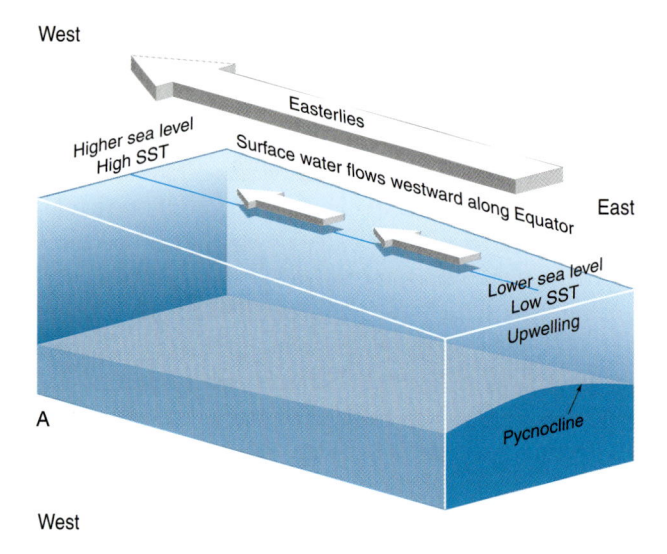

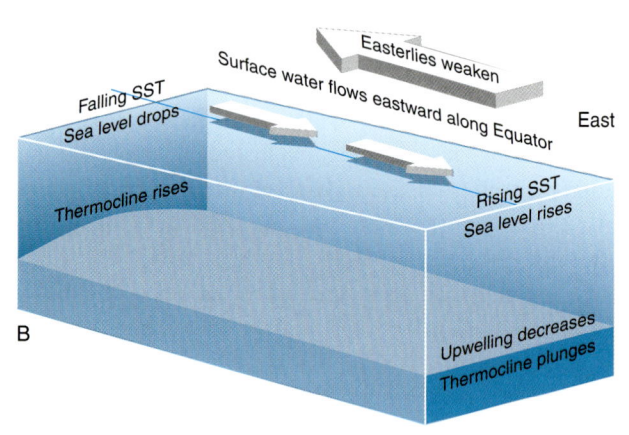

FIGURE 11.4
Block diagram showing ocean/atmosphere conditions in the tropical Pacific during (A) neutral conditions, and (B) El Niño.

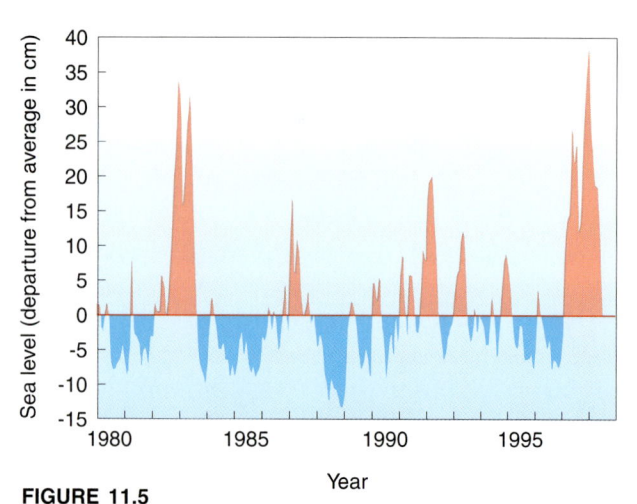

FIGURE 11.5
Sea level record at Galapagos in the eastern tropical Pacific based on tide gauge records and expressed in cm as departure from the long-term average. Relatively high sea levels correspond to El Niño episodes. [From NOAA PMEL]

microscopic algae that supply coral with oxygen and some organic compounds produced through photosynthesis. Recall from Chapter 10 that without zooxanthallae, coral polyps have little pigmentation and appear nearly transparent on the coral's white skeleton, a condition referred to as *coral bleaching*. Excessive bleaching can kill coral polyps, destroying the habitat for a variety of marine life. Coral bleaching also has serious economic impact on local fisheries and tourism. Extensive coral bleaching was reported during the 1997-98 El Niño.

During El Niño, lower than usual SST in the western tropical Pacific and higher than usual SST in the central and eastern tropical Pacific coupled with the change in trade wind circulation give rise to anomalous weather patterns in the tropics and subtropics. Long-term average winds blow onshore over Indonesia so that rainfall is normally abundant, but during El Niño, prevailing winds are offshore and the weather of Indonesia is dry. El Niño-related droughts may also grip India, eastern Australia, northeastern Brazil, and southern Africa. Meanwhile, warmer than usual surface waters off the west coast of South America spur convection and heavy rainfall along the normally arid coastal plain, causing flash flooding. Wetter conditions tend to occur in southern Brazil, Uruguay, and equatorial East Africa.

Apparently, El Niño also influences the intensity, frequency, and spatial distribution of tropical cyclones (e.g., hurricanes). Stronger than usual winds aloft (the subtropical jet stream) tend to inhibit the development of tropical cyclones over the Atlantic Basin (Chapter 8). Those few that do develop are usually weaker and shorter lived than usual. In the Pacific and Indian Oceans, changes in SST that accompany El Niño appear to alter the intensity and spatial distribution of tropical cyclones rather than their frequency. Because of the extensive area of warmer water over the eastern tropical Pacific, hurricanes that form there can travel farther north and west.

El Niño usually brings dry weather to the Hawaiian Islands. Almost all of Hawaii's major droughts during the 20th century coincided with an El Niño event. As part of the usual atmospheric circulation changes that take place during El Niño, the North Pacific subtropical anticyclone shifts so that the associated region of descending air moves closer to the Islands. The greater than usual frequency of sinking air over Hawaii is responsible for a persistent dry weather pattern.

El Niño also has a ripple effect on the weather and climate of middle latitudes, especially in winter. A linkage between changes in atmospheric circulation occurring in widely separated regions of the globe, often thousands of kilometers apart, is known as a **teleconnection**. What causes these teleconnections? Latent heat released into the atmosphere during deep convection and the buildup of thunderstorms in the tropical troposphere is one of the major controls of the planetary-scale circulation. Changes in the location of these heat sources alter wind and weather patterns worldwide. Higher than usual SST over the central and eastern tropical Pacific during El Niño heats and destabilizes the troposphere. Deep convection generates towering thunderstorms that help drive atmospheric circulation, governing the course of jet streams, storm tracks, and moisture transport at higher latitudes.

Events in the ocean and atmosphere are somewhat analogous to the effect of large boulders on a swiftly flowing stream. Boulders in the streambed induce a train of turbulent eddies that extends downstream. Moving the boulders also displaces the train of eddies. Thunderstorm clouds building high into the tropical troposphere deflect the upper air winds in much the same manner as boulders in a stream; that is, a shift in location of the principal area of convection eastward over the tropical Pacific redirects the atmospheric circulation.

During typical El Niño winters, prevailing storm tracks bring abundant rainfall and cooler than usual conditions to the Gulf Coast states, from Texas to Florida. Over the northern U.S. and Canada, prevailing winds tend to blow from west to east so that extremely cold air masses tend to move eastward across the Arctic and northern Canada. Persistence of this circulation pattern prevents exceptionally cold air masses from invading the northern U.S so that mild weather prevails over much of Canada, Alaska, and parts of the northern U.S. West-to-east flow in the westerlies also diminishes the usual spring contrast between warm, humid air masses moving northeastward from the Gulf of Mexico and cold, dry air masses sweeping southeastward from Canada. Consequently, fewer severe thunderstorms and tornadoes break out in the Ohio and Tennessee River Valleys.

Although some weather extremes almost always accompany El Niño, no two events are exactly the same because El Niño is only one of many factors that influence inter-annual variations in climate. In southern California, for example, heavy winter rains (snows at higher elevations) have occurred during some but not all El Niño events. Record heavy rainfall in Southern California during January 1995 was linked to a shift of the jet stream (and storm track) south of its usual position over the eastern Pacific. In that case, a change in atmospheric circulation associated with El Niño was the culprit. Whereas the 1982-83 El Niño brought severe drought to eastern Australia, dry conditions in Australia during the 1997-98 El Niño were far less severe.

THE 1997-98 EL NIÑO

The 1997-98 El Niño rivaled its 1982-83 predecessor as the most intense of the 20th century (Figure 11.6). Weather extremes and disease outbreaks (related to rainfall and standing water, such as malaria and cholera) worldwide claimed an estimated 22,000 lives and caused $36 billion in economic losses. This El Niño developed rapidly with the trade winds weakening and eventually reversing direction in the western tropical Pacific in early 1997. Equatorial upwelling ceased during the Northern Hemisphere summer of 1997. A pool of exceptionally warm surface waters (SST greater than 29 °C or 84 °F) migrated eastward from the western tropical Pacific. During the Northern Hemisphere fall of 1997, SSTs over the eastern tropical Pacific were at least 5 Celsius degrees (9 Fahrenheit degrees) above the long-term average. Warming was rapid in the eastern tropical Pacific with SST setting new record highs each successive month from June through December 1997. By late 1997, the pycnocline essentially flattened across the tropical Pacific basin, rising some 20 to 40 m (65 to 130 ft) in the west and falling more than 90 m (300 ft) in the east. Somewhat lower surface water temperatures in the west and much higher than usual surface water temperatures in the east weakened the usual east-west SST gradient. By further reducing the east-west pressure gradient, the weaker SST gradient contributed to additional slackening of the trade winds.

The 1997-98 El Niño came to an abrupt end in mid-May 1998. Trade winds strengthened rapidly, upwelling resumed along the equator and off the northwest coast of South America, and SST over the eastern tropical Pacific plummeted in response to upwelling of very cold water. At one location along the equator (125 degrees W), the SST fell 8 Celsius degrees (14 Fahrenheit degrees) in only four weeks.

LA NIÑA, THE COLD PHASE

Sometimes, but not always, La Niña follows El Niño. **La Niña** is a period of unusually strong trade winds and exceptionally vigorous upwelling in the eastern tropical Pacific. Like its warm counterpart, La Niña tends to persist for 12 to 18 months. During La Niña, SST anomalies (departures from the long-term average) are essentially opposite those observed during El Niño; that is, surface waters are colder than usual over the central and eastern tropical Pacific and somewhat warmer than usual over the western tropical Pacific. SST anomalies over the eastern tropical Pacific typically have a greater magnitude during El Niño than during La Niña. SST usually rise about 5 to 6 degrees Celsius (9 to 11 Fahrenheit degrees) above the long-term average during an intense El Niño and drop 2 to 3 Celsius degrees (3.6 to 5.4 Fahrenheit degrees) below the long-term average during a strong La Niña.

Accompanying La Niña are worldwide weather extremes that are often opposite those observed during El Niño. And as with El Niño, the most consistent mid-latitude teleconnections appear in winter. In the tropical Pacific, lower than usual SST inhibit rainfall in the east and higher than usual SST enhance rainfall in the west, including Indonesia, Malaysia, and northern Australia during the Northern Hemisphere winter and the Philippines during the Northern Hemisphere summer. Elsewhere around the globe, the Indian monsoon rainfall (in summer) tends to be heavier than average (especially in northwest India) and wet conditions prevail over southeastern Africa and northern Brazil (during the Northern Hemisphere winter). Southern Brazil to central Argentina experience a dry winter. Finally, during La Niña, weak winds aloft favor tropical cyclone formation over the Atlantic.

Across middle latitudes of the Northern Hemisphere, westerlies tend to be more *meridional* (i.e., more north-south) during La Niña; that is, prevailing winds tend to encircle the globe from west to east in great north-south loops. These winds steer cold air masses toward the southeast and warm air masses toward the northeast. Occasionally, a meridional flow pattern becomes so extreme that a broad pool of rotating air essentially separates from the main current and remains over the same geographical area for many weeks to months. (This is somewhat analogous to a whirlpool that forms in a swiftly flowing stream.) A pool of air rotating in a clockwise direction (viewed from above) is known as a *cutoff anticyclone* (or *high*). For as long as a cutoff high persists over the same area, the weather remains dry and the probability of drought increases. Just such an atmospheric circulation pattern was responsible for the severe summer drought that afflicted the central U.S. during the La Niña year of 1988.

In spring, a more meridional flow pattern in the westerlies increases the likelihood of severe thunderstorms and tornadoes across the central U.S. Such a flow pattern brings together air masses that contrast in temperature and humidity, a key ingredient for severe weather development. Also in the U.S., La Niña tends to be accompanied by be-

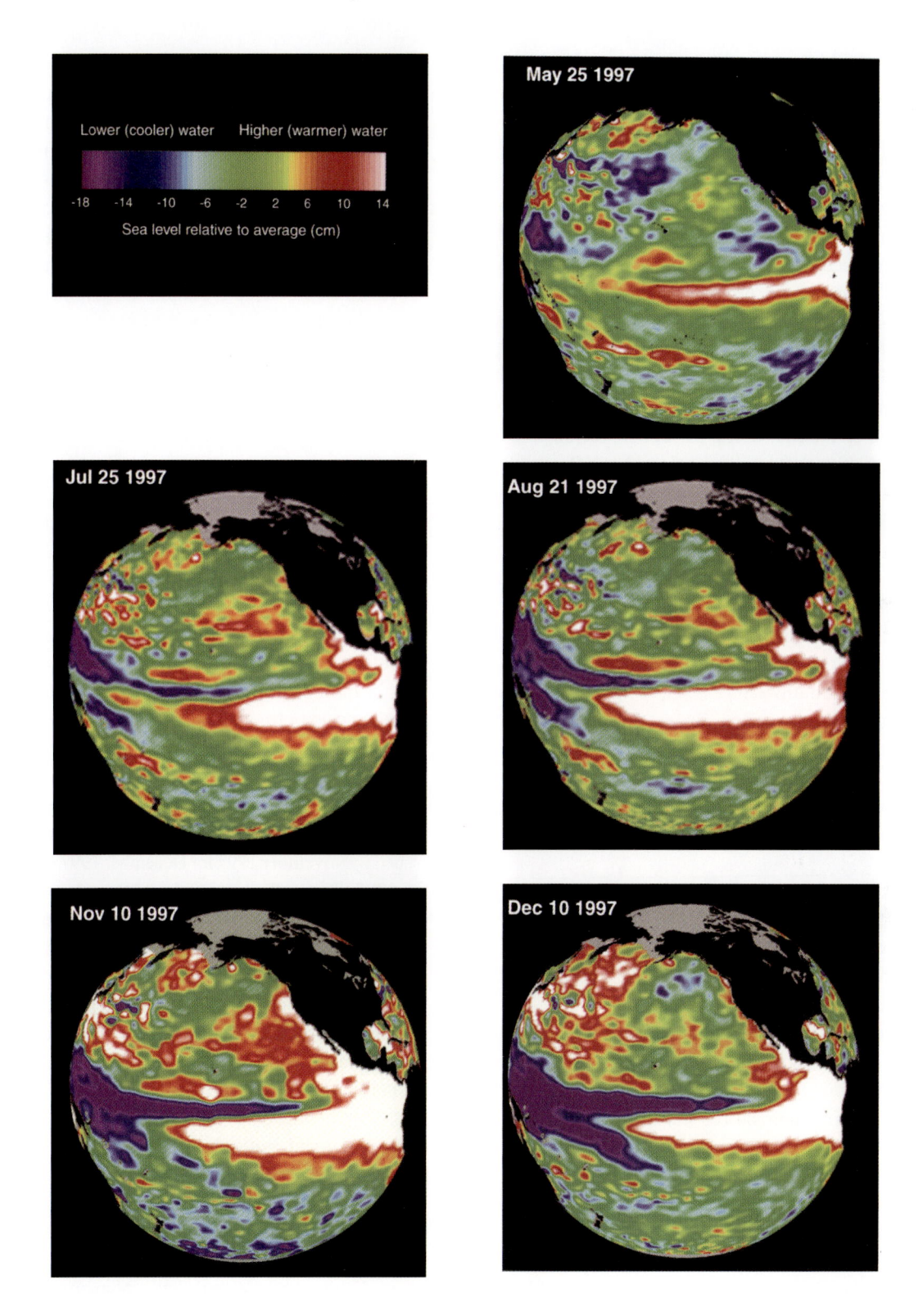

FIGURE 11.6
Evolution of the 1997-98 El Niño derived from changes in ocean surface height (compared to long-term average) as measured by the TOPEX/Poseidon satellite. On the color scale, whites and reds indicate elevated areas (and warmer than normal water). In the white areas, the sea surface is 14 to 32 cm (6 to 13 in.) above normal; in the red areas it is about 10 cm (4 in.) above normal. Green indicates normal sea level whereas purple corresponds to areas that are at least 18 cm (7 in.) below normal sea level (and colder than normal water). [NASA Goddard Space Flight Center]

low average winter precipitation and mild temperatures in a band from the Southwest, through the central and southern Rockies, and eastward to the Gulf Coast. Winter in the Pacific Northwest tends to be cool and wet. Lower than usual winter temperatures also occur in the northern Intermountain West and north-central states.

Changes in the atmospheric circulation pattern associated with a recent La Niña were likely responsible for a persistent cut-off high that brought drought to southwestern Asia, from Iran eastward to western Pakistan. From 1998 to 2001, Afghanistan, near the center of the dry belt, experienced its most severe drought in 50 years, compounding the human misery brought on by decades of political instability and civil strife. Heidi M. Cullen, a climatologist with the National Center for Atmospheric Research (NCAR) in Boulder, CO proposes that changes in the atmospheric circulation pattern were linked to exceptionally low SSTs in the central tropical Pacific and unusually high SSTs in the western tropical Pacific. These La Niña conditions brought heavy winter rains to the eastern Indian Ocean (e.g., Malaysia), a shift in the location of the subtropical jet stream, and development of a cut-off high over southwestern Asia. The cut-off high blocked the usual west to east movement of storm systems bringing an extended period of dry weather to portions of southwestern Asia. Cullen cautions, however, that not all La Niña events bring drought to the Iran-Afghanistan-Pakistan region.

PREDICTING AND MONITORING EL NIÑO AND LA NIÑA

Scientists have developed numerical models that simulate El Niño and La Niña. These models approximate oceanic processes that alter sea surface temperatures, and the atmospheric response, including convection, clouds, and winds. Forecasters rely on two basic types of numerical models to predict the onset, evolution, and decay of El Niño (or La Niña): empirical (or statistical) models and dynamical models. An *empirical model* compares current and evolving oceanic and atmospheric conditions with comparable observational data for periods preceding El Niño (or La Niña) episodes over the prior 40 years. A match or at least some resemblance between past and present conditions is the basis for a prediction. A *dynamical model* consists of a series of mathematical equations that simulates interactions or coupling among atmosphere, ocean, and land. These so-called *coupled* models are more sophisticated than empirical models and run on supercomputers.

Two of the earliest El Niño models were developed at Scripps Institution of Oceanography in La Jolla, CA and at Columbia University's Lamont-Doherty Earth Observatory in Palisades, NY. The Scripps model is empirical whereas the Lamont-Doherty model is a coupled dynamical model. The Lamont-Doherty model successfully forecasted the 1986-87 El Niño and both models gave advance warning of the 1991-92 El Niño. But neither model successfully predicted the rapid development and intensity of the 1997-98 El Niño.

An assessment of the performance of 12 models (6 empirical and 6 dynamical) in predicting the onset, evolution, and demise of the 1997-98 El Niño was somewhat disappointing. The dynamical model operated by the Climate Prediction Center (CPC) of NOAA's National Centers for Environmental Prediction (NCEP) in Camp Springs, MD performed well in predicting the onset of the 1997-98 El Niño. But the NCEP and all other models greatly underestimated (by about 50%) the magnitude of sea surface warming in the eastern tropical Pacific and demonstrated little skill in forecasting the decay of El Niño. The NCEP model predicted gradual cooling in the eastern tropical Pacific through 1998 when in fact sea-surface temperatures plunged in May 1998.

Official forecasts of an impending El Niño appeared in the spring of 1997 (after warming of the eastern tropical Pacific had already begun). Model output strongly influenced the Climate Prediction Center's winter outlook issued in November 1997 for December 1997 through February 1998. That outlook correctly predicted heavier than usual precipitation across the southern U.S. and anomalous warmth over the northern one-third of the nation. Predictions of unusually dry conditions in Indonesia, northern South America, and southern Africa were verified. Above average rainfall predicted for Peru, east Africa, and northern Argentina was also correct. But the extreme drought expected for northeast Australia did not materialize.

Reliable observational data from the tropical Pacific Ocean and atmosphere are essential for detecting a developing El Niño or La Niña, and for initializing numerical models. The accuracy of dynamical models depends not only on how well their component equations simulate the coupled ocean/atmosphere/land system, but also reliable observational data. Numerical models use those initial conditions as a starting point for predicting future states of the ocean and atmosphere and for verifying the model predictions and results. Data are also assimilated into model runs to correct or "nudge" the model as events evolve.

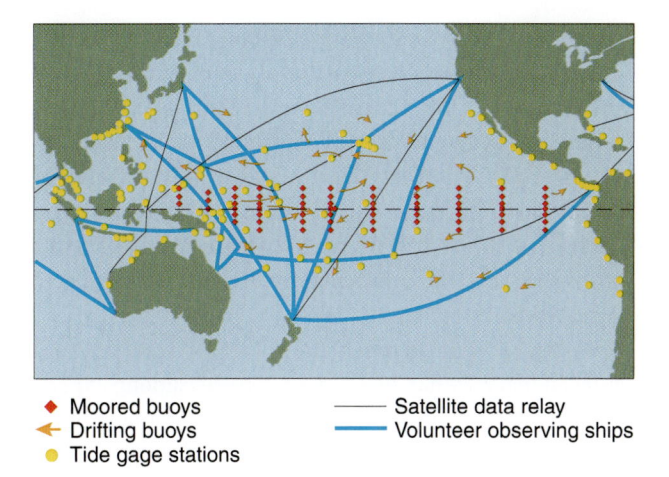

FIGURE 11.7
Components of the ENSO Observing System provide advance warning and monitor the development and decay of El Niño and La Niña events. [NOAA, Pacific Marine Environmental Laboratory, Seattle, WA]

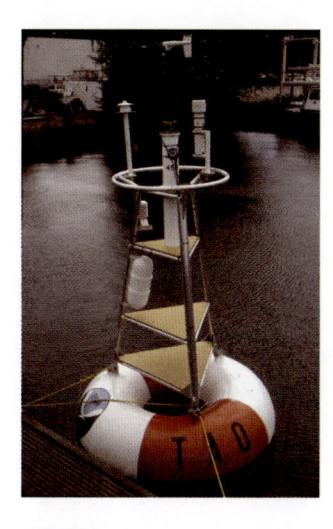

- Moored buoys
- Drifting buoys
- Tide gage stations
- Satellite data relay
- Volunteer observing ships

FIGURE 11.8
An instrumented TAO moored buoy photographed at NOAA's Pacific Marine Environmental Laboratory in Seattle, WA. An array of similar moored buoys gathers oceanic and atmospheric data from the tropical Pacific as part of the ENSO Observing System. [Photo by J.M. Moran]

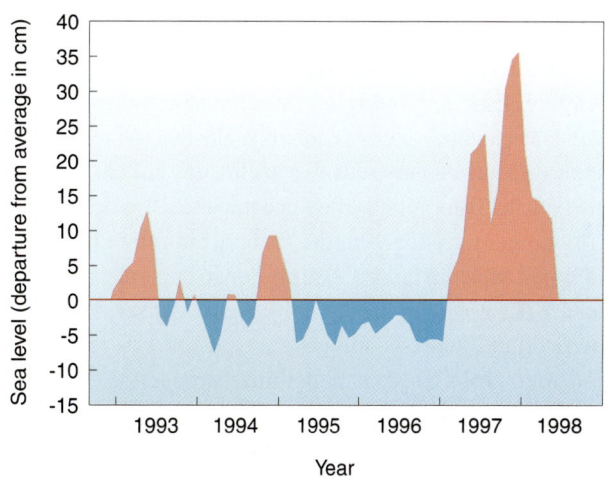

FIGURE 11.9
Sea level record at a location along the equator in the eastern tropical Pacific derived from measurements made by the TOPEX/Poseidon satellite. Sea level is expressed in cm as departure from the long-term average. Relatively high sea levels correspond to El Niño episodes. [From NOAA PMEL]

A major reason why dynamical models performed well in detecting the onset of the 1997-98 El Niño was the increasing amount of ocean/atmosphere observational data from the tropical Pacific. Those data were obtained by monitoring systems deployed as part of the 10-year (1985-94) international *Tropical Ocean Global Atmosphere (TOGA)* study. TOGA was aimed at improving understanding, detection, and prediction of ENSO-related variability. An important product of TOGA, the *ENSO Observing System,* was fully operational by December 1994. The ENSO Observing System consists of an array of moored and drifting buoys, island and coastal tide gauges, ship-based measurements, and satellites (Figure 11.7).

One component of the ENSO Observing System is the *TAO (Tropical Atmosphere/Ocean)* array of moored buoys (small, unmanned, instrumented platforms) in the tropical Pacific Ocean (Figure 11.8). Buoys are strategically placed over an area bounded by 8 degrees N, 8 degrees S, 95 degrees W, and 137 degrees E. This instrument array, renamed TAO/TRITON in 2000, presently consists of approximately 70 deep-sea moorings that measure several atmospheric variables (air temperature, wind, relative humidity) as well as oceanic parameters (sea surface and subsurface temperatures at 10 depths in the upper 500 m or 1650 ft). Several newer moorings also have salinity sensors, along with additional meteorological sensors. Five moorings along the equator also measure ocean current velocity using Subsurface Acoustic Doppler Current Profilers. Observational data are telemetered to NOAA's Pacific Marine Environmental Laboratory (PMEL) in Seattle, WA via a NOAA polar-orbiting satellite. Those data are available on the Internet in near real-time.

Remote sensing by satellite plays an important role in providing early warning of an evolving El Niño or La Niña. NOAA and NASA satellites monitor cloud cover and map sea surface temperatures. The TOPEX/Poseidon satellite, a joint mission of NASA and the Centre National d'Etudes Spatiales in France, launched in 1992, provides images of ocean surface topography (Figures 11.6 and 11.9). Radar (microwave) altimeters onboard the satellite (orbiting at an altitude of 1340 km or 830 mi) bounce microwaves off the ocean surface to obtain precise measurements of the distance between the satellite and the sea surface. These data are combined with data from the Global Positioning System (GPS) locations, to produce images of sea surface height. Elevated topography (hills) indicates warmer than usual water whereas areas of low topography (valleys) indicate colder than usual water. Such images can be used to calculate surface ocean currents through the geostrophic assumption. In December 2001, NASA and its French counterpart launched Jason 1, which is designed to replace TOPEX/Poseidon and promises more accurate ocean height measurements. For more on the technique used to remotely determine ocean surface elevation, refer to the second Essay in Chapter 7.

Another satellite mission that is helping scientists detect the onset and follow the evolution of El Niño and La Niña is the joint U.S.-Japanese Tropical Rainfall Measuring Mission (TRMM) launched in November 1997 (Figure 11.3). The TRMM satellite uses active radar plus passive microwave energy sensors to monitor clouds, precipitation, and radiation processes over the area between 40 degrees N and 40 degrees S in the Pacific Ocean. During the 1997-1998 El Niño, sensors on the TRMM satellite recorded a reduction in precipitation in the western tropical Pacific and an increase in precipitation east of 150 degrees W longitude. An important feature of the satellite is its TRMM Microwave Imager (TMI), an instrument that "sees through" clouds and measures sea-surface temperatures. The TMI uses the microwave energy emitted by the sea surface to characterize its IR emission spectrum and determine its radiation temperature. Microwaves pass through clouds with little attenuation but are strongly absorbed and reflected by rainfall so that TMI can measure SST only during fair weather.

Increasingly accurate predictions of El Niño and La Niña are enabling humankind to better cope with the impacts of short-term (inter-annual) climate variability. Such forecasts allow informed strategic planning in agriculture, fisheries, and water resource management. Consider an example: Peru's economy, like that of most developing nations, is very sensitive to climate. In Peru, El Niño is bad for fishing and often accompanied by destructive flooding. La Niña benefits fishing but may also mean drought and crop failure. Warning of an impending El Niño or La Niña prior to the start of the growing season allows agricultural interests and government officials to consult on what crops to plant to optimize production. For example, if the forecast calls for El Niño, rice is favored over cotton because rice thrives during a wet growing season whereas cotton is more drought-tolerant and is more suitable during La Niña. Also, in anticipation of heavy rains that are likely to accompany a full-blown El Niño, water resource managers can direct the gradual draw down of reservoirs to reduce flooding.

FREQUENCY OF EL NIÑO AND LA NIÑA

El Niño can be expected about once every 3 to 7 years and to last from 12 to 18 months. An index that was commonly used to identify El Niño and La Niña is the departure of SST from long-term average in the area bounded by latitudes 5 degrees N and 5 degrees S, and longitudes 90 degrees W and 150 degrees W (Figure 11.10). Large positive values of this index indicate El Niño whereas large negative values define La Niña.

On 30 September 2003, NOAA scientists announced that they had reached a consensus among experts in the federal government and academia for a new index that forms the basis for operational definitions of El Niño and La Niña. Sea surface temperatures for the index are drawn from an area of the tropical Pacific that encompasses the equatorial cold tongue and is shifted somewhat to the west of the region represented by Figure 11.10. This newly designated key area of the tropical Pacific is bounded by longitude 120 degrees W and 170 degrees W, latitude 5 degrees N and 5 degrees S. El Niño is characterized by a positive sea surface temperature departure from normal (based on the period 1971-2000) greater than or equal to 0.5 Celsius degree, averaged over three consecutive months. La Niña is characterized by a negative sea surface temperature departure from normal (based on the period 1971-2000) greater than or equal to 0.5 Celsius degree, averaged over three consecutive months.

Although El Niño does not always give way to La Niña, La Niña appears more likely to follow an intense El Niño rather than a weak one. This may be explained by the fact that during an intense El Niño, the global water cycle conveys a great amount of heat out of the eastern tropical Pacific and into higher latitudes. Unusually cold water situated just below the warm surface water is poised to well up to the surface as soon as the trade winds

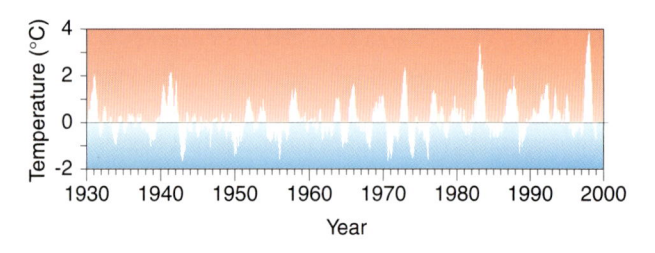

FIGURE 11.10
Sea-surface temperature anomalies (departures from long-term averages) for the area in the tropical Pacific Ocean between 5 degrees N and 5 degrees S latitude and from 90 degrees W to 150 degrees W longitude. Warm anomalies (greater than about 0.5 Celsius degree) generally indicate El Niño whereas cold anomalies (more negative than about −0.5 Celsius degree) generally indicate La Niña. [Adapted from NOAA PMEL]

strengthen. For example, a significant La Niña followed the intense 1997-98 El Niño.

During the second half of the 20th century, El Niño conditions prevailed 31% of the time and La Niña occurred 23% of the time. The rest of the time neutral or near-neutral conditions prevailed. Thirteen El Niño episodes have occurred since 1950; the most recent one was in 2002-03. Prior El Niño episodes took place in 1997-98, 1991-95, 1986-87, 1982-83, 1976-77, 1972-73, 1969, 1965-66, 1963, 1957-58, 1953, and 1951. During the 1980s and 1990s, La Niña was less frequent than El Niño. The most recent La Niña was in 1998-2000; others occurred in 1995-96, 1988-89, 1973-75, 1970-71, 1964, 1955-56, and 1949-1950. While all El Niño episodes share some common characteristics as described earlier in this chapter, they differ from one another in duration, intensity, and the way they develop. The same can be said of La Niña.

El Niño and La Niña are not just recent phenomena. In fact, documentary, archeological, and geological evidence indicates that ENSO has operated over at least tens of thousands of years. For more on this topic, refer to this chapter's second Essay.

North Atlantic Oscillation

Research on El Niño has spurred interest in other regular oscillations that involve the interaction of ocean and atmosphere and impact short-term climate variability. These include the North Atlantic Oscillation (NAO), the Arctic Oscillation (AO), and Pacific Decadal Oscillation (PDO). In general, NAO, AO, and PDO affect more restricted geographical areas and operate over longer time periods than either El Niño or La Niña.

Over the North Atlantic, the time-averaged planetary-scale atmospheric circulation features subpolar low pressure near Iceland (the *Icelandic low*) and a massive subtropical anticyclone centered near 30 degrees N that stretches from Bermuda to near the Azores (Chapter 5). The **North Atlantic Oscillation (NAO)** refers to a seesaw variation in air pressure between Iceland and the Azores. When air pressure is higher than the long-term average over the Azores, it is lower than the long-term average over Iceland and vice versa. The air pressure gradient between the Bermuda-Azores subtropical anticyclone and the Icelandic low drives winds that steer storms from west to east across the North Atlantic.

The North Atlantic Oscillation influences precipitation patterns and winter temperatures over eastern North America and much of Europe and North Africa. The so-called **NAO Index** is directly proportional to the strength of the North Atlantic air pressure gradient, i.e., the difference in sea level air pressure between the Bermuda-Azores high and the Icelandic low. When the NAO Index is relatively high, stronger than usual winter winds blow across the North Atlantic moving cold air masses over eastern Canada and the U.S. so that winters tend to be colder than usual in that region. But cold air masses modify considerably as they move over the relatively mild ocean surface, warming and becoming more humid so that winters are milder and wetter than usual downstream over Europe. Meanwhile, dry winters prevail in the Mediterranean region. On the other hand, when the NAO Index is low, steering winds over the North Atlantic shift southward so that winters are colder than usual over northern Europe and wet and mild conditions prevail from the Mediterranean eastward into the Middle East. The eastern U.S. and Canada tend to experience relatively mild winters while winters in the southeast U.S. are colder than usual.

The NAO Index varies significantly from one year to the next and from decade to decade and is much less regular than the ENSO cycle. The NAO Index was generally low in the 1950s and 1960s and high from about the mid-1970s through the 1990s (with the exception of 1997-98).

Changes in winter moisture supply associated with NAO have had varied impacts in Europe and North Africa. In recent decades of relatively high NAO-Index, wetter winters have increased hydroelectric power potential in the Scandinavian nations, lengthened the growing season over northern Eurasia, but also diminished the snow cover for winter recreation. Meanwhile, a moisture deficit has been the problem in the Iberian Peninsula, the watershed of the Tigris and Euphrates Rivers, and the Sahel of North Africa.

Arctic Oscillation

Related to the North Atlantic Oscillation, the **Arctic Oscillation (AO)** is a seesaw variation in air pressure between the North Pole and the margins of the polar region. Associated changes in the horizontal air pressure gradient alter the speed of the band of winds aloft (the *polar vortex*) that blow counterclockwise (viewed from above) around the Arctic. Strengthening and weakening of these polar winds impact winter weather in middle latitudes and contribute to climate variability and changes in ocean circulation.

The Arctic Oscillation features a negative and positive phase. During its negative phase, the pressure gradient is weaker and the polar vortex circulation is not as strong as usual. This allows bitterly cold Arctic air masses to more frequently move out of their source regions in the far north and plunge southeastward into middle latitudes. This brings colder than usual winter weather to most of the U.S., Northern Europe, Russia, China, and Japan. Heavy lake-effect snows are more frequent and nor'easters are more likely along the Eastern Seaboard (Chapter 8). But when the Arctic Oscillation is in its positive phase, the pressure gradient is steeper and winds encircling the Arctic are stronger. These stronger winds act as a dam to impede the southward flow of Arctic air. The mid-latitude westerlies also strengthen and blow more directly from west to east, flooding much of the U.S. with relatively mild air from off the Pacific Ocean (instead of frigid Arctic air from Canada). Winter temperatures are milder than usual (especially east of the Rocky Mountains) and major snowstorms are less likely. Meanwhile, Alaska, Scotland, and Scandinavia are snowier than usual and California and Spain are drier.

Although during any winter the Arctic Oscillation can shift many times between its negative and positive phases, extended periods occur when either the negative or positive phase dominates the winter season. In the 1960s, the negative phase of the Arctic Oscillation dominated. Since then, the positive phase has been more frequent, a trend that is consistent with observed climate fluctuations in middle latitudes (e.g., less frequent episodes of extreme cold and major snowstorms). Furthermore, during this recent episode of dominantly positive AO phase, winds have been delivering warmer than usual air and ocean water into the Arctic. As we will see in Chapter 12, this may at least help explain the recent shrinkage of Arctic ice cover.

Pacific Decadal Oscillation

The **Pacific Decadal Oscillation (PDO)** is a long-lived variation in climate over the North Pacific and North America. Sea surface temperatures fluctuate between the north central Pacific and the west coast of North America. During a PDO *warm phase*, SST are lower than usual over the broad central interior of the North Pacific and above average in a narrow strip along the coasts of Alaska, western Canada, and the Pacific Northwest. In an interesting parallel to what happens off the coast of Ecuador and Chile during El Niño, the layer of relatively warm surface waters off the Pacific Northwest Coast significantly reduces upwelling of nutrient-rich bottom water. Populations of phytoplankton and zooplankton plummet and juvenile salmon migrating to coastal waters from streams and rivers starve (Chapter 10). On the other hand, during a PDO *cold phase*, SST are higher in the North Pacific interior and lower along the coast associated with the return of nutrients, primary production, and salmon.

Key to the climatic impact of PDO is the strength of the subpolar *Aleutian low*, which prevails through the winter off the Alaskan coast. During a PDO *warm phase*, the Aleutian low is well developed and its strong counter-clockwise winds steer mild and relatively dry air masses into the Pacific Northwest. Winters tend to be mild and dry and water supplies suffer from reduced mountain snow pack. But during a PDO *cold phase*, the Aleutian cyclone is weaker so that cold, moist air masses more frequently invade the Pacific Northwest. Winters are colder and wetter, and the mountain snow pack is thicker. PDO phases tend to last for 20 to 30 years. Cold phases persisted from 1890-1924 and again from 1947-1976 whereas warm phases prevailed from 1925-1946 and 1977 through the 1990s.

Conclusions

El Niño and La Niña involve interactions between the tropical Pacific Ocean and atmosphere. These phenomena underscore the importance of the flux of heat and moisture between the ocean and atmosphere. Changes in these fluxes during El Niño and La Niña have relatively short-term (one to two year) impacts on ocean circulation and the weather and climate in various parts of the world with implications for marine life, fisheries, and hydrologic budgets on land (e.g., drought, flooding rains). Short-term fluctuations in climate induced by El Niño and La Niña as well as the NAO, AAO, and PDO are superimposed on much longer period variations in climate that have longer lasting impacts on the Earth system. The ocean plays a central role in these long-term climate changes. In the next chapter, we examine the nature of climate change, causes of climate change, and possible implications of those changes for the ocean.

Basic Understandings

- Many factors working together shape the climate of any locality. Controls of climate consist of (1) latitude, (2) elevation, (3) topography, (4) proximity to large bodies of water, (5) Earth's surface characteristics, (6) long-term average atmospheric circulation, and

(7) prevailing ocean circulation. Climate responds in periodic fashion to the diurnal and seasonal variations in incoming solar radiation. On the time scales of millions of years, all climate controls are variable. On shorter time scales, for all practical purposes, the first four climate controls are essentially fixed and exert regular and predictable influences on climate.

- The ocean is a major player in Earth's climate system operating on time scales of days to millennia and spatial scales from local to global. The ocean influences the radiational heating and cooling of the planet.
- Ocean currents strongly influence climate. Cold surface currents are heat sinks; they chill and stabilize the overlying air, thereby increasing the frequency of sea fogs and reducing the likelihood of thunderstorms. Relatively warm surface currents are heat sources; they supply heat and moisture to the overlying air, destabilizing the air, thereby energizing storm systems. Furthermore, the global oceanic conveyor belt contributes to the poleward transport of heat.
- In 1924, Sir Gilbert Walker discovered the southern oscillation, a seesaw variation in air pressure between the western and central tropical Pacific. In the mid-1960s, Jacob Bjerknes found that an El Niño episode begins when the air pressure gradient across the tropical Pacific begins to weaken.
- Over the long-term average (neutral conditions), southerly and southwesterly winds along the west coast of South America drive warm surface waters westward (via Ekman transport), away from the coast. Departing warm surface waters are replaced by cold, nutrient-rich water that wells up from below the pycnocline (upwelling) fueling high biological production. Upwelling also occurs along the equator, east of the international dateline.
- During neutral conditions, relatively cool surface waters in the central and eastern tropical Pacific chill the overlying air and suppress convection so that rainfall is light in that region and along the adjacent western coastal plain of South America. Meanwhile, over the western tropical Pacific, relatively warm surface waters heat the overlying air, strengthening convection and giving rise to heavy rainfall. Higher air pressure in the east and lower air pressure in the west result in trade winds blowing from the east piling up a thick layer of warm surface water in the tropical western Pacific.
- At the onset of El Niño, air pressure falls over the eastern tropical Pacific and rises over the western tropical Pacific as part of the southern oscillation. The air pressure gradient across the tropical Pacific weakens and the trade winds slacken and may even reverse direction west of the international dateline. This allows the thick layer of warm surface water piled up in the western tropical Pacific to slosh eastward along the equator into the eastern tropical Pacific, swamping coastal upwelling.
- As El Niño evolves, sea-surface temperatures decline, sea level falls, and depth of the pycnocline decreases in the western tropical Pacific. Meanwhile, sea-surface temperatures rise, sea level climbs, and depth of the pycnocline increases in the eastern tropical Pacific. Conditions are drier than usual in the western tropical Pacific and wetter than usual in the central tropical Pacific.
- A linkage between changes in atmospheric circulation occurring in widely separated regions of the world is known as a teleconnection. Through teleconnections, El Niño and La Niña have ripple effects on the weather and climate of middle latitudes, especially in winter. During El Niño, prevailing westerlies tend to blow more directly from west to east so that winters are milder than normal in western Canada and across parts of the northern U.S. and wet and cool along the Gulf Coast.
- Although some weather extremes almost always accompany El Niño, no two events are exactly the same because El Niño is only one of many factors that influence short-term variations in climate. The 1982-83 and 1997-98 El Niño events were the two most intense of the 20th century.
- La Niña is a period of unusually strong trade winds and exceptionally vigorous upwelling in the eastern tropical Pacific. During La Niña, sea-surface temperature anomalies are essentially opposite those observed during El Niño. Accompanying La Niña are worldwide weather extremes that are often opposite those observed during El Niño.
- Scientists employ two types of numerical models to predict the evolution of El Niño or La Niña: empirical (or statistical) models and dynamical models. An empirical model bases predictions on records of past ocean/atmosphere conditions. A dynamical model consists of a series of mathematical equations that simulates interactions among ocean, atmosphere, and land. Dynamical models tend to outperform empirical models.
- The accuracy of dynamical models in predicting the evolution of El Niño or La Niña depends not only on

- how well their constituent equations simulate the coupled ocean/atmosphere/land system, but also the reliability of observational data used to initialize, correct, and verify the model.
- The ENSO Observing System in the tropical Pacific consists of an array of instrumented moored buoys, current meters, and satellites. The system became fully operational in December 1994. Radar (microwave) altimeters onboard the TOPEX/Poseidon and Jason 1 satellites accurately measure changes in ocean surface elevation that accompany El Niño or La Niña.
- El Niño occurs about once every 3 to 7 years. La Niña appears more likely to follow an intense El Niño than a weak one.
- Other quasi-regular oscillations involving the ocean and atmosphere that impact climate include the North Atlantic Oscillation, Arctic Oscillation, and the Pacific Decadal Oscillation.
- The North Atlantic Oscillation (NAO) refers to a seesaw variation in air pressure between Iceland and the Azores. When air pressure is higher than the long-term average over the Azores, it is lower than the long-term average over Iceland and vice versa. The North Atlantic Oscillation influences precipitation patterns and winter temperatures over eastern North America and much of Europe and North Africa.
- The Arctic Oscillation (AO) is a seesaw variation in air pressure between the North Pole and the margins of the polar region. Associated changes in the horizontal air pressure gradient alter the speed of the band of winds aloft that encircle the Arctic. Strengthening and weakening of these polar winds impact winter weather in middle latitudes and contribute to climate variability and changes in ocean circulation.
- The Pacific Decadal Oscillation (PDO) is a long-lived variation in climate over the North Pacific and North America. Sea surface temperatures fluctuate between the north central Pacific and along the west coast of North America.

ESSAY: Sea Surface Temperature and Drought in Sub-Saharan Africa

Perhaps nowhere in the world has prolonged drought caused more human misery than in sub-Saharan Africa, much of which is known as the Sahel. The Sahel is the transition zone in West Africa between the Sahara Desert to the north and the rainforest of the Guinea coast to the south. As shown on the map, the Sahel includes all or part of Mauritania, Senegal, Mali, Burkina Faso, Niger, and Chad. These are the poorest nations on Earth. Low average annual rainfall, considerable year-to-year variability in rainfall, and prolonged droughts has brought considerable hardship to residents of the Sahel. The people of the Sahel are particularly vulnerable to the effects of drought because most of them depend on agriculture for their livelihood. Droughts forced them off their lands that, even in the best of times, are marginal for survival of crops and livestock. They migrated to cities in search of food and work, and many ended up in refugee camps. In the worst of times, horrible scenes of starving children, emaciated livestock, and withered crops were televised to a worldwide audience.

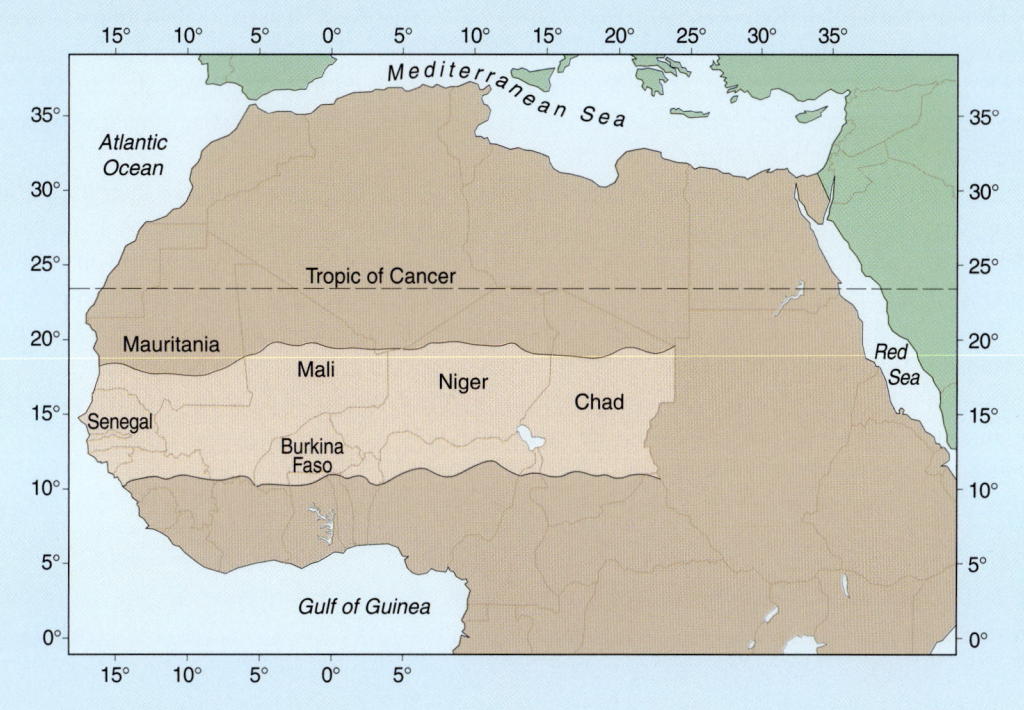

FIGURE
Changes in sea surface temperature perhaps coupled with human activity may explain the occurrence of multi-decadal droughts in the Sahel of North Africa (shaded area), one of the most impoverished regions of the world.

Drought in the Sahel is the product of interactions involving the atmosphere, ocean, land, and people. The Sahel has a semiarid climate that features a rainy "high sun" season (Northern Hemisphere summer) and a dry "low sun" season (Northern Hemisphere winter). The average length of the rainy season and mean annual rainfall increases from north to south across the Sahel. At the northern edge of the Sahel (about 18 degrees N), the rainy season usually does not begin until June and may last only a month or two; mean annual rainfall typically is less than 100 mm (about 4 in.). In the extreme southern Sahel (about 10 degrees N), the rainy season is underway as early as April and persists up to 5 or 6 months; mean annual rainfall exceeds 500 mm (about 20 in.). Shifts between the dry season and rainy season are linked to changes in prevailing winds associated with the location of the Intertropical Convergence Zone (ITCZ). As the ITCZ follows the sun, its northward surge in spring triggers rainfall, and its southward shift in fall brings the rainy season to an end. During the rainy season, surface winds blow from the south and southwest, transporting humid air from over both the Atlantic and Indian Oceans. During the dry season, the Sahel is under the eastern flank of the Bermuda-Azores subtropical high and surface winds blow from the dry north and northeast.

In a summer when the ITCZ does not move as far north as usual, or arrives late, or shifts southward early, rainfall is below average. A succession of such summers means drought. In the 20th century, the people of the Sahel endured three major long-term droughts: 1910-1914, 1940-1944, and 1970-1985. Rainfall during the 1961-1990 period was 20% to 40% lower than it was during the prior three decades, constituting the greatest 30-year anomaly in precipitation recorded anywhere in the world. Drought and famine claimed over 600,000 lives in 1972-1975 and again in 1984-1985. The reconstructed long-term climate record (based on lake-level fluctuations and historical accounts of landscape changes) indicates that droughts lasting 1 or 2 decades are the norm in the Sahel.

Some scientists argue that the principal reason for drought in the Sahel is changes in atmospheric circulation patterns (i.e., seasonal shifts of the moisture-bearing ITCZ) that are linked to anomalies in sea-surface temperature (SST). Initial studies showed that when SST is higher than normal in the eastern tropical Atlantic, south of the equator and southwest of West Africa (e.g., the Gulf of Guinea), rainfall is below the long-term average in the Sahel. In 2003 scientists at the International Research Institute for Climate Prediction at Palisades, NY announced the results of a NASA global climate model study of rainfall in the Sahel covering the period 1930 to 2000. They found that rising sea-surface temperature in the Indian Ocean was more important than changes in Atlantic SST in simulating long-term trends in Sahel rainfall. They also found that El Niño and La Niña accounted for much of the year-to-year variability in Sahel rainfall. More recently, runs of another global climate model developed at the Max Planck Institute for Meteorology in Germany found a similar dependence of decadal trends in Sahel rainfall on Indian Ocean SST.

Human activity (i.e., overgrazing, conversion of woodland to agriculture) may also contribute to persistent drought in sub-Saharan Africa by reinforcing dry conditions. Overgrazing and deforestation alter the regional radiation balance by denuding the soil's vegetative cover and degrading the quality and moisture holding capacity of soils. Soil albedo increases, evaporation decreases, convection weakens, and rainfall is less likely. What has happened in parts of the Sahel may be *desertification*, the name applied to the conversion of arable land to desert due to some combination of climate change and human mismanagement of the land.

Analysis of drought in the Sahel is a good example of how the Earth system perspective provides valuable insights on the environmental factors that contribute to climate variability.

ESSAY: El Niño in the Past

Lengthy records of past El Niño episodes would provide us with some perspective on more recent events. However, in most areas of the world, reliable instrument-derived weather records that might point to past El Niño episodes extend back only to about the mid 19th century. For information on prior occurrences of El Niño, scientists must rely on *proxy climatic data*, that is, information inferred from documentary (e.g., logs, diaries), geological (e.g., cores of ocean bottom sediments and glacial ice), or biological (e.g., tree rings, tropical Pacific corals) indicators of climate. Consider a few examples. Written records enumerate agricultural losses due to prolonged drought in India and Southeast Asia. Sediment cores extracted from a lake bottom indicate an episode of heavy rains in normally arid coastal Peru. These events occurring at the same time are consistent with a strong El Niño.

Geological evidence indicates that El Niño was occurring at least as far back as late in the Pleistocene Ice Age (1.7 million to 10,500 years ago). The El Niño signal shows up in a 4000-year record of lake sediments extracted from Glacial Lake Hitchcock, which occupied the Connecticut River valley during the waning phase of the last ice age. The layers of lake sediments chronicle short-term climatic fluctuations spanning the period from about 17,500 to 13,500 years ago (during the final retreat of glacial ice from New England), and apparently resolve both intense and weak El Niño episodes (with periods of 2.5 to 5 years).

In late 2002, Geoffrey Seltzer of Syracuse University and his colleagues reported on their analysis of two 8-m (26-ft) sediment cores extracted from the bottom of Laguna Pallcacocha high in the Andes of southern Ecuador. Rainfall governs the amount of sediment delivered to the lake but rainfall associated with a weak El Niño is not likely to reach the 4200-m (13,800-ft) high lake. Hence, they interpreted anomalously thick layers of silt in the cores as indicating heavy rainfall associated with an intense El Niño event. The lake sediment cores span the past 12,000 years and indicate that between 12,000 and 7000 years ago, strong El Niño episodes occurred five or fewer times per century. The frequency of intense El Niño episodes then increased and peaked in the 9th century A.D. when they occurred about every three years.

Using documentary evidence from explorers and early settlers of the coast of northwest South America, scientists have been able to reconstruct the chronology of El Niño events back to the year 1525 A.D. For example, El Niño conditions prevailed during 1531-32, when Francisco Pizarro conquered the Incas. Although heavy rains impeded his advance, his horses were well fed by the unusually lush vegetation. Records of the maximum discharge of the Nile River extending back to 622 A.D. are another valuable source of information on past El Niño episodes. The summer monsoon rains in the Ethiopian highlands, near the headwaters of the Nile, decrease during El Niño.

The long-term reconstructed climate record suggests that El Niño has occurred regularly since the late Pleistocene—albeit with at least one significant lull from 12,000 to 7000 years ago. However, El Niño intensity has varied over thousands of years (i.e., the tempo is roughly the same but the beat has alternately intensified and weakened). We now appear to be in a period of particularly intense El Niño events.

CHAPTER 12

THE OCEAN AND CLIMATE CHANGE

Case-in-Point
Driving Question
The Climate Record
    Marine Sediments and Climate
    Other Proxy Climatic Data Sources
    Geologic Time
    Past Two Million Years
    Instrument-Based Temperature Trends
Lessons of the Climatic Past
Possible Causes of Climate Change
    Climate and Plate Tectonics
    Climate and Solar Variability
    Climate and Volcanoes
    Climate and Earth's Surface Properties
    Climate and Human Activity
The Climate Future
    Global Climate Models
    Enhanced Greenhouse Effect and Global Warming
Impact of Climate Change on the Ocean
    Sea Level Fluctuations
    Arctic Sea Ice Cover
    Marine Life
Conclusions
Basic Understandings
ESSAY: Climate Rhythms in Glacial Ice Cores
ESSAY: The Drying of the Mediterranean Sea
ESSAY: Sea Level Rise and Saltwater Intrusion

Small ice melting, seen from the bridge of the Nathaniel B. Palmer–December 1998. [Courtesy of NOAA NESDIS, ORA]

Case-in-Point

People and other animals living in the Arctic face an uncertain future because of a recent warming trend. The Inuit people (also called Eskimos) live around the Arctic Ocean, in Greenland, Canada, Alaska, and Siberia. This is one of Earth's coldest regions and among the most remote and inhospitable for humans. Here, people still depend largely on hunting seals, caribou, and polar bears, herding reindeer, and gathering berries and other foods from the land.

    Warming is melting the *permafrost* (permanently frozen ground) weakening the foundations for homes and other structures. Warming—perhaps coupled with a change in ocean circulation—is also shrinking the Arctic sea ice cover restricting the Inuit's access to traditional hunting and gathering grounds. Warming means more summer precipitation falling as rain instead of snow. Rainwater freezes, forming a glaze several centimeters thick covering the ground that prevents caribou, reindeer, and musk ox (all hunted by the Inuit) from reaching the lichens they normally eat. Furthermore, rainwater and higher temperatures promote the growth of toxic molds on the lichens so that these animals avoid them and eventually must migrate to other areas to feed, or face death by starvation.

Another food source for the Inuit is seriously threatened. Polar bears, the Arctic's largest land predators, face possible extinction within this century as the Arctic sea ice cover shrinks. These animals depend on sea ice floes as platforms from which they hunt seals, their principal food source. They also travel across the sea ice to reach their dens on land. In short, they are uniquely adapted to living on Arctic sea ice and are often considered to be marine mammals because they spend so much time in the ice-infested water. The Arctic Ocean could be nearly ice-free during summer by the middle of the century.

Polar bears live on the sea ice and in coastal areas of Alaska, northern Canada, northern Russia, Greenland, and Norway. Effects of the shrinking sea ice cover are especially serious in Hudson Bay in northern Canada, which is near the southern limit of the polar bears' range. They scavenge for food in garbage dumps around settlements when the lack of sea ice prevents them from hunting seals. Some bears have come to depend on humans for food and are further threatened because residents may be forced to kill nuisance bears.

Driving Question:

How and why does climate change and how does the ocean participate in and respond to climate change?

In this chapter, we examine the instrument-derived and reconstructed climate record. One of our primary goals is to learn more about how climate changes through time. These lessons of the climate past are useful in establishing a perspective on the present climate and how climate might change in the future. Two of the most obvious lessons of the climate past are: (1) climate changes over a broad range of temporal scales, from years to decades to centuries to millennia, and (2) many forces working together are responsible for climate change.

The ocean is a central factor in our investigation of changes in Earth's climate. Deep-sea sediment cores are a source of information on past climates extending back hundreds of thousands of years. The ocean's low albedo, great thermal inertia, and role as a sink and source of heat energy and gases (especially water vapor and carbon dioxide) inhibit wild swings in climate. At the same time, the ocean responds to global climate change. Warming elevates sea level by raising sea surface temperatures (expanding sea water) and melting glaciers (whose runoff flows into the ocean).

We begin by examining the climate record for what it tells us about climate behavior. We then discuss the many factors that influence the variation of climate through time and finally summarize what is known about the climate future particularly as it relates to the world ocean.

The Climate Record

Climate varies not only from one place to another but also with time. Examining the record of the climatic past reveals some basic understandings regarding the nature of climate behavior and provides a useful perspective on current and future climate. In most places, however, the reliable instrument-based record of past weather and climate is limited to not much more than a century or so. For information on earlier fluctuations in climate, scientists rely on reconstructions of climate based on historical documents (e.g., personal diaries, ship's logs, weather-sensitive economic data) and longer-term geological and biological evidence such as bedrock type, fossil plants and animals, pollen, tree growth rings, glacial ice cores, and deep-sea sediment cores. In this section, we first consider some of the non-instrumental sources of information on past climates (*proxy climatic data sources*). We then summarized the climate record from geologic time through to the present.

MARINE SEDIMENTS AND CLIMATE

Cores taken from marine sediments that blanket the ocean floor yield a continuous record of sedimentation going back many hundreds of thousands and in places, millions of years (Figure 12.1). Much of what we know about the climatic fluctuations of the Pleistocene Ice Age is based on analysis of the shell and skeletal remains of microscopic marine organisms that are extracted from deep-sea sediment cores. Identification of these organism plus **oxygen isotope analysis** of

FIGURE 12.1
Scientists onboard the drill ship *JOIDES Resolution* have split open and are about to examine a sediment core recovered from the sea floor. Sediment and bedrock cores extracted from the ocean floor provide valuable information on the geologic and climatic past. [Ocean Drilling Program]

their remains enable scientists to distinguish between cold and mild climatic episodes of the past.

Scientists use a special property of water to reconstruct large-scale climatic fluctuations of the Pleistocene Ice Age. A water molecule (H_2O) is composed of either of two stable isotopes of oxygen, O^{16} or O^{18}. (Isotopes of a single element differ in atomic mass based on the number of neutrons in the nucleus of the atom.) The lighter isotope (O^{16}) is much more abundant than the heavier isotope (O^{18}); only one O^{18} exists for every thousand or so O^{16}. Nonetheless, small but significant variations are measured in the amount of light oxygen compared to heavy oxygen circulating in the global water cycle. These variations have important implications for past fluctuations in glacial ice volume on the planet.

On average, water molecules containing the lighter O^{16} isotope move slightly faster than water molecules containing the heavier O^{18} isotope and evaporate more readily. Water molecules that evaporate from the ocean (or land) are enhanced in light oxygen compared to heavy oxygen. The amount of O^{16} compared to O^{18} is also greater in cloud particles and rain and snow versus liquid water on Earth's surface. Most precipitation either falls back into the ocean directly or drains from land back to the sea, replenishing the ocean's supply of light oxygen and maintaining a relatively constant average ratio of O^{16} to O^{18}. However, geographical variations in the oxygen isotope ratio of seawater arise because of differences in precipitation amounts and evaporation rates. Seawater has more O^{18} at subtropical latitudes where evaporation exceeds precipitation and less in middle latitudes where rainfall is heavier.

During a glacial climatic episode, snow that accumulates on land is converted to ice. Heavy water molecules condense and precipitate slightly more readily than light water molecules so that moisture plumes moving from the tropics to high latitudes lose heavy oxygen along the way. Hence, snow falling in Canada has less O^{18} than rain falling in the tropics. The result is that growing ice sheets sequester more and more light oxygen while ocean water ends up with less and less O^{16} proportionately. With a shift to an interglacial climate, ice sheets melt and O^{16}-rich melt water drains back into the ocean increasing the amount of O^{16} compared to O^{18}. Hence, over time the proportion of light to heavy oxygen in ocean water decreases with increasing glacial ice cover.

Organic sediments on the ocean floor record fluctuations in the oxygen isotope ratio of seawater. Marine organisms living in the sunlit surface waters such as foraminifera build their shells from calcium carbonate ($CaCO_3$) that is dissolved in seawater. Shells formed during warmer interglacial climatic episodes contain more light oxygen than those formed during colder glacial climatic episodes. When these organisms die, their shells settle to the ocean bottom and mix with other marine sediments (Chapter 4). From specially outfitted deep-sea drilling ships, scientists extract cores from an undisturbed sequence of ocean bottom sediments, with the youngest sediments at the top of the core and the oldest sediments at the bottom. In the laboratory, the core is split open, and shells are extracted and analyzed for their oxygen isotope ratio. Variations in oxygen isotope ratio document changes in the planet's glacial ice volume.

Oxygen isotope analysis of deep-sea sediment cores indicates that the Pleistocene was punctuated by numerous abrupt changes between numerous glacial and interglacial climatic episodes. Oxygen isotope analysis has also been applied to ice layers within cores extracted from the Greenland and Antarctic ice sheets. These analyses confirm the abrupt change behavior of climate back hundreds of thousands of years. For more information on climatic inferences drawn from glacial ice cores, refer to this chapter's first Essay.

OTHER PROXY CLIMATIC DATA SOURCES

On land, much has been learned about past climate from study of pollen records, tree growth rings, and glacial ice cores. Pollen is a valuable source of information on late Ice Age vegetation and climate, especially over the past 15,000 years. *Pollen* is the tiny dust-like fertilizing component of a seed plant that is dispersed by the wind and accumulates on the bottom of lakes (and in other depositional environments) along with other organic and inorganic sediments. Scientists use a corer to extract a sediment column (core) that chronicles past changes in pollen (and therefore, vegetation). Back at the laboratory, pollen grains are separated from the other sediment in the core, identified as to source vegetation, and counted to determine the dominant vegetation type. Implications for past climate are drawn assuming that (1) the pollen is of local origin and (2) climate largely controls vegetation (and pollen) type. Changes in dominant pollen type in a core signal changes in nearby vegetation, likely in response to climate change.

Tree growth rings can yield a year-to-year record of past climate variations stretching back many thousands of years. Each spring/summer, living trees add a growth-ring whose thickness and density depend on growing season weather. A small hollow drill is used to extract a tree's growth-ring record. Although other environmental factors (e.g., soil type, drainage) can be important, tree growth rings are especially sensitive to moisture stress and have been used to reconstruct lengthy chronologies of drought prior to the era of instrument-based records.

Ice cores extracted from glaciers yield a record of past seasonal snowfall preserved as thin layers of ice (Figure 12.2). Cores extracted from the Greenland and Antarctic ice sheets provide detailed decadal-scale climatic information going back roughly 420,000 years. Using the oxygen isotope technique, scientists can distinguish between cold and mild episodes in the past and through chemical analysis of tiny bubbles trapped in the ice, determine the chemical composition of ancient air. This latter information is important in tracking long-term trends in the concentration of greenhouse gases.

GEOLOGIC TIME

Throughout most of the approximately 4.5 billion years that constitute *geologic time*, information on climate is generally unreliable. Descriptions of early climates are suspect because of lengthy gaps in the proxy climatic record and difficulties in determining the timing of events and correlating events that occurred in widely separated locations. Furthermore, plate tectonics complicates climate reconstruction efforts that focus on periods spanning hundreds of million of years (Chapter 2). Nonetheless, the available evidence supports some general conclusions regarding the climate over geologic time. For convenience of study, the geologic past and its climate record is subdivided using the **geologic time scale**, a standard division of Earth history into eons, eras, periods, and epochs based on large-scale geological events (Figure 12.3).

Geologic evidence indicates an interval of extreme climatic fluctuations about 570 million years ago, corresponding to the transition between Precambrian and Phanerozoic Eons. Along Namibia's Skeleton Coast rock layers that formed in tropical seas directly overlie glacial deposits. Some scientists interpret these rock sequences as indicating abrupt climatic changes between extreme cold and tropical heat. During as many as four cold episodes, each lasting perhaps 10 million years, the continents were encased in glacial ice and the ocean froze to a depth of more than 1000 m (3300 ft). At the close of each cold episode, temperatures rose rapidly, and within only a few centuries, all the ice melted.

Global warming appears to have persisted through much of the Mesozoic Era, from about 245 million to 70 million years ago. At the boundary between the Triassic and Jurassic Periods, the global mean temperature rose perhaps 3 to 4 Celsius degrees (5.5 to 7 Fahrenheit degrees), causing a major extinction of animals and substantial changes in vegetation. At peak warming during the Cretaceous Period, the global mean temperature was perhaps 6 to 8 Celsius degrees (11 to 14 Fahrenheit degrees) higher than now. Subtropical plants and animals lived as far north as 60 degrees N and dinosaurs roamed what is now the North Slope of Alaska. Adding to the geological processes impacting climate, a great meteorite impact about 65 million years ago threw huge quantities of dust into the atmosphere, blocking sunlight, and causing cooling that apparently was

FIGURE 12.2
This 6-m (20-ft) long ice core was extracted from the Greenland ice sheet and is a source of information on past variations in climate and atmospheric composition. Ice layers are made cloudy by dust particles that settled out of the atmosphere. [Photo by Mark Twickler, University of New Hampshire; National Geophysical Data Center]

Chapter 12 THE OCEAN AND CLIMATE CHANGE 283

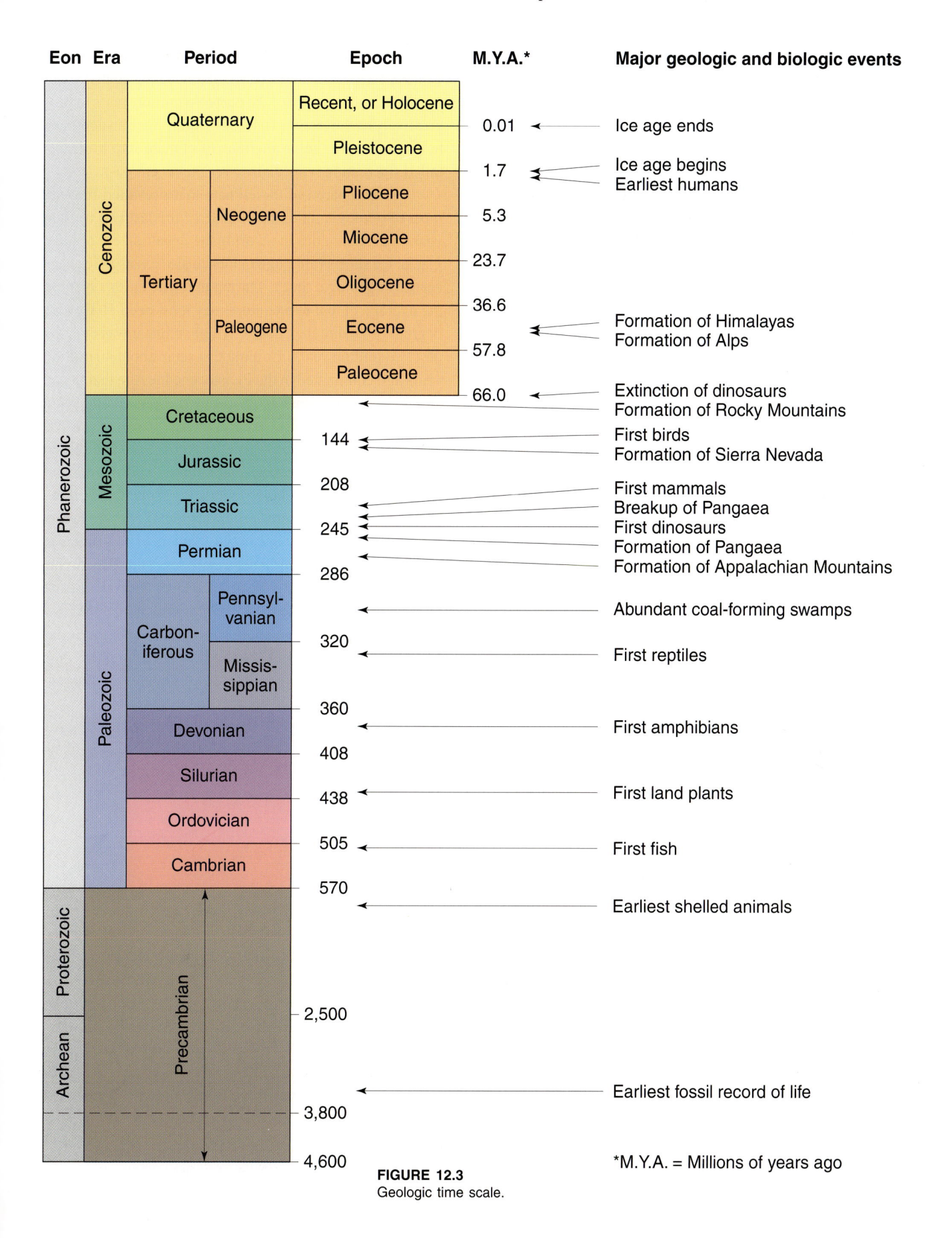

Eon	Era	Period	Epoch	M.Y.A.*	Major geologic and biologic events
Phanerozoic	Cenozoic	Quaternary	Recent, or Holocene	0.01	Ice age ends
			Pleistocene	1.7	Ice age begins / Earliest humans
		Tertiary (Neogene)	Pliocene	5.3	
			Miocene	23.7	
		Tertiary (Paleogene)	Oligocene	36.6	
			Eocene	57.8	Formation of Himalayas / Formation of Alps
			Paleocene	66.0	Extinction of dinosaurs / Formation of Rocky Mountains
	Mesozoic	Cretaceous		144	First birds / Formation of Sierra Nevada
		Jurassic		208	
		Triassic		245	First mammals / Breakup of Pangaea / First dinosaurs / Formation of Pangaea / Formation of Appalachian Mountains
	Paleozoic	Permian		286	
		Carboniferous (Pennsylvanian)			Abundant coal-forming swamps
		Carboniferous (Mississippian)		320	First reptiles
		Devonian		360	First amphibians
		Silurian		408	
		Ordovician		438	First land plants
		Cambrian		505	First fish
				570	Earliest shelled animals
Proterozoic		Precambrian		2,500	
Archean				3,800	Earliest fossil record of life
				4,600	

FIGURE 12.3
Geologic time scale.

*M.Y.A. = Millions of years ago

responsible for the demise of the dinosaurs (refer to the first Essay in Chapter 1). Extinction of the dinosaurs was followed by an explosion in the population of mammals.

The Cenozoic Era was a time of great climatic fluctuations. About 55 million years ago, methane (CH_4) released from deep-sea sediments escaped into the ocean and then into the atmosphere where it enhanced the greenhouse effect, causing an already warm planet to become even warmer. But by 40 million years ago, Earth's climate began shifting toward colder, drier, and more variable conditions setting the stage for the Pleistocene Ice Age. According to research findings of W.F. Ruddiman of Lamont-Doherty Geological Observatory of Columbia University and J.E. Kutzbach of the University of Wisconsin-Madison, mountain building may be the principal explanation for this change in Earth's climate, specifically the rise of the Colorado Plateau, Tibetan Plateau, and Himalayan Mountains. Prominent mountain ranges influence the geographical distribution of clouds and precipitation and can alter the planetary-scale circulation. Furthermore, mountain building may alter the global carbon cycle. Enhanced weathering of bedrock exposed in mountain ranges sequesters more atmospheric carbon dioxide in sediments thereby weakening the natural greenhouse effect.

In the American West, the region from the California Sierras to the Rockies, known as the Colorado Plateau, has an average elevation of 1500 to 2500 m (5000 to 8200 ft). Although mountain building began about 40 million years ago, about half of the total uplift took place between 10 and 5 million years ago. The Tibetan Plateau and Himalayan Mountains of southern Asia cover an area of more than 2 million square km (0.8 million square mi) and have an average elevation of more than 4500 m (14,700 ft). About half of total Himalayan uplift took place over the past 10 million years. These plateaus diverted the planetary-scale westerlies into a more meridional pattern, increasing the north-south exchange of air masses and influencing the climate over a broad region of the globe. Also, seasonal heating and cooling of the plateaus causes deeper low pressure to develop in summer and high pressure in winter, contributing to a more discernable monsoon-type circulation over southern Asia.

PAST TWO MILLION YEARS

Over the past two million years, plate tectonics was not a major factor in climatic fluctuations. For example, assuming a spreading rate of 4 cm (1.6 in.) per year, in two million years, the Atlantic Basin would spread a total distance of 80 km (50 mi), not very significant in a climatic sense. For all practical purposes, mountain ranges, continents, and ocean basins were essentially as they are today. Climate varies over a wide range of time scales so that viewing the climatic record of the past two million years in progressively narrower time frames is useful. Such an approach helps to resolve the oscillations of climate into more detailed fluctuations especially over the recent past.

Compared to the climate that prevailed through most of geologic time, the climate of the last two million years was anomalous in favoring the development of huge glacial ice sheets (although evidence also exists of ice ages earlier in geologic time). During much of Earth's history, the average global temperature may have been 10 Celsius degrees (18 Fahrenheit degrees) higher than it was over the past two million years. Cooling that set in about 40 million years ago culminated in the Pleistocene Ice Age that began about 1.7 million years ago and ended about 10,500 years ago.

During the Pleistocene Ice Age the climate shifted numerous times between glacial climates and interglacial climates. A **glacial climate** favors the thickening and expansion of glaciers whereas an **interglacial climate** favors the thinning and retreat of existing glaciers or no glaciers at all. During major glacial climatic episodes of the Pleistocene, the Laurentide ice sheet developed over central Canada and spread westward to the Rocky Mountains, eastward to the Atlantic Ocean, and southward over the northern tier states of the United States (Figure 12.4). At about the same time, mountain glaciers in the Rockies coalesced into the

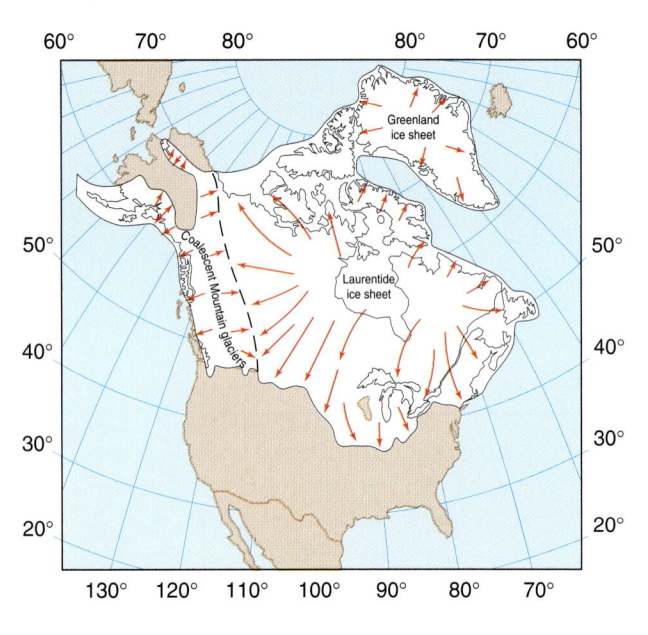

FIGURE 12.4
Extent of glacial ice cover over North America about 18,000 to 20,000 years ago, the time of the last glacial maximum.

Cordilleran ice sheet, a relatively thin ice sheet covered the Arctic Archipelago, and an ice sheet much smaller than the Laurentide developed over northwest Europe including the British Isles and Scandinavia. The vast quantity of water locked up in these ice sheets caused sea level to drop by 113 to 135 m (370 to 443 ft), exposing portions of the continental shelf, including a land bridge linking Siberia and North America.

The Laurentide and European ice sheets thinned and retreated, and may even have disappeared entirely, during relatively mild interglacial climatic episodes, which typically lasted about 10,000 years. Throughout these interglacials, however, glacial ice persisted over most of Antarctica and Greenland as it still does today.

During glacial climatic episodes, temperatures were lower than they are today but the cooling was not geographically uniform. A variety of geologic evidence indicates that during the Pleistocene, temperature fluctuations between major glacial and interglacial climatic episodes typically amounted to as much as 5 Celsius degrees (9 Fahrenheit degrees) in the tropics, 6 to 8 Celsius degrees (11 to 14 Fahrenheit degrees) at middle latitudes, and 10 Celsius degrees (18 Fahrenheit degrees) or more at high latitudes. An increase in the magnitude of a climatic change with increasing latitude is known as **polar amplification**, indicating that polar areas are subject to greater changes in climate.

Oxygen isotope analysis of the deep-sea sediment cores shows numerous fluctuations between major glacial and interglacial climatic episodes over the past 600,000 years (Figure 12.5A). Shifting focus to the past 160,000 years, resolution of the climate record improves. The temperature curve in Figure 12.5B is based on analysis of an ice core extracted from the Antarctic ice sheet at Vostok. A relatively mild interglacial episode, referred to as the *Eemian*, began about 127,000 years ago and persisted for about 7000 years. In some localities, temperatures may have been 1 to 2 Celsius degrees (2 to 4 Fahrenheit degrees) higher than during the warmest portion of the present interglacial (the *Holocene*). The Eemian interglacial was followed by numerous fluctuations between glacial and interglacial episodes. The last major glacial climatic episode began about 27,000 years ago and reached its peak about 18,000 to 20,000 years ago when glacial ice cover over North America was about as extensive as it had ever been (Figure 12.4).

The global mean temperature 18,000 to 20,000 years ago was about 4 to 6 Celsius degrees (7.2 to 10.8 Fahrenheit degrees) lower than at present. A general warming trend (and rise in sea level from melting ice) followed the last glacial maximum, punctuated by relatively brief returns to

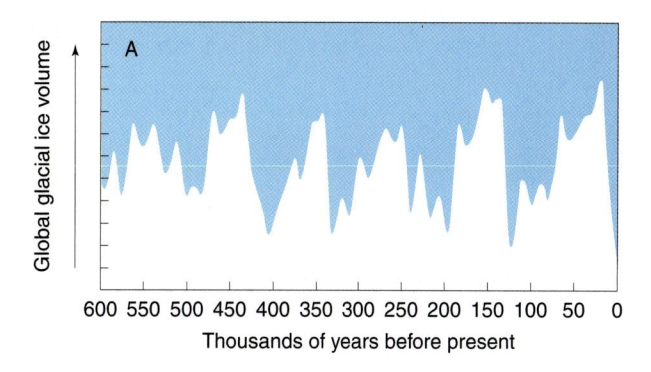

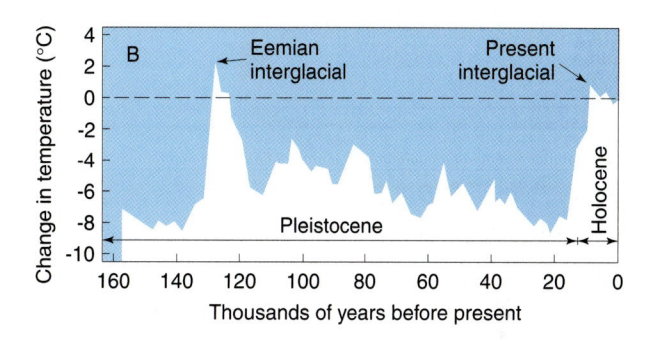

FIGURE 12.5
Reconstructed records of (A) the variation in global glacial ice volume over the past 600,000 years based on analysis of the oxygen isotope ratio of shells in deep-sea sediment cores, and (B) temperature variation over the past 160,000 years derived from oxygen isotope analysis of an ice core extracted from the Antarctic ice sheet at Vostok and expressed as a departure in Celsius degrees from the 1900 global mean temperature. [Compiled by R.S. Bradley and J.A. Eddy from J. Jousel et al., *Nature* 329(1987):403-408 and reported in *EarthQuest* 5, No. 1 (1991).]

glacial climatic episodes (Figure 12.6). A notable example is the cold period from about 11,000 to 10,000 years ago known as the *Younger Dryas* (named for the polar wildflower, *Dryas octopetala*, that reappeared in portions of Europe at the time). The return of glacial climatic conditions triggered short-lived re-advances of remnant ice sheets in North America, Scotland, and Scandinavia. Glacial ice finally withdrew from the Great Lakes region about 10,500 years ago ushering in the present interglacial, the **Holocene**.

Events surrounding the *Younger Dryas* point to the role of large-scale ocean thermohaline circulation in influencing climate. About 11,000 years ago, movement of glacial ice lobes diverted melt water from the Mississippi River to the St. Lawrence River and into the North Atlantic. With the input of freshwater, North Atlantic surface waters became less saline and eventually were not sufficiently dense to sink and form deep water. This stopped the conveyor belt circulation, which in turn halted the flow of warm

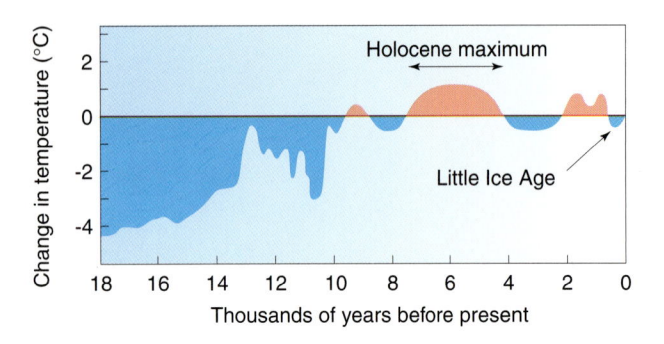

FIGURE 12.6
Reconstructed temperature variation over the past 18,000 years based on a variety of proxy climatic indicators and expressed as a departure in Celsius degrees from the 1900 global mean temperature. [Compiled by R.S. Bradley and J.A. Eddy based on J.T. Houghton et al. (eds.), *Climate Change: The IPCC Assessment*, Cambridge University Press, U.K., 1990, and reported in *EarthQuest* 5, No. 1 (1991).]

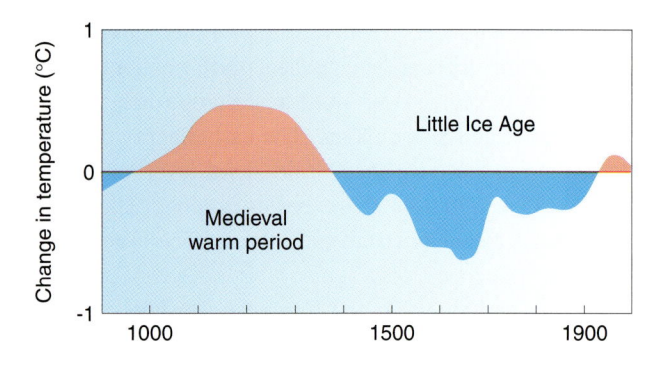

FIGURE 12.7
Reconstructed temperature variation over the past 1000 years based on analysis of historical documents and expressed as a departure in Celsius degrees from the 1900 global mean temperature. [Adapted from J.T. Houghton et al. (eds.), *Climate Change: The IPCC Assessment*, Cambridge University Press, U.K., 1990, p. 202.]

water into the North Atlantic causing a marked cooling of lands surrounding the North Atlantic. This was the beginning of the *Younger Dryas*. At the onset of the Holocene (about 10,500 years ago), a warming trend caused glaciers to retreat sufficiently that fresh melt water was diverted from the St. Lawrence back to the Mississippi River. North Atlantic surface waters become saltier and denser, deep water formed, and the conveyor belt circulation was back in operation, warming the lands surrounding the North Atlantic. This signaled the end of the *Younger Dryas*.

Although the Laurentide ice sheet was melting and would disappear almost entirely by about 5500 years ago, the Holocene has been an epoch of spatially and temporally variable temperature and precipitation. Ice cores extracted from the Greenland ice sheet and sediment cores taken from the bottom of the North Atlantic reveal that the overall post-glacial warming trend was interrupted by abrupt millennial-scale fluctuations in climate. Post-glacial warming during the Holocene gave way to a cold episode about 8200 years ago; significant cooling also occurred between 3100 and 2400 years ago. On the other hand, at times during the mid-Holocene (classically known as the *Hypsithermal*) mean annual global temperature was perhaps 1 Celsius degree (2 Fahrenheit degrees) higher than it was in 1900, the warmest in more than 110,000 years, that is, since the Eemian interglacial. A pollen-based climate reconstruction indicates that 6000 years ago, July mean temperatures were about 2 Celsius degrees (3.6 Fahrenheit degrees) higher than now over most of Europe.

A generalized temperature curve for the past 1000 years, derived mostly from historical documents, is shown in Figure 12.7. The most notable features of this record are (1) the **Medieval Warm Period** from about 950 to 1250 AD and (2) the cooling that followed, from about 1400 to 1850 AD, a period now known as the **Little Ice Age**. The Medieval Warm Period and the Little Ice Age were not episodes of sustained warming and cooling, respectively. On the contrary, sediment and glacial ice core records plus historical evidence indicate decadal fluctuations in temperature and precipitation. During much of the Medieval Warm Period, the global temperature averaged about 0.5 Celsius degree (0.9 Fahrenheit degree) higher than in 1900. The first Norse settlements appeared along the southern coast of Greenland and vineyards thrived in the British Isles.

Independent lines of evidence confirm that the Little Ice Age was a relatively cool period in many regions with mean annual global temperatures perhaps 0.5 Celsius degree (0.9 Fahrenheit degree) lower than it was in 1900. Sea ice cover expanded, mountain glaciers advanced, growing seasons shortened, and erratic harvests caused much hardship for many people, including the end of the Norse settlements in Greenland.

INSTRUMENT-BASED TEMPERATURE TRENDS

Invention of weather instruments, establishment of weather observational networks around the world, and standardized methods of observation and record-keeping made the climatic record much more detailed and dependable. The most reliable temperature records date from the late 1800s and the birth of the predecessor to today's World Meteorological Organization (WMO) and U.S. National Weather Service (NWS). Examination of temperature trends over the past 120 years or so is instructive as to the short-term vari-

Chapter 12 THE OCEAN AND CLIMATE CHANGE

ability of climate.

Plotted in Figure 12.8 are variations from 1880 to 2000 in (1) global (land and ocean) mean annual temperature, (2) global mean sea surface temperature, and (3) global mean land surface temperature. In all three cases, the temperature is expressed as a departure (in degrees Celsius) from the long-term (120 years) period average. As expected, these temperature time series, assembled by NOAA's National Climatic Data Center (NCDC), indicate greater year-to-year variability over land than ocean. The trend in global mean temperature is generally upward from 1880 until about 1940, downward or steady from 1940 to about 1970, and upward again through the 1990s. The total temperature fluctuation amounts to only about ±0.4 Celsius degrees (±0.7 Fahrenheit degrees) about the period average. Note that this temperature record for the globe as a whole is not representative (in direction or magnitude) of all locations worldwide; that is, the trend was amplified or reversed, or both, in specific regions.

In late 1999, the World Meteorological Organization reported that the global mean annual temperature was about 0.7 Celsius degree (1.3 Fahrenheit degrees) higher at the end of the 20th century than at the close of the 19th century. In Spring 2000, the NCDC reported that global warming accelerated during the final quarter of the 20th century. Since 1976, global mean annual temperature climbed at the rate of 0.2 to 0.3 Celsius degrees (0.4 to 0.5 Fahrenheit degrees) per decade. The seven warmest years in the instrument-based record occurred in the 1990s with 1998 being the warmest. During 1997 and 1998, record high global mean temperatures were set for 16 consecutive months.

Furthermore, climate reconstructions (mostly in the Northern Hemisphere) indicate that the 20th century was the warmest in 1000 years. For example, analysis of ice cores extracted from a glacier at 7163 m (23,500 ft) in the Himalayan Mountains revealed that the 1990s and the last half of the 20th century were the warmest of any equivalent period in a millennium.

Some critics question the integrity of large-scale (hemispheric or global) temperature records. They cite as potential sources of error: (1) improved sophistication and reliability of weather instruments through the period of record, (2) changes in location and exposure of instruments at most long-term weather stations, (3) huge gaps in monitoring networks, especially over the ocean, and (4) the warming influence of urbanization (*urban heat islands*). By careful statistical interrogation of available data, however, global-scale temperature trends have been confirmed to a reasonable level of confidence.

While the general consensus in the scientific community holds that a global-scale warming trend has prevailed since the end of the

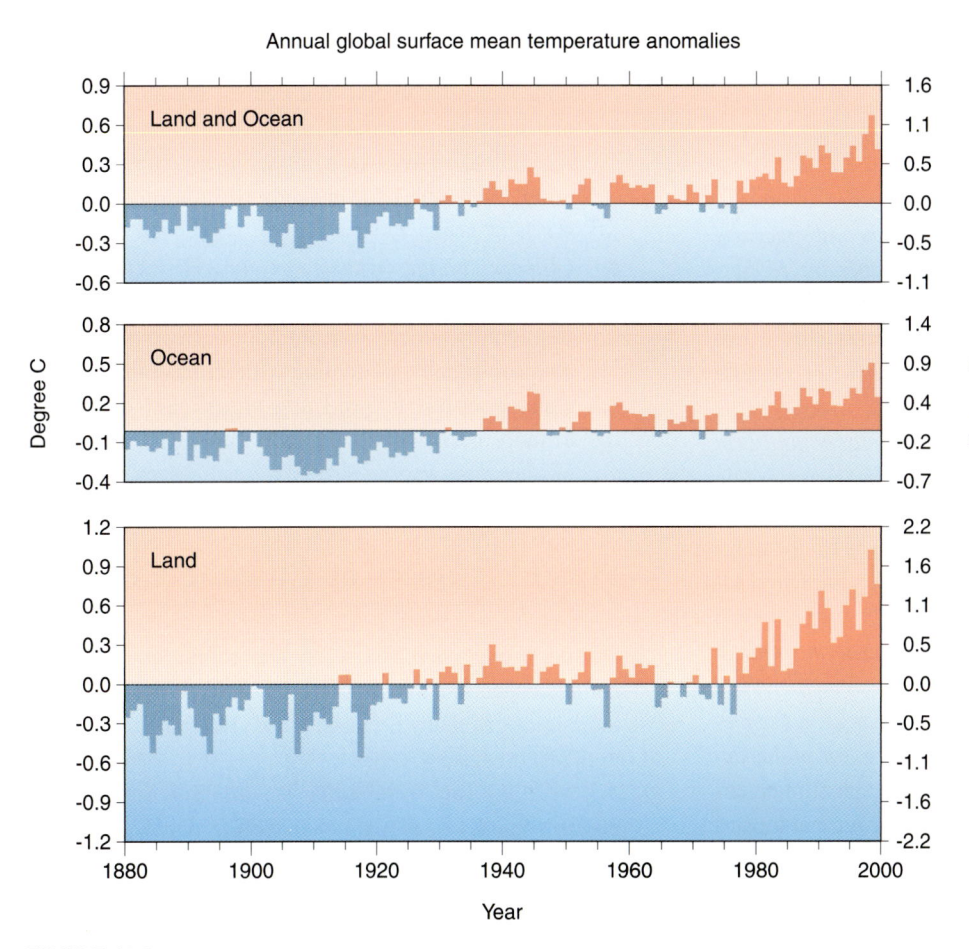

FIGURE 12.8
Instrument-derived trends in mean annual global (land and ocean), sea-surface, and land temperatures and expressed as departures in degrees Celsius and degrees Fahrenheit from the 120-year period average. [NOAA, National Climatic Data Center]

Little Ice Age, agreement is not universal as to the cause of the warming or how long it might continue. The most popular explanation for the warming trend is the steady build-up of carbon dioxide in the atmosphere and the consequent enhancement of the natural greenhouse effect (Chapter 5). Continued rise in the concentration of atmospheric CO_2 (and other infrared-absorbing gases) may mean global warming throughout this century and perhaps beyond. However, the climate system is complex and not completely understood and global climate models require further development. As we will see later in this chapter, many factors interact to shape the changing climate and the possibility exists for changes in Earth's climate system that might compensate to some extent for the influence of rising levels of greenhouse gases.

Lessons of the Climate Past

What does the climatic record tell us about the behavior of climate through time? The following are some of the lessons of the climatic record. These lessons are useful in assessing prospects for the climate future and the possible impacts of climate change.

1. *Climate is inherently variable over a broad spectrum of time scales ranging from years to decades, to centuries, to millennia.* Variability is an endemic characteristic of climate. The question for the future is not *whether* the climate will change but *how* the climate will change.
2. *Variations in climate are geographically nonuniform in both sign (direction) and magnitude.* Large-scale trends in climate are not necessarily duplicated at particular locations although the tendency is for the magnitude of temperature changes to amplify with increasing latitude (*polar amplification*).
3. *Climate change may consist of a long-term trend in the various climatic elements (e.g., mean temperature or average precipitation) and/or a change in the frequency of extreme weather events (e.g., drought, excessive cold).* Recall from Chapter 5 that climate encompasses mean values plus extremes. A trend toward warmer or cooler, wetter or drier conditions may or may not be accompanied by a change in frequency of weather extremes. On the other hand, a climatic regime featuring relatively little change in mean temperature or precipitation through time could be accompanied by an increase or decrease in frequency of weather extremes.
4. *Climate change tends to be abrupt rather than gradual.* In the context of the climate record, *abrupt* is a relative term. If the time of transition between climatic episodes is much shorter than the duration of the episodes, then the transition is considered to be relatively abrupt. Analysis of cores extracted from the Greenland ice sheet indicate that cold and warm climatic episodes, each lasting about 1000 years, were punctuated by abrupt change over periods as brief as a single decade. The abrupt-change nature of climate would test the resilience of society to respond effectively to climate change.
5. *Only a few cyclical variations can be discerned from the long-term climate record.* Reliable cycles include diurnal and seasonal changes in incoming solar radiation (the forcing) and temperature (the response). This means simply that days are usually warmer than nights and summers are warmer than winters. Quasi-regular variations in climate include El Niño (occurring about every 3 to 7 years), Holocene millennial-scale fluctuations identified in glacial ice cores, and the major glacial-interglacial climatic shifts of the Pleistocene Ice Age unlocked from deep-sea sediment cores and operating over tens of thousands to hundreds of thousands of years.
6. *Climate change impacts society.* History recounts numerous instances when climate change significantly impacted society. Although modern societies may be more capable of dealing with climate change than early peoples, a rapid and significant change in climate is likely to seriously impact all sectors of modern society.

Possible Causes of Climatic Change

No simple explanation exists for why climate changes. The complex spectrum of climate variability is a response to the interactions of many processes, both internal and external to the Earth system.

One way to organize our thinking on the many possible causes of climate change is to match a possible cause (or *forcing*) with a specific climatic oscillation (or *response*), based on similar periods of oscillation (Figure 12.9). For example, changes in continents and ocean basins due to plate tectonics might explain long-term climate changes over hundreds of millions of years. Systematic changes in Earth's orbit about the sun may account for

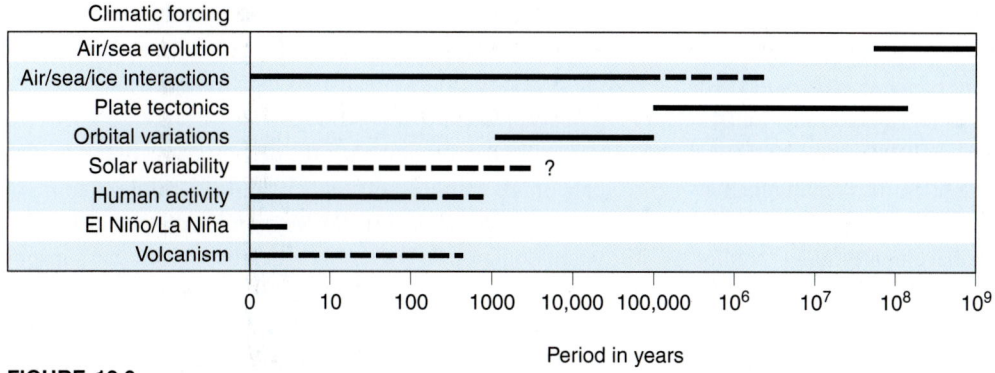

FIGURE 12.9
The various causes of climate change operate over a range of time frames.

climatic shifts of the order of 10,000 to 100,000 years. Changes in sunspot number and variations in the sun's energy output may be associated with climatic fluctuations of decades to centuries; volcanic eruptions, El Niño, or La Niña may account for climatic fluctuations lasting several months or a few years. But matching some forcing mechanism with a climate response based on similar periods is no guarantee of a real physical relationship.

Another way to think about the possible causes of climate change is in terms of *global radiative equilibrium* (Chapter 5). Energy entering the Earth system (i.e., absorbed solar radiation) must ultimately equal energy leaving the system (i.e., infrared radiation emitted to space). Any change in either energy input or energy output will shift the Earth system to a new equilibrium and change the planet's climate. A number of factors can alter global radiative equilibrium, including changes in solar energy output and its receipt, volcanic eruptions, human activity, and changes in Earth's surface properties.

In Chapter 11, we described the influence of El Niño and La Niña on inter-annual climatic variability. In this section, we summarize the many other factors that may contribute to climate change over a range of time scales.

CLIMATE AND PLATE TECTONICS

Plate tectonics likely has operated on the planet for perhaps 2.5 billion years and influenced climate in many ways. Recall from Chapters 1 and 2 that the solid outer skin of the planet is divided into a dozen gigantic rigid plates (and many smaller ones) that drift very slowly over the face of the planet. As plates move, continents move, ocean basins open and close, and tectonic stresses build mountain ranges and volcanoes. Plate movements are so slow compared to the span of human existence that we can consider topography and the geographical distribution of the ocean and continents as essentially fixed controls of climate. Over the vast expanse of geologic time, however, plate tectonics was a major player in large-scale climate change.

Changes in the location of continents (*continental drift*) altered the local and regional radiation balance and the response of air temperature. Continental drift explains such seemingly anomalous finds as glacial deposits in the Sahara Desert, fossil tropical plants in Greenland, and coal in Antarctica. These discoveries reflect climatic conditions millions to hundreds of millions of years ago when landmasses were situated at different latitudes than they are today. Furthermore, mountain ranges rose and eroded away, altering atmospheric circulation and the patterns of clouds and precipitation.

Opening and closing of ocean basins changed the course of heat-transporting ocean currents and the global conveyor belt. About 100 million years ago, the *Tethys Sea* separated Africa and Eurasia and Central America was submerged so that warm water currents flowed around the equator connecting what are now the Pacific, Atlantic, and Indian Oceans. About 40 million years ago, diverging plates separated Antarctica from Australia. As plate movements continued, about 30 million years ago South America moved away from Antarctica, opening the Drake Passage between Antarctica and the southern tip of South America. The Drake Passage permitted a circumpolar current to flow around Antarctica. In addition, these plate movements eventually left the Antarctic continent situated over the South Pole. By blocking the transport of heat from the tropics to the Southern Ocean, the Antarctic circumpolar current probably led to the formation of the Antarctic ice sheets about 17 million years ago. About 20 million years ago, the movement of Saudi Arabia northward against Asia sealed off the Tethys Sea, forming the Mediterranean Sea. About 3 million years ago, volcanic eruptions in what is now Central America formed a narrow isthmus of land, blocking the equatorial currents that previously flowed from the Atlantic through the Caribbean and into the equatorial Pacific Ocean. For another example of how plate tectonics may have played a role in climate change, refer to this chapter's second Essay.

CLIMATE AND SOLAR VARIABILITY

Fluctuations in the sun's energy output, sunspots, or regular variations in Earth's orbital parameters, so-called Milankovitch cycles, are external factors that can alter Earth's climate. Satellite measurements in the 1980s and 1990s confirmed long-held suspicions in the scientific community that the sun's total energy output at all wavelengths varies with time. Furthermore, numerical global climate models predict that only a 1% change in the sun's energy output could significantly alter the mean temperature of the Earth-atmosphere system.

Changes in solar energy output are apparently related to sunspot number. A **sunspot** is a dark blotch on the face of the sun, typically thousands of kilometers across that develops where an intense magnetic field suppresses the flow of gases transporting heat from the sun's interior. A sunspot appears dark because its temperature is about 400 to 1800 Celsius degrees (720 to 3240 Fahrenheit degrees) lower than the surrounding surface of the sun, the *photosphere*. A sunspot typically lasts only a few days, but the rate of sunspot generation is such that the number of sunspots varies systematically. The time between successive sunspot maxima or minima averages about 11 years with a range of 10 to 12 years (Figure 12.10). Also, the strong magnetic field associated with sunspots exhibits an approximate 22-year oscillation in polarity (the *double sunspot cycle*). Sunspot number reached a maximum in 1989, a minimum in 1996, and a maximum in 2000.

Satellite monitoring reveals that the sun's energy output varies directly with sunspot number; that is, a slightly brighter sun has more sunspots whereas a slightly dimmer sun has fewer sunspots. The variation in total solar energy output through one 11-year sunspot cycle amounts to less than 0.1%, with much of that taking place in the ultraviolet portion of the solar spectrum. A brighter sun is associated with more sunspots because of a concurrent increase in bright areas, known as *faculae*, which appear near sunspots on the photosphere. Faculae dominate sunspots and the sun brightens. More sunspots may contribute to a warmer global climate and fewer sunspots may translate into a colder global climate.

How reasonable is the proposed link between global climate and sunspot number? In 1893, while examining sunspot records at the Old Royal Observatory at Greenwich, England, E. Walter Maunder discovered that sunspot activity greatly diminished during the 70-year period between 1645 and 1715, now referred to as the **Maunder minimum** (Figure 12.10). The scientific community for the most part ignored Maunder's finding until the 1970s when John A. Eddy of the University Corporation for Atmospheric Research (UCAR) in Boulder, CO reinvestigated his work. Eddy pointed out that the Maunder minimum plus a prior period of reduced sunspot number, called the *Spörer minimum* (1400 to 1510 AD), occurred about the same time as relatively cold phases of the Little Ice Age in Western Europe. Furthermore, the Medieval Warm Period coincided with an interval of heightened sunspot activity between about 1100 and 1250 AD.

Skeptics dismiss the significance of the match between the Maunder minimum and a cooler climate. They argue that relatively cold episodes occurred in Europe just prior to and after the Maunder minimum (i.e., 1605-1615, 1805-1815), and relatively cool conditions did not persist throughout the Maunder minimum and were not global in

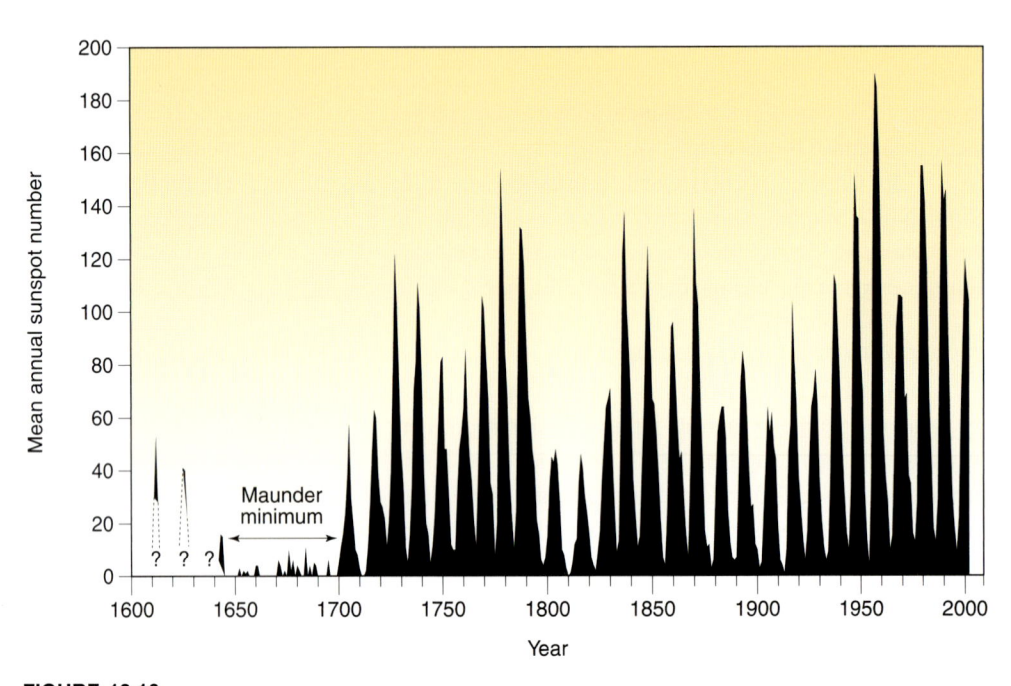

FIGURE 12.10
Variation in mean annual sunspot number since the early 17th century. [National Geophysical Data Center]

extent. Furthermore, some scientists argue that a 0.1% variation in the sun's total energy output during the 11-year sunspot cycle is much too weak to significantly impact Earth's climate. Other scientists counter that certain mechanisms operating within the Earth-atmosphere system could amplify changes in total solar output, making the slight brightening and dimming of the sun an important player in Earth's climate system. These variations are responsible for seasonal and latitudinal changes in solar radiation received by planet Earth.

Milankovitch cycles are regular variations in the precession and tilt of Earth's rotational axis and the eccentricity of its orbit about the sun (Figure 12.11). Named for the Serbian astrophysicist Milutin Milankovitch (1879-1958) who studied them extensively in the 1920s and 1930s, these long-term, quasi-rhythmic changes in Earth-sun geometry are caused by gravitational influences exerted on Earth by other large planets, the moon, and the sun. Combined, Milankovitch cycles drive climatic fluctuations operating over tens of thousands to hundreds of thousands of years. Milankovich cycles were likely responsible for the major advances and recessions of the Laurentide ice sheet over North America during the Pleistocene.

Over a period of about 23,000 years, Earth's spin axis describes a complete circle, much like the wobble of a spinning top (Figure 12.11A). These variations are responsible for seasonal and latitudinal changes in solar radiation received by planet Earth. This precession cycle changes the dates of *perihelion*, when Earth is closest to the sun, and *aphelion*, when Earth is farthest from the sun, increasing the summer-to-winter seasonal contrast in one hemisphere and decreasing it in the other. Currently, perihelion is in early January and aphelion is in early July. In about 10,000 years, those dates will be reversed (perihelion in July and aphelion in January) and the seasonal contrast will be greater than it is now in the Northern Hemisphere (i.e., colder winters and warmer summers) and less in the Southern Hemisphere (i.e., milder winters and cooler summers).

The tilt of Earth's spin axis changes from 22.1 degrees to 24.5 degrees and then back to 22.1 degrees over a period of about 41,000 years, the consequence of long-period changes in the orientation of Earth's orbital plane with respect to its spin axis (Figure 12.11B). (Presently the tilt is 23.5 degrees.) As axial tilt increases, winters become colder and summers become warmer in both hemispheres. Earth's axial tilt has been slowly decreasing for about 10,000 years and will continue to do so for the next 10,000 years.

The shape of the Earth's orbit changes from ellip-

A. Changes in axial precession ~23,000 years

B. Changes in axis tilt ~41,100 years

C. Changes in eccentricity ~100,000 years

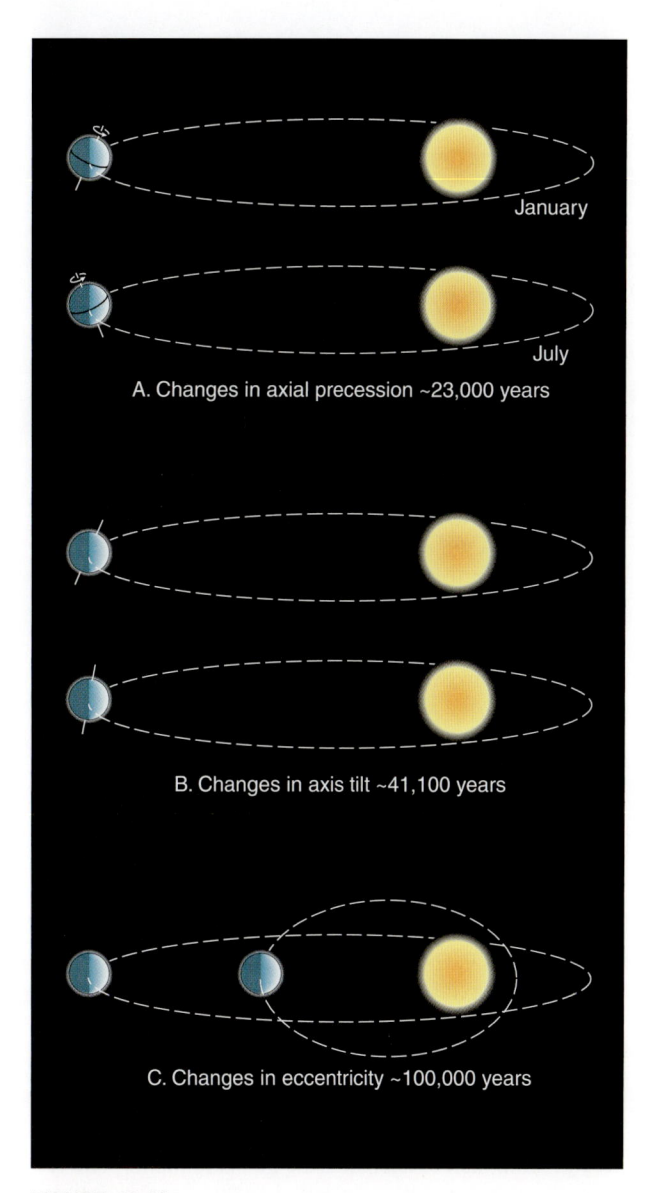

FIGURE 12.11
Milankovitch cycles likely explain the large-scale fluctuations of Earth's glacial ice cover during the Pleistocene. Note that diagrams greatly exaggerate changes in Earth-sun geometry.

tical (high eccentricity) to nearly circular (low eccentricity) in an irregular cycle of 90,000 to 100,000 years (Figure 12.11C). Variation in orbital eccentricity alters the distance between Earth and sun at aphelion and perihelion, thereby changing the amount of solar radiation received by the planet at those times of the year. When Earth's orbit is highly elliptical, the amount of radiation received at perihelion is significantly greater than at aphelion.

Milankovitch cycles do not alter appreciably the total amount of solar energy received by the Earth-atmosphere system annually, but they do change significantly the latitudinal and seasonal distribution of incoming solar radiation. Milankovitch developed a numerical model based on the three cycles that calculated latitudinal differences in incoming solar radiation and the corresponding surface temperature for the time span of 600,000 years prior to the year 1800. He proposed that glacial climatic episodes began when Earth-sun geometry favored an extended period of increased solar radiation in winter and decreased solar radiation in summer at subpolar latitudes. More intense winter radiation at these latitudes translates into somewhat higher temperatures, higher humidity, and more snowfall. Weaker solar radiation in summer means that some of the winter snow cover, especially north of 60 degrees N, would survive summer, and a succession of many such cool summers would favor formation of a glacier. This process, according to Milankovitch, was the origin of the Laurentide ice sheet. At other times, Earth-sun geometry favored enhanced solar radiation in summer at higher latitudes, triggering an interglacial climate and shrinkage of the Laurentide ice sheet.

Milankovitch's ideas on the cause of large-scale fluctuations in the planet's glacial ice cover during the Pleistocene were largely ignored for 80 years. Then in 1976, analysis of deep-sea sediment cores revealed regular shifts between glacial and interglacial climatic episodes over the past 450,000 years that closely correspond to periodic changes in Earth's orbital parameters (i.e., precession, axial tilt, and eccentricity). This discovery strongly argues for Milankovitch cycles as the principal forcing of the regular large-scale waxing and waning of the planet's glacial ice cover during the Pleistocene. However, Milankovitch cycles do not account for the period of climate quiescence prior to the Pleistocene. Apparently, Milankovitch cycles (which likely operated throughout Earth history) were not effective in initiating continental-scale glaciation until plate tectonics provided the appropriate boundary conditions, that is, landmasses at high latitudes and mountain ranges in place—conditions not fully achieved until onset of the Pleistocene.

CLIMATE AND VOLCANOES

The idea that volcanic eruptions influence climate has been around for more than two centuries. Benjamin Franklin (1706-1790) proposed that eruption of Iceland's Laki volcano in the summer of 1783 was responsible for the severe winter of 1783-84. The unusually cool summer of 1816 (the so-called *year without a summer* in New England) followed the violent eruption of Tambora, an Indonesian volcano, in the spring of 1815. Several relatively cold years occurred on the heels of the 1883 eruption of Krakatau. Is the relationship between volcanic eruptions and cooling real or coincidental?

Only explosive volcanic eruptions that are rich in sulfur dioxide are likely to impact global or hemispheric climate and then only for a few years at most. A violent volcanic eruption can send sulfur dioxide (SO_2) high into the *stratosphere* (the atmospheric layer above the troposphere). In the stratosphere, sulfur dioxide combines with water vapor to form tiny droplets of sulfuric acid (H_2SO_4) and sulfate particles, collectively called sulfurous aerosols. The small size of **sulfurous aerosols** (averaging about 0.1 micrometer in diameter) coupled with the absence of precipitation in the stratosphere, allow sulfurous aerosols to remain suspended in the stratosphere for many months to perhaps a year or longer before they cycle to Earth's surface. Successive volcanic eruptions have produced a sulfurous aerosol veil in the stratosphere at altitudes of about 15 to 25 km (9 to 16 mi). Sulfur dioxide emissions from clusters of volcanic eruptions temporarily thicken the stratospheric aerosol veil.

Sulfurous aerosols absorb both incoming solar radiation and outgoing infrared radiation warming the lower stratosphere (especially in the tropics). Sulfurous aerosols also reflect solar radiation to space. For example, NASA scientists reported that in the months following the June 1991 eruption of Mount Pinatubo (on Luzon Island in the Philippines), satellite sensors detected a 3.8% increase in the amount of solar radiation reflected to space (Figure 12.12). These interactions reduce the amount of solar radiation reaching the troposphere and Earth's surface, translating into lower air temperatures.

A violent sulfur-rich volcanic eruption is unlikely to lower the mean hemispheric or global temperature by more than 1.0 Celsius degree (1.8 Fahrenheit degrees) although the magnitude of local and regional temperature change may be greater (Figure 12.13). The 1963 eruption of Agung in Bali lowered the mean temperature of the Northern Hemisphere an estimated 0.3 Celsius degree (0.5 Fahrenheit degree) for a year or two. The violent eruption of the Mexican volcano El Chichón in March-April 1982 may have produced hemispheric cooling of about 0.2 Celsius degree (0.4 Fahrenheit degree). The 1991 eruption of Mount Pinatubo injected an estimated 20 megatons of sulfur dioxide into the stratosphere, the most massive stratospheric volcanic aerosol cloud of the 20th century. Cooling associated with the Mount Pinatubo eruption temporarily interrupted the post-1970s global warming trend. From 1991 to 1992, the global mean annual temperature dropped 0.6 Celsius degree (1.1 Fahrenheit degrees).

Chapter 12 THE OCEAN AND CLIMATE CHANGE 293

FIGURE 12.12
The June 1991 explosive eruption of Mount Pinatubo (in the Philippines) was rich in sulfur dioxide. The resulting sulfurous aerosol veil in the stratosphere caused cooling at the Earth's surface, interrupting the post-1970s global warming trend for a few years. [U.S. Geological Survey photo]

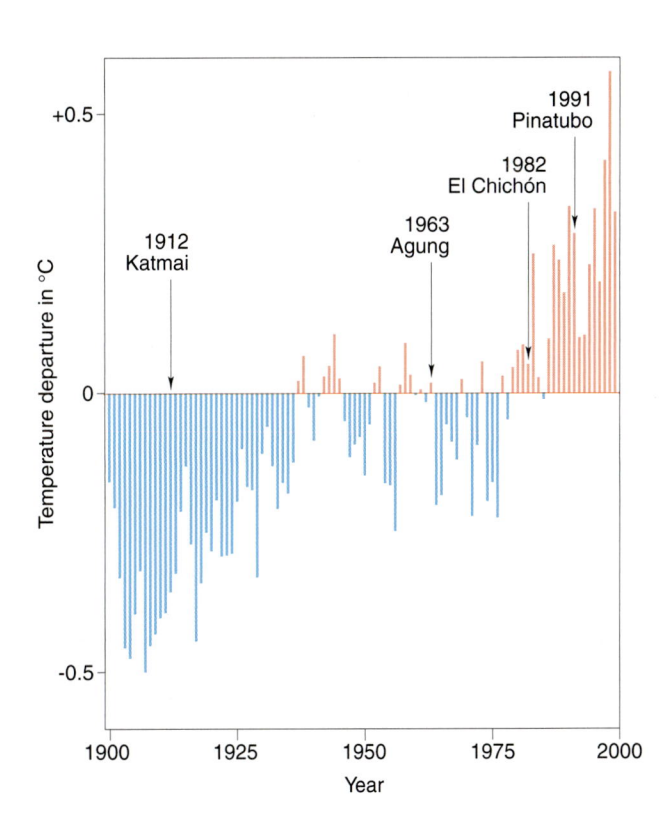

FIGURE 12.13
Large-scale cooling often followed massive volcanic eruptions that emitted sulfur dioxide into the stratosphere.

CLIMATE AND EARTH'S SURFACE PROPERTIES

Earth's surface, which is mostly ocean water, is the prime absorber of solar radiation (Chapter 5). Any change in the physical properties of Earth's water or land surfaces or in the relative distribution of ocean, land, and ice may affect Earth's radiation balance and climate.

Variations in mean regional snow cover may contribute to climate change because an extensive snow cover has a refrigerating effect on the atmosphere. Fresh-fallen snow typically reflects 80% or more of incident solar radiation, thereby substantially reducing the amount of solar heating and lowering the daily maximum temperature. Snow is also an excellent emitter of infrared radiation, so heat is efficiently radiated to space, especially on nights when the sky is clear. Because of this radiational feedback, a snow cover tends to be self-sustaining. An unusually extensive winter snow cover favors persistence of an episode of cold weather. On the other hand, less than the usual extent of winter snow cover raises average air temperatures.

Whereas changes in regional snow cover might impact climate over the short-term (seasonal), changes in Earth's sea ice or glacial ice coverage are likely to have longer-lasting effects on climate. Sea ice (formed from the freezing of seawater) covers an average area of about 25 million square km (9.6 million square mi), about the area of the North American continent. Terrestrial ice sheets, ice caps, and mountain glaciers cover a total area of about 15 million square km (5.8 million square mi), roughly 10% of the land area of the planet. Ice (especially snow-covered ice) is much more reflective of incident solar radiation than either the ocean or snow-free land so that any change in glacial or sea ice coverage would affect climate. As we will see later in this chapter, the Arctic climate is particularly sensitive to changes in sea ice cover.

Changes in ocean circulation and sea-surface temperatures contribute to large-scale climate change. As described in detail in Chapter 11, changes in sea-surface temperature patterns accompanying El Niño and La Niña significantly influence interannual climate variability. Ocean circulation includes warm and cold surface currents and the deep-ocean conveyor belt that transports heat throughout the world. Regular changes in the strength of these conveyor belts may explain millennial-scale (1400 to 1500 year) cycles over the past 10,000 years. A strong conveyor belt brings a relatively mild climate (for the latitude) to the North Atlantic and Western Europe whereas a weakening of the conveyor belt triggers cooling. Evidence shows that the actual turning of the conveyor belt on and off can happen as quickly as a decade or less.

CLIMATE AND HUMAN ACTIVITY

Human activity may contribute to climate change. Humans modify the landscape (e.g., urbanization, clear-cutting of forests) and thereby alter radiational properties of Earth's surface. Cities are slightly warmer than the surrounding countryside. Combustion of fossil fuels (i.e., coal, oil, and natural gas) alters concentrations of certain key gaseous and aerosol components of the atmosphere. Of these human impacts on the environment, the final one is most likely to impact climate on a hemispheric or global scale.

Many scientists as well as many public policy makers are concerned about the possible global climate impact of the steadily rising concentrations of atmospheric carbon dioxide (CO_2) and other infrared-absorbing gases. Higher levels of these gases appear likely to enhance the greenhouse effect and could be contributing to warming on a global scale. (Recall from Chapter 5, however, that water vapor is the chief greenhouse gas.)

In 1957, systematic monitoring of atmospheric carbon dioxide began at NOAA's Mauna Loa Observatory in Hawaii under the direction of Charles D. Keeling of Scripps Institution of Oceanography. The observatory is situated on the slope of a volcano 3400 m (11,200 ft) above sea level in the middle of the Pacific Ocean—sufficiently distant from major sources of air pollution that carbon dioxide levels are considered representative of at least the Northern Hemisphere. Also since 1957, atmospheric CO_2 has been monitored at the South Pole station of the U.S. Antarctic Program and that record closely parallels the one at Mauna Loa. The Mauna Loa and South Pole records both show a sustained increase in average annual atmospheric carbon dioxide concentration from about 316 ppmv (parts per million by volume) in 1959 to 379 ppmv in 2004 (Figure 12.14). Superimposed on this upward trend is an annual carbon dioxide cycle caused by seasonal changes in Northern Hemisphere vegetation. Levels of carbon dioxide fall during the growing season to a minimum in October, recover in winter, and reach a maximum in May.

The upward trend in atmospheric carbon dioxide was underway long before Keeling's monitoring and appears likely to continue well into the future (Figure 12.14). Humankind's contribution to the buildup of atmospheric CO_2 began roughly three centuries ago with the clearing of land for agriculture and settlement. Land clearing contributes CO_2 to the atmosphere via burning of vegetation, decay of wood residue, and reduced photosynthetic removal of carbon dioxide from the atmosphere. By the middle of the 19th century, growing dependency on coal burning associated with the beginnings of the Industrial Revolution

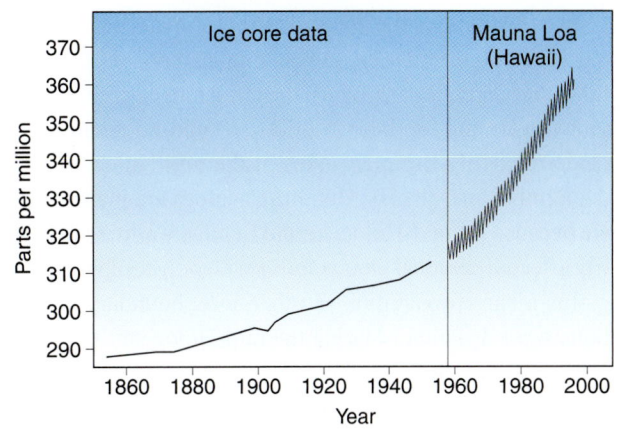

FIGURE 12.14
Reconstructed and measured trend in atmospheric carbon dioxide concentration. The record since 1957 is based on measurements made at the Mauna Loa Observatory in Hawaii. [Source of Mauna Loa data is C.D. Keeling et al., Scripps Institution of Oceanography, University of California, La Jolla, CA.]

triggered a more rapid rise in CO_2 concentration. Carbon dioxide is a byproduct of the burning of coal and other fossil fuels. The concentration of atmospheric CO_2 is now about 31% higher than it was in the pre-industrial era. Fossil fuel combustion accounts for roughly 75% of the increase in atmospheric carbon dioxide while deforestation (and other land clearing) is likely responsible for the balance. With continued growth in fossil fuel combustion, the atmospheric carbon dioxide concentration could top 550 ppmv (double the pre-industrial level) by the end of the 21st century. As noted in Chapter 9, the ocean is an important player in governing the amount of carbon dioxide in the atmosphere. From the beginning of the Industrial Revolution, scientists were able to estimate readily the amount of CO_2 released to the atmosphere by human activity. But in measuring the actual amount of carbon dioxide in the atmosphere, about 50% was missing. As it turns out, the ocean takes up about half of the carbon dioxide of anthropogenic origin (via photosynthesis and cold water absorbing CO_2 and sinking). Some anthropogenic carbon dioxide is transported with the oceanic conveyor belt and may be sequestered for 1000 years before returning to the air/sea interface.

Furthermore, rising levels of other infrared-absorbing gases (e.g., methane and nitrous oxide) could also enhance the greenhouse effect. Methane's concentration in the atmosphere has increased by an estimated 1060 ppb (parts per billion) (151%) since 1750. Methane (CH_4) is the product of organic decay in the absence of oxygen and its concentration may be rising because of more rice cultivation, cattle, landfills, and/or termites, all sources of methane. The atmospheric concentration of nitrous oxide (N_2O) has increased by an estimated 46 ppb (17%) since 1750 likely because of industrial air pollution. Although occurring in extremely low concentrations, methane and nitrous oxide are very efficient absorbers of infrared radiation and are probably contributing to global warming.

Atmospheric aerosols that are byproducts of human activity apparently have the opposite effect on temperatures at Earth's surface as greenhouse gases. These aerosols vary in size, shape, and chemical composition. Larger aerosols tend to settle out of the atmosphere quickly whereas smaller ones may remain suspended for many days to weeks and can be transported thousands of kilometers by the wind, possibly impacting large-scale climate. Perhaps 90% of these aerosols are byproducts of fossil fuel burning in the Northern Hemisphere.

Sulfur oxides emitted from power plant smokestacks and boiler vent pipes combine with water vapor in the air to produce tiny droplets of sulfuric acid and sulfate particles. These *sulfurous aerosols* appear to raise the atmosphere's albedo directly by reflecting sunlight to space and indirectly by acting as condensation nuclei that spur cloud development. Greater reflectivity cools the lower atmosphere. Sulfurous aerosols in the troposphere have a shorter-term impact on climate than carbon dioxide and other greenhouse gases. Rain and snow wash sulfurous aerosols from the atmosphere so that the residence time of these substances is typically only a few days. On the other hand, a CO_2 molecule typically resides in the atmosphere for upwards of three-quarters of a century before being cycled out by natural processes operating as part of the global carbon cycle.

The Climate Future

What does the climate future hold? Scientists attempt to answer this question primarily by relying on global climate models that run on supercomputers.

GLOBAL CLIMATE MODELS

A **global climate model** is a simulation of Earth's climate system. One type of global climate model consists of dozens of mathematical equations that describe the physical interactions among the various components of the climate system, that is, the atmosphere, ocean, land, cryosphere, and biosphere. A global climate model differs from numerical models used for weather forecasting in that it predicts broad regions of expected positive and negative temperature and precipitation *anomalies* (departures from long-term averages) over much longer time scales, and the mean location of circulation features such as jet streams and principal storm tracks.

Global climate models, for example, are used to predict the potential climatic impacts of rising levels of atmospheric carbon dioxide (or other greenhouse gases). Using current boundary conditions (i.e., climate controls), a global climate model simulates the present climate. Then, holding all other variables constant, the concentration of carbon dioxide (or another greenhouse gas) is elevated and the model is run to a new equilibrium state. By comparing the new climate state with the present climate, scientists deduce the impact of an enhanced greenhouse effect on patterns of temperature and precipitation.

Most modelers agree that global climate models are in need of considerable refinement. Today's models may not adequately simulate the role of small-scale weather systems (e.g., thunderstorms) or accurately portray local and regional conditions and may miss important feedback processes. A major uncertainty is the net feedback of clouds. Clouds cause both cooling (by reflecting sunlight to space) and warming (by absorbing and emitting to Earth's surface outgoing infrared radiation). The cooling effect prevails with an increase in low cloud cover whereas the warming effect prevails with an increase in high cloud cover.

Problems with global climate models stem in part from limited spatial resolution of the models. Today's models partition the global atmosphere into a three-dimensional grid of boxes with each box typically having an area of 250 square km (155 square mi) and a thickness of 1 km (0.6 mi). Limited spatial resolution in climate models stems from limited computational speed. Although today's supercomputers can perform 10 to 50 billion operations per second, the complexity of the climate system means that simulation of climate change over a century requires months of computing time. Much greater resolution of global climate models will come with development of faster supercomputers.

Another approach to forecasting future climate is to identify the various factors that may have contributed to past fluctuations in climate and to extrapolate their influence into the future. Atmospheric scientists have probed climate records in search of (1) regular cycles that might be extended into the future and (2) analogs that might provide clues as to how the climate in specific regions responds to global-scale climate change. Few of the quasi-regular oscillations appearing in the climate record have much practical value for climate forecasting at least over the next century.

Although the record of the climate past yields much useful information on how climate behaves through time, the search for analogs in the climate record for future global warming has been fruitless. Proposed analogs include relatively warm episodes of the mid-Holocene and the Eemian interglacial. But those analogs are inappropriate because the mid-Holocene and Eemian warming primarily affected seasonal temperatures with only a slight rise in global mean temperature. Furthermore, boundary conditions were different. During the mid-Holocene, sea level was lower, ice sheets were more extensive, and the dates of perihelion and aphelion were different than they are now or will be over the next several centuries. Although the level of atmospheric CO_2 trended upward during the mid-Holocene, the rate of increase (about 0.5 ppmv per century) was far lower than at present (more than 60 ppmv per century). Pre-Pleistocene analogs are also inappropriate because of the absence of ice sheets and significant differences in topography and land-ocean distribution.

ENHANCED GREENHOUSE EFFECT AND GLOBAL WARMING

Over the next 10 to 20 thousand years, Milankovitch cycles favor a return to Ice Age conditions. What about climate in the near term—over the next century or so? If all other controls of climate remain fixed, rising concentrations of atmospheric carbon dioxide and other greenhouse gases are likely to cause global warming to continue throughout this century.

How much warming might accompany a doubling of atmospheric CO_2? Based on the most recent runs of global climate models, the *Intergovernmental Panel on Climate Change (IPCC)* in 2000 revised upward its earlier projections of the magnitude of global warming that could accompany an enhanced greenhouse effect. Various models project that the globally averaged surface temperature will rise by 1.4 to 5.8 Celsius degrees (2.5 to 10.4 Fahrenheit degrees) over the period 1990 to 2100. Recall, however, that climate change is geographically non-uniform (in both magnitude and direction) so that this projected rise in global mean annual temperature is not necessarily representative of what might happen everywhere. For example, *polar amplification* suggests that global warming will be greater at higher latitudes. In any event, enhancement of the natural greenhouse effect could cause a climate change that would be greater in magnitude than any previous climate change over the past 10,000 years.

Models of global warming based on an enhanced greenhouse effect assume that all other climate controls remain constant. Whether that actually happens is not known. Comparing the post-1957 trend in atmospheric CO_2 to the trend in mean annual global temperature strongly suggests that recent climate has been shaped by many interacting factors. The rapid rise in CO_2 concentration was not accompanied by a consistent rise in global mean temperature over the same period. Recall, for example, that sulfurous aerosols from the June 1991 eruption of Mount Pinatubo apparently were responsible for significant global-scale cooling the following year. Also as noted earlier, El Niño and La Niña influence climate in some areas of the globe over periods of 12 to 18 months. Furthermore, analysis of tiny air bubbles in cores extracted from the Greenland and Antarctic ice sheets indicate that during the Pleistocene Ice Age, atmospheric CO_2 varied between about 260 and 280 ppmv. Although CO_2 was consistently higher during milder interglacial climatic episodes than during colder glacial climatic episodes, fluctuations in CO_2 levels are out of phase with reconstructed variations in climate. For example, at the beginning of the last major glacial climatic episode, the decline in CO_2 concentration significantly lagged cooling in the Antarctic. Hence, fluctuations in atmospheric CO_2 may have been a response to large-scale climate oscillations rather than a cause of those oscillations.

Impact of Climate Change on the Ocean

As we have seen, since the Little Ice Age ended in the mid 19th century, the global mean temperature has generally trended upward. In recent years, that warming trend has accelerated—possibly the result of the build up of greenhouse gases. If the warming trend continues, how might this climate change impact the ocean? Higher sea level is one consequence that has worldwide ramifications. In addition, warmer conditions in the Arctic may significantly reduce the sea ice cover with important implications for Earth's climate system. Furthermore, global warming is likely to impact marine life.

SEA LEVEL FLUCTUATIONS

Climate change, which was responsible for the waxing and waning of Earth's glacial ice cover during the Pleistocene Ice Age, also caused sea level to alternately fall and rise. Geological evidence such as drowned beaches and river valleys, and submarine canyons attests to periods when sea level was much lower than at present. Scientists estimate that during the last glacial maximum, about 18,000 to 20,000 years ago, mean sea level was 113 to 135 m (370 to 443 ft) lower than it is today. More than 90% of this drop in sea level was due to a change in the global water cycle brought about by a colder climate. As noted in Chapter 1, the total amount of water on the planet is essentially constant—an assumption that likely holds throughout recent geologic time. Practically all the water locked in glaciers came from the ocean via the global water cycle. During glacial climatic episodes, glaciers on land thicken and expand, and the volume of water in the ocean basins decreases (i.e., sea level falls). Conversely, during interglacial climatic episodes, glaciers on land thin and retreat, and the volume of water in the ocean basins increases (i.e., sea level rises). Furthermore, perhaps 7% or 8% of the sea level drop during the last glacial maximum was due to a decline in ocean temperature resulting in an increase in water density and contraction of the ocean water. Recall from Chapter 3 that seawater always contracts when its temperature drops and expands when its temperature rises.

The waxing and waning of glaciers plus ocean temperature fluctuations are two factors that govern **eustasy**, the global variation in sea level brought about by a change in the volume of water in the ocean basins. Another factor that contributes to changes in sea-level change and can be important over geologic time is alterations in the size of the ocean basins brought about by tectonic movements (Chapter 2). Continuation of the global warming trend appears likely to cause sea level to rise in response to melting of land-based polar ice sheets and mountain glaciers, and thermal expansion of seawater. Thermal expansion of warming oceans probably will be the greater contributor to sea level rise in the 21st century.

How did sea level respond to the warming trend of the 20th century? For most of the century, coastal tide gauges were the principal source of data on sea level change. Care must be taken to exclude (or adjust) those tide gauge records that might be influenced by tectonic uplift or subsidence or post-glacial rebound. Since 1993, sea level has also been measured remotely using satellite-borne microwave altimeters (Figure 12.15). (Refer to our description of TOPEX/Poseidon and its successor Jason 1 in Chapter 11). Between January 1993 and December 2000, TOPEX/Poseidon measured a global mean sea level rise of 2.5 ± 0.2 mm per year. In total, mean sea level rose about 18 cm (7.1 in.) during the 20th century. Most mountain glaciers have been shrinking since the mid 20th century; portions of the Greenland ice sheet have shown recent signs of accelerated melting; and portions of the ocean are warming. But

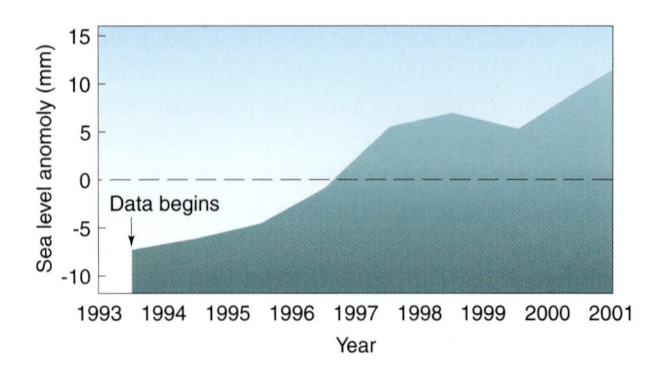

FIGURE 12.15
Yearly averaged sea level variation based on TOPEX/Poseidon measurements for 1993 to 2000. [Adapted from C. Cabanes *et al.*, "Sea Level Rise During Past 40 Years Determined from Satellite and in Situ Observations," *Science* 294(2001):840.]

how much of the recent rise in sea level was due to melting of glacial ice and how much was due to thermal expansion of the warming ocean is not known.

Amplification of the warming trend at higher latitudes would increase the threat to the Greenland and Antarctic ice sheets. About 90% of the planet's glacial ice blankets Antarctica and melting could cause a considerable rise in sea level. How likely is this to happen? Separate ice sheets cover West Antarctica and East Antarctica. The West Antarctic ice sheet sits on a former ocean bottom and is mostly below sea level whereas the East Antarctic ice sheet is situated on a continent and is well above sea level. While geological evidence suggests that the East Antarctic ice sheet has been stable for the past 15 million years, the West Antarctic sheet has undergone episodes of rapid disintegration and may have completely melted at least once in the past 600,000 years. More than 30 years ago, recognition of the relative instability of the West Antarctic ice sheet prompted some scientists to speculate that ice streams flowing from the interior of the glacier to the Ross and Ronne ice shelves might cause a total collapse of the ice sheet in a few centuries or less. Such a catastrophic event would greatly accelerate the rate of sea level rise. (Complete disintegration of the West Antarctic ice sheet would raise sea level by about 5 m or 16 ft.)

In 2001, concerns about the possible disintegration of the West Antarctic ice sheet were alleviated with the discovery that new snowfall is keeping pace with the loss of ice from bergs breaking off the Ross Ice Shelf. In early 2002, scientists at the California Institute of Technology and the University of California-Santa Cruz reported that based on satellite analysis of flow measurements of the Ross ice streams, the West Antarctic ice sheet appears to be thickening. Meanwhile, the region of the ice sheet that feeds the Thwaites and Pine Island glaciers is thinning. These glaciers transport ice directly into the ocean (rather than adding to an ice shelf) and are responsible for about 10% of the average annual rise in sea level (Figure 1.3). Nonetheless, the consensus of scientific opinion today is that the West Antarctic ice sheet is unlikely to suddenly accelerate the current rise in sea level. According to some experts, long-term gradual shrinkage of the West Antarctic ice could raise sea level at a rate of about 1 m (3.3 ft) per 500 years.

Melting (and the rate of sea level rise) could accelerate if the Antarctic ice sheets begin to feel the effects of global warming. Over most of Antarctica, the mean air temperature has fluctuated very little over the past 50 years. The Antarctic Peninsula is the only part of the continent that has shown significant warming with summer mean temperatures rising more than 2 Celsius degrees (3.6 Fahrenheit degrees) over the past half century. This warming has been accompanied by considerable ice breakup along the coast.

NASA research results released in 2000 concluded that while the central interior of the Greenland ice sheet showed no sign of thinning, about 70% of the margin was thinning substantially. Two separate research teams—one using a Global Positioning System (GPS) to monitor ice flow and the other relying on an airborne laser altimeter to measure ice thickness—reached the same conclusion based on observations made between 1993 and 1999. The maximum melting rate at the margin was about 1 m per year. An estimated 50 cubic km (12 cubic mi) of Greenland's ice melts each year, enough to raise sea level by 0.13 mm annually.

More recently, scientists at the University of Colorado reported that during the summer of 2002, surface melting on the Greenland ice sheet encompassed an area of about 695,000 square km (265,000 square mi)—about 9% greater than observed in 24 years of monitoring by satellite. In addition, melting in the northern and northeastern portion of the ice sheet occurred at elevations as high as 2000 m (6550 ft) where normally temperatures are too low for any melting. It appears that winter snowfall is insufficient to offset the summer melt so that overall the Greenland ice sheet is shrinking. Furthermore, by reducing frictional resistance within the ice mass, the greater supply of melt water accelerates the flow of Greenland ice streams into the ocean and increases the rate of rise of sea level.

According to Roger G. Barry, director of the National Snow and Ice Data Center at the University of Colorado at Boulder, the rate of melting of most of the world's mountain glaciers accelerated after the mid-1900s and especially since the mid-1970s. Some mountain glaciers have disappeared entirely. For example, today only 26 glaciers remain of the 150 glaciers present in Montana's Glacier National Park a century ago. Barry estimates that runoff from melting mountain glaciers contributes about 0.4 mm to the annual rise in sea level.

In Spring 2000, NOAA scientists reported that the combined Atlantic, Pacific, and Indian Oceans warmed significantly between 1955 and 1995. The greatest warming occurred in the upper 300 m (980 ft) of the ocean and amounted to 0.31 Celsius degree (0.56 Fahrenheit degree). Water in the upper 3000 m (9850 ft) warmed by an average 0.06 Celsius degree (0.11 Fahrenheit degree). In February 2002, researchers at Scripps Institution of Oceanography reported that temperatures at mid-depths (between 700 m and 1100 m) in the Southern Ocean rose 0.17 Celsius degree between the 1950s and 1980s. Although these magnitudes of temperature change may sound trivial, recall that water has an unusually high specific heat so that even a very small change in temperature of such vast volumes of water represents a tremendous heat input into the ocean. Sequestering of vast quantities of heat in the ocean may help explain why global warming during the 20th century was less than some climate models predicted based on the concurrent buildup of greenhouse gases. That is, heating of the ocean partially offset warming of the lower atmosphere. This finding also underscores the importance of the ocean's moderating influence on global climate change.

According to IPCC estimates, melting glaciers combined with thermal expansion of ocean water could raise mean sea level somewhere in the range of 9 to 80 cm (4 to 30 in.) over the period 1990 to 2100. Higher sea level would accelerate coastal erosion by wave action, inundate wetlands, estuaries and some islands, and make low-lying coastal plains more vulnerable to storm surges (Chapter 8). Rising sea level would disrupt coastal ecosystems and could threaten historical, cultural, and recreational resources (Figure 12.16). In some coastal areas, higher sea level is likely to exacerbate the problem of saltwater intrusion into groundwater. For more on this problem, refer to this chapter's third Essay.

According to a 1997 report by the U.S. Office of Science and Technology Policy, a 50-cm (20-in.) rise in sea level would result in a substantial loss of coastal land, especially along the U.S. Gulf and southern Atlantic coasts. Particularly vulnerable is South Florida where one-third of the Everglades is less than 30 cm (12 in.) above sea level. Globally, a 50-cm (20-in.) rise in sea level would double the number of people at risk from storm surges from about 45 million at present to more than 90 million, not counting any additional population growth in the coastal zone.

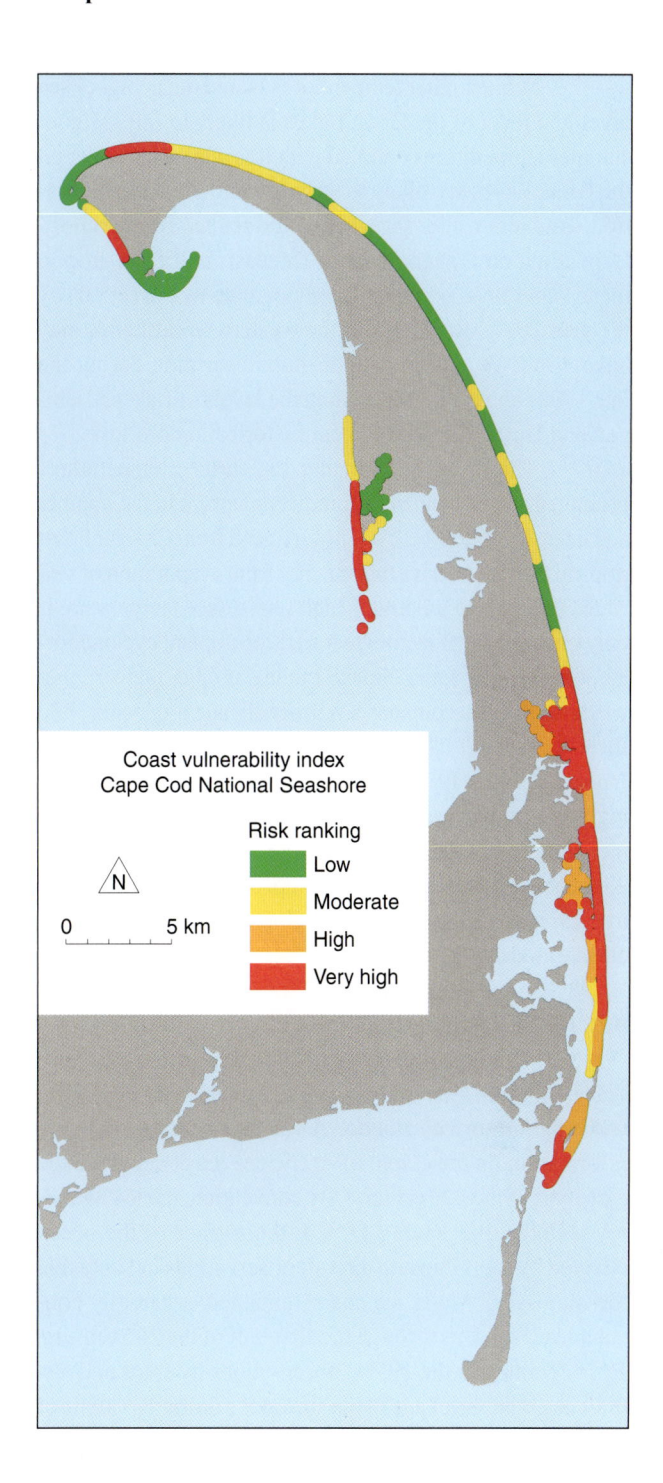

FIGURE 12.16
Map of the U.S. Geological Survey's Coastal Vulnerability Index (CVI) for Cape Cod National Seashore, MA showing the vulnerability of the coast to changes in sea level. The CVI is based on tidal range, wave height, coastal slope, historic shoreline change rates, geomorphology, and historical rates of relative sea-level change due to eustatic sea-level rise and tectonic uplift or subsidence. [USGS Fact Sheet FS-095-02, September 2002]

While higher temperatures would mean higher sea level, the level of the Great Lakes is likely to fall. Higher summer temperatures coupled with less winter ice cover on the Great Lakes are likely to translate into greater evaporation. And less winter snowfall would reduce spring runoff. Depending on the model used, forecasts call for a drop in mean water level on Lake Michigan of up to 2 m (6.5 ft) by the year 2070. Residents of the western Great Lakes may have previewed the impact of global warming during the late 1990s and early 2000s when the levels of Lakes Michigan and Huron dropped to near historical record lows.

The prospect of higher sea surface temperatures prompted some scientists to predict an upturn in the number and intensity of tropical cyclones. (Recall from Chapter 8 that tropical cyclones derive their energy from evaporation of seawater.). However, relatively high sea surface temperature is only one of several factors required for tropical cyclone formation. In fact, some climate models predict that stronger winds aloft will accompany a warmer climate producing wind shear that would inhibit tropical cyclone development. In January 2001, the IPCC concluded that in spite of global warming, there is no evidence that top wind speeds or rainfall had increased in tropical cyclones during the second half of the 20th century. Resolving the likely impact of global warming on tropical cyclones awaits the development of more realistic climate models.

ARCTIC SEA ICE COVER

Although melting of floating sea ice does not raise sea level, it can alter climate in a significant way by greatly reducing the surface albedo. A variety of sources provide information on the extent of Arctic sea ice cover since the early part of the 20th century. By convention, the ice boundary is defined as having 10% of the surface of the ocean covered by ice. Ship and aircraft observations indicate that the multi-year Arctic ice cover remained essentially constant in all seasons through the first half of the 20th century. But beginning in the 1950s, observations by ships and aircraft detected shrinkage in the summer minimum extent of ice while the winter maximum remained nearly constant. By the mid-1970s, satellite surveillance, submarines, and ice-cores were finding a decline in the winter maximum as well.

Norwegian researchers reported that the area covered by multi-year ice in the Arctic decreased by about 14% between 1978 and 1998. Their findings were based on satellite monitoring of the spectrum of microwave energy emitted by the ice. Comparing ice-thickness measurements made by U.S. Navy submarines from 1958 to 1976 with those made during the *Scientific Ice Expedition* program in 1993, 1996, and 1997, University of Washington scientists found that ice had thinned from an average thickness of 3.1 m (10 ft) to an average thickness of 1.8 m (6 ft). Thinning at the rate of 15% per decade translates into a total volumetric ice loss of about 40% in three decades. (Upward-looking acoustic sounders were used to map ice depth.)

From analysis of satellite data, scientists from the Cooperative Institute for Research in Environmental Sciences (CIRES) at the University of Colorado reported that in 2002 the extent of Arctic sea ice was the lowest in the satellite record, likely the lowest since the early 1950s, and perhaps the lowest in several centuries. In September 2002, sea ice covered about 5.3 million square km (2 million square mi) compared to the long-term average of about 6.3 million square km (2.4 million square mi). If current trends continue, by 2050 there may be a 20% reduction in mean annual sea ice in the Arctic and little or no ice in summer.

Shrinkage of Arctic sea ice raises concerns about a possible ice-albedo feedback mechanism that would accelerate melting of sea-ice and amplify warming. Sea ice insulates the overlying air from warmer seawater and reflects much more incident solar radiation than ocean water. Snow-covered sea ice has an albedo of about 85% whereas ice-free Arctic Ocean water has an average albedo of only about 7%. As sea ice cover shrinks, the greater area of ice-free ocean waters will absorb more solar radiation, sea-surface temperature will rise, and more ice will melt. This positive feedback mechanism could rapidly reduce the sea ice cover and greatly alter the flux of heat energy and moisture between the ocean and atmosphere with possible ramifications for global climate.

Less sea ice cover on the Arctic Ocean is likely to increase the humidity of the overlying air leading to more cloudiness. As pointed out earlier in this chapter, clouds cause both cooling (by reflecting sunlight to space) and warming (by absorbing and emitting to Earth's surface outgoing infrared radiation). During the long dark polar winter, clouds would have a warming effect. In summer, the impact of a greater cloud cover depends on the height of the clouds. Cooling would prevail with an increase in low cloud cover whereas warming would likely accompany an increase in high cloud cover.

Shrinkage of the Arctic sea ice may be the direct consequence of higher air temperatures or indirectly the result of changes in ocean circulation (i.e., greater input of warmer Atlantic water under the Arctic ice). Some scientists argue that higher air and ocean temperatures in the Arctic are natural variations in climate associated with the *Arctic Oscilla-*

tion (AO) and that the ice-cover will return to normal after the AO changes phase. As noted in Chapter 11, during the present AO phase winds have been delivering warmer than usual air and seawater into the Arctic.

Mean annual air temperatures in the Arctic climbed by about 0.5 Celsius degree (0.9 Fahrenheit degree) per decade over the past thirty years. Proxy climatic data indicate that present temperatures (especially in winter and spring) may be at their highest level in four centuries. In response, mountain glaciers in Alaska are shrinking at historically unprecedented rates, *permafrost* (permanently frozen ground) is beginning to thaw, and freshwater runoff into the ocean has increased. More winter precipitation combined with an increased flow of groundwater (due to melting permafrost) is likely responsible for a 7% increase in the discharge of six major Eurasian rivers into the Arctic Ocean since the 1930s.

Input of more fresh surface water (from rivers and melting glacial ice) would impact the oceanic conveyor belt by reducing the salinity and thus the density of surface ocean waters. Recall from Chapter 6 that as part of the conveyor belt, relatively dense cold, salty water sinks at high latitudes of the Atlantic Ocean. As pointed out earlier in this chapter, changes in strength of the oceanic conveyor belt affect air temperatures over the North Atlantic and Western Europe. Any alteration of the oceanic conveyor belt is likely to impact the poleward transport of heat (Chapter 5). Paradoxically, warming at high latitudes could lead to regional cooling due to a weakening of the oceanic conveyor belt.

MARINE LIFE

How might climate change impact marine life? As described in Chapters 9 and 10, marine animals and plants are components of ecosystems. An *ecosystem* is a fundamental subdivision of the Earth system in which organisms depend for their survival on other organisms and the physical and chemical constituents of their surroundings. Within and between ecosystems, materials and energy flow from one organism to another via food webs. Climate change could alter the physical and chemical conditions in the ocean, perhaps exceeding the tolerance limits of organisms. If these organisms are unable to avoid these stressful conditions, they may perish. For example, global warming could raise the sea surface temperature sufficiently to threaten coral. As noted in Chapter 10, coral polyps are sensitive to small changes in water temperature and prolonged periods of exposure to excessively warm waters can lead to coral bleaching and death. Because of the interdependency of organisms (e.g., for food, habitat), loss of one species can have a disruptive effect on food webs and the entire ecosystem.

Organisms are particularly vulnerable to environmental change that affects a limiting factor. A *limiting factor* is an essential resource that is in lowest supply compared to what is required by the organism. For example, a climate change may indirectly affect the supply of an essential nutrient. Concurrent shifts in atmospheric circulation may suppress upwelling in some locations or reduce the transport of micronutrients to portions of the open ocean. Furthermore, rising sea level is likely to alter marine habitats especially in the coastal zone. In some cases—where the coastline has been developed for roads and buildings—flooding may eliminate marine habitats entirely causing the demise of organisms that are dependent on those habitats.

Consider, for example, the fate of salt marshes already threatened by development. Higher temperatures will increase the rate of evaporation from the soil surface thereby elevating the soil salinity. More saline soils will cut biological productivity and may exceed the tolerance limits of plants. Rising sea level threatens to drown salt marshes. If plants cannot shift inland to higher ground, their fate is sealed and the marsh may be lost. Loss of salt marshes, in turn, will make the coastline more vulnerable to flooding and erosion by storm waves. Furthermore the filtration function of salt marshes will be lost so that higher levels of pollutants and nutrients will enter estuaries with runoff.

A key consideration in the impact of climate change on marine organisms is the rate of that change. Marine ecosystems can more readily adjust to gradual rather than abrupt changes in the ocean environment brought on by global scale climate fluctuations. But as noted earlier in this chapter, the climate change due to an enhanced greenhouse effect could be without precedent in the past 10,000 years. If this accelerated global warming materializes, considerable disruption of marine ecosystems may occur.

Conclusions

Climate varies not only spatially but also with time. In most places, the reliable instrument-based climatic record extends back only 120 years or so. For information on climate prior to this period, scientists must rely on a variety of documentary, geological, and biological proxy climatic indicators. Although the climate record loses detail, continuity, and reliability with increasing time before present, it is evident that climate is inherently variable and changes over a wide spectrum of time scales.

The interaction of many factors is responsible for the inherent variability of climate. Although we can isolate specific climatic controls that are internal or external to the Earth-atmosphere-ocean system, our understanding of how these controls interact is far from complete. This state of the art limits the ability of scientists to forecast the climatic future. We should therefore be wary of simplistic scenarios of the climatic future, for not enough is known about the causes of climate change.

Continued research on climate is needed. It is reasonable to assume that physical laws govern climatic change; that is, variations in climate are not arbitrary, random events. As scientists more fully comprehend the laws regulating climatic change and especially the role played by the ocean in Earth's climate system, their ability to predict the climatic future will improve. Meanwhile, trends in climate must be monitored closely, especially in view of the potential wide ranging impacts of climate change.

Basic Understandings

- For information on climate prior to the reliable instrument-based era, scientists rely on climatic inferences drawn from historical documents, and geological/biological proxy climatic evidence such as bedrock, fossils, pollen, tree growth rings, glacial ice cores, and deep-sea sediment cores.
- Much of what is known about the climatic fluctuations of the Pleistocene Ice Age is based on analysis of the shell and skeletal remains of microscopic marine organisms that are extracted from deep-sea sediment cores. Identification of the organism plus oxygen isotope analysis of remains enable scientists to distinguish between cold and mild climatic episodes of the past.
- Plate tectonics complicate climate reconstruction of periods spanning hundreds of millions of years. In the context of geologic time, topography and the geographical distribution of continents and the ocean are variable controls of climate.
- By 40 million years ago, global climate began shifting toward cooler, drier, and more variable conditions. Scientists have implicated tectonic forces and the building of the Colorado Plateau, Tibetan Plateau, and Himalayan Mountains as the principal causes of this climatic change.
- Cooling culminated in the Pleistocene Ice Age about 1.7 million years ago. During the Pleistocene, the climate shifted numerous times between glacial climatic episodes (favoring expansion of glaciers) and interglacial climatic episodes (favoring shrinkage of glaciers).
- Cooling during the Pleistocene was geographically non-uniform in magnitude with maximum cooling at high latitudes and minimum cooling in the tropics. This latitudinal variation in temperature change is known as *polar amplification*.
- Notable post-glacial climatic episodes were the relatively warm mid-Holocene, the Medieval Warm Period, from about 950 to 1250 AD, and the Little Ice Age, from about 1400 to 1850 AD. In all cases, the temperature change was geographically non-uniform in magnitude.
- The instrument-based record of global mean temperature indicates a gradual warming trend since 1880, interrupted by cooling from about 1940 to 1970. The warming trend appears to have accelerated in the 1990s.
- Analysis of the climate record reveals many useful observations about the temporal behavior of climate. Climate is inherently variable; climate change is geographically non-uniform in direction and magnitude; climate change may involve changes in mean values of temperature or precipitation as well as changes in the frequency of weather extremes; climate change tends to be abrupt rather than gradual; the climate record contains few cycles that are sufficiently reliable to permit forecasting climate over the next century; and climate change impacts society.
- Matching a particular forcing mechanism with a climate response based on similar periods of oscillation is no guarantee of a real physical relationship. Factors that could alter the global radiative equilibrium and change Earth's climate include fluctuations in solar energy output, volcanic eruptions, changes in Earth's surface properties, and certain human activities.
- Changes in the sun's total energy output are apparently related to sunspot activity. Solar output varies directly and minutely with sunspot number; that is, a slightly brighter sun has more sunspots (because of a concurrent increase in faculae, bright areas), and a slightly dimmer sun exhibits fewer sunspots.
- Milankovitch cycles drive climatic oscillations operating over tens of thousands to hundreds of thousands of years and were likely responsible for the major advances and recessions of the Laurentide ice sheet over North America during the Pleistocene. They consist of regular variations in precession and tilt of Earth's rotational axis and the eccentricity of its orbit about the sun. These same cycles also show up in deep-sea sediment cores that date from the Pleistocene Ice Age.

- Only violent volcanic eruptions rich in sulfur dioxide gases are likely to impact hemispheric or global-scale climate. Such eruptions are unlikely to lower the mean annual global surface temperature by more than about 1.0 Celsius degree (1.8 Fahrenheit degrees) for a year or two.
- Earth's surface, which is mostly ocean water, is the prime absorber of solar radiation so that any change in the physical properties of water or land surfaces or in the relative distribution of ocean and land may impact the global radiation balance and climate.
- Human activity may impact global-scale climate by elevating the concentration of greenhouse gases (causing warming) or sulfurous aerosols (causing cooling). The current upward trend in atmospheric carbon dioxide is primarily due to burning of fossil fuels and to a lesser extent the clearing of vegetation.
- Global climate models consist of a series of mathematical equations that describes the physical laws that govern the interactions among the various components of the climate system. They are used to predict broad regions of expected positive and negative temperature and precipitation anomalies.
- The climate record does not contain reliable cycles that can be used to forecast the climate over a period of several centuries. Also, no satisfactory analogs exist in the climate record that would allow scientists to predict regional responses to future global-scale climate change.
- Current global climate models predict that significant global warming will accompany a doubling of atmospheric CO_2 (possible by the close of this century). Warming may be greater in magnitude than any other prior climate change during the history of civilization.
- Projections of the magnitude of future global warming arising from CO_2-enhancement of the greenhouse effect assumes that all other controls of climate remain constant. To what extent this will happen is not known.
- Global warming is likely to cause sea level to rise in response to melting of polar ice sheets and mountain glaciers plus thermal expansion of seawater. Higher sea level would accelerate coastal erosion, inundate wetlands, estuaries and some islands, and make low-lying coastal plains more vulnerable to storm surges. Rising sea level would disrupt coastal ecosystems and could threaten historical, cultural, and recreational resources.
- Recent shrinkage of Arctic sea ice cover raises concerns about a possible ice-albedo feedback mechanism that would accelerate melting of sea-ice and amplify warming in the Arctic. Sea ice insulates the overlying air from warmer seawater and reflects much more incident solar radiation than ocean water. If sea ice cover shrinks, the greater area of ice-free ocean waters will absorb more solar radiation, sea surface temperatures will rise, and more ice will melt. This positive feedback mechanism could rapidly reduce the sea ice cover and greatly alter the flux of heat energy and moisture between the ocean and atmosphere with possible ramifications for global climate.
- Global warming could disrupt the functioning of marine ecosystems by exceeding tolerance limits or destroying habitats.

ESSAY: Climate Rhythms in Glacial Ice Cores

The Pleistocene Epoch, the most recent of Earth's Ice Ages, began about 1.7 million years ago and ended about 10,500 years ago. Since then, conditions in the present Epoch (the Holocene) have been reasonably mild, with relatively minor temperature fluctuations compared to the Pleistocene. In an effort to better understand climate change, scientists are collecting and analyzing climatic data from the Pleistocene Ice Age with an eye toward predicting future climate (especially in view of the exploding world population and the influence of climate on energy demand and supplies of food and fresh water). An important source of data on the climate as well as the chemical composition of air during the Pleistocene is ice cores extracted from the polar ice sheets.

In 1988, Soviet and French scientists reported on their analysis of a 2200-m (7200-ft) ice core extracted at Vostok station on the East Antarctic ice sheet. The ice core spanned 160,000 years. Oxygen isotope analysis yielded a temperature record and chemical analysis of trapped air bubbles revealed trends in the greenhouse gases carbon dioxide and methane. In the mid-1990s, drilling at Vostok recovered a 3100-m (10,170-ft) ice core spanning the past 420,000 years. More recently, in 2004, the European Project for Ice Coring in Antarctica (EPICA) extracted an ice core from East Antarctica representing a time interval of 740,000 years. During the summers of 1991-93, two independent scientific teams, one American and the other European, drilled into the thickest portion of the Greenland ice sheet. The two drill sites were located within 30 km (19 mi) of each other, about 650 km (403 mi) north of the Arctic Circle. Both cores were about 3000 m (9850 ft) in length and spanned a time interval of roughly 200,000 years.

Ice cores from both Greenland and Antarctica clearly reveal an approximately 100,000 year Ice Age cycle consisting of cold glacial climatic episodes (e.g., the Wisconsinan stage) sandwiched between mild interglacial climatic episodes (e.g., the Holocene). Perhaps 16 of these long-term cycles operated over the 1.7 million years of the Pleistocene Epoch. As discussed elsewhere in this chapter, evidence from deep-sea sediment cores indicates that regular variations in Earth-sun geometry (the Milankovitch cycles) drive this approximately 100,000-year glacial/interglacial cycle.

The Greenland and Antarctic ice core records correlate well both in terms of magnitude of temperature change and the timing of events indicating that the 100,000-year Ice Age cycles were globally synchronous. However, comparison of the Greenland and Antarctic ice core data over the most recent Ice Age cycle (i.e., from about 142,000 years ago to 10,500 years ago) reveals marked differences between the Southern and Northern Hemisphere. Whereas the Antarctic record is reasonably smooth and "calm," the Greenland record shows numerous abrupt and drastic flip-flops between glacial and interglacial climatic episodes. Temperatures changed as much as 7 Celsius degrees (12.6 Fahrenheit degrees) over periods of decades or less (in some cases in only 3 years.) These abrupt temperature changes—having two basic periods of 2000 to 3000 years and 7000 to 12,000 years—occurred during the Wisconsinan stage but not during the subsequent Holocene Epoch. The periods of these temperature fluctuations are much shorter than those of the Milankovitch cycles and hence are probably unrelated to changes in Earth-sun geometry.

The most likely explanation for these short-term abrupt changes in temperature is the weakening and strengthening of the oceanic conveyor belt (Chapter 6). For example, as discussed elsewhere in this chapter, the turning on and off of the conveyor belt may explain the occurrence of the *Younger Dryas* cold episode 11,000 to 10,000 years ago. The *Younger Dryas* began abruptly when a sudden influx of fresh water discharged by the St. Lawrence River into the North Atlantic prevented the formation of North Atlantic Deep Water (NADW), shutting down (or diverting) the conveyor belt. With the shutdown of the conveyor belt, temperatures in the North Atlantic and the surrounding lands plunged. The *Younger Dryas* ended just as abruptly as it began when the input of fresh water into the North Atlantic decreased, formation of NADW resumed, and the conveyor belt restarted. The geographic pattern of the *Younger Dryas* climatic impacts (e.g., little in western North America) and only a muted response in the Antarctic ice core record suggest that *Younger Dryas* was not part of the larger ice age variability driven by the Milankovich cycles. Rather, the *Younger Dryas* was a regional shorter-term climatic fluctuation linked to changes in the Atlantic conveyor belt circulation.

Changes in the conveyor belt circulation that delineate the *Younger Dryas* also occurred at other times. Scientists have interpreted certain layers of lithogenous sediment in cores extracted from the floor of the North Atlantic as materials released during melting of fleets of icebergs. These icebergs surged or slid off glaciated North America and floated out onto the North Atlantic every 2000 to 3000 years as the climate flip-flopped between warm and cold episodes. Melting of the icebergs freshened the North Atlantic surface waters turning off or diverting the conveyor belt. With colder conditions and fewer icebergs, freshening of the surface waters ceased, the water became salty again due to wind-driven evaporation, and the conveyor belt resumed. After two or three of these events, an even greater discharge of icebergs occurred at intervals of 7000 to 12,000 years. The smaller, shorter, more frequent events are called *Dansgaard-Oeschger events* (named for the paleoclimatologists Willi Dansgaard and Hans Oeschger) or "flickers" because of their relatively short period. The Greenland ice core record contains some 23 Dansgaard-Oeschger events during the period from 110,000 to 15,000 years ago. The larger, longer, and less frequent events are known as *Heinrich events* (discussed in the first Essay of Chapter 4). Flickers and Heinrich events occurred during both glacial and interglacial times and are evident in the temperature record reconstructed from Greenland ice cores.

ESSAY: The Drying of the Mediterranean Sea

The continents of Africa, Europe, and Asia surround the Mediterranean Sea, which is connected to the Atlantic Ocean through the narrow Strait of Gibraltar. The Mediterranean is a remnant of the once vast *Tethys Sea* but was nearly squeezed shut during the Oligocene Epoch, 23 to 33 million years ago. The Mediterranean continues to be tectonically active and to shrink slowly in size as the African plate pushes northward and subducts under the Eurasian plate. This subduction has caused folding and uplift of marine sedimentary rocks to form the Alps along with volcanic activity along the northern edge of the Mediterranean. The Mediterranean is about 3900 km (2400 mi) long and 1600 km (1000 mi) wide. The average water depth is 1500 m (4900 ft) with a maximum depth of 5150 m (16,900 ft) off the south coast of Greece. A bathymetric sill in the Strait of Gibraltar separates the Mediterranean basin from the Atlantic Ocean basin; the maximum water depth over the sill is only 284 m (931 ft) at a point where the Strait is about 30 km (18.6 mi) wide.

In the 1960s, William Ryan of the Lamont-Doherty Earth Observatory at Columbia University made a remarkable discovery regarding the composition of sediments on the floor of the Mediterranean basin. While sailing in the Mediterranean on the *R/V Chain* from Woods Hole Oceanographic Institution, Ryan was using a new continuous seismic profiler that could penetrate into sea bottom sediments. The acoustic return revealed a reflecting layer 100 to 200 m (325 to 650 ft) beneath the sea bottom, which was whimsically labeled the "mysterious layer," or M-layer. Through subsequent years of collecting M-layer data, it soon became obvious that the layer was ubiquitous in the Mediterranean and was deposited after the deep basin of the Mediterranean had already formed and had almost the same bathymetry as today.

In 1972, Ryan and Kenneth Hsü (a Swiss geologist) onboard the drill ship *Glomar Challenger* in the western Mediterranean brought up the first cores of the M-layer. The cores resembled marble and were labeled "the pillars of Atlantis." The core material turned out to be anhydrite (calcium sulfate, $CaSO_4$) and stromatolites (mats of the remains of sediment-trapping cyanobacteria). They dated from the late Miocene Epoch, about 5 million years old. These two types of sediment core material were extremely unusual in that anhydrites form only in hot, dry deserts where salty groundwater close to the desert surface evaporates causing calcium sulfate to precipitate as a solid. How is it possible for evaporites to form beneath 200 m (650 ft) of marine sediments on the bottom of the Mediterranean Sea? Stromatolites also appear out of place on the bottom of the present Mediterranean. Today, stromatolites form in broad, intertidal mud flats in the Bahamas and in salty bays in Western Australia and require light for photosynthesis. What are they doing on the bottom of the Mediterranean, and why are they associated with desert evaporites?

Determination of the M-layer composition points to one of the most extraordinary events in the history of the planet Earth. The atmosphere, ocean, plate tectonics, and glaciation combined to drastically alter the climate of the Mediterranean Sea. About 6 million years ago, the Mediterranean Sea became isolated from the Atlantic Ocean and almost completely dried up. For some 1 to 2 million years, the basin bottom consisted of deserts, salty lakes, and salt marshes. Then the water refilled the basin pouring over the sill at the Strait of Gibraltar as a spectacular waterfall lasting about 100 years.

Several hypotheses seek to explain the drying up of the Mediterranean, most of which invoke tectonic plate movement or large-scale glaciation, plus an extended period of evaporation. The tectonic hypothesis suggests that the entire basin was uplifted and the Mediterranean Sea emptied into the Atlantic, and later dropped back down and refilled. This cycle may have happened not once, but several times. The driving force for uplift would be subduction of the African plate under the Eurasian plate, which has caused similar uplift in the Alps. Another hypothesis attributes the drying to large-scale glaciation that caused sea level to fall below the sill at Gibraltar, cutting off input of water from the Atlantic. This is a reasonable hypothesis considering that the sill at Gibraltar has a maximum depth of only 284 m (931 ft). When the ice sheets melted, sea level rose above the sill and refilled the Mediterranean basin. Other hypotheses propose that either the Strait of Gibraltar was squeezed closed or parts of the Atlantic Ocean floor were deformed and uplifted forming a gate that alternately closed and opened the Strait of Gibraltar.

Evaporation of the isolated Mediterranean Sea is readily explained by reconstructions of the late Miocene climate. The rock record shows that at the time, the climate in the Mediterranean region probably was as hot and arid as it is today with little input of water by rivers and streams—the type of climate that would evaporate the Mediterranean waters leaving behind anhydrite deposits.

The most reasonable explanation for the drying of the Mediterranean Sea is likely some combination of factors. In the late Miocene, tectonic stresses (causing moderate regional uplift) combined with falling sea level dropped the Atlantic Ocean below the sill at Gibraltar, cutting off the Mediterranean from the Atlantic. Over the subsequent 1000 years, the waters of the Mediterranean evaporated, eventually exposing the bottom of the basin where deserts formed among shallow hypersaline lakes, salt marshes, and salt flats. With the loss of the Mediterranean Sea, the region probably became cooler and even more arid. In the deeper eastern Mediterranean basin, a system of salty lakes, similar to North America's Great Salt Lake, was fed by salty overflow from the Black Sea to the east. Such conditions can produce both anhydrites and stromatolites together at the bottom of the Mediterranean basin. Alternating layers of biogenous sediments and evaporites suggest that over the next two million years, the Mediterranean probably partial refilled and dried up between 8 and 40 times. Finally about 4.5 million years ago, tectonic forces relaxed dropping Gibraltar and/or sea level rose causing the Atlantic waters to spill over the Gibraltar sill, refilling the basin. Hsü calculated that the flow of water over the Gibraltar sill was about 1000 times greater than Niagara Falls and yet took a century to fill the Mediterranean basin.

ESSAY: Sea Level Rise and Saltwater Intrusion

Rising sea level threatens to exacerbate the problem of saltwater intrusion in certain low-lying coastal areas. *Saltwater intrusion* is the movement of saltwater into subsurface zones previously occupied by fresh groundwater and is most common along flat coastal plains (e.g., Florida and southeastern Georgia) and islands. Water that completely fills the openings (e.g., pore space, fractures) in sediment and rock is known as *groundwater*. An *aquifer* consists of porous and permeable earth material that is saturated with water and will yield water in usable quantities to a well or spring. In general, groundwater under land is fresh whereas groundwater under the ocean is salty. Fresh groundwater is the single most important source of potable water for mankind.

In coastal areas where fresh groundwater is situated adjacent to salty marine groundwater, excessive withdrawal of fresh groundwater can allow salt water to migrate inland. As saltwater replaces fresh water in an aquifer, coastal wells begin to deliver saltwater. One of the most serious consequences of saltwater intrusion is contamination of fresh water sources so that they cannot be used for domestic purposes or irrigating crops.

The potential for saltwater intrusion is greatest in fresh water aquifers that are hydraulically connected to seawater. In such a system, a transition zone develops between fresh groundwater and salty groundwater. Saltwater is denser than fresh water so that a higher hydraulic head of fresh water is needed to balance the hydraulic head of saltwater and keep the interface offshore. Because the dry land surface is higher than the ocean bottom, this is usually the case. But when large amounts of water are pumped from a fresh-water coastal aquifer (without replacement), the accompanying change in hydraulic gradient encourages the flow of saltwater toward wells. That is, the interface migrates inland and upward and well water turns salty. Rising sea level increases the potential for saltwater intrusion by increasing the hydraulic head of saltwater.

Islands are particularly vulnerable to saltwater intrusion. On an island subsurface freshwater floats like a lens on the denser underlying saltwater. Excess pumping of wells on an island or rising sea level causes upward movement of saltwater in wells. Increasing withdrawal of groundwater that has accompanied the development of Hilton Head Island, SC is causing the subsurface saltwater wedge to gradually migrate toward the freshwater aquifer. The U.S. Geological Survey reports that salty groundwater is approaching the northeast side of the island at about 30 to 43 m (100 to 140 ft) per year and according to model predictions, saltwater could contaminate most of the island's wells by 2032.

What can be done to prevent or alleviate salt-water intrusion? Reducing the rate of groundwater withdrawal and relocating wells so that they are farther apart will help. Artificially recharging the freshwater aquifer (using injection wells) diminishes seawater encroachment. Another strategy is to drill extraction wells to remove seawater from an aquifer before it has a chance to migrate to producing wells. Prior to employing any of these strategies, however, it is important to determine the geologic and hydrologic properties of the aquifer and to install monitoring wells to locate the subsurface salt/fresh water transition zone.

CHAPTER A

THE FUTURE OF OCEAN SCIENCE

Investigating the Ocean
    Voyages of Exploration
    Challenger Expedition (1872-1876)
Modern Ocean Studies
    Technological Innovations
    Remote Sensing
    Scientific Ocean Drilling
Emerging Ocean-Sensing Technologies
    Autonomous Instrumented Platforms and Vehicles
    Ocean Floor Observatories
    Animal-Borne Instruments
    Computers and Numerical Models
Challenges in Ocean-Sensing Technologies
Conclusions
Basic Understandings

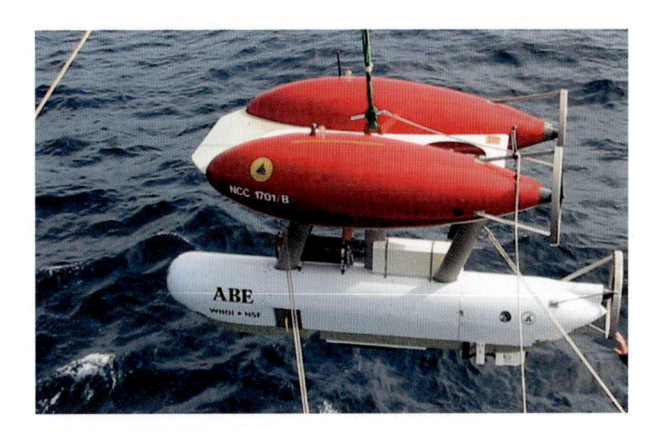

The *Autonomous Benthic Explorer* (ABE), operated by the Woods Hole Oceanographic Institution. [NOAA photo]

The first systematic scientific observations of the ocean were made in the late 19th century from the decks of a converted sailing warship. A hundred years later, Earth-orbiting satellites were routinely providing ocean scientists with a unique perspective of the ocean surface that is impossible to obtain from ships at sea. Prior to remote sensing by satellite, scientists mapped the ocean by combining data from ship-borne observations often separated by thousands of kilometers in distance and decades in time. Today, sensors onboard satellites provide ocean data instantaneously and can monitor the entire surface of the planet in just a few hours. While surface and deep ocean data still must be collected to calibrate and verify remotely sensed observations, using these modern sources of higher resolution data, ocean scientists can observe changes in major ocean currents and sea surface temperature, map sea surface topography, and detect early signs of a developing El Niño or La Niña.

Throughout history, the ocean has commanded a prominent position in mythology, religion, and literature as a mysterious and threatening place. Today, we continue to regard the ocean with awe and treat it with respect. In this chapter, we examine humankind's efforts to learn more about the ocean, that is, to map the ocean floor, measure the properties of seawater, and monitor marine life as we seek to understand ocean's role in the Earth system.

Humankind's ability to investigate the ocean depends on the availability of appropriate observing systems. The vastness of the ocean continues to challenge the available observational technologies. For one, satellite-borne sensors are limited to observing mostly surface waters. Even though remote sensing by satellite is developing rapidly, the ocean—especially the deep ocean—remains undersampled so that many of its properties and processes are still poorly understood. Scientists too frequently are limited by their inability to sample and measure the properties of seawater and marine life. As one renowned ocean scientist put it, our scientific models of the ocean often have a "curious, dreamlike quality." This chapter summarizes humankind's efforts to investigate the ocean scientifically—the basic goal of oceanography. In this chapter, we also show where progress is being made, the challenges that remain, and how ocean scientists are seeking to answer questions that have so far gone unanswered. We begin by summarizing the history of human exploration of the ocean.

Investigating the Ocean

From the beginning of human existence, people have been fascinated by the ocean and the mystery of what lies beyond the sea-surface horizon. From tentative forays into local waters onboard crude boats, people eventually developed ocean-going vessels capable of circumnavigating the globe.

VOYAGES OF EXPLORATION

The earliest migrations of modern humans (*Homo sapiens*) out of Africa began around 50,000 years ago. Human remains dating from 40,000 years ago are found in Australia. These early migrants apparently knew about fishing and using boats, which they needed to cross the sea separating Australia and Southeast Asia. Further evidence of human migrations via the ocean is the widespread distribution of human populations. Humans had populated all the continents except Antarctica thousands of years before Europeans began their era of ocean exploration in the 15th century. Accounts of voyages and shipwrecks 2000 years ago are found in the Bible and other works of comparable antiquity.

Little is contained in surviving written documents or charts about these ancient ocean explorations for a number of reasons. Many ancient societies had no written language. Most sailors were illiterate and thus unable to document their travels, although oral accounts were passed down from one generation to the next. Perhaps more importantly, knowledge of the sea routes that they explored were trade secrets, too valuable to be shared by the nations or companies that supported the expeditions.

One documented ancient ocean-exploring voyage was the royal trade expedition organized by the Egyptian queen and pharaoh Hatshepsut around 1450 BC. From northern Egypt, the expedition journeyed through the Red Sea, and southward along the African coast to what is now Mozambique. Scenes from the expedition are displayed at Hatshepsut's funerary temple in southern Egypt near Luxor. The purpose of the expedition was to bring back some of the valuable resources of the region including timber, gold, ivory, and wild animals (e.g., monkeys and baboons). Commercial traffic likely had been conducted for centuries through this part of the Indian Ocean but no written records of such voyages have been found. The Egyptians developed sailing ships around 4000 BC, some 2500 years earlier, but probably used them only in the eastern Mediterranean, near the mouth of the Nile River. Polynesians were expert sailors who discovered and settled widely scattered islands in the Pacific from about 2000 BC, when they left New Guinea, to 400 AD when they first settled in the Hawaiian Islands.

For some time prior to about 1200 BC, Mycenae in Greece was the center of a great civilization that controlled trade in the Aegean Sea and much of the Mediterranean. Following the decline and fall of Mycenae (at least partially due to a prolonged drought), the Phoenicians, from the eastern end of the Mediterranean Sea (now modern Lebanon), dominated sea trade for about a thousand years. They established many prosperous colonies along the shores of the Mediterranean, in Sicily, Spain, North Africa, and beyond the Strait of Gibraltar into the Atlantic and down the Red Sea into the Indian Ocean. Mythic accounts of Phoenician expeditions indicate that one expedition reached Cornwall in southwestern England (then called the *Island of Tin*). Another expedition sailed along the West African coast to the mouth of Niger River.

After the fall of the Phoenician city of Tyre to the Assyrians in 640 BC, the Greeks reestablished their dominance of Mediterranean and Black Sea trade routes. They developed more reliable methods of navigation at sea. To estimate latitude, the Greeks used a calculation based on length of daylight corrected for the time of year. About 450 BC, the Greek Herodotus published an accurate map of the world according to the Greeks. Another Greek astronomer and geographer, Pytheas, sailed out of the Mediterranean, around England and possibly even reached Iceland and the Baltic Sea around 325 BC. He also was the first to determine latitude by measuring the angular distance above the horizon of the North Star (Polaris), a technique that is still used today. About 300 BC, Alexander the Great founded a library at Alexandria, which served as a repository for scrolls from ships and land caravans. For the next 600 years, the library was a maritime studies center, fostering advances in science and celestial navigation by many Greek and Egyptian scholars. About 150 AD, the geographer Ptolemy produced an accurate map of the world known to the Romans that influenced geographic thinking for centuries.

The Vikings set sail from Scandinavia around 750 AD and explored and colonized lands bordering the North Atlantic, including Iceland (810 AD), Greenland (980 AD), and Newfoundland (1000 AD). The relatively mild conditions of the Medieval Warm Period (about 950 to 1250 AD) and reduced sea ice cover apparently made their voyages possible. The full extent of Viking colonization of North America during this period is much debated today. Vikings who had settled along the coast of Greenland were unable (or chose not) to adapt to deteriorating climatic conditions that heralded the Little Ice Age (Chapter 12). Their increasing inability to grow European style crops and livestock and to communicate with their homeland by sea (due to more widespread sea ice cover in the North Atlantic) eventually led to the collapse of the Viking settlements in Greenland.

Around 1420 AD, during the Ming Dynasty (1368-1644), the Chinese embarked on seven major expeditions involving hundreds of ships technically advanced for the day (Chinese junks), using magnetic compasses. The purpose of these expeditions was to map unknown regions, obtain treasures and exotic animals for the Forbidden City (walled section of Peking containing royal palaces), and extend Chinese influence throughout the region. Some claim that the Chinese circumnavigated the world and may even have reached North America. Economic and political pressures at home terminated these voyages and began the long period of Chinese isolation.

In the 15th century, the Portuguese began the European *Age of Discovery*, an era of ocean exploration and colonization of the Americas, India, Asia, Australia, and New Zealand. In 1420, Prince Henry the Navigator (1394-1460) founded a naval observatory in Portugal, the first school for teaching navigation, astronomy, and cartography. Vasco da Gama (c. 1469-1524) sailed around Africa, reaching India in 1498 to open a lucrative trade in spices. In 1519-1522, ships under the command of Ferdinand Magellan (c.1480-1521) were credited with being the first to sail around the world, although he died in the Philippines in 1521 after reaching the eastern edge of the then known world. His crew completed the voyage back to Spain.

English ships played a major role in exploring the ocean. Captain James Cook (1728-1779) of the British Royal Navy used the latest navigational tools (precise clocks known as chronometers) to determine longitude accurately. During three major voyages conducted between 1768 and 1780, he mapped the Southern Ocean as well as many other parts of the Pacific, "discovering" Australia, New Zealand, and the Hawaiian Islands. He was the first to circumnavigate Earth at high latitudes and sailed as far south as about 70 degrees S but did not sight Antarctica. The chief objective of Cook's voyages was not science, but to establish a British presence in the South Seas. Nonetheless, some valuable scientific information was acquired. During their exploration of New Zealand, the Great Barrier Reef, Tonga, and the Easter Islands, Cook and his crew collected samples of terrestrial plants and animals, marine life, and the ocean bottom. On his search for a Northwest Passage, Cook prepared charts of the West Coast of North America that remained very useful until World War II.

Two early English explorers, Sir John Ross and his nephew Sir James Clark Ross, were interested in conditions in the deep ocean. In 1818, Sir John Ross obtained a bottom sample at a depth greater than 1900 m (6200 ft) off the coast of Greenland. Later, Sir James Clark Ross, discoverer of the Ross Sea and Victoria Land in Antarctica, reached an ocean depth of almost 4900 m (16,000 ft) in the South Atlantic using a long rope attached to a weight. They discovered abundant marine animals living on the ocean bottom at great depths of water—an extreme environment previously believed to be devoid of all life. According to the then widely accepted **azoic hypothesis**, deep-ocean waters below 300 fathoms (550 m or 1800 ft) were barren, lacking sufficient dissolved oxygen to support marine life. The hypothesis originated in the mid-1800s with Sir Edward Forbes (1815-54), at one time considered the father of deep-sea biology. Forbes spent more than a year on the Royal Navy ship HMS *Beacon* in the eastern Mediterranean. Dredging the sea bottom at depths as great as 230 fathoms (425 m or 1380 ft), he found that the number of marine species decreased with increasing depth. By simple extrapolation, Forbes concluded that no life existed below 300 fathoms. However, discovery of live animals on the deep-ocean floor during several British ocean expeditions debunked the azoic hypothesis and spurred scientists' interest in further exploration of the ocean basins.

From 1831 to 1836, the HMS *Beagle* undertook a voyage to study the natural science of Galapagos Islands (in the equatorial Pacific off Ecuador) as well as many other locations. Charles Darwin (1809-1882) was onboard as a naturalist. From his observations came his first major published work, *Structure and Distribution of Coral Reefs* (1842). In it, Darwin correctly argued that the form and structure of reefs and atolls develop because they are living organisms growing upward in an effort to remain in the photic zone as compensation for the sinking sea floor (Chapters 2 and 10). Darwin also wrote on several other marine subjects including barnacle biology and fossils, all of which was overshadowed by his revolutionary work, *On the Origin of Species* (1859).

In 1838-42, the United States launched its Exploring Expedition. Had it not been for the contentious personality of its leader, Lt. Charles Wilkes, USN, this voyage might have been as famous as those of Cook or the Challenger Expedition (discussed below). The *Vincennes* was the flagship of a six-vessel fleet. Goals included showing the flag, charting, whale watching, gathering geological specimens, and general scientific observations. Exploration and charting of the east Antarctic coast confirmed that Antarctica was a continent. The many specimens and artifacts collected during the voyage formed the nucleus of the recently established Smithsonian Institution in Washington, DC. Wilkes and his scientific staff produced a 19-volume final report.

Matthew Fontaine Maury (1806-73) began his naval career as a midshipman onboard the U.S. sloop-of-war *Falmouth*, serving as navigation officer. After being injured in a stagecoach accident in 1839, Maury was unable to return to sea and was assigned desk-duty at home. In 1842, he was appointed Superintendent of the Navy's Depot of Charts and Instruments (later the U.S. Naval Observatory and Hydrographical Office) in Washington, DC. Maury discovered that existing ocean navigation charts were out of date, inaccurate, and supplied little information on winds and currents. He decided to revise the charts. Maury and his staff poured over naval logbooks stored at the Depot to glean any useful navigational information. Captains of naval vessels were ordered to supply Maury with navigational, hydrographic, and meteorological data. Later, Maury supplied captains of merchant ships with special logbooks for recording similar observations; the logbooks were mailed to the Depot at the end of a voyage. Out of this work came more accurate and useful navigation charts of winds and currents in the world ocean. The first of Maury's "Wind and Current Charts" was issued in 1847. With these charts, sea captains could take advantage of favorable currents and winds and thereby significantly reduce sailing time on many routes. For his pioneering work on ocean navigation, Maury became known as the "pathfinder of the seas." In 1855, Maury published *The Physical Geography of the Sea*, the first textbook on modern oceanography. With the outbreak of the Civil War, Maury resigned his position in 1861 and joined the Confederate Navy.

CHALLENGER EXPEDITION (1872-1876)

The Scottish oceanographer Charles Wyville Thomson (1830-82) led the first voyage dedicated exclusively to marine science. The **Challenger Expedition** was underway in December 1872 and ended in May 1876 having covered 127,500 km (76,500 mi), three times the circumference of the planet. Scientists sampled every ocean basin except the Arctic, probed the ocean and the seafloor to depths as great as 9000 m (29,500 ft), and sailed as far south as 61 degrees S before being turned back by sea ice. The Challenger Expedition laid the foundation for modern ocean science.

The Royal Society of London funded and organized the expedition while the Royal Navy provided the vessel, the HMS *Challenger*, a three-masted, square rigged wooden warship, 68.5 m (226 ft) long. The Royal Navy also supplied the captain and crew. In a forerunner of modern oceanographic practice, the captain ran the ship while a board of six chief scientists was in charge of the scientific enterprise. Thomson and his colleague Sir John Murray are credited with coining the term *oceanography*. On the ship's deck, scientific gear, a laboratory, and facilities for storage of samples took the place of all but two of the ship's original 19 guns (Figure A.1). *Challenger* was equipped with an auxiliary steam engine for maneuvering while taking samples and scientific observations.

Many of the Challenger Expedition's contributions are still used today; an example is the global map of marine sediment distribution (Figure 4.5). Other important contributions of the Challenger Expedition include the first systematic map of major ocean currents and water temperatures, a map of ocean bottom features (e.g., mid-ocean ridges), discovery of the *Challenger Deep* in the Mariana Trench in the Pacific, finding manganese nodules at the bottom of the North Atlantic, and documentation of diverse forms of marine plants and animals, especially microscopic plankton. All told, the Challenger Expedition discovered 4017 previously unknown species of marine organisms. The 50-volume report of the expedition's scientific findings, the *Challenger Report*, was written and published between 1880 and 1895 by Sir John Murray. This report formed the foundation for the science of oceanography and is still highly valued today. In short, the Challenger Expedition marked the beginning of the modern study of the world ocean.

Some of the instruments used to collect samples on the Challenger Expedition differ little from those in use today: towed nets to gather plankton and larger organisms, bottles for collecting water samples, and coring tubes for retrieving seafloor sediment samples. Since the time of *Challenger*, winches for lowering and retrieving instruments have improved greatly and electronic instruments have largely replaced mercury thermometers and other mechanical devices. Navigation, originally done by astronomical observations, is now largely based on the Global Positioning System (described in the second Essay of Chapter 6). Furthermore, satellites and electronic computers have greatly accelerated data collection, relay, and analysis.

Chapter A THE FUTURE OF OCEAN SCIENCE   313

FIGURE A.1
Dredging and sounding equipment onboard the HMS *Challenger*. The 1872-76 Challenger Expedition laid the foundation for modern ocean science. [NOAA Photo Library]

Modern Ocean Studies

The Challenger Expedition was a model for modern scientific study of the ocean carried out throughout the 20th century and into the present century. In this section, we examine modern initiatives aimed at developing a greater understanding of the ocean, its properties, processes, and life forms. We open with an historical perspective of technological innovations that enabled scientists to more extensively probe the sea.

TECHNOLOGICAL INNOVATIONS

The USS *Albatross* of the U.S. Fish Commission sailed throughout the world ocean from 1887 to 1925, collecting hundreds of marine species using specially designed nets and benthic dredges (Figure A.2). The first major oceanographic expedition of the 20th century was the German Atlantic Ocean Expedition of 1925-27. The Institute of Marine Research in Berlin organized the expedition and the German Navy supplied the ship, the *R/V Meteor* (the first of several German oceanographic research ships to bear that name). The primary mission of the voyage was to gather data on the physical properties of the Atlantic Ocean using optical, acoustic, and electronic equipment. Probably the most important innovation was use of an acoustic echo sounder to nearly instantly measure and record the depth and profile of the sea bottom. These observations drastically changed the prevailing view of the ocean floor as monotonously flat to a place of considerable topographic relief—information that would later help support the theory of sea floor spreading and plate tectonics (Chapter 2).

Since its founding in 1902, the Scripps Institution of Oceanography in La Jolla, CA has spurred the advance of oceanographic research. In 1937, the Institution's schooner, *E.W. Scripps*, began a broad research program on physical, biological, and chemical oceanography in Pacific waters off Southern California. Work at Scripps eventually led to publication of the well-known book on ocean science, *The Oceans*, in 1942.

National defense needs during World War II (1939-1945) and the subsequent Cold War (1945-1990) spurred advances in ocean science and technology by the industrialized nations of the West and the Soviet Block. Many nations supported multi-purpose exploration of spe-

FIGURE A.2
The USS *Albatross* of the United States Fish Commission carried out many expeditions throughout the world from 1887 to 1925 and discovered hundreds of marine species using surface townets and benthic trawls and dredges. [NOAA Photo Library]

cific areas of the ocean using ship-borne observations and backed development of observing instruments for ocean studies. Some of these studies were also aimed at improving long-range weather forecasting and detection of missile-launching submarines. The latter led to a better understanding of sound propagation in the ocean. Enormous amounts of data were collected and analyzed. Eventually these data were declassified and are now available for peacetime scientific research.

Paralleling the increasing use of ships (both military and civilian) for ocean research was the design and development of piloted submersible vehicles that enabled scientists to observe directly the ocean depths. The *bathysphere*, developed in the early 1930s by William Beebe and Otis Barton of the New York Zoological Society, was a hollow steel ball about 1.5 m (5 ft) in diameter, equipped with oxygen tanks, chemicals to absorb carbon dioxide, and portholes and a searchlight for viewing. A cable from a ship lowered and raised the bathysphere into and out of the ocean. In 1934 just off the Bermuda coast, Beebe and Barton descended in the bathysphere to a depth of 923 m (3028 ft). They reported seeing fish and invertebrates previously unknown.

More maneuverable than the bathysphere for exploration of the deep ocean was the *bathyscaph*, designed by Auguste Piccard (1884-1962) in the late 1930s and tested in 1948. The bathyscaph was not tethered to a ship. It consisted of a thick-walled cabin suspended under a tank (float) containing less-dense-than-water petroleum. Ballast tanks attached to the float kept the bathyscaph afloat while on the surface. Flooding the ballast tanks caused the bathyscaph to dive whereas systematically releasing weights (iron pellets) slowed the descent or allowed the vehicle to surface. In the early 1950s, Piccard and his son Jacques built a new and improved version of the bathyscaph, the *Trieste*, which reached a maximum depth of 3167 m (10,392 ft) in the Mediterranean Sea. Facing high maintenance and operational costs, the Piccards sold the Trieste to the U.S. Navy in 1958. On 23 January 1960, the bathyscaph *Trieste*, piloted by Jacques Piccard and U.S. Navy Lieutenant Don Walsh, dove to a new world record depth of 10,912 m (35,800 ft) very close to the bottom of one of the deepest places in the ocean, the Challenger Deep in the Mariana Trench—in the Pacific Ocean about 325 km (200 mi) southwest of Guam.

The 3-passenger (pilot plus two scientific observers) research submersible *Alvin* was built in 1964 and subsequently re-built and technologically upgraded about every three years (Figure A.3). It is one of only five deep-sea

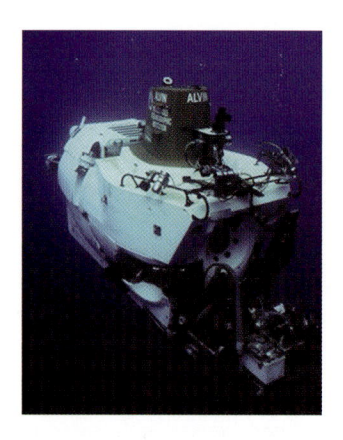

FIGURE A.3
On 12 April 2004, the research submersible *Alvin* made its 4000th dive since christening in 1964. The dive took place on the East Pacific Rise off the coast of Mexico to a maximum depth of 2500 m (8200 ft). Operated by the Woods Hole Oceanographic Institution (WHOI), *Alvin* transports a pilot and two passengers to ocean depths as great as 4500 m (14,764 ft). [WHOI photo by Rod Catanach]

research submersibles in the world. The National Science Foundation (NSF), NOAA, and the U.S. Navy own and fund *Alvin*. Operated by the Woods Hole Oceanographic Institution (WHOI) in Falmouth, MA, *Alvin* is considered the workhorse of submersibles, averaging 175 dives per year (Figure A.4). In April 2004, *Alvin* made its 4000th dive. *Alvin* made possible the discovery of many important deep-sea features, such as hydrothermal vents and associated unique biological communities (Chapters 2 and 9)

FIGURE A.4
Aft view of the Research Vessel *Atlantis* moored at the Woods Hole Oceanographic Institution in Massachusetts. The 83.5-m (274-ft) *Atlantis* is the support ship for the submersible *Alvin*. [NOAA photo]

as well as locating the wreck of the RMS *Titanic* and many other ships. However, *Alvin's* maximum depth range of about 4500 m (14,764 ft) is far short of the greatest ocean depths.

In the late 1980s, scientists and engineers at the Woods Hole Oceanographic Institution designed and built *JASON*, a **remotely operated vehicle (ROV)** capable of reaching a depth of 6000 m (19,680 ft). Cables (transmitting power and data) tether a ROV to a surface ship. *JASON* was used to explore and photograph hydrothermal vents along mid-ocean ridges and survey old shipwrecks among other activities. *JASON* was retired in 2001 and replaced by *JASON II* the following year. *JASON II* is a significant technological upgrade with greater maneuverability and sampling capacity than its predecessor. *JASON II* has the latest in robotics and is equipped with two mechanical arms that can reach twice as far and lift five times as much material as the single arm on the first *JASON*. *JASON II* has more power for brighter lighting and is equipped with digital cameras. Designed for routine operation at depths as great as 6500 m (21,320 ft), *JASON II* can remain on the ocean floor for days at a time. Both *Alvin* and *JASON II* are components of the National Deep Submergence Laboratory at WHOI, the only facility of its type in the nation.

In recent years, Japan and France have built research submersibles capable of diving to depths of 6500 m (21,325 ft). Also, the U.S. Navy has dedicated a nuclear powered submarine, the *RN-1 Deep Submergence Craft*, to scientific research. While it can dive to a maximum depth of only 700 m (2300 ft), the submarine can remain underwater for 30 days with a crew of 7 and is well suited for scientific voyages under the Arctic sea ice.

A major impetus for development of ocean instrumentation came from the offshore oil industry, which played a leading role in the design of remotely operated vehicles (Figure A.5). Since 1970, more than 1000 ROVs have been built primarily for operational needs such as servicing oil pipelines on the ocean bottom, retrieving lost objects, and even investigating wrecked aircraft and ships on the ocean floor.

ROVs and Autonomous Underwater Vehicles (AUV) (discussed below) have greatly increased our knowledge of mid- and deep-water marine life. They are equipped with robotic arms and traps that can capture fragile, gelatinous organisms that would never survive being caught in traditional nets. As we saw in Chapters 9 and 10, these organisms play an important role in marine ecosystems. In early 2003, scientists from the Monterey Bay Aquarium Research Institute announced that using a ROV equipped with video cameras, they had discovered an entirely new

FIGURE A.5
Launch of the ROV *Innovator* with the Apache oil platform in the background. Sample buckets are attached to the sides. [Photo by Jeremy Potter, NOAA Office of Ocean Exploration]

species of red jellyfish. Nicknamed "Big Red," these animals are up to one meter (3.3 ft) in diameter and live at depths of 650 to 1500 m (2100 to 4900 ft). The fact that scientists overlooked such a large animal until now underscores how little is known about the deep ocean and how modern technology enables scientists to expand our understanding.

The need to study large, remote areas of the ocean has prompted groups of nations to cooperate in international scientific studies. The high cost of conducting research in the ocean and operating expensive research vessels are strong incentives for international cooperation. An early example was the scientific initiatives underway during 1957, the International Geophysical Year (IGY). Among studies begun during the IGY was regular monitoring of carbon dioxide levels in the atmosphere, an important component of global climate change studies (Chapter 12). The United Nations Educational, Scientific, and Cultural Organization (UNESCO) sponsored the International Decade of Ocean Exploration (IDOE) in the 1970s. IDOE organized the first modern systematic surveys of ocean currents and the chemical composition of seawater. IDOE also initiated field studies that eventually led to our present ability to monitor the evolution of El Niño and La Niña (Chapter 11). Other programs were eventually organized to marshal international support for decade-long studies of the ocean, primarily focused on specific processes such as the carbon cycle.

REMOTE SENSING

Some of the most significant advances in the study of the ocean and its basins came in the final three decades of the 20th century with the development and application of remote sensing by high altitude aircraft and satellites (Chapter 1). Throughout this book, we use various images processed from data acquired by sensors onboard Earth-orbiting satellites. These space platforms provide a unique global perspective of the ocean, its properties and processes. Initially, ocean scientists were limited to using satellite sensors to study only the ocean/atmosphere interface because radiation at most wavelengths cannot penetrate to great depths in the ocean. But in recent years, they have developed a variety of techniques that permit them to draw inferences about the state of the ocean at some depth. Satellite sensors remotely measure such ocean parameters as sea surface temperature, sea-level fluctuations, currents, eddies, sea ice extent, and biological productivity.

Aquarius, a NASA satellite planned for launch in 2006 or 2007, is expected to provide the first global salinity maps for ocean surface waters. Remote sensing of salinity is a major achievement in itself, but combining global satellite observations of temperature and salinity will likely yield a better understanding of the processes involved in ENSO and longer-term changes in ocean circulation. It is also possible that such data will improve long-range weather and short-term climate forecasts based on a better understanding of the Pacific Decadal Oscillation (PDO), North Atlantic Oscillation (NAO), the Arctic Oscillation (AO), and other ocean-atmosphere interactions (Chapter 11).

Paralleling advances in remote sensing by satellite was the development of electronic computers, powerful and fast enough to store and analyze vast quantities of observational data. Earth-orbiting satellites also dramatically improved communications capabilities at sea. Today, data collected in the most remote reaches of the ocean can be sent nearly instantaneously to laboratories for analysis almost anywhere in the world. Other satellite-based technologies that greatly aid ocean studies include the Global Positioning System (GPS) that enables un-piloted instrumented platforms to report their locations accurately and communications satellites that relay data to computers for analysis.

Remote sensing refers to acquisition of data on the properties of some object without the sensor being in direct contact with the object. Remote sensing involves not only Earth-orbiting satellites, but also certain automated observing platforms. For example, arrays of underwater microphones (originally intended for submarine detection) are used to measure water temperatures over great distances within the ocean basins based on sound propagation (Chapter 3). Such measurements are unobtainable using ships alone. These observations are the basis for detecting and following the movements of water masses through the ocean, locating submarine volcanic eruptions, and tracking migrating whales.

SCIENTIFIC OCEAN DRILLING

Sampling marine sediments and the ocean crust for scientific purposes began in the early 1960s when engineers demonstrated that small drilling vessels and barges equipped with oil-field drilling rigs could be used to sample the deep-ocean bottom. The success of these experiments led to the **Deep Sea Drilling Program (DSDP)**, which operated from 1968 to 1983, using the drill ship *Glomar Challenger*. A technique known as *dynamical positioning* permitted the ship to drill in deep waters without anchoring. Powerful computer-controlled propulsion units kept the ship within a fixed distance of the spot where the drill penetrated the sea floor. The computer responds to an acoustic signal emitted by a beacon positioned on the ocean bottom. In 1983, a larger and more capable drill ship, the JOIDES *Resolution*, using essentially the same technology continued deep-sea drilling in the **Ocean Drilling Program (ODP)**. ODP operated between 1983 and 2003 and was supported by 22 nations. Newer drilling ships use GPS, rather than sea floor acoustic beacons, to fix the precise locations of the drill hole and the drill ship.

Over the past 40 years, drilling up to 1700 holes annually, scientists have recovered from the ocean floor rock and sediment cores that if laid end-to-end would have a total length of about 160 km (100 mi). Many important scientific discoveries came out of the DSDP and ODP in what is now the world's largest internationally supported Earth science program. Among the many important findings of ocean drilling was verification of sea-floor spreading from analysis of rock samples recovered from the bottom of the North Atlantic (Chapter 2). Scientists recovered thick deposits of salt underlying marine sediments in the Gulf of Mexico and the Mediterranean Sea (as described in the second Essay of Chapter 12). These deposits formed when the basins were isolated from the rest of the ocean and their waters almost completely evaporated. Furthermore, deep-sea sediment cores yield a record of climate variations as far back as about 190 million years ago.

In October 2003, the **Integrated Ocean Drilling Program (IODP)** began. Sponsored initially by the U.S. and Japan, the IODP may eventually involve as many as 20 other nations. The goal of the new drilling program is to use two drill ships plus specialized drilling platforms rather than a single general-purpose ship and drill more and deeper holes on the ocean floor. The original IODP plan called for an upgraded *JOIDES Resolution* or a new similar drill ship but because of budget constraints, the U.S. sponsor (the National Science Foundation) decided to continue using an unmodified *JOIDES Resolution* at least for the foreseeable future.

By late 2006, the *Chikyu*, a new drill ship built and operated by Japan, will join the U.S. ship. The *Chikyu* is much larger than earlier drill ships, technologically more advanced, and equipped with specialized equipment that will minimize environmental hazards (e.g., petroleum blowouts). Lacking such capabilities, DSDP and ODP drilling sites were restricted to the deep-ocean floor far from shore where the chance of encountering oil or natural gas deposits is minimal. *Chikyu's* first project will be to investigate earthquake mechanisms by boring into the Nankai Trough subduction zone located offshore of Honshu, Japan—a location that was off limits to the *JOIDES Resolution* because of the potential for a petroleum blowout. Another option with IODP is the use of specialized drilling platforms for areas where drill ships could not safely or efficiently operate, such as in the Arctic multi-year sea ice.

With its unique capabilities, IODP promises to add significantly to scientific knowledge. IODP scientists can explore the oldest rocks in the ocean basins, those occurring near continental margins (Chapter 2). Gas hydrate deposits, a potential source of natural gas, can be safely sampled. The deep biosphere contained in marine sediments and oceanic crust (home to perhaps two-thirds of the planet's total microbial population) can be explored at depths below that reached by DSDP and ODP drill holes; some of these microbes may play a role in biotechnology. The new drilling program will also add to our understanding of global climate change and earthquake generation.

Emerging Ocean-Sensing Technologies

Sending a well-equipped research vessel to sea for an extended period of time with a complete crew of sailors, technicians, and scientists is very costly. While seagoing oceanographers will not be replaced, new microelectronic instruments, better computers, and improved communications and positioning systems developed in the late 1990s have made possible new types of un-piloted ocean observing systems. These innovations have greatly expanded the range of ocean observations into areas not usually traversed by ships. We consider some of these developments in this section.

AUTONOMOUS INSTRUMENTED PLATFORMS AND VEHICLES

The need to understand the subsurface ocean on all temporal and spatial scales has led to the development of new types of research platforms. They are designed to overcome the limitations of oceanographic ships, including their slow speeds, high costs, uncomfortable working conditions, and limited supply versus demand. A variety of such platforms is available and more are under development. Available now are instrumented buoys, ARGO floats, Slocum gliders, and Autonomous Underwater Vehicles.

Buoys are small floating un-piloted platforms, typically several meters in diameter that are moored at fixed locations in the ocean (such as the buoys positioned across the equatorial Pacific to monitor El Niño and La Niña; see Chapter 11). Sensors on these buoys take continual observations of the lower atmosphere and upper-ocean. Data are then transmitted to polar-orbiting satellites for transmission to computers ashore. These platforms are sources of data at locations considered critical for improving weather and climate prediction. Many buoys have lines suspended below them with sensors attached at various depths to continually measure water temperature, currents, and salinity. A few of these buoys also have cables that connect to sensors placed on the sea floor (discussed later in this chapter). Other smaller *drifting buoys* move with ocean currents and provide information on their speed and direction as well as changes in subsurface ocean properties.

Other autonomous platforms, known as *ARGO floats*, are cylindrical devices equipped with sensors that augment satellite-based observations of the ocean surface as well as buoy observations of the upper ocean. As described in detail in the third Essay of Chapter 6, ARGO floats obtain profiles of ocean temperature and salinity to depths as great as 2000 m (6600 ft). A drawback of ARGO floats is the absence of a propulsion system so that they can only drift with the ocean currents.

A more recent type of float, the *Slocum glider*, covers greater distances (a range of 1500 km or 930 mi) and has a longer sampling life span than the ARGO float. Plus it is designed to follow a prescribed trajectory. The Slocum glider is named for New England Sea Captain Joshua Slocum (1844-1909) who was the first person to sail solo around the world. (He completed the 65,000-km (40,000-mi) voyage in three years and two months (1895-98) in his 11-m (36-ft) long sailboat.) The 1.5-m (5-ft) long winged torpedo-shaped glider is highly maneuverable, sinking and rising through the ocean by changing its buoyancy. As it sinks and rises, the glider's wings and rudder control horizontal movements so the vehicle moves independently of the ocean currents. At a horizontal speed of about 0.5 m per sec (1 nautical mi per hr), the glider follows a saw-tooth sampling pattern as it descends to a predetermined depth, measuring temperature, conductivity (salinity), and other water properties at various depths. At regular intervals, the glider surfaces, determines its position by GPS, and relays the stored observational data via satellite.

The Slocum glider employs the same buoyancy changing technique used in the ARGO float. Hydraulic oil is pumped between an internal reservoir and an external bladder thereby changing the density of the vehicle by changing its total volume. To ascend, oil is pumped to the bladder that expands and thereby increases the volume and lowers the density of the glider. To sink, oil is pumped back to the internal reservoir, the bladder shrinks, and the overall density of the glider increases.

Until recently, the oil pump that controlled buoyancy was powered by batteries, limiting the range of the glider to the lifetime of the batteries. But scientists at Woods Hole Oceanographic Institution developed a novel engine powered by the temperature difference between surface- and deep-ocean waters that promises to extend the range of the Slocum glider. At the heart of the engine is a tube containing a wax that solidifies and shrinks at about 10 °C (50 °F). The wax melts and expands when the glider is in relatively warm surface waters but solidifies and shrinks when the glider enters the colder deep-ocean waters. Expansion and shrinkage of the wax controls the movement of a device that distributes the oil between the internal reservoir and external bladder. While the wax tube does not replace the glider batteries, it reduces the drain on them.

Gliders have been used to carry instruments that analyze and take samples of toxic algal blooms in the Gulf of Mexico. Scientists at the Mote Marine Laboratory in Sarasota, FL developed a winged, robotic-controlled platform that is released at the ocean surface and glides slowly to the bottom taking either observations or water samples. The device provides a relatively inexpensive platform to work within a limited area to obtain background data on the ecosystem, detect and sample a red tide should one develop, and bring back water and plankton samples to be analyzed in the lab.

Autonomous, powered vehicles augment the capabilities of research submersibles and ROVs and move through ocean waters faster than gliders. They are called **Autonomous Underwater Vehicles (AUVs)** because they are un-piloted, remotely controlled, and do not rely on a

FIGURE A.6
Launch of *ABE*, an autonomous underwater vehicle (AUV) equipped with sensors to measure temperature, conductivity, magnetics, and multibeam bathymetry. [NOAA photo]

cable tethering them to a mother ship, as do ROVs (Figure A.6). These vehicles carry sensors that measure ocean water properties (e.g., temperature, salinity, depth, chlorophyll, light, dissolved oxygen) along trajectories that can be preset or controlled while the AUV is underway. The operational range of an AUV is restricted by the limited amount of power their batteries can provide for propulsion and instruments. They must return to the mother ship or shore-based facility for recharging. AUVs also have limited space and power for electronic instrument packages. On the plus side, AUVs can observe the ocean in places and under conditions where research ships and other instrumented platforms cannot. AUVs are especially suited for studies under ice (such as the Antarctic ice shelves) and can be equipped with sonar to determine the thickness of the ice-cover above. With side-scan sonar, they can efficiently and economically map the ocean floor in considerable detail (Figure A.7). They can probe coastal waters at the edge of the continental shelf as well as specific locations within submarine canyons. AUVs are also useful in highly energetic ocean systems such as current rings (Chapter 6). In the future, hydrogen-powered fuel cells promise greater range and more power for AUVs.

Autonomous, ultra-light, un-piloted aircraft, once primarily used for military surveillance are also promising platforms for ocean studies. Groups of such solar-powered aircraft (called drones), can fly for days or weeks over thousands of kilometers of ocean surface and observe the development of tropical cyclones, monitor sea surface temperature, salinity, and ocean color, or collect atmospheric aerosols. GPS tracks the aircraft and polar-orbiting satel-

FIGURE A.7
Recovery of a side-scan sonar towfish used to map the Pearl Harbor Defense Area. [NOAA photo by J. Smith and C. Kelley]

lites relay data back to computers ashore. Drones are especially useful in coastal and estuarine areas where satellite sensors do not provide adequate resolution for study of complex environmental processes.

OCEAN FLOOR OBSERVATORIES

Ocean floor observatories are a relatively new but powerful means of studying ocean waters, marine life, and the sea floor. An **ocean floor observatory** is a facility that can perform experiments, collect data, and instantly communicate observations to data networks and scientists worldwide. The first ocean floor observatories stored data that were retrieved during infrequent visits by surface ships or submersibles. With the new generation of ocean-floor observatories, data collection can continue indefinitely as underwater cables transmit the data to networks on land for distribution. Here we consider three types of ocean floor observatories: deep-ocean-floor installations to record earthquakes worldwide, a seafloor volcanic observatory on an active submerged volcano, and a continental-shelf observatory in relatively shallow waters.

Detailed studies of seismic waves from earthquakes worldwide require fairly uniformly spaced observations. Previously most observations came from instruments, called *seismometers*, on land. The only seismometers within the ocean basins were located on islands and unfortunately large portions of the ocean such as the North Pacific have few islands. About 20 deep-ocean-floor seismic observatories are being planned to fill these gaps. Seismic instruments are installed in holes drilled into the sea floor. At some observatories ships will recover the instrument packages and retrieve their recorded data for later analysis. Others will use acoustic transmission of data to listening stations. Still others will be connected to networks on land by submarine cables providing instantaneous distribution of seismic data. These data are essential to improving our understanding of interactions between oceanic crust and the deeper parts of Earth's geosphere. These data also will aid investigations of some of the longer-term fluctuations in the Earth system such as ocean basin opening and closing (Chapter 2).

Ocean-floor seismic stations placed in a 1000-m (3300-ft) deep bore hole near Japan monitor earthquakes occurring near the Japan Trench where the Pacific Plate is subducting under the Eurasian Plate. When the plates slip past one another, seismic energy is released causing destructive earthquakes, which, along with associated tsunamis (Chapter 7), can cause considerable property damage and great loss of life on the islands of Japan. Plans are to have the Japanese drill ship, part of the Integrated Ocean Drilling Project (IODP), install additional seismic stations. One hoped-for outcome of better earthquake surveillance is improved earthquake prediction. Initially these observatories will use replaceable data-storage units to record data but eventually fiber-optic cable will link the observatories to data networks ashore.

The *Hawaiian Underwater Geo-Observatory (HUGO)* was situated at the summit of the Loihi seamount south of the Big Island of Hawaii and operated from 1997 to 2002. Loihi is the youngest volcano in the Hawaiian chain and, like the other volcanoes of the Big Island, is located over a hot spot (Chapter 2). Although its summit is more than 900 m (3000 ft) below sea level, Loihi rises more than 2700 m (9000 ft) above the sea floor—taller than Mount St. Helens was above sea level prior to its explosive eruption in 1980. HUGO was equipped with seismometers, an underwater microphone, and bottom pressure sensor. These instruments monitored seismic activity, volcanic eruptions, and hydrothermal venting. A submarine fiber-optic cable linked the observatory to a shore station on Hawaii and provided scientists with data from the seamount. However, the harsh conditions at the seamount disrupted the data flow so that as of this writing, HUGO is no longer operating.

A shallow-water station off New Jersey's central coast—consisting of two observatories—provides near real-time data for coastal managers, students, teachers, fishers, swimmers, surfers, and beachcombers. The undersea station, known as *LEO-15* for *Long-term Ecosystem Observatory*, is located in 15 m (50 ft) of water, 9 km (5.6 mi) offshore on the mid-continental shelf. Observatory instruments monitor ocean currents, water temperature, salinity, light and chlorophyll levels, wave height and period, sediment transport, and phytoplankton blooms. Other data sources include a tower for atmospheric observations and Doppler radar. A small autonomous underwater vehicle is programmed to gather data from areas surrounding the station. Data gathered by undersea instruments are transmitted by cable to the Rutgers Marine Field Station in Tuckerton, NJ.

ANIMAL-BORNE INSTRUMENTS

An innovative approach to obtaining observations in remote areas of the ocean uses small, electronic instrument packages attached to free-swimming marine animals. These electronic devices, called *tags*, are attached to large, long-lived animals having strong homing instincts, such as seals and whales. (Homing is important because it means that these animals will repeat ocean transects rather than just remaining in one locale.) Some instruments report via satellite, providing information on the animals' movements as well as the marine environment in which they live. In some cases, the data are recovered when the animals return to their home locations. In this way, ocean scientists obtain data from areas outside the major shipping lanes and infrequently sampled by ocean survey ships.

The most straightforward application of this technique tracks animals as they migrate and at the same time obtains other data such as water temperature at various depths as the animals dive. With development of smaller solid-state electronic components, the sophistication and communication range of tags increases. Among the animals that have been studied using this technique are seals, whales, tuna, sharks, swordfish, and sea turtles. With even smaller tags, it is likely that many more animals such as sea birds will be used to gather ocean data in this way.

Instrument-bearing large marine animals were used in the world's first *Census of Marine Life*, begun in the late 1990s. The Census, a decade-long international effort, was organized to determine the regional distribution and abundance of marine organisms, as well as their relationships to each other and to their ocean habitat. These instruments were used in studies of large animals such as seals, whales, and tuna that range over vast and remote areas of the ocean. One early application of instrumented animals in the Census involved elephant seals in the northeastern Pacific, tagged by scientists from the Southwest Center of NOAA's National Marine Fisheries Service in La Jolla, CA. Elephant seals live in large rookeries on islands offshore from central California, where they breed and raise their pups. Information from the tags revealed that these animals leave the rookeries on regular foraging trips lasting 2 to 9 months. However, males and females follow different migration routes. Males travel along the Alaskan and Aleutian continental shelf and slope whereas females remain in the northeastern Pacific west of California. Elephant seals continually dive in search of food; an average dive lasts about 20 minutes and typically reaches depths of about 400 to 600 m (1300 to 2000 ft) but can be as deep as 1600 m (5200 ft). When elephant seals return to the surface, their electronic tags communicate data to polar-orbiting satellites for transmission to scientists.

The northern elephant seal's cousins, southern elephant seals, live in the Southern Ocean. These animals routinely swim thousands of kilometers in the vicinity of Antarctica and off the southern tip of South America. The feeding migrations of the males are the longest known marine mammal migration, except for whales. If similarly instrumented, they would provide ocean data on a regular basis in a remote and poorly known portion of the ocean.

Instruments have also been attached to wild white (beluga) whales living in the Arctic and nearby high-latitude waters where ocean data are scarce. In winter, beluga whales often enter ice-covered waters. Tagged beluga whales permit studies of oceanic processes such as sea-ice formation in northern fjords. In one study, oceanographers from the Norwegian Polar Institute captured two beluga whales from the Storfjorden, Svalbard, Arctic fjord and fitted them with tags. The small instrument package was inserted into the whale's blubber so it would not interfere with the animal's normal activities. These instruments allowed scientists to determine the whales' location and record salinity and temperature at one-second intervals as the whales ascended from their bottom feeding. The data was downloaded via satellite when the whales surfaced for air. One discovery that came out of this experiment, reported in 2003, was the presence of a thin layer of relatively warm North Atlantic water immediately under the ice. Previously, ocean scientists had assumed that the water beneath the ice was uniformly cold. This discovery may have important implications for understanding recent trends in the extent and thickness of Arctic sea-ice cover (Chapter 12).

COMPUTERS AND NUMERICAL MODELS

Whereas past advances in ocean science have depended heavily on sampling platforms and observing instruments, future advances will increasingly depend on access to the world's largest and fastest computers (so-called supercomputers). New high-performance computers will be necessary to assimilate the flood of observational data and to run coupled ocean-atmosphere numerical models. In this way, ocean science is following the lead of atmospheric science where even routine weather predictions require the largest and fastest supercomputers available. NOAA put the latest version of such a computer into operation early in 2003 in Gaithersburg, MD. NOAA's National Weather Service uses computer output to generate a variety of forecast and other guidance products, including some that deal with ocean conditions (e.g., sea-ice extent, tracks of tropical cyclones). Such computers permit more accurate, longer-range forecasts.

One of the major achievements in the application of coupled air-sea numerical models is the prediction of the onset and evolution of El Niño and La Niña months in advance of the event. As noted in Chapter 11, observational data acquired by an array of moored and drifting buoys in the tropical Pacific along with satellite and tide gauge data are used to initialize these numerical models. Model-based predictions are very helpful for the many government agencies responsible for dealing with the potential impacts of extreme weather associated with El Niño and La Niña.

Challenges in Ocean-Sensing Technologies

The principal benefit of newly available ocean sensing technology is the promise of a better understanding of how the ocean functions as part of the Earth system and hence, better prediction capabilities. For example, more complete knowledge of the role of the ocean in Earth's climate system is key to answering questions concerning global climate change (Chapter 12). New observing systems are planned to monitor the global ocean over a broad spectrum of temporal and spatial scales. Such observational data help scientists more accurately model ocean properties and processes, enabling them to predict, for example, how Earth's climate system is likely to respond to increased concentrations of greenhouse gases. But despite recent advances in ocean-sensing capabilities, many problems remain to be solved. Some of these problems and possible solutions are considered in this section.

The coastal zone is perhaps an area where substantial advances in ocean-observing technology are most needed, in part because this is where the actual and potential impacts of human population pressures are most severe (Chapter 8). Some of the observing techniques or platforms discussed in this chapter cannot be employed close to the shoreline. Yet this is the most heavily used part of the ocean and one of the most dynamic regions in the Earth system where the hydrosphere, atmosphere, biosphere, and geosphere interact. For example, the coastal zone is where storm surges and tsunamis have their greatest impact.

One emerging technology that may be useful in the coastal zone is autonomous, remotely controlled ultralight aircraft capable of flying low and making observations over extended periods. Because the most dramatic alterations of the coastal zone occur during infrequent intense storms, such aircraft in their present form cannot be the only answer. Probably an older but expensive technology will be used in the near future: instrumented towers and shallow-water sea-floor observatories.

Better technologies for studying marine life are needed. As we have learned, most of the organisms living in the ocean are one-celled microbes, but many sampling techniques are designed for studying only those organisms that can be caught in nets, from large plankton up to the size of fish (Figure A.8). Some scientists are using long

FIGURE A.8
Crew members of the NOAA ship *McArthur* inspect a set of bongo nets used to determine the type and concentration of marine plankton. The *McArthur* conducts a range of oceanographic research missions mostly in marine sanctuaries along the U.S. West Coast. [NOAA photo]

tubes to pump water from depth over long periods of time. In this way, they sample and filter large volumes of water to gather sufficient numbers of microbes for study. Many of the new techniques to study marine microbes will probably come from advances in biotechnology. One example is the use of new techniques to study archaea, a little understood but extremely abundant organism that often occupies extreme marine environments (Chapter 9). Archaea cannot be seen by standard microscopes and cannot be grown in laboratories using conventional microbiological techniques. However, techniques used to sequence the human genome in the early 21st century may assist in the study of marine life, to identify and assess its abundance. These data will aid in identifying organisms and mechanisms responsible for sequestering carbon in the ocean and in this way add to our understanding of the ocean's role in the global carbon cycle.

New microchips are needed to perform chemical analyses of compounds in seawater, including routine monitoring of essential nutrients, the fertilizers of the sea, as well as toxins released by harmful algal blooms (Chapter 9). This promising technology must overcome one of the oldest problems known to sailors, that is, **bio-fouling**. Organisms rapidly colonize any object placed in ocean water (especially warm ocean water) unless the object is first treated by toxic chemicals. First come bacteria and other marine microbes, then diatoms, and eventually mussels and barnacles (as any boat owner knows). Such communities will also coat and impair the functioning of any chemical-sensing chip in the ocean unless the bio-fouling problem is solved.

Better knowledge of the ocean requires more accurate maps of the ocean floor. Mapping the deep-ocean floor, begun by the Challenger Expedition, is still incomplete. Exploration of Earth's neighbors in space has produced better maps of their surfaces than we have of Earth's ocean bottom. Obtaining such maps of the ocean floor is an expensive and daunting task. The best sea-floor maps are made by oceanographic ships equipped with GPS and state-of-the-art bottom-mapping equipment that surveys broad swaths of ocean bottom along the ship's track. NOAA's *National Ocean Survey* ships are equipped to make such maps, but about 125 ship-years would be required to complete the task of mapping the world ocean bottom at a cost of about $1 billion. Because of high cost, to date only about 10% of the sea floor has been mapped by modern surveying techniques.

As noted earlier, an economically attractive alternative for mapping the ocean floor uses autonomous underwater vehicles (AUVs) equipped with side-scan sonar.

Another approach is to rely on new satellite-borne radar altimeters. These instruments map the shape of the sea surface, which is strongly influenced by Earth's gravitational field. Ocean floor topography affects the local gravity field and hence, the shape of the sea surface (as discussed in the second Essay of Chapter 7). One revolutionary new global ocean-bottom map was made by combining satellite altimetry data, depth soundings from numerous ships, and newly declassified military data. This new map revealed a level of detail previously unavailable, including the discovery of previously unknown seafloor features. Engineers estimate that about 6 years would be required to develop an improved radar altimeter to be flown on a polar-orbiting satellite with non-repeating orbits. For best results, satellite data would be combined with ships' measurements to refine and provide ground-truth for mapping.

More complete mapping of the sea floor would yield numerous benefits. These maps will permit refinement of plate-tectonic models. More precise depiction of sea-floor roughness would permit better understanding of mixing processes in the deep ocean due to tidal currents flowing over features such as volcanic ridges or seamounts. Commercial applications of deep-ocean maps include determining the optimum sites for laying optical cables for communication networks and managing fishery resources in Exclusive Economic Zones (EEZ) (Chapter 4) and the deep ocean. Tsunami prediction and hazard models would also benefit from more accurate ocean-floor maps.

A technique also exists for obtaining detailed, three-dimensional images of specific areas of the seafloor. This instrument, called *multibeam sonar*, operates from a ship somewhat like a depth sounder, but combines information received from dozens or even hundreds of sonar beams simultaneously. Images obtained this way are so detailed that in 20 m (65 ft) of water objects as small as a lobster trap can be easily seen. For objects at greater depths, less detail can be resolved, but the images still represent a remarkable step forward in identifying seafloor habitats. For example, fishing grounds can be surveyed so that gear (for example, scallop dredges) is deployed more carefully to avoid damage to the environment.

Conclusions

Humankind's fascination with the sea extends back thousands of years. Sailing vessels tapped the energy of the winds and currents enabling humans to explore, discover, settle, and exploit the resources of new lands. With the development of reliable methods of navigation, ocean-going vessels provided access to all continents and commercially important trade routes were established. At first an object of mystery and curiosity, the ocean eventually became the subject of intense scientific scrutiny. Beginning with the Challenger Expedition of the 1870s, understanding of the ocean's properties and processes rapidly grew during the 20th century and into the present century. This new knowledge was partially a product of peacetime applications of technologies originally developed for national defense. It was also the result of innovative technologies either adapted from other fields (e.g., deep-sea drilling) or developed specifically for probing the ocean (e.g., *Alvin*, ARGO floats, Slocum glider).

Although many challenges need to be overcome and many questions remain to be answered, humankind's understanding of the role of the ocean in the Earth's system is progressing at an encouraging pace. Rock and sediment cores extracted from the ocean floor provided evidence of sea-floor spreading and confirmed the hypothesis of plate tectonics, revolutionizing our understanding of Earth's large-scale geological processes. Deep-sea sediment cores unlocked the secrets of the climatic fluctuations of the Pleistocene Ice Age and may provide information that is key to understanding future climate. A variety of piloted and unpiloted vehicles provide near real-time data on the physical, chemical, biological, and geological characteristics of the ocean. And sensors onboard Earth-orbiting satellites acquire a flood of valuable data from vast stretches of the ocean.

Basic Understandings

- The earliest humans to migrate from Southeast Asia to Australia around 40,000 years ago apparently knew about fishing and relied on boats for transport. Few written documents or charts refer to ancient ocean explorations because many ancient societies had no written language, most sailors were illiterate, and sea routes were trade secrets.
- The Egyptians developed sailing ships around 4000 BC and by about 1500 BC a royal trade expedition traveled from northern Egypt, through the Red Sea, and southward along the African coast to what is now Mozambique. The Phoenicians dominated trade in the Mediterranean for about 1000 years following the demise of Mycenae in Greece about 1200 BC. After the fall of the Phoenician city of Tyre in 640 BC, the Greeks reestablished control of Mediterranean and Black Sea trade routes and contributed significantly to the development of navigation techniques. The Vikings set out from Scandinavia around 750 AD and later explored and colonized Iceland, Greenland, and Newfoundland.
- In the 15th century, the Portuguese began the European *Age of Discovery*, a time of ocean exploration and colonization of the Americas, India, Asia, Australia, and New Zealand. Vasco da Gama sailed around Africa and reached India; ships commanded by Ferdinand Magellan were the first to circumnavigate the globe; and James Cook mapped the Southern Ocean and many islands in the Pacific Ocean.
- In 1842, Matthew Fontaine Maury was appointed Superintendent of the Navy's Depot of Charts and Instruments in Washington, DC. Maury found that existing ocean navigation charts were out of date, inaccurate, and supplied little information on winds and currents. By thorough scrutiny of logbooks maintained by the captains of both naval and merchant ships, Maury and his staff developed more accurate and useful navigation charts of winds and currents in the world ocean.
- The Challenger Expedition was the first voyage intended exclusively for marine scientific purposes. Between December 1872 and May 1876, the HMS *Challenger* circumnavigated the globe, sampling all ocean basins except the Arctic, probing the ocean and seafloor to depths of 9000 m (29,500 ft), and laying the foundation for modern ocean science.
- National defense needs during World War II and the subsequent Cold War spurred ocean studies and technology development by the industrialized nations of the West and the Soviet Block. Paralleling the increasing use of military and civilian ships for ocean research was the development of piloted submersible vehicles that enabled scientists to observe directly the ocean depths. These vehicles included the bathysphere in the 1930s and the bathyscaph in 1950s and 1960s. *Alvin*, the workhorse of submersibles, has operated since 1964 and made possible the discovery of many deep-sea features including hydrothermal vents.
- Another impetus for development of ocean instrumentation came from the offshore oil industry, which played a leading role in the design of remotely operated vehicles (ROVs). These un-piloted craft are tethered to a ship and can spend more time on the ocean bottom than a piloted vehicle.
- The need to study large, remote areas of the ocean prompted nations to cooperate in international scientific investigations. Examples include the International

Geophysical Year (IGY) in 1957 and the United Nation's International Decade of Ocean Exploration (IDOE) in the 1970s.
- Earth-orbiting satellites provide a unique global perspective of the ocean. Initially, satellite sensors were limited to monitoring processes operating at the ocean/atmosphere interface, but recently satellite-based methods have been developed to infer the state of the ocean at some depth. Paralleling advances in remote sensing by satellite was development of electronic computers capable of analyzing vast quantities of observational data using coupled ocean-atmosphere numerical models.
- Sampling marine sediments and ocean crust in deep waters for scientific purposes began in the early 1960s, using oilfield drilling rigs. The success of these experiments led to the Deep Sea Drilling Program (1968-83), the Ocean Drilling Program (1983-2003), and the Integrated Ocean Drilling Program (2003-). Convincing evidence for sea-floor spreading and plate tectonics is one of the major outcomes of these drilling programs.
- New un-piloted ocean observing systems include instrumented buoys (both moored and drifting), ARGO floats that profile temperature and salinity in the upper 2000 m (6600 ft) of the ocean, the Slocum glider that gathers ocean data along a predetermined trajectory, and Autonomous Underwater Vehicles that are highly maneuverable and can observe the ocean in places where other instrumented platforms cannot.
- An ocean floor observatory is a facility designed for long-term experiments and data gathering (e.g., seismic and volcanic activity). An example is the Long-term Ecosystem Observatory off the central coast of New Jersey.
- An innovative approach to observing remote areas of the ocean uses small, electronic instrument packages (tags) that are attached to free-swimming marine animals having strong homing instincts such as seals and whales. The tags provide information on the animals' movements as well as the environment in which they live.
- Whereas past advances in ocean science have depended heavily on sampling platforms and observing instruments, future advances will increasingly depend on access to supercomputers. Supercomputers are necessary to assimilate the flood of observational data and to run coupled ocean-atmosphere numerical models.
- Despite recent advances in ocean-sensing capabilities, many technological challenges remain to be solved. Needed are observing methods suited to the dynamic conditions of the coastal zone, better technologies for studying marine life, a solution to the problem of biofouling, and more accurate maps of the sea floor.

CHAPTER B

OCEAN STEWARDSHIP

Stewardship of Ocean Life
Fisheries and Sustainable Exploitation
    Overfishing
    Maximum Sustainable Yield
    Ecologically Sustainable Yield
    Bycatch
    Restoring Fisheries
    Habitat Destruction and Restoration
    Recreational Fisheries
Protecting Endangered Marine Species
    Sea Turtles
    Whales
    Water Birds
Mariculture
Marine Exotic Species
Conclusions
Basic Understandings

Artificial reefs can increase productivity of sandy bottoms.
[Courtesy of NOAA /OAR/National Undersea Research Program]

In early 2003, two Canadian fisheries scientists released the results of their 10-year study, which showed that commercial fishing has reduced the populations of as many as 90% of the world's largest fish species to such small numbers that their populations may not survive. Even more surprisingly, this population decline has happened in only 50 years, the result of intensive, highly industrialized fishing practices.

The rapid decline in the populations of species such as cod, marlin, swordfish, and halibut has a severe impact on marine ecosystems because these fish occupy a high level in marine food webs; most of them are **top predators** in the ecosystem. These large fish are slow growing and take a relatively long time (decades) to reach reproductive maturity. Once they are adults, the older fish tend to be the most fecund (fertile). Hence, removal of the largest fish of a species reduces the reproductive success of the population as a whole.

In this chapter we examine some examples of adverse human impacts on marine fisheries and ecosystems along with some of the steps that are being taken to protect marine species from extinction. Such efforts are part of **stewardship**, action taken by society to protect the ocean and its resources for now and the future.

Stewardship of Ocean Life

Components of the Earth system are highly interdependent so that disturbing one part of the Earth system can impact other parts—perhaps negatively. Thus, stewardship of the ocean and its resources involves responsibly managing all resources to benefit present and future generations. However, despite continuing advances in ocean science and technology (Chapter A), less than 5% of the ocean bottom has been explored or mapped to the same resolution as the surface of Earth's neighboring planets, Mars and Venus. For this reason, we often lack the basic understandings needed to effectively manage and protect marine living resources and their associated ecosystems.

In the United States, the *National Oceanic and Atmospheric Administration (NOAA)* is the principal government agency charged with stewardship responsibilities for the nation's marine environment and living resources. These responsibilities include maintaining sustainable fisheries in U.S. waters, and protecting and restoring the populations of endangered marine animals, such as sea turtles and whales. NOAA also protects and maintains the viability of the nation's coastal zone. Effective stewardship requires consideration of both the biotic and abiotic components of marine ecosystems to protect living organisms, their habitats, and population sustaining interactions (e.g., predator-prey relationships). Stewardship of ecosystem components also includes managing their exploitation by commercial and recreational fishers for the benefit of people who make their living in these industries or use them for recreation.

One concept that is central to exercising stewardship is **sustainability**, defined by the United Nations' *World Commission on Environment and Development* as "developments that meet the needs of the present without compromising the ability of future generations to meet their own needs." Achieving sustainability requires effectively balancing environmental issues with social and economic concerns. Sustainability involves intergenerational equity in managing resources and the environment; that is, we need to consider the needs of future generations as well as our own. It also includes the concept of maintaining a level of critical ecosystem function and biodiversity.

Fisheries and Sustainable Exploitation

Worldwide, fisheries supply about 19% of all the animal protein consumed by people and the fishing industry employs more than 200 million people, 95% of whom live in developing nations. In coastal developing nations—especially in the tropics—almost all the protein intake comes from near shore waters. These fishers use traditional methods (*artisanal fisheries*) and account for about half the world's marine fish production. Most of the fish they catch is eaten directly by humans. In developed countries, fishers employ technologies intended to increase the fish catch and a large amount of commercial-fish production is processed into fishmeal to feed pets and livestock, and for use in fertilizers.

In this section, we investigate overfishing, the contributing factors and the implications of overfishing for survival of top predator fish and the viability of associated marine ecosystems. We consider the advantages of a more holistic, ecologically based approach to fisheries management in order to achieve sustainable yields. Following this are discussions of fisheries and habitat restoration and recreational fisheries.

OVERFISHING

An estimated 25% to 30% of all exploited fish are overfished. **Overfishing** occurs when a fish species is taken at a rate that exceeds the maximum catch that would allow reproduction to replace the population. The two types of overfishing are known as recruitment overfishing and growth overfishing. *Recruitment overfishing* involves taking adult fish in such high numbers that too few survive to replenish the breeding stock. A 2003 study of global fisheries spanning the past 50 years showed that within 15 years of the opening of a new fishery, 80% of the largest fish are taken. The largest fish in a population not only have the greatest commercial value, they are also the most valuable ecologically because the largest fish in a species are the most fecund. *Growth overfishing* takes place when fish are taken too small, before the animals can grow to a size that would produce the maximum yield. For example, growth overfishing may have reduced the population of Chesapeake Bay crabs by as much as 30%.

For more than 10,000 years, humans have taken food from the sea and for most of this time the ocean's bounty of life was considered limitless. However, even early on there were signs that this view of an inexhaustible food resource from the sea was seriously flawed. In fact, the first known human induced collapse of a marine stock took place about 3000 years ago along the Peruvian coast. Fishers continued to harvest shellfish even after a natural disaster had greatly reduced shellfish populations.

In the years since the Industrial Revolution, a variety of technological innovations enabled fishers from developed nations to greatly increase the fish catch. These innovations included the steam and internal combustion engines that powered fishing vessels, extending their range from coastal areas to well offshore. These engines also powered heavier fishing gear such as trawls. In addition, refrigeration and factory trawlers made possible the preservation and processing of more fish. In the early 20th century, large, diesel-powered fishing boats permitted fishing on larger scales throughout the ocean's depths. In the last half of the 20th century, freezing to preserve freshness, improved fish-locating techniques (e.g., acoustic fish finders), and precision navigational technologies (GPS) have aided exploitation of fish stocks worldwide, and from all depths in the ocean.

Economics is an important factor in overfishing. Governments worldwide spend billions of dollars annually subsidizing their commercial fishing industry. Consequently, many fisheries scientists and economists believe that the world fishing capacity is too large. They estimate that the world fishing fleet is about 2.5 times larger than needed to catch the ocean's sustainable fish production. Furthermore, worldwide government subsidies—totaling about $100 billion each year—support an industry that produces about $70 billion worth of fish annually. (Annual subsidies amounting to between $15 billion and $30 billion compensate fishers whose income has been lost because of government restrictions on fisheries.) In addition, trends in price structure can sometimes mask a fishery that is overexploited. As the supply of a commercially valuable fish declines, its price rises. Rising prices, in turn, makes the fishery more attractive for fishers, hastening the decline in population and perhaps the collapse of the fishery.

Government subsidies, greater fishing efficiency, coupled with the growing demand for fish by the soaring human population drove an increase in total fish catch through the 1950s into the 1980s. The United Nations' *Food and Agriculture Organization (FAO)* reported that the total annual fish catch rose from 19.2 million metric tons (21.2 million tons) in 1950 to 88.6 million metric tons (97.6 million tons) in 1988. (Complicating matters during this period, the People's Republic of China routinely over-reported their fish catch, significantly distorting available statistics on global fish production.) Because the rate of fish production during this period exceeded human population growth rates, marine fisheries were thought to be the solution to the problem of feeding the world's burgeoning human population. At the time, the principal fishery-management approach was to expand the areas fished and exploit newly discovered fish populations. Because many marine fishes are fecund, ocean fisheries were assumed inexhaustible.

As fishers became more adept at taking fish in large numbers, the frequency of stock collapses increased, forcing fishery managers to impose quotas on commercially desirable fish, limit fishing time at sea, or close fisheries altogether. The first modern day fishery collapse occurred in 1971-72 in the Peruvian anchovy fishery, then the world's largest. Overfishing coupled with the effects of an intense El Niño apparently caused the collapse (Chapter 11). Total marine fish production peaked in the late 1980s; by then, discovery of new fisheries could no longer compensate for the decline in production. The global fish catch has been declining since the late 1980s by about 700,000 metric tons (770,000 tons) per year.

Overfishing continues as shown by the recent collapse of the cod fishery off Atlantic Canada and New England, which was closed in the early 1990s and has not reopened (see the Case-in-Point at the beginning of Chapter 10). In the early years of the 21st century, cod fisheries bordering Northern Europe and in the Baltic Sea also showed signs of stress and possible pending collapse. Overfishing has led to the serial depletion of fish populations as each newly discovered stock is fished out and the fishing fleet moves on to the next stock or fishing area. Many fisheries continue to be at risk of collapse and extinction. In 2000, the United Nation's Food and Agriculture Organization (FAO) estimated that one quarter of the world's commercial fisheries were overexploited or depleted and about half of the ocean's fishing areas were "fully exploited." Expansion of production is possible in only about one quarter of the world's fish stocks. According to NOAA's *National Marine Fisheries Service (NMFS)*, more than 60% of the 200 most commercially valuable fish stocks are either overfished or fished to the limit. One result is that many marine fishes in fish markets come primarily from fish farms. Wild salmon, for example, are increasingly rare in the market place, replaced by those raised in pens (see the section on mariculture later in this chapter).

Sharks are an example of an over-exploited, top predator with a relatively low reproductive capacity. Sharks are slow to mature sexually, some requiring 15 years or longer. Dusky sharks first breed when they are 20 to 25 years old. Sharks bear live young and most have long gestation periods. The spiny dogfish shark, used in British fish and chips, has a 22-month gestation period. Thus, many sharks have annual replacement rates of only 3% to 4%. As their population declines, their slow reproduction rates mean that shark species may risk extinction. Even if the fishery is closed, shark populations are slow to recover. For example, in the 1960s, overfishing caused the collapse of the porbeagle fishery in the northwest Atlantic. Almost three decades passed before these sharks recovered. When the fishery reopened, it took only about three years for the stock to be overfished again.

Between 1986 and 2000, the numbers of sharks caught in the northwestern Atlantic on the long lines used to catch tuna and swordfish diminished significantly. Over the 15-year period, hammerhead shark numbers were down by approximately 89% while white and thresher shark populations declined by about 75%. The thresher shark population may have collapsed; that is, they became extinct for commercial purposes. Some areas of the ocean had no white sharks at all. Only the mako shark did not show a decline.

Overall, the rates of shark declines in the northwestern Atlantic indicate that the shark fishery there is not sustainable.

Sharks are taken for their meat; some are killed accidentally by collisions with fishing boats whereas others are caught along with target fish and then dumped overboard. The most egregious threat to shark populations, however, is catching them just for their fins. Demand for pricey Asian shark-fin soups fuels the demand for fins that bring a much higher price than shark meat. Crews engaged in "shark finning" typically slice off the fin of a captured shark and then toss the rest of the animal overboard, a tremendous waste of food. Worldwide, fishers take an estimated 100 million sharks each year for their fins. In response to this practice, the U.S. Shark Finning Prohibition Act has been in effect since March 2002. This law bans U.S. vessels anywhere in the world ocean from possessing shark fins unless the rest of the shark carcass is also onboard. The same regulation applies to foreign vessels fishing in the U.S. *Exclusive Economic Zone (EEZ)*, 370 km (200 nautical mi) or more offshore (Chapter 4). Whereas this law is well intentioned, trade in shark fins continues unregulated throughout most of the world ocean. In August 1992, the U.S. Coast Guard stopped and boarded the *King Diamond II* out of Honolulu in waters off of Acapulco, Mexico. Law-enforcement officials found 29 metric tons (32 tons) of fins onboard but no other shark remains. According to a NOAA official, this finding would indicate that about 30,000 sharks were taken and almost 0.6 million kg (1.3 million lbs.) of shark meat was disposed of at sea.

Fisheries scientists estimate that about 8% of the ocean's *primary production* is now harvested in the global fisheries (Chapter 9). That is, 8% of the total global fixation of carbon by phytoplankton is being removed from the ocean in the form of fish. This percentage is somewhat misleading because most of the ocean's area is relatively unproductive. Considering only the commercially significant areas of the ocean, the situation is much more serious. Fisheries remove about one quarter of the primary production in upwelling zones and the tropical ocean shelf areas. About 35% of the primary production of temperate coastal shelf seas is removed in this way. Based on these findings, many scientists believe that the fish catch is already near the limit of the ocean's fish yield. This view is further supported by the apparent decline in global marine fisheries early in the 21st century. Humans are now the top predators in many—perhaps most—of the world's major marine ecosystems. Unlike other top predators, however, humans return essentially nothing useful to the ecosystems from which they take their fish.

MAXIMUM SUSTAINABLE YIELD

To prevent overfishing of a fish stock, fishery managers typically set catch quotas. The goal is to adhere to the **maximum sustainable yield** of the fish stock; that is, limits are set on fish catches (*total allowable catch*) so that stocks are maintained at a level that will ensure the long-term viability of the target species. Quotas must take into consideration the life span, population growth rate, and reproductive capacity of a particular fish species. A fishery is sustainable if it can be fished indefinitely at reasonable levels while maintaining the ecosystem (function, structure, and diversity) on which the fishery depends, and the integrity of the habitat essential to the fish species. As we have seen, in cases where a fishery has collapsed, the fishery may be closed in order to allow its population to recover.

The maximum sustainable yield is based on a model that simulates the population growth of an exploited fish species. One model consists of a sigmoid (*S*-shaped) growth curve as shown in Figure B.1A. According to this model, population growth of a fish species is slow at first and then more rapid. Eventually, as the number of fish becomes so large that available food resources begin to become limiting, population growth again slows and eventually levels off at or near the carrying capacity. The **carrying capacity** is the maximum population that can be sustained by the resources of the marine habitat. In the case of fisheries, the carrying capacity is the maximum population that would exist in the absence of commercial fishing.

Note in Figure B.1A that the population growth rate is low when the population is small or when the population nears the carrying capacity. At some intermediate size, the fish population has the greatest potential for growth and reproduction. For most fish species, reducing the fish population to about half its natural size (i.e., the carrying capacity) makes available more resources per individual and produces the maximum sustainable yield (Figure B.1B). If the fish population does not drop below this intermediate size, the fish population can be sustained by limiting the catch to no more than the annual population growth rate. Today, the populations of most fish stocks are well below that which would produce the maximum sustainable yield.

ECOLOGICALLY SUSTAINABLE YIELD

In recent years, fisheries managers have begun to reassess the traditional approach to sustaining the populations of exploited fish stocks. This reassessment stems from the realization that fish species are components of ecosystems in which they interact with other organisms (e.g., as food sources) and the physical/chemi-

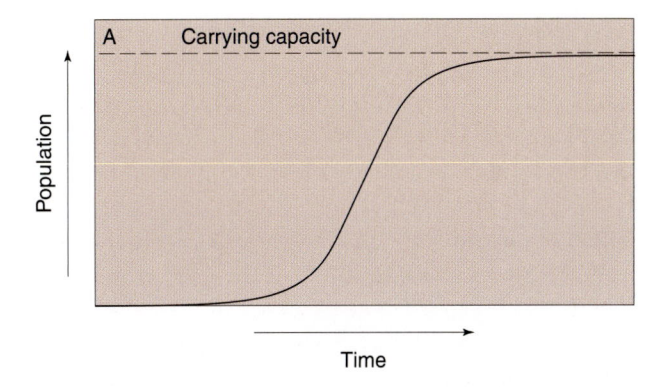

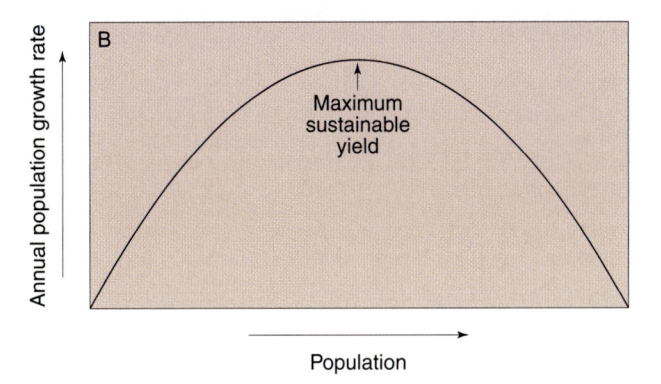

FIGURE B.1
The maximum sustainable yield of fish is based on a model (A) that simulates the growth of the population of a fish species. The maximum population, the carrying capacity, is determined by the availability of resources in the marine habitat. The annual population growth rate (B) peaks at some intermediate population size; this is the maximum sustainable yield.

cal environment (e.g., for habitat). A decline in the population of a particular species of fish may have a ripple effect on the ecosystem, altering its biotic composition and reducing its stability. This is especially the case when fishers decimate the populations of commercially attractive fish such as cod, tuna, swordfish and other species that are at the top of food webs (top predators). This ecosystem perspective argues for a more holistic approach to fisheries management.

As increasing fishing pressures cause populations of top predators to decline, organisms (e.g., other fish) at progressively lower trophic levels begin to dominate food webs in marine ecosystems. Fishers, in turn, must target fish populations at these lower trophic levels. Daniel Pauly of the University of British Columbia coined the phrase *"fishing down the food web"* for this shift in targeted fishes induced by overfishing of top predators. From an economic perspective, ecosystems that are depleted of top predators and dominated by disproportionate numbers of small fish, crustaceans, and phytoplankton are less valuable.

A study of fish catches over the period 1950 to 1994 conducted by the United Nations' Food and Agriculture Organization (FAO) confirmed fishing down the food web. Researchers discovered a gradual shift away from food webs dominated by high trophic level fish (e.g., cod, haddock) to food webs dominated by low trophic level invertebrates and plankton-feeding fish. The FAO found that the rate of fishing down the food web was about 0.1 trophic levels per decade—likely an underestimate because of limited data from fisheries in the tropics.

At first targeting fishes at lower trophic levels may be accompanied by an increase in total catch. With depletion of top predators, their prey populations (smaller fish) increase in numbers. However, the greater total catch is short-lived as populations of the competitors of the top predators increase. Furthermore, smaller fish and other organisms at lower trophic levels undergo much more rapid population fluctuations than do organisms at the higher trophic levels. This implies a less reliable food supply for the limited number of remaining top predators and a less stable food web; that is, top predators are more vulnerable to environmental change (including climate change) that impact their prey.

Protection of fisheries requires a thorough understanding of the entire ecosystem, not just the targeted commercial or recreational fish species. This understanding is the basis for achieving an **ecologically sustainable yield**, that is, the yield that a marine ecosystem can sustain without undergoing an undesirable change in state. New management concepts and associated policies must take into account all the species that interact with the targeted species and the habitats they depend on. Even reducing fish catches by 40% may not be enough to protect fish stocks and their ecosystems while allowing both to recover from past overfishing. Thus the goal of sustainable fisheries may be quite different from, and even incompatible with the ecologically sustainable goals of maintaining natural populations of fish and other members of the ecosystem. Furthermore, sustainability will require more extensive monitoring of the marine environment in which these ecosystems function to detect changes in conditions that might significantly alter the system. In short, humans will not only continue to be the dominant predator in marine ecosystems, but will also be responsible for managing, protecting, and restoring them if they are damaged by human activities or climate change.

BYCATCH

Many commercial fishers accidentally catch undersized fish or unwanted species in their nets or hook lines; these are discarded, either dead or dying, because they cannot be sold. **Bycatch** refers to fish and other marine animals that are taken in addition to the target species. Today, about one-third of the annual commercial fish catch worldwide is discarded as bycatch.

Bycatch is a major threat to many endangered species, such as sea turtles, dolphins, and other marine mammals. For example, sea turtles die in large numbers each year after swallowing floating hooks intended to catch swordfish, or after they become tangled in shrimp nets. Ironically, the long-line technique was developed as a substitute for nets that were previously used to catch tuna but that also killed large numbers of dolphins each year.

In the late 1950s, fishers in the eastern tropical Pacific discovered that yellowtail tuna aggregated below schools of dolphins. Since then, the dominant method of catching tuna has been to set nets around the dolphin schools to capture the tuna beneath. For some time, on average more than 350,000 dolphins died each year when they became trapped in the fishing nets. More recently fishers have used special nets that allow dolphins to escape unharmed. Since 1998, about 2000 dolphins were killed each year in these nets, a 99% reduction in mortality. Despite this advance in protecting dolphins, three dolphin species are still listed as depleted (i.e., two species of spotted dolphin and one species of spinner dolphin). NOAA's *National Marine Fisheries Service* continues to work with other nations fishing in the eastern tropical Pacific to further reduce dolphin mortality.

Trawling for shrimp (also called prawns) is an especially large and little recognized contributor to the bycatch problem worldwide (Figure B.2). Nets used to capture shrimp are dragged along the shallow sea bottom where they also catch all kinds of fishes, shellfish, and other animals, including sea turtles. An estimated 150,000 sea turtles are killed each year in shrimp-trawl nets. As noted later in this chapter, however, these turtle deaths can be reduced or avoided by using excluder devices that allow them to escape shrimp nets unharmed.

Worldwide, shrimp trawling results in bycatches of 5 to 20 kg (11 to 44 lb.) of unwanted animals and plants for every 1.0 kg (2.2 lb.) of shrimp caught. (In the past, local artisinal fishers would use unwanted fish for their own consumption.) Much shrimp trawling is conducted in shallow tropical waters and the catch is exported to European or North American food markets. The European Union is

FIGURE B.2
A double-rigged shrimp trawler with one net up and the other being brought aboard. Trawling for shrimp is an especially large contributor to bycatch. [NOAA Photo Library, Photo by Robert K. Brigham]

the world's largest shrimp importer, much of it coming from tropical waters near developing countries. Shrimp fishing is also heavily subsidized by the European Union, which pays nearly half the operating expenses for shrimp trawlers working off the West African (Guinea-Bissau) coast. Trawling for shrimp accounts for about one-third of the world's total bycatch while representing only about 2% of the global fish catch. Trawl nets also damage the sea bottom (Chapter 10).

Buying farmed shrimp does little good for the protection of marine fisheries. Each kilogram of farmed shrimp produced requires feeding them two kilograms of fishmeal processed from wild-fish species. Furthermore, large areas of mangrove swamps are destroyed to construct shrimp ponds, which have a short productive life before disease outbreaks necessitate building new ones and abandoning the old ones.

Part of the bycatch problem might be alleviated by changing or rewriting governmental regulations to relax quotas, permitting accidental catches to be landed legally. Success in this effort may be elusive because a policy that is too relaxed on accidental catches may actually encourage such "accidents." These regulations would only apply within national waters. In open-ocean waters, beyond the boundaries of Exclusive Economic Zones (EEZ), international cooperation would be required. This policy is important for marine mammals and sea turtles.

RESTORING FISHERIES

Few marine fisheries have been restored after experiencing severe depletion. The North Atlantic herring fishery temporarily recovered after fishing ceased during World Wars I and II. Another example is striped bass, also known as rockfish, which recovered from near collapse in the 1980s when the states of Maryland and Virginia declared moratoriums on taking the species in Chesapeake Bay.

To establish sustainable fisheries for other species will require new strategies. Earlier in this chapter, we described the concept of ecologically sustainable yield. The goal of the ecosystem-based approach to fisheries management is to maintain or re-establish marine ecosystems that are home to fisheries. This would require protection of seafloor habitats (e.g., by phasing out destructive trawling practices) and expanding the areas of the world ocean designated as protected areas, called *marine reserves*, where fishing is prohibited. In these so-called *no-take zones*, fishes can grow to maturity and reach their optimal reproductive potential. The largest individual fishes are most desirable because they are the most prolific producers of eggs and sperm. For example, a single female red snapper weighing 12.5 kg (27.5 lb.) produces about the same number of eggs as more than 200 smaller female red snappers each weighing 1.2 kg (2.6 lb.). The resulting increased fish production could be taken in the waters surrounding marine reserves where fishing is permitted. However, existing no-take zones are small and few in number; together they cover only 0.01% of the area of the world ocean.

A marine reserve has been proposed for the Channel Islands Marine Sanctuary off Southern California. This is one of 13 marine sanctuaries established since 1972 and administered by the NOAA National Ocean Service's *National Marine Sanctuary Program* in U.S. coastal waters and the U.S. Great Lakes. In a similar vein, New York's Bronx Zoo operates a sanctuary for penguins and other sea birds on two uninhabited rocky islands near the Falkland (or Malvinas) Islands, off Argentina in the South Atlantic. (For more on marine sanctuaries and marine reserves, refer to the first Essay in Chapter 10.)

Other strategies proposed to help re-establish or protect fisheries include reducing the capacity of the global fishing fleet. As noted earlier, the world fishing fleet is estimated to be about 2.5 times larger than needed to catch the ocean's sustainable fish production. Related to this strategy is the proposed elimination of government subsidies for unprofitable fishing operations. Furthermore, greater effort is needed to deal with the problems associated with bycatch and the wasteful practice of disposing of bycatch at sea.

HABITAT DESTRUCTION AND RESTORATION

In addition to overfishing, habitat destruction has also caused depletion or destruction of major fish stocks. Since steam-powered fishing boats began dragging heavy steel nets over the ocean bottom to catch bottom-feeding fishes (e.g., cod and haddock), benthic ecosystems and habitats have been heavily damaged. For decades the damage was ignored and fishers spoke of "plowing" the sea bottom to improve fish catches, much like farmers cultivate the land. Some fisheries scientists, however, argue that the effects of trawling the sea floor is more like clear cutting forests to improve hunting for rabbits or deer. This activity deprives young fishes of places to hide from predators and food for the adults. Due to widespread damage to shallow sea bottoms, cod and other long-lived, bottom-feeding fishes have experienced severe population declines. Consequently, some commercial fishing operations have shifted to rapidly growing and reproducing open-water organisms, such as squid, which do not require intact benthic ecosystems and habitats for their survival.

Habitat destruction is especially serious for oysters in coastal and estuarine waters. For instance, in the 17th century, when Europeans first sailed into Chesapeake Bay, they encountered vast oyster reefs that stood above sea level at low tide. In fact, oyster reefs were hazards to ships navigating the Bay. Native Americans and early colonists gathered oysters by hand and easily retrieved them from the shallow Bay bottom. The extensive oyster reefs also provided shelter for many fishes and other species in the crevices between oyster shells.

In the early 20th century, powerboats replaced sailing vessels (called *skipjacks*) and powered winches permitted oysters to be harvested from all depths in Chesapeake Bay. Oyster production peaked in the 1880s and then declined. Not only were the oysters overfished, oyster reefs were also destroyed in the process. Without the large reefs, the remaining oysters lived on the muddy Bay bottom where they were exposed to more predators and pathogens. Two oyster diseases, accidentally introduced with non-native oysters, further contributed to their decline. In the early 20th century, commercial oyster production slipped to historically low levels and "oystering" as a traditional way of life for watermen on the Bay was threatened in both Maryland and Virginia.

Native oysters spawned and raised in hatcheries were placed on oyster beds to grow. Unfortunately, few reached marketable size before dying from disease, especially in Virginia's high-salinity waters, nearest the ocean. In an attempt to restore the Virginia oyster industry, sterile

Asian oysters were introduced. This began on a trial basis in 2003 to test the marketability of mariculture of the non-native oyster species. These oysters are larger, faster growing, and more resistant to the diseases that killed the Bay's native oysters. They are also better adapted to living in the muddy Chesapeake Bay waters than were the native oysters.

Other efforts to restore the oyster fishery in Chesapeake Bay focus on physically rebuilding oyster reefs that stand above the muddy bottom, much as they had before Europeans arrived. These reef-building efforts, led by the Virginia Marine Resources Commission, continue but are limited by cost and the scarcity of appropriate materials for oyster attachment. Oyster larvae preferentially attach to clean oyster shells, which are available in only small quantities from oyster canneries. Shells from fossil reefs are also used but are less suitable as a substrate for juvenile oysters. Mud settling on the shells or growths of benthic organisms inhibits attachment of the larvae.

Destruction of oyster populations and efforts to restore the native populations are not restricted to the Chesapeake Bay. Tomales Bay, along the California coast about 70 km (43 mi) north of San Francisco, was originally home to a small, slow-growing, native oyster. Considered a delicacy, it was essentially destroyed as a commercial fishery in the late 1800s due to local market demand during the Gold Rush days. Efforts to restore this native oyster in the Bay use floats with mesh bags to keep the oysters off the bottom so that they can grow faster and avoid predation. Here part of the objective is to restore the natural filtering capacity of Tomales Bay's waters, similar to the function that oysters served in Chesapeake Bay.

Sea grass beds in Chesapeake Bay provide another important habitat for shallow-water organisms such as blue crabs. Young crabs can hide in grass beds to avoid being eaten by predators, including fish, when they are especially vulnerable after shedding their hard shells and before the new, soft shells harden. Trawling to catch blue crabs cut wide swaths and decimated sea grasses in Chesapeake Bay. Input of excess levels of nutrients from farms, cities, and industry also played a role by promoting algal growth in the water and on the sea grass blades, depriving the plants of the light they need for photosynthesis. Efforts to restore the beds include regulation and reduction of anthropogenic nutrient discharges and by planting new sea grass beds. The goal is to restore about 46,000 hectares (114,000 acres) of the nearly 240,000 hectares (600,000 acres) of submerged aquatic vegetation that originally grew on the Bay bottom providing food for ducks and habitat for many species of small fishes and invertebrates. In 1978, before restoration efforts began, only 16,000 hectares (40,000 acres) of sea grasses remained.

RECREATIONAL FISHERIES

Recreational fisheries in the coastal ocean and in lakes are also important activities worldwide. They involve large numbers of people who catch fish for sport rather than for profit. This type of fishing is part of the tourism and recreational industry, the world's largest and fastest growing economic activity. In the U.S., more than 17 million people engage in recreational fishing each year and spend more than $25 billion; the industry employs about 288,000 people. For some fish species, the value and the amount of the recreational catch equal or exceed that of the commercial fishery. Striped bass, blue fish, and flounder in the Mid-Atlantic Region are examples of such prized fishes. More than half the fishes caught by recreational fishers are released alive; unfortunately many die due to the stress of "catch and release."

Important differences exist between recreational and commercial fisheries. Most recreational fishers prefer to fish near home; hence, sport fish populations are most heavily fished near major urban areas. Also recreational fisheries do not damage fish habitat as much as some commercial fishing practices. Furthermore, when catches of popular recreational fishes decline, fisheries management agencies often resort to fish hatcheries to replenish depleted populations. The small fishes are released into recreational waters where they grow to a size that will satisfy the sport fishing demand. Lake trout is one example of such a hatchery-raised fish released to many North American lakes; another is salmon on the Pacific coast.

Despite the large numbers of people involved and their associated expenditures, data for recreational fisheries are not as readily available as for commercial fisheries. Based on anecdotal accounts, recreational fisheries have followed some of the same exploitation patterns as commercial fisheries. They expand into new territories as populations of the large-bodied, slow growing but highly prized fishes decline. Like commercial fisheries, the sports fisheries shift to less desirable species and often end up with smaller, fast-growing fish. Collapse of recreational fisheries may be the culmination of a slow population decline over several decades. In time, the expectations of sport fishers change as memories of the original preferred species fade.

Protecting Endangered Marine Species

Many large marine animals are protected by government regulations because their populations have been greatly depleted by commercial fishing or other human activities over the past century or so. In 1973, the U.S. *Endangered Species Act* became law providing for the protection of endangered and threatened species (and their habitats). Species determined to be in imminent danger of extinction throughout a significant portion of their range are listed as *endangered*. Species are listed as *threatened* if they are likely to become endangered in the foreseeable future. NOAA's National Marine Fisheries Service (NMFS) has authority for listing (most) marine species as threatened or endangered.

The *Convention on International Trade in Endangered Species of Wild Fauna and Flora (CITES)* controls international commerce in endangered species and their products. Species listed in Appendix I of CITES are threatened with extinction; trade in these species is prohibited. Species listed in Appendix II of CITES may become threatened and their trade is restricted. At present, 67 nations are signatory to CITES. In spite of CITES, many marine species including sea turtles, whales, and water birds remain threatened or endangered.

SEA TURTLES

The large graceful marine turtles, called *sea turtles*, that inhabit mostly tropical and subtropical ocean waters are also found occasionally in bays and estuaries where they are most likely to interact with humans. While they resemble their more familiar terrestrial relatives, tortoises and freshwater turtles, sea turtles have special adaptations for the marine environment. Instead of legs, they have flippers for swimming and their shells are lighter and more streamlined to reduce water resistance. All sea turtles, except leatherbacks, have hard upper shells (*carapace*) covered with large scale-like structures (*scutes*) and a hard lower shell (*plastron*). Unlike most other animals, the carapace in turtles incorporates their backbone and ribs. The leatherback sea turtle has boney plates under the leathery skin on its back.

Sea turtles breathe air but can hold their breath for long periods when they dive or rest on the sea bottom. While swimming, they remain near the ocean surface and come up to breath every few minutes. Some species draw water into their mouth and throat via their nasal passages and oxygen is extracted by the pharynx (similar to a gill). Like all reptiles, sea turtles are cold blooded; that is, their body temperature is the same as the surrounding water. They migrate seasonally, remaining in waters having their desired temperature range and may die if they enter waters that are too cold for their metabolic needs. One species, the leatherback turtle, apparently can create some body heat, permitting them to withstand the cold waters off Canada and Iceland, where they spend the summer. Sea turtles live in the ocean their entire lives except for adult females who come ashore during nesting season to lay eggs in beach sand (Figure B.3).

FIGURE B.3
A sea turtle excavates a nest on a sandy beach on Jobos Bay, Puerto Rico. [NOAA National Estuarine Reserve Collection]

Seven species of sea turtles are recognized worldwide. The populations of all of them are listed as either endangered or threatened. Technically, the flatback sea turtle is neither threatened nor endangered—it nests on the remote beaches of northern Australia. But because it resembles the green sea turtle, trade in its products are prohibited by CITES. Many human activities threaten sea-turtle populations: (1) commercial harvesting of adult turtles (e.g., for food, hides, oil) and the poaching of eggs (believed by some to be an aphrodisiac) from nests, (2) development projects (e.g., seawalls) along beaches used by turtles to nest, (3) bycatch in commercial fishing nets, especially shrimp nets, (4) collisions with ships, including recreational boats, (5) ingestion of plastic litter, especially bags, and (6) marine pollution.

Green sea turtles occur throughout the world ocean between about 35 degrees S and 35 degrees N and are considered the most commercially valuable of all sea turtles. Hawksbills prefer warmer tropical waters and have a more limited range than the other sea turtles. The shells of hawksbills (" tortoiseshell") are used for jewelry, the principal reason for its endangered status. Kemp's (Atlantic) ridley, the rarest of sea turtles, lives in the Gulf of Mexico and along the U.S. East Coast. The only known nesting beach for this species of sea turtle is Rancho Nuevo, Mexico. The olive (Pacific) ridley is found mostly in the warm waters of the Pacific but also occurs in the Atlantic and Indian Oceans. It nests along the west coast of Mexico; its status is endangered along the Pacific coast of Mexico and threatened elsewhere in its range. The loggerhead is found throughout the world with the second greatest population along the U.S. southeast coast. Loggerheads nest on beaches from Texas to New Jersey.

Leatherback turtles are the largest sea turtles (and one of the largest of all marine reptiles); they can reach a length of 2 m (6.5 ft) and weigh up to 590 kg (1300 lb.). Leatherbacks have a different lifestyle from other sea turtles, living most of their lives in deep-ocean waters, between about 50 degrees N and 50 degrees S. In a single year, they can migrate from South America to Nova Scotia. They feed primarily on jellyfish, but can dive to depths of almost a kilometer in search of other food. These long-lived animals take 10 to 20 years to reach sexual maturity. Like all sea turtles, leatherbacks are threatened by poachers who take their eggs from nests on beaches. Their greatest threat, however, comes from long-line fishing for tuna and swordfish. Long-lines have lengths up to 60 km (37 mi) and each line has thousands of floating hooks. Leatherbacks are caught when they take the bait and swallow the hook or they become entangled in the lines. One scientist calculated that on any given day up to four million hooks are in the ocean for tuna and swordfish.

The population of sea turtles declined by nearly 90% between 1980 and 2000. Unless this trend is reversed, they may be extinct by 2030. Because they live primarily in open-ocean waters, protecting them will require international action. There are several ways in which the threats to sea turtles by commercial fishing can be reduced. One is the use of a special trap door in fishing nets that are used to trawl the sea floor for shrimp. These low-cost devices allow turtles to escape the net easily without impeding the shrimp catch. Also, reducing the time a trawl net remains on the bottom lowers the death rate among trapped turtles. Another approach is to declare areas with high seasonal populations of turtles off limits to fishing at these times.

In the United States, NOAA's *Office of Protected Resources* within the National Marine Fisheries Service oversees activities to protect sea turtles. These efforts involve training commercial fishers to use turtle exclusion devices. Because sea turtles nest on land, the U.S. Fish and Wildlife Service co-ordinates conservation efforts on beaches.

WHALES

Hunting whales began early in human history. The first whalers used small boats and hunted close to shore, as some coastal-dwelling native peoples still do, primarily around the Arctic Ocean. Around 1000 AD, Basque whalers living in northern Spain began hunting in the North Atlantic. In the 16th century, these whalers used larger vessels to reach North America where they established whaling stations in Newfoundland. Later, English and Dutch whalers joined them to hunt whales in the Arctic Ocean. These early whalers hunted the now nearly extinct North Atlantic right whale, a large (45 metric ton or 50 ton), surface feeding whale. It was called the "right whale" because the animal is so easy to take, that is, it swims slowly, was easy to kill, and floated after death so it could be towed ashore for processing.

Whales provide meat for humans (especially in Japan) and for animal feed. Before petroleum became available for oil lamps in the second half of the 19th century, whale oil was used for that purpose. Other uses for whale products included baleen for corset stays, hoop skirts, and umbrellas. Spermaceti, a valuable oily substance from sperm whales, was used to make cosmetics and candles.

When Northern Hemisphere whale stocks declined markedly early in the 20th century, commercial whalers using larger steam-powered vessels moved into the untapped Southern Ocean to exploit the large whale stocks living there. Norwegians dominated commercial whaling worldwide. Initially, whales were taken to factories ashore for processing. After 1924, factory ships were developed to process whales at sea. The increased efficiency permitted whale catches to rise from 12,000 in 1910 to 40,000 in 1940.

Like fishers, whalers focused on a single species until its numbers were so reduced that they had to hunt another (usually smaller) species. Whales are social animals and some use protected inlets for breeding, making their populations especially vulnerable to whalers, who used small boats and hand harpoons to take the whales in these close quarters. In the early 1900s, the huge blue whales were preferred and in 1930-31, 29,000 of these animals were taken. Blue whales were finally protected in 1965 but by then their numbers had dropped to an estimated 6000 in the entire world ocean. After blue whales (135 metric tons or 150 tons) were depleted, whalers took fin whales (45 metric tons or 50 tons) and then the smaller sei whales. Today, Japanese and Norwegian whalers still take the much smaller minke whales (5 metric tons or 6 tons). This is another example of serial depletion and harvesting down the food web.

To protect the world's remaining whale stocks, in 1986 the *International Whaling Commission (IWC)* instituted a moratorium on commercial whaling. The IWC, founded in 1946 and now consisting of 54 member nations, is one of the few international regulatory bodies setting policies to protect marine animals. It regulates whaling to help conserve stocks, which were greatly depleted during centuries of uncontrolled commercial whaling. However, Japan, Norway, and Iceland objected to the moratorium and continued commercial whaling. These countries claim that the populations of smaller whales, such as the minke whale, are now at healthy levels and can sustain whaling. Another argument is that whales eat many commercially valuable fishes and must be hunted to protect these fish stocks.

Japan claims that whales are taken for research purposes only, which is permitted under the moratorium. Japan took about 700 whales in the Southern Ocean and North Pacific in 2002. Japan defends its continued whaling on cultural, historic and economical grounds. In Japan, a government agency sponsors the country's "scientific whaling" operations that take 500 to 700 whales each year. The agency sells whale meat not used for research and much of it ends up in Japanese restaurants where whale meat is considered a delicacy. Many Japanese consider the moratorium, strongly supported in the United States and many other nations, as a form of cultural imperialism. Cultural differences such as this make international conservation policies extremely difficult to negotiate and enforce.

Norway resumed commercial whaling in 1993, defying the IWC's moratorium. In 2002, Norwegian whalers took 634 whales. Most of the meat is eaten locally but small amounts were exported to Iceland before it too resumed whaling. Because no market now exists for whale fat (blubber), it is either burned or dumped at sea. Claiming that whale stocks had recovered sufficiently to permit research whaling, in 2002 Iceland announced plans to resume whaling before 2006, taking 100 minke, 100 fin and 50 Sei whales over a two-year period. Iceland intends to study the effect of whales on ecosystems. Many countries and organizations object to whaling, even for research purposes, arguing that it is actually commercial whaling and has little to do with scientific research.

The IWC allowed indigenous people (Inuits) in Alaska and Siberia to conduct small whale hunts for cultural reasons and for food as they had done traditionally. In 2002, their request to take 280 bowhead whales was denied by the IWC. Interestingly, opposition to the Inuit request was led by Japan, Norway and Iceland.

Whale watching is an important activity of *ecotourism*, one of the fastest growing sectors of the tourism industry. Iceland's decision to resume whaling was controversial among its citizens at least partially because whale watching is a significant part of that nation's tourism industry, earning Iceland an estimated $8 million per year. Whale watching has become so popular in many places such as Seattle, WA that there is concern that the tour boats are making so much noise that it disturbs the whales, disrupting their normal behavior.

Even after decades of protection, many whale stocks have still not rebounded to their original numbers. Some whales are endangered, among them the North Atlantic right whale; fewer than about 550 animals remain. Other endangered whales include the blue, bowhead, finback, Sei, sperm, and humpback (Figure B.4). A few whales, such as the California gray whale, have recovered to roughly historic levels; their major breeding ground in Baja California is protected by Mexico, which has banned whaling in its coastal waters thus creating the world's largest whale reserve.

Marine mammals and other animals, including fishes, may be adversely affected by other environmental factors, including increasing noise levels in the ocean. Noise levels in the ocean have been increasing since the beginning of the Industrial Revolution from a variety of activities including ships, petroleum exploration and production, and naval exercises. The ability of mammals to use sound pulses for communication, locating prey, and possibly even to kill prey has been known for many years. Excess noise levels at sea apparently may cause whales to beach them-

FIGURE B.4
The humpback whale is one of many marine mammals listed as endangered species. [NOAA Photo Library]

selves on shore where they subsequently die. Sound pulses used in acoustic tomography to map water temperatures over entire ocean basins (Chapter 3) have also been challenged as potentially harmful, although these signals are of much lower strength.

WATER BIRDS

Restoration of some water-bird populations is among the few successes resulting from changed environmental policies and regulations in the late 20th century. Populations of brown pelicans were greatly depleted by commercial hunters for their feathers to decorate ladies hats and dresses in the late 19th and early 20th centuries. Brown pelicans are large, fish-eating coastal birds. In 1903, President Theodore Roosevelt designated Florida's Pelican Islands as the nation's first wildlife reserve. In 1918 the Migratory Bird Treaty gave them further protection from hunting.

Commercial fishers often killed pelicans by raiding their nests because they were thought to be voracious predators of commercially valuable fish stocks. As a result, pelican populations declined markedly. After 1940 their numbers were further reduced by the widespread use of DDT (dichlorodiphenyltrichloroethane) and other persistent organic pesticides. These compounds accumulated in the fatty tissues of the fish they ate and in the birds themselves. DDT caused pelicans eggs (and the eggs of other birds) to have such thin shells that they broke before hatching.

Brown pelican populations began recovering after 1972 when the U.S. Environmental Protection Agency banned the manufacture and use of DDT (Figure B.5). By 1985, brown pelican populations on the U.S. Atlantic coast had recovered to such an extent that they were removed from the list of endangered species. Since then the birds have expanded their range, moving northward as far as the Chesapeake Bay. However, populations of brown pelicans living on the Pacific and Gulf coasts as well as in Central and South America remain endangered. DDT is still manufactured and used in some developing countries to control mosquitoes and combat malaria.

North American ospreys had a similar history during the 20th century. They too are large fish-eating birds (sometimes called fish hawks). They live on rivers, lakes, and estuaries across much of North America in summer and winter in South America. Like brown pelicans, osprey populations were decimated by the egg thinning caused by DDT accumulation in their tissues. The 1972 DDT-ban greatly helped them to recover. Ospreys build their nests in dead trees near the water. They are very resourceful, using duck blinds, power poles, or channel markers as bases

FIGURE B.5
The brown pelican population on the U.S. Atlantic coast has recovered to such an extent that they were removed from the endangered species list in 1985. [NOAA Photo Library]

for their nests. For many years, U.S. Coast Guard personnel destroyed their nests on navigational aids, thus limiting the number of available nesting sites. After the Coast Guard reversed that policy, the number of ospreys rebounded dramatically. Now they are commonly seen and heard near coasts, lakes, and rivers.

Mariculture

Wild fish stocks are increasingly stressed by overfishing and even collapsing, such as the once-plentiful cod in the North Atlantic. **Mariculture**, industrial farming of fish and shellfish in the ocean, is a growing industry worldwide. For many varieties of fish and shellfish (e.g., salmon, catfish, trout, mussels, oysters, and clams) farmed fish and shellfish dominate retail markets. Most of these oceanic fish farms are large-scale operations. They have huge floating pens, usually located in protected inlets, holding hundreds of thousands to millions of animals (Figure B.6). Even marine algae are farmed in Asia, but our focus here is on fish and shellfish.

The simplest mariculture operations involve filter-feeding mollusks (i.e., mussels, oysters and clams). They are usually grown in mesh bags suspended in seawater from floating platforms. Because these animals feed by filtering plankton from the water, they do not require additional food. As they grow, mollusks must be tended to remove algae that grow on the bags and limit the circulation of oxygenated water. Periodically the bags must be replaced as the animals grow.

FIGURE B.6
Moi (Pacific threadfin) inside an offshore cage in Hawaii. An example of mariculture. [NOAA Photo Library]

All fish-farming operations release large amounts of fecal matter that settle to the bottom. Decay of these organic deposits can deplete near-bottom waters of dissolved oxygen if the currents are not sufficiently strong to disperse them. Some particularly large mussel, clam and oyster farms are situated in inlets along the northwest coast of Galicia in Spain but this fish farming has destroyed the original ecosystems.

The salmon-farming industry has grown dramatically since the 1960s when it began off Norway. Today, it dominates the world market for salmon, which is supplied by large farms in cold-water inlets in the coastal waters of Norway, Scotland, Atlantic Canada, British Columbia, and Chile. Wild salmon, while highly prized for its flavor, is now rare, but farmed salmon is abundant, inexpensive, and available year-round. Atlantic salmon is the preferred species for these fish farms because it grows faster, is easier to handle, and more animals can be grown in a single pen than other salmon species. The fish are grown in square steel-net pens (about 30 m or 100 ft on a side) suspended in the water from floating platforms.

Fish farms can add to water pollution problems. Carnivorous fish must be fed a diet of fish oils and fishmeal; some of it sinks uneaten to the bottom along with fecal matter. Critics of fish farms compare them to floating pig farms or cattle feed lots on land in terms of the amount of waste material produced. About 1.1 kg (2.4 lb.) of wild fishmeal is required to grow 0.5 kg (1.0 lb.) of salmon. In addition, the fish must be inoculated against diseases that would otherwise kill them because they live in such crowded conditions, unlike the widely dispersed wild species. Pesticides are also added to their food to control infestations of fish lice. Toxic chemicals are used to treat the nets to prevent algal growth that would otherwise clog them. Finally the salmon must be fed pigments with their food to make their flesh salmon-pink rather than a less appealing pale gray. In the wild, the salmon's normal color comes from eating pink krill, a variety of large zooplankton.

Salmon farms affect wild salmon and may be responsible for reducing native salmon populations in areas where the farms are located. Captive fish escape when their pens are damaged by storms or attacked by sea lions. They may breed with the wild fish, reducing their genetic variability, or even replace them entirely. Wild fish coming close to the pens may also become infested by sea lice, which otherwise occur in low concentrations in the scattered low-density wild populations.

Less complex are freshwater, catfish farms, often grown in converted rice fields. These herbivorous fish can feed on terrestrial grains rather than fish, thus reducing the pressure on wild stocks of anchovies, sardines, and mackerel—ingredients of fishmeal. Furthermore, catfish are more efficient than salmon in converting feed to fish proteins.

Marine Exotic Species

Exotic species, sometimes called *alien species*, are animals and plants introduced into ecosystems, usually by humans. Some introductions are intentional (new organisms used for mariculture) whereas others are unintentional (brought in ships' ballast waters). Introduction of exotic species have been common for centuries, especially in coastal waters. In most cases the introduced organisms do not survive because they are not well adapted to in the physical/chemical conditions of their new environment or they are unable to compete in the new ecosystem. But sometimes an exotic species finds its new environment to be favorable; lacking competitors or diseases and their population expands rapidly. Such introduced species often have no predators to control its numbers. Also they often are outside the range of many of the diseases they normally encounter in their native habitat. Thus, they grow larger than normal, reproduce faster than usual, and take over—perhaps destroying entire ecosystems. This situation develops in both coastal ocean waters and lakes. (Examples in the Great Lakes are lamprey eels and zebra mussels.)

One example of a marine exotic species is the Asian carp, a large (50 kg or 110 lbs.) fast-growing fish imported from China to the southern United States by fish farmers. Each female produces millions of eggs each year so that they can quickly overwhelm native fish species. Asian carp consumes up to 40% of their body weight daily, eating vegetation, phytoplankton, mussels and fishes, thereby decimating ecosystems. In the early 1990s, Asian carp were accidentally released into the lower Mississippi River system when floodwaters broke through the dikes around fishponds holding the carp. Since then the carp has taken over large stretches of the Mississippi, destroying commercial fisheries on the Lower Mississippi as they migrate northward about 80 km (50 mi) per year to cooler waters, which they prefer. By 2003, they had reached Chicago where a canal connects the Mississippi River system to Lake Michigan and to all the other Great Lakes.

The *International Joint Commission (IJC)*, the U.S.-Canadian organization established to protect the Great Lakes that the two nations share, has proposed various schemes to prevent the further spread of Asian carp. Among these schemes is the use of an existing electrical barrier in the canal or building a new, physical barrier. Keeping Asian carp out of the Great Lakes will be difficult, however, as they have already been found in a fountain in Toronto, Ontario and in Lake Erie.

New exotic organisms may also result from subtle exchanges of genetic materials. The tall marsh grass (*Phragmites*) commonly found in the mid-Atlantic salt marshes has become a widespread and invasive species that crowds out other types of marine grasses. For many thousands of years prior to about 1910, the native form of *Phragmites* grew along with other marine grasses. Beginning in the 1970s, the plant became far more intrusive. Apparently some genetic material, perhaps only a small root fragment of a closely related grass, was brought to North America from Europe or Asia. The genetic material from this exotic variety became dominant and has supplanted the native, less invasive form. Now only plants having genetic material from the Asian form are common.

Exotic marine organisms can also cause health problems among humans. In 1993, a new strain of bacteria, native to Bangladesh and India, appeared in Peru's coastal waters and later spread throughout Central and South America. The bacteria were apparently introduced by ballast water discharged from a ship. Carried by coastal plankton, this strain is now endemic in the region and causes severe cholera-like symptoms when ingested by humans.

Conclusions

For thousands of years humans have looked to the ocean as an important source of food. In fact, the belief was widespread that the sea held an almost limitless bounty of food and inspired some to speculate that marine fisheries could meet the needs of Earth's burgeoning human population well into the future. A variety of technological innovations enabled fishers to become more and more efficient at taking the most desirable top predator fishes and as those fish stocks diminished, they typically moved on to another fishery and repeated the process. However, by the late 1980s, the world fish catch peaked and has been declining ever since. Overfishing forced quotas and the closure of some fisheries but not before some species came to the brink of extinction. Overfishing and its associated problems (e.g., habitat destruction, bycatch) spurred fisheries managers to reassess their management schemes. The move is to shift from the traditional method of managing fisheries for maximum sustainable yield to a more holistic, ecologically based approach that emphasizes the sustainability of the marine ecosystem (structure and function) that is home to the fishery.

Basic Understandings

- Stewardship of the ocean and its resources involves responsibly managing all resources to benefit present and future generations. In the United States, the National Oceanic and Atmospheric Administration (NOAA) is the principal government agency with stewardship responsibilities for the nation's marine environment and living resources.

- Overfishing occurs when a fish species is taken at a rate that exceeds the maximum catch that would allow reproduction to replace the population. Recruitment overfishing involves taking adult fish in such high numbers that too few survive to replenish the breeding stock. Growth overfishing takes place when fish are taken too small, before the animals can grow to a size that would produce the maximum yield.

- A variety of technological innovations enabled fishers to greatly increase the fish catch. These innovations include new engines to power larger fishing vessels, refrigeration, factory trawlers, fish-locating equipment, and GPS. In addition, governments worldwide spend billions of dollars annually subsidizing their commercial fishing industry—often spending more than the fishery is worth.

- As fishers became more adept at taking fish in large numbers, the frequency of stock collapses increased,

forcing fishery managers to impose quotas on commercially desirable fish, limit fishing time at sea, or close fisheries altogether.
- Sharks are an example of an over-exploited, top predator with a relatively low reproductive capacity. Sharks are taken for their meat, but some are killed accidentally by collisions with fishing boats whereas others are unintentionally caught along with target fish and then dumped overboard. The most egregious threat to shark populations, however, is catching them just for their fins and dumping the carcasses overboard (known as shark finning).
- Many commercial fishers accidentally catch undersized fish or unwanted species; this so-called bycatch is discarded, either dead or dying, because it cannot be sold. Bycatch is a major threat to many endangered species, including sea turtles, dolphins, and other marine mammals.
- To prevent overfishing of a fish stock, fishery managers typically set catch quotas. The goal is to adhere to the maximum sustainable yield of the fish stock; that is, limits are set on fish catches so that stocks are maintained at a level that will preserve the long-term viability of the target species.
- As increasing fishing pressures cause populations of top predators to decline, organisms at progressively lower trophic levels begin to dominate the food webs of marine ecosystems. Fishers, in turn, must target fish populations at these lower trophic levels. This shift in targeted fishes triggered by overfishing of top predators is described as "fishing down the food web."
- Protection of fisheries requires a thorough understanding of the entire ecosystem, not just the targeted commercial or recreational fish species. This understanding is the basis for achieving an ecologically sustainable yield, that is, the yield that a marine ecosystem can sustain without undergoing an undesirable change in state.
- The ecosystem-based approach to fisheries management would require protection of seafloor habitats (e.g., by phasing out destructive trawling practices) and expanding the areas of the world ocean designated as protected areas, called *marine reserves*, where fishing is prohibited.
- Recreational fisheries and commercial fisheries differ in important ways. Most recreational fishers prefer to fish near home; hence, sport fish populations are most heavily fished near major urban areas. Recreational fisheries do not damage fish habitat as much as some commercial fishing practices. Also, when catches of popular recreational fishes decline, fisheries management agencies often resort to fish hatcheries to replenish depleted populations.
- Around the world, sea-turtle populations are threatened by many human activities including (1) commercial over-harvesting of adult turtles and illegal poaching eggs from beaches, (2) development along beaches used for nesting, (3) bycatch in commercial fishing nets, especially shrimp nets, (4) collisions with ships, including recreational boats, (5) ingestion of plastic litter, especially plastic bags mistaken for jellyfish, and (6) marine pollution.
- Like fishers, whalers focused on a single species until its numbers were so reduced that they had to hunt another (usually smaller) species. To protect the world's remaining whale stocks, in 1986 the International Whaling Commission (IWC) declared a moratorium on commercial whaling. However, Japan, Norway, and Iceland objected to the moratorium and continued commercial whaling, arguing that the populations of smaller whales (e.g., the minke whale) are now at healthy levels and can sustain whaling, and that whales eat many commercially valuable fishes.
- Restoration of some water-bird populations is among the few successes resulting from changed environmental policies and regulations in the late 20th century. North American brown pelicans and ospreys are examples.
- Mariculture, industrial farming of fish and shellfish in the ocean, is a growing industry worldwide. For many varieties of fish and shellfish (e.g., salmon, catfish, trout, mussels, oysters, and clams) farmed fish and shellfish dominate retail markets.
- Marine exotic species are animals or plants introduced into ecosystems, usually by humans. Some introductions are intentional (new organisms used for mariculture) whereas others are unintentional (brought in ships' ballast waters). In some cases, an exotic species finds its new environment to be favorable; lacking competitors or diseases and their population expands rapidly. Thus, they grow larger than normal, reproduce faster than usual, and take over—perhaps destroying entire ecosystems.

CHAPTER C

OCEAN PROBLEMS AND POLICY

Milestones in Ocean Governance
    Freedom of the Seas
    Antarctic Treaty
    Exclusive Economic Zones
Human Impact in the Coastal Zone
    Population Trends
    Environmental Pollution
Oil Spills
Dams and Marine Habitats
Restoring Chesapeake Bay
Waste Disposal in the Ocean
Deep-Ocean Carbon Storage
Obstacles to Ocean Policy Making
Conclusions
Basic Understandings

September 1992. Timbalier Bay, Terrebonne Parish, Louisiana. An oil rig blew out spewing crude oil into Timbalier Bay. The rig is surrounded a boom to contain the oil. [NOAA photo]

This chapter is concerned with some of the effects of human activities on the ocean, its ecosystems and habitats, and how society is attempting to minimize those impacts, remediate past damage, and avoid future problems. Earlier in this book, we presented many examples of how humans impact the ocean and the coastal zone. In Chapter 8, we described how humankind has attempted to stabilize the coast by constructing seawalls, jetties, and groins, often to no avail. In Chapter B, we discussed overfishing and endangered marine species and the need for an ecosystem perspective in fisheries management. In large measure, efforts at ocean stewardship arise from humankind's desire to regulate exploitation of ocean resources and space to meet human needs and to achieve *sustainable development*. While the United States is the locale of many of the issues considered in this book, vast areas of the ocean are in international waters and marine ecosystems adhere to no political boundaries. Hence, in dealing with human impacts on the ocean, a global perspective with international cooperation is essential.

The first commission to develop a national ocean policy for the United States was the *Commission on Marine Science, Engineering and Resources*, more commonly referred to as *Stratton Commission* after its Chair, Julius A. Stratton (1901-94), then also Chair of the Ford Foundation. The Commission worked for two years and presented its final report, *Our Nation and The Sea, A Plan for National Action*, in January 1969. Among other recommendations, it called for creation of the *National Oceanic and Atmospheric Administration (NOAA)* by combining existing government agencies responsible for the ocean and atmosphere. Recognizing that the ocean plays an integral role in the economic, environmental, and security interests of the United States, a second *U.S. Commission on Ocean Policy* was established to make recommendations to the President and Congress for a coordinated and comprehensive national ocean policy. It began work in 2001 and is expected to issue its final report at the end of summer 2004. Among expected recommendations is an overhaul of ocean policy including a shift to ecosystem-based management of marine life (Chapter B), creation of a National Oceans Council within the Executive Branch of government, and doubling of federal funding of ocean research.

Other nations, including Australia and Canada, have developed ocean policies for their Exclusive Economic Zones (EEZs) (Chapter 4). Such policies are tailored to the interests and priorities of the individual countries. No international ocean policy has yet been developed to deal with open-ocean areas outside the various EEZs.

This chapter addresses several ways whereby human activities impact the orderly functioning of the coastal and open ocean environment. Among the topics we cover are oil spills, the effect of dams on marine habitats, and the problem of waste disposal in ocean waters. We begin with a brief summary of important milestones in the governance of the ocean.

Milestones in Ocean Governance

On land, humans have long divided space and resources into two categories: private property and public property held in trust for citizens of the nation in which these resources are found. The history of ocean governance, however, differs dramatically from that of the land. For most of human history, the open ocean has been beyond the direct management of coastal nations or any other governments, just like outer space. As the human population soared and human exploitation of ocean resources expanded, regulations and institutions to govern the ocean have developed.

FREEDOM OF THE SEAS

In 1609, the Dutch legal scholar Hugo Grotius (1585-1645), a pioneer natural rights theorist, published his influential treatise, *The Freedom of the Seas*. In it, he argued that the ocean was open with free access to all nations; that is, no nation or group of nations has the right to monopolize either its use or access to it. For Grotius, liberty of the sea was key to maintaining communication among peoples and nations. Along with this doctrine of free access, the general consensus was that the sea was limitless and its resources inexhaustible. There was no need to control access if marine resources were essentially unlimited and therefore virtually immune to any possible damage caused by human activities.

Partitioning of the coastal ocean began in the late 1700s, when the new United States of America claimed a 5-km (3-mi) "territorial sea" off its coast. Such a narrow coastal fringe could be guarded and controlled by the military cannons of the day. Authority over these territorial seas was assigned to the individual states. In 1945, the U.S. extended its national claims on coastal resources to the outer edge of the continental shelf. A major reason for this policy change was the expansion of offshore oil and natural gas production out on the shallow continental shelf around the Gulf of Mexico and off the California coast. To this day, the U.S. considers waters outside national territorial waters to be international waters, open to all. With a few exceptions, no effective international agreements exist to control the exploitation of the ocean and its resources in international waters.

ANTARCTIC TREATY

In the last half of the 20th century, Antarctica and the Southern Ocean came under the protection of the *Antarctic Treaty System*. The Antarctic Treaty System is a complex of arrangements that regulate relations among nations near Antarctica and those conducting research and exploration there. In 1959, the Antarctic Treaty System was adopted by the dozen nations that engaged in scientific research in the Antarctic region during the International Geophysical Year (1957-58). The Treaty has since expanded to 45 nations sponsoring research in Antarctica and/or the Southern Ocean. In 2003, these countries represented about two-thirds of the world's population. The Antarctic Treaty System ensures that no national claim of territorial rights in Antarctica or the surrounding waters will ever be legally recognized and that Antarctica forever would be used exclusively for peaceful purposes.

The Antarctic Treaty System applies to waters, ice shelves, and islands in the Southern Ocean, that is, south of 60 degrees S (Chapter 2). This is the approximate latitude of the *Antarctic convergence* where cold Antarctic waters from the south meet warmer waters from the north (Chapter 6). The Antarctic convergence acts as a biological barrier so that the Southern Ocean is essentially a separate ecosystem. The *Convention on the Conservation of Antarctic Marine Living Resources (CAMLR)* was adopted in 1982 as part of the Antarctic Treaty System. CAMLR was established in response to international concerns that increased catches of krill (a large shrimp-like zooplanktonic organism) in the Southern Ocean could threaten krill populations as well as the large whales, other marine mammals, birds, and fishes that feed primarily on krill and associated high seas fisheries.

EXCLUSIVE ECONOMIC ZONES

In 1976, the United States extended its jurisdiction over some marine fisheries. The *Magnuson-Stevens Fisheries and Conservation Act* (also known as the *Magnuson Act*) extended the jurisdiction of the nation's fisheries-management seaward from the shore to 330 km (200 mi), waters that were formerly heavily fished by foreign vessels. The Magnuson Act also established regional fishery management councils to manage the fishery resources in each region of the U.S. In 1996, the *Sustainable Fisheries Act* amended the Magnuson Act imposing strict new mandates to halt overfishing, rebuild overfished stocks, reduce bycatch, and protect essential fish habitat.

In 1982, the *United Nations Convention on the Law of the Sea* expanded the narrow territorial seas to 20 km (12 nautical mi) from the shoreline and authorized nations to establish *Exclusive Economic Zones (EEZs)*. The U.S. established its EEZ in 1983, extending federal jurisdiction from the seaward limits of the state-controlled territorial seas to 370 km (200 nautical mi) from the coastline. In 1996, EEZs were extended to the edge of the continental shelf when the shelf edge was farther than 370 km (200 nautical mi). Within their EEZs, each coastal nation has the same rights and responsibilities that they exercise over their land areas. Migratory fishes, which may cross national boundaries, and range beyond the limits of the EEZs, are regulated under the *Convention on Straddling Stocks*. Cooperation on the conservation of such stocks is the responsibility of those nations where the fish occur or where they are taken.

At the beginning of the 21st century, about two-thirds of the ocean was part of the **ocean commons** and beyond the control of any coastal nation. The ocean commons essentially is free to all for unlimited exploitation. Little or no regulation exists over the ocean commons, except for a few specialized treaties such as the 1972 *London Dumping Convention* that bans the discharge of wastes from land by ships or aircraft. For this reason, commercial whaling was uncontrolled until banned by the International Whaling Commission in 1986 (Chapter B). Unregulated commercial fishing also led to depletion of some open-ocean fish populations. An example is the overfishing and eventual collapse of the Canadian cod fishery in the early 1990s (Chapter 10).

As the global human population continues to increase and growing demands for natural resources exceed supplies on land, ocean resources are likely to be increasingly exploited. In the future, such pressures are likely to drive human governance over more ocean space and resources. Innovative technologies could be used to enforce international regulations. As noted in Chapter A, various technologies now provide continuous surveillance of the ocean surface and interior. Some experts argue that all ships, planes, and submarines operating in, on, and over the ocean can be tracked, and their activities monitored in various ways. For example, acoustic methods can detect whether fishing boats have nets in the water and can even identify the types of nets in use.

Future governance of ocean resource exploitation may apply the *Precautionary Principle*: "When an action causes a threat to human health or the environment, precautionary measures should be taken even if some cause-and-effect relationships are not fully established scientifically." The 1992 United Nations Conference on Environment and Development held in Rio de Janeiro, Brazil adopted the Precautionary Principle. It was also included in the Rio Declaration's "Agenda 21," which was adopted by 178 nations in June 1992, and later ratified by the United States. The precautionary approach helps to overcome the enormous barrier to action posed by the inevitable scientific uncertainty about cause-effect relationships in complex systems. This is especially important in the ocean where many important sub-systems and processes, and their interactions are still poorly understood.

Human Impact in the Coastal Zone

As noted in Chapter 8, the majority of the human population lives in the coastal zone where they impact and are impacted by the ocean (and large lakes). It is in the coastal zone that stewardship over the ocean is most urgent and, in many cases, most challenging. People living in the coastal zone are vulnerable to certain natural hazards (e.g., flooding due to storm surge). They also impact the natural functioning of the coastal zone where the land, ocean, atmosphere, and biosphere interact.

POPULATION TRENDS

About 80% of the land on Earth is now inhabited or otherwise used by human activities such as agriculture, pasturing animals, growing trees, or mining. Satellite images taken at night of lighted settlements combined with census data (base year 1990) show that half the world's people live within about 30 km (19 mi) of the coastline and nearly 75% within 50 km (31 mi). Population densities in the Earth's coastal zone are three times those of the overall Earth average. The most densely populated coastal zones are in Europe and Asia. Most of these coastal dwellers live in rural areas and small- to medium-sized cities, rather than large cities.

In Chapter 8, we described how human habitation of the coastal-zone exposes people to many natural hazards including storm surges and tsunamis. The hazard risk is especially great for people living in low-lying coastal plains less than 100 m (330 ft) above sea level. The principal problems are expected to be in low-lying delta areas, such as the Netherlands in Western Europe, Bangladesh in South Asia, and the Mississippi River delta of North America. In these areas, storm surge flooding can extend inland 100 km (62 mi) or more.

Population growth in the coastal zone during the first half of the 21st century is expected to take place in small- and mid-sized communities in developing countries. Older mega-cities, which are mostly located at shorelines and near large rivers, are expected to grow more slowly. A significant fraction of this coastal-zone population growth will likely come from people moving from inland agricultural communities to small coastal communities, filling in the space between large metropolitan areas, such as the Boston-New York-Washington, DC corridor in the United States or the Tokyo-Osaka corridor in Japan. Consequently, a larger fraction of the world's population will impact the coastal ocean and risk exposure to coastal hazards.

ENVIRONMENTAL POLLUTION

Increasing density of the human population in the coastal zone is likely to further exacerbate environmental pollution. **Pollution** is defined as an intentional or unintentional disturbance of the environment that adversely affects the wellbeing of organisms (including humans) directly or the natural processes upon which they depend. The disturbance might involve the alteration of a biogeochemical cycle as when burning of fossil fuels changes the global carbon cycle and elevates the concentration of atmospheric carbon dioxide (Chapter 12). In other cases, toxic and hazardous materials may be dumped into reservoirs where they may not normally occur. As described in the Case-in-Point of Chapter 4, disposal of mercury waste in Minamata Bay, Japan contaminated fish and shellfish that were subsequently consumed by humans who developed adverse health effects. In the U.S., all along the Mississippi River, agricultural runoff is carried along and eventually dumped into the Gulf of Mexico off Louisiana. This results in excessive primary production off the Mississippi River delta. The subsequent decomposition of this production causes annual oxygen-deficient "dead zones" on the ocean bottom.

Inevitably, all organisms (including humans) disturb their environment by exploiting and utilizing resources, as well as producing waste products. Through the years, humans have been particularly pervasive in disturbing their environment so that many areas of the world suffer from air and water pollution. In addition, natural physical forces such as hurricanes, floods, volcanic eruptions, and earthquakes disturb the Earth system. Anyone who has survived a hurricane or destructive earthquake would agree that humankind is unable to prevent such catastrophes. As noted in Chapter 8, even efforts to lessen the impact by structural methods (e.g., building sea walls to protect against storm surge) have limitations and often fail. We can, however, reduce the toll of lives lost and property damage by planning for such disasters and avoiding habitation of areas that are particularly prone to natural hazards.

On the other hand, we can do something about disturbances for which humans are responsible. For example, we can reduce the input of excess nutrients from agricultural fields into estuaries and other coastal waters, as well as fresh water lakes and systems, and stop the dumping of untreated sewage and industrial waste into rivers and streams that ultimately empty into the ocean. Most long-term efforts to control or prevent *harmful algal blooms (HAB)* center upon reducing the amount of nutrients in the water that stimulate algal blooms in general, specifically by reducing nitrogen and phosphorus discharges from wastewater treatment plants or runoff from farms and factories (Chapter 9).

If we are responsible for disturbances that reduce the quality of the coastal marine environment, then corrective action is needed. However, not everyone agrees on whether action is needed or on what specific steps should be taken even when steps are agreed upon as needed. Some people argue that any disturbance that harms plants, animals, or humans in any way is unacceptable. But how seriously must organisms and/or the ecosystems be impaired before the disturbance is considered unacceptable? Are we concerned about all species or just a select group of species or the ecosystem-species interaction as in ocean fisheries? What are the political and economic ramifications of our choices? There are often no easy answers to these questions.

A better understanding of the problem of pollution in the coastal zone and elsewhere in the ocean is based on the concepts of carrying capacity, assimilative capacity, and limiting factors. Waste disposal in ocean waters is an ongoing problem that can only be exacerbated by human population growth in the coastal zone. The human **carrying capacity** of a region is based on the maximum rate of resource consumption and waste discharge that is sustainable indefinitely without progressively impairing ecosystem productivity and integrity. **Assimilative capacity** is the amount of waste that an ecosystem can assimilate (via decomposition by bacteria, fungi, and invertebrates) without damaging ecosystem functions or building up waste products to levels that may cause unwanted or harmful impacts on living organisms (including humans). The growth and wellbeing of an organism is limited by the essential resource (e.g., nutrients, dissolved oxygen) that is in lowest supply relative to what is required. This most deficient resource is known as the **limiting factor** (Chapter 9).

Carrying capacity and assimilative capacity are especially important considerations for islands, which are often very densely populated and lack adequate facilities to treat the wastes of large and seasonally variable populations of tourists. For example, this is a serious problem in many popular beach resorts around the Mediterranean Sea.

Oil Spills

Releases of oil, hazardous substances, and other pollutants continue to threaten fragile coastal ecosystems. According to the Office of Response and Restoration of NOAA's National Ocean Service, about 200 maritime accidents collectively spill about 7.6 million liters (2 million gal) of oil each year into the coastal zone. More than 700 hazardous waste sites contaminate the U.S. coast. Contaminants transported to the coastal environment by rivers and streams can kill fish and birds, destroy habitats, disrupt marine food webs, and close recreational beaches. In this section, we consider the causes and impacts of oil spills in coastal waters.

Oil is a major industrial commodity shipped in huge quantities across the ocean every day. For many reasons (shipwreck, collisions, pumping of ships' bilges, war, and terrorism), oil is spilled accidentally or intentionally into the ocean on a daily basis. Oil routinely enters the ocean from storm sewer and wastewater discharges. In addition, a large but mostly unknown quantity of oil enters the ocean naturally, leaking through the seafloor into the overlying water. All of these discharges adversely affect the ocean and its ecosystems with the exception of an unknown number of chemosynthetic ecosystems. Consider the impacts on the ocean of some of the larger oil spills of the late 20th and 21st centuries.

One of the world's most well-known oil spills occurred in the coastal waters of Alaska in 1989. Because of navigational errors, early on 24 March 1989, the supertanker *Exxon Valdez* ran aground on a reef in Prince William Sound, ripping open the tanker's hull. Eventually nearly 42 million liters (11 million gal) of crude oil spilled into the water, creating the largest oil spill in U.S. history. The spill occurred in one of the world's most bountiful coastal areas home to diverse and productive ecosystems.

During the next several weeks, in part due to lack of man's planning, technology and ability to control and corral the leaking oil, strong currents and storm winds moved the spilled oil as a slick through the Sound out onto the open North Pacific coast. Some of the oil moved westward onto beaches along about 2400 km (1500 mi.) of the Alaska Peninsula. The Exxon Corporation, the ship's owner, conducted a massive, $2 billion shoreline-cleanup. Thousands of people washed beaches with hot and cold water, removed oiled deposits, and spread chemical fertilizers to spur bacterial growth and decomposition of petroleum residues (Figure C.1). Naturally occurring oil-decomposing bacteria bloomed in the near shore waters after the spill and helped to clean shorelines. Smaller cleanup crews worked during subsequent summers, removing oily sediments and recording the condition of the treated beaches.

The spilled oil devastated many communities of marine organisms, which were covered by oil and thereby exposed to many toxic constituents. Organisms living attached to the rocky upper and middle parts of the intertidal zone were severely impacted. In such areas, most seaweeds, and attached and burrowing organism perished. Weathered oil that sank to the bottom below low tide levels killed eelgrass beds and the animals living in them. The spilled oil also decimated sea birds; an estimated half-million birds representing 90 species probably died. Sea birds that live mostly on the water's surface were especially vulnerable, including common murres and harlequin ducks. About 250 bald eagles were apparently killed, either directly by the

FIGURE C.1
Cleanup in the Alaska Prince William Sound area following the 1989 *Exxon Valdez* oil spill [EXXON VALDEZ Oil Spill Trustee Council, NOAA Photo Library].

spill or later when they ate oil-coated fish and oil-contaminated carcasses.

Marine mammal populations near the oil spill were also affected. About 2800 sea otters were probably killed immediately when their fur coats were oiled, thus depriving them of their fur's insulation. Furthermore, oil slicks that spread from the spill contaminated many prime haulout-areas used by hundreds of harbor seals, just before pupping season began. Large marine mammals, such as humpback whales, apparently were little affected. However, nearly two-dozen killer whales from a resident pod were missing and presumed dead.

Among commercially exploited fishes in Prince William Sound, herring and salmon were most seriously affected. Billions of salmon died. For many fish species, the cause of death will never be known. Their corpses likely sank to the bottom, were washed out to sea, or were eaten by scavengers. Unfortunately, many of the scavengers that ate them probably also died from the oil-contaminated meat.

A federal-state council of resource agencies planned and implemented a program of natural resource damage assessment. Together with the U.S. Environmental Protection Agency, the council developed plans for restoring Prince William Sound and its ecosystems. With the massive cleanup efforts and more than a decade of natural healing, many of the resources of Prince William Sound have recovered or are still recovering. Unfortunately, some ecosystems may take much longer to recover and some may never return to their original state.

When Prince William Sound was surveyed nearly 14 years after the spill and cleanup, the results were mixed. Six species had nearly fully recovered: murres, black oystercatchers, bald eagles, river otters, and two salmon species. Oil was still present under rocks on beaches. Many species had not recovered to pre-spill levels, including herring (a major food source in local ecosystems), several species of seabirds (loons, cormorants, ducks), harbor seals, and killer whales.

Some ecologists question whether humans can restore such a system that has been so extensively disturbed. All parties agree that more money is needed for restoration activities. Exxon settled claims for damages to the marine environment amounting to $1 billion. In addition, the company was fined billions of dollars for punitive damages. Much of the money already spent has gone into purchasing wildlife areas and studying the long-term effects of the spill.

A much smaller oil spill occurred off Cape Cod, MA on 15 December 1976, following the wreck of an oil barge. Much of the spilled oil (about 700,000 liters or 185,000 gal) was buried in the sediment deposits of a nearby small salt marsh. After nearly three decades, the oil remained in the same location, essentially unaltered. Under certain circumstances, oil can persist and remain toxic for decades. One explanation is that, in the absence of dissolved oxygen in the interstitial waters of the marsh sediments, bacteria were relatively ineffective at degrading the oil.

One of the world's largest oil spills, this time not an accident, occurred in February 1991 when retreating Iraqi armies released 6 to 8 million barrels of oil into the shallow Persian Gulf waters, forming a slick that coated nearly 1600 km (1000 mi) of shoreline. Cleanup cost about $700 million. When these coastal areas were surveyed in 1998, marine organisms appeared to have recovered significantly. Shrimp catches had returned to pre-spill levels and coral reefs appeared to be healthy.

Another large oil spill occurred in mid-November 2002 off Galicia, Spain's northwestern province. Waves whipped by strong storm winds caused the old, single-hulled tanker *Prestige* to split in two and sink about 250 km (150 mi) off shore. The wreck slowly discharged at least 11,000 metric tons (10,000 tons) of bunker oil (a mix of different grades of petroleum) through holes in the damaged hull. The spill immediately killed sea birds (including the locally endangered murre population) and threatened beaches, rich fishing grounds, as well as mariculture operations along the Spanish and Portuguese coasts. Mariculture of mussels is Galicia's largest industry and is particularly vulnerable to oil spills that would smother the filter-feeding organisms. Fearing toxic effects, the Spanish government banned fishing and shellfish harvesting along a 300-km (186-mi) stretch of impacted coastline. The oil slick eventually reached French Atlantic beaches coating surfaces with a gooey mess having the consistency of molasses. Economic losses and the cost of clean-up operations are expected to top $145 million. Another 74,000 metric tons (67,000 tons) of the oil went down with the ship and could adversely impact the seafloor environment for an extended period of time. The economic losses and possible cost of clean-up operations of the oil that went down with the ship (e.g., pumping out the sunken ship) are unknown.

The European Union subsequently passed legislation to ban such single-hulled tankers older than 23 years from its ports, and seeks to join with the International Maritime Organization (IMO) to institute such bans worldwide. In 1990, the United States banned single-hulled tankers older than 23 years from operating in its waters and ports after 2010. Use of double-hulled ships in place of single-hulled ships greatly reduces the risk of spills associated with oil transport on the high seas. A double-hulled ship is essentially a ship within a ship and costs about 15% to 25% more to build. According to the U.S. Coast Guard, double-hulled ships could eliminate an estimated 95% of all oil spills, but probably would not prevent spills in the case of major accidents, such as the one involving the *Exxon Valdez* where both hulls would still have been punctured.

Much of the oil entering the coastal ocean comes from untreated storm-sewer discharges. To deal with these discharges, engineers have developed ways to use natural processes to treat these waters at low costs. The initial surge of runoff from a paved road during a heavy rain event carries the greatest pollutant loads. Unburned fuel from cars and trucks along with accumulated grease from road surfaces flows into storm drains soon after the rain starts. This water is diverted into gravel-lined trenches where suspended particles settle into the gravel and the water infiltrates the underlying soil. Bacteria in the trenches decompose hydrocarbons while metal-rich particles from brake pads and particles from tires are trapped in soils and can be periodically removed for later disposal in a sanitary landfill. Runoff from later in the rainfall event contains fewer pollutants and flows through drainage channels directly into the ocean. Storm waters flowing through wetlands are also filtered naturally and at low cost.

Dams and Marine Habitats

A dam is a barrier constructed across a watercourse that impounds water in an upstream reservoir (Figure C.2). Worldwide about 15% of the Earth system's renewable freshwater supply is stored in reservoirs behind dams (about 6000 cubic km of water). Almost 3000 of these reservoirs have a combined storage capacity of more than 94 billion liters (25 billion gal); this is equivalent to all the water in Lakes Michigan and Ontario.

Most dams and their associated reservoirs are multipurpose structures; they are used for flood control, for recreation (e.g., boating, swimming, fishing), to generate hydroelectric power, to supply water for irrigation, and to regulate water levels for navigation, as well for water for municipalities. No reliable statistics exist on the number of dams worldwide. In the United States, the best estimate is 2 million dams, 75,000 of which are taller than about 2 m (7 ft). The Hoover Dam, constructed on the Colorado River in the 1930s, marked the beginning of the era of so-called *superdams*, dams that are more than 150 m (500 ft) high and

FIGURE C.2
Aerial view of DeGray Dam along the Caddo River showing a closeup view of the dam and powerhouse. Arkadelphia, AR. [U.S. Army Corps of Engineers Digital Visual Library]

capable of storing one or more years of the average annual flow of a river. Today, more than 100 superdams operate worldwide.

Dam building is a big business worldwide, especially in developing nations. During the 1990s, dam construction costs amounted to nearly $40 billion dollars annually. In 2003, nearly 1700 dams were under construction—nearly 500 in Brazil and more than 700 in India. Dam construction has received major funding from the World Bank and the International Monetary Fund. Now, however, the environmental and human costs of dams are more widely recognized and in many developed nations such as the United States, emphasis has shifted from dam construction to floodplain management in which the uses of the floodplain are regulated to protect lives and property and preserve ecosystems. About 25% of all U.S. dams are more than 50 years old. Unfortunately, many have been abandoned, are no longer maintained, and pose safety hazards to downstream communities.

Dams cause major dislocations among native and indigenous peoples who are forced to move from areas flooded by reservoirs, losing their homes, farms, and communities. For instance, flooding associated with the world's largest superdam, the Three Gorges Dam constructed during the 1990s and the early 21st century, displaced 1.3 million Chinese people. This project dammed the Yangtze River, China's largest river, to generate electricity and to control flooding. The 700-km (400-mi) long reservoir flooded 60,000 hectares (150,000 acres) inundating 160 towns, many villages, and 1600 factories. Untreated wastes discharged from upstream cities and factories will flow slowly through the reservoir, creating local pollution problems and exacerbating the already severe pollution problems in Shanghai, downstream from the dam.

Prior to the dam, the Yangtze was among the world's largest sediment-transporting rivers (Chapter 4). After the dam's completion, most of the sediment carried by the river will accumulate in the reservoir. Over time, these deposits reduce the reservoir's storage capacity, thus diminishing its flood-control and hydroelectric generating capabilities over time. Loss of the sediment-load previously carried downstream in the river waters will cause increased erosion along the river below the dam. Furthermore, the loss of sediment previously deposited in the Yangtze delta region during floods will halt the regular renewal of this fertile agricultural land. This situation is similar to what happened to the Nile Delta after the Aswan High Dam was completed (Chapter 4).

Dams have caused major changes in coastal river and ocean ecosystems. The U.S. government extensively dammed the Columbia and Snake Rivers during the past century to produce hydroelectric power, control flooding, and allow river-barge traffic throughout the region; 14 dams were constructed on the Columbia River and 13 on the Snake River. In addition, about 4000 smaller non-federal dams were built in Oregon and Washington to provide electrical power, and supply water for factories, farms, and cities in the region. Damming transformed these once cold, fast-flowing rivers and streams into a series of large lakes with warmer, sluggishly flowing waters. The effects of dams on *anadromous fish* populations have been especially severe (Chapter 10); salmon populations plummeted from an estimated 16 million before the rivers were dammed to 300,000 by early in the 21st century. Some salmon stocks are now nearly extinct. (Overfishing and climate change may be responsible for some part of this population decline.)

Dams block adult salmon returning from the Pacific to spawn on upstream gravel bars and sandy riverbanks, eliminating much of the original spawning grounds for these fish populations. Fish ladders were built on the dams to allow fish to swim past the dams but they did not work well. Furthermore, hydroelectric turbines kill many of the newly hatched salmon as they attempt to swim to the sea. Like a food processor, turbine blades grind up the young fish and the extreme pressure changes and the turbulence they encounter from the spinning blades kill an estimated 15% at each dam. Overall about 60% to 70% of the juvenile salmon never reach the ocean.

The most radical proposal to protect and restore wild salmon stocks calls for removal or partial breaching of some of the large hydroelectric dams. Some water would flow through a breached dam, permitting fish to swim downstream without encountering the turbines. The problem is that either removal or breaching of dams would reduce the amount of power generated. The region has already experienced electrical shortages and further loss of generating capacity would only exacerbate energy problems. The hotly debated alternative of removing dams has been proposed for the lowermost four dams on the Snake River.

In the United States, removal of even small dams is a relatively recent strategy. Hundreds of small dams were removed in the 1990s although few have been studied enough to demonstrate conclusively the environmental benefits of dam removals. One of the success stories involves the breaching of the Edwards Dam on the Kennebec River in Augusta, ME on 1 July 1999. For the first time in 162 years, the river flowed freely to the ocean. At the time, the Kennebec was the largest river in the nation to have a dam removed and the Edwards Dam is still among the largest dams ever removed. Built in 1837, the Edwards Dam decimated local fish populations, first by flooding critical habitat and second by preventing anadromous fishes that migrate from the ocean from reaching their upstream spawning grounds. Removing the Edwards Dam reopened about 30 km (19 mi) of fish spawning and nursery habitat in the Kennebec River. Environmental improvements in Merrymeeting Bay (at the river mouth), the largest freshwater tidal complex on the Atlantic coast north of the Chesapeake Bay, were reported soon after the removal. Populations of 10 species of anadromous fishes exhibited varying levels of recovery; among them were alewives, American shad, Atlantic salmon, striped bass, and Atlantic sturgeon. Some of the fish populations will require releases of hatchery-raised young fish to restore their populations to pre-dam levels. Birds have also increased in number and water quality has improved. The recovery of the river spurred revitalization of the Augusta's waterfront and other river-side communities as tourists, recreational fishers, and boaters returned to the area.

Dam removals can also cause problems. For example, the upper Hudson River in New York has several low dams, which previously formed large, slowly flowing pools where PCB-contaminated sediments released by manufacturing plants accumulated over several decades. Before 1973, most of these contaminated sediments were sequestered at the bottom of the reservoir behind a dam at Ft. Edward. Concern over a possible failure due to the dam's age and poor structural condition led to its removal in 1973. Before another dam could be built in its place, major floods in 1974 and 1976 moved the contaminated sediments downstream where they were deposited behind the next dam. Plans call for dredging the contaminants and storing them in an enclosed structure to prevent them from further movement and polluting the river downstream.

A smaller dam on the Patapsco River, a tributary of the Chesapeake Bay near Baltimore MD, was repaired rather than removed. In this case, the dam was needed to retain contaminated sediment deposited behind it. These deposits had high chromium concentrations discharged by earlier industrial operations. Many small dams near older manufacturing plants and in mining districts face similar problems involving contaminated sediment deposits.

Restoring Chesapeake Bay

Since 1983, the Chesapeake Bay estuary has been the focus of the nation's largest federal-state environmental restoration project, the *Chesapeake Bay Program*. The Bay Program seeks to stop further degradation of the Bay, restore its water quality, and rebuild its fisheries. Total costs for the Bay's restoration are estimated at approximately $20 billion.

The Chesapeake Bay region was among earliest settled parts of the United States; European colonists arrived at the beginning of the 17th century. By 2000, the Bay's *watershed*, the geographical area drained by rivers flowing into the Bay, was home to about 16 million people (Figure C.3). Some 150 rivers and streams drain an approximately 166,000 square km (64,000 square mi) area encompassing portions of New York, Pennsylvania, West Virginia, Delaware, Maryland, Virginia, and the District of Columbia. Forests cover about 60% of the Bay's watershed; farms and pastures cover about 30%; cities, suburbs and highways cover the remaining 10%. Point and non-point (area) sources within the watershed discharge wastes into the Bay primarily through rivers that empty into the Bay. A **point source** of pollution is a discernible conduit, such as pipes, chimneys, ditches, channels, sewers, tunnels or vessels, which transport contaminants. A **non-point source** of pollution is a broad area of the landscape, such as agricultural fields or parking lots. In addition, some airborne pollutants (e.g., ozone, nitrogen oxides), primarily from fossil fuel fired electric power plants in the Ohio River

FIGURE C.3
Since 1983, the Chesapeake Bay estuary has been the focus of the nation's largest federal-state environmental restoration project, the Chesapeake Bay Program. [U.S. Geological Survey Photo]

Valley, also reach the Bay and settle or are washed by rain and snow into its waters.

Chesapeake Bay is generally shallow with deeper dredged shipping channels leading into port facilities at Baltimore, MD to the north and Norfolk, VA near the Bay's mouth. As described in Chapter 8, it is a moderately stratified (partially mixed) estuary with several smaller, tributary estuaries emptying into it. The northern end of the Bay is essentially the estuary of Susquehanna River, which drains portions of Pennsylvania and New York; it contributes about half the river water flowing into the Bay. The Potomac and James Rivers flowing from the west each contribute about one quarter of the total river inflow. An unknown amount of groundwater flows into the Bay, primarily from its eastern side. Sediments enter the Bay directly from the Susquehanna River, whereas sediments transported by the Potomac and James Rivers are mostly trapped in their lower reaches and do not reach the Bay. Sands from the continental shelf enter the Bay with the flow of seawater through its mouth.

Partners in the Chesapeake Bay Program include political jurisdictions (District of Columbia, Maryland, Pennsylvania, and Virginia), more than 20 federal agencies, and a tri-state legislative body. The U.S. Environmental Protection Agency (EPA) represents the federal government. The federal partners are in ten departments having land-holdings in the watershed and numerous agencies having responsibilities or programs in the Bay or its watershed, among them the National Oceanic and Atmospheric Administration (NOAA).

Chesapeake Bay was the first U.S. estuary selected for such restoration and protection. Research on the Bay's environmental problems conducted in the 1970s had indicated that three problems required immediate attention: Over-enrichment of nutrients in Bay waters, continued losses of underwater sea-grass beds, and pollution by toxic chemicals. The Bay Program focused on restoring the Bay's living resources, especially its finfish and shellfish stocks, and underwater grasses. After more than two decades, the accomplishments of the Chesapeake Bay Program are modest. Communications among federal and state agencies on a regional level have improved. Plans for fisheries management and habitat restoration have been formulated and adopted. The expansion of underwater grasses was mapped. Goals were set for reducing the amounts of nutrients (primarily phosphate and nitrate compounds) entering the Bay from land-based sources. Numerical models were developed to describe the amounts and effects of nutrients discharged to the Bay. And a monitoring program was set up for the Bay's water quality.

Unfortunately, the Bay Program could not prevent the demise of the oyster and blue crab fisheries, the last two commercial fisheries in the Bay. Phosphate concentrations in Bay waters declined, but this was primarily due to government policies dating to the early 1970s banning phosphates in detergents. Although reduced runoff from land did cut nitrogen levels in Bay waters, the Program's goal of a 40% reduction by 2000 was not met. About 25% to 30% of the nitrogen entering the Bay and its watershed is airborne material emitted by motor vehicles and coal-fired electric power plants. On the plus side, the Chesapeake Bay Program has helped to prevent environmental conditions in the Bay from deteriorating as rapidly as they were prior to the 1970s.

The Bay's striped bass population was rescued from near collapse by a short moratorium on fishing in the mid- and late 1980s; the fishery was declared fully restored in 1995 and is again a major recreational fishery in the region. Maryland initially imposed the moratorium followed by Virginia but for a much shorter period. This is an example of successful policy action by individual states.

The efforts of *non-governmental organizations (NGOs)* have also helped to improve Bay conditions. NGOs educate and organize residents in volunteer conservation programs. They try to garner public support to influence government policies about environmental issues. The *Chesapeake Bay Foundation (CBF)*, the largest NGO dedicated to restoring the Bay, worked to prevent pollutants from entering the Bay as well as to restore the Bay's natural water-filtering mechanisms: forests, wetlands, underwater grasses, and oysters. Volunteers carry out these activities. Educational programs for children and adults as well as efforts to build an environmental ethic in the watershed have been important factors in the ongoing effort to restore the Bay.

Waste Disposal in the Ocean

The global human population will continue to grow rapidly during the next century. Population growth will increase pressure for more fresh water, food, energy, consumer products, and living space. It will be increasingly important to find new and acceptable ways to dispose of waste generated by human activities while maintaining a satisfactory level of environmental quality both on land and in the ocean. The rate of waste production is likely to keep pace with or perhaps outpace population growth. Disposing of waste in an environmentally acceptable manner is likely to be most challenging in the coastal zone where the human population is growing at a much more rapid rate. Sanitary landfills are filling rapidly and high-tech incinerators are expensive to operate. Recycling of discarded materials is among the strategies to reduce the amount of material entering the waste stream. Some cities encourage and even now require recycling of paper, metals, glass, and plastics. Management of toxic and hazardous waste is particularly challenging. Treaties banning the manufacture of some persistent chemicals, such as DDT, PCBs, and CFCs have been most successful in developed nations but have had less success in developing countries. Discharges of nutrients and sediments eroded from the land will likely continue worldwide.

The ocean is viewed as one possible solution to the problems of waste disposal as terrestrial options are used up. International treaties and national regulations prohibit ocean disposal of most waste. However, the ocean has been used for disposal of a great variety of wastes since humans began navigating and fishing its waters. Shipwrecks, abandoned petroleum production platforms, and other debris from human activities are common on the ocean bottom, especially near land and in shipping lanes. Several countries have directly discharged wastes at sea, including materials dredged from coastal waterways, solids (sewage sludge) from wastewater treatment plants, various industrial and construction wastes, and radioactive materials.

Deliberate discharge of these substances from ships, planes, and platforms is regulated under the *Convention on the Prevention of Marine Pollution by Dumping of Waste and Other Matter*, more commonly known as the *London Dumping Convention*, administered by the United Nations International Maritime Organization (IMO). The treaty was first established in 1972 and strengthened in 1996. Dumping of some materials was banned entirely (e.g., radioactive wastes); others were regulated (e.g., dredged materials, sewage sludge, mining wastes).

Given the population pressures on land, some people feel that waste disposal in the ocean is an attractive alternative for many waste disposal operations. The ocean is vast and the deep-ocean floor (more than 3000 m or 10,000 ft below the sea surface) covers about half of Earth's surface. Much of this deep-ocean bottom is a featureless, sediment-covered plain, seemingly monotonous physically, chemically, and biologically. Unfortunately, our knowledge of this realm is very limited. Thus, it is easy to assume that part of this region could be used for human purposes without interfering with other uses of the sea. For example, the argument continues, dumping barge-loads of crushed rock from Indonesian gold-mine operations down a submarine canyon off New Guinea would have few, if any, adverse effects on the nearby ocean bottom. This region naturally receives huge amounts of river-borne sediment.

Other people argue that we should use the deep ocean for disposal of nuclear wastes, or carbon dioxide from industrial chimneys (in liquid or solid form). The ocean is viewed as a potential receptacle for wastes that would be considered undesirable in a land disposal site. But most scientists argue that we do not yet know nearly enough about the chemistry, circulation processes, and marine ecosystems in the deep ocean to be sure that such disposal would be safe.

In addition to scientific concerns, ethical issues are involved in ocean waste disposal. Many individuals and organizations refuse to consider using the ocean in this way. They consider it immoral to pollute the ocean and oppose all forms of waste disposal at sea. This is especially true for disposal of radioactive materials. Disposal of these wastes pose extremely difficult problems for the industries and government agencies that produce them. Radioactive wastes (e.g., from nuclear power plants) emit ionizing radiation that poses serious health hazards. As a general rule, radioactive isotopes must be kept isolated for a period equivalent to 10 half-lives before they can be safely released into the environment. (One *half-life* is the amount of time required for the radioactivity emanating from a particular radioactive substance to decrease by one-half.) For some high-level radioactive isotopes, this means isolation for hundreds of thousands of years. Because human institutions and structures rarely survive intact for such long periods, the materials must be safely isolated. Isolation through burial in geologically stable regions is usually the preferred disposal option.

Land disposal sites proposed for radioactive wastes are bitterly opposed because of the risk of escape of radioactive materials to the groundwater or air. The "NIMBY" (not in my back yard) argument is often heard when this issue is debated. Sealing radioactive wastes in special containers and then burying them in the deep-ocean floor is one option that has been proposed. This is hypothesized to keep the radioactive waste out of direct human contact for thousands, perhaps millions, of years. Other schemes would place the radioactive-waste containers in subduction zones so that they would be drawn down into Earth's mantle and presumably, out of contact with humans (Chapter 2).

Essentially these disposal plans are resisted because we know so little about the deep ocean. The basic argument is that it is unwise to put the most dangerous waste materials where they may be subjected to processes not fully understood. Furthermore, retrieving such waste from the sea floor might be difficult if not impossible if they had to be relocated. The deep ocean appears to be even more complex, with greater biodiversity, than many ecosystems on land. Deep-ocean currents are apparently stronger than previously thought, and the ocean conveyor belt has a time scale of ~1000 years, which is far less than the half-life of most radioactive wastes. Furthermore, new fishing techniques permit commercial fishing in deep-ocean waters so that wastes might be accidentally introduced into commercial fish catches, or trawled up.

Deep-Ocean Carbon Storage

As we have learned, levels of the greenhouse gas carbon dioxide have been rising since the middle of the 19th century and may be contributing to global climate change (Chapters 5 and 12). Burning fossil fuels and deforestation are primarily responsible for the increase in atmospheric CO_2.

Schemes to remove carbon dioxide from the atmosphere or to store it directly without releasing it to the atmosphere have been proposed.

Replanting tropical forests has been proposed as a low-cost way to remove atmospheric carbon dioxide. Plantation costs are low in many developing countries in the tropics and rain forest vegetation grows rapidly, thus storing carbon for decades to centuries. Unfortunately, such projects do not provide enough carbon storage over the long term to be effective. Growth slows as trees mature, so the carbon uptake also diminishes. Furthermore, forests are either harvested or burned due to natural causes (e.g., lightning) every 30 to 50 years. As populations grow and more agricultural land is needed, forests will probably shrink, rather than expand.

Another way to store carbon dioxide is in rock formations below the sea floor. This technique has been used at some offshore natural-gas production facilities where carbon dioxide is separated from methane. The methane, the chief component of natural gas, is piped ashore and the carbon dioxide is injected back into formations below the facility (rather than being vented to the atmosphere). Using this technique, Norway's state-controlled oil company has disposed of about 1 million tons of carbon dioxide per year since 1996 at its natural gas operation in the North Sea. Other options for sequestration of carbon dioxide include injection into aquifers, deep coal seams, or depleted oil reservoir rock.

Fertilization of ocean surface waters has been proposed as a means of increasing the ocean's uptake of carbon dioxide from the atmosphere. Recall from Chapter 9 that broad areas of the subtropical open-ocean receive abundant sunlight and have relatively high concentrations of nitrogen and phosphorus compounds, all essential ingredients for photosynthesis by planktonic algae. Yet, primary production is meager in these areas. In the late 1980s and 1990s, ocean scientists demonstrated that these waters are unproductive because they lack essential micronutrients, such as iron (refer to the first Essay in Chapter 9). Events of iron-rich dust particles blown from the land and settling into the ocean's surface layer trigger infrequent pulses of short-lived algal blooms. Scientists propose releasing a soluble form of iron into these waters to artificially stimulate longer-term algal blooms, which take up carbon dioxide from the atmosphere that would not otherwise be used in photosynthesis. When the algal cells die, some sink through the surface layer and are eaten or decomposed in the deep ocean or are deposited on the deep-ocean floor via the physical and biological pumps (Chapter 9). In any event, once the algal cells sink through the surface layer, the carbon they contain is removed from contact with the atmosphere for centuries to perhaps millennia.

Researchers continue investigating the feasibility of this approach and any possible adverse effects on marine ecosystems. Field experiments conducted in the Southern Ocean and in Pacific equatorial waters did show temporary increases in phytoplankton populations and in primary production following the introduction of iron. But as yet no proof exists that such a method could be economically used on a large scale to bring about long-term removal of carbon from surface-ocean waters and the overlying atmosphere. Furthermore, open-ocean and deep-ocean ecosystems are too poorly known to accurately predict possible consequences of such iron releases. More field experiments are planned in other areas of the ocean and over greater temporal and spatial scales to identify possible adverse effects. For example, too much phytoplankton production might deplete the dissolved oxygen as the organic matter subsequently decomposed.

Another option is to inject carbon dioxide directly into deep-ocean waters for long-term storage. The Italian energy expert E. Marchetti first proposed this idea in 1977. Marchetti proposed to separate CO_2 from the emissions of coal-fired power plants and pump it into the outgoing Mediterranean waters off Gibraltar. These relatively dense waters sink into the Atlantic carrying the carbon dioxide with it. Model studies of such injections showed that the gas must be injected at depths greater than 3000 m (9800 ft) to keep the carbon dioxide in the ocean for more than a few centuries. On the other hand, liquid carbon dioxide injected at depths below 4500 m (14,800 ft) would form lakes occupying topographic depressions on the seafloor. A thin layer of ice-like hydrate would form at the lake surface thereby slowing the escape of CO_2.

The deep-ocean injection method has several drawbacks. CO_2 injections at shallow depths were less effective. The scarcity of carbon dioxide sources near suitably deep waters limits the feasibility of this approach. Injections into the deep Pacific were more successful than those into the deep Atlantic, which has more vigorous deep-ocean currents. Furthermore, direct injection of carbon dioxide into ocean waters is expected to cause the waters to become more acidic with unknown effects on organisms.

Many scientists and conservationists feel that it would be more practical, economical, and less ecologically risky, to focus on reducing anthropogenic carbon dioxide emissions in the first place. Much of the technology to do this exists and is being used in several industrialized nations of Europe. In North America, however, the majority of the people do not support this approach and the measures it would require, such as increasing the fuel efficiency of automobiles and reducing reliance on fossil fuels for generation of electricity.

Obstacles to Ocean Policy Making

We have considered examples of policies designed to protect the ocean and its resources in this and the previous chapter. They include local, regional, national and international policies, forged by local, state and national governments, and through international treaties and organizations. However, the agreements required to forge an acceptable environmental policy at any level of government are usually complex and fraught with obstacles and controversy. A number of problems arise repeatedly. Issues are very often divisive and the opposing sides hold rigid views. For example, fishers rarely believe the warnings of scientists and government agencies that fish stocks are declining. They fish with more determination, further hastening the decline, until draconian measures, such as a complete moratorium on fishing, are necessary.

Often, such as in the efforts to restore and protect the Chesapeake Bay, multiple levels of government are involved. The Bay Program involved agencies in three states, the District of Columbia, the federal government and many of its agencies. Political differences sometimes get in the way of forging agreements on the policies needed to achieve common goals. Various governments and organizations must respond to competing interest groups and all parties must be invited to negotiations even when they have different agendas. Diverse sectors of society may blame each other for the damage done while failing to make progress toward repairing it. The agricultural industry in rural areas surrounding Chesapeake Bay feels that it is unfairly blamed for nutrient pollution in the Bay and argues that the cities and industries should bear more of the blame and costs of cleanup because of their sewage and other waste discharges.

At the international level, the problems are even more complex. Industrialized nations possessing advanced technology and many resources to combat pollution dominate the Northern Hemisphere. These nations have small population growth rates (if any) and an educated, affluent society that can push governments to take action to combat environmental problems. In many cases the environmental damage was done a century or more ago and is now forgotten or at least partially repaired. Developing nations are trying to improve their economies and reduce poverty by establishing industries that will create jobs. Unfortunately, the rush to industrialize and the huge population pressures in many developing countries have resulted in enormous pollution problems and damage to ecosystems. For example, the use of dynamite or cyanide to kill coral reef fish for easier harvesting causes severe harm to the reef ecosystem. The practice is banned in much of the world, but is still commonly practiced in some developing countries. DDT is still used in some areas as a cheap and effective means of combating the mosquito that carries malaria (Chapter B).

The attitude of leaders of many developing countries is that economic development must be their top priority if they are to alleviate widespread poverty. Developed countries blame them for many pollution and environmental problems. In return the developing countries point to the damage done in the past by industrialized countries (e.g. whaling, over fishing, strip mining, clear-cut logging, pollution) and demand financial and technological assistance if they are to take action on their own environmental problems. This conflict frequently frustrates, or greatly delays, attempts to develop international policies to protect the ocean and marine life.

Conclusions

Humans have long assumed that the vast ocean holds a bounty of virtually limitless resources that are there for the taking by almost anyone. They have also assumed that the ocean's ability to assimilate waste was limitless so that the ocean was considered an appropriate repository for a wide variety of wastes ranging from sewage sludge to construction debris. In more recent times, people have begun to reevaluate these assumptions as they become more aware of issues such as overfishing, accelerated shoreline erosion, pollution, harmful algal blooms, oxygen-deficient "dead zones," and exotic species invasions.

Guided by an Earth system perspective and the fundamental understanding that marine life depends on the orderly functioning of ecosystems, scientists and public policy makers have joined ranks to rectify past practices that disrupt marine ecosystems and to formulate new policies that promise to safeguard the ecological integrity of the ocean and coastal zone for future generations. This stewardship effort is becoming more important as the human population density in the coastal zone continues to increase. Individual nations have taken more control over human activities in their EEZs (e.g., closing stressed fisheries) and the international community is beginning to take steps to further regulate human activities in the ocean commons.

In order for ocean policy to be appropriate and effective, however, more resources need to be directed toward developing a more complete understanding of the properties and processes of the ocean.

Basic Understandings

- The history of ocean governance differs dramatically from that of the land. For most of human history, the open ocean has been beyond the direct management of coastal nations or any other governments. As the human population soared and exploitation of ocean resources expanded, regulations and institutions to govern human activities in the ocean have been developed.
- Partitioning of the coastal ocean began in the late 1700s, when the new United States of America claimed a 5-km (3-mi) "territorial sea" off its coasts. In 1945, the U.S. extended its claims on coastal resources to the outer edge of the continental shelf—principally because of the expansion of offshore oil and natural gas production out on the shallow continental shelf. With a few exceptions, at present, effective international agreements still do not exist to control the exploitation of the ocean and its resources in international waters.
- n 1959, Antarctica and the Southern Ocean came under the protection of the Antarctic Treaty System ensuring no national claim of territorial rights in Antarctica or the surrounding waters will ever be legally recognized and that Antarctica would be forever used exclusively for scientific and other peaceful purposes.
- The U.S. established its Exclusive Economic Zone (EEZ) in 1983, extending federal jurisdiction from the seaward limits of the state-controlled territorial seas (5 km or 3 nautical mi) to 370 km (200 nautical mi) from the coastline. In 1996, EEZs were extended to the edge of the continental shelf in the cases when the shelf edge was farther than 370 km (200 nautical mi). Within their EEZs, each coastal nation has the same rights and responsibilities that they exercise over their land areas.
- Future governance of ocean resource exploitation may be guided by the Precautionary Principle: "When an action causes a threat to human health or the environment, precautionary measures should be taken even if some cause-and-effect relationships are not fully established scientifically." The 1992 United Nations Conference on Environment and Development held in Rio de Janeiro, Brazil adopted this principle.
- Increasing density of the human population in the coastal zone is likely to further exacerbate the problem of environmental pollution. Pollution is defined as an intentional or unintentional disturbance of the environment that adversely affects the wellbeing of an organism (including humans) directly or the natural processes (e.g., ecosystem) upon which it depends.
- A better understanding of the pollution problem in the coastal zone and elsewhere in the ocean is based on application of the concepts of carrying capacity, assimilative capacity, and limiting factors. The human carrying capacity of a region is based on the maximum rate of resource consumption and waste discharge that is sustainable indefinitely without progressively impairing ecosystem productivity and integrity. Assimilative capacity is the amount of waste that an ecosystem can decompose without damaging ecosystem functions or building up waste products to levels that may cause unwanted or harmful impacts on living organisms (including humans). The growth and wellbeing of an organism is limited by the essential resource that is in lowest supply relative to what is required; this most deficient resource is known as the limiting factor.
- For many reasons (shipwreck, collisions, pumping of ships' bilges, war, and terrorism), oil is accidentally or deliberately spilled into the ocean. Some oil enters the ocean routinely from storm sewer discharges into the sea and an unknown quantity of oil seeps into the ocean naturally, leaking through the sea floor into the overlying water. The vast majority of these petroleum discharges adversely affect the ocean and its ecosystems, although some chemosynthetic bacteria and systems thrive on hydrocarbons.
- Dams have caused major changes in coastal river and ocean ecosystems. The effects of dams on *anadromous fish* populations have been especially severe so to the point that some salmon stocks are now nearly extinct. For example, dams in the Columbia River system block adult salmon returning from the Pacific to spawn on upstream gravel bars and sandy riverbanks, eliminating much of the original spawning grounds for these fish populations. Fish ladders were built on the dams to allow fish to swim past the dams but they did not work well. Furthermore, hydroelectric turbines kill many of the newly hatched salmon as they attempt to swim to the sea. The most radical proposal to protect

and restore wild salmon stocks calls for removal or partial breaching of some large dams.
- Since 1983, Chesapeake Bay has been the focus of the nation's largest federal-state environmental restoration project, the Chesapeake Bay Program. The Bay Program seeks to stop further degradation of the Bay, restore its water quality, and rebuild its fisheries. Chesapeake Bay was the first U.S. estuary selected for restoration and protection.
- Within the Chesapeake Bay watershed, numerous point and non-point sources discharge wastes into the Bay waters, primarily through rivers that empty into the Bay. A point source of pollution is a discernible conduit, such as pipes, chimneys, ditches, channels, sewers, tunnels or vessels, which transport contaminants. A non-point source of pollution is a broad area of the landscape, such as agricultural fields or parking lots.
- The vast ocean is viewed as a possible solution to the growing problem of waste disposal as terrestrial options are used up. But most scientists argue that we do not yet know nearly enough about the chemistry, circulation processes, and marine ecosystems in the deep ocean to be sure that such disposal would be safe. Treaties and national regulations prohibit ocean disposal of most waste. Deliberate discharge of wastes from ships, planes, and platforms is regulated under the London Dumping Convention of 1972, administered by the United Nations International Maritime Organization (IMO).
- To head off global climate change due to a build up of atmospheric carbon dioxide (a greenhouse gas), many schemes exist to remove carbon dioxide from the atmosphere or to store it directly without releasing it to the atmosphere. One possibility is to store CO_2 in rock formations below the sea floor. This technique has been used at some offshore natural-gas production facilities where carbon dioxide is separated from methane. Fertilization of ocean surface waters (e.g., iron as a micronutrient) has been proposed as a means of increasing the ocean's uptake of carbon dioxide from the atmosphere and letting nature pump it to the ocean bottom via the biological and physical "pumps". Another option is to inject carbon dioxide directly into deep-ocean waters for long-term storage. Thus far, none of these proposals have been embraced as a solution.
- Agreements required to forge an acceptable national and/or international ocean environmental policy at any level of government are usually complex and fraught with obstacles and controversy.

APPENDIX I

CONVERSION FACTORS

	Multiply	By	To obtain
LENGTH	inches (in.)	2.54	centimeters (cm)
	centimeters	0.3937	inches
	centimeters	10000	micrometers (m)
	feet (ft)	0.3048	meters (m)
	feet	6	fathoms (fm)
	meters	3.281	feet
	meters	0.546	fathom
	statute (land) miles (mi)	1.6093	kilometers (km)
	statute miles	0.869	nautical miles (n. mi.)
	kilometers	0.6214	statute miles
	kilometers	0.54	nautical miles
	nautical miles	1.852	kilometers
	nautical miles	1.1516	statute miles (mi)
	kilometers	3281	feet
	feet	0.0003048	kilometers
	fathoms	1.8288	meters
	fathoms	0.1666	feet
SPEED	miles per hour (mph)	1.6093	kilometers per hour (kph)
	miles per hour	1.1516	knots (kts)
	miles per hour	0.447	meters per second (m/s)
	knots	0.869	miles per hour
	knots	0.5148	meters per second
	knots	1.852	kilometers per hour
	kilometers per hour	0.6213	miles per hour
	kilometers per hour	0.540	knots
	meters per second	1.944	knots
	meters per second	2.237	miles per hour
WEIGHTS AND MASS	ounces (oz)	28.35	grams (g)
	grams	0.0353	ounces
	pounds (lb)	0.4536	kilograms (kg)
	kilograms	1000	grams

Appendix I CONVERSION FACTORS

	kilograms	2.205	pounds
	tons	0.9072	metric tons
	metric tons	1.102	tons
	metric tons	1000	kilograms
DENSITY	grams per cubic centimeter (g/cm^3)	1000	kilograms per cubic meter (kg/m^3)
	pounds per cubic foot (lb/ft^3)	16.02	kilograms per cubic meter
	ounces per cubic inch (oz/in^3)	1.73	grams per cubic centimeter
FLOW	cubic kilometers per day (km^3/d)	11 574	cubic meters per second (m^3/s)
	cubic meters per second	35.32	cubic feet per second (ft^3/s)
	cubic meters per second	70.03	acre-feet per day (A-ft/d)
AREA	acres (A)	0.4047	hectares (ha)
	square yards (yd^2)	0.8361	square meters (m^2)
	square miles (mi^2)	640	acres
	square miles	2.590	square kilometers (km^2)
	hectares	0.010	square kilometers
	hectares	2.471	acres
	hectares	10 000	square meters
	square kilometers	1 000 000	square meters
	square kilometers	0.292	square nautical miles (n. mi^2)
	square meters	1.196	square yards
VOLUME	fluid ounces (fl oz)	0.0296	liters (L)
	gallons (gal)	3.785	liters
	liters	0.2642	gallons
	liters	33.815	fluid ounces
	acre-feet (A-ft)	1233.5	cubic meters (m^3)
	cubic meters	1000	litres
	cubic meters	35.32	cubic feet (ft^3)
	cubic kilometers (km^3)	1 000 000 000	cubic meters
	cubic kilometers	0.157	cubic nautical miles (n. mi^3)
	cubic miles (mi^3)	4.168	cubic kilometers
PRESSURE/ FORCE	pounds force (lb)	4.448	newtons (N)
	newtons	0.2248	pounds
	millimeters of mercury at 0°C	133.32	pascals (Pa; N per m^2)
	pounds per square inch (psi)	6.895	kilopascals (kPa; 1000 pascals)
	pascals	0.0075	millimeters of mercury at 0 °C
	kilopascals	0.1450	pounds per square inch
	bars	1000	millibars (mb)
	bars	100000	pascals
	bars	0.9869	atmospheres (atm)
ENERGY	joules (J)	0.2389	calories (cal)
	kilocalories (kcal)	1000	calories
	joules	1.0	watt-seconds (W-sec)
	kilojoules (kJ)	1000	joules
	calories	4.186	joules
	calories	0.00397	Btu (British thermal units)
	Btu	252	calories

POWER			
	joules per second	1.0	watts (W)
	kilowatts (kW)	1000	watts
	megawatts (MW)	1000	kilowatts
	kilocalories per minute	69.93	watts
	watts	0.00134	horsepower (hp)
	kilowatts	56.87	Btu per minute
	Btu per minute	0.235	horsepower

Appendix I CONVERSION FACTORS

APPENDIX II

OCEAN TIMELINE

ca. 4000 B.C.E.	Egyptians developed shipbuilding and ocean piloting capabilities, as well as trading on the Nile.
ca. 3800 B.C.E.	First maps showing water (river charts).
ca. 2000-500 B.C.E.	The Polynesians voyaged across the Pacific Ocean and settled all the major islands, first inhabiting Tonga, Samoa, ca. 1000 B.C.E. Micronesian stick chart used.
ca. 900 B.C.E.	Greeks first use the term *okeanos*, a mythical god, to describe the great river that flowed in a circle around the Earth and the root of the present word ocean.
ca. 800 B.C.E.	First graphic aids to marine navigation.
ca. 600 B.C.E.	The Greek Pythagoreans assumed a spherical Earth.
ca. 600 B.C.E.	Thales of Miletus, a Greek natural philosopher, developed the origin of science as we define it today and he is alleged to have written a book on navigation.
ca. 1200-850 B.C.E.	Phoenicians explored entire Mediterranean Sea and sailed into the Atlantic to West Africa and to Cornwall, England for trading.
ca. 450 B.C.E.	The Greek Herodotus compiled a map of the known world centered on the Mediterranean region.
ca. 325 B.C.E.	The Greek Pytheas explored the coasts of England, Norway, and perhaps Iceland. He developed a means of determining latitude from the angular distance of the North Star and proposed a connection between the phases of the moon and the tides.
ca. 325 B.C.E.	The great Greek philosopher, Aristotle, published *Meteorologica,* which described the geography of the Greek world, and *Historia Animalium* the first known treatise on marine biology—a catalogue of marine organisms
ca. 300 B.C.E.	Library founded at Alexandria by Alexander the Great, which served as the repository for scrolls from ships and land caravans; for the next 600 years, it was a maritime studies center, fostering advances in science, celestial navigation by a variety of Greek and Egyptian scholars.
ca. 276-192 B.C.E.	The Greek Eratosthenes, a scholar and librarian at Alexandria, calculated the circumference of a spherical Earth with remarkable accuracy using trigonometry noting the specific angle of sunlight that occurred at Alexandria and Syene (now called Aswan), Egypt. He also invented latitude and longitude lines based on landmarks.
ca. 200 B.C.E.	Chinese invent first magnetic compass.

ca. 127 B.C.E.	Hipparchus of Nicaea, a Greek mathematician, developed a 360 degree system and arranged latitude and longitude in regular grid by degrees.
ca. 54 B.C.E. - AD. 30	The Roman Seneca devised the hydrologic cycle to show that, despite the inflow of river water, the level of the ocean remained stable because of evaporation.
ca. AD 150	The Greek-Egyptian scientist, Claudius Ptolemy, added minutes and seconds to the latitude-longitude system and compiled a map of the Roman World, but erred in estimating the Earth's circumference, which remained uncorrected for hundreds of years.
AD 415	Alexandrian library with an estimated 700,000 scrolls destroyed and the last librarian, Hypatia, murdered by religious mob; Earth considered flat again.
AD 500	Hawaii colonized by Polynesians.
AD 673--735	The English monk Bede published *De Temporum Ratione,* which discussed the lunar control of the tides and recognized monthly tidal variations and the effect of wind drag on tidal height.
AD 780	Viking raids begin.
AD 982	Eric the Red, a Norse chieftain, completed the first transatlantic voyage, reaching present-day Baffin Island, Canada.
AD 995	Leif Ericson, son of Eric the Red, established the settlement of Vinland in what is now Newfoundland, Canada.
1000	Norwegian colonies in North America.
1050	The astrolabe, a navigation instrument used to measure the height of celestial bodies above the horizon, first arrived in Europe from the East.
1405	Admiral Cheng Ho, the commander of the Treasure Fleet of the Dragon Throne began his voyages for Emperor Chu Ti, ultimately ruling the South Pacific and Indian Ocean until 1433.
1420	Prince Henry the Navigator of Portugal founded a naval observatory and the first school for teaching navigation, astronomy, and cartography.
1452-1519	Leonardo da Vinci, the famous Italian scientist, observed, recorded and interpreted characteristics of currents and waves and noted that fossils In Italian mountains implied that sea level had been higher in the ancient past.
1460	Prince Henry the Navigator dies.
1492	On his first voyage, Christopher Columbus rediscovered North America, sailing to the islands of the West Indies.
1497–1499	Vasco da Gama, a Portuguese navigator, journeyed to India by sea by rounding the Cape of Good Hope with four ships, thereby establishing the first European all water trade route between Europe and India.
1500	The Portuguese navigator and explorer, Pedro Alvares Cabral, traveled westward in search of a route to India, but reached and explored Brazil.
1513	Juan Ponce de Leon described the swift and powerful Florida Current.
1513-1518	Vasco de Nunez de Balboa crossed the Isthmus of Panama and sailed in the Pacific Ocean.
1515	Peter Martyr proposed an origin for the Gulf Stream.
1519-1522	Ferdinand Magellan embarked on the first circumnavigation of the globe; the crew led by Sebastian del Cano completed the voyage. On the voyage, named the Pacific Ocean.

1569	Geradus Mercator, the Flemish mapmaker, constructed a map projection of the world that was adapted to navigational charts.
1609	Hugo Grotius, a Dutch legal scholar, published *Mare Liberum,* the foundation for all modern law of the sea.
1674	The British scientist Robert Boyle investigated the relation among temperature, salinity, and pressure of seawater with depth and reported his findings in *Observations and Experiments on the Saltiness of the Sea.*
1687	Isaac Newton published *Principia Mathematica*, which includes an explanation of the operation of gravity and that tides are the result of gravitational attraction of the moon and sun.
1725	Count Luigi Marsigli from Bologna, Italy compiled *Histoire Physique de la Mer*, the first book pertaining entirely to oceanography, examining the geological formation of ocean basins along with the marine life that lived in the sea.
1735	George Hadley proposed an explanation for the trade wind regime that involved a consideration of the Earth's rotation.
1740	Leonhard Euler, the noted Swiss mathematician, developed an approximation of the three-body problem to calculate the magnitude of the lunar and solar attractive forces that generate ocean tides.
1742	Anders Celsius invents a temperature scale.
1758	Carolus Linnaeus publishes tenth edition of *Systema Naturae*, in which biological nomenclature is formalized.
1760	John Harrison invented the Number Four chronometer in his quest to solve the longitude problem.
1760	Joseph Black proposed the concept of specific heat.
1768-1780	Captain James Cook commanded three major ocean voyages gathering extensive data on geography, geology, biota, currents, tides, and water temperatures of all of the principal oceans.
1769 or 1770	The American scientist, Benjamin Franklin, published the first chart of the Gulf Stream (by Folger), which was used by ships to cross the North Atlantic Ocean faster.
1779	James Cook dies in Hawaii.
1802	Nathaniel Bowditch of Massachusetts published the *New American Practical Navigator,* a navigational resource that continues to be revised and published to this day.
1805	Admiral Sir Francis Beaufort of the Royal Navy developed his 12-category wind force scale to aid mariners estimate wind speed; this scale was later modified.
1807	President Thomas Jefferson mandated coastal charting of the entire United States and established the U.S. Coast and Geodetic Survey.
1817-1818	Sir John Ross ventured into the Arctic Ocean to explore Baffin Island, where he sounded the bottom successfully (took the first deep-water and sediment samples) and recovered starfish and mud worms from a depth of 1.8 km.
1820	Alexander Marcet, a London physician, noted that the proportion of the chemical ingredients in seawater is unvarying in all ocean basins.
1831-1836	The epic five-year journey of Charles Darwin aboard the British research ship HMS *Beagle* led to a theory of atoll formation and later the theory of evolution by natural selection.

1835	Gaspard Gustav de Coriolis, a French mathematican, publishes first papers on an object's horizontal motion across the surface of the rotating Earth.
1836	William Henry Harvey, an Irish botanist, devises a taxonomy of seaweeds.
1838-1842	Departure of the United States Exploring Expedition led by Lieutenant Charles Wilkes on the flagship *Vincennes*. Six vessels mapping coastal areas, collected specimens.
1839-1843	Sir James Clark Ross led an expedition to Antarctica, recovering samples of deep-sea benthos down to a depth of 4.9 km.
1841, 1854	Sir Edward Forbes published *The History of British Star-Fishes* (1841) and then his *Distribution of Marine Life* (1854), in which he argued that sea life cannot exist below a depth of about 600 m (the so called azoic hypothesis).
1847	Hans Christian Oersted observes plankton.
1851	The first telegraph cable was laid across the Straits of Dover, stimulating new technologies to protect, raise and lower cables to the sea floor.
1853	The first international conference convened in Brussels by Lt. Matthew Fontaine Maury, USN for the purpose of establishing a uniform means of meteorological observations at sea.
1855	Lt. Matthew Fontaine Maury, USN compiled and standardized the wind and current data recorded in U.S. ship logs and summarized his findings in *The Physical Geography of the Sea*, credited as the first textbook of modern oceanography.
1859	Darwin's *Origin of Species* published.
1865	Fr. Pietro Angelo Secchi, a papal scientific advisor, perfected the Secchi disc for determination of the transparency of Mediterranean Sea water to sunlight.
1868-1870	Charles Wyville Thomson, aboard HMS *Lightning* and HMS *Porcupine*, made the first series of deep-sea temperature measurements and collected marine organisms from great depths, disproving Forbes' azoic hypothesis.
1870s	William Ferrel, an American meteorologist, drew attention to the deflecting effect of the Earth's rotation on ocean currents (the Coriolis effect).
1871	The U.S. Fish Commission was established with a modern laboratory at Woods Hole, MA.
1872-1876	Under the leadership of Charles Wyville Thomson, HMS *Challenger* conducted worldwide scientific expeditions, collecting data and specimens that were later analyzed in the more than 50 volumes of the *Challenger Reports*.
1872	Departure of *Challenger* Expedition.
1873	Charles Wyville Thomson published a general oceanography book called the *Depths of the Sea*.
1877-1880	Alexander Agassiz, an American naturalist, founded the first U.S. marine station, the Anderson School of Natural History, on Penikese Island, Buzzards Bay, MA.
1880	William Dittmar determines major salts in seawater.
1884-1901	USS *Albatross* was designed and constructed specifically to conduct scientific research at sea and undertook numerous oceanographic cruises.
1880s	Development of the civil time zones and the ultimate specification of the Greenwich Mean Time (GMT) as the world-wide reference.
1888	The Marine Biological Laboratory was established at Woods Hole, MA.

1890	Alfred Thayer Mahan completes *The Influence of Sea Power upon History*.
1891	Sir John Murray and Alphonse Renard classify marine sediments.
1893–1896	The Norwegian Fridtjof Nansen while aboard the *Fram*, which had a reinforced hull for use in sea ice, studied the circulation pattern of the Arctic Ocean and confirmed that a northern continent did not exist.
1902	Danish scientists with government backing established the International Council for the Exploration of the Sea (ICES) to investigate oceanographic conditions that affect North Atlantic fisheries. Council representatives were from Great Britain, Germany, Sweden, Norway, Denmark, Holland, and the Soviet Union.
1903	The Friday Harbor Oceanographic Laboratory was established at the University of Washington in Seattle.
1903	The Laboratory that become the Scripps Institution of Biological Research, and later the Scripps Institution of Oceanography, was founded in San Diego.
1905	Marconi perfects wireless telegraphy, which becomes important for communication on the ocean.
1905	The Swedish oceanographer, V.W. Ekman, published his classic work on the cause of the spiral nature of wind driven ocean currents near the surface.
1906	The first Sonar type listening device was invented by Lewis Nixon in order to detect icebergs.
	Prince Albert I of Monaco establishes the Muscée Océanographique.
1907	Bertram Boltwood calculates age of Earth by radioactive decay.
1911	Roald Amundsen was first at South Pole.
1912	Following the sinking of the *Titanic*, the International Ice Patrol formed to monitor icebergs.
	German meteorologist Alfred Wegener proposed his theory of continental drift in his Frankfurt lectures.
	Scripps Institution allied with the University of California.
1918	Vilhelm Bjerknes and colleagues at the Bergen (Norway) school formulate theories of air masses, atmospheric fronts, and midlatitude storm systems.
1921	International Hydrographic Bureau founded.
1925-1927	A German expedition aboard the research vessel *Meteor* studied the physical oceanography of the Atlantic Ocean, using an echo sounder extensively for the first time.
1930	The Woods Hole Oceanographic Institution was established on the southwestern shore of Cape Cod, MA.
1931	*Atlantis* launched.
1932	The International Whaling Commission was organized to collect data on whale species and to enforce voluntary regulations on whaling.
1937	*E.W. Scripps* launched.
1942	Harald Sverdrup, Richard Fleming, and Martin Johnson published the first modern reference text, *The Oceans,* which is still consulted today.
1943	Jacques Cousteau and Emile Gagnan invent the scuba regulator and tank combination, the "aqualung" for diving underwater.

1949	Maurice Ewing formed the Lamont (later changed to Lamont Doherty) Geological Observatory at Columbia University in New York.
1957-1958	The International Geophysical Year (IGY) was organized as an international effort to coordinate geophysical investigations of the Earth, including the ocean.
1958	The first U.S. nuclear submarine, USS *Nautilus*, also made the first submerged transit of the Arctic ice pack, passing the geographic North Pole.
1959-1965	The International Indian Ocean Expedition was established under United Nations auspices to intensively investigate Indian Ocean oceanography.
1960	The bathyscaphe *Trieste* carrying Jacques Piccard and Don Walsh reached the bottom of the deepest (Mariana) trench (10,915 m, or 35,801 ft).
1960s	Jacob Bjerknes developed concept of teleconnections involving El Niño events and the Southern Oscillation.
1962	Rachel Carson's book *Silent Spring* initiates the U.S. environmental movement.
1966	The U.S. Congress adopted the Sea Grant College and Programs Act to provide nonmilitary funding for marine science education and research.
1968	*Glomar Challenger* returns first cores, indicating the age of the Earth's crust. The cores support theories of plate tectonics.
	The U.S. National Science Foundation organized the Deep Sea Drilling project (DSDP) to core through the sediments and crust of the ocean bottom.
1969	Santa Barbara, CA oil well blowout captures national attention.
1970s	The United Nations initiated the International Decade of Ocean, Exploration (IDOE) to improve scientific knowledge of the ocean.
1970	The U.S. government created the National Oceanic and Atmospheric Administration (NOAA) to oversee and coordinate government activities related to oceanography and meteorology.
	John Tuzo Wilson, a Canadian geologist, proposes a cyclic model of tectonic revolution.
1972	The Geochemical Ocean Sections Study (GEOSECS) was organized to study seawater chemistry and investigate ocean circulation and mixing, and the biogeochemical recycling of chemical substances.
	Marine Sanctuaries Program in NOAA was initiated.
1974	Project FAMOUS (French-American Mid-Ocean Undersea Study) maps and samples the Mid-Atlantic Ridge, a zone of seafloor spreading.
1977	The submersible *Alvin* finds hydrothermal vents in the Galapagos rift.
1978	*Seasat-A*, the first satellite dedicated to the remote sensing of the oceans, was launched.
1982	The UN Convention on the Law of the Sea established.
1985	The scientific research vessel *JOIDES Resolution* replaces *Glomar Challenger* in Deep Sea Drilling Project.
	Tropical Ocean-Global Atmosphere (TOGA) commences a 10-year experiment in the equatorial Pacific, with the deployment of the TOGA Tropical Atmosphere Ocean (TAO) Array.
	R. D. Ballard locates wreck of *Titanic*.

1990	The *JOIDES Resolution* drilling vessel retrieved a sediment sample estimated to be 170 million years old.
1991	JOI researchers bore to a depth of 2 km (1.24 mi) beneath the seafloor near the Galapagos Islands.
1992	The U.S.-French *TOPEX/Poseidon* satellite launched to continuously monitor global ocean topography.
	The Tropical Ocean Global Atmosphere Coupled Ocean-Atmosphere Response Experiment (TOGA COARE) was conducted in western Pacific.
1994	The Law of the Sea Treaty entered into force.
1995	*Keiko*, a small remotely controlled Japanese submersible, sets a new depth record, reaching 10,978 m (36,008 ft) in the Challenger Deep.
1998	UN declares the "Year of the Oceans" to increase awareness of the importance of the ocean.
	Galileo spacecraft finds possible evidence of an ocean on Jupiter's moon Europa.
2000	The first ARGO free-drifting profiling floats deployed.

Appendix II OCEAN TIMELINE

GLOSSARY

A

absorption The process through which incident radiant energy is retained by a substance. The absorbed radiation is then converted to another form of energy (e.g., heat).

abyssal plain The flat surface of the sea floor, usually at the base of a *continental rise* and formed by the deposition of sediments that obscure the preexisting topography.

acid rain Precipitation with a pH of less than 5.6.

acoustic tomography A method of monitoring the changes in ocean temperature by measuring the speed of sound through the ocean.

adaptation A genetically controlled trait or characteristic that enhances an organism's chance for survival and reproduction in its environment.

adaptive coloration Camouflage whereby an organism's color pattern closely matches its background substrate.

aerosols Minute solid and liquid particles which are suspended in the atmosphere.

air mass A large widespread volume of air, the thermal, moisture, and stability properties of which are characteristic of its source region and modified as it moves away from its source.

air pressure The force exerted per unit of area by the atmosphere as a consequence of gravitational attraction upon the molecules in a column of air lying directly above a specified location.

air pressure gradient A change in air pressure from one place to another.

albedo The ratio of the amount of electromagnetic radiation reflected by a body to the amount incident to it; commonly expressed as a percentage. Usually, albedo refers to radiation in the visible range or to the full spectrum of solar radiation.

anadromous fishes Fishes which travel from salt water to fresh water or up rivers to spawn. They include salmon, shad, sturgeon and striped bass.

Arctic Oscillation (AO) A seesaw variation in air pressure between the North Pole and the margins of the polar region. Changes in the horizontal air pressure gradient alter the speed of the polar vortex.

artificial beach nourishment A process whereby sand dredged offshore is moved to badly eroded beaches with an aim to re-establishing the natural balance between sediment inputs and outputs on the beach.

assimilative capacity The amount of waste that an ecosystem can assimilate (via decomposition by bacteria, fungi, and invertebrates) without damaging ecosystem functions or building up waste products to levels that may cause unwanted or harmful impacts on living organisms (including humans).

asthenosphere A region of the upper mantle between 200 and 400 km deep which exhibits plastic-like behavior and readily deforms in response to stress. Magma may be generated here.

astronomical tides The periodic rise and fall of sea level resulting from the gravitational interaction and motions of the moon, sun, and Earth.

atmosphere A relatively thin envelope of gases and suspended particles surrounding the Earth and held there by gravity.

atmospheric window Wavelength bands in which atmospheric constituents absorb very little or no electromagnetic radiation.

atoll A series of coral reefs surrounding a *lagoon* that remains when a volcanic island sinks beneath the waves or erodes away.

Autonomous Underwater Vehicles (AUVs) Un-piloted, remotely controlled powered vehicles that are not tethered to a mother ship. They carry sensors that measure ocean water properties (e.g., temperature, salinity, dissolved oxygen).

azoic hypothesis The erroneous belief that deep-ocean waters (below 300 fathoms or 550 m) lacked sufficient dissolved oxygen to support marine life.

B

barrier island An elongated, narrow accumulation of sand oriented parallel to a shoreline, but separated from the mainland by a *lagoon*, *estuary*, or bay.

beach A deposit of unconsolidated sediment (usually sand and gravel), extending landward from low tide to a change in topography or where permanent vegetation begins.

beach sediment budget The sum of all sediment outputs and inputs on a beach.

benthic zone One of the two basic subdivisions of the marine biome which includes the sea floor and bottom dwelling organisms.

benthos A collective term for marine organisms which live on or near the sea floor, or *benthic zone*.

berm A platform of sand, flat-topped and sloping steeply seaward, formed near the mean high-water mark.

bioaccumulation The process by which persistent materials (that resist chemical, physical, or biological breakdown) gradually become increasingly concentrated in living tissue as one organism consumes another within a food web.

bio-fouling The rapid colonization by organisms of any object placed in ocean water.

biogenous sediment Marine sediment formed from the excretions, secretions, and remains of organisms.

biogeochemical cycle Pathways along which solid, liquid, and gaseous materials flow among the various reservoirs of the Earth system.

biological pump The process whereby carbon cycles through the ocean as organic matter decomposes.

bioluminescence Production of light by living organisms. Light is generally the product of a chemical reaction that takes place in specialized cells or organs.

bioturbation The churning and stirring of sediment deposits by benthic animals in the course of feeding.

Bowen ratio For any moist surface, the ratio of heat energy used for sensible heating (conduction and convection) to the heat energy used for latent heating (evaporation of water or sublimation of snow or ice).

breakwater A long, narrow offshore structure, usually constructed of large blocks of rock or concrete, oriented parallel to the shoreline and intended to provide calm waters for docking boats and to protect beaches from erosion.

buffer A substance that causes chemical equilibrium.

bycatch Fish and other marine animals that are caught in addition to the target species.

C

calcareous oozes Deep-sea pelagic sediment containing at least 30% calcareous (calcium carbonate) skeletal remains by weight.

calving The breaking away of a mass of ice from the leading edge of a glacier forming icebergs upon entering the ocean.

capillary waves Small ocean waves with a wavelength of less than 1.7 cm (0.7 in.).

carbonate compensation depth The depth of the ocean below which material composed of calcium carbonate ($CaCO_3$) dissolves and does not accumulate.

carrying capacity The maximum population of a species that can be sustained by the resources of the habitat.

cartilaginous fishes Also called *elasmobranchs*, these generally primitive fishes lack true bones and their skeletons consist of cartilage. Examples include sharks, skates, and rays.

catadromous fishes Fish that breed in the open ocean, but spend their adult lives in fresh water. An example is the American eel.

celerity The rate at which a surface wave progresses outward in still water from the point where the water was disturbed.

cellular respiration The process whereby food is broken down liberating energy for maintenance, growth and reproduction, while releasing carbon dioxide, water and heat energy to the environment.

Challenger Expedition The first voyage dedicated exclusively to marine exploration. Conducted from December 1872 to May 1876, scientists sampled every ocean basin except the Arctic.

chemosnythesis The process whereby marine organisms in the absence of sunlight derive energy from substances such as hydrogen sulfide (H_2S).

climate The weather of some locality averaged over some specific interval of time (e.g., 30 years) plus extremes in weather observed during the same period or during the entire period of record.

coast A strip of land of indefinite width that extends from the low-tide line inland to the first major change in landform features; transitional between land and ocean.

coastline The farthest inland extent of storm waves, in some cases marked by sand dunes or wave-cut cliffs.

coccolithophores Any of numerous minute marine single-celled photosynthesizing organisms covered with calcium carbonate plates.

cold-core rings Eddies having a core of relatively cold water that break off from an ocean surface current; viewed from above in the Northern Hemisphere, they rotate in a counterclockwise direction.

compensation depth The ocean depth below which no *net primary production* occurs, usually where the light level diminishes to about 1% of what it is at the surface.

condensation The process whereby water changes phase from vapor to liquid.

constructive wave interference The result when the crests of sets of waves coincide to form a wave of greater height.

consumer An organism that is unable to manufacture its food form nonliving materials but is dependent on the energy stored in other living things; also known as *heterotrophs*.

continental crust The outermost part of the lithosphere overlying the mantle and forming the solid surface of Earth's landmasses; mostly granite in composition.

continental drift The slow movement of landmasses as part of *tectonic plates*. The continents of today were once a single landmass (Pangaea) that broke apart with the various fragments moving over the surface of the planet.

continental rise The gently sloping area of the sea floor beyond the base of the *continental shelf*.

continental shelf The submerged zone between the shoreline and *continental slope* where the sea floor slopes seaward at less than one degree.

continental slope The relatively steep downward sloping area from the shelf-slope break to the more gently sloping *continental rise* or directly into an *ocean trench*.

convergent plate boundary The boundary between two tectonic plates that are moving toward one another.

copepods A diverse group of tiny crustaceans covered with an exoskeleton made of chitin.

coral atoll A ring-shaped island surrounding a seawater *lagoon*.

coral bleaching The whitening of coral colonies due to the loss of symbiotic zooxanthellae from the tissues of coral polyps; may be triggered by a rise in seawater temperature.

coral reef A calcareous reef in relatively shallow, tropical seas composed of a thin veneer of living coral growing on older layers of dead coral or volcanic rock.

Coriolis effect An apparent force relative to Earth's surface due to Earth's rotation that causes deflection of moving objects to the right in the Northern Hemisphere and to the left in the Southern Hemisphere.

cosmogenous sediment Particles from outer space, often originating from meteorite fragments.

cotidal line A line plotted on a chart connecting points at which high water (high tide) occurs simultaneously. Lines show the time lapse, in lunar-hour intervals, between the Moon's passage over a reference meridian (usually the prime meridian) and the succeeding high water at a specified location.

countershading Protective coloration found in fish whereby their dorsal (back) side is a dark color making it difficult to see them from above and their ventral (underbelly) side is light making it difficult to see them from below.

cryosphere The frozen portion of the *hydrosphere*, encompassing glacial ice, icebergs, sea ice, and the ice in permafrost.

D

decomposers *Consumers*, usually microscopic, that feed on dead organic matter, either on the ocean bottom or in the water column, thus aiding in recycling nutrients.

deep layer Dark, cold nearly isothermal ocean water below the *pycnocline*; accounts for most of the ocean's mass.

Deep Sea Drilling Program (DSDP) Using the drill ship *Glomar Challenger*, this project sampled marine sediments and oceanic crust for scientific purposes from 1968 to 1983.

deep-water wave A wave on the surface of a body of water whose depth is more than one-half the wavelength.

delta A landform at the mouth of a river produced by the sudden dissipation of a stream's velocity and the resulting deposition of the river's sediment load in the shape of the Greek letter *delta*.

density Mass per unit volume.

deposition The process whereby water changes directly from vapor to solid (ice crystals) without first becoming liquid.

destructive wave interference When the troughs of sets of waves coincide with the crests of another set of waves producing waves of reduced height.

diatoms A diverse group of minute shell-covered phytoplanktonic marine organisms having silica exoskeletons.

dinoflagellates Any of a class of single-cell marine planktonic organisms having characteristics of both plants and animals.

diurnal inequality The difference in heights between the two successive high waters (high tides) or of the two successive low waters (low tides) of a tidal day.

diurnal tide An *astronomical tide* with only one high water (high tide) and one low water (low tide) occurring each lunar (tidal) day.

divergent plate boundary The zone between tectonic plates that are pulling apart with magma and new crust moving in to fill the gap; most often occurring at a mid oceanic ridge.

downwelling Downward motion of surface water, especially along the western side of an ocean basin, caused by *Ekman transport* onshore.

E

Earth-atmosphere system The interaction of processes operating at the Earth's surface with those of the overlying atmosphere.

ebb tides Tidal currents flowing seaward with falling sea levels.

ecological efficiency The fraction of the total energy input of a system that is transformed into work or some other usable form of energy.

ecologically sustainable yield The maximum catch that a marine ecosystem can sustain without undergoing an undesirable change in state.

ecosystem Communities of plants and animals that interact with one another, together with the physical conditions and chemical constituents in a specific geographical area.

Ekman spiral In response to a steady wind blowing over the ocean surface, water at increasing depths moves in directions more and more to the right (in the Northern Hemisphere) until at about 100 m depth the water is moving in a direction opposite to that of the wind; the spiral is in the opposite direction in the Southern Hemisphere.

Ekman transport The net transport of water due to the *Ekman spiral*; 90 degrees to the right of the surface wind in the Northern Hemisphere and 90 degrees to the left of the surface wind in the Southern Hemisphere.

El Niño Anomalous warming of ocean surface waters in the eastern tropical Pacific; accompanied by suppression of upwelling off the coasts of Ecuador and northern Peru and along the equator east of the international dateline. Typically lasting for 12 to 18 months and occurring every 3 to 7 years, El Niño is accompanied by weather extremes in various parts of the world.

elasmobranchs Cartilaginous fishes including sharks, skates and rays. More primitive than bony fishes, their skeletons consist entirely of cartilage.

electromagnetic radiation Energy in the form of waves that have both electrical and magnetic properties; these waves can travel through gases, liquids, and solids and require no physical medium. All objects emit all forms of electromagnetic radiation, although each object emits its peak radiation at a certain wavelength within the electromagnetic spectrum. Forms of electromagnetic radiation include gamma rays, x-rays, ultraviolet, visible light, infrared, microwaves, and radio waves.

electromagnetic spectrum The various forms of *electromagnetic radiation* arranged and distinguished by type by their wavelength (or frequency).

ENSO A contraction for El Niño/southern oscillation. The term for the coupled ocean-atmosphere interactions in the tropical Pacific characterized by episodes of anomalously high sea surface temperatures in the equatorial and tropical eastern Pacific. It is associated with large-scale swings in surface air pressure between the western and eastern tropical Pacific and is the most prominent source of inter-annual variability in weather and climate around the world.

epifauna Creatures that live on the surface rather than within marine sediments on the ocean floor.

erosion The removal and transport of sediments by running water, glaciers, or wind.

estuary The portion of a river affected by ocean tides; the semi-enclosed region in the vicinity of a river mouth, in which the freshwater of the river mixes with the saltwater of the ocean.

euphausiid Larger members of the zooplankton community, including krill.

eustasy A condition of world-wide sea level and its fluctuations which may be caused by change in the global water cycle (e.g., the waxing and waning of Earth's glacial ice sheets).

evaporation The process whereby water changes phase from a liquid to a vapor.

evaporative cooling The cooling a surface, such as Earth's surface, experiences as water evaporates, absorbing heat and transferring that heat to the atmosphere via water vapor. Evaporation of water requires the latent heat of vaporization.

exclusive economic zone (EEZ) The jurisdictional zone initially established in 1983 for marine resources; the inner boundary of that zone is coterminous with the seaward boundary of coastal nations and extends seaward 370 km (200 nautical miles).

exotic species Animals and plants introduced into an ecosystem usually by humans intentionally or unintentionally.

extratropical cyclone Any synoptic-scale storm system that is not a *tropical cyclone*, usually referring only to the migratory low pressure systems of middle and high latitudes.

eye The roughly circular area of comparatively light winds and fair weather found at the center of an intense tropical cyclone (i.e., a hurricane).

eyewall The organized ring of intense thunderstorms surrounding the eye of a *tropical cyclone*, typically a hurricane.

F

fecal pellets An organic excrement found especially in marine sediment deposits and made up of the undigested organic matter secreted by animals (mainly invertebrates).

fetch The distance the wind blows over a continuous water surface.

filter feeders Zooplankton that use tiny hairs (called *cilia*) or mucous-covered surfaces to capture food particles suspended in the water.

fjord A narrow inlet or arm of the sea bordered by steep cliffs; formed when the post-glacial rise in sea level flooded a glacially eroded river valley.

float An instrument package that obtains vertical profiles of water temperature and salinity and is used to study deep-ocean currents.

flood tides Tidal currents directed toward land, causing water levels to rise in harbors and rivers.

food chain A sequence of feeding relationships among organisms whereby energy and biomass is transferred from one trophic (feeding) level to the next higher trophic level (i.e., from autotrophs to heterotrophs).

food web A complex of feeding relationships consisting of interacting food chains in an ecosystem.

G

geologic time scale A standard division of time on Earth into eons, eras, periods, and epochs based on occurrence of large-scale geological events; spans millions to billions of years in the past.

geostationary satellite A satellite which revolves around Earth at the same rate and in the same direction as the planet rotates so that it always remains over the same point on the equator and monitors the same field of view. The satellite orbits Earth at an altitude of about 36,000 km (22,300 mi).

geostrophic flow Horizontal movement of surface water parallel to ocean height contours and arising from a balance between the pressure gradient force and the Coriolis force.

glacial climate A climate which favors the thickening and expansion of glaciers.

glacier A mass of ice that flows internally under the influence of gravity.

global climate model A simulation of Earth's climate system usually consisting of a series of mathematical equations and used to predict climatic anomalies.

global oceanic conveyor belt Large-scale vertical and horizontal movement of ocean water that transports heat and salt within the various ocean basins; an important contributor to poleward heat transport.

global radiative equilibrium The balance between net incoming solar radiation and infrared radiation emitted to space by the Earth-atmosphere system.

global water cycle The ceaseless movement of water among its various reservoirs on a planetary scale.

greenhouse effect Heating of Earth's surface and lower atmosphere as a consequence of differences in atmospheric transparency to electromagnetic radiation. The atmosphere is nearly transparent to incoming solar radiation, but much less transparent to outgoing infrared radiation. Terrestrial infrared radiation is absorbed and radiated principally by water vapor and, to a lesser extent, by carbon dioxide and other trace gases, thereby slowing the loss of heat to space by the Earth-atmosphere system and significantly elevating the average temperature of Earth's surface.

greenhouse gases Those atmospheric gases that absorb appreciable terrestrial infrared radiation and contribute to the greenhouse effect in the Earth-atmosphere system. The principal greenhouse gas is water vapor; others are carbon dioxide, ozone, methane, and nitrous oxide.

groin A low artificial structure often composed of rock rubble that is built perpendicular to the shoreline to trap littoral drift; similar to a jetty but smaller.

gross primary production The total amount of carbon fixed into organic matter through *photosynthesis* in a given unit of time; usually expressed in units of grams of carbon per square m per day or year.

guyot A flat-topped seamount of volcanic origin rising more than 1 km above the sea floor.

gyre Refers to the nearly circular motion of surface ocean currents in each of the major ocean basins centered under subtropical high-pressure systems. Viewed from above the subtropical gyres rotate clockwise in the Northern Hemisphere and counterclockwise in the Southern Hemisphere.

H

HNLC regions Portions of the ocean featuring *H*igh *N*utrients and *L*ow *C*hlorophyll. Biological production is low even though surface waters have relatively high concentrations of nutrients (nitrogen and phosphorus compounds).

halocline A layer of water characterized by a relatively large change in salinity with increasing depth.

heat A form of energy transferred between systems in response to a difference in temperature. Heat is always transferred from a warmer system to a colder system.

high A dome of air that exerts relatively high surface air pressure compared to the surrounding air and is usually associated with fair weather; same as an anticyclone.

Holocene The time interval since the end of the Pleistocene glaciation (about 10,500 years ago) that represents the current interglacial.

horse latitudes A nautical term describing the latitude belts over the ocean at approximately 30 to 35 degrees N and S where winds are usually light or the air is calm and the weather is hot and dry. These zones coincide with the location of the subtropical anticyclones.

hot spot A long-lived source of magma caused by rising plumes of hot material originating deep in the mantle. As a tectonic plate moves over a hot spot, magma may break through the crust and form a volcano.

hurricane An intense tropical cyclone originating over tropical ocean waters, usually in late summer or early fall, and with a sustained wind speed of 119 km (74 mi) per hour or higher.

hydrogen bonding An attractive force whereby a positively charged (hydrogen) pole of a water molecule attracts the negatively charged (oxygen) pole of another water molecule.

hydrogenous sediment Particles which are chemically precipitated from seawater.

hydrosphere One of the major interacting subsystems of the Earth system which includes water in all three phases (ice, liquid, and vapor) that continually cycles from one reservoir to another within that system (i.e., the global water cycle).

hydrothermal vent A site where ocean water enters fractures in newly formed oceanic crust, is heated, and then surfaces again through fractures; usually located near oceanic ridge systems.

I

infauna Creatures that burrow into and live within marine sediments on the ocean floor.

Integrated Ocean Drilling Project (IODP) A program of ocean floor exploration underway in October 2003 with the goal of using two drill ships plus specialized drilling platforms to drill more and deeper holes in the ocean floor.

interglacial climate A climate which favors the thinning and retreat of exiting glaciers or no glaciers at all.

internal tides Waves generated by tides well below the sea surface and occurring at tidal frequencies.

internal waves Waves that form and propagate well below the sea surface along the boundary between layers of water that differ in density.

intertidal zone The shore area between high and low tides.

intertropical convergence zone (ITCZ) A narrow, discontinuous belt of convective clouds and thunderstorms paralleling the equator and marking the convergence of the trade winds of the two hemispheres; this zone shifts north and south seasonally.

island arc A curved chain of volcanic islands that parallels a deep-sea trench where oceanic lithosphere is subducted causing volcanism and producing volcanic islands.

J

jetty A breakwater oriented perpendicular to the shoreline and extending seaward up to a kilometer or more; intended to protect a harbor or tidal inlet from filling by littoral drift.

K

kelp Includes various species of brown algae which grow to enormous size, found in cool waters worldwide, especially in coastal upwelling zones.

krill Shrimp-like crustaceans that are the major food source for whales and other organisms in the Southern Ocean.

L

La Niña An episode of strong trade winds and unusually low sea surface temperatures in the central and eastern tropical Pacific. Essentially the opposite of *El Niño*, La Niña is accompanied by weather extremes in various parts of the world.

lagoon A partially enclosed shallow stretch of seawater separating a barrier island from the mainland.

latent heat The heat energy that is used to change the phase of water but not the temperature of the water. Hence the term *latent*, meaning hidden.

latent heat of condensation Heat released to the environment during the change in phase of water from vapor to liquid.

latent heat of deposition Heat released to the environment during the change in phase of water from vapor to solid (ice).

latent heat of fusion Heat released to the environment when water changes phase from liquid to solid.

latent heat of sublimation Heat absorbed from the environment when ice or snow vaporizes.

latent heat of vaporization Heat absorbed from the environment when water changes phase from liquid to vapor.

latent heating The transfer of heat energy from one place to another as a consequence of the phase changes of water.

law of the minimum The growth and well being of an organism is limited by the essential resource that is in lowest supply relative to what is required by the organism.

limiting factor The most deficient of the essential resources an organism requires for growth and well being.

lithification The process whereby sediment is converted to sedimentary rock involving compaction and/or cementation; part of the rock cycle.

lithogenous sediment Marine sediment formed mostly by the weathering and erosion of pre-existing rock.

lithosphere The outer, rigid part of the Earth, consisting of the upper mantle, oceanic crust and continental crust.

Little Ice Age A relatively cool interval during the Holocene interglacial, from about 1400 to 1850 AD, when average temperatures were lower in many areas, and alpine glaciers advanced down mountain valleys. The Little Ice Age followed the *Medieval Warm Period*.

littoral drift The sediment transported by a longshore current along a coast, either nourishing or cutting back beaches.

longshore current The component of water motion parallel to the shore and in the surf zone; the consequence of waves breaking at an angle to the shore.

low An area of relatively low atmospheric pressure often associated with clouds and precipitation; same as a cyclone.

M

magma Hot molten rock material formed deep in the crust or upper mantle which wells up and migrates along rock fissures; called lava when it flows onto Earth's surface.

manganese nodules Irregularly shaped, sooty black or brown nodules on the sea floor which contain a high concentration of manganese and iron.

mangrove Any of various coastal or aquatic salt-tolerant trees that form large colonies in swamps or shallow water.

mangrove swamp A marshy coastal wetland area with a dense growth of mangroves and other tropical plant species which tolerate salt water flooding.

mariculture Industrial farming of fish and shellfish in the ocean.

Maunder minimum The period of greatly diminished sunspot activity between 1645 and 1715 AD identified by and named for E. Walter Maunder.

maximum sustainable yield The maximum catch of a fish species that will ensure the long-term viability of its population.

Medieval Warm Period A relatively mild episode of the *Holocene* from about 950 to 1250 AD.

mesoscale systems Weather phenomena that are so small that they influence atmospheric conditions over only a portion of a city or county; includes thunderstorms and sea breezes. These systems have dimensions of 1 to 100 km (1 to 60 mi) and last from hours to a day or so.

microbial loop A micro-food chain that works within (or along side) the classical food chain. The smallest organisms, heterotrophic bacteria, use dissolved inorganic material directly as carbon and energy sources.

microscale system Weather phenomena representing the smallest spatial subdivision of atmospheric circulation, such as a weak tornado. These systems have dimensions of 1 m to 1 km (3 ft to less than a mile) and last from seconds to an hour or so.

Milankovitch cycles Systematic variations in the precession and tilt of the Earth's rotational axis and the eccentricity of its orbit about the sun; affects the seasonal and latitudinal distribution of incoming solar radiation and influences climate fluctuations operating over tens of thousands to hundreds of thousands of years.

mixed layer The surface layer of the ocean that is mixed by the action of waves and tides so that the waters are nearly isothermal and isohaline; underlain by the pycnocline at low and middle latitudes.

mixed tide An *astronomical tide* with two high waters (high tides) and two low waters (low tides) occurring during a tidal day and having a marked diurnal inequality.

mixotrophs Marine organisms having characteristics of both plants (autotrophs) and animals (heterotrophs).

mud flat A nearly level area of fine silt near the shore and an intertidal habitat for submerged aquatic vegetation and benthic animals.

N

NAO Index A measure of the strength of the horizontal air pressure gradient between Iceland and the Azores in the North Atlantic; influences the climate of eastern North America and much of Europe and North Africa; varies from year to year and decade to decade.

neap tides *Astronomical tides* that have the least monthly *tidal range* occurring at the first and third quarter phases of the moon.

nekton Pelagic animals that are free-swimmers such as fish, adult squid, turtles, and marine mammals.

neritic deposits River-borne lithogenous sediment which settles along the continental margin.

neritic zone Also called the *coastal zone*, the area seaward from the shore to the *continental shelf* break at a depth of about 200 m (650 ft); includes the *intertidal zone*.

net primary production The amount of organic matter produced by living organisms within a given volume or area in a given time, minus that which is consumed through *cellular respiration* by the organisms.

net production The amount of organic matter produced during photosynthesis that exceeds the amount consumed in the process of *cellular respiration*.

neutrally stable system A system that, following a disturbance, does not return to its initial state and is easily mixed.

Newton's first law of motion An object in constant straight-line motion or a rest remains that way unless acted upon by an unbalanced force.

non-point source A broad area of the landscape that yields contaminants to the air or waterways.

nor'easter Common contraction for northeaster, an intense *extratropical cyclone* that tracks along the East Coast of North America and is named for the direction from which its most destructive winds blow.

North Atlantic Oscillation (NAO) A seesaw variation in air pressure between Iceland (the Icelandic low) and the Azores (the Bermuda-Azores subtropical high); influences the climate of eastern North America and much of Europe and North Africa over periods up to decades.

O

ocean basin A large topographic depression on the ocean floor.

ocean commons The vast portion of the ocean that is beyond the control of any coastal nation and essentially free to all for unlimited exploitation.

Ocean Drilling Program (ODP) A project that obtained samples of ocean crust and sediment using the drill ship JOIDES *Resolution*; operated from 1983 to 2003.

ocean floor observatory A long-term facility on the sea floor designed to collect data, perform experiments, and communicate information to scientists onshore.

ocean trench A long narrow deep depression in the ocean floor caused by subduction of tectonic plates.

oceanic crust The outermost part of the lithosphere overlying the mantle and lying beneath the ocean; mostly composed of the fine-grained ferromagnesian igneous rock known as basalt.

overfishing Occurs when a fish species is taken at a rate that exceeds the maximum catch that would allow reproduction to replace the population.

oxygen isotope analysis A technique used to identify climatic fluctuations of the past by examining the ratio of light and heavy isotopes of oxygen (O^{16} and O^{18}) found, for example, in shells extracted from deep-sea sediment cores.

P

Pacific Decadal Oscillation (PDO) A long-lived variation in climate over the North Pacific and North America in which sea surface temperatures fluctuate between the north central Pacific and the west coast of North America; linked to changes in strength of the Aleutian low.

partial tides Harmonic components that comprise the *astronomical tide* at any point. The periods of partial tides are derived from the tidal forces of the moon and sun.

partially mixed estuary A partially isolated body of water where freshwater from rivers and streams mix with seawater to the extent that stratification is weak and the *salinity* typically varies by less than 10 parts per thousand from bottom to top. Mixing is greater than in a *salt-wedge estuary* but not as great as in a *well-mixed estuary*.

particulate organic carbon (POC) The carbon contained in organic particles that sink out of the ocean's surface layer and is then consumed by zooplankton or decomposed by bacteria and converted back to dissolved inorganic carbon.

pelagic deposit Fine-grained sediments that accumulate over time on the deep-ocean floor.

pelagic zone The open-ocean environment, divided into the *neritic zone* (seaward to a depth of 200 m) and the oceanic zone (depth greater than 200 m).

photic zone The upper sunlit layer of the ocean where photosynthesis takes place.

photosynthesis The process whereby autotrophs use light energy from the sun to combine carbon dioxide from the atmosphere with water to produce sugar, a form of carbohydrate that contains a relatively large amount of energy.

physical pump The physical process whereby carbon dioxide sinks deeply in the cold ocean water at high latitudes and is sequestered in the deep ocean in the conveyor belt circulation.

phytoplankton Microscopic unicellular algae and photosynthetic bacteria found in the ocean and responsible for considerable biological production.

pinnipeds Marine mammals which have distinctive swimming flippers; species include seals, walruses and sea lions.

planetary albedo The fraction (or percent) of incident solar radiation that is scattered and reflected back to space by the Earth-atmosphere system; satellite sensors indicate a planetary albedo of about 31%.

planetary-scale systems Weather phenomena operating at the largest spatial scale of atmospheric circulation; includes the global wind belts (i.e., trade winds, westerlies, polar easterlies) and semipermanent pressure systems (e.g., subtropical anticyclones).

plankton Plant and animal life found floating or drifting in the ocean and used as food by nearly all marine animals.

plate tectonics The process by which the massive plates of the lithosphere are slowly driven across the face of the globe by huge convection currents in the Earth's mantle.

point source A discernible conduit such as a pipe or chimney that transports contaminants to a waterway or the atmosphere.

polar amplification An increase in the magnitude of a climatic change with increasing latitude.

polar-orbiting satellite Satellites in relatively low-altitude orbits that pass near the north and south geographical poles. Earth rotates through the plane of the satellite's orbit, which is at an altitude of about 800 to 1000 km (500 to 620 mi).

poleward heat transport A meridional flow of latent and sensible heat from tropical to middle and high latitudes in response to latitudinal imbalances in radiational heating and cooling; brought about by air mass exchange, storms, and ocean circulation.

pollution Intentional or unintentional disturbance of the environment that adversely affects the wellbeing of organisms (including humans) directly or the natural processes upon which they depend.

precipitation Water in frozen or unfrozen forms (rain, snow, drizzle, ice pellets, hail) that falls from clouds and reaches the Earth's surface.

pressure gradient force A three-dimensional force operating in the atmosphere that accelerates air away from regions of high pressure and toward regions of low pressure in response to an air pressure gradient.

primary production The amount of organic matter synthesized by simple organisms from inorganic substances.

principle of constant proportions The major constituents of seawater occur in the same relative concentrations throughout the ocean system.

producers Simple marine organisms also called autotrophs that manufacture the food they need from the physical environment.

pycnocline Layer of ocean water in which density increases rapidly with depth (due to vertical changes in temperature and/or salinity); in low and middle latitudes situated between the mixed layer and deep layer.

R

recreational fisheries Fish that are caught for sport rather than profit.

red tide A discoloration of surface ocean waters usually in the coastal zone caused by a high concentration of microscopic organisms (e.g., dinoflagellates).

reflection The process whereby a portion of the radiation striking the interface between two different media (e.g., atmosphere and ocean) is redirected such that the angle of reflection equals the angle of incidence.

refraction The bending of a wave in response to changing wave speed.

remote sensing Acquisition of data on the properties of some object without the sensor being in direct contact with the object.

remotely operated vehicle (ROV) A data-gathering submersible tethered to a ship by cables that transmit power and data.

resonance A buildup of amplitude in a physical system when the frequency of an applied force is close to the natural frequency of the system.

rings Large turbulent rotating warm-core and cold-core eddies that break off from the relatively swift western boundary currents (e.g., the Gulf Stream).

rock cycle A sequence of events involving the formation, alteration, destruction, and reformation of igneous, sedimentary, and metamorphic rocks as a result of such processes as erosion, transportation, deposition, lithification, metamorphism, and melting.

S

Saffir-Simpson Hurricane Intensity Scale Hurricane intensity scale based on central pressure, and specifying a range of wind speed, height of storm surge, and damage potential; 1 is minimal, 5 is most intense.

salinity A measure of the quantity of dissolved salt in seawater.

salt marsh Coastal wetlands consisting of salt-tolerant grasses regularly covered with seawater.

salt-wedge estuary An estuary where river inflow is swift and tidal currents are weak; the denser high-salinity seawater forms a distinct layer beneath the low-salinity river water.

sand spit A finger-like ridge of sand or gravel that projects from the shore into the ocean.

scattering The process by which small particles suspended in a medium such as air or water diffuse a portion of the incident radiation in all directions.

scientific method A systematic form of inquiry that involves observation, gathering data, critical thinking, and formulating and testing hypotheses.

schooling The characteristic behavior of many fish species to swim together in organized groups to reduce predation.

sea The state of the surface of the ocean with regard to waves or swells.

sea-floor spreading The divergence of adjacent plates on the ocean bottom; occurs along the oceanic ridge system.

sea wave An oscillation on the ocean surface that propagates along the interface between the atmosphere and the ocean.

seamounts A structure of volcanic origin rising more than 1 km above the sea floor.

seawall A concrete embankment intended to protect beaches, roads, buildings, and shoreline cliffs from erosion by storm waves.

seawater The water of the ocean, distinguished from freshwater by its higher salinity.

secondary production The organic material produced in the growth of consumers (heterotrophs).

sediment Particles of organic or inorganic origin that are transported from their place of origin and deposited by wind, water, or ice; typically in unconsolidated form.

seiche A rhythmic oscillation of water in an enclosed basin or partially enclosed coastal inlet; a type of standing wave.

semi-diurnal tide An *astronomical tide* with two high waters (high tides) and two low waters (low tides) occurring during a tidal day and having a small diurnal inequality.

sensible heating Transport of heat from one location or object to another through conduction, convection, or both, which brings about temperature changes.

shallow-water waves Waves in water shallower than the *wave-base* (half the wave length).

shore Land exposed at low tide up to the coastline.

shoreline The boundary line between a water body and land, usually taken at mean high tide.

siliceous ooze Pelagic deposit made from the shells of silica-secreting organisms.

slack water Periods of little or no horizontal water movement occurring between *ebb* and *flood tides*.

SLOSH (Sea, Lake, and Overland Surges from Hurricanes) A numerical model which accurately predicts the location and height of a *storm surge*.

SOFAR channel S*O*und *F*ixing *A*nd *R*anging. A zone at an ocean depth of about 100 m (3300 ft) where the speed of sound is at a minimum value.

solar altitude The angle of the sun above the horizon; varies from 0 degree (horizon) to 90 degrees (zenith).

sorting The dynamic process by which sedimentary particles are separated by size; well-sorted sediment has a narrow range of particle size whereas poorly-sorted sediment has a broad range of particle size.

southern oscillation Opposing swings of surface air pressure between the western and central tropical Pacific Ocean; associated with intense *El Niño* events.

specific heat The amount of heat required to raise the temperature of 1 gram of a substance by 1 Celsius degree.

spring bloom A dramatic increase in phytoplankton populations in the ocean which occurs as a consequence of more sunlight and abundant nutrients.

spring tide An *astronomical tide* occurring twice each month at or near the times of new moon and full moon when the gravitational pull of the sun reinforces that of the moon, and having an unusually large or increased *tidal range*.

stable system A system which tends to persist in its current state without changing and, following a disturbance, tends to return to its original state or condition.

stewardship Action taken by society to protect the ocean and its resources for now and the future.

storm surge An abnormal local rise in sea level accompanying a *tropical cyclone* or other intense storm system and whose height is the difference between the observed level of the sea surface and the level that would have occurred in the absence of the storm.

subduction The process whereby the leading edge of one tectonic plate on the ocean bottom descends beneath the margin of another plate and then into the mantle.

subduction zone A long, narrow zone at a convergent plate boundary where an oceanic plate descends beneath another plate, either oceanic or continental.

sublimation The process by which water changes directly from a solid to a vapor without first becoming liquid.

submarine canyon A steep-sided canyon below sea level in the continental shelf and slope.

submarine fan A cone-shaped sedimentary deposit that accumulates on the continental slope and rise.

subpolar low A region of low pressure found in both the Northern and Southern Hemispheres where the middle latitude westerlies converge with the polar easterlies.

subtropical high One of the massive semipermanent anticyclonic systems centered over the ocean basins near 30 degrees N and S.

sulfurous aerosols The tiny droplets of sulfuric acid (H_2SO_4) and sulfate particles which form in the stratosphere when sulfur dioxide combines with moisture.

sunspot A dark blotch on the face of the sun, typically thousands of kilometers across that develops where an intense magnetic field suppresses the flow of gases transporting heat from the sun's interior.

surf A nearly continuous train of waves breaking along a shore.

surface tension The attraction between molecules at or near the surface of a liquid.

sustainability Actions that meet present needs without compromising the ability of future generations to meet their own needs.

swash Intermittent landward flow of water across a beach following the breaking of a wave.

swell Large long-period ocean waves that radiate away from the region where they were generated by strong storm winds.

symbiotic relationship A mutually beneficial association between organisms.

synoptic-scale systems Weather phenomena operating at the continental or oceanic spatial scale, including migrating *tropical* and *extratropical cyclones*. These systems have dimensions of 100 to 10,000 km (60 to 6000 mi) and last from days to a week or so.

system An interacting set of components that behave in an orderly way according to the laws of physics, chemistry, geology, and biology.

T

tektite Black fragments of glass formed from rock which has been liquefied when a meteor strikes the Earth.

teleconnection A linkage between weather changes occurring in widely separated regions of the globe.

teleost Bony fishes which have skeletons and scales.

temperature A measure of the average kinetic energy of the individual atoms or molecules composing a substance.

tephra Sediments of volcanic origin which accumulate in the ocean.

terminal velocity Constant downward-directed motion of a particle within a fluid due to a balance between gravity (directed downward) and fluid resistance (directed upward).

thermal inertia Resistance to a change in temperature.

thermocline A layer of water in which the temperature decreases rapidly with increasing depth (e.g., between the warmer mixed layer and the colder, deep layer in a thermally stratified ocean).

thermohaline circulation Subsurface movement of water masses caused by density contrasts arising from differences in temperature and salinity.

tidal currents Alternating horizontal movements of water accompanying the rise and fall of *astronomical tides*.

tidal day The moon-based day based on the interval of time between two successive passes of the moon over a meridian (approximately 24 hours, 50 minutes).

tidal period The elapsed time between successive high tides or successive low tides.

tidal range The difference in height between consecutive high water (high tide) and low water (low tide).

tide pool A volume of water left behind in a rock basin or other intertidal depression by an ebbing tide.

top predators Organisms that occupy a high trophic level in a marine food web.

trade winds The wind system, occupying most of the tropics, which blows outward from the *subtropical highs* toward the equatorial trough or *intertropical convergence zone*; a major component of the planetary-scale circulation of the atmosphere. Trade winds blow from the northeast in the Northern Hemisphere and from the southeast in the Southern Hemisphere.

transform plate boundary The region where adjacent *tectonic plates* slide laterally past one another.

transitional wave A wave entering water having a depth of between one-twentieth and one-half of the wavelength.

transpiration The process whereby water that is taken up from the soil by plant roots eventually escapes as vapor through the tiny pores (stomates) on the surface of green leaves.

trophic level The feeding position of an organism within a *food chain* or *food web*.

tropical cyclone Generic term for a non-frontal synoptic-scale *cyclone* originating over warm tropical or subtropical ocean waters with cyclonic surface wind circulation (e.g., hurricane, tropical storm).

tropical depression A *tropical cyclone* in which the sustained surface wind is at least 37 km (23 mi) per hour but less than 63 km (39 mi) per hour; an early stage in the development of a *hurricane*.

tropical disturbance A discrete system of organized convection in the tropics or subtropics with a detectable center of low air pressure; the initial stage in the development of a *hurricane*.

tropical storm A *tropical cyclone* having a sustained surface wind speed of 63 to 118 km (39-73 mi) per hour.

troposphere The lowest thermal layer of the atmosphere; where the atmosphere interfaces with the ocean, cryosphere, lithosphere, and biosphere and where most weather takes place.

tsunami A rapidly propagating shallow-water ocean wave that develops when a submarine earthquake, landslide or volcanic eruption disturbs deep ocean water; known to build to tremendous wave height in coastal areas.

turbidites Sedimentary deposits produced by turbidity currents.

turbidity current A sediment-water mixture denser than normal seawater that flows downslope to the deep-sea floor.

turbidity maximum The area in an estuary where sediments are most concentrated.

twilight zone The intermediate zone of the ocean, below the photic zone, but above the greater depths where there is less biological activity.

U

unstable system A system which following a disturbance tends not to return to its original state or condition.

upwelling Upward circulation of cold, nutrient-rich bottom water toward the ocean surface; may occur along the coast or equator.

V

vertical migration The daily migration of zooplankton from deep waters to the surface zone to feed on phytoplankton.

W

warm-core rings Eddies having a core of relatively warm water that break off from an ocean surface current; viewed from above rotate in a clockwise direction in the Northern Hemisphere.

water mass A large, homogenous volume of ocean water featuring a characteristic range of temperature and salinity.

wave A regular oscillation that occurs in a solid, liquid, or gaseous medium as energy is transmitted through that medium.

wave-base A depth of about one-half wavelength, where the diameter of the orbits of water particles in waves is essentially zero; the depth below which water is not affected by surface waves.

wave crest The highest point reached by an oscillating water surface.

wave frequency The number of waves passing a fixed point over an interval of time.

wave height A measurement of the vertical distance between wave crest and wave trough.

wave period A measurement of the time needed for two successive wavelengths to pass a fixed point.

wave trough The lowest point in an oscillating water surface.

wavelength The distance between successive wave crests (or equivalently, wave troughs).

weather The state of the atmosphere at some place and time, described in terms of such variables as temperature, precipitation, cloud cover, and wind speed.

weathering The physical disintegration, chemical decomposition, or solution of exposed rock which takes place where the lithosphere (mainly the crust) interfaces with the other Earth subsystems.

well-mixed estuary An *estuary* where strong tidal currents dominate the inflow from rivers and thoroughly mix the fresh water and saltwater.

western boundary currents Generic term for a relatively strong and narrow flow of ocean water (current) that runs along the western edge of a major ocean basin; the Gulf Stream is an example.

wetland Low-lying flat areas that are covered by water or have soils that are saturated with water for at least part of the year.

wind-waves Sea waves that are the produced as the kinetic energy of the wind is transferred to surface waters.

Wilson cycles Cycles of ocean basin spreading and closing operating over hundreds of millions of years.

Z

zooplankton Single-celled and multi-cellular animal plankton that drift passively with ocean currents.

INDEX

A

absorption, 103
abyss, 231
abyssal plain, 35
abyssopelagic zone, 231
acid rain, 14, 65
acid snow, 65
Acoustic Doppler Current Profiler, 269
acoustic fish finders, 68
acoustic tomography, 18, 70, 138
active continental margin, 175
adaptation, 232, 234-236
adaptive coloration, 234
Advanced Very High-Resolution Radiometer (AVHRR), 217
aerosols, 7
Afghanistan drought, 268
Africa,
    Sub-Saharan drought, 275-276
Age of Discovery, 311
Agenda 21, 343
Agulhas Current, 133, 135, 142, 151, 152
Agung, Bali, 292, 293
ahermatypic corals, 242
air density, 112-113
air masses, 111-112
    exchange of, 111-112
    source region, 111
air pressure, 112-113
air pressure gradient, 113
air pressure gradient force, 112-113
Alaska, 345-346
Alaska gyre, 132
albedo, 102, 103
    of the ocean, 103-104
    planetary, 104
Aleutian low, 116, 132, 272

Alexander the Great, 310
algae, 19, 237
algal bloom, 19-20
alien species, 337-338
Alvin, 43, 314-315
Amazon River, 162, 184
American eel, 247
amnesic shellfish poisoning, 203
anadromous fish, 11, 247, 348
Andes Mountains, 41
anemometer, 168
angiosperms, 238
animal-borne instruments, 320
Antarctic Bottom Water, 140
Antarctic Circumpolar Current, 131, 132
Antarctic Convergence, 342
Antarctic Deep Water, 140
Antarctic Intermediate Water, 140, 141
Antarctic Treaty System, 342
Antarctic ozone hole, 122-123
Antarctica, 5, 6, 74, 298, 304, 342
anticyclone (high), 113
anti-esuarine circulation, 185
antinode, 156
APEX, 148-149
aphelion, 291
apogee, 160
Aquarius, 316
aquifer, 308
archaea, 213-214, 223, 322
Arctic Bottom Water, 140
Arctic Oscillation (AO), 271-272, 300-301
Arctic sea ice cover, 280, 300-301
    albedo of, 300
    shrinkage of, 300-301
ARGO, 138, 149, 317

Array for Real-time Geostrophic Oceanography (ARGO), 138, 149, 317
artificial beach nourishment, 182-183
artisanal fisheries, 326
Asian carp, 337-338
assimilative capacity, 14, 344-345
asteroid, 23-24
asthenosphere, 32
astronomical tides, 157-164
Aswan High Dam, 87
atmosphere, 5-7
    circulation of, 114-118, 259
    composition of, 6-7
    evolution of, 25-27
    scales of motion, 114
    temperature profile, 6
atmospheric aerosols, 7
atmospheric fixation, 26
atmospheric window, 105-106
atoll, 43, 242, 243
Autonomous Benthic Explorer (ABE), 309
Autonomous Profiling Explorer (APEX), 148-149
Autonomous Underwater Vehicles (AUV), 318
autotrophs, 10, 122, 205-207
Azam, Farooq, 214
azoic hypothesis, 311

B

backwash, 177
Bacon, Francis, 36
bacteria, 10, 206
Baja California, 335
bald eagles, 345-346
baleen whale, 235-236
Bangladesh, 186
barnacles, 237
barrier islands, 179-180
barrier reef, 240-241, 242
Barringer Crater, AZ, 23
Barry, Roger G., 298
Barton, Otis, 314
bathypelagic zone, 231
bathyscaph, 314
bathysphere, 314
bathythermograph (BT), 148
Battisti, David, 260
Bay of Fundy, Nova Scotia, 161-162

beach drift, 177
beach sediment budget, 179
beach sediment, 177, 179
beaches, 177-179
Beaufort, Francis, 168
Beaufort Scale, 168
Beebe, William, 314
benthic organisms, 236-244
    selective feeders, 243
    unselective feeders, 243
benthic ecosystems, 231, 243-244
benthic zone, 231, 236, 243-244
benthos, 231, 236
Bering Sea, 213
berm, 177
Bermuda, 218
Bermuda Triangle, 152
Bermuda-Azores subtropical high, 132, 271, 275
bio-fouling, 322
bioaccumulation, 209-210
biogenous sediment, 83-85
biogeochemical cycles, 11-12, 26, 33-34
biological fixation, 26
biological pump, 215-217
bioluminescence, 234-235
biomass, 10, 209
biosphere, 9-11
bioturbation, 243
Bjerknes, Jacob, 261
Black, Joseph, 58
black smokers, 43
Blackwater National Wildlife Refuge, 198
blue whales, 235
bomb (cyclone), 193
bottom waters, 139
Bowen ratio, 110
bowhead whales, 335
Brazil Current, 131, 133
breaker, 155
breakwater, 182
Briggs, Derek, 96
brown mud, 88
brown pelicans, 336
buffer, 65
buoy, 168, 269, 317
Burgess Shale, 95-96
bycatch, 330

C

calcareous oozes, 88-89
California Current, 131, 134, 188-189, 260
California gray whale, 335
Callisto, 2
calorie, 57
calving, 83
Cambrian explosion, 95
Canada, Atlantic fishery, 229-230
Canary Current, 131, 132, 134
Cape Cod National Seashore, 299
Cape Hatteras Lighthouse, 196-197
Cape Mendacino Fracture Zone, 40
capillary waves, 153
carapace, 333
carbon cycle, 11-12, 26, 215-217
carbon dioxide, 7, 25-26, 64, 105, 106, 259, 288, 294-295, 352-353
carbon fixation, 207
carbonate compensation depth (CCD), 88-89
carbonate rock, 12
Carboniferous Period, 12
carnivores, 10, 205
carrying capacity, 328, 329, 344
cartilaginous fishes, 245
cast, 139
catadromous fishes, 247
celerity, 154
cellular respiration, 10, 64, 209
Celsius, Anders, 57
Celsius temperature scale, 57
Cenozoic Era, 284
Census of Marine Life, 320
Center for Operational Oceanographic Products and Services (CO-OPS), 162
central waters, 139
Challenger Expedition (1872-1876), 312, 313
Challenger Deep, 312, 314
Channel Islands National Marine Sanctuary, 254, 331
chemosynthesis, 10, 44, 206-207, 211, 244
Chesapeake Bay, 184, 198, 238-239, 349-350
    blue crab fisheries, 239
    oyster fisheries, 331-332
    striped bass population, 350
Chesapeake Bay Foundation, 350
Chesapeake Bay Program, 349
Chesapeake Marsh Restoration/Nutria Control Project, 198
Chicxulub crater, 23
Chikyu, 49, 317
Chisso Chemical Plant, 78

chlorofluorocarbons (CFCs), 122-123
chlorophyll, 226
cilia, 207
ciliates, 214
circumpolar vortex, 122-123
climate, 99, 258
    and human activity, 294-295
    and the ocean, 259-261, 297-300
    continental, 59-60, 101
    glacial, 281, 284
    interglacial, 281, 284
    maritime, 59-60, 101
climate change
    and Earth's surface, 294
    and Earth's surface properties, 294
    and marine life, 301
    and plate tectonics, 289
    and solar variability, 290-292
    and volcanoes, 292-293
    causes of, 288-295
    impact on ocean, 297-300
climate controls, 259
climate future, 295-297
Climate Prediction Center (CPC), 268
climate record, 280-288
    lessons of, 288
climate system, 258-261
cloud condensation nuclei, 108
clouds, 14, 109-110
coal, 12
Coale, Kenneth, 221
coast, 174
coastal armor, 181-183
coastal storms, 185-186
coastal zone, 231, 321, 343-345
    evacuation, 191-192
    human population trends, 343-344
    management, 193-194
    natural hazards of, 185-193, 343-344
    pollution in, 344-345
Coastal Zone Color Scanner (CZCS), 226
Coastal Zone Management Program (CZMP), 194
coastal zone management, 193-194
coastline,
    features, 177-183
    formation, 175-177
    human alterations, 181-183
coccolithophores, 83, 84, 206
coccoliths, 83, 84

cod, 229-230, 327
cold-core rings, 134
collection bottles, 148
Colorado Plateau, 284
Columbia River, 348
comets, 25
commercial fish production, 326-332
Commission on Marine Science, Engineering and Resources, 341
Commission on Ocean Policy, 341
Common Water, 140-141, 142
compensation depth, 211
conceptual model, 19
condensation, 13-14, 58
conduction of heat, 108-109
constructive wave interference, 152, 154
consumers, 10, 205, 207-208
continental climate, 59-60, 101
continental crust, 32, 42
continental drift, 37, 289
continental margins, 34-35
continental rise, 34, 35
continental shelf, 34-35
continental slope, 34, 35
convection, 108-109
Convention on International Trade in Endangered Species of Wild Fauna & Flora (CITES), 333
Convention on Straddling Stocks, 343
Convention on the Conservation of Antarctic Marine Living Resources (CAMLR), 342
Convention on the Prevention of Marine Pollution by Dumping of Waste & Other Matter, 343, 351
convergent plate boundary, 40-42
Cook, Captain James, 48, 311
copepods, 207
coral atoll, 43, 242, 243
coral bleaching, 242, 264-265, 301
coral reefs, 83, 240-243
corals, 240-243
    ahermatypic, 242
    hermatypic, 242
Coriolis deflection, 113-114, 185, 187
Coriolis effect, 113-114, 129-130, 133, 200
Coriolis, Gaspard Gustav de, 113
Cornwall, England, 310
cosmogenous sediment, 86
cotidal line, 161
countershading, 234
coupled model, 268

crust, 8, 32, 42
Crutzen, P.J., 122
cryosphere, 5
Cullen, Heidi M., 268
cumulonimbus clouds, 109-110
cumulus clouds, 109
Curie, Pierre, 37
Curie temperature, 37
current rings, 134-135
currents
    equatorial, 132-133
    ring, 134-135
    tidal, 162
    western boundary, 133-134
    wind-driven, 130-137
cutoff anticyclone, 266
cyanobacteria, 206, 207
cycling rate, 11
cyclone (low), 112-113, 185-193

D

dams, 247, 347-349
    and dislocation of human populations, 347-348
    impact on ecosystems, 348
    removal of, 348-349
DART, 165-166
Dansgaard-Oeschger events, 305
Dansgaard, Willi, 305
Darwin, Charles, 43, 242, 311
    On the Origin of Species, 311
    Structure and Distribution of Coral Reefs, 311
    voyage on the *HMS Beagle,* 43, 311
Davis, R.E., 193
daylight, 101, 102
DDT (dichlorodiphenyltrichloroethane), 336
Dead Sea, 53
dead zone, 19-20, 344
decibar, 68
decomposers, 10, 208
deep layer, 127, 128
Deep-Ocean Assessment and Reporting of Tsunamis (DART), 165-166
deep-ocean carbon storage, 352-353
deep scattering layer (DSL), 234
Deep-Sea Drilling Program (DSDP), 48, 316
deep-sea mud,
    brown mud, 88
    red clay, 88

deep sea sediment cores, 48-49, 280-281
deep-water waves, 154-155
deep waters, 139
deeps, 41, 312, 314
delta, 87, 180-181
demersal fishes, 246
density, 66
    freshwater, 66
    saltwater, 66-67
deposition, 14, 58
desalination, 73
desertification, 276
destructive wave interference, 154
dentritivores, 208
detritus, 208
Deuterium (D), 25
diatoms, 85, 206, 232
Dietz, Robert S., 37
dinoflagellates, 203-204, 206, 207
dinosaurs, extinction of, 23-24
dissolved gases, 63-64
dissolved organic carbon (DOC), 214
dissolved oxygen, 64
distillation, 14, 73
distributary, 87
Dittmar, William, 60-61
diurnal inequality, 160
diurnal tide, 159, 160
divergent plate boundary, 38-40
Dolan, R., 193
doldrums, 132
dolphins, 330
dominant wave period, 168
domoic acid, 203-204
double-hulled ships, 347
double sunspot cycle, 290
downwelling, 135-137
drainage basin, 15
Drake Passage, 31
drifting buoys, 317
drones, 318-319, 321
drought, 268, 275-276
Dry Tortugas Islands, 254
dugongs, 248
dust (wind-blown), 81-83
dynamic model of tides, 161
dyamical model, 268
dynamical positioning, 316
Earth-atmosphere system, 98

Earth Science Enterprise, 226
Earth system, 2, 3-12
    origin of, 25-26
earthquakes, 37, 38, 44
East Antarctica ice sheet, 298
East Australia Current, 131, 133
East Greenland Current, 131
East Pacific Rise, 39
ebb tide, 162
echo sounder, 49, 68
ecological efficiency, 209
ecologically sustainable yield, 328-329
ecosystem, 10-11, 204, 205, 301
    based management, 255

E

eddies, 154
Eddy, John A., 290
Edmund Fitzgerald, 152, 201
Edwards Dam, 348
Eemian interglacial, 285
Ekman spiral, 129
Ekman transport, 128-130, 136, 187, 200, 262
Ekman, V. Walfrid, 129
El Chichón volcano, 292, 293
El Niño, 137, 148, 241, 242, 249, 264-266, 277
    frequency of, 270-271
    historical perspective, 261-262, 277
    observing, 148
    predicting, 268-270
El Niño of 1972-73, 262
El Niño of 1982-83, 257-258
El Niño of 1997-98, 266, 267
elasmobranchs, 245
electromagnetic radiation, 15, 16
electromagnetic spectrum, 15, 16
electron, 54
elephant seals, 320
Emperor Seamounts, 42
empirical model, 268
ENSO, 261
ENSO Observing System, 269
endangered species, 333-336
Endangered Species Act, 333
energy, 11
epifauna, 243
epipelagic zone, 231
equatorial countercurrent, 131, 132

Hudson River, 348-349
HUGO, 319
humpback whales, 335
Hurricane Andrew, 188, 192
Hurricane Beulah, 190
Hurricane Camille, 186
Hurricane Elena, 192
Hurricane Floyd, 192
Hurricane Hugo, 183, 192
Hurricane Iniki, 189
Hurricane Isabel, 197
hurricanes, 186-187
    evacuation from, 191-192
    hazards, 189-191
    life cycle, 189
    where and when, 187-189
hydrogen bonding, 55, 153
hydrogenous sediment, 85-86
hydrosphere, 3-5
hydrothermal circulation, 43-44, 244
hydrothermal mineral deposits, 91
hydrothermal vent, 43-44, 244
hypolimnion, 201
Hypsithermal, 286

I

ice cores, 282, 304-305
ice floe, 74
ice keel, 75
ice lattice, 55
ice sheets, 5
    East Antarctica, 298
    Greenland, 298, 304
    Laurentide, 5, 94, 284-285, 292
    West Antarctica, 298
ice shelves, 5
ice worms, 225
icebergs, 5, 6, 74, 83, 94
Iceland, 40, 50
    whaling, 335
Icelandic low, 116, 132, 271
igneous rock, 32, 33, 81
iguanas, 249
Industrial Revolution, 294-295
inertial circle, 200
infauna, 243
infiltration, 15
infrared radiation, 16, 18, 105-106

infrared window, 105, 106
instrument-based climate record, 286-288
Integrated Ocean Drilling Program (IODP), 49, 316-317, 319
interglacial climate, 281, 284
Intergovernmental Panel on Climate Change (IPCC), 296, 299
intermediate waters, 139
internal energy, 56-57
internal geological processes, 8-9
internal tides, 163-164
internal waves, 164
International Date Line, 120
International Decade of Ocean Exploration (IDOE), 315
International Geophysical Year of 1957-58, 315, 342
International Great Lakes Datum (IGLD), 201-202
International Joint Commission (IJC), 338
International Maritime Organization (IMO), 346, 351
International Whaling Commission (IWC), 335, 343
intertidal zone, 175, 231, 236-238
intertropical convergence zone (ITCZ), 115, 132, 260-261, 275-276
Inuit people, 278-279, 335
ion, 60
iron fertilization, 221-222
island arc, 41
Island of Tin, 310
isomotic fish, 245
isotacy, 32

J

Japanese whaling industry, 335
Jason1 (satellite), 170, 270
JASON, 315
JASON II, 315
Jeddah, Saudi Arabia, 73
jellyfish, 232, 235
jet, 200
jet stream, 116
jetty, 182
JOIDES, 48
JOIDES *Resolution*, 48, 49, 316
Joint Global Ocean Flux Study (JGOFS), 217-218
Joint Oceanographic Institutions for Deep Earth Sampling (JOIDES), 48
joule, 57
Juan de Fuca plate, 175
Juan de Fuca Ridge, 175
Julie N., 145
Jupiter, 2, 25

K

katabatic wind, 75
Keeling, Charles D., 294
kelp, 240
kelp forest, 240
Kelvin waves, 200
Kennebec River, 348
keystone species, 208
kinetic energy, 56, 79, 127, 152, 153
King Diamond II, 328
Kircher, Athanasius, 125
Kiritimati (Christmas Island), 258
Krakatoa, eruption of, 29-30, 165, 292
krill, 207-208, 342
Kuroshio Current, 131, 133
Kutzbach, J. E., 284

L

La Niña, 137, 148, 266, 268
    frequency of, 270-271
    historical perspective, 261-262
    observing, 148
    predicting, 268-270
Labrador Current, 131
lag concentrate, 90-91
lagoon, 180
Laguna Pallcacocha, Ecuador, 277
Lake Geneva, Switzerland, 156
Lake Huron, 199
Lake Michigan, 199
Lake Superior, 199
Laki volcano, 292
Lamont-Doherty Earth Observatory, 268
land breeze, 115
laser bathymetry, 49
latent heat, 56, 57-58
latent heat of condensation, 58
latent heat of deposition, 58
latent heat of fusion, 57
latent heat of sublimation, 58
latent heat of vaporization, 58
latent heating, 108, 260
latitude, 120
Laurasia, 37
Laurentide ice sheet, 5, 94, 284-285, 292
lava, 9, 32
law of conservation of matter, 11
law of energy conservation, 11

law of the minimum, 205
Law of the Sea (LOS) Treaty, 91, 343
lead, 74-75
Leeuwenhoek, Antony van, 214
length of daylight, 101, 102
LEO-15, 319-320
light, 15, 16, 104-105
limestone, 12
limiting factor, 205, 301, 345
lithification, 89
lithogenous sediment, 80-83
lithosphere, 8, 32-33
Little Ice Age, 98, 142, 286, 290
littoral drift, 177, 179
littoral zone, 175, 231, 236-238
living fossil, 233
location at sea, 120, 127, 146-147, 316
Loihi seamount, 319
London Dumping Convention, 343, 351
Long-term Ecosystem Observatory, 319-320
longitude, 120
longshore current, 177, 179
low (cyclone), 112-113, 185-193

M

M-layer, 306
MacAyeal, Doug, 94
Mackenzie delta (Canada), 223
macroalgae, 236
Magellan, Ferdinand, 311
magma, 9, 32, 33, 42
magnetic anomalies, 37, 38
magnetic reversals, 37-38
Magnuson-Stevens Fisheries and Conservation Act, 342-343
manatees, 248
manganese nodules, 85-86, 91
mangroves, 181
mangrove swamp, 239-240
mantle, 8
mantle plume, 50-51
Marchetti, E., 352
Marianas Deep, 41
Marianas Trench, 41
mariculture, 336-337
marine ecosystems, 204-218
    microbial, 213-215
    observations and models, 217-218
    processes, 210-215
    structure of, 205-215

marine exotic species, 337-338
marine habitats, 230-234
marine iguanas, 249
marine life,
    adaptations, 232, 234-236
    and climate change, 301
    requirements for, 204-205
Marine Mammal Protection Act, 248
marine mammals, 247-248
Marine Protection, Research and Sanctuaries Act of 1972, 253, 254
marine reptiles, 248-249
marine reserves, 242-243, 254, 331
marine sanctuaries, 242-243, 253-254
marine sediments, 77-92
    and climate record, 280-281
    biogenous, 83-85
    classification of, 80-86
    cosmogenous, 86
    deposits of, 86-89
    glaciomarine, 83
    hydrogenous, 85-86
    lithogenous, 80-83
    phosphatic, 85
marine snow, 83, 216
marine volcanism, 42-43, 50-51
maritime climate, 59-60, 97-98, 101
Mars, 1-2
Mars Global Surveyor, 1
Mars Odyssey, 1
marsh grass (*Phragmites*), 338
Martin, John H., 221
mass continuity, 136
Matthews, D. H., 38
Mauna Loa Observatory, 294-295
Maunder minimum, 290
Maunder, E. Walter, 290
Maury, Matthew Fontaine, 126-127, 312
maximum sustainable yield, 328, 329
McConnell, R.G., 96
McIsaac, G.F., 19-20
McMurdo Station, Antarctica, 123
Medieval Warm Period, 286, 290
Mediterranean Sea, 41, 53, 62, 87, 140, 185, 306-307, 352
Mediterranean Intermediate Water, 140
menhaden, 239
mercury, 77-78
meridional westerlies, 266
mermaid's purses, 245

Merrymeeting Bay, 348
mesopelagic zone, 231
mesoscale systems, 114, 115
mesosphere, 6
Mesozoic Era, 282-284
metamorphic rock, 32, 33
meteorite, 25
methane, 295
methane hydrate, 223
methane ice, 223
methyl mercury, 77-78
microbial loop, 214-215
micrometer, 99
micronutrients, 205
microplankton, 207
microscale systems (atmospheric), 114
Mid-Atlantic Ridge, 39, 40, 44
Mid-Indian Ridge, 39
Migratory Bird Treaty, 336
Milankovitch cycles, 291-292, 304
Milankovitch, Milutin, 291
mineralization, 91
mineral resources, 90-91
minerals, 32
Minamata, Japan, 77-78
Minamata Disease, 77-78
Ming Dynasty (China), 311
Mississippi River, 19-20
mistral winds, 140
mixed layer, 127
mixed tide, 159, 160
mixotrophs, 206
model, 18-19
Molina, M.J., 122
mollusks, 336
monsoon, 141
Monterey Bay, CA, 254
Montreal, Canada, 203
moon, and tides, 158-159
Morgan, W.J., 50
Morris, Simon Conway, 96
Moses Project, 174
Mote Marine Laboratory (Florida), 318
Mount Pinatubo, 30, 292, 293
Mount St. Helens, 30
mud flat, 237
multi-year ice, 68, 74
multibeam sonar, 322
Murray, Sir John, 312

Mycenae, Greece, 310

N

Namibia, 282
Nankai Trough, 49, 224
Nansen, Fridtjof, 129, 164
NAO Index, 271
National Aeronautics and Space Administration (NASA), 1, 269
National Centers for Environmental Prediction (NCEP), 268
National Climatic Data Center (NCDC), 287
National Data Buoy Center (NDBC), 168
National Deep Submergence Laboratory, 315
National Estuaries Research Reserve System (NERRS), 194
National Hurricane Center, 191
National Marine Fisheries Service, 230, 248, 327, 330, 333
National Ocean Survey, 322
National Oceanic and Atmospheric Administration (NOAA), 326, 341
    Atlantic Oceanographic and Meteorological Laboratory, 148, 149
    Coastal Change Analysis Program, 145
    Deep-Ocean Assessment & Reporting of Tsunamis (DART), 165-166
    National Marine Fisheries Service (NMFS), 230, 248, 327, 330, 333
    National Marine Sanctuary Program, 253, 331
    National Ocean Service CO-OPS, 162
    National Ocean Survey, 322
    National Weather Service, 286
    Office of Protected Resources, 248, 334
    Office of Response and Restoration, 345
    Pacific Marine Environmental Laboratory, 148
    Physical Oceanographic Real Time Services (PORTS), 162
National Ozone Expeditions, 122
National Park Service, 196
National Science Foundation (NSF), 49
National Tsunami Hazard Mitigation Program, 165-166
National Weather Service (NWS), 286
natural gas, generation of, 90
nautilus, 233
Navistar satellites, 146-147
neap tides, 160
nebula, 25
nekton, 231, 233-234
neritic deposits, 86-88
neritic zone, 231
net primary production, 211
net production, 211

neutral conditions (tropical Pacific), 262-264
neutrally stable system, 128
New England, 230
new production, 211
Newfoundland,
    cod industry, 230
Newton, Sir Isaac, 158
Newton's first law of motion, 79-80, 113, 158, 200
nilas, 74
Nile River delta, 87
nitrogen cycle, 12, 26
nitrous oxide, 295
no-take zones, 254, 331
node, 156
non-governmental organizations (NGOs), 350
non-point source, 349
nonferromagnesium silicate minerals, 81
nonrenewable resources, 34
nor'easter, 192-193
North Atlantic Current, 131, 142
North Atlantic Deep Water, 140, 141-142, 304
North Atlantic Equatorial Current, 131, 132
North Atlantic Intermediate Water, 140
North Atlantic Oscillation (NAO), 271
North Atlantic subtropical gyre, 131-132
North East Water, 75
North Equatorial Current, 131
North Pacific Intermediate Water, 140, 141
North Sea, 230, 352
North Water, 75
Northridge, CA, 157
Norwegian Current, 131-132
Norwegian Polar Institute, 320
Norwegian whaling industry, 335
numerical model, 19, 320-321
nutria, 198
nutrients, 213

O

ocean, 3-5
    and carbon storage, 352-353
    and climate, 297-301
    and nuclear waste, 351
    and waste disposal, 350-351
    and water cycle, 12-15
    as a heat sink, 112
    as a heat source, 112
    circulation of, 125-142

color, 17, 226-227
forces acting on, 128-130
geography of, 31
modeling, 18-19
profiling, 148-149
remote sensing of, 15-18
ridge system, 39-40
structure of, 127-128
ocean basin, 35-36
and plate tectonics, 36-46
ocean bottom (or floor)
mapping of, 322
probing, 48-49
profile of, 34-36
Ocean City, MD, 182, 193
ocean currents, 125-142
ocean commons, 343
ocean drilling, 316-317
Ocean Drilling Program (ODP), 48-49, 223, 316
ocean floor observatories, 319-320
ocean policy, 342-343
ocean trench, 41
ocean water,
properties of, 53-70
oceanic conveyor belt, 98, 112, 141-142, 260, 289, 304-305
oceanic crust, 32, 42
oceanic life zones, 231
oceanography, 312
Oeschger, Hans, 305
oil, formation of, 90
oil spills, 145, 345-347
Okushiri, Japan, 164
oligotrophic waters, 231
omnivores, 10, 205
On the Origin of Species, 311
oolith, 85
ospreys, 336
oozes, 88-89
Our Nation and the Sea, A Plan for National Action, 341
Outer Banks (North Carolina), 73, 196-197
outgassing, 25
overfishing, 326-328
growth, 326
recruitment, 326
oxygen, origin of, 26
oxygen isotope analysis, 280-281, 304
oysters, 331-332
ozone, 6, 7, 26, 121-123

P

Pacific Decadal Oscillation (PDO), 272
Pacific Marine Environmental Laboratory (PMEL), 269
Pacific Ring of Fire, 41, 46
Pacific Subarctic Water, 139, 140, 141
Pacific Tsunami Warning Center, 165
pack ice, 5, 74
Padre Island, TX, 179-180
PALACE, 148-149
pancake ice, 74
Pangaea, 9, 37, 39
Panthalassa, 46
partial tides, 162
partially mixed estuary, 184
particulate organic carbon (POC), 216
passive continental margin, 175
Patapsco River, 349
Pauly, Daniel, 329
pelagic deposit, 88-89
pelagic fish, 233-234
pelagic zone, 231, 232-234
pelagos, 231
Pelican Islands, Florida, 336
pelicans, 249
penguins, 249-250
perigee, 160
perihelion, 291
permafrost, 5, 279-280, 301
Peru-Chile Trench, 41
Peru Current, 131, 132
Peruvian anchovy fishery, 327
pH, 64-65
Phoenicians, 310
phosphorescence, 234
photic zone, 64, 104-105, 211-213
photocytes, 234
photodissociation, 12
photophores, 234
photosphere, 290
photosynthesis, 10, 26, 64, 104, 210-211
phragmites, 338
physical model, 19
Physical Oceanographic Real Time Services (PORTS), 162
physical pump, 215
phytoplankton, 205-206, 232
Piccard, Auguste, 314
Piccard, Jacques, 314
pillow basalt, 38, 39

Pine Island Glacier, 6
pinnipeds, 248
Pizarro, Francisco, 277
placer deposits, 90-91
planetary albedo, 104
planetary-scale circulation, 114, 115-116
planetary-scale systems, 114, 115-116
plankton, 231, 232-233
plastron, 333
plate boundaries, 38-42
plate tectonics, 8-9, 29-31, 36-46
    and climate, 282
    evidence for, 36-38
Pleistocene Ice Age, 26, 35, 94, 175-176, 183, 199-200, 277, 281, 297, 304
point source, 349
polar amplification, 285, 296
polar bears, 248, 280
polar easterlies, 114, 115
polar front, 115, 116
polar-orbiting satellite, 17
polar vortex, 122-123
Polaris, 99
polders, 198
poleward heat transport, 111-112
pollen, 281-282
pollution, 344-345
polychlorinated biphenyls (PCBs), 209, 210
Polynesians, 310
polynyas, 75
polyps (coral), 241, 242
Port Valdez, AK, 183
Portuguese man-of-war, 232-233
practical salinity units (psu), 61
Precautionary Principle, 343
precipitation, 14
pressure, 56, 67-68
pressure gradient force, 112-113
pressure ridge, 75
Prestige, 346
primary production, 205, 328
prime meridian, 120
Prince William Sound, AK, 345-346
principle of constant proportions, 61
prochlorococus, 214
producers, 10, 205-207
Profiling Autonomous Lagrangian Circulation (PALACE), 148-149
progressive wave, 156

prokaryotes, 214
protista, 206
protozoa, 214
proxy climatic data, 277, 280
pteropods, 83, 85
Ptolemy, 120
Puget Sound, 184
pycnocline, 127, 128, 164, 263
Pytheas, 310

Q

quartz, 81
Queen Elizabeth 2, 151
QuikSCAT, 169
radar altimeter, 170
radiolaria, 85
radiation, 15
ram ventilation, 245
recreational fisheries, 332
recruitment overfishing, 326
red clay, 88
Red Sea, 141
Red Sea Intermediate Water, 140, 141
red snapper, 331
red tide, 83, 203-204, 318
reference ellipsoid, 170
reflection, 102-104
refraction (wave), 176-177
regenerated production, 211
remineralization, 216-217
remote sensing, 15-18, 315-316
    active, 17
    passive, 15
remotely operated vehicle (ROV), 315
Resen, Bishop, 125
reservoir rock, 90
residence time, 11, 199
resonance, 156-157, 161-162
resources, of the seafloor, 90-91, 92
respiration, 10, 64, 209
reverse estuaries, 185
reserve osmosis, 73
reversing thermometers, 148
rift valley, 39-40, 44
rings, 134-135
rip current, 179
rocks, 32
rock cycle, 33-34, 89

Rocky Mountains, 41-42
rogue waves, 151-152
Ronne Ice Shelf, 298
Roosevelt, Theodore, 336
Ross, Sir James Clark, 311
Ross, Sir John, 311
Ross ice shelf, 298
Ross Sea, 311
Rowland, F.S., 122
Ruddiman, W. F., 284
runoff, 15
runoff component, 15
Rutgers Marine Field Station, 320
R/V Chain, 306
R/V Meteor, 313
R.V. Melville, 221
Ryan, William, 306

S

Saffir, H. S., 190
Saffir-Simpson Hurricane Intensity Scale, 190-191
Sahara Desert, 81, 82
Sahel (Africa), 275, 276
salinity, 60-63
salmon, 247, 337, 348
salmon farming, 337
salt marsh, 181, 183, 198, 239, 301, 338
saltwater intrusion, 308
salt-wedge estuary, 184
San Andreas fault, 42
San Francisco Bay, 184
sand dunes, 179-180
sand spit, 180
Santa Barbara, CA, 182
Santorini, 30
Sargasso Sea, 132, 134, 237
sargassum, 237
satellite remote sensing, 15-18
scattering, 102, 103
scatterometer, 168-169
schooling, 247
Scientific Committee of Oceanic Research (SCOR), 217
Scientific Ice Expedition, 300
scientific method, 36
Scotch Cap Lighthouse, AK, 164
Scripp's Institution of Oceanography, 268, 313
scutes, 333
sea, 154

sea breeze, 115
sea fog, 260
sea grass, 238-239, 332
sea ice, 68, 69, 74-75, 294
    extent of, 69
    first-year ice, 68, 74
    multi-year ice, 68, 74
    new ice, 74
    old ice, 74
    physical properties of, 68, 69, 74-75
    second-year ice, 74
    terminology, 74-75
    young ice, 74
sea level fluctuations, 267, 297-300
    and saltwater intrusion, 308
sea lily, 244
sea lion, 248
sea otters, 240, 346
sea salts, 14, 60-63
sea snakes, 249
sea state, 168
sea turtles, 229, 249, 333-334
    carapace, 333
    flatback, 333
    green, 333
    hawksbill, 333
    Kemp's (Atlantic) ridley, 333-334
    leatherback, 248, 334
    loggerhead, 334
    olive (Pacific) ridley, 334
    plastron, 333
    scutes, 333
sea urchins, 240
sea wave, 152
sea-floor spreading, 37
Sea-viewing Wide Field-of-view Sensor (SeaWiFS), 9, 217, 226-227
seabirds, 249-250
seafloor resources, 90-92
Seager, Richard, 260
seals, 247-248
seamounts, 42
SEASAT-A Satellite Scatterometer (SASS), 169
seasons, 99-101
SeaStar satellite, 226
seawall, 182
seawater, 60
    alkalinity of, 64-65
    chemical properties of, 60-65

conservative properties of, 61
dissolved gases, 63-64
non-conservative properties of, 61
physical properties of, 65-70
pH of, 64-65
standard, 61
seaweed, 237
SeaWiFS, 9, 217, 226-227
secondary production, 205
sediment, 8, 77-92
size classification, 78-79
sedimentary rock, 32, 33, 89-90
seiche, 156-157
seismometer, 319
Seltzer, Geoffrey, 277
semi-diurnal tide, 159-160
sensible heating, 108-110, 260
Severn River, 162
shallow-water waves, 155
shark fin soup, 328
sharks, 245, 327-328
basking, 245
finning (for soup), 328
mako, 327
overfishing of, 327-328
porbeagle, 327
spiny dogfish, 327
thresher, 327
tiger, 245
viviparous reproduction of, 245
shore, 175
shoreline, 175
shrimp, 235
shrimp trawling, 330
significant wave height, 168
siliceous ooze, 89
silicon-oxygen tetrahedron, 80-81
Simpson, R. H., 190
Skeleton Coast, Namibia, 282
slack water, 162
Slocum, Josiah, 317
Slocum glider, 317-318
SLOSH (Sea, Lake, and Overland Surges from Hurricanes), 186, 191
slush, 74
Snake River, 348
snow cover, 294
snow ice, 74
soil, 8

SOFAR channel, 68, 70
Sojourner, 1
solar altitude, 99-101
solar radiation, 99-105, 259
budget of, 102-104
and the ocean, 104-105
solar still, 73
solar variability, and climate, 290-292
solstices, 99
solubility pump, 215
SONAR (Sound Navigation and Ranging), 68
sorting, 78-79
sound, 235, 335-336
transmission of, 18, 68, 70
sounding, 139
South Atlantic gyre, 131, 132
South Equatorial Current, 131, 132-133
Southern Ocean, 31, 311
southern oscillation, 261
southern oscillation index (SOI), 261, 262
specific heat, 58-59
specific heat of water, 58-59
spermaceti, 334
Spörer minimum, 290
sport fisheries, 332
spring bloom, 212
spring tides, 160
spring turnover, 201
squall line, 193
squid, 233, 235
stability, 128
stable system, 128
standard atmospheric pressure, 112
standard seawater, 61
standing wave, 156
state of the sea, 168-169
Steele, John, 213-214
stewardship (of resources), 325
storm-sewer discharge, 347
storm surge, 185-186
Strait of Bab el Mandeb, 141
Strait of Gilbralter, 140, 306
strategic retreat, 183
stratosphere, 6, 26, 121-123
stratospheric ozone shield, 6, 26, 103
and marine life, 121-123
Stratton, Julius A., 341
Stratton Commission, 341
stromatolites, 306

subduction, 37
subduction zone, 41
sublimation, 13, 58
submarine canyon, 35
submarine fans, 35
submarine landslides, 224
sub-polar gyres, 132
subpolar lows, 116
Sub-Saharan Africa, 275-276
Subsurface Acoustic Doppler Current Profilers, 269
subtropical gyres, 131-132
subtropical highs, 114, 115, 116
Suess, Eduard, 37
sulfurous aerosols, 30, 292, 295
sunspots, 290
superdams, 347
surf, 155
surf zone, 179
surface geological processes, 8
surface tension, 153
Susquehanna River, 184, 349
sustainability, 326
sustainable development, 341
Sustainable Fisheries Act of 1996, 343
sustainable yield, 328-329
sustained wind speed, 186
swash, 177
swell, 154
swordfish, 325
symbiotic relationship, 241
synoptic-scale weather systems, 114, 116-118
    highs, 113, 116-117
    lows, 113, 116-118
system, 3

T

tags, 320
tai-fung, 187
Tamano, 145
Tambora, eruption of, 30, 292
TAO/TRITON, 148, 269
Taylor, Frank B., 37
tectonic processes, 8-9, 29-31, 36-46
tektite, 86
teleconnections, 265
teleosts, 246-247
temperature, 56-57
temperature gradients, 7

tephra, 80
Tethys Sea, 44, 289, 306
thermal inertia, 59, 259
terminal velocity, 79-80
thermocline, 127, 128, 201
thermohaline circulation, 137, 141
thermosphere, 6
Thomson, Charles Wyville, 312
threatened species, 333
Three Gorges Dam, 347-348
Thunder Bay National Marine Sanctuary and Underwater Preserve, 253-254
Tibetan Plateau, 284
tidal bore, 162
tidal currents, 162
tidal day, 159
tidal period, 157
tidal power, 171
tidal range, 157
tidal wave, 157, 165
tide (astronomical), 157-164
    diurnal, 159, 160
    ebb, 162
    equatorial, 160
    flood, 162
    fortnightly, 160
    in ocean basins, 160-162
    internal, 163-164
    mixed, 159, 160
    neap, 160
    observing, 162-163
    open-ocean, 163-164
    partial, 162
    predicting, 162-163
    semi-diurnal, 159-160
    spring, 160
    tropic, 160
tide cracks, 74
tide generating forces, 158-159
tide pool, 238
tide prediction, 162-163
Titanic, RMS, 5
Tomales Bay, CA, 332
top predators, 234, 325
TOPEX/POSEIDON, 163, 170, 267, 269, 270, 297, 298
tornado, 190
total allowable catch, 328
toxic algal blooms, 83
trace elements, 205, 213

trade winds, 114, 115, 132
trajectory analysis, 145
transform plate boundary, 42
transitional wave, 155
transpiration, 13
tree growth rings, 282
trench, 41
Trieste, 314
TRMM Microwave Imager (TMI), 270
trophic level, 10
Tropic of Cancer, 101. 159
Tropic of Capricorn, 101, 159
tropic tides, 160
tropical cyclone, 185, 186-192, 300
    and El Niño, 265
    evacuation, 191-192
    hazards, 189-191
    life cycle, 189
    where and when, 187-189
tropical depression, 189
tropical disturbance, 189
Tropical Atmosphere/Ocean (TAO) array, 269
Tropical Ocean Global Atmosphere (TOGA), 148, 269
Tropical Prediction Center/National Hurricane Center, 191
Tropical Rainfall Measuring Mission (TRMM), 263, 270
tropical storm, 189
troposphere, 6
tsunami, 23, 30, 164-166, 224
Tsunami Warning Centers, 165
tubeworms, 244
tuna, 330
turbidites, 87, 88
turbidity current, 35, 87-88, 95
turbidity maximum, 185
Twain, Mark, 139
twilight zone, 216, 234
typhoon, 187

U

UN Conference on Environment & Development, 343
UN Convention on the Law of the Sea, 91, 343
UN Food & Agriculture Organization (FAO), 327, 328
UN International Maritime Organization (IMO), 346, 351
UNESCO World Heritage Site, 96
U.S. Coast Guard, 336
U.S. Commission on Ocean Policy, 341
U.S. Environmental Protection Agency, 336, 346
U.S. Fish Commission, 313

U.S. Fish and Wildlife Service, 334
U.S.-Japanese Tropical Rainfall Measuring Mission (TRMM), 263, 270
U.S. Naval Observatory & Hydrological Office, 312
USS Albatross, 313
USS Monitor, 253
USS Ramapo, 152
universal solvent, 54, 60
unstable system, 128
upwash, 177
upwelling, 135-137, 212, 262-263
urban heat island, 287
UV radiation, 121-123
UVB, 121-122

V

Venice, Italy, 173-174
vertical migration, 234
Victoria Land, Antarctica, 311
Vikings, 310-311
Vincennes, 312
Vine, F. J., 38
viruses, 214
visible light, 15, 16, 104-105
visible window, 105, 106
volcanic eruptions, 9, 29-30
    and climate, 30
Vostok, Antarctica, 304

W

Walcott, Charles D., 96
Walker Circulation, 264
Walker, Sir Gilbert, 261, 264
Walsh, Don, 314
warm-core rings, 134, 135
water
    origin of, 25-26
    phases, of, 55-56, 57-58
    as a solvent, 60
    specific heat of, 58-59
water birds, 336
water cycle, 12-15, 26
water mass, 139-141
water molecule, 54-55
water vapor, 7
watershed, 15
wave, 152-157

    capillary, 153
    deep-water, 154-155
    forced, 154
    generation, 153-154
    internal, 164
    progressive, 156
    rogue, 151-152
    sea, 152
    shallow-water, 155
    standing, 156
    transitional, 155
    tsunami, 23, 30, 164-165, 224
wave crest, 152, 153
wave frequency, 153
wave height, 152, 153
wave period, 152-153
wave power index, 193
wave refraction, 176-177
wave trough, 152, 153
wave-base, 153
wavelength, 15, 16, 152, 153
weather, 98-99
weathering, 8, 32, 80, 81, 89
Weddell Sea, 74
Wegener, Alfred, 37
well-mixed estuary, 184
Wentworth Classification, 78-79
West Antarctica ice sheet, 298
West Coast and Alaska Tsunami Warning Center, 165
westerlies, 114, 115, 116
western boundary currents, 133-134
wetlands, 87
whale oil,
whale watching, 335
whales, 334-336
    blue, 334
    humpback, 335
    hunting of, 334
    right, 334
    white (beluga), 320
whaling, commercial, 335
White Cliffs of Dover, 206
Whittington, Harry, 96
Wilkes, Lt. Charles, 312
Wilson cycles, 44-46
    declining stage, 44-45
    embryonic stage, 44, 45
    juvenile stage, 44, 45
    mature stage, 44, 45
    suturing stage, 44, 46
    terminal stage, 45, 46
Wilson, J. Tuzo, 44
wind belts, 114, 115
wind shear, 188
wind-blown dust, 81-83
wind-driven currents, 130-137
wind-driven waves, 152-157
wind waves, generation of, 153-154
Woods Hole Oceanographic Institution, 315
World Meteorological Organization (WMO), 74, 286

Y

Yangtze River, 348
Year without a summer, 30, 292
Yellowstone National Park, 9, 50
Younger Dryas, 285-286, 304
Yucatan Peninsula, 23

Z

zooplankton, 207-208, 232
zooplankton bloom, 212
zooxanthellae, 241-242